MARTIN—Nonlinear Operators and Differential Equations in Banach Spaces

MELZAK—Companion to Concrete Mathematics

MELZAK—Mathematical Ideas, Modeling, and Applications (Volume II of Companion to Concrete Mathematics)

MEYER—Introduction to Mathematical Fluid Dynamics

MORSE—Variational Analysis: Critical Extremals and Sturmian Extensions

NAYFEH and MOOK—Nonlinear Oscillations

NAYFEH—Perturbation Methods

ODEN and REDDY—An Introduction to the Mathematical Theory of Finite Elements

PAGE—Topological Uniform Structures

PASSMAN—The Algebraic Structure of Group Rings

PRENTER—Splines and Variational Methods

RIBENBOIM—Algebraic Numbers

RICHTMYER and MORTON—Difference Methods for Initial-Value Problems, 2nd Edition

RIVLIN—The Chebyshev Polynomials

RUDIN—Fourier Analysis on Groups

SAMELSON—An Introduction to Linear Algebra

SIEGEL—Topics in Complex Function Theory
Volume 1—Elliptic Functions and Uniformization Theory
Volume 2—Automorphic Functions and Abelian Integrals
Volume 3—Abelian Functions and Modular Functions of Several Variables

STOKER—Differential Geometry

STOKER—Nonlinear Vibrations in Mechanical and Electrical Systems

STOKER—Water Waves

WHITHAM—Linear and Nonlinear Waves

WOUK—A Course of Applied Functional Analysis

YOSIDA—Lectures on Differential and Integral Equations

TOPOLOGICAL UNIFORM STRUCTURES

TOPOLOGICAL UNIFORM STRUCTURES

WARREN PAGE

City University of New York

A WILEY-INTERSCIENCE PUBLICATION

JOHN WILEY & SONS, New York • Chichester • Brisbane • Toronto

Library of Congress Cataloging in Publication Data:

Page, Warren, 1939–
Topological uniform structures.

"A Wiley-Interscience publication."
Bibliography: p.
Includes index.
1. Uniform spaces. I. Title.

QA611.25.P33 514′.3202 78–930
ISBN 0-471-02231-4

Printed in the United States of America

10 9 8 7 6 5 4 3 2 1

To

Esther Page

and

the memory of Irving Page.

They taught me hope, faith, and perseverance.

Preface

This book aims to acquaint the reader with a slice of mathematics that is interesting, meaningful, and in the mainstream of contemporary mathematical developments. Admittedly a number of excellent sources cover, in part, uniform spaces, topological groups, topological vector spaces, topological algebras, and abstract harmonic analysis. I believe, however, that this is the first known text to give a thorough and fully detailed account of all these topics. It also contains a section on vector-valued measure spaces and a plethora of problems and examples that have not yet appeared in book form. The aforementioned topological structures are worth studying (particularly for the neophyte mathematician), since they are the juncture points at which many significant mathematical ideas come into contact.

Most advanced undergraduate and beginning graduate students, constrained by time and compartmentalized course requirements, do not yet fully appreciate the true essence of mathematics as a structured, coherent, and harmonious whole. This, to a large extent, motivated my style and presentation. The two most salient features of this work are the following:

(1) The overall unifying theme of topologies compatible with increasingly enriched algebraic structures.

(2) Its attempt to gradually deepen the reader's appreciation of the rich interplay among diverse areas of mathematics.

The interested reader can get a better overview by quickly scanning the chapter perspectives and section summaries. Briefly, (1) may be described by following the order of topics covered, subject to the requirement that each structure's algebraic operations be continuous. What emerges from (2) is a number of striking results that combine and interlace algebraic, topological, and measure-theoretic properties associated with the structure under consideration.

A book of this nature could easily become encyclopedic were it not for the decision to omit topics. This is always heartrending, even more so when such topics fit so well and clamor for attention. Given freer rein or an expanded future edition, I would include a section on proximity spaces (as structures layered between topological and uniform spaces) and some material on topological fields and generalized topological structures (i.e., topological spaces with continuous generalised internal and external operations). The second theme would also be enhanced by a fuller treatment of representation theory and an increased cross-pollination with our Gelfand theory approach to harmonic analysis.

While writing this text I gave readability and motivation prime consideration. The pace is leisurely and the contents, highly illustrative. Each section begins with a brief summary advertising the salient points to appear. The theory is intended to unfold smoothly through a good deal of commentary and an abundance of examples. Furthermore, the remarks (which follow most theorems and examples) are chock full of useful information, counterexamples, and other points worth noting. In the notation a remark is referred to as $u \cdot v^+$ or $u \cdot v^-$, depending on whether it follows or precedes item v of Section u. Finally, each section ends with a set of problems (more than 220 each with several subparts in all). Some test understanding and give additional insights. A large number, however, come from findings in recent literature and research papers. Their arrangement and emphasis I hope will broaden the student's perspective and whet his appetite for delving further into some of them.

This work is reasonably self-contained and accessible to students with a background in elementary analysis, linear algebra, and point set topology. At the same time it covers a good amount of advanced material without going off into the purple deep. My intent was to design this text with a maximum of flexibility. Accordingly, it is a set of interconnected chapters, each of which is a minibook standing at the threshold of several advanced texts in these areas. A few to which these chapters are well geared are *Uniform Spaces* (Cech [16], Isbell [45]); *Topological Groups* (Pontryagin [80], Montgomery and Zippin [64]); *Topological Vector Spaces* (Köthe [55], Treves [98], Edwards [28]); *Topological Algebras* (Naimark [69], Narici [72], Rickart [82]) and *Abstract Harmonic Analysis* (Hewitt and Ross [41], Sugiura [94], and Rudin [85], followed by Dunkl and Ramirez [27]). In essence, this text is well suited to a variety of one- or two-semester courses as outlined in A Guide to This Book. I feel that it will also prove useful as a reference source and as a source of seminar material at the advanced undergraduate and beginning graduate level.

Finally, I record here my profound gratitude to Professor George Bachman, who first triggered my impulse to write this book. His valuable advice, imaginative ideas, and timely infusions of enthusiasm helped me carry it to completion. I am also indebted to Professor Paul Sally for his insightful comments and constructive suggestions during the early stages of this work.

WARREN PAGE

New York, New York
February 1978

A Guide to this Book

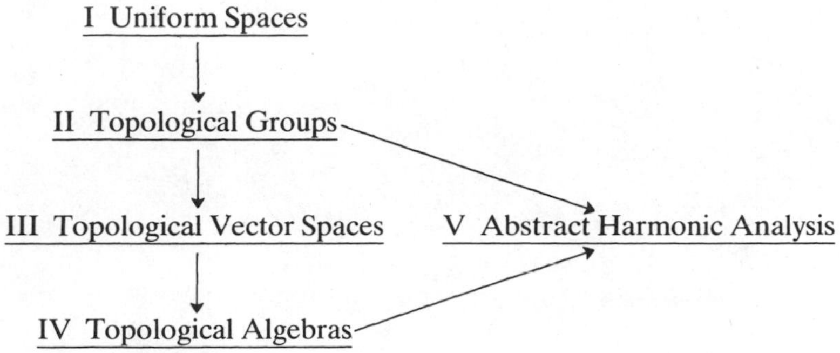

One Year Courses	Chapters/Sections
Topological Uniform Structures	I–IV
Elements of Modern Analysis	II–V

One Semester Courses	Chapters/Sections
Topology and Groups	I, II
Topological Groups and Harmonic Analysis	II, 22 & 23, V
Topological Vector Spaces	III
Topological Algebras	13 & 14, IV, 27 & 28

Contents

Symbols and Notations

iff if and only if
$\forall$ for each, for every
$\aleph$ cardinality
$\delta_{\alpha,\beta}$ kronecker delta
$\mathbb{F}$ field of scalars
$\mathbb{R}$ real numbers
$\mathbb{C}$ complex numbers
$\mathbb{I}$ integers
$\mathbb{N}$ natural numbers
$\mathbb{N}_0 = \mathbb{N} \cup \{0\}$
[K] vector subspace spanned by K
$\mathbb{1}$ identity mapping
$\mathfrak{e}$ evaluation mapping, 55
$\mathscr{B}_x$ and $\mathscr{N}_x$, 363
$\mathscr{P}(X)$ = power set of X, 363
$\mathrm{Catg}_i(X)$, 364
$\mathbb{A}_i$ and $\mathscr{B}_{ao}$, 364
$\mathscr{A} < \mathscr{B}$, $\mathscr{A} \overset{*}{<} \mathscr{B}$, and $\mathscr{A} \overset{b}{<} \mathscr{B}$, 365–366
$\mathrm{St}(E, \mathscr{A})$, $\mathscr{A}^b$, and $\mathscr{A} \wedge \mathscr{B}$, 366
$\sum_{\mathscr{A}} x_\alpha$, 367
fip, 367
$\mathscr{F}(\mathscr{B})$, $\mathscr{F} \to x$, and $\mathscr{F} \infty x$, 367
$\mathscr{F}\langle X \rangle$, 38
f_{Re} and f_{Im}, 372
$f^{\pm}$ and $|f|$, 372
χ_E characteristic function of E
$\mu \times \nu$ product measure, 374
$\mu^{\pm}$ and $|\mu|$, 375
$\nu \ll \mu$ and $\nu \perp \mu$, 376
$\mathscr{C}$, $\mathscr{C}_\delta$, $\mathfrak{C}$ and $\mathscr{O}$, $\mathscr{O}_\delta$, $\mathfrak{O}$, 377
$\mathfrak{R}_\sigma(\mathscr{C})$, $\mathscr{R}_\sigma(\mathscr{C}_\delta)$, and $\mathfrak{R}_\sigma$, 377

Δ, $A \circ B$, and A^{-1}, 2
$\mathscr{B}_d$ and U_ε, 2
$\mathscr{C}_{U_\varepsilon}$, 16
$\mathscr{U}(X)$, $\mathbb{S}(X)$, $\mathscr{Q}(X)$, and $\mathscr{L}(X)$, 3
$t \cdot \mathbb{S}(X)$, 6
$\mathscr{U}_{tb}(X)$, 39
$\mathscr{S}_{\mathrm{sym}}$, 4
$\mathscr{K}_2(X, \mathfrak{T})$, 50
$\mathfrak{m}$ and $\mathfrak{i}$, 67
${}_g\mathfrak{m}$ and $\mathfrak{m}_g$, 68
$\mathfrak{s}_\alpha$ and $\mathfrak{s}$, 133 and 134
$\mathbb{M}_n(\mathbb{F})$ and $G_n(\mathbb{F})$, 68
$f \circ g$ composition map
$\chi_d \to \chi$ and $\chi_d \infty \chi$, 367
$S'(\mathbb{C})$ = circle (torus) group, 88
$G \overset{t}{=} H \otimes K$, 89
$H \otimes K$, 88
$\mathrm{gr}(f)$ graph of f, 93
G' = character group, 102
$\hat{f}$ and $\tilde{f}$, 111 and 118
I_G Haar integral, 112
μ_G Haar measure, 112
μ'_G extended Haar measure, 121
$\Delta: G \to \mathbb{R}$ modular function, 117
$T(G)$ and $T(X)$, 67 and 133, 273
$N(f)$ support of $f \in C_0(\Omega)$, 369
K_b, K_c, and K_{bc}, 135
$S_{\varepsilon,p}(0)$ and $\bar{S}_{\varepsilon,p}(0)$, 154
p_K Minkowski functional, 155
(E_B, p_B), 157
$X_p = (X/p^{-1}(0), \|\,\|_p)$, 167

Spaces

Uniformities

$\mathcal{U}_{\mathrm{d}}$ $\mathcal{U}_{\mathfrak{g}}$, $\mathcal{U}_{\mathfrak{D}}$, and $\mathcal{U}_{<}$, 4
${}_G\mathfrak{U}$ $\mathfrak{U}_G$, and $\mathcal{U}$ for topological groups, 72
$m\mathcal{S}$ uniform modification of a semiuniformity, 14
$\sup_{\mathfrak{a}} \mathcal{U}_\alpha$ and $\inf_{\mathfrak{a}} \mathcal{U}_\alpha$, 14
$\mathcal{U}_F$ the fine uniformity, 15
$\mathcal{U}_c$ the coarse uniformity, 15
$\mathcal{U}_{\beta(X)}$ the Cech uniformity, 43
$(X/R, \mathcal{U}_{\mathfrak{g}(\nu)})$ quotient uniform space, 27
$\mathfrak{c}_{\mathcal{U}}$ covering uniformity determined by $\mathcal{U}$, 17
$\mathfrak{c}_\rho$ and $\mathfrak{c}_<$, 19
$\mathfrak{c}_F$ the fine covering uniformity, 21
$\mathfrak{c}_c$ the coarse covering uniformity, 21
$\sup_{\mathcal{A}} \mathfrak{c}_\alpha$ and $\inf_{\mathcal{A}} \mathfrak{c}_\alpha$, 21
$\mathcal{U}_{\mathcal{P}(\mathcal{F})}$ and $\mathfrak{c}_{\mathcal{P}(\mathcal{F})}$ projective limit uniformities, 22
$\mathcal{U}_{\mathfrak{g}(\mathcal{F})}\{\mathcal{U}_{\mathfrak{g}'(\mathcal{F})}\}$ and $\mathfrak{c}_{\mathfrak{g}(\mathcal{F})}$ inductive limit uniformities, 26 {168}
$\mathcal{U}_\pi$ and $\mathfrak{c}_\pi$ product uniformities, 25
$\mathcal{U}_E$ and $\mathfrak{c}_E$ induced subspace uniformities, 24
$\mathcal{U}_{\mathfrak{G}}$ and $\mathfrak{c}_{\mathfrak{G}}$ uniformities determined by a gage $\mathfrak{G}$, 33

Topologies

$\mathfrak{T}_{\mathcal{U}}$ and $\mathfrak{T}_{\mathfrak{c}}$ uniform space topology, 5 and 18
$\mathfrak{T}_{\mathcal{P}(\mathcal{F})}$ projective limit topology, 22
$\mathfrak{T}_{\mathfrak{g}(\mathcal{F})}$ inductive limit topology, 26
$\mathfrak{T}_{\mathfrak{g}(\P)}$ internal inductive limit topology, 176, 184
$\mathfrak{T}(\mathcal{P})$ topology determined by a family $\mathcal{P}$ of seminorms, 154
$\mathfrak{T}(\mathfrak{G})$ topology determined by a gage $\mathfrak{G}$, 31
$\mathfrak{T}_{\mathcal{S}}$ polar topology, 207
$\mathfrak{T}_{F_c}$ finest compatible locally convex topology, 156
$\mathfrak{T}_{co}$ compact open topology, 65
$\mathfrak{T}_g$ the Gelfand topology on $\mathfrak{M}$, 302
$\mathfrak{T}_{hk}$ the hull-kernel topology on $\mathfrak{M}$, 302
$\mathfrak{T}_{hk}$ the hull-kernel topology on $\mathfrak{M}$, 307
$\sigma(X, Y)$, $\beta(X, Y)$, and $\tau(X, Y)$, 208–209
$\kappa(X, X')$, $\kappa_c(X, X')$, $\kappa_{bc}(X, X')$, and $\lambda(X, X')$, 216
$\beta^*(X, X')$ and $\kappa^*_{bc}(X, X')$, 217

CHAPTER I

Uniform Spaces

Perspective. A mapping between pseudometric spaces (X, d) and (Y, ρ) is uniformly continuous if $\rho\{f(x_1), f(x_2)\} < \varepsilon$ whenever $d(x_1, x_2) < \delta(\varepsilon)$ that is, images under f of points $\delta(\varepsilon)$-close in X are ε-close in Y. Since orders of nearness are not defined for a general topological space, the notion of uniform continuity is meaningless here. The same, of course, can be said about size-dependent notions such as total boundedness. All this motivates generalizing the concept of pseudometric space. We do so by introducing topology-generating structures (called uniformities) on a set X, both in terms of subsets of $X \times X$ (section 1) and in terms of coverings of X (section 2). The nexus between these two types of uniformity is spelled out in detail. Just about everything else that follows serves to define, develop, and illustrate the basic properties of uniform spaces.

1. Entourage Uniformities

Let X be a set and suppose $\mathcal{Q}$ is an index set such that for each $\alpha \in \mathcal{Q}$ and $x \in X$ there is a subset $u_\alpha(x) \subset X$. Assume furthermore that $\mathcal{U} = \{u_\alpha(x) : (\alpha, x) \in \mathcal{Q} \times X\}$ satisfies $N_1 - N_4$ of Appendix t.1. Then $\{u_\alpha(x) : x \in X\}$ may be used to define "α-sized" nbds in X (relative to the topology $\mathfrak{T}_{\mathcal{U}}$ on X having $\mathcal{U}$ as a nbd base). Simply define $x, x' \in X$ to be α-close iff $x' \in u_\alpha(x)$. If Y also has a topology $\mathfrak{T}_{\mathcal{V}}$ defined by an index set $\mathfrak{B}$, then $f : (X, \mathfrak{T}_{\mathcal{U}}) \to (Y, \mathfrak{T}_{\mathcal{V}})$ may be considered uniformly continuous in the sense that for each $\beta \in \mathfrak{B}$ there exists an $\alpha \in \mathcal{Q}$ such that $f(u_\alpha(x)) \subset v_\beta(f(x))\ \forall x \in X$.

Example 1.1. If d is a pseudometric on a set X and $U_\varepsilon = \{(x, x') \in X \times X : d(x, x') < \varepsilon\}$, then $U_\varepsilon[x] = \{x' \in X : (x, x') \in U_\varepsilon\}$ is the "ε-sized" set $S_\varepsilon(x) = \{x' \in X : d(x', x) < \varepsilon\}$. Clearly, $\mathcal{U} = \{U_\varepsilon[x] : x \in X \text{ and } \varepsilon > 0\}$ satisfies $N_1 - N_4$. If (Y, ρ) is also a pseudometric space, then $f : (X, d) \to (Y, \rho)$ is uniformly continuous iff $f\{S_\delta(x)\} \subset S_\varepsilon\{f(x)\}\ \forall x \in X$.

The preceding example suggests that one approach to generalizing (pseudo)metric spaces might be to consider families $\mathcal{U} \subset X \times X$ for which the corresponding U-sized sets $\mathfrak{N} = \{U[x] : x \in X \text{ and } U \in \mathcal{U}\}$, where $U[x] = \{y \in X : (x, y) \in U\}$ define a nbd system for a topology on X. Such indeed is the case, as we shall see shortly.

DEFINITION 1.2. Let X be a set and let $\Delta = \{(x, x) : x \in X\}$. One calls $A \subset X \times X$ symmetric if $A = A^{-1}$, where $A^{-1} = \{(x, y) : (y, x) \in A\}$. For $A, B \subset X \times X$ we define $A \circ B = \{(x, y) : (x, z) \in A \text{ and } (z, y) \in B \text{ for some } z \in X\}$.

Some elementary properties of "$\circ$" are summarized in

THEOREM 1.3. Let $A, B, C \subset X \times X$. Then

(i) $A \circ \Delta = \Delta \circ A = A$;
(ii) $A \circ (B \circ C) = (A \circ B) \circ C$ (with common value denoted $A \circ B \circ C$);
(iii) $A^{n+1} = A^n \circ A$ and $A^n \circ A^m = A^{n+m}$;
(iv) $(A \circ B)^{-1} = B^{-1} \circ A^{-1}$;
(v) $(A^n)^{-1} = (A^{-1})^n$. However, $A^{n+1} \circ A^{-1} \neq A^n$ in general, since it is not always true that $A \circ A^{-1} = \Delta$.
(vi) If $A \subset B$, then $A^{-1} \subset B^{-1}$ and $A^n \subset B^n$. Moreover, $A^n \subset A^{n+m}$ whenever $\Delta \subset A$.
(vii) If A is symmetric, so is A^n $\forall n \in \mathbb{N}$.

Proof. Left as a simple exercise.

Example 1.4. If (X, d) is a (pseudo)metric space and $\mathcal{B}_d = \{U_\varepsilon : \varepsilon > 0\}$, then

(i) $\Delta \in U_\varepsilon$ $\forall U_\varepsilon \in \mathcal{B}_d$;
(ii) $U_\varepsilon^{-1} = U_\varepsilon$ $\forall U_\varepsilon \in \mathcal{B}_d$;
(iii) $\forall U_\varepsilon \in \mathcal{B}_d$, there is a $U_{\varepsilon/2} \in \mathcal{B}_d$ such that $U_{\varepsilon/2}^2 \subset U_\varepsilon$;
(iv) $\forall U_{\varepsilon_1}, U_{\varepsilon_2} \in \mathcal{B}_d$, there is a $U_{\varepsilon_3} \in \mathcal{B}_d$ satisfying $U_{\varepsilon_3} \subset U_{\varepsilon_1} \cap U_{\varepsilon_2}$. (Note that $d(x, y) \leq \min\{\varepsilon_1, \varepsilon_2\}$ if $d(x, y) < \varepsilon_1$ and $d(x, y) < \varepsilon_2$.)

Remark. Properties (i), (ii), and (iii) are the respective analogs of reflexivity, symmetry, and the triangle inequality.

THEOREM 1.5. Let $\mathcal{B}$ be a filterbase on $X \times X$ such that for each $U \in \mathcal{B}$ (i) $\Delta \in U$, (ii) $V^{-1} \subset U$ for some $V \in \mathcal{B}$, and (iii) $V^2 \subset U$ for some $V \in \mathcal{B}$. Then the filter $\mathcal{F}(\mathcal{B})$ determined by $\mathcal{B}$ satisfies

(i)′ $\Delta \in U \; \forall U \in \mathfrak{F}(\mathcal{B})$;
(ii)′ $U^{-1} \in \mathfrak{F}(\mathcal{B}) \; \forall U \in \mathfrak{F}(\mathcal{B})$;
(iii)′ $\forall U \in \mathfrak{F}(\mathcal{B})$, there is a $W \in \mathfrak{F}(\mathcal{B})$ such that $W^2 \subset U$.

Conversely, suppose $\mathfrak{F}$ is a filter on $X \times X$ satisfying (i)′–(iii)′. Then conditions (i)–(iii) are satisfied by every filterbase $\mathcal{B}$ which determines $\mathfrak{F}$.

Proof. (i)′ is obvious. To prove (ii)′ and (iii)′, consider any $U \in \mathfrak{F}(\mathcal{B})$. Then $W \subset U$ for some $W \in \mathcal{B}$. Using (ii), one has $V \in \mathcal{B} \subset \mathfrak{F}(\mathcal{B})$ satisfying $V^{-1} \subset W \subset U$ and $V \subset U^{-1}$; that is, $U^{-1} \in \mathfrak{F}(\mathcal{B})$. Moreover, there is [by (ii)] a $V \in \mathcal{B}$ such that $V^2 \subset W \subset U$.

Conversely, suppose that $\mathfrak{F}$ satisfies (i)′–(iii)′ and let $\mathcal{B}$ be a filterbase which determines $\mathfrak{F}$. Since $\mathcal{B} \subset \mathfrak{F}(\mathcal{B})$, (i) holds. If $U \in \mathcal{B}$, then $U^{-1} \in \mathfrak{F}(\mathcal{B})$ and $V \subset U^{-1}$ for some $V \in \mathcal{B}$. Thus $V^{-1} \subset U$ and (ii) is satisfied. Finally let $U \in \mathcal{B} \subset \mathfrak{F}(\mathcal{B})$ and let $W^2 \subset U$. Since $W \in \mathfrak{F}(\mathcal{B})$, one has $V \subset W$ for some $V \in \mathcal{B}$. In particular, $V^2 \subset W^2 \subset U$ and (iii) is established. ∎

Remark. By Theorem 1.3 conditions (ii)′ and (iii)′ are together equivalent to
(iv)′ $\forall U \in \mathfrak{F}(\mathcal{B})$, there is a $W \in \mathfrak{F}(\mathcal{B})$ satisfying $W \circ W^{-1} \subset U$.

We turn now to some special subcollections of $X \times X$.

DEFINITION 1.6. A filter $\mathcal{U}$ of subsets on $X \times X$ is called a

(a) semiuniformity if $\Delta \in U$ and $U^{-1} \in \mathcal{U} \; \forall U \in \mathcal{U}$;

(b) quasi-uniformity if $\Delta \in U$ and there is a $V \in \mathcal{U}$ such that $V^2 \subset U \; \forall U \in \mathcal{U}$;

(c) uniformity if $\mathcal{U}$ is both a semiuniformity and a quasi-uniformity;

(d) local uniformity if it is a semiuniformity and $\forall U \in \mathcal{U}$ and $x \in X$, there is a $V \in \mathcal{U}$ such that $V^2[x] \subset U[x]$.

$\mathcal{U}$ is said to be separated if $\bigcap_{\mathcal{U}} U = \Delta$. The members $U \in \mathcal{U}$ are called entourages. Furthermore, $x, y \in X$ are termed U-close ($U \in \mathcal{U}$) whenever $(x, y) \in U$.

Notation. $\mathfrak{U}(X)$ $\{\mathfrak{S}(X), \mathfrak{Q}(X), \mathfrak{L}(X)\}$ denote the class of all uniformities {semiuniformities, quasi-uniformities, local uniformities} on $X \times X$. We use the expression "type uniformity" as the generic term representing any of the above filters. Unless specified otherwise, similar definitions are assumed for the other type uniformities above.

DEFINITION 1.7. A pair $(X, \mathcal{U})$ consisting of a set X and a uniformity $\mathcal{U}$ on $X \times X$ is called a uniform space. If $\mathcal{U}_1$ and $\mathcal{U}_2$ are both uniformities for X, we say that $\mathcal{U}_1$ is coarser or weaker than $\mathcal{U}_2$ (and $\mathcal{U}_2$ is finer or stronger than $\mathcal{U}_1$) if every entourage of $\mathcal{U}_1$ is also a member of $\mathcal{U}_2$.

DEFINITION 1.8. A base for a uniformity $\mathscr{W}$ is a filter base $\mathscr{B}$ such that $W \in \mathscr{W}$ iff W contains some $B \in \mathscr{B}$. (Since a base $\mathscr{B}$ for $\mathscr{W}$ completely determines $\mathscr{W}$, we frequently refer to $\mathscr{B}$ as a base.)

Remark. Note that the symmetric entourages $\mathscr{S}_{\text{sym}}$ of a semiuniformity $\mathscr{S}$ constitutes a base for $\mathscr{S}$.

Two bases $\mathscr{B}_1$, $\mathscr{B}_2$ are equivalent if $\forall B_1 \in \mathscr{B}_1$ there is a $B_2 \in \mathscr{B}_2$ such that $B_2 \subset B_1$, and $\forall B_2' \in \mathscr{B}_2$ some $B_1' \in \mathscr{B}_1$ satisfies $B_1' \subset B_2'$. (We sometimes refer to this condition by saying that $\mathscr{B}_1$, $\mathscr{B}_2$ internest.)

Example 1.9. (a) The indiscrete uniformity $\mathscr{U}_{\mathfrak{I}} = X \times X$ is the weakest uniformity for a set X, whereas the discrete uniformity $\mathscr{U}_{\mathfrak{D}} = \{U \subset X \times X : \Delta \in U\}$ is the strongest uniformity for X. A base for $\mathscr{U}_{\mathfrak{I}}$ is given by $\{X \times X\}$. Clearly, $\{\Delta\}$ is a base for $\mathscr{U}_{\mathfrak{D}}$.

(b) A pseudometric ρ on a set X determines a base $\mathscr{B}_\rho$ for a uniformity $\mathscr{U}_\rho$, where $\mathscr{B}_\rho = \{U_\varepsilon : \varepsilon > 0\}$ (cf Example 1.1). $\mathscr{U}_\rho$ is called the uniformity determined by ρ. For the special case in which $X = \mathbb{R}$ and $d_1(x, y) = |x - y|$, one calls $\mathscr{U}_{d_1}$ the standard (natural, usual) uniformity for $\mathbb{R}$. (See Example 1.28 for analogous results concerning quasi-uniformities and semiuniformities.)

(c) Let $(X, <)$ be an ordered set. For each $x_0 \in X$, let $B_{x_0} = \{(x, y) : x, y > x_0\}$ and define $U_{x_0} = \Delta \cup B_{x_0}$. Then $B_< = \{U_x : x \in X\}$ is the base for a uniformity $U_<$ on $X \times X$.

(d) Given an ordered set $(X, <)$, let $\mathscr{B} = \{(x, y) : y \geq x\}$ be the "upper triangle" in $X \times X$. Then the filter determined by $\mathscr{B}$ is a quasi-uniformity but not a semiuniformity (therefore neither a local uniformity nor a uniformity).

(e) If $X = \{0, 1, 2\}$, the filter determined by $\mathscr{B} = \{\Delta \cup (0, 1) \cup (1, 0) \cup (1, 2) \cup (2, 1)\}$ is a semiuniformity, but not a quasi-uniformity (nor a uniformity) for X.

Remark. A locally uniform space need not be a uniform space. This will be clear after Theorem 1.22.

THEOREM 1.10. Let $\mathscr{W}$ be a filter on $X \times X$ such that

$$\Delta \in W \; \forall W \in \mathscr{W}.$$

(i) There exists a unique topology $\mathfrak{T}_{\mathscr{W}}$ on X such that $\mathscr{W}_{x_0} = \{W[x_0] : W \in \mathscr{W}\}$ is a base for the $\mathfrak{T}_{\mathscr{W}}$-nbds of $x_0 \in X$. One defines $u \in \mathfrak{T}_{\mathscr{W}}$ iff $\forall x \in u$ there is a $W \in \mathscr{W}$ satisfying $W[x] \subset u$.

(ii) Suppose further that $\forall W \in \mathscr{W}$ and $x \in X$ there is a $\hat{W} \in \mathscr{W}$ such that $\hat{\mathscr{W}}^2[x] \subset W[x]$. Then each $\mathscr{W}_{x_0}$ is precisely the collection of $\mathfrak{T}_{\mathscr{W}}$-nbds of $x_0 \in X$.

Proof. Fix any $x_0 \in X$. The properties of $\mathscr{W}$ assure that $\mathscr{W}_{x_0}$ satisfies N_1 and N_2 of Appendix t.1. To establish that N_3 holds, suppose $W[x_0) \subset w$ for $W \in \mathscr{W}$. Then $W \subset U = (x_0 \times w) \cup \bigcup_{x \neq x_0} (x \times W[x])$ and (since $U \in \mathscr{W}$) there follows $U[x_0] = w \in \mathscr{W}_{x_0}$. This then proves N_3 and the first part of the theorem. The condition in (ii) assures that $\hat{W}[y] \subset W[x_0]\ \forall y \in \hat{W}[x_0]$ (since $y \in \hat{W}[x_0]$ implies that $\hat{W}[y] \subset \hat{W}^2[x_0]$). Therefore N_4 and the second part of the theorem follow. ■

COROLLARY 1.11. If $\mathscr{B}$ is a base for a type uniformity $\mathscr{W} \subset X \times X$, then $\{B[x] : B \in \mathscr{B}\}$ is a nbd base for $\mathfrak{T}_{\mathscr{W}}$ on X.

Theorem 1.10 can be used to introduce

DEFINITION 1.12. The topology $\mathfrak{T}_{\mathscr{U}}$ for a uniform space $(X, \mathscr{U})$ is called the uniform topology on X. Analogous definitions apply to each type uniformity in Definition 1.6.

Unless specified otherwise, $(X, \mathscr{W})$ will always be assumed to carry $\mathfrak{T}_{\mathscr{W}}$.

Different type uniformities may determine the same topology as $\mathscr{U}_{\mathfrak{D}}$ and $\mathscr{U}_{\prec}$ in Example 1.19 each determine the discrete topology on X. Such is also true for uniformities of the same type. For instance, equivalent metrics determine the same uniform topology but not necessarily the same uniformity as demonstrated by

Example 1.13. On $X = [1, \infty)$, let $d_1(x, y) = |x - y|$ and let $\rho(x, y) = |1/x - 1/y|$. Then $h : (X, d_1) \underset{x \rightsquigarrow 1/x}{\rightarrow} (X, \rho)$ is a homeomorphism. However $\mathscr{B}_{d_1}$ and $\mathscr{B}_\rho$ are not equivalent (and so $\mathscr{U}_{d_1} \neq \mathscr{U}_\rho$) on $X \times X$, since they do not internest (Definition 1.9).† In fact, the ε-entourages ($\varepsilon > 0$) of $\mathscr{B}_{d_1}$ and $\mathscr{B}_\rho$ are given by

$$U_{\varepsilon, d_1} = \left\{ (x, y) : \begin{matrix} x > 1 \\ x - \varepsilon < y < x + \varepsilon \end{matrix} \right\}$$

and

$$U_{\varepsilon, \rho} \quad \left\{ (x, y) : \begin{matrix} x > 1 \\ \dfrac{x}{1 - \varepsilon x} < y < \dfrac{x}{1 + \varepsilon x} \end{matrix} \right\}$$

† In general, equivalence "in the small" (viz., d_1 and ρ on X or $\mathfrak{T}_{\mathscr{U}_{d_1}} = \mathfrak{T}_{\mathscr{U}_\rho}$ on X) does not imply equivalence "in the large" (viz., $\mathscr{B}_{d_1}$ and $\mathscr{B}_\rho$ on $X \times X$).

respectively. Thus,

$$U_{\varepsilon,d_1}[x]=(x-\varepsilon,\, x+\varepsilon) \quad \text{and} \quad U_{\varepsilon,\rho}[x]=\left(\frac{x}{1+\varepsilon x}, \frac{x}{1-\varepsilon x}\right) \ \forall x \in X$$

A semiuniform space $(X, \mathscr{S})$ need not be locally uniform so that property N_4 of Appendix t.1 need not hold. Thus, although every $\mathfrak{T}_{\mathscr{S}}$-nbd of $x_0 \in X$ is a member of $\mathscr{S}_{x_0}=\{S[x_0]: S \in \mathscr{S}\}$, the converse is not necessarily true and this complicates matters considerably. Put another way, $\xi_{\mathscr{S}}: \mathfrak{P}(X) \longrightarrow \mathfrak{P}(X)$, $A \rightsquigarrow \{x: S[x] \cap A \neq \varnothing \ \forall S \in \mathscr{S}\}$ satisfies $(K_1)-(K_3)$, but not necessarily (K_4), in Appendix t.2. Therefore $\xi_{\mathscr{S}}(A) \subset \bar{A}^{\mathfrak{T}_{\mathscr{S}}}$ (without equality being assured) $\forall$ nonclosed subset $A \subset X$. Such complications, of course, do not occur for quasi- or locally uniform spaces $(X, \mathscr{W})$, since (ii) of Theorem 1.10 is satisfied and $\xi_{\mathscr{W}}$ is precisely $\mathfrak{T}_{\mathscr{W}}$-closure.

Despite the above storm warnings, a little caution and the use of $\xi_{\mathscr{S}}$ instead of $\mathfrak{T}_{\mathscr{S}}$ enables us to proceed. The passage will be made smoother by introducing

Definition 1.14. By a topological semiuniformity $\mathscr{S}$ for X (abbreviated, $t \cdot$ semiuniformity) is meant an $\mathscr{S} \in \mathfrak{S}(X)$ such that $\xi_{\mathscr{S}}=\mathfrak{T}_{\mathscr{S}}$-closure on X. The class of $t \cdot$ semiuniformities on $X \times X$ is denoted $t \cdot \mathscr{S}(X)$. Note that $\mathfrak{Q}(X) \subset \mathfrak{L}(X) \subset t \cdot \mathfrak{S}(X)$.

A simple calculation shows that

$$W \circ M \circ V = \bigcup_{(x,y) \in M} W^{-1}[x] \times V[y] \quad \forall W, V, M \subset X \times Y.$$

If $(X, \mathscr{S})$ and $(Y, \mathscr{R})$ are semiuniform spaces, define

$$(\xi_{\mathscr{S}} \times \xi_{\mathscr{R}})(M)=\{(x, y) \in X \times Y: (S[x] \times R[y]) \cap M \neq \varnothing \ \forall (S, R) \in \mathscr{S} \times \mathscr{R}\}.$$

We can now go on to

Theorem 1.15. Let $M \subset X \times Y$, where $(X, \mathscr{W})$ and $(Y, \mathscr{V})$ are quasi-{semi-} uniform spaces. Then

(i) $\operatorname{Int} M=\{(x, y) \in M: W[x] \times V[y] \subset M$ for some $(W, V) \in \mathscr{W} \times \mathscr{V}\}$,

(ii) $\bar{M}= \bigcap_{\mathscr{W} \times \mathscr{V}} W \circ M \circ V^{-1} \left\{(\xi_{\mathscr{W}} \times \xi_{\mathscr{V}})(M)= \bigcap_{\mathscr{W} \times \mathscr{V}} W \circ M \circ V^{-1}\right\}$.

Proof. That (i) holds is clear, since $\operatorname{Int} M$ consists of all interior points of M. For (ii) just note that $(x_0, y_0) \in (\xi_{\mathscr{W}} \times \xi_{\mathscr{V}})(M)$ iff

$$(x_0, y_0) \in \bigcup_{M} W^{-1}[x] \times V[y] \ \forall (W, V) \in \mathscr{W} \times \mathfrak{B}:$$

that is, $(x_0, y_0) \in W \circ M \circ V^{-1}\ \forall (W, V) \in \mathscr{W} \times \mathscr{V}$. If $\mathscr{W}$, $\mathscr{V}$ are quasi-uniformities, then $\bar{M} = (\xi_{\mathscr{W}} \times \xi_{\mathscr{V}})(M)$ so that (ii) also holds. ■

COROLLARY 1.16. Let $M \subset X \times X$, where $(X, \mathscr{V})$ is a quasi {semi-} uniform space.

(i) $\operatorname{Int} M = \{(x, y) \in M : V[x] \times V[y] \subset M \text{ for some } V \in \mathscr{V}\}$,

(ii) $\bar{M} = \bigcap_{\mathscr{V}} V \circ M \circ V^{-1} \left\{ \xi_{\mathscr{V}} \times \xi_{\mathscr{V}})(M) = \bigcap_{\mathscr{V}} V \circ M \circ V^{-1} = \bigcap_{V \in \mathscr{S}_{\text{sym}}} V \circ M \circ V \right\}$.

Proof. If $V_1[x] \times V_2[y] \subset M$ for $V_1, V_2 \in \mathscr{V}$, then $V = V_1 \cap V_2 \in \mathscr{V}$ satisfies $V[x] \times V[y] \subset M$. Clearly, $(\xi_{\mathscr{V}} \times \xi_{\mathscr{V}})(M) = \bigcap_{\mathscr{V} \times \mathscr{V}} V_1 \circ M \circ V_2^{-1} \subset \bigcap_{\mathscr{V}} V \circ M \circ V^{-1}$. On the other hand, $(V_1 \cap V_2) \circ M \circ (V_1 \cap V_2)^{-1} \subset V_1 \circ M \circ V_2^{-1}\ \forall V_1, V_2 \in \mathscr{V}$ assures that $\bigcap_{\mathscr{V}} V \circ M \circ V^{-1} \subset (\xi_{\mathscr{V}} \times \xi_{\mathscr{V}})(M)$. Thus $(\xi_{\mathscr{V}} \times \xi_{\mathscr{V}})(M) = \bigcap_{\mathscr{V}} V \circ M \circ V^{-1}$ which (by Definition 1.8$^+$) equals $\bigcap_{V \in \mathscr{S}_{\text{sym}}} V \circ M \circ V$. ■

Corollary 1.16 can be used to illustrate the added scope a uniformity has over its semiuniform or quasi-uniform component.

COROLLARY 1.17. The symmetric open (closed) entourages of a uniformity $\mathscr{U}$ constitutes a base for $\mathscr{U}$.

Proof. Given any $U \in \mathscr{U}$, choose a symmetric $V \in \mathscr{U}$ such that $V^3 \subset U$. Then $\bar{V} \subset U$ and $V \subset \operatorname{Int} U$; therefore $\operatorname{Int} V \subset U$. Specifically, $\bar{V} = \bigcap_{\mathscr{U}} W \circ V \circ W^{-1} \subset V \circ V \circ V^{-1} = V^3$ and $U \supset V^3 = \bigcup_{(x,y) \in V} V[x] \times V[y]$ by Theorem 1.15$^-$. It remains only to note that $\operatorname{Int} V$ and $\bar{V}$ are also symmetric for symmetric $V \in \mathscr{U}$. ■

COROLLARY 1.18. Let $A \subset X$, where $(X, \mathscr{V})$ is a quasi- {semi-} uniform space. Then $\bar{A} = \bigcap_{\mathscr{V}} V^{-1}[A] \left\{ \xi_{\mathscr{V}}(A) = \bigcap_{\mathscr{V}} V^{-1}[A] = \bigcap_{\mathscr{V}_{\text{sym}}} V[A] \right\}$ and $\operatorname{Int} A = \{a \in A : V[a] \subset A \text{ for some } V \in \mathscr{V}\}$.

Proof. Clearly $a \in \xi_{\mathscr{V}}(A)$ iff $V[a] \cap A \neq \varnothing\ \forall V \in \mathscr{V}$ iff $a \in V^{-1}[A]\ \forall V \in \mathscr{V}$, which means $a = \bigcap_{\mathscr{V}} V^{-1}[A]$. For the second assertion fix $x_0 \in X$. Then $s_{x_0} : (X, \mathfrak{T}_{\mathscr{V}}) \to (x_0 \times X, x_0 \times \mathfrak{T}_{\mathscr{V}})$, $x \rightsquigarrow (x_0, x)$ is a homeomorphism and $\operatorname{Int} A = s_{x_0}^{-1}\{x_0 \times \operatorname{Int} A\} = s_{x_0}^{-1}\{\operatorname{Int}(x_0 \times A)\} = \{a \in A : V[a] \subset A \text{ for some } V \in \mathscr{V}\}$. ■

Corollary 1.18 provides the basis for

THEOREM 1.19. If $(X, \mathcal{V})$ is a quasi-{semi-} uniform space, then (i)–(v) {(i)′ and (ii)} are equivalent:

(i) $\mathfrak{T}_{\mathcal{V}}$ is T_k $(k=0,1,2)$;
(i)′ $\mathfrak{T}_{\mathcal{V}}$ is T_k $(k=0,1)$;
(ii) $\Delta=\bigcap_{\mathcal{V}} V\left(=\bigcap_{\mathcal{B}} B, \text{ where } \mathcal{B} \text{ is a base for } \mathcal{V}\right)$;
(iii) Δ is closed in $X\times X$;
(iv) $\mathfrak{T}_{\mathcal{V}}$-compact subsets of X are closed;
(v) for each pair of disjoint compact sets $K_i\neq\varnothing$, $(i=1,2)$ there exist open sets $U_i\supset K_i$ such that $K_i\cap U_j=\varnothing$ for $i\neq j$ $(i,j=1,2)$.

Proof. Suppose $\mathfrak{T}_{\mathcal{V}}$ is T_0. If $x\neq y$, some $\mathfrak{T}_{\mathcal{V}}$-nbd of y does not contain x. In particular $x\notin V_0[y]$ and $(x,y)\notin V_0$ for some $V_0\in\mathcal{V}$ so that $\bigcap_{\mathcal{V}} V=\Delta$. This, in turn, implies that $\xi_{\mathcal{V}}(x)=\{x\}\ \forall x\in X$; that is, $\mathfrak{T}_{\mathcal{V}}$ is T_1 (therefore T_0) if $\mathcal{V}$ is a semiuniformity, and $\{\bar{x}\}$ is the intersection of all $\mathfrak{T}_{\mathcal{V}}$-nbds; that is, $\mathfrak{T}_{\mathcal{V}}$ is T_2 (therefore T_0) if $\mathcal{V}$ is a quasi-uniformity. This then establishes that (i) and (ii) {(i)′ and (ii)} are equivalent. Assume now that $\mathcal{V}$ is a quasi-uniformity for X. Then (iii) is equivalent to (i) (specifically, to $\mathfrak{T}_{\mathcal{V}}$ being T_2) and (since compact subsets in a T_2 space are closed) (i)$\Leftrightarrow$(iv). Clearly, (v)$\Rightarrow$(iv) (each point in the complement of a compact set K is an interior point of $X-K$) and it remains only to verify that (iv)$\Rightarrow$(v). For each $x\in K_2$ let $u_x\subset X-K_1$ be an open nbd of x. Since $\{u_x : x\in K_2\}$ covers K_2, some open set $u_2=\bigcup_{i=1}^{N} u_{x_i}$ covering K_2 is disjoint from K_1. A similar argument on K_1 completes the proof. ■

DEFINITION 1.20. A topological space $(X, \mathfrak{T})$ is said to be uniformizible {quasi-uniformizible, $t\cdot$semiuniformizible, locally uniformizible} if $\mathfrak{T}_{\mathcal{W}}=\mathfrak{T}$ for some uniformity {quasi-uniformity, $t\cdot$semiuniformity, local uniformity} $\mathcal{W}$ on $X\times X$.

The above is also characterized by saying that $\mathcal{W}$ is compatible with $\mathfrak{T}$.

Remark. Note that $(X,\mathfrak{T})$ is $t\cdot$semiuniformizible iff $\xi_{\mathcal{S}}=\mathfrak{T}$-closure on X for some semiuniformity $\mathcal{S}\in X\times X$. Every $\mathcal{S}\in t\cdot\mathbb{S}(X)$ obviously has $\xi_{\mathcal{S}}$ as its $\mathfrak{T}_{\mathcal{S}}$-closure on X. If $\xi_{\mathcal{S}}=\mathfrak{T}$-closure for $\mathcal{S}\in\mathbb{S}(X)$, then $\xi_{\mathcal{S}}\{\xi_{\mathcal{S}}(A)\}=\xi_{\mathcal{S}}(A)\ \forall A\subset x$ and (Appendix t.2) $\xi_{\mathcal{S}}=\mathfrak{T}_{\mathcal{S}}-$closure; that is, $\mathfrak{T}_{\mathcal{S}}=\mathfrak{T}$ for $\mathcal{S}\in t\cdot\mathbb{S}(X)$.

If $\mathcal{S}$, $\mathcal{R}$ are $t\cdot$semiuniformities in Definition 1.14, then $\bar{M}=(\xi_{\mathcal{S}}\times\xi_{\mathcal{R}})(M)$ in the product topology on $X\times Y$. Thus quasi-uniformities and $t\cdot$semiuniformities have the same properties in Theorems 1.15 and 1.19. The distinguishing feature of such uniformities is encompassed in

THEOREM 1.21. A topological space $(X, \mathfrak{T})$ is

(i) quasi-uniformizible;
(ii) locally uniformizible iff it is regular;
(iii) $t\cdot$semiuniformizible iff $x \in \overline{\{y\}} \Leftrightarrow y \in \overline{\{x\}}\ \forall x, y \in X$;
(iv) uniformizible iff it is completely regular.

Proof. (i) For $A \subset X$ define $Q_A = (A \times A) \cup [(X-A) \times X]$ and observe that $\Delta \in Q_A = Q_A^2$. To be sure $Q_A \subset Q_A^2$ by Theorem 1.3(vi). On the other hand, $(a, y) \in Q_A^2$ implies $y \in A$ [otherwise (a, z) and (z, y) belong to Q_A and $z \in A \cap X - A$ for some $z \in X$] and $Q_A^2 \subset Q_A$ follows. If $\mathfrak{N}$ is a $\mathfrak{T}$-nbd base, then $\mathscr{B} = \left\{ \bigcap_{i=1}^{n} \mathfrak{Q}_{A_i} : A_i \in \mathfrak{N} \right\}$ is the base for a quasi-uniformity $\mathfrak{Q}$ compatible with $\mathfrak{T}$. Verification: For $u \in \mathfrak{T}$ and $x \in u$ one has $u = Q_u[x] \in \mathfrak{T}_{\mathfrak{Q}}$, from which follows $\mathfrak{T} \subset \mathfrak{T}_{\mathfrak{Q}}$. If $v \in \mathfrak{T}_{\mathfrak{Q}}$ and $x \in v$, then $x \in \left\{ \bigcap_{i=1}^{n} Q_{A_i} \right\}[x] \subset v$ for $\{A_1, \ldots, A_n\} \in \mathfrak{N}$. Since $\left(\bigcap_{i=1}^{n} Q_{A_i} \right)[x] = \bigcap_{i=1}^{n} A_i \in \mathfrak{N}$, we have $v \in \mathfrak{T}$ and $\mathfrak{T}_{\mathfrak{Q}} \subset \mathfrak{T}$ as required.

(ii) Suppose $(X, \mathfrak{L})$ is a locally uniform space. If $L \in \mathfrak{L}$, then $L[x]$ is a $\mathfrak{T}_{\mathfrak{L}}$-nbd of $x \in X$ and $L_0^2[x] \subset L[x]$ for some $L_0 \in \mathfrak{L}$. In particular, (Corollary 1.18) $\overline{L_0[x]} = \bigcap_{L' \in \mathfrak{L}} L'[L_0[x]] \subset L_0^2[x] \subset L[x]$, thereby proving that $\mathfrak{T}_{\mathfrak{L}}$ is regular. Conversely, suppose $(X, \mathfrak{T})$ is regular and let $\mathscr{B}$ denote the collection of all sets of the form $\bigcup_{\mathscr{A}} (A_\alpha \times A_\alpha)$ for $\mathfrak{T}$-open covers $\{A_\alpha : \alpha \in \mathscr{A}\}$ of X. Then $\Delta \in B = B^{-1}\ \forall B \in \mathscr{B}$. If $L = \bigcup_{\mathscr{A}} (A_\alpha \times A_\alpha)$ is a member of $\mathscr{B}$ and $x \in X$, then $L[x] = \bigcup_{x \in A_\alpha} A_\alpha \in \mathfrak{T}$. Let $u, v \in \mathfrak{T}$ satisfy $x \in \bar{v} \subset u \subset \bar{u} \subset L[x]$. Then $L_0 = (u \times u) \cup \{L[x] - v \times L[x] - v\} \cup \{X - \bar{u} \times X - \bar{u}\}$ is a member of $\mathscr{B}$ satisfying $L_0^2[x] = L[u] \subset L[x]$. The filter $\mathfrak{L}$ determined by $\mathscr{B}$ is therefore a local uniformity, which we now show to be compatible with $\mathfrak{T}$. Clearly, $L[x] \in \mathfrak{T}\ \forall L \in \mathfrak{L}$ assures that $\mathfrak{T}_{\mathfrak{L}} \subset \mathfrak{T}$. For the reverse inclusion consider any $u \in \mathfrak{T}$ and $x \in u$. Pick $v, w \in \mathfrak{T}$ such that $x \in w \subset \bar{w} \subset v \subset \bar{v} \subset u$ and let $L = (v \times v) \cup \{u - \bar{w} \times u - \bar{w}\} \cup \{X - \bar{v} \times X - \bar{v}\}$. Then $L \in \mathfrak{L}$ and $L[x] = v \subset u$, thus yielding $\mathfrak{T} \subset \mathfrak{T}_{\mathfrak{L}}$.

(iii) If $\mathfrak{T}_{\mathscr{S}} = \mathfrak{T}$ for some $\mathscr{S} \in t\cdot\mathfrak{S}(X)$, then (Corollary 1.17) $x \in \overline{\{y\}} = \bigcap_{\mathscr{S}_{\text{sym}}} S[y]$ iff $y \in \bigcap_{\mathscr{S}_{\text{sym}}} S[x] = \overline{\{x\}}$. Assume, conversely, that $x \in \overline{\{y\}}$ iff $y \in \overline{\{x\}}\ \forall x, y \in X$. Let u_x be a $\mathfrak{T}$-nbd of $x\ \forall x \in X$, take $R = \bigcup_X (x \times u_x)$, and let $S = R \cup R^{-1}$. The collection $\mathscr{B}$ of all such $S \subset X \times X$ is the base for an $\mathscr{S} \in \mathfrak{S}(X)$, which we claim determines the $\mathfrak{T}$-closure operation. That $\bar{A}^{\mathfrak{T}} \subset \xi_{\mathscr{S}}(A)$ is clear, since every $S[x]$ $(S \in \mathscr{B})$ is some $u_x \in \mathfrak{T}$. Now fix $x \in X$ and let

$w \in \mathfrak{T}$ contain x. Choose a cover of X by $\mathfrak{T}$-nbds $\{v_y(y): y \in X\}$ such that $v_x \subset w$ and $x \in X - v_y$ for $y \notin \{\bar{x}\}$. If $R = \bigcup_X (y \times v_y)$, then $S = R \cup R^{-1}$ belongs to $\mathscr{B}$ and $v_x = S[x]$. (Surely, $v_x \subset S[x]$. If $y \in S[x] - v_x = R[x] \cup R^{-1}[x] - v_x = R^{-1}[x]$, then $x \in R[y] = v_y$ and $y \in \{\bar{x}\}$ so that $x \in \{\bar{y}\}$ and $y \in v_x$, which is impossible.) The conclusion $S[x] \subset w$ further establishes that $\xi_{\mathscr{S}}(A) \subset \bar{A}^{\mathfrak{T}}\ \forall A \subset X$.

(iv) The proof will be encompassed in Corollary 4.3. ■

Remark. The local uniformity determined by the $\mathfrak{T}$-nbds of a regular, noncompletely regular space $(X, \mathfrak{T})$ cannot be a uniformity [Example 1.9(c)].

Remark. The proofs of (ii), (iii) are attributed to G. D. Richardson (unpublished) and Cech [16], respectively.

Uniform Type Continuity

Every mapping $f: X \to Y$ induces a new mapping

$$f \times f: \underset{(x_1, x_2) \rightsquigarrow (f(x_1), f(x_2))}{X \times X \longrightarrow Y \times Y}.$$

We use this notation in

DEFINITION 1.22. A mapping $f: (X, \mathscr{U}) \to (Y, \mathscr{V})$ between uniform spaces $(X, \mathscr{U})$ and $(Y, \mathscr{V})$ is called uniformly continuous if any of the following equivalent conditions hold:

(i) For each $V \in \mathscr{V}$ there is a $U \in \mathscr{U}$ such that $(f \times f)U \subset V$; that is $(f(x_1), f(x_2)) \subset V\ \forall (x_1, x_2) \in U$.

(ii) $\{(x_1, x_2): (f(x_1), f(x_2)) \subset V\} = (f \times f)^{-1} V \in \mathscr{U}\ \forall V \in \mathscr{V}$.

(iii) $(f \times f)^{-1} \mathscr{V} = \{(f \times f)^{-1} V: V \in \mathscr{V}\} \subset \mathscr{U}$.

(iv) Given any $V \in \mathscr{V}$, there is a $U \in \mathscr{U}$ satisfying $f\{U[x]\} \subset V[f(x)]\ \forall x \in X$.

One calls f above a unimorphism if it is bijective and both f, f^{-1} are uniformly continuous.

Analogous definitions are assumed for each pair of uniformities of the same category in Definition 1.6. For instance, f is semiuniformly continuous if the above holds for semiuniformities $\mathscr{U}, \mathscr{V}$. By a semiunimorphism is meant a bijection f for which f, f^{-1} are semiuniformly continuous.

Remark. The composition of two uniform-type continuous functions $f: (X, \mathscr{W}) \to (Y, \mathscr{V})$ and $g: (Y, \mathscr{V}) \to (z, \mathscr{U})$ have the same uniform type continuity since $(gf \times gf)^{-1} U = (f \times f)^{-1}\{(g \times g)^{-1} U\} \in \mathscr{W}\ \forall U \in \mathscr{U}$.

THEOREM 1.23. Every quasi- $\{t \cdot \text{semi-}\}$ uniformly continuous function $f: (X, \mathscr{W}) \to (Y, \mathscr{V})$ is $\mathfrak{T}_{\mathscr{W}} - \mathfrak{T}_{\mathscr{V}}$ continuous.

Proof. Given any $x_0 \in X$ and $\mathfrak{T}_{\mathscr{V}}$-base nbd $V[f(x_0)]$ of $f(x_0) \in Y$, one has $f\{W[x_0]\} \subset V[f(x_0)]$ for some $W \in \mathscr{W}$. Since $W[x_0]$ is a $\mathfrak{T}_{\mathscr{W}}$-nbd of x_0, the continuity of f at $x_0 \in X$ is proved. ■

Remark. The converse of Theorem 1.23 is not true in general. The homeomorphism $h: ([0, \infty), d_1) \to ([0, \infty), d_1)$, $x \rightsquigarrow x^3$, is not uniformly continuous.

The following point is important. Although continuity is a topologically dependent concept, uniform type continuity of $f: (X, \mathscr{W}) \to (Y, \mathscr{V})$ depends only on the particular entourages of $\mathscr{W}$, $\mathscr{V}$ and not on the topologies $\mathfrak{T}_{\mathscr{W}}$, $\mathfrak{T}_{\mathscr{V}}$.

Example 1.24. Let $(X, <)$ be an ordered set. For each pair $x, y \in X$ define $\rho(x, x) = 0$ and $\rho(x, y) = 1$ if $x \neq y$. Then (Example 1.9) $\mathfrak{T}_{\mathscr{U}_<} = \mathfrak{T}_{\mathscr{U}_\rho} = \mathfrak{D}$ on X and $\mathbb{1}: (X, U_<) \to (X, U_\rho) = (X, \mathfrak{D})$ is a nonuniformly continuous homeomorphism. (The nonuniform continuity of $\mathbb{1}$ follows from the fact that for $\varepsilon < 1$, $U_\varepsilon = \Delta \in U_{\mathfrak{D}} - U_<$ and no $U \in U_<$ satisfies $U[x] = \mathbb{1}\{U[x]\} \subset U_\varepsilon[x] = x\ \forall x \in X$.)

Corollary 1.17 affords the following partial converse to Theorem 1.23.

THEOREM 1.25. Suppose $(Y, \mathscr{V})$ is a uniform space and $(X, \mathscr{W})$ is a compact uniform space (i.e., $(X, \mathfrak{T}_{\mathscr{W}})$ is compact). Then every $\mathfrak{T}_{\mathscr{W}} - \mathfrak{T}_{\mathscr{V}}$ continuous function $f: (X, \mathscr{W}) \to (Y, \mathscr{V})$ is uniformly continuous.

Proof. Given any $V \in \mathscr{V}$, choose a symmetric $\tilde{V} \in \mathscr{V}$ such that $\tilde{V}^2 \subset V$. For each $x \in X$ there is some $W_x \in \mathscr{W}$ satisfying $f\{W_x[x]\} \subset \tilde{V}[f(x)]$. Choose symmetric open entourages $\tilde{W}_x \in W$ satisfying $\tilde{W}_x^2 \subset W_x\ \forall x \in X$. Then $X = \bigcup_{i=1}^{n} \tilde{W}_{x_i}[x_i]$ and $\tilde{W} = \bigcap_{i=1}^{n} \tilde{W}_{x_i} \in \mathscr{W}$ for some finite set $\{x_1, \ldots, x_n\}$. We claim that $(f \times f)\tilde{W} \subset V$, from which follows f's uniform continuity. Clearly, $(x, y) \in \tilde{W}$ implies $(x, y) \in \tilde{W}_{x_i}\ \forall i = 1, 2, \ldots, n$. Since y belongs to some such $\tilde{W}_{x_{i'}}[x_{i'}] \subset W_{x_{i'}}[x_{i'}]$, we have $(y, x_{i'}) \in \tilde{W}_{x_{i'}}$ and $(x, x_{i'}) \in \tilde{W}^2_{x_{i'}} \subset W_{x_{i'}}$; therefore $x \in W_{x_{i'}}[x_{i'}]$. Thus $(f \times f)(x, y) \in f\{W_{x_{i'}}[x_{i'}]\} \times f\{W_{x_{i'}}[x_{i'}]\} \subset V[f(x_{i'})] \times V[f(x_{i'})] \subset V$ as asserted. ■

An especially useful consequence of Theorem 1.25 is

THEOREM 1.26. There exists exactly one uniformity $\mathscr{U}_F$ compatible with a compact, completely regular space $(X, \mathfrak{T})$. If $(X, \mathfrak{T})$ is compact and T_2, then $\mathscr{U}_F$ has {open sets $V \subset X \times X: \Delta \in V$} as a base.

Proof. If $\mathfrak{T}_{\mathcal{W}} = \mathfrak{T} = \mathfrak{T}_{\mathcal{V}}$ for $\mathcal{W}, \mathcal{V} \in \mathfrak{U}(X)$, the homeomorphism $\mathbb{1}: (X, \mathcal{W}) \to (X, \mathcal{V})$ is a unimorphism. This, together with Theorem 1.21, establishes the existence of $\mathcal{U}_F$. If $(X, \mathfrak{T})$ is compact and T_2, so is $X \times X$ in the product topology. Therefore for any open entourage $U \in \mathcal{U}_F$ there is an open set $V \subset X \times X$ satisfying $\Delta \in V \subset U$. We claim that $V \in \mathcal{U}_F$. Verification is based on the fact (Corollary 1.17, Theorem 1.19) that $\bigcap_{\mathcal{B}} B = \Delta \in V$, where $\mathcal{B}$ denotes the collection of closed symmetric entourages of $\mathcal{U}_F$. Clearly $V \notin \mathcal{U}_F$ implies $\bigcap_{i=1}^{n} \{U_i \cap (X \times X - V)\} = \left(\bigcap_{i=1}^{n} U_i\right) \cap (X \times X - V) \neq \varnothing$ for all finite sets $\{U_1, \ldots, U_n\} \in \mathcal{B}$. Since $\{U \cap (X \times X - V): U \in \mathcal{B}\}$ is a collection of closed sets with fip (Appendix t.7), one has $\varnothing \neq \bigcap_{\mathcal{B}} \{U \cap (X \times X - V)\} = \Delta \cap (X \times X - V)$—which is impossible. ∎

Remark. Compactness cannot be replaced by local compactness. For instance, $(\mathbb{R}, d_1)$ has more than one compatible uniformity.

Remark. The converse of Theorem 1.26 is clearly not true, since $\mathcal{U}_F = \mathcal{U}_{\mathfrak{I}}$ for the indiscrete space $(X, \mathfrak{I})$.

Definition 1.22, Theorem 1.23, partially extend (pseudo)metric-dependent concepts to more general topological spaces. Quasi-uniform {semiuniform} spaces lack the generalized symmetry {triangle inequality} property and so cannot be considered the most natural generalization of a (pseudo)metric space (see problems 1A, B). Uniform spaces, however, are well suited for this purpose (Example 1.1) and it is for this reason that we emphasize such spaces in the sections that follow.

Problems

1A. Pseudometric Generalizations

A real-valued, nonnegative function ρ on $X \times X$ satisfying $\rho(x, x) = 0$ is called a semipseudometric (spm) if $\rho(x, y) = \rho(y, x)\ \forall x, y \in X$; and a quasi-pseudometric (qpm) if $\rho(x, z) \leq \rho(x, y) + \rho(y, z)\ \forall x, y, z \in X$. Thus ρ is a pseudometric if it is both an spm and qpm on X.

(a) Let $W_\varepsilon = \{(x, y) \in X \times X : \rho(x, y) < \varepsilon\}$ and take $\mathcal{W}_\rho = \{W_\varepsilon : \varepsilon > 0\}$. If ρ is an spm, then $\mathcal{W}_\rho[x_0] = \{W_\varepsilon[x_0] : \varepsilon > 0\}$ is a base for the ρ-nbds of $x_0 \in X$. In all other cases $\mathcal{W}_\rho[x_0]$ is a ρ-nbd base at $x_0 \in X$.

(b) If ρ is an spm {qpm} on X, then $\mathcal{W}_\rho$ is the base for a semiuniformity {quasi-uniformity} on $X \times X$ compatible with $\mathfrak{T}(\rho)$.

1B. Quasi-Uniform Spaces

(a) If $\mathfrak{Q}$ is a quasi-uniformity for X, so is $\mathfrak{Q}^{-1}=\{Q^{-1}:Q\in\mathfrak{Q}\}$. Suppose $\mathcal{B}$ is a base for $\mathfrak{Q}$ on $X\times X$. Then $\mathcal{B}^{-1}$ is a base for $\mathfrak{Q}^{-1}$. Moreover, $\mathcal{B}^n=\{B^n:B\in\mathcal{B}\}[\mathcal{B}^{-n}=\{B^{-n}:B\in\mathcal{B}\}]$ is a base for $\mathfrak{Q}$ [for $\mathfrak{Q}^{-1}$] $\forall n\in\mathbb{N}$.

(b) Let $\mathfrak{Q}_1, \mathfrak{Q}_2\in\mathfrak{Q}(X)$. Then $M\subset X\times X$ is $\mathfrak{T}_{\mathfrak{Q}_1}\times\mathfrak{T}_{\mathfrak{Q}_2}$-open (closed) iff M^{-1} is $\mathfrak{T}_{\mathfrak{Q}_2}\times\mathfrak{T}_{\mathfrak{Q}_1}$-open (closed). [*Hint*: $r(x, y)=(y, x)$ defines a homeomorphism between $(X\times X, \mathfrak{T}_{\mathfrak{Q}_1}\times\mathfrak{T}_{\mathfrak{Q}_2})$ and $(X\times X, \mathfrak{T}_{\mathfrak{Q}_2}\times\mathfrak{T}_{\mathfrak{Q}_1})$.]

(c) Let $Q\in\mathfrak{Q}$, where $\mathfrak{Q}\in\mathfrak{Q}(X)$. Then Int $Q\in\mathfrak{Q}$, where interior is taken relative to $\mathfrak{T}_{\mathfrak{Q}^{-1}}\times\mathfrak{T}_{\mathfrak{Q}}$ on $X\times X$. In particular, the $\mathfrak{T}_{\mathfrak{Q}^{-1}}\times\mathfrak{T}_{\mathfrak{Q}}$-open entourages of $\mathfrak{Q}$ form a base for $\mathfrak{Q}$. [*Hint*: Consider Theorem 1.15 and (b) above.]

(d) The $\mathfrak{T}_{\mathfrak{Q}}\times\mathfrak{T}_{\mathfrak{Q}^{-1}}$-closed entourages of a quasi-uniformity $\mathfrak{Q}$ form a base for $\mathfrak{Q}$.

(e) If $\mathfrak{T}_{\mathfrak{Q}}\subset\mathfrak{T}_{\mathfrak{Q}^{-1}}$ on X for $\mathfrak{Q}\in\mathfrak{Q}(X)$, then $(X, \mathfrak{T}_{\mathfrak{Q}^{-1}})$ is completely regular; that is, uniformizible. [*Hint*: Show that $\mathfrak{T}_{\mathfrak{Q}^{-1}}=\mathfrak{T}_{\mathcal{W}}$, where $\mathcal{W}=\sup\{\mathfrak{Q}, \mathfrak{Q}^{-1}\}$.] This does not necessarily mean that $\mathfrak{Q}$ is a uniformity for X. Why not?

(f) Show for $\mathfrak{Q}_1, \mathfrak{Q}_2\in\mathfrak{Q}(X)$, that $\mathfrak{T}_{\mathfrak{Q}_1}=\mathfrak{T}_{\mathfrak{Q}_2}$ on X does not imply $\mathfrak{T}_{\mathfrak{Q}^{-1}}=\mathfrak{T}_{\mathfrak{Q}^{-1}}$ on X.

(g) A topological space $(X, \mathfrak{T})$ is quasi-pseudometrizible if there is a quasi-pseudometric ρ on X whose topology $\mathfrak{T}(\rho)$ coincides with $\mathfrak{T}$. Every quasi-pseudometric space is quasi-uniformizible (Problem 1A). Prove that a quasi-uniform space $(X, \mathfrak{Q})$ is quasi-pseuodometrizible if (therefore iff!) $\mathfrak{Q}$ has a countable base.

(h) In view of Theorem 1.21(i), does (g) assert that a topological space $(X, \mathfrak{T})$ is quasi-pseudometrizible iff $\mathfrak{T}$ is 1°?

(i) A quasi-pseudometric space (X, ρ) is T_0 iff $\rho(x, y)+\rho(y, x)=0\Rightarrow x=y$ $\forall x, y\in X$. [Compare this with Corollary 1.19 using problem 1A(a).]

1C. Locally Uniform Spaces

Assume throughout that $(X, \mathfrak{L})$ is a locally uniform space.

(a) Suppose $L\subset\mathcal{S}$, where $\mathcal{S}\in\mathfrak{S}(X)$ satisfies $\mathfrak{T}_{\mathfrak{L}}=\mathfrak{T}_{\mathcal{S}}$ on X. Then $\mathcal{S}\in\mathfrak{L}(X)$.

(b) For each $n\in\mathbb{N}$, confirm that $\mathfrak{N}^n=\{N\in X\times X: L^n\subset N$ for some $L\in\mathfrak{L}\}$ is a local uniformity for X such that $\mathfrak{T}_{\mathfrak{L}}=\mathfrak{T}_{\mathfrak{N}^n}$ on X.

(c) $\mathcal{B}'=\{\bar{L}:L\in\mathfrak{L}\}$ is the base for some $\mathfrak{L}'\in\mathfrak{L}(X)$ such that $\mathfrak{L}'\subset\mathfrak{L}$ and $\mathfrak{T}_{\mathfrak{L}'}=\mathfrak{T}_{\mathfrak{L}}$ on X. [*Hint*: Assume that the entourages of $\mathfrak{L}$ are symmetric. Then $\mathcal{B}'$ is the base for an $\mathcal{S}\in\mathfrak{S}(X)$. For any $\bar{L}$ and $x\in X$ let $L_0\in\mathfrak{L}$ satisfy $L_0^6[x]\subset L[x]$. Then $\bar{L}_0=\bigcap_{\mathcal{S}}\{S\circ L\circ S\}[x]\subset L_0^3[x]$ and $\bar{L}_0^2[x]\subset L_0^6[x]\subset\bar{L}[x]$ so that $\mathcal{S}\in\mathfrak{L}(X)$. Show that $\mathfrak{L}^3\subset\mathcal{S}\subset\mathfrak{L}$ (therefore $\mathfrak{T}_{\mathcal{S}}=\mathfrak{T}_{\mathfrak{L}^3}=\mathfrak{T}_{\mathfrak{L}}$).]

(d) $\mathcal{B}''=\{\text{Int } L:L\in\mathfrak{L}\}$ is the base for an $\mathfrak{L}''\in\mathfrak{L}(X)$ such that $\mathfrak{L}\subset\mathfrak{L}''$ and $\mathfrak{T}_{\mathfrak{L}}=\mathfrak{T}_{\mathfrak{L}''}$ on X.

(e) $\mathfrak{L}$ is called a nbd local uniformity (NLU) for X if every $L\in\mathfrak{L}$ is a $\mathfrak{T}_{\mathfrak{L}}\times\mathfrak{T}_{\mathfrak{L}}-$ nbd of $\Delta\in X\times X$. Show that there is a finest NLU $\mathfrak{L}_0$ on $X\times X$ such that $\mathfrak{L}_0\subset\mathfrak{L}$ and $\mathfrak{T}_{\mathfrak{L}_0}=\mathfrak{T}_{\mathfrak{L}}$ on X.

(f) Every pseudometric space (being uniformizible) is locally uniformizible. prove that $(X, \mathfrak{L})$ is pseudometrizible (i.e., $\mathfrak{T}_{\mathfrak{L}} = \mathfrak{T}(\rho)$ for a pseudometric ρ on X) if $\mathfrak{L}$ has a countable base.

(g) Two local uniformities $\mathfrak{L}_1, \mathfrak{L}_2 \in \mathfrak{L}(X)$ are said to be weakly equivalent if $\mathfrak{L}_1^n \subset \mathfrak{L}_2$ and $\mathfrak{L}_2^m \subset \mathfrak{L}_1$ for some $m, n \in \mathbb{N}$. (Thus every local uniformity $\mathfrak{L}_0$ is weakly equivalent to $\mathfrak{L}_0^n$ $\forall n \in \mathbb{N}$.) Verify that weakly equivalent local uniformities determine the same topology on X. Show further that $\mathfrak{L}$ is weakly equivalent to a local uniformity $\mathfrak{L}_0$ having an open (closed) base $\mathscr{B}_0$. [*Hint*: let $\mathfrak{L}_0$ be determined by $\mathscr{B}_0 = \{\text{Int } L^3 : L \in \mathfrak{L}\}$ in the first case and use $\mathscr{B}_0 = \{\bar{L} : L \in \mathfrak{L}\}$ for the second case.]

1D. Uniform Modification of a Semiuniformity

Assume throughout that $(X, \mathscr{S})$ is a semiuniform space.

(a) There is a largest uniformity $m\mathscr{S}$ for X (called the uniform modification of $\mathscr{S}$) satisfying $m\mathscr{S} \subset \mathscr{S}$. An entourage $S \subset \mathscr{S}$ belongs to $m\mathscr{S}$ iff there is a sequence $\{S_n\} \in \mathscr{S}$ such that $S_1 \subset S$ and $S_{n+1}^2 \subset S_n$ $\forall n \in \mathbb{N}$. [*Hint*: If $\mathscr{W} = \{$uniformities $\mathscr{U}$ for $X : \mathscr{U} \subset \mathscr{S}\}$, then $\left\{\bigcap_{i=1}^{n} U_{\alpha_i} : U_{\alpha_i} \in \mathscr{U}_{\alpha_i} \in \mathscr{W}\right\}$ is the base for a uniformity $m\mathscr{S} \subset \mathscr{S}$. For $\{S_n\} \in \mathscr{S}$ above, $(S_{n+1}^{-1})^2 \subset S_n^{-1}$ $\forall n$ and $\{U_n = S_n \cap S_n^{-1} : n \in \mathbb{N}\} \in \mathscr{S}$ satisfies $U_1 \subset S$ and $U_{n+1}^2 \subset U_n$ $\forall n \in \mathbb{N}$.]

(b) Let $\mathscr{U}$ be a uniformity for X. Then $\mathscr{U} = m\mathscr{S}$ iff for any uniform space $(Y, \mathscr{V})$, semiuniform continuity of $f : (X, \mathscr{S}) \to (Y, \mathscr{V})$ is equivalent to uniform continuity of $f : (X, \mathscr{U}) \to (Y, \mathscr{V})$.

(c) If $f : (X_1, \mathscr{S}_1) \to (X_2, \mathscr{S}_2)$ is semiuniformly continuous, then $f : (X_1, m\mathscr{S}_1) \to (X_2, m\mathscr{S}_2)$ is uniformly continuous.

Remark. A full treatment of $m\mathscr{S}$ is given in Cech [16].

1E. More on Uniform Type Continuity

(a) Verify that conditions (i)–(iv) in Definition 1.22 are equivalent.

(b) If $\mathscr{B}_2$ is a base (subbase) for $\mathscr{U}_2$, then Definition 1.22(ii) is equivalent to requiring that $(f \times f)^{-1}\mathscr{B}_2 \subset \mathscr{U}_1$.

(c) Let $f : (X_1, \mathfrak{Q}_1) \to (X_2, \mathfrak{Q}_2)$ be quasi-uniformly continuous. Then $f : (X_1, \mathfrak{Q}_1^{-1}) \to (X_2, \mathfrak{Q}_2^{-1})$ and $f : (X_1 \bar{\bar{\mathfrak{Q}}}_1) \to (X_2, \bar{\bar{\mathfrak{Q}}}_2)$ are quasi-uniformly continuous. [$\bar{\bar{\mathfrak{Q}}}_i$ denotes $\sup\{\mathfrak{Q}_i, \mathfrak{Q}_i^{-1}\}$, $(i = 1, 2)$.] Suppose $\mathfrak{Q}_1$ is a uniformity for X_1. Then $f : (X_1, \mathfrak{Q}_1) \to (X_2, \mathfrak{Q}_2)$ is quasi-uniformly continuous iff $f : (X_1, \bar{\bar{\mathfrak{Q}}}_1) \to (X_2, \bar{\bar{\mathfrak{Q}}}_2)$ is.

1F. Supremum and Infimum of Uniformities

(a) The coarsest uniformity for X which is finer than each member of $\{\mathscr{U}_\alpha \in \mathfrak{U}(X) : \alpha \in \mathfrak{A}\}$ is called the $\sup_{\mathfrak{A}} \mathscr{U}_\alpha$. Prove the following:

(i) If $\mathscr{B}_\alpha$ is a base for $\mathscr{U}_\alpha$ $\forall \alpha \in \mathfrak{A}$, then $\left\{\bigcap_{i=1}^{n} U_{\alpha_i} : U_{\alpha_i} \in \mathscr{U}_{\alpha_i} \text{ and } \alpha_i \in \mathfrak{A}\right\}$ is a base for $\sup_{\mathfrak{A}} \mathscr{U}_\alpha \in \mathfrak{U}(X)$.

(ii) $\mathfrak{T}_{\sup \mathscr{U}_\alpha} = \sup_{\mathfrak{a}} \{\mathfrak{T}_{\mathscr{U}_\alpha}\}$ (i.e., the uniform topology determined by $\sup_{\mathfrak{a}} \mathscr{U}_\alpha$ is the coarsest topology finer than every $\mathfrak{T}_{\mathscr{U}_\alpha}$ on X.)

(b) If $\{\mathscr{W}_\alpha : \alpha \in \mathfrak{a}\}$ is a collection of {semi-} uniformities for X, define $\inf_{\mathfrak{a}} \mathscr{W}_\alpha$ as the finest {semi-} uniformity for X which is coarser than each $\mathscr{W}_\alpha$ $(\alpha \in \mathfrak{a})$.

(i) prove that $\inf_{\mathfrak{a}} \{\mathscr{S}_\alpha \in \mathfrak{S}(X)\} = \bigcap_{\mathfrak{a}} \mathscr{S}_\alpha$ on $X \times X$. Show, by counterexample, that $\bigcap_{\mathfrak{a}} \{\mathscr{U}_\alpha \in \mathfrak{U}(X)\}$ need not belong to $\mathfrak{U}(X)$ and conclude that

$$\inf_{\mathfrak{a}} \{\mathscr{U}_\alpha \in \mathfrak{U}(X)\} \neq \bigcap_{\mathfrak{a}} \mathscr{U}_\alpha.$$

(ii) If $\mathscr{S}_\alpha \in \mathfrak{S}(X)\ \forall \alpha \in \mathfrak{a}$, then $m \bigcap_{\mathfrak{a}} \mathscr{S}_\alpha = m \inf_{\mathfrak{a}} \mathscr{S}_\alpha = \inf_{\mathfrak{a}} \{m\mathscr{S}_\alpha\}$ in $\mathfrak{U}(X)$. In particular, $\inf_{\mathfrak{a}} \{\mathscr{U}_\alpha \in \mathfrak{U}(X)\} = m \bigcap_{\mathfrak{a}} \mathscr{U}_\alpha$ in $\mathfrak{U}(X)$.

1G. The Fine Uniformity, $\mathscr{U}_F$

(a) For every topological space $(X, \mathfrak{T})$ there is a finest uniformity $\mathscr{U}_F$ for X such that $\mathfrak{T}_{\mathscr{U}_F} \subset \mathfrak{T}$ on X. Furthermore, $\mathfrak{T}_{\mathscr{U}_F} = \mathfrak{T}$ iff $(X, \mathfrak{T})$ is uniformizible. [*Hint*: The collection $\mathfrak{U} = \{\mathscr{U} \in \mathfrak{U}(X) : \mathfrak{T}_{\mathscr{U}} \subset \mathfrak{T}\}$ contains $\mathscr{U}_{\mathfrak{I}}$ and is therefore nonempty. Define $\mathscr{U}_F = \sup_{\mathfrak{U}} \mathscr{U}$ so that $\mathscr{U} \subset \mathscr{U}_F \in \mathfrak{U}(X)$ and $\mathfrak{T}_{\mathscr{U}} \subset \mathfrak{T}_{\mathscr{U}_F} \subset \mathfrak{T}\ \forall \mathscr{U} \in \mathfrak{U}$. (Use Problem 1F and Corollary 1.17.)]

(b) if $\{\mathscr{U}_\alpha \in \mathfrak{U}(X) : \alpha \in \mathfrak{a}\}$ is compatible with $(X, \mathfrak{T})$, then $\mathfrak{T}_{\sup \mathscr{U}_\alpha} = \mathfrak{T}$ on X.

(c) A "fine" uniformity may have few or many members. Show that $\mathscr{U}_F = \{X \times X\}$ for $(X, \mathfrak{I})$, whereas $\mathscr{U}_F = \mathscr{U}_{\mathfrak{D}}$ for $(X, \mathfrak{D})$.

(d) If $(X, \mathfrak{T})$ is uniformizible, a base for U_F is given by $\mathscr{B}_F = \{$open sets $E \subset X \times X$; there is a sequence $\{E_n\}$ of open sets, each containing Δ, such that $E_1 = E$ and $E_{n+1}^2 \subset E_n\ \forall n \in \mathbb{N}\}$.

(c) For a paracompact, uniformizible space $\mathscr{B}_F = \{$open sets $E \subset X \times X : \Delta \in E\}$. [*Hint*: This is more easily proved via problem 2D(c).]

(f) Using (d) above, prove that a continuous function from a fine uniform space [meaning $\mathscr{U}_F \in \mathfrak{U}(X)$ is compatible with $(X, \mathfrak{T})$] to a uniform space is uniformly continuous. Note that the converse of (f) is false, since every $f : (X, \mathscr{U}) \to (X, \mathscr{U}_{\mathfrak{I}})$ must be uniformly continuous.

1H. The Coarse Uniformity, $\mathscr{U}_c$

The coarsest uniformity compatible with a uniformizible space $(X, \mathfrak{T})$ is called the coarse uniformity for $(X, \mathfrak{T})$, denoted $\mathscr{U}_c$. The coarse uniformity does not always exist (cf Theorem 6.17).

(a) There exists a coarsest $t \cdot$ semiuniformity $\mathscr{S}_c$ for X such that $\mathfrak{T}_{\mathscr{S}_c} \subset \mathfrak{T}$. Furthermore, $\mathfrak{T}_{\mathscr{S}_c} = \mathfrak{T}$ iff $(X, \mathfrak{T})$ is semiuniformizible. [*Hint*: Let $\nu = \{\mathscr{S} \in t \cdot \mathfrak{S}(X) : \mathfrak{T} \subset \mathfrak{T}_{\mathscr{S}}\}$ and define $\mathscr{S}_c = \bigcap_{\nu} \mathscr{S} = \inf_{\nu} \mathscr{S}$ (Problem 1F). Then $\mathfrak{T}_{\mathscr{S}_c} = \inf_{\nu} \mathfrak{T}_{\mathscr{S}} = \bigcap_{\nu} \mathfrak{T}_{\mathscr{S}}$ yields $\mathscr{S}_c \in t \cdot \mathfrak{S}(X)$ and $\mathfrak{T} \subset \mathfrak{T}_{\mathscr{S}_c} \subset \mathfrak{T}_{\mathscr{S}}\ \forall \mathscr{S} \in \nu$.]

(b) If $\mathfrak{T}_{\mathscr{S}} = \mathfrak{T}$ for some $\mathscr{S} \in t \cdot \mathbb{S}(X)$, then $\mathscr{B}_c = \{\langle S; F\rangle = S \cup (X-F)\times(X-F): S \in \mathscr{S}$ and finite subsets $F \subset X\}$ is a base for $\mathscr{S}_c$. [*Hint*: Show that $\mathscr{B}_c$ is the base for some $\mathscr{W} \in t \cdot \mathbb{S}(X)$ compatible with $\mathfrak{T}$ on X. Suppose next that $\mathfrak{T} = \mathfrak{T}_{\mathscr{V}}$ for $\mathscr{V} \in t \cdot \mathbb{S}(X)$ and let $W_1 = \langle W, F\rangle \in \mathscr{B}_c$ for some symmetric $W \in \mathscr{W}$ and finite set $F \subset X$. Clearly, $W_1 = X \times X \in \mathscr{V}$ for $F = \emptyset$. If $F \neq \emptyset$, there exist {symmetric $V_x \in \mathscr{V}: x \in F\}$ satisfying $V_x[x] \subset W[x] \forall x \in F$. Establish that $V = \bigcap_F V_x \in \mathscr{V}$ is symmetric and contained in W_1.]

(c) $m\mathscr{S}_c \subset \mathscr{U}\ \forall U \in \mathfrak{U}(X)$ compatible with $(X, \mathfrak{T}_{m\mathscr{S}_c})$. In particular, $m\mathscr{S}_c$ is the coarse uniformity for $(X, \mathfrak{T}_{m\mathscr{S}_c})$.

(d) $\mathscr{S}_c = \mathscr{U}_c$ if $(X, \mathfrak{T})$ is uniformizible and $\mathscr{S}_c \in \mathfrak{U}(X)$.

1I. More on the Fine and Coarse Semiuniformity

The finest $t \cdot$ semiuniformity compatible with a semiuniformizible space $(X, \mathfrak{T})$ is called the fine semiuniformity for $(X, \mathfrak{T})$, denoted $\mathscr{S}_F$.

(a) If $(X, \mathfrak{F})$ is semiuniformizible, $\mathscr{S}_F = \{S \subset X \times X: (S \cap S^{-1})[x]$ is a $\mathfrak{T}$-nbd of $x\ \forall x \in X\}$.

(b) If $\mathscr{S}_F$ is the fine semiuniformity for a uniformizible space $(X, \mathfrak{T})$, then $\mathscr{U}_F = m\mathscr{S}_F$. [*Hint*: $\mathscr{S}_F \supset m\mathscr{S}_F \supset m\mathscr{U} = \mathscr{U}\ \forall U \in \mathfrak{U}(X)$ compatible with $\mathfrak{T}$ so that $m\mathscr{S}_F$ is compatible with $\mathfrak{T}$.]

(c) The analog of (b) for coarse uniformities does not generally hold. [*Hint*: $\mathscr{S}_c \supset m\mathscr{S}_c = \mathscr{U}_c$ implies $\mathscr{S}_c = \mathscr{U}_c$ (i.e., $m\mathscr{S}_c = \mathscr{S}_c$). In fact, $m\mathscr{S}_c = \mathscr{S}_c$ iff $m\mathscr{S}_c = \mathscr{U}_c$ (Problem 1H(d).]

References for Further Study

On Semiuniform Spaces: Cech [16].
On Quasiuniform Spaces: Murdeshwar and Naimpally [65]; Naimpally [70].
On Locally Uniform Spaces: Williams [106].
On Uniform Spaces: Bourbaki [12]; Cech [16]; Isbell [45]; Tukey [99]; Weil [104].

2. Covering Uniformities

Let ρ be a pseudometric on a set X. Then (Example 1.1) every entourage $U_\varepsilon \in \mathscr{U}_\rho$ determines an ε-sized cover $\mathscr{C}_{U_\varepsilon} = \{U_\varepsilon[x] = S_\varepsilon(x): x \in X\}$ of X and, conversely, every ε-sized cover $\{S_\varepsilon(x): x \in X\}$ determines a symmetric $\bigcup_X S_\varepsilon(x) \times S_\varepsilon(x) \in \mathscr{U}_\rho$. This interplay between coverings and entourages extends to all uniform spaces. Our goal here is to define uniform spaces and to prove some of their elementary properties in terms of coverings, most of which are the analog of results proved earlier via entourages.

DEFINITION 2.1. A collection $\mathfrak{c}$ of coverings of a set X is called a covering uniformity for X, denoted $\mathfrak{c} \in \mathfrak{U}(X)$ if

(C_1) $\mathscr{C} < \mathscr{C}'$ and $\mathscr{C} \in \mathfrak{c}$ implies $\mathscr{C}' \in \mathfrak{c}$

(C_2) $\forall \mathscr{C}_1, \mathscr{C}_2 \in \mathfrak{c}$, there is a $\mathscr{C}_3 \in \mathfrak{c}$ such that $\mathscr{C}_3 \overset{*}{<} \mathscr{C}_1 \wedge \mathscr{C}_2$.

A base for $\mathfrak{c} \in \mathfrak{U}(X)$ is a subcollection $\mathfrak{b}_\mathfrak{c}$ of $\mathfrak{c}$ such that every $\mathscr{C} \in \mathfrak{c}$ is refined by some $\mathscr{C}_0 \in \mathfrak{b}_\mathfrak{c}$.

Remark. (C_2) may be replaced by $(C_2)': \forall \mathscr{C}_1, \mathscr{C}_2 \in \mathfrak{c}$, there is a $\mathscr{C}_3 \in \mathfrak{c}$ such that $\mathscr{C}_3 \overset{b}{<} \mathscr{C}_1 \wedge \mathscr{C}_2$. (Clearly, $(C_2) \Rightarrow (C_2)'$. On the other hand, $\mathscr{C} \overset{b}{<} \mathscr{C}_1 \wedge \mathscr{C}_2$ and $\mathscr{C}_3 \overset{b}{<} \mathscr{C} \wedge \mathscr{C} < \mathscr{C}$ imply $\mathscr{C}_3 \overset{*}{<} \mathscr{C}_1 \wedge \mathscr{C}_2$ (Appendix t.5).

Remark. Note that a collection $\mathfrak{b}$ of covers of X is the base for a covering uniformity $\mathfrak{c} = \{\mathscr{C}: \mathscr{C}_0 < \mathscr{C}$ for some $\mathscr{C}_0 \in \mathfrak{b}\}$ iff $\mathfrak{b}$ satisfies $(C_2)\{(C_2)'\}$.

DEFINITION 2.2. A cover $\mathscr{C}$ of a uniform space $(X, \mathscr{U})$ is said to be uniform if $\{U[x]: x \in X\} = \mathscr{C}_U < \mathscr{C}$ for some $U \in \mathscr{U}$. The collection of $(X, \mathscr{U})$-uniform covers is denoted $\mathfrak{c}_\mathscr{U}$.

The exact relationship of entourage and covering uniformities for a set X is given in

THEOREM 2.3. $\underset{\mathscr{U} \rightsquigarrow \mathfrak{c}_\mathscr{U}}{\phi: \mathfrak{U}(X) \to \mathfrak{U}(X)}$ is a bijection.

Proof. Clearly, (C_1) holds for $\mathfrak{c}_\mathscr{U}$. We verify that (C_2) holds so that $\mathfrak{c}_\mathscr{U} \in \mathfrak{U}(X)$. Given $\mathscr{C}_i \in \mathfrak{c}$ $(i = 1, 2)$, let $U_i \in \mathscr{U}$ satisfy $\mathscr{C}_{\mathscr{U}_i} < \mathscr{C}_i$ $(i = 1, 2)$ and pick symmetric entourages $V, W \in \mathscr{U}$ such that $W^4 \subset V^2 \subset U_1 \cap U_2$. Then $St(x, \mathscr{C}_V) \subset U_1[x] \cap U_2[x] \in \mathscr{C}_1 \wedge \mathscr{C}_2$ $\forall x \in X$; that is, $\mathscr{C}_V \overset{b}{<} \mathscr{C}_1 \wedge \mathscr{C}_2$. A similar argument yields $\mathscr{C}_W \overset{b}{<} \mathscr{C}_V$, from which it follows (Appendix t.5) that $\mathscr{C}_W \overset{*}{<} \mathscr{C}_1 \wedge \mathscr{C}_2$ as required (surely, $\mathscr{C}_W \in \mathfrak{c}$!). To prove ϕ surjective, consider any $\mathfrak{c} \in \mathfrak{U}(X)$ and let $\mathscr{B}_\mathfrak{c} = \{U_\mathscr{C} = \bigcup_{E \in \mathscr{C}} E \times E: \mathscr{C} \in \mathfrak{c}\}$. Clearly, $\Delta \in U_\mathscr{C} = U_\mathscr{C}^{-1} \in \mathscr{B}_\mathfrak{c}$ $\forall \mathscr{C} \in \mathfrak{c}$. For any $\mathscr{C}_1, \mathscr{C}_2 \in \mathfrak{c}$, one has [by (C_2)] a $\mathscr{C}_3 < \mathscr{C}_1 \wedge \mathscr{C}_2$ and $U_{\mathscr{C}_3} \subset U_{\mathscr{C}_1 \wedge \mathscr{C}_2} = U_{\mathscr{C}_1} \cap U_{\mathscr{C}_2}$. The quasi-uniform property of $\mathscr{B}_\mathfrak{c}$ can also be shown to hold. For any $U_\mathscr{C} \in \mathscr{B}_\mathfrak{c}$ let $\tilde{\mathscr{C}} \in \mathfrak{c}$ be a barycentric refinement of $\mathscr{C}$ [note that $\tilde{\mathscr{C}}$ exists by (C_2) and Appendix t.5]. Then $(x_1, x_2) \in U_{\tilde{\mathscr{C}}}^2$ iff $(x_i, z) \in U_{\tilde{\mathscr{C}}}$; that is, $(x_i, z) \in \tilde{E}_i \in \tilde{\mathscr{C}}$ $(i = 1, 2)$ for some $z \in X$. Therefore $(x_1, x_2) \in \tilde{E}_1 \times \tilde{E}_2 \subset St(z, \tilde{\mathscr{C}}) \times St(z, \tilde{\mathscr{C}}) \subset E \times E \subset U_\mathscr{C}$ for some $E \in \mathscr{C}$, which means that $U_{\tilde{\mathscr{C}}}^2 \subset U_\mathscr{C}$. We have thus established that $\mathscr{B}_\mathfrak{c}$ is the base for a uniformity $\mathscr{U}_\mathfrak{c}$ on $X \times X$ and now show that $\mathfrak{c} = \mathfrak{c}_{\mathscr{U}_\mathfrak{c}}$. Clearly, $\mathfrak{c} \subset \mathfrak{c}_{\mathscr{U}\mathfrak{c}}$, since $\mathscr{C}_{\mathscr{U}_\mathfrak{c}} = \mathscr{C} < \mathscr{C}$ $\forall \mathscr{C} \in \mathfrak{c}$. For any $\mathscr{O} \in \mathfrak{c}_{\mathscr{U}_\mathfrak{c}}$ there is (by definition) a $U \in \mathscr{U}_\mathfrak{c}$ such that $\mathscr{C}_U < \mathscr{O}$ and (since $\mathscr{B}_\mathfrak{c}$ is a base for $\mathscr{U}_\mathfrak{c}$) a $\mathscr{C}' \in \mathfrak{c}$ such that $U_{\mathscr{C}'} \subset U$. Therefore $\mathscr{C}' = \mathscr{C}_{U_{\mathscr{C}'}} < \mathscr{C}_U < \mathscr{O}$ and $\mathscr{O} \in \mathfrak{c}$ by property (C_1). To prove ϕ

injective suppose $\mathfrak{c}_{\mathcal{U}} = \mathfrak{c}_{\mathcal{V}}$ on X. If $U \in \mathcal{U}$, then $\mathcal{C}_U \in \mathfrak{c}_{\mathcal{U}} = \mathfrak{c}_{\mathcal{V}}$ and $\mathcal{C}_V \overset{b}{<} \mathcal{C}_U$ for some $V \in \mathcal{V}$. Therefore (as above) $V^2 \subset U$ and $U \in \mathcal{V}$ assures that $\mathcal{U} \subset \mathcal{V}$. The reverse inclusion follows similarly. ■

Remark. Suppose $\mathfrak{c}$ is a collection of covers of X such that (C_0): $\forall \mathcal{C}_1, \mathcal{C}_2 \in \mathfrak{c}$, there is a $\mathcal{C}_3 \in \mathfrak{c}$ satisfying $\mathcal{C}_3 < \mathcal{C}_1 \wedge \mathcal{C}_2$. Then $\mathcal{B}_{\mathfrak{c}}$ is the base for a semiuniformity $\mathcal{S}_{\mathfrak{c}}$ on $X \times X$ such that $\mathfrak{c}$ coincides with the collection of "semiuniform" covers of $(X, \mathcal{S}_{\mathfrak{c}})$.

For future reference it will be convenient to include

COROLLARY 2.4. If $(X, \mathcal{U})$ is a uniform space, then
(i) $\mathcal{C}_V \overset{b}{<} \mathcal{C}_U \; \forall V, U \in \mathcal{U}$ such that $V \circ V^{-1} \subset U$,
(ii) $U^2_{\tilde{\mathcal{C}}} \subset U_{\mathcal{C}} \; \forall \tilde{\mathcal{C}}, \mathcal{C} \in \mathfrak{c}_{\mathcal{U}}$ such that $\tilde{\mathcal{C}} \overset{b}{<} \mathcal{C}$.

Proof. $V \circ V^{-1} \subset U$ implies $St(x, \mathcal{C}_V) \subset U[x] \; \forall x \in X$. ■

COROLLARY 2.5. If $\mathfrak{b}$ is a base for $\mathfrak{c} \in \mathfrak{U}(X)$, then $\mathcal{B}_{\mathfrak{b}} = \{U_{\mathcal{C}} : \mathcal{C} \in \mathfrak{b}\}$ is a base for $\mathcal{U}_{\mathfrak{c}}$. If $\mathcal{B}$ is a base for $\mathcal{U} \in \mathfrak{U}(X)$, then $\mathfrak{b}_{\mathcal{B}} = \{\mathcal{C}_U : U \in \mathcal{B}\}$ is a base for $\mathfrak{c}_{\mathcal{U}}$.

Proof. If $U \in \mathcal{U}_{\mathfrak{c}}$, then $U_{\mathcal{C}} \subset U$ for some $\mathcal{C} \in \mathfrak{c}$. Since $\mathfrak{b}$ is a base for $\mathfrak{c}$, some $\tilde{\mathcal{C}} \in \mathfrak{b}$ is a barycentric refinement of $\mathcal{C}$. Therefore $U_{\tilde{\mathcal{C}}} \subset U^2_{\tilde{\mathcal{C}}} \subset U_{\mathcal{C}} \subset U$ and $U_{\tilde{\mathcal{C}}} \in \mathcal{B}_{\mathfrak{b}}$ completes our first assertion of the corollary. For the second claim let $\mathcal{C} \in \mathfrak{c}_{\mathcal{U}}$. Then $\mathcal{C}_U < \mathcal{C}$ for some $U \in \mathcal{U}$ and (since $\mathcal{B}$ is a base for $\mathcal{U}$) there is a symmetric $V \in \mathcal{B}$ such that $V^2 \subset U$, which means $\mathcal{C}_V \overset{b}{<} \mathcal{C}_U < \mathcal{C}$. ■

If $\mathfrak{c} \in \mathfrak{U}(X)$, call $E \subset X$ a nbd of $x \in X$ if $St(x, \mathcal{C}) \subset E$ for some $\mathcal{C} \in \mathfrak{c}$. Then $\eta_x = \{\text{nbds of } x \in X\}$ satisfies $(N_1)-(N_4)$ of Appendix t.1. We verify (N_4). Suppose $St(x, \mathcal{C}) \subset E \in \eta_x$ and let $\tilde{\mathcal{C}} \in \mathfrak{c}$ satisfy $\tilde{\mathcal{C}} \overset{b}{<} \mathcal{C}$. Then $F = St(x, \tilde{\mathcal{C}}) \in \eta_x$ and $St(y, \tilde{\mathcal{C}}) \subset St(x, \mathcal{C}) \; \forall y \in F$ so that [by (N_3)] $St\ (x, \mathcal{C}) \in \eta_y \; \forall y \in F$. It makes sense now to introduce

DEFINITION 2.6. The uniform topology on X determined by $\mathfrak{c} \in \mathfrak{U}(X)$ is the unique topology $\mathfrak{T}_{\mathfrak{c}} = \{E : E \text{ is a nbd of each } x \in E\}$.

The question which naturally arises concerns the nexus between $\mathfrak{T}_{\mathfrak{c}}$ and $\mathfrak{T}_{\mathcal{U}_{\mathfrak{c}}}$ for $\mathfrak{c} \in \mathfrak{U}(X)$. This is easily answered in

THEOREM 2.7. Let $\mathfrak{c} \in \mathfrak{U}(X)$. Then $\mathfrak{T}_{\mathfrak{c}} = \mathfrak{T}_{\mathcal{U}_{\mathfrak{c}}}$ on X. If $\mathfrak{b}$ is a base for $\mathfrak{c}$, then $\eta_{\mathfrak{b}} = \{St(x, \mathcal{C}) : \mathcal{C} \in \mathfrak{b}, x \in X\}$ is a base for $\mathfrak{T}_{\mathfrak{c}}$.

Proof. It suffices, for the first assertion, to demonstrate that η_x and $(\mathcal{U}_{\mathfrak{c}})_x = \{U[x] : U \in \mathcal{U}_{\mathfrak{c}}\}$ internest at each $x \in X$. Surely $U[x] \subset$

$St(x, \mathscr{C}_U)\ \forall U \in \mathscr{U}_{\mathfrak{c}}$. On the other hand, for $U[x] \in (\mathscr{U}_{\mathfrak{c}})_x$ and $V \in \mathscr{U}_{\mathfrak{c}}$ such that $V \circ V^{-1} \subset U$ one has $St(x, \mathscr{C}_V) \subset U[x]$. The second assertion should be clear. ■

The counterpart to Corollary 1.17 is

COROLLARY 2.8. The $\mathfrak{T}_{\mathfrak{c}}$-open (closed) covers in $\mathfrak{c} \in \mathfrak{U}(X)$ constitute a base for $\mathfrak{c}$.

Proof. For any $\mathscr{C} \in \mathfrak{c}$ there is a symmetric $\mathfrak{T}_{\mathscr{U}_{\mathfrak{c}}} \times \mathfrak{T}_{\mathscr{U}_{\mathfrak{c}}}$-open (closed) entourage $V \in \mathscr{U}_{\mathfrak{c}}$ such that $V^2 \subset U_{\mathscr{C}}$ (Corollary 1.18) and $\mathscr{C}_V < \mathscr{C}_{U_{\mathscr{C}}} = \mathscr{C}$. Since V is $\mathfrak{T}_{\mathscr{U}_{\mathfrak{c}}} \times \mathfrak{T}_{\mathscr{U}_{\mathfrak{c}}}$-open (closed), the member sets $\{V[x]: x \in X\}$ of $\mathscr{C}_V$ are all $\mathfrak{T}_{\mathscr{U}_{\mathfrak{c}}} = \mathfrak{T}_{\mathfrak{c}}$-open (closed). ■

Example 2.9. Refer to Example 1.9.

(a) $\mathfrak{c} = \{X\} \in \mathfrak{U}(X)$ is compatible with $(X, \mathfrak{I})$ and $\mathscr{B}_{\mathfrak{c}} = \{U_X = X \times X\}$ is a base for $\mathscr{U}_{\mathfrak{I}} \in \mathfrak{U}(X)$. Also, $\mathfrak{c} = \{\{x\}: x \in X\} \in \mathfrak{U}(X)$ is compatible with $(X, \mathfrak{D})$ and $\mathscr{B}_{\mathfrak{c}} = \{\Delta\}$ is a base for $\mathscr{U}_{\mathfrak{D}} \in \mathfrak{U}(X)$.

(b) If (X, ρ) is a pseudometric space, then $\mathfrak{b}_\rho = \{\mathscr{C}_\varepsilon = \{S_\varepsilon(X): x \in X\}: \varepsilon > 0\}$ is a base for $\mathfrak{c}_\rho \in \mathfrak{U}(X)$ and $\mathscr{B}_{\mathfrak{b}_\rho} = \{U_{\mathscr{C}_\varepsilon} = U_\varepsilon : \varepsilon > 0\} = \mathscr{B}_\rho$ is a base for $\mathscr{U}_{\mathfrak{c}_\rho} = \mathscr{U}_\rho \in \mathfrak{U}(X)$. Furthermore, $\mathfrak{T}_{\mathfrak{c}_\rho} = \mathfrak{T}_{\mathscr{U}_\rho} = \mathfrak{T}(\rho)$ on X.

(c) Let $(X, >)$ be a directed set, define $B_r = \{x \in X: x > r\}\ \forall r \in X$, and let $\mathscr{C}_r$ be the cover of X consisting of sets B_r and singleton sets $\{x\}$, where $x \leq r$. Then $\mathfrak{b} = \{\mathscr{C}_r: r \in X\}$ is the base for a covering uniformity $\mathfrak{c}_<$ for X such that $\mathscr{B}_{\mathfrak{c}_<}$ is a base for $\mathscr{U}_<$. [To verify that (C_2) holds, let $\mathscr{C}_r$ and $\mathscr{C}_{r'}$ belong to $\mathfrak{b}$ and take $r > r_2 > r_1$. Then $\mathscr{C}_r^* = \mathscr{C}_r < \mathscr{C}_{r_1} \wedge \mathscr{C}_{r_2} = \mathscr{C}_{r_2}$.]

Remark. A uniform cover in $(X, \mathscr{U}_\rho)$ need not consist of ε-sized sets. The member sets in $\mathscr{C}_{U_\varepsilon}$ are given by

$$U_\varepsilon[x] = \begin{cases} x, & x < \varepsilon \\ [\varepsilon, \infty), & x > \varepsilon \end{cases} \quad \text{for } (X, <) = (\mathbb{R}, \leq)$$

In terms of coverings, Definition 1.23 becomes

DEFINITION 2.10. A mapping between two uniform spaces $f: (X, \mathfrak{c}) \to (X', \mathfrak{c}')$ is said to be uniformly continuous if any of the following equivalent conditions hold:

(i) For each $\mathscr{C}' \in \mathfrak{c}'$ there is a $\mathscr{C} \in \mathfrak{c}$ such that $\mathscr{C} < f^{-1}(\mathscr{C}') = \{f^{-1}(E'): E' \in \mathscr{C}'\}$.

(ii) $f^{-1}(\mathfrak{c}') \subset \mathfrak{c}$, where $f^{-1}(\mathfrak{c}') = \{f^{-1}(\mathscr{C}'): \mathscr{C}' \in \mathfrak{c}'\}$.

(iii) $f^{-1}(\mathfrak{b}') \subset \mathfrak{c}$, where $\mathfrak{b}'$ is a base for $\mathfrak{c}'$.

THEOREM 2.11. If $(X, \mathfrak{c}')$ is a uniform space and $(X, \mathfrak{c})$ is a compact uniform space, then every $\mathfrak{T}_{\mathfrak{c}}-\mathfrak{T}_{\mathfrak{c}'}$ continuous function $f:(X, \mathfrak{c})\to(X', \mathfrak{c}')$ is uniformly continuous.

Proof. (Theorem 2.7) $f:(X, \mathfrak{T}_{\mathcal{U}_{\mathfrak{c}}})\to(X', \mathfrak{T}_{\mathcal{U}_{\mathfrak{c}'}})$ is continuous, therefore uniformly continuous (Theorem 1.25), and this is equivalent to f being $\mathfrak{c}-\mathfrak{c}'$ uniformly continuous [Problem 2B(b)]. ■

We conclude with the counterpart to Theorem 1.26.

THEOREM 2.12. A compact, completely regular space $(X, \mathfrak{T})$ has a unique compatible covering uniformity $\mathfrak{c}_F\in\mathfrak{U}(X)$. If $(X, \mathfrak{T})$ is compact and T_2, then $\mathfrak{c}_F$ has {all open covers} as a base.

Proof. The existence of $\mathfrak{c}_F$ follows from Theorem 2.11 as did $\mathcal{U}_F$ in Theorem 1.26. Since $\mathfrak{T}_{\mathcal{U}_{\mathfrak{c}_F}}=\mathfrak{T}_{\mathfrak{c}_F}=\mathfrak{T}$, one has $\mathfrak{U}_{\mathfrak{c}_F}=\mathcal{U}_F$ (Theorem 1.26) and the conclusion that $\mathfrak{b}_0=\{\text{all open covers of } X\}\subset\mathfrak{c}_F$. To be sure, $\mathscr{C}_0\in\mathfrak{b}_0$ implies $U_{\mathscr{C}_0}\in\mathscr{B}_0\subset\mathcal{U}_F=\mathfrak{U}_{\mathfrak{c}_F}$ so that $\mathscr{C}_0=\mathscr{C}_{U_{\mathscr{C}_0}}\in\mathfrak{c}_F$. Finally, consider any $\mathscr{C}\in\mathfrak{c}_F$. Since $U_{\mathscr{C}}\in\mathfrak{U}_{\mathfrak{c}_F}=\mathcal{U}_F$, there is a $V\in\mathscr{B}_0$ such that $V\circ V^{-1}\subset U_{\mathscr{C}}$ and it follows that $\mathscr{C}_V\in\mathfrak{b}_0$ and $\mathscr{C}_V<\mathscr{C}$. ■

PROBLEMS

2A. Separation Properties

If $\mathfrak{c}\in\mathfrak{U}(X)$, then $\mathfrak{T}_{\mathfrak{c}}$ is regular and the following are equivalent:

(a) $\mathfrak{T}_{\mathfrak{c}}$ has property T_k $(k=0, 1, 2, 3)$
(b) $x\neq y$ implies $St(x, \mathscr{C})\cap St(y, \mathscr{C})=\varnothing$ for some $\mathscr{C}\in\mathfrak{c}$
(c) $x\neq y$ implies $x\notin St(y, \mathscr{C})$ for some $\mathscr{C}\in\mathfrak{c}$
(d) $\bigcap_{\mathfrak{c}} St(x, \mathscr{C})=\{x\}\ \forall x\in X$.

$\mathfrak{c}$ is said to be separated if any of the above conditions hold.

2B. Uniform Continuity

(a) Every uniformly continuous function $f:(X, \mathfrak{c})\to(X', \mathfrak{c}')$ is $\mathfrak{T}_{\mathfrak{c}}-\mathfrak{T}_{\mathfrak{c}'}$ continuous.

(b) $f:(X, \mathfrak{c})\to(X', \mathfrak{c}')$ is uniformly continuous iff it is $\mathcal{U}_{\mathfrak{c}}-\mathcal{U}_{\mathfrak{c}'}$ uniformly continuous.

(c) $f:(X, \mathcal{U})\to(Y, \mathcal{V})$ is uniformly continuous iff it is $\mathfrak{c}_{\mathcal{U}}-\mathfrak{c}_{\mathcal{V}}$ uniformly continuous.

(d) If $f:(X, \mathfrak{c}) \to (X', \mathfrak{c}')$ is uniformly continuous on each subspace $(G_i, G_i \cap \mathfrak{c})$ of a finite cover $\mathcal{G} = \{G_1, \ldots, G_n\} \in \mathfrak{c}$, then f is uniformly continuous on X. [*Hint*: If $\mathscr{C}' \in \mathfrak{c}'$, then $G_i \cap f^{-1}(\mathscr{C}') = G_i \cap \mathscr{C}_i$ for covers $\mathscr{C}_i \in \mathfrak{c}$ $(i = 1, 2, \ldots, n)$ and $\mathscr{C} = \bigwedge_{i=1}^{n} \mathscr{C}_i \in \mathfrak{c}$ satisfies $\mathcal{G} \wedge \mathscr{C} < f^{-1}(\mathscr{C}')$.]

(e) [Example 2.9, (b)]. Let $I = [0, 1]$ *carry* $\mathfrak{c}_{d_1}$, where d_1 is the usual metric on $\mathbb{R}$. Given subsets A, B of a uniform space $(X, \mathfrak{c})$, there is a uniformly continuous $f:(X, \mathfrak{c}) \to (I, \mathfrak{c}_{d_1})$ with $f(A) = 0$ and $f(B) = 1$ iff $St(A, \mathscr{C}) \cap B = \varnothing$ for some $\mathscr{C} \in \mathfrak{c}$.

2C. Supremum and Infimum Covering Uniformities

(a) The coarsest covering uniformity for X finer than each member of $\{\mathfrak{c}_\alpha \in \mathcal{U}(X): \alpha \in \mathcal{Q}\}$ is called the $\sup_{\mathcal{Q}} \mathfrak{c}_\alpha$. Show that $\left\{ \bigwedge_{i=1}^{n} \mathscr{C}_{\alpha_i} : \mathscr{C}_{\alpha_i} \in \mathfrak{c}_{\alpha_i}, \alpha_i \in \mathcal{Q} \right\}$ is a base for $\sup_{\mathcal{Q}} \mathfrak{c}_\alpha$. Furthermore, $\mathfrak{T}_{\sup \mathfrak{c}_\alpha} = \sup_{\mathcal{Q}} \mathfrak{T}_{\mathfrak{c}_\alpha}$ on X.

(b) The finest covering uniformity coarser than each member of $\{\mathfrak{c}_\alpha \in \mathcal{U}(X): \alpha \in \mathcal{Q}\}$ is termed the $\inf_{\mathcal{Q}} \mathfrak{c}_\alpha$. See how far you can parallel Problem 1F(b) in terms of coverings.

2D. The Fine Covering Uniformity, $\mathfrak{c}_F$

(a) For every topological space $(X, \mathfrak{T})$ there is a finest covering uniformity $\mathfrak{c}_F$ for X such that $\mathfrak{T}_{\mathfrak{c}_F} \subset \mathfrak{T}$. Moreover, $\mathfrak{T}_{\mathfrak{c}_F} = \mathfrak{T}$ iff $(X, \mathfrak{T})$ is uniformizable. [*Hint*: Let $\mathfrak{u} = \{\mathfrak{c} \in \mathcal{U}(X): \mathfrak{T}_{\mathfrak{c}} \subset \mathfrak{T}\}$ and define $\mathfrak{c}_F = \sup_{\mathfrak{u}} \mathfrak{c}$. For any $\mathfrak{T}_{\mathfrak{c}_F}$-nbd $St(x, \mathscr{C})$ of $x \in X$ there are open covers $\mathscr{C}_{\alpha_i} \in \mathfrak{c}_{\alpha_i} \in \mathfrak{u}$ such that $\bigwedge_{i=1}^{n} \mathscr{C}_{\alpha_i} < \mathscr{C}$. Proceed by verifying that $St\left(x, \bigwedge_{i=1}^{n} \mathscr{C}_{\alpha_i}\right) = \bigcap_{i=1}^{n} St(x, \mathscr{C}_{\alpha_i}) \subset St(x, \mathscr{C})$.]

(b) If $(X, \mathfrak{T})$ is uniformizible, a base for $\mathfrak{c}_F$ is given by $\mathfrak{b}_F = \{$open covers $\mathcal{O}$ of X: there is a sequence of open covers $\{\mathcal{O}_n\}$ such that $\mathcal{O}_1 \overset{*}{<} \mathcal{O}$ and $\mathcal{O}_{n+1} \overset{*}{<} \mathcal{O}_n$ $\forall n \in \mathbb{N}\}$. [*Hint*: Given any $\mathscr{C} \in \mathfrak{c}_F$, choose an $\mathcal{O} \in \mathfrak{b}_F \subset \mathfrak{c}_F$ such that $\mathcal{O} < \mathscr{C}$ and let $\tilde{\mathcal{O}}_1 \in \mathfrak{c}_F$ satisfy $\tilde{\mathcal{O}}_1 \overset{*}{<} \mathcal{O} \wedge \mathcal{O} < \mathcal{O}$. Since $\tilde{\mathcal{O}}_1 \in \mathfrak{c}_F$, there is an open $\mathcal{O}_1 \in \mathfrak{c}_F$ such that $\mathcal{O}_1 < \tilde{\mathcal{O}}_1$. Thus $\mathcal{O}_1 \overset{*}{<} \mathcal{O}$. The same argument produces an $\tilde{\mathcal{O}}_2 \in \mathfrak{c}_F$ and $\mathcal{O}_2 \in \mathfrak{c}_F$ such that $\mathcal{O}_2 < \tilde{\mathcal{O}}_2 < \mathcal{O}_1 \wedge \mathcal{O}_1 < \mathcal{O}_1$ and $\mathcal{O}_2 \overset{*}{<} \mathcal{O}_1$. Proceed!]

(c) If $(X, \mathfrak{T})$ in (b) is also paracompact, then $\mathfrak{b}_F = \{$open covers of $X\}$. [*Hint*: Every open cover has an open nbd finite refinement, therefore an open star refinement (Appendix t.5).]

(d) Use (b) above to prove that a continuous function from a fine covering uniform space to a covering uniform space is uniformly continuous. [*Hint*: If $f:(X, \mathfrak{c}) \to (X', \mathfrak{c}')$ and $\mathcal{Q}, \mathscr{B} \in \mathfrak{c}'$ satisfy $\mathcal{Q} \overset{*}{<} \mathscr{B}$, then $f^{-1}(\mathcal{Q}) \overset{*}{<} f^{-1}(\mathscr{B})$ since $St(A, \mathcal{Q}) \subset B \in \mathscr{B}$ implies $St(f^{-1}(A), f^{-1}(\mathcal{Q})) \subset f^{-1}(\mathscr{B})$.]

2E. The Coarse Covering Uniformity, $\mathfrak{c}_c$

The coarsest covering uniformity compatible with a uniformizible space $(X, \mathfrak{T})$ is called the coarse covering uniformity, denoted $\mathfrak{c}_c$. Show (Problem 1H) that $\mathfrak{c}_c$ need not exist. What relationship exists between $\mathcal{U}_c$ and $\mathfrak{c}_c$ when they do exist?

2F. Lebesgue Covering Theorem Generalized

Recall that for each open cover $\mathcal{O}$ of a compact subset K of a pseudometric space (X, ρ) there is a $\lambda_{\mathcal{O}} > 0$ (called the Lebesgue number of $\mathcal{O}$) such that $\{S_{\rho,\lambda_{\mathcal{O}}}(x) : x \in K\} < \mathcal{O}$.

(a) Let $\mathcal{O}$ be an open cover of K, where K is a compact subset of a uniform space $(X, \mathfrak{c})$ $\{(X, \mathcal{U})\}$. Then $\mathcal{G} \cap K < \mathcal{O}$ for some $\mathcal{G} \in \mathfrak{c}$. $\{\mathcal{O}$ is a $\mathcal{U}_K = \mathcal{U} \cap (K \times K)$-uniform covering.$\}$

(b) The above remains true if "compact" is replaced by "paracompact." [This is involved. Kelley ([50] 156) provides enough to proceed.]

3. Projective and Inductive Limits

A family of mappings $\mathcal{F} = \{f_\alpha : X \to Y_\alpha : \alpha \in \mathcal{Q}\}$ from a set X to topological spaces Y_α determines a weakest topology $\mathfrak{T}_{\mathcal{P}\{\mathcal{F}\}}$ on X (called the projective limit) under which every $f_\alpha \in \mathcal{F}$ is continuous. By definition, $f_\alpha^{-1}(u_\alpha)$ (u_α is a Y_α-nbd) are the $\mathfrak{T}_{\mathcal{P}\{\mathcal{F}\}}$-subbase sets. At the other extreme $\mathcal{G} = \{g_\alpha : Y_\alpha \to X : \alpha \in \mathcal{Q}\}$ generates a finest topology $\mathfrak{T}_{\mathcal{I}\{\mathcal{G}\}}$ on X (called the inductive limit) under which every $g_\alpha \in \mathcal{G}$ is continuous. Here $\mathfrak{T}_{\mathcal{I}\{\mathcal{G}\}}$-nbds are of the form $\{u \subset X : g_\alpha^{-1}(u)$ is a Y_α-nbd $\forall \alpha \in \mathcal{Q}\}$. This section extends these notions to uniform spaces. Accordingly, subspaces and products of uniform spaces are treated as projective limits and quotient uniform spaces are viewed as inductive limits.

Projective Limits

Substituting "uniform space" for "topological space" and "uniformly continuous" for "continuous" leads to

DEFINITION 3.1. The projective limit uniformity generated by $\mathcal{F} = \{f_\alpha : X \to (Y_\alpha, \xi_\alpha) : \xi_\alpha \in \mathfrak{U}(Y_\alpha)\, \forall \alpha \in \mathcal{Q}\}$ is the weakest uniformity for X under which every $f_\alpha \in \mathcal{F}$ is uniformly continuous. This is denoted $\mathcal{U}_{\mathcal{P}(\mathcal{F})}$ $\{\mathfrak{c}_{\mathcal{P}(\mathcal{F})}\}$ when all $\xi_\alpha = \mathcal{U}_\alpha$ $\{\xi_\alpha = \mathcal{C}_\alpha\}$ are entourage $\{$covering$\}$ uniformities for Y_α.

$\mathcal{S}_{\mathcal{P}(\mathcal{F})}$ and $\mathcal{Q}_{\mathcal{P}(\mathcal{F})}$ are defined similarly, using entourage filters $\xi_\alpha \in \mathfrak{S}(Y_\alpha)$ and $\xi_\alpha \in \mathfrak{Q}(Y_\alpha)\, \forall \alpha \in \mathcal{Q}$.

THEOREM 3.2. For $\mathcal{F} = \{f_\alpha : X \to (Y_\alpha, \xi_\alpha) : \xi_\alpha \in \mathfrak{U}(Y_\alpha)\, \forall \alpha \in \mathcal{Q}\}$

(i) $\left\{ \bigcap_{i=1}^{n} (f_{\alpha_i} \times f_{\alpha_i})^{-1} U_{\alpha_i} : U_{\alpha_i} \in \mathcal{U}_{\alpha_i}, \alpha_i \in \mathcal{Q} \right\}$ is a base for $U_{\mathcal{P}(\mathcal{F})} \in \mathfrak{U}(X)$,

(ii) $\left\{ \bigwedge_{i=1}^{n} f_{\alpha_i}^{-1}(\mathcal{G}_{\alpha_i}) : \mathcal{G}_{\alpha_i} \in \mathfrak{c}_{\alpha_i}, \alpha \in \mathcal{Q} \right\}$ is a base for $\mathfrak{c}_{\mathcal{P}(\mathcal{F})} \in \mathfrak{U}(X)$,

(iii) $\mathfrak{T}_{U_{\mathcal{P}(\mathcal{F})}} = \mathfrak{T}_{\mathcal{P}(\mathcal{F})} = \mathfrak{T}_{\mathfrak{c}_{\mathcal{P}(\mathcal{F})}}$ on X.

Proof. (i) Every member of the filterbase

$$\mathscr{B} = \left\{ \bigcap_{i=1}^{n} (f_{\alpha_i} \times f_{\alpha_i})^{-1} U_{\alpha_i} : U_{\alpha_i} \in \mathscr{U}_{\alpha_i}, \alpha_i \in \mathfrak{A} \right\}$$

contains $\Delta \in X \times X$. Consider any $B = \bigcap_{i=1}^{n} (f_{\alpha_i} \times f_{\alpha_i})^{-1} U_{\alpha_i}$ in $\mathscr{B}$. Since each $\mathscr{U}_{\alpha_i} \in \mathfrak{U}(Y_{\alpha_i}) \in \mathfrak{S}(Y_{\alpha_i})$, it follows that $B^{-1} = \bigcap_{i=1}^{n} \{(f_{\alpha_i} \times f_{\alpha_i})^{-1} U_{\alpha_i}\}^{-1} = \bigcap_{i=1}^{n} (f_{\alpha_i} \times f_{\alpha_i})^{-1} U_{\alpha_i}^{-1} \in \mathscr{B}$. Furthermore, $\mathfrak{U}(Y_{\alpha_i}) \subset \mathfrak{Q}(Y_{\alpha_i})$ so that there exist $V_{\alpha_i} \in \mathscr{U}_{\alpha_i}$ satisfying $V_{\alpha_i}^2 \subset U_{\alpha_i}$, and this assures that $\left\{ \bigcap_{i=1}^{n} (f_{\alpha_i} \times f_{\alpha_i})^{-1} V_{\alpha_i} \right\}^2 \subset \bigcap_{i=1}^{n} \{(f_{\alpha_i} \times f_{\alpha_i})^{-1} V_{\alpha_i}\}^2 \subset \bigcap_{i=1}^{n} (f_{\alpha_i} \times f_{\alpha_i})^{-1} U_{\alpha_i} = B$. Thus $\mathscr{B}$ is the base for some $\mathscr{U} \in \mathfrak{U}(X)$, which (by definition) is $\mathscr{U}_{\mathscr{P}(\mathfrak{F})}$. Indeed, if $\mathscr{W} \in \mathfrak{U}(X)$ and every $f_\alpha \in \mathfrak{F}$ is $\mathscr{W}$-uniformly continuous, then $(f_\alpha \times f_\alpha)^{-1} U_\alpha \in \mathscr{W}\ \forall \alpha \in \mathfrak{A}$ and $U_\alpha \in \mathscr{U}_\alpha$ so that $\mathscr{U} \subset \mathscr{W}$.

(ii) First note that $f_\alpha^{-1}\{St(f_\alpha(x), \mathcal{G}_\alpha)\} = St(x, f_\alpha^{-1}(\mathcal{G}_\alpha))\ \forall x \in X,\ f_\alpha \in \mathfrak{F}$, and $\mathcal{G}_\alpha \in \mathfrak{c}_\alpha$. For any $\mathscr{A} = \bigwedge_{i=1}^{n} f_{\alpha_i}^{-1}(\mathcal{G}_{\alpha_i})$ and $\mathscr{A}' = \bigwedge_{j=1}^{m} f_{\beta_j}^{-1}(\mathcal{G}_{\beta_j})$ belonging to $\mathfrak{b} = \left\{ \bigwedge_{i=1}^{n} f_{\alpha_i}^{-1}(\mathcal{G}_{\alpha_i}) : \mathcal{G}_{\alpha_i} \in \mathfrak{c}_{\alpha_i}, \alpha_i \in \mathfrak{A} \right\}$, let $\tilde{\mathcal{G}}_{\alpha_i} \in \mathfrak{c}_{\alpha_i}$ and $\tilde{\mathcal{G}}_{\beta_j} \in \mathfrak{c}_{\beta_j}$ satisfy $\tilde{\mathcal{G}}_{\alpha_i} \overset{b}{<} \mathcal{G}_{\alpha_i}$ ($\forall i = 1, 2, \ldots, n$) and $\tilde{\mathcal{G}}_{\beta_j} \overset{b}{<} \mathcal{G}_{\beta_j}$ ($\forall j = 1, 2, \ldots, m$), respectively. Then $\mathscr{A}'' = \bigwedge_{i=1}^{n} f_{\alpha_i}^{-1}(\tilde{\mathcal{G}}_{\alpha_i}) \wedge \bigwedge_{j=1}^{m} f_{\beta_j}^{-1}(\tilde{\mathcal{G}}_{\beta_j}) \overset{b}{<} \mathscr{A} \wedge \mathscr{A}'$ and $(C_2)'$ of Definition 2.1 holds. If $\mathfrak{c}$ is the covering uniformity for X determined by $\mathfrak{b}$, then every $f_\alpha \in \mathfrak{F}$ is $\mathfrak{c}-\mathfrak{c}_\alpha$ uniformly continuous. Any other $\mathfrak{c} \in \mathfrak{U}(X)$ for which all $f_\alpha \in \mathfrak{F}$ are uniformly continuous necessarily contains every $f_\alpha^{-1}(\mathcal{G}_\alpha)$ and so also contains $\mathfrak{c}$. Thus $\mathfrak{c} = \mathfrak{c}_{\mathscr{P}(\mathfrak{F})}$.

(iii) Theorems 1.23 and 2.11 and the definition of $\mathfrak{T}_{\mathscr{P}(\mathfrak{F})}$ assures that $\mathfrak{T}_{\mathscr{P}(\mathfrak{F})} \subset \mathfrak{T}_{\mathscr{U}_{\mathscr{P}(\mathfrak{F})}}, \mathfrak{T}_{\mathfrak{c}_{\mathscr{P}(\mathfrak{F})}}$. The reverse inclusion also holds since every $\mathfrak{T}_{\mathscr{U}_{\mathscr{P}(\mathfrak{F})}}$-subbase nbd $\{(f_\alpha \times f_\alpha)^{-1} U_\alpha\}[x_0] = f_\alpha^{-1}\{U_\alpha[f_\alpha(x_0)]\}$ and every $\mathfrak{T}_{\mathfrak{c}_{\mathscr{P}(\mathfrak{F})}}$-subbase nbd $St(x_0, f_\alpha^{-1}(\mathcal{G}_\alpha)) = f_\alpha^{-1}\{St(f_\alpha(x_0), \mathcal{G}_\alpha))$ is a $\mathfrak{T}_{\mathscr{P}(\mathfrak{F})}$-nbd of $x_0 \in X$. These assertions follow from the $\mathfrak{T}_{\mathscr{P}(\mathfrak{F})}$-continuity of every $f_\alpha \in \mathfrak{F}$. ■

Remark. Both (i), (iii) hold for $\mathfrak{Q}_{\mathscr{P}(\mathfrak{F})}$ and (i) holds for $\mathscr{S}_{\mathscr{P}(\mathfrak{F})}$.

This is a good place (cf Corollary 4.3) to include

COROLLARY 3.3. The projective limit of uniformizable (completely regular) spaces is uniformizible (completely regular).

THEOREM 3.4. The following are equivalent for $\mathfrak{F} = \{f_\alpha : X \to (Y_\alpha, \mathscr{U}_\alpha);\ \mathscr{U}_\alpha \in \mathfrak{U}(Y_\alpha)\ \forall \alpha \in \mathfrak{A}\}$ and $\mathscr{U} \in \mathfrak{U}(X)$:

(i) $\mathcal{U} = \mathcal{U}_{\mathcal{P}(\mathfrak{F})}$,

(ii) $f:(Z, \mathcal{W}) \to (X, \mathfrak{U})$ is uniformly continuous for any uniform space $(Z, \mathcal{W})$ iff every $f_\alpha f:(Z, \mathcal{W}) \to (Y_\alpha, \mathcal{U}_\alpha)$ is uniformly continuous.

Proof. (i)$\Rightarrow$(ii). Suppose $U \in \mathcal{U}_{\mathcal{P}(\mathfrak{F})} = \mathcal{U}$ so that $\bigcap_{i=1}^{n} (f_{\alpha_i} \times f_{\alpha_i})^{-1} U_{\alpha_i} \subset U$ for $\{U_{\alpha_i} \in \mathcal{U}_{\alpha_i} : i = 1, 2, \ldots, n\}$. If all $f_\alpha f$ are uniformly continuous, then $W = \bigcap_{i=1}^{n} (f_{\alpha_i} f \times f_{\alpha_i} f)^{-1} U_{\alpha_i} \in \mathcal{W}$ and so

$$(f \times f)^{-1} U \supset (f \times f)^{-1} \left\{ \bigcap_{i=1}^{n} (f_{\alpha_i} \times f_{\alpha_i})^{-1} U_{\alpha_i} \right\} = W$$

belongs to $\mathcal{W}$. Thus f is uniformly continuous. The reverse implication holds by Definition 1.23$^+$.

(ii)$\Rightarrow$(i). Assuming (ii), take $f = \mathbb{1}:(X, \mathcal{U}) \to (X, \mathcal{U})$. Then every $f_\alpha \mathbb{1} = f_\alpha:(X, \mathcal{U}) \to (Y_\alpha, \mathcal{U}_\alpha)$ is uniformly continuous and $\mathcal{U}_{\mathcal{P}(\mathfrak{F})} \subset \mathcal{U}$. On the other hand, the uniform continuity of each $f_\alpha = f_\alpha \mathbb{1}:(X, \mathcal{U}_{\mathcal{P}(\mathfrak{F})}) \to (X, \mathcal{U}) \to (Y_\alpha, \mathcal{U}_\alpha)$ assures that of $\mathbb{1}$, which means that $\mathcal{U} \subset \mathcal{U}_{\mathcal{P}(\mathfrak{F})}$. ■

Remark. The analogs of Theorem 3.4 and Corollary 3.5 below also hold for $\mathcal{S}_{\mathcal{P}(\mathfrak{F})}$ and $\mathcal{Q}_{\mathcal{P}(\mathfrak{F})}$.

COROLLARY 3.5. Hold $\mathfrak{F}$ as above. For each $\alpha \in \mathfrak{A}$, let $\mathcal{B}(\alpha)$ be a nonempty index set with a corresponding collection $\mathcal{G}_{\mathcal{B}(\alpha)} = \{g_\beta : Y_\alpha \to (Z_\beta, \mathcal{W}_\beta); \mathcal{W}_\beta \in \mathfrak{U}(Z_\beta)\ \forall \beta \in \mathcal{B}(\alpha)\}$. Suppose, further, that $\mathcal{U}_\alpha = \mathcal{U}_{\mathcal{P}\{\mathcal{G}_{\mathcal{B}(\alpha)}\}}\ \forall \alpha \in \mathfrak{A}$. Then $\mathcal{U}_{\mathcal{P}(\mathfrak{F})} = \mathcal{U}_{\mathcal{P}\{\mathcal{G}\mathfrak{F}\}}$, where $\mathcal{G}\mathfrak{F} = \{g_\beta f_\alpha : \alpha \in \mathfrak{A}, \beta \in \mathcal{B}(\alpha)\}$.

Example 3.6. Let E be a subset of a set X.

(a) If $\mathcal{U} \in \mathfrak{U}(X)$, then $\mathcal{U}_E = \{U \cap (E \times E) : U \in \mathcal{U}\}$ is a uniformity for E (called the relative uniformity for E). Since $\{U \cap (E \times E)\}[x] = U[x] \cap E\ \forall x \in X$ and $U \in \mathcal{U}$, the uniform topology $\mathfrak{T}_{\mathcal{U}_E}$ on E and the induced relative topology $\mathfrak{T}_{\mathcal{U}}$ on E coincide. Let $\mathbb{1}$ be the (into) identity mapping $E \to (X, \mathcal{U})$. Then $\mathcal{U}_E = \mathcal{U}_{\mathcal{P}(\mathbb{1})}$ is compatible with $\mathfrak{T}_{\mathcal{U}_E} = \mathfrak{T}_{\mathcal{P}(\mathbb{1})}$ on E.

(b) If $\mathfrak{c} \in \mathfrak{U}(X)$, then $\mathfrak{c} \cap E = \{\mathcal{G} \cap E : \mathcal{G} \in \mathfrak{c}\}$ constitutes a base for a covering uniformity $\mathfrak{c}_E$ for E (called the relative covering uniformity for E). As above, $\mathfrak{c}_E = \mathfrak{c}_{\mathcal{P}(\mathbb{1})}$ is compatible with $\mathfrak{T}_{\mathfrak{c}_E} = \mathfrak{T}_{\mathcal{P}(\mathbb{1})}$ on E.

Remark. If $\mathcal{B}$, $\mathcal{B}'$ are equivalent bases for some $\mathcal{U} \in \mathfrak{U}(X)$, then $\mathcal{B}_E$, $\mathcal{B}_{E'}$ are equivalent bases for $\mathcal{U}_E \in \mathfrak{U}(E)$. The converse however, is false. (Simply take $E = \{x\}$ in Example 1.13.)

Example 3.7. Let $X = \prod_{\mathfrak{A}} X_\alpha$, where each $(X_\alpha, \mathfrak{T}_\alpha)$ $(\alpha \in \mathfrak{A})$ is a topological space. Recall that the product topology $\mathfrak{T}_\Pi$ on X is precisely the projective limit topology generated by the family of projections $\pi = \{\pi_\alpha : X \underset{\{x\} \rightsquigarrow x_\alpha}{\to} X_\alpha ; \alpha \in \mathcal{A}\}$. If each $\mathfrak{T}_\alpha$ $(\alpha \in \mathcal{A})$ is uniformizible, then (Corollary 3.3) $\mathfrak{T}_\Pi$ is uniformizible.

(a) Suppose that $\mathcal{U}_\alpha \in \mathfrak{U}(X_\alpha)\ \forall \alpha \in \mathcal{A}$. One then defines $\mathcal{U}_{\mathcal{P}(\pi)}$ as the product uniformity for X, denoted $\mathcal{U}_\Pi$. This is reasonable since $\mathcal{U}_\Pi$ is compatible with $\mathfrak{T}_\Pi$ on X. Subbase entourages of $\mathcal{U}_\Pi$ are of the form $(\pi_\alpha \times \pi_\alpha)^{-1} U_\alpha = \{(\{x\}, \{y\}) \in X \times X : (x_\alpha, y_\alpha) \in U_\alpha \in \mathcal{U}_\alpha\} (\alpha \in \mathcal{A})$.

(b) For each $\alpha \in \mathcal{A}$, assume that $\mathfrak{c}_\alpha$ is a covering uniformity for X_α. Then $\mathfrak{c}_{\mathcal{P}(\pi)}$ is called the product covering uniformity for X, denoted $\mathfrak{c}_\pi$. By definition, subbase entourages for $\mathfrak{c}_\pi$ are of the form $\pi_\alpha^{-1}(\mathcal{G}_\alpha) = \{\prod_{\mathcal{A}} A_\beta : A_\beta \in \mathcal{G}_\alpha \text{ if } \beta = \alpha, \text{ and } A_\beta = X_\beta, \text{ if } \beta \neq \alpha\}$. [As above, $\mathfrak{c}_\pi$ is compatible with $\mathfrak{T}_\Pi = \mathfrak{T}_{\mathcal{P}(\pi)}$.]

(c) Given a family of surjections $\mathcal{F} = \{f_\alpha : X \to Y_\alpha ; \alpha \in \mathcal{A}\}$, suppose $\mathcal{G}_\alpha$ is a filter in Y_α such that every element of $\mathcal{B} = \left\{\bigcap_{i=1}^{n} f_{\alpha_i}^{-1}(G_{\alpha_i}) : G_{\alpha_i} \in \mathcal{G}_{\alpha_i}, \alpha_i \in \mathcal{A}\right\}$ is nonempty. Then $\mathcal{B}$ is the base for the smallest filter $\mathcal{G}_{\mathcal{P}(\mathcal{F})}$ in X which contains $f_\alpha^{-1}(\mathcal{G}_\alpha) = \{f_\alpha^{-1}(G_\alpha) : G_\alpha \in \mathcal{G}_\alpha\}$ and satisfies $f_\alpha(\mathcal{G}_{\mathcal{P}(\mathcal{F})}) = \mathcal{G}_\alpha\ \forall \alpha \in \mathcal{A}$. This filter $\mathcal{G}_{\mathcal{P}(\mathcal{F})}$, when it exists, is called the projective limit of $\{\mathcal{G}_\alpha : \alpha \in \mathcal{A}\}$ by $\mathcal{F}$.

(d) The projective limit of $\{\mathcal{G}_\alpha \subset Y_\alpha : \alpha \in \mathcal{A}\}$ generated by $\pi = \{\pi_\beta : X = \prod_{\mathcal{A}} X_\alpha \to X_\beta ; \beta \in \mathcal{A}\}$ always exists and is called the product filter of $\{\mathcal{G}_\alpha : \alpha \in \mathcal{A}\}$, denoted $\mathcal{G}_\pi$. Existence is clear since

$$\pi_\beta^{-1}(\mathcal{G}_\beta) = \mathcal{G}_\beta \times \prod_{\alpha \neq \beta} X_\alpha \quad \text{and} \quad \bigcap_{i=1}^{n} \pi_{\alpha_i}^{-1}(G_{\alpha_i}) = \prod_{i=1}^{n} G_{\alpha_i} \times \prod_{\mathcal{A} - \{\alpha_1, \ldots, \alpha_n\}} X_\alpha \neq \varnothing.$$

Example 3.8. (a) Let $C(X; \mathbb{R})$ denote the real-valued continuous functions on a topological space $(X, \mathfrak{T})$, the usual assumption being that $\mathbb{R}$ carries $\mathcal{U}_{d_1}$ (Example 1.9). Then $\mathcal{U}_{\mathcal{P}\{C(X;\mathbb{R})\}}$ is compatible with the projective limit topology $\mathfrak{T}_{\mathcal{P}\{C(X;\mathbb{R})\}}$. If $\mathfrak{T}$ is completely regular, $\mathfrak{T} = \mathfrak{T}_{\mathcal{P}\{C(X;\mathbb{R})\}}$— in which case, $(X, \mathfrak{T})$ is uniformizible with compatible uniformity $U_{\mathcal{P}\{C(X;\mathbb{R})\}}$.

(b) It is easily demonstrated that the induced uniformity $\mathcal{U}_{d_1}$ for $X = [0, \infty)$ is strictly coarser than $\mathcal{U}_F \in \mathfrak{U}(X)$ (Problem 1G). Both $\mathcal{U}_{d_1}$ and $\mathcal{U}_{\mathcal{P}\{C(X;\mathbb{R})\}}$ are compatible with the metric topology $\mathfrak{T}(d_1)$ on X. Although $\mathcal{U}_{d_1} \subset \mathcal{U}_{\mathcal{P}\{C(X;\mathbb{R})\}}$ on $X \times X$, equality fails since $(X, \mathcal{U}_{d_1}) \underset{x \rightsquigarrow x^3}{\to} (X, \mathcal{U}_{d_1})$ is a nonuniformly continuous homeomorphism. Note further that $\mathfrak{U}_{\mathcal{P}\{C(X;\mathbb{R})\}} \subset \mathcal{U}_F$ by Problem 1I(f).

Remark. For results concerning uniformities under which every continuous function is uniformly continuous see Atsuji [4], [5].

Inductive Limits

The counterpart to Definition 3.1 is given by

DEFINITION 3.9. The inductive limit uniformity generated by a family of surjections $\mathfrak{F}=\{f_\alpha:(X_\alpha,\xi_\alpha)\to Y:\xi_\alpha\in\mathfrak{U}(X_\alpha)\ \forall\alpha\in\mathfrak{A}\}$ is the finest uniformity for Y under which every $f_\alpha\in\mathfrak{F}$ is uniformly continuous. This is denoted $\mathcal{U}_{\mathfrak{g}(\mathfrak{F})}$ $\{\mathfrak{c}_{\mathfrak{g}(\mathfrak{F})}\}$ when all $\xi_\alpha=\mathcal{U}_\alpha$ $\{\xi_\alpha=\mathfrak{c}_\sigma\}$ are entourage {covering} uniformities for Y_α.

Analogous definitions are assumed for $\mathcal{S}_{\mathfrak{g}(\mathfrak{F})}$ and $\mathfrak{Q}_{\mathfrak{g}(\mathfrak{F})}$.

It is best to begin by dispelling some enticing conjectures.

Example 3.10. (a) If each $\xi_\alpha=\mathcal{S}_\alpha\in\mathfrak{S}(X_\alpha)$ in $\mathfrak{F}$ above, then $\mathcal{S}_{\mathfrak{g}(\mathfrak{F})}=\{S\subset Y\times Y:\Delta\subset S$ and $(f_\alpha\times f_\alpha)^{-1}S\in\mathcal{S}_\alpha\ \forall\alpha\in\mathfrak{A}\}$. Note that $\mathcal{S}_{\mathfrak{g}(f)}$ is $(f\times f)\mathcal{S}=\{(f\times f)S:S\in\mathcal{S}\}$ for a single surjection $f:(X,\mathcal{S})\to Y$.

(b) The corresponding statements for $\mathfrak{Q}_{\mathfrak{g}(\mathfrak{F})}$ (therefore $\mathcal{U}_{\mathfrak{g}(\mathfrak{F})}$) do not hold in general. For instance, let $\mathcal{U}$ be the uniformity for $X=\{0,1,2,3\}$ having base $\{U\}$, where $U=\Delta\cup\{(0,1),(1,0),(2,3),(3,2)\}$ and define $f:X\to Y=\{0,1,2\}$ by $f(0)=0$, $f(1)=f(2)=1$, and $f(3)=2$. Then $(f\times f)U=\Delta\cup\{(0,1),(1,0),(1,2),(2,1)\}$ and $\{(f\times f)W\}^2=Y\times Y\not\subset(f\times f)U$ for any $W\in\mathcal{U}$. Consequently, $(f\times f)\mathcal{U}\notin\mathfrak{Q}(Y)$ and $\mathfrak{Q}_{\mathfrak{g}(f)}\neq(f\times f)\mathcal{U}=\{Q\subset Y\times Y:\Delta\in Q$ and $(f\times f)^{-1}\mathfrak{Q}\in\mathcal{U}\}$.

(c) In contrast with Theorem 3.2 no topology $\mathfrak{T}_{\mathcal{S}_{\mathfrak{g}(\mathfrak{F})}}$, $\mathfrak{T}_{\mathfrak{Q}_{\mathfrak{g}(\mathfrak{F})}}$, $\mathfrak{T}_{\mathcal{U}_{\mathfrak{g}(\mathfrak{F})}}=\mathfrak{T}_{\mathfrak{c}_{\mathfrak{g}(\mathfrak{F})}}$ need be compatible with the inductive limit topology $\mathfrak{T}_{\mathfrak{g}(\mathfrak{F})}$. To be sure, let $\mathcal{U}$ be the uniformity for $X=\{0,1,2\}$ having base $\{U_1,U_2\}$, where $U_1=\Delta\cup\{(0,2),(2,0)\}$ and $U_2=U_1\cup\{0,1),(1,0)\}$. If $f:X\to Y=\{0,1\}$ is defined by $f(0)=f(1)=0$ and $f(2)=1$, then $\mathfrak{T}_{\mathfrak{g}(f)}=\{\varnothing,\{0\},Y\}$ (clearly, $\{0\}\in\mathfrak{T}_{\mathfrak{g}(f)}$, since $f^{-1}(0)=\{0,1\}=U_2[1]\in\mathfrak{T}_{\mathcal{U}}$ on X). However, $\mathfrak{S}(Y)=\mathfrak{Q}(Y)=\mathfrak{U}(Y)=Y\times Y$ so that $\mathfrak{T}_{\mathcal{S}_{\mathfrak{g}(f)}}=\mathfrak{T}_{\mathfrak{Q}_{\mathfrak{g}(f)}}=\mathfrak{T}_{\mathcal{U}_{\mathfrak{g}(f)}}=\mathfrak{g}$ (the indiscrete topology) is strictly weaker than $\mathfrak{T}_{\mathfrak{g}(f)}$.

Remark. That $\mathcal{U}_{\mathfrak{g}(\mathfrak{F})}$ is not always compatible with $\mathfrak{T}_{\mathfrak{g}(\mathfrak{F})}$ also follows by noting that $\mathfrak{T}_{\mathfrak{g}(f)}$ above is not completely regular; that is, uniformizible (Theorem 1.21), since $\mathfrak{T}_{\mathfrak{g}(f)}$ is connected and $\aleph(Y)>1$ [Theorem 8.24(ii)].

The duals of Theorem 3.4, Corollary 3.5 are

THEOREM 3.11. *The following are equivalent for surjections* $\mathfrak{F}=\{f_\alpha:(X_\alpha,\mathcal{U}_\alpha)\to Y:\mathcal{U}_\alpha\in\mathfrak{U}(X_\alpha)\ \forall\alpha\in\mathfrak{A}\}$ *and* $\mathcal{U}\in\mathfrak{U}(X)$:

(i) $\mathcal{U} = \mathcal{U}_{\mathfrak{I}(\mathfrak{F})}$,

(ii) $f:(Y, \mathcal{U}) \to (Z, \mathcal{W})$ is uniformly continuous for any uniform space $(Z, \mathcal{W})$ iff every $ff_\alpha:(X_\alpha, \mathcal{U}_\alpha) \to (Z, \mathcal{W})$ is uniformly continuous.

Proof. (i) $\Rightarrow$ (ii) follows from Definition 1.23$^+$ and the fact that $(f_\alpha \times f_\alpha)^{-1}\{(f \times f)^{-1}W\} = (f_\alpha f \times f_\alpha f)^{-1}W \; \forall W \in \mathcal{W}$. For (ii) $\Rightarrow$ (i) consider $f = \mathbb{1}:(Y, \mathcal{U}) \to (Y, \mathcal{U}_{\mathfrak{I}(\mathfrak{F})})$ and $\mathbb{1}:(Y, \mathcal{U}) \to (Y, \mathcal{U})$. ■

Remark. The analogs of Theorem 3.11 and Corollary 3.12 hold for $\mathcal{S}_{\mathfrak{I}(\mathfrak{F})}$ and $\mathfrak{Q}_{\mathfrak{I}(\mathfrak{F})}$.

COROLLARY 3.12. Given a family of surjections $\mathcal{G} = \{g_\alpha:(Y_\alpha, \mathcal{V}_\alpha) \to Z: V_\alpha \in \mathfrak{U}(Y_\alpha) \forall \alpha \in \mathfrak{A}\}$, suppose there is an index set $\mathcal{B}(\alpha)$ and corresponding family of surjections $\mathfrak{F}_{\mathcal{B}(\alpha)} = \{f_\beta: (X_\beta, \mathcal{U}_\beta) \to Y_\alpha: \mathcal{U}_\beta \in \mathfrak{U}(X_\beta) \forall \beta \in \mathcal{B}(\alpha)\} \forall \alpha \in \mathfrak{A}$. If each $\mathcal{V}_\alpha = \mathcal{V}_{\mathfrak{I}\{\mathfrak{F}_{\mathcal{B}(\alpha)}\}}$, then $\mathcal{U}_{\mathfrak{I}(\mathcal{G})} = \mathcal{U}_{\mathfrak{I}(\mathcal{G}\mathfrak{F})}$ on $X \times X$.

Sometimes $\mathcal{U}_{\mathfrak{P}(\mathfrak{F})}$ and $\mathcal{U}_{\mathfrak{I}(\mathfrak{F})}$ occur simultaneously.

Example 3.13. If $f:(X, \mathcal{U}) \to (Y, \mathcal{V})$ is a unimorphism, then $\mathcal{U} = \mathcal{U}_{\mathfrak{P}(\mathfrak{F})}$ and $\mathcal{V} = \mathcal{V}_{\mathfrak{I}(f)}$. Verification: If $V \in \mathcal{V}_{\mathfrak{I}(f)}$, then $(f \times f)^{-1}V \in \mathcal{U}$ and (since f^{-1} is $\mathcal{V} - \mathcal{U}$ uniformly continuous) $V = (f^{-1} \times f^{-1})^{-1}\{(f \times f)^{-1}V\} \in \mathcal{V}$. Thus $\mathcal{V} = \mathcal{V}_{\mathfrak{I}(f)}$. If $U \in \mathcal{U}$, then $(f \times f)U \in \mathcal{V}_{\mathfrak{I}(f)} = \mathcal{V}$ and $U = (f \times f)^{-1}\{(f \times f)U\} \in \mathcal{U}_{\mathfrak{P}(f)}$, which demonstrates that $\mathcal{U} = \mathcal{U}_{\mathfrak{P}(f)}$.

Quotient uniformities represent one important application of inductive limits.

DEFINITION 3.14. Let $\nu:(X, \mathcal{U}) \to X/\mathfrak{R}$ be the canonical mapping of a uniform space $(X, \mathcal{U})$ onto the quotient set (of equivalence classes) determined by an equivalence relation $\mathfrak{R}$ on X. The pair $(X/\mathfrak{R}, \mathcal{U}_{\mathfrak{I}(\nu)})$ is called the quotient uniform space determined by $\mathfrak{R}$.

Example 3.15. Every surjection f from a set X onto a set Y determines an equivalence relation $\mathfrak{R}_f$ on X by taking $(x_1, x_2) \in \mathfrak{R}_f$ iff $f(x_1) = f(x_2)$. The corresponding canonical mapping $\nu: X \underset{x \rightsquigarrow f^{-1}\{f(x)\}}{\longrightarrow} X/\mathfrak{R}_f$ induces a bijection $F_f: X/\mathfrak{R}_f \underset{f^{-1}\{f(x)\} \rightsquigarrow f(x)}{\longrightarrow} Y$ such that $f = F_f \nu$.

Suppose both X, Y are topological {uniform spaces}. Then (Theorem 3.11) f is continuous {uniformly continuous} iff F_f is continuous {uniformly continuous}. In fact (Corollary 3.12), $\mathfrak{T}_{\mathfrak{I}(f)} = \mathfrak{T}_{\mathfrak{I}(F_f)}$ on $Y\{\mathcal{U}_{\mathfrak{I}(f)} = \mathcal{U}_{\mathfrak{I}(F_f)}$ on $Y \times Y\}$. Actually, F_f is a homeomorphism {unimorphism} iff Y carries $\mathfrak{T}_{\mathfrak{I}(F_f)}\{\mathcal{U}_{\mathfrak{I}(F_f)}\}$. We verify this for uniformities (the case for topological spaces is left as a simple exercise). If Y carries $\mathcal{U}_{\mathfrak{I}(F_f)}$ (so that F_f is uniformly continuous) and $V \in \mathcal{U}_{\mathcal{I}(\nu)} \in \mathfrak{U}(X/\mathfrak{R}_f)$, then $(F_f^{-1} \times F_f^{-1})^{-1} V = (F_f \times F_f)V \in \mathcal{U}_{\mathcal{I}(F_f)}$ which shows that F_f^{-1} is $\mathcal{U}_{\mathcal{I}(F_f)} - \mathcal{U}_{\mathcal{I}(\nu)}$ uniformly continuous. The converse follows by Example 3.13.

Every uniform space $(X, \mathcal{U})$ determines an equivalence relation $\mathfrak{R} = \bigcap_{\mathcal{U}} U$ on X. If $(\nu(x), \nu(y)) \in (\nu \times \nu)\{\mathfrak{R} \circ U \circ \mathfrak{R}\}$, then $(x', y') \in \mathfrak{R} \circ U \circ \mathfrak{R}$ for some $x', y' \in X$ such that $\nu(x) = \nu(x')$ and $\nu(y) = \nu(y')$. Accordingly, $(x, x') \in \mathfrak{R}$ and $(x, y') \in \mathfrak{R}^2 \circ U \circ \mathfrak{R} = \mathfrak{R} \circ U \circ \mathfrak{R}$ so that $(x, y) \in \mathfrak{R} \circ U \circ \mathfrak{R}^2 = \mathfrak{R} \circ U \circ \mathfrak{R}$. This sets the stage for

THEOREM 3.16. For each $\mathcal{U} \in \mathfrak{U}(X)$ there is a unique separated uniformity $\mathcal{U}_{\mathcal{I}(\nu)} \in \mathfrak{U}(X/\mathfrak{R})$ compatible with $\mathfrak{T}_{\mathcal{I}(\nu)}$. Furthermore, $\nu: (X, \mathcal{U}) \to (X/\mathfrak{R}, \mathcal{U}_{\mathcal{I}(\nu)})$ is a uniformly continuous, open, closed surjection.

Proof. We first show that $(\nu \times \nu)U = (\nu \times \nu)\{\mathfrak{R} \circ U \circ \mathfrak{R}\}\ \forall U \in \mathcal{U}$. This yields $U \subset (\nu \times \nu)^{-1}\{(\nu \times \nu)U\} = \mathfrak{R} \circ U \circ \mathfrak{R} \subset U^3\ \forall U \in \mathcal{U}$, from which we conclude that $\mathcal{U}_{\mathcal{P}(\nu)} = \mathcal{U}$ has $\{\mathfrak{R} \circ U \circ \mathfrak{R}: U \in \mathcal{U}\}$ as a base. Clearly, $U \subset \mathfrak{R} \circ U \circ \mathfrak{R}$ implies $(\nu \times \nu)U \subset (\nu \times \nu)\{\mathfrak{R} \circ U \circ \mathfrak{R}\}$. If $(x, y) \in \mathfrak{R} \circ U \circ \mathfrak{R}$, then $\mathfrak{R}\{x\} = \mathfrak{R}\{x'\}$ and $\mathfrak{R}\{y\} = \mathfrak{R}\{y'\}$ for some $(x', y') \in U$. Therefore $(\nu \times \nu)(x, y) = \mathfrak{R}\{x'\} = \mathfrak{R}\{y'\} \subset \bigcup_{(x', y') \in U} \mathfrak{R}\{x'\} \times \mathfrak{R}\{y'\} = (\nu \times \nu)U$ yields $(\nu \times \nu)\{\mathfrak{R} \circ U \circ \mathfrak{R}\} \subset (\nu \times \nu)U$ as required. Next, note that $\mathfrak{T}_{\mathcal{U}_{\mathcal{I}(\nu)}} \subset \mathfrak{T}_{\mathcal{I}(\nu)}$ by Theorem 1.23. Suppose, conversely, that $E \subset X/\mathfrak{R}$ is $\mathfrak{T}_{\mathcal{I}(\nu)}$-open and let $\nu(x_0) \in E$. Since $\nu^{-1}(E)$ is $\mathfrak{T}_{\mathcal{P}(\nu)} = \mathfrak{T}_{\mathcal{U}_{\mathcal{P}(\nu)}}$-open, some $W \in \mathcal{U}_{\mathcal{I}(\nu)}$ satisfies $\{(\nu \times \nu)^{-1} W\}\{x_0\} \subset \nu^{-1}(E)$ and $\nu(x_0) \in W\{\nu(x_0)\} \subset E$. Consequently, E is $\mathfrak{T}_{\mathcal{U}_{\mathcal{I}(\nu)}}$-open and $\mathfrak{T}_{\mathcal{I}(\nu)} \subset \mathfrak{T}_{\mathcal{U}_{\mathcal{I}(\nu)}}$ follows. With regard to separation, $(\nu(x), \nu(y)) \in \bigcap_{\mathcal{U}_{\mathcal{I}(\nu)}} V = \bigcap_{\mathcal{U}} (\nu \times \nu)U = \bigcap_{\mathcal{U}} (\nu \times \nu)\{\mathfrak{R} \circ U \circ \mathfrak{R}\}$ implies $(x, y) \in \bigcap_{\mathcal{U}} \mathfrak{R} \circ U \circ \mathfrak{R} = \bigcap_{\mathcal{U}} U^3 = \mathfrak{R}$ and $\nu(x) = \nu(y)$. Therefore $\bigcap_{\mathcal{U}_{\mathcal{I}(\nu)}} V = \Delta \in X/\mathfrak{R} \times X/\mathfrak{R}$. It remains only to verify that ν is an open, closed mapping. If $E \subset X$ is $\mathfrak{T}_{\mathcal{P}(\nu)}$-open and $\nu(x_0) \in \nu(E)$, then $\nu(x_0) = \nu(x_e)$ for some $x_e \in E$. Thus $x_e \in \nu^{-1}(w) \subset E$ and $\nu(x_0) \in w \subset \nu(E)$ for some $\mathfrak{T}_{\mathcal{I}(\nu)}$-nbd $w \subset X/\mathfrak{R}$. In particular, $\nu(E) \in \mathfrak{T}_{\mathcal{I}(\nu)}$ and ν is open. Finally, suppose $E \subset X$ is $\mathfrak{T}_{\mathcal{P}(\nu)}$-closed. If $\nu(x_0) \notin \nu(E)$, then $x_0 \notin E$ and $U\{x_0\} \cap E = \emptyset$ for some $U \in \mathcal{U}$. Therefore $(\mathfrak{R} \circ V \circ \mathfrak{R})\{x_0\} \cap E = \emptyset$ for some $V \in \mathcal{U}$, which means $\{(\nu \times \nu)V\}\{\nu(x_0)\} \cap \nu(E) = \{(\nu \times \nu)(\mathfrak{R} \circ V \circ \mathfrak{R})\}\{\nu(x_0)\} \cap \nu(E) = \emptyset$ and $\nu(x_0) \notin \overline{\nu(E)}$, the conclusion being that $\nu(E) = \overline{\nu(E)}$ and ν is closed. ■

Remark. One calls $(X/\mathfrak{R}, \mathcal{U}_{\mathcal{I}(\nu)})$ the separated uniform space associated with $(X, \mathcal{U})$.

Example 3.17. A pseudometric ρ on a set X generates an equivalence relation $\mathfrak{R}_\rho$ on X by $(x, y) \in \mathfrak{R}_\rho$ iff $\rho(x, y) = 0$. This, in turn, determines a metric $d_\rho(\{x\},\{y\})=\rho(x, y)$ on the quotient set $X/\mathfrak{R}_\rho$ of equivalence classes $\{z\}=\{z'\in X: (z, z')\in\mathfrak{R}_\rho\}$; but $\mathfrak{R}_\rho$ is precisely $\bigcap_{\varepsilon>0} U_{\varepsilon,\rho}$ for $\mathcal{U}_\rho\in\mathfrak{U}(X)$. Thus $X/\mathfrak{R}_\rho$ carries two uniformities $\mathcal{U}_{\mathfrak{g}(\nu)}$ and $\mathcal{U}_{d_\rho}$, which, in fact, are the same since $(\nu\times\nu)^{-1}U_{\varepsilon,d_\rho}=U_{\varepsilon,\rho}\ \forall\varepsilon>0$. It follows, therefor, that $\nu:(X, \mathfrak{T}(\rho))\to(X/\mathfrak{R}_\rho, \mathfrak{T}_{\mathfrak{g}(\nu)}=\mathfrak{T}(d_\rho))$ is a continuous, open, closed surjection.

PROBLEMS

3A. Uniformities for Product Spaces

Assume that $U_\alpha\in\mathfrak{U}(X_\alpha)\ \forall\alpha\in\mathfrak{A}$ and let $X=\prod_{\mathfrak{A}} X_\alpha$ carry U_Π:

(a) Why is it not necessarily true that $(\pi_{\alpha_0}\times\pi_{\alpha_0})^{-1}U_{\alpha_0}=U_{\alpha_0}\times\prod_{\mathfrak{A}-\alpha_0}(X_\alpha\times X_\alpha)$? If $j: X\times X\to\prod_{\mathfrak{A}}(X_\alpha\times X_\alpha)$, then

$(\{x_\alpha\},\{y_\alpha\})\rightsquigarrow\{x_\alpha,y_\alpha\}$

$(\pi_{\alpha_0}\times\pi_{\alpha_0})^{-1}U_{\alpha_0}=j^{-1}\{U_{\alpha_0}\times\prod_{\mathfrak{A}-\alpha_0}(X_\alpha\times X_\alpha)\}\ \forall\alpha_0\in\mathfrak{A}$ and $U_\alpha\in\mathcal{U}_\alpha$.

(b) Show that $\mathcal{U}_\alpha=\mathcal{U}_{\mathfrak{g}(\pi_\alpha)}$ on $X_\alpha\times X_\alpha\ \forall\alpha\in\mathfrak{A}$.

(c) Fix $\gamma=\{\gamma_\alpha\}\in X$. For each $\alpha_0\in\mathfrak{A}$, define $j_{\alpha_0}:(X_{\alpha_0}, \mathcal{U}_{\alpha_0})\to X$ by $j_{\alpha_0}(x_{\alpha_0})=\{\gamma_\alpha+(x_{\alpha_0}-\gamma_{\alpha_0})\delta_{\alpha,\alpha_0}\}$ and let $J=\{j_{\alpha_0}:\alpha_0\in\mathfrak{A}\}$. Then $\mathcal{U}_\Pi\subset\mathcal{U}_{\mathfrak{g}(J)}$ on $X\times X$.

(d) Why is $\mathcal{U}_\Pi$ separated iff every $\mathcal{U}_\alpha$ $(\alpha\in\mathfrak{A})$ is separated?

3B. Products of Extremal Uniformities

(a) If X, Y are compact and T_2, then $\mathcal{U}_\Pi$ is the unique uniformity compatible with the product topology on $X\times Y$.

(b) Suppose X, Y are discrete. Then $\mathcal{U}_\Pi=\mathcal{U}_{\mathfrak{D}}\in\mathfrak{U}(X\times Y)$.

(c) Give [in contrast to (a), (b) above] an example for each of the following situations:

(i) $\mathcal{U}_\Pi$ is not a product of fine uniformities.
(ii) $\mathcal{U}_\Pi$ is not a product of coarse uniformities.
(iii) $\mathcal{U}_\Pi$ is not $\mathcal{U}_{\mathcal{P}\{C(X\times Y;\mathbb{R})\}}$ for $(X, \mathcal{U}_{\mathcal{P}\{C(X;\mathbb{R})\}})\times(Y, \mathcal{U}_{\mathcal{P}\{C(Y;\mathbb{R})\}})$.

Remark. Products of extremal uniformities are further considered in Hager [6], [37].

3C. Inductive Limits and Uniform Modification

(Problem 1D). Given a family of surjections $\mathfrak{F}=\{f_\alpha:(X_\alpha, \mathcal{S}_\alpha)\to Y;\ \mathcal{S}_\alpha\in\mathfrak{S}(X_\alpha)\ \forall\alpha\in\mathfrak{A}\}$, prove

(i) $m\mathcal{S}_{\mathfrak{g}(\mathfrak{F})}\subset\mathcal{U}_{\mathfrak{g}(m\mathfrak{F})}$ on $Y\times Y$, where $m\mathfrak{F}=\{f_\alpha:(X_\alpha, m\mathcal{S}_\alpha)\to Y;\alpha\in\mathfrak{A}\}$,
(ii) $\mathcal{U}_{\mathfrak{g}(\mathfrak{F})}=m\mathcal{S}_{\mathfrak{g}(\mathfrak{F})}$ if $m\mathfrak{F}=\mathfrak{F}$.

3D. Supremum and Infimum as Limits

(a) Show that $\sup_{\mathcal{A}} \{\xi_\alpha \in \mathfrak{U}(X_\alpha) : \alpha \in \mathcal{A}\} = \mathcal{U}_{\mathcal{P}(\mathbb{1})}$, where $\mathbb{1} = \{\mathbb{1}_\alpha = \mathbb{1} : X \to (X_\alpha, \xi_\alpha) ; \alpha \in \mathcal{A}\}$.

(b) Given a family of surjections $\mathfrak{F} = \{f_\alpha : (X_\alpha, \mathcal{S}_\alpha) \to Y ; \mathcal{S}_\alpha \in \mathbb{S}(X_\alpha)\ \forall \alpha \in \mathcal{A}\}$, one has $\mathcal{U}_{\mathfrak{I}(\mathfrak{F})} = \inf_{\mathcal{A}} \{m\mathcal{S}_{\mathfrak{I}(f_\alpha)}\} \in \mathfrak{U}(X)$. [*Hint*: $\mathcal{S}_{\mathfrak{I}(\mathfrak{F})} = \bigcap_{\mathcal{A}} \mathcal{S}_{\mathfrak{I}(f_\alpha)} = \inf_{\mathcal{A}} \mathcal{S}_\alpha$ and Problem 1F(b) applies.]

(c) Verify that $\inf_{\mathcal{A}} \{\mathcal{U}_\alpha \in \mathfrak{U}(X) : \alpha \in \mathcal{A}\} = \mathcal{U}_{\mathfrak{I}(\mathbb{1})}$, where $\mathbb{1} = \{\mathbb{1}_\alpha = \mathbb{1} : (X_\alpha, \mathcal{U}_\alpha) \to X ; \alpha \in \mathcal{A}\}$. [*Hint*: Use (b) and the fact that $m\mathcal{S}_{\mathfrak{I}(\mathbb{1}_\alpha)} = \mathcal{U}_{\mathfrak{I}(\mathbb{1}_\alpha)}\ \forall \alpha \in \mathcal{A}$.]

4. Uniformities and Gages

The key feature of this section is the proved equivalence of complete regularity, uniformizibility, and being gage (i.e., the topology is determined by a family of pseudometrics). Concepts described in terms of uniformities are shown to be described via gages and conversely.

Gage Spaces

Let $\mathcal{G} = \{\rho_\alpha : \alpha \in \mathcal{A}\}$ be a collection of pseudometrics on a set X. One can easily verify that $\mathcal{S} = \{S_{\varepsilon,\rho_\alpha}(x) : x \in X,\ \rho_\alpha \in \mathcal{G},\ \varepsilon > 0\}$ is a subbase $\left(\text{and } \left\{\bigcap_{i=1}^{n} S_{\varepsilon_i,\rho_{\alpha_i}}(x) : S_{\varepsilon_i,\rho_{\alpha_i}}(x) \in \mathcal{S}\right\} \text{ is a base}\right)$ for a topology $\mathfrak{T}(\mathcal{G})$ on X. For each finite index set $a = \{\alpha_1, \alpha_2, \ldots, \alpha_n\} \in \mathcal{A}$ the corresponding pseudometric $\rho_a(x, y) = \max_{1 \le i \le n} \rho_{\alpha_i}(x, y)$ has $S_{\varepsilon,\rho_a}(x) = \bigcap_{i=1}^{n} S_{\varepsilon,\rho_{\alpha_i}}(x)$. Therefore $\{S_{\varepsilon,\rho_a}(x) : x \in X,\ \varepsilon > 0 \text{ and finite } a \subset \mathcal{A}\}$ is also a base for $\mathfrak{T}(\mathcal{G})$.

For each $\rho_\alpha \in \mathcal{G}$ the collection $\mathfrak{b}_{\rho_\alpha} = \{\mathcal{C}_{\varepsilon,\rho_\alpha} = \{S_{\varepsilon,\rho_\alpha}(x) : x \in X\} : \varepsilon > 0\}$ is a base for a covering† uniformity $\mathfrak{c}_{\rho_\alpha}$ compatible with $\mathfrak{T}(\rho_\alpha)$ on X [Example 2.9(b)]. Furthermore, (Problem 2C) the collection $\{\mathfrak{b}_{\rho_\alpha} : \alpha \in \mathcal{A}\}$ is a subbase for $\mathfrak{c}_{\mathcal{G}} = \sup_{\mathcal{A}} \mathfrak{c}_{\rho_\alpha} \in \mathfrak{U}(X)$ which has $\mathfrak{T}_{\mathfrak{c}_{\mathcal{G}}} = \sup_{\mathcal{A}} \mathfrak{T}_{\mathfrak{c}_{\rho_\alpha}} = \sup_{\mathcal{A}} \mathfrak{T}(\rho_\alpha)$ on X.

An especially interesting feature is the fact that $\mathfrak{T}(\mathcal{G}) = \mathfrak{T}_{\mathfrak{c}_{\mathcal{G}}}$ on X, since both topologies share the same subbase sets $S_{\varepsilon,\rho_\alpha}(x)$. This essentially proves the easier part of the next theorem once we introduce

† We take the liberty of using either (entourage or covering) approach to the problem at hand. This free movement between uniformities (e.g., Theorem 4.8) allows greater flexibility but more importantly keeps both related concepts operational for the reader. It is usually a good (and certainly instructive!) exercise to prove a verified result from the dual viewpoint. For instance, each $\mathcal{B}_{\rho_\alpha}$ is a base for $\mathcal{U}_{\rho_\alpha}$ (Example 1.10) and $\{\mathcal{B}_{\rho_\alpha} : \alpha \in \mathcal{A}\}$ is a subbase for $\mathcal{U}_{\mathcal{G}} = \sup_{\mathcal{A}} \mathcal{U}_{\rho_\alpha} \in \mathfrak{U}(X)$ with $\mathfrak{T}_{\mathcal{U}_{\mathcal{G}}} = \sup_{\mathcal{A}} \mathfrak{T}_{\mathcal{U}_{\rho_\alpha}} = \sup_{\mathcal{A}} \mathfrak{T}(\rho_\alpha)$ [Problem 1F(a)]. Furthermore, $\mathfrak{T}(\mathcal{G}) = \mathfrak{T}_{\mathcal{U}_{\mathcal{G}}}$ since each $U_{\varepsilon,\rho_\alpha}[x] = S_{\varepsilon,\rho_\alpha}(x)$.

DEFINITION 4.1. A topological space $(X, \mathfrak{T})$ is called a gage space if there is a collection $\mathcal{G}$ of pseudometrics on X [called a gage for $(X, \mathfrak{T})$] such that $\mathfrak{T} = \mathfrak{T}(\mathcal{G})$. Note therefore that $(X, \mathfrak{T})$ is (pseudo)metrizible iff it admits a gage consisting of a single (pseudo)metric.

THEOREM 4.2. A topological space $(X, \mathfrak{T})$ is uniformizible iff it is a gage space.

Proof. If $\mathcal{G}$ is a gage for $(X, \mathfrak{T})$, then $\mathfrak{T} = \mathfrak{T}(\mathcal{G}) = \mathfrak{T}_{\mathfrak{c}_{\mathcal{G}}}$ assures that $(X, \mathfrak{T})$ is uniformizible with compatible uniformity $\mathfrak{c}_{\mathcal{G}}$. Suppose, conversely, that $\mathfrak{c} \in \mathfrak{U}(X)$ is compatible with $\mathfrak{T}$ and let $\mathfrak{b}_0$ denote the base of open coverings in $\mathfrak{c}$ (Corollary 2.8). Our first objective is to show that each $\mathscr{C} \in \mathfrak{b}_0$ determines a pseudometric $\rho_{\mathscr{C}}$ on X. Let $\mathcal{O}_0 = \{X\}$, take $\mathcal{O}_1 = \mathscr{C}$, and define inductively a sequence of open covers $\{\mathcal{O}_n\} \in \mathfrak{b}_0$ satisfying $\mathcal{O}_{n+1} \overset{*}{<} \mathcal{O}_n \; \forall n \in \mathbb{N}$. For each pair $x, y \in X$, write $\lambda(x, y) = \inf\{2^{-n} : x \in St(y, \mathcal{O}_n)\}$ and define $\rho_{\mathscr{C}}(x, y) = \inf\left\{\sum_{i=1}^{k} \lambda(x_{i-1}, x_i) \text{: finite sets } \{x_0, x_1, \ldots, x_k\} \subset X \text{ such that } x_0 = x \text{ and } x_k = y\right\}$. Then $\rho_{\mathscr{C}}$ is a pseudometric on X and $\rho_{\mathscr{C}}(x, y) \leq \lambda(x, y) \; \forall x, y \in X$ implies $St(y, \mathcal{O}_n) \subset S_{2^{-n+1}, \rho_{\mathscr{C}}}(y) \; \forall y \in X$. Let $\mathfrak{T}(\mathcal{G})$ denote the topology on X determined by $\mathcal{G} = \{\rho_{\mathscr{C}} : \mathscr{C} \in \mathfrak{b}_0\}$. Surely, $\mathfrak{T}(\mathcal{G}) \subset \mathfrak{T}$ since every $\mathfrak{T}(\mathcal{G})$-subbase nbd of $y \in X$ contains a $\mathfrak{T}_{\mathfrak{c}}$-base nbd of y). The reverse inclusion will follow by showing that $S_{2^{-n}, \rho_{\mathscr{C}}}(y) \subset St(y, \mathcal{O}_n) \; \forall y \in X$. It suffices for this to prove that $x_0 \in St(x_k, \mathcal{O}_n)$ for every $\{x_0, x_1, \ldots, x_k\} \subset X$ satisfying $\sum_{i=1}^{k} \lambda(x_{i-1}, x_i) < 2^{-n}$. (Indeed, $\rho_{\mathscr{C}}(x, y) < 2^{-n}$ implies that $\sum_{i=1}^{k} \lambda(x_{i-1}, x_i) < 2^{-n}$ for some $\{x = x_0, x_1, \ldots, x_{k-1}, x_k = y\} \subset X$.) We proceed by induction on k. If $k = 1$, then $\lambda(x_0, x_1) < 2^{-n}$ assures that $x_0 \in St(x_1, \mathcal{O}_m)$ for some $m > n$. Since $m \geq n+1$ and $\mathcal{O}_{n+1} \overset{*}{<} \mathcal{O}_n$, it follows that $x_0 \in St(x_1, \mathcal{O}_n)$. Assume now that $x_0 \in St(x_k, \mathcal{O}_n)$ for every $\{x_0, x_1, \ldots, x_k\}$ with $k \leq N$ such that $\sum_{i=1}^{k} \lambda(x_{i-1}, x_i) < 2^{-n}$. If $\sum_{i=1}^{N+1} \lambda(x_{i-1}, x_i) < 2^{-n}$, let m be the largest integer $(1 \leq m \leq N)$ for which $\sum_{i=1}^{m} \lambda(x_{i-1}, x_i) \leq 2^{-(n+1)}$. Then $\lambda(x_m, x_{m+1}) < 2^{-n}$ so that both x_m and x_{m+1} belong to some common open set $E_{n+1} \in \mathcal{O}_{n+1}$. By inductive hypothesis one also has $x_0 \in St(x_m, \mathcal{O}_{n+1}) \subset St(E_{n+1}, \mathcal{O}_{n+1})$. Since $\sum_{i=1}^{m+1} \lambda(x_{i-1}, x_i) > 2^{-(n+1)}$), it follows that $\sum_{i=m+2}^{N+1} \lambda(x_{i-1}, x_i) = \sum_{i=1}^{N+1} \lambda(x_{i-1}, x_i) - \sum_{i=1}^{m+1} \lambda(x_{i-1}, x_i) < 2^{-(n+1)}$ and (again by hypothesis) $x_{m+1} \in St(x_{N+1}, \mathcal{O}_{n+1})$. Thus $x_{m+1}, x_{N+1} \in \tilde{E}_{n+1}$ for some $\tilde{E}_{n+1} \in \mathcal{O}_{n+1}$. Specifically, $x_{N+1} \in \tilde{E}_{n+1} \subset St(E_{n+1}, \mathcal{O}_{n+1})$. Since both $x_0, x_{N+1} \in St(E_{n+1}, \mathcal{O}_{n+1}) \subset F_n$ for some $F_n \in \mathcal{O}_n$, we have $x_0 \in St(x_{N+1}, \mathcal{O}_n)$. ■

Our next statement confirms that the topological spaces which admit compatible uniformities and gages are the already familiar completely regular spaces.

COROLLARY 4.3. The following properties are equivalent for a topological space $(X, \mathfrak{T})$:

(i) $(X, \mathfrak{T})$ is a gage space;
(ii) $(X, \mathfrak{T})$ is uniformizible;
(iii) $(X, \mathfrak{T})$ is completely regular.

Proof. (i)$\Leftrightarrow$(ii) by Theorem 4.2 and (iii)$\Rightarrow$(ii) by Example 3.8(a). If $\mathfrak{T}=\mathfrak{T}(\mathcal{G})$ for a gage $\mathcal{G}$ and $x_0 \in X-C$ for a closed set $C \subset X$, then $S_{\varepsilon,\rho}(x_0) \subset X-C$ for some $\rho \in \mathcal{G}$ and $\varepsilon>0$. Accordingly, $f(x_0)=0$ and $f(C)=\{1\}$ for $f \in C(X;\mathbb{R})$, defined by $f(x)=\min\{1, [\rho(x_0, x)]/\varepsilon\}$, and we have (i)$\Rightarrow$(iii) as required. ■

Quite understandably the reader may be perturbed. Since (i)$\Rightarrow$(ii) and (ii)$\Leftrightarrow$(iii), the gory details needed to confirm that (ii)$\Rightarrow$(i) seem unnecessary; there are many simple proofs (cf Problem 4E) that (iii)$\Rightarrow$(i). A partial answer to this lies in the following common misconception (cf Problem 4A): If $\xi \in \mathfrak{U}(X)$ is (pseudo)metrizible; that is, $\xi=\xi_\rho$ for some (pseudo)metric ρ on X, then $\mathfrak{T}_\xi=\mathfrak{T}(\rho)$ and so $(X, \mathfrak{T})$ is (pseudo)metrizible. The converse, however, is false. If $(X, \mathfrak{T})$ is (pseudo)metrizible and $\xi \in \mathfrak{U}(X)$ is compatible with $\mathfrak{T}=\mathfrak{T}(\rho)$, it need not be true that ξ is (pseudo)metrizible! (See Problem 4A.) The details of Theorem 4.2 therefore are needed for

COROLLARY 4.4 A uniformity for a set X is pseudometrizible iff it has a countable base.

Proof. If $\mathfrak{c}=\mathfrak{c}_\rho$ for some pseudometric ρ on X, then $\{\mathscr{C}_{2^{-n}}: n \in \mathbb{N}\}$ is a base for $\mathfrak{c}$ (Example 2.9). Suppose, conversely, that $\mathfrak{c} \in \mathfrak{U}(X)$ has a countable base $\mathfrak{b}=\{\mathcal{O}, \mathcal{O}_2, \ldots\}$ whose coverings we can assume are open and satisfy $\mathcal{O}_n \overset{*}{<} \mathcal{O}_{n-1}\ \forall n \in \mathbb{N}$ (Corollary 2.8). Let (Theorem 4.2) ρ_n denote the corresponding pseudometric on X associated with $\mathcal{O}_n \in \mathfrak{b}$. Then $\rho(x, y)=\sum_{n=1}^{\infty} \min\{\rho_n(x, y), 2^{-n}\}$ is a pseudometric on X and we need only show that $\mathfrak{c}=\mathfrak{c}_\rho$. From $S_{\varepsilon,\rho}(x) \subset \bigcap_{n=1}^{\infty} S_{\varepsilon,\rho_n}(x)\ \forall x \in X$, follows $\mathscr{C}_{\varepsilon,\rho} < \mathscr{C}_{\varepsilon,\rho_n}\ \forall n \in \mathbb{N}$. If $\mathcal{O}_n \in \mathfrak{b}$, then (see Theorem 4.2) $S_{2^{-(n+1)},\rho_n}(x) \subset St(x, \mathcal{O}_{n+1})\ \forall x \in X$ and $\mathscr{C}_{2^{-(n+1)},\rho} < \mathscr{C}_{2^{-(n+1)},n} < \{\mathcal{O}_n\} \overset{b}{<} \mathcal{O}_n$. Thus $\mathcal{O}_n \in \mathfrak{c}$ and $\mathfrak{c} \subset \mathfrak{c}_\rho$. For the reverse inclusion let $\varepsilon>0$ be given and choose N sufficiently large to ensure that $\sum_{n=N+1}^{\infty} 2^{-n} <$

$\varepsilon/4$. Since (again see Theorem 4.2) $St(x, \mathcal{O}_n) \subset S_{2^{-n+1},\rho_n}(x)\ \forall x \in X$, one has $\mathcal{O}_n < \{\mathcal{O}_n\} \overset{b}{<} \mathscr{C}_{2^{-n+1},\rho_n}$ so that $\mathscr{C}_{2^{-n+1},\rho_n} \in \mathfrak{c}\ \forall n$. Since $2^{-m+1} < 2^{-n+1}(\varepsilon/4)$ for large enough m, one also obtains $\mathscr{C}_{2^{-n+1}(\varepsilon/4),\rho_n} \in \mathfrak{c}\ \forall n$. Now $\mathscr{H} = \bigwedge_{n=1}^{N} \mathscr{C}_{2^{-n+1}(\varepsilon/4),\rho_n} \in \mathfrak{c}$ and a little computation reveals that $\mathscr{H} < \mathscr{C}_{\varepsilon,\rho}$. This, of course, establishes that $\mathscr{C}_{\varepsilon,\rho} \in \mathfrak{c}$ and $\mathfrak{c}_\rho \subset \mathfrak{c}$. ■

COROLLARY 4.5. A uniformity is metrizible iff it is separated and has a countable base.

Proof. Because of Problem 2A, only sufficiency requires proof. Let $\mathfrak{c} \in \mathfrak{U}(X)$ be separated and have a countable base. If $\mathfrak{c} = \mathfrak{c}_\rho$, where ρ is the pseudometric on X defined in Corollary 4.4, then $\rho(x, y) = 0$ implies $y \in \bigcap_{\varepsilon>0} S_{\varepsilon,\rho}(x)$. Thus $y \in \bigcap_{\mathscr{C} \in \mathfrak{c}} St(x, \mathscr{C}) = \{x\}$ and ρ is a metric on X. ■

Uniformly Continuous Pseudometrics

By definition, a pseudometric $\rho: X \times X \to (\mathbb{R}, \mathscr{U}_{d_1})$ is $\mathscr{U}_{\mathscr{P}(\rho)}$-uniformly continuous, therefore $\mathscr{W}$-uniformly continuous $\forall W \in \mathfrak{U}(X \times X)$ finer than $U_{\mathscr{P}(\rho)}$. Another criterion for uniform continuity is

THEOREM 4.6. A pseudometric $\rho: (X \times X, \mathscr{U} \times \mathscr{U}) \to (\mathbb{R}, \mathscr{U}_{d_1})$ is uniformly continuous iff $\mathscr{U}_\rho \subset \mathscr{U}$ on $X \times X$.

Proof. Recall (Theorem 3.2) that a base for $\mathscr{U} \times \mathscr{U}$ consists of entourages $\pi_1^{-1}(U) \cap \pi_2^{-1}(U) = \{(x, y), (u, v)) \in (X \times X) \times (X \times X) : \langle x, u \rangle \in \mathscr{U}$ and $\langle y, v \rangle \in \mathscr{U}\}$ $(U \in \mathscr{U})$. Suppose ρ is $\mathscr{U} \times \mathscr{U}$-uniformly continuous and let $\varepsilon > 0$. Then $\pi_1^{-1}(U) \cap \pi_2^{-1}(U) \subset (\rho \times \rho)^{-1} V_\varepsilon$ for some $U \in \mathscr{U}$; that is, $|\rho(x, y) - \rho(u, v)| < \varepsilon\ \forall \langle x, u \rangle,\ \langle y, v \rangle \in U$. For $\langle u, v \rangle = \langle y, y \rangle$ this means $\rho(x, y) < \varepsilon\ \forall (x, y) \in U$. Thus $U \subset U_\varepsilon$ and it follows that $\mathscr{U}_\rho \subset \mathscr{U}$. Conversely, $\mathscr{U}_\rho \subset \mathscr{U}$ implies (by an application of the triangle inequality) that $W \subset (\rho \times \rho)^{-1} V_\varepsilon$ for $W = \{(x, y), (u, v)) \in (X \times X) \times (X \times X) : (x, u) \in U_{\varepsilon/2}$ and $(y, v) \in U_{\varepsilon/2}\} \in \mathscr{U}_\rho \times \mathscr{U}_\rho \subset \mathscr{U} \times \mathscr{U}$. Therefore $(\rho \times \rho)^{-1} V_\varepsilon \in \mathscr{U} \times \mathscr{U}$ and ρ is $\mathscr{U} \times \mathscr{U}$-uniformly continuous. ■

The reasoning above confirms that every pseudometric $\rho: X \times X \to (\mathbb{R}, \mathscr{U}_{d_1})$ is $\mathscr{U}_\rho \times \mathscr{U}_\rho$-uniformly continuous. This, in fact, is a special case of

COROLLARY 4.7 Let $\mathcal{G}$ be any collection of pseudometrics on X. Then $\mathscr{U}_{\mathcal{G}}$ (equivalently $\mathfrak{c}_{\mathcal{G}}$) is the smallest uniformity for X such that each $\rho \in \mathcal{G}$ is $\mathscr{U}_{\mathcal{G}} \times \mathscr{U}_{\mathcal{G}}$-uniformly continuous.

Proof. $\mathcal{U}_\rho \subset \mathcal{U}_\mathcal{G} = \sup_\mathcal{G} \mathcal{U}_\rho \ \forall \rho \in \mathcal{G}$. ■

We show next that every uniformity $\mathcal{U}$ is determined by its collection of $\mathcal{U}\times\mathcal{U}$-uniformly continuous pseudometrics (this collection is called the gage of $\mathcal{U}$).

THEOREM 4.8. If $(X, \mathcal{U})$ is a uniform space and $\mathcal{G}$ is the gage of $\mathcal{U}$, then $\mathcal{U}_\mathcal{G} = \mathcal{U}$.

Proof. Clearly $\mathcal{U}_\mathcal{G} \subset \mathcal{U}$ by Corollary 4.7. Now consider any symmetric $U \in \mathcal{U}$. Since $\mathscr{C}_U \in \mathfrak{c}_\mathcal{U}$, there is an open cover $\mathcal{O}_1 \in \mathfrak{c}_\mathcal{U}$ such that $\mathcal{O}_1 \overset{*}{<} \mathscr{C}_U$. If ρ denotes the $\mathcal{O}_1$-corresponding pseudometric on X (Theorem 4.2), then $S_{2^{-1},\rho}(x) \subset St(x, \mathcal{O}_1) \ \forall x \in X$ and it follows that $\mathscr{C}_{1/2,\rho} < \{\mathcal{O}_1\} \overset{b}{<} \mathscr{C}_U$; that is, $\mathscr{C}_{1/2,\rho} \overset{b}{<} \mathscr{C}_U$. Thus (Corollary 2.4) $U_{1/2,\rho} \subset \bigcup_X S_{1/2,\rho}(x) \times S_{1/2,\rho}(x) \subset \left\{\bigcup_X S_{1/2,\rho}(x) \times S_{1/2,\rho}(x)\right\}^2 \subset U_{\mathfrak{c}_U} = U$; but ρ is $\mathcal{U}_\rho \times \mathcal{U}_\rho$-uniformly continuous and $\mathcal{U}_\rho \times \mathcal{U}_\rho \subset \mathcal{U}_\mathcal{G} \times \mathcal{U}_\mathcal{G} \subset \mathcal{U} \times \mathcal{U}$. Therefore ρ is $\mathcal{U}\times\mathcal{U}$-uniformly continuous and belongs to $\mathcal{G}$ so that $U_{1/2,\rho} \in \mathcal{U}_\mathcal{G}$ and $\mathcal{U} \subset \mathcal{U}_\mathcal{G}$ follows. ■

Our first example demonstrates the regrettable fact that the type of symmetry for uniformities in Theorem 4.8 does not carry over to gages in general.

Example 4.9. Let $\mathcal{G}$ denote a collection of pseudometrics on X and let $\tilde{\mathcal{G}}$ be the gage of $\mathcal{U}_\mathcal{G} = \sup_\mathcal{G} \mathcal{U}_\rho$; that is, $\tilde{\mathcal{G}}$ is the collection of all $\mathcal{U}_\mathcal{G} \times \mathcal{U}_\mathcal{G}$-uniformly continuous pseudometrics on X.

(a) Although $\mathcal{G} \subset \tilde{\mathcal{G}}$ (Corollary 4.7), equality need not prevail. For instance, the pseudometric $q(x, y) = \rho_1(x, y) + \rho_2(x, y)$ does not belong to $\mathcal{G} = \{\rho_1, \rho_2\}$. However, $q \in \tilde{\mathcal{G}}$ since $\mathcal{U}_\mathcal{G} = \sup\{\mathcal{U}_{\rho_1}, \mathcal{U}_{\rho_2}\} = \mathcal{U}_q$ (by virtue of $\mathcal{U}_{\rho_1}, \mathcal{U}_{\rho_2} \subset \mathcal{U}_q$ and $U_{\varepsilon/2,\rho_1} \cap U_{\varepsilon/2,\rho_2} \subset U_{\varepsilon,q} \ \forall \varepsilon > 0$).

Remark. Note (Theorem 4.8) that $\mathcal{U}_\mathcal{G} = \mathcal{U}_{\tilde{\mathcal{G}}}$ on $X \times X$. One calls $\tilde{\mathcal{G}}$ the gage generated by $\mathcal{G}$. (See Problem 4C.)

(b) Let $\mathcal{G}^+ = \{\rho_\lambda = \max_\lambda \rho_\alpha$: finite index sets λ and $\rho_\alpha \in \mathcal{G} \ \forall \alpha \in \lambda\}$. From $\mathcal{G} \subset \mathcal{G}^+$ follows $\mathcal{U}_\mathcal{G} \subset \mathcal{U}_{\mathcal{G}^+}$. One also has $\mathcal{U}_{\mathcal{G}^+} \subset \mathcal{U}_\mathcal{G}$, since $U_{\varepsilon,\rho_\lambda} = \bigcap_\lambda U_{\varepsilon,\rho_\alpha}$. Therefore $\mathcal{U}_\mathcal{G} = \mathcal{U}_{\mathcal{G}^+} = \mathcal{U}_{\tilde{\mathcal{G}}}$ and $\mathcal{G}^+ \subset \tilde{\mathcal{G}}$. Note also that $\{U_{\varepsilon,\rho_\lambda} : \rho_\lambda \in \mathcal{G}^+, \varepsilon > 0\}$ is a base for $\mathcal{U}_\mathcal{G}$.

The essence of Theorem 4.8 is that every concept defined in terms of uniformities can be described via gages and conversely. For a glimpse into the nature of such dual results, consider

Example 4.10. Let $\mathcal{U}_{\mathcal{G}}$ be the uniformity determined by a collection $\mathcal{G}$ of pseudometrics on X.

(a) $\mathcal{U}_{\mathcal{G}}$ is separated iff $x \neq y$ implies $\rho(x, y) > 0$ for some $\rho \in \mathcal{G}$. [One says that $\mathcal{G}$ is separating when $\mathcal{U}_{\mathcal{G}}$ is separated. This, of course, is equivalent to $\mathfrak{T}(\mathcal{G})$ being T_2.] The condition is necessary by Problem 1F(a), Theorem 1.19. It is also sufficient, since $\rho_0(x, y) = \varepsilon > 0$ for $\rho_0 \in \mathcal{G}$ implies $(x, y) \notin U_{\varepsilon/2, \rho_0}$ and $(x, y) \notin \bigcap_{\mathcal{G}, \varepsilon > 0} U_{\varepsilon, \rho}$.

(b) The $\mathfrak{T}_{\mathcal{U}_{\mathcal{G}}}$–closure of $A \subset X$ is $\bar{A} = \{x \in X : \rho(x, A) = 0 \ \forall \rho \in \mathcal{G}\}$. To be sure, $x \in \bar{A}$ iff $U[x] \cap A \neq \varnothing \ \forall U \in \mathcal{U}_{\mathcal{G}}$; that is, $S_{\varepsilon,\rho}(x) \cap A \neq \varnothing \ \forall \varepsilon > 0$ and $\rho \in \mathcal{G}$, which is equivalent to $\rho(x, A) = 0 \ \forall \rho \in \mathcal{G}$.

(c) The $\mathfrak{T}_{\mathcal{U}_{\mathcal{G}}}$-interior of $A \subset X$ is $\operatorname{Int} A = \{x \in A : S_{\varepsilon, \rho_\lambda}(x) \subset A$ for some $\varepsilon > 0$ and $\rho \in \mathcal{G}^+\}$. This is simply the note in Example 4.9(b) combined with Corollary 1.18.

(d) A filter $\mathcal{K}$ on X is $\mathcal{U}_{\mathcal{G}}$-Cauchy iff $\forall \varepsilon > 0$ and $\rho \in \mathcal{G}$ there is a set $K \in \mathcal{K}$ such that $\sup_{K \times K} \rho(x, y) < \varepsilon$. The condition is necessary since $U_{\varepsilon,\rho} \in \mathcal{U}_{\mathcal{G}}$ and $K \times K \subset U_{\varepsilon,\rho}$ iff $\sup_{K \times K} \rho(x, y) < \varepsilon$. If $U = \bigcup_{i=1}^{n} U_{\varepsilon_i, \rho_{\alpha_i}}$ $(\rho_{\alpha_i} \in \mathcal{G})$ and $\sup_{K_{\alpha_i} \times K_{\alpha_i}} \rho_{\alpha_i}(x, y) < \varepsilon = \min_{1 \le i \le n} \varepsilon_i$, then $K \times K \subset U$ for $K = \bigcap_{i=1}^{n} K_{\alpha_i} \in \mathcal{K}$.

(e) Suppose $\mathcal{U}_{\mathcal{H}} \in \mathfrak{U}(Y)$ is determined by a collection $\mathcal{H}$ of pseudometrics on Y. Then $f : (X, \mathcal{U}_{\mathcal{G}}) \to (Y, \mathcal{U}_{\mathcal{H}})$ is uniformly continuous iff $(f \times f)^{-1} U_{\varepsilon,d} \in \mathcal{U}_{\mathcal{G}} \ \forall \varepsilon > 0$ and $d \in \mathcal{H}$. Let $d \circ f \times f$ be the pseudometric $\underset{(x_1, x_2) \rightsquigarrow d\{f(x_1) f(x_2)\}}{X \times X \longrightarrow \mathbb{R}}$. Then f is uniformly continuous iff $\forall d \in \mathcal{H}$, some $\bigcap_{i=1}^{n} U_{\delta_i, \rho_i} \subset (f \times f)^{-1} U_{\varepsilon,d} = U_{\varepsilon, d \circ (f \times f)}$ $(\rho_i \in \mathcal{G})$, and this is equivalent to $d \circ (f \times f) \in \mathcal{G} \ \forall d \in \mathcal{H}$.

Problems

4A. (Pseudo)Metrizible Uniformities and Topologies

(a) If $\mathcal{U} = \mathcal{U}_\rho$ for some (pseudo)metric ρ on X, then $\mathfrak{T}_{\mathcal{U}} = \mathfrak{T}(\rho)$; that is, $\mathfrak{T}_{\mathcal{U}}$ is determined by ρ.

(b) Suppose that $(X, \mathfrak{T})$ is (pseudo)metrizible and $\mathcal{U} \in \mathfrak{U}(X)$ is compatible with $\mathfrak{T}(\rho)$. It need not be true that $\mathcal{U} = \mathcal{U}_\rho$. [*Hint*: Consider Examples 1.10 and 1.25.]

(c) Extend the statement in (b) by showing that a $\mathcal{U} \in \mathfrak{U}(X)$ compatible with $\mathfrak{T} = \mathfrak{T}(\rho)$ need not even be (pseudo)metrizible. [*Hint*: Take X as the set of all countable ordinals in the hint above.]

4B. The Uniformity $\mathcal{U}_{\mathcal{G}}$

Let $\mathcal{G}$ be a collection of pseudometrics on a set X.

(a) If $\hat{\rho}(x, y) = \sup_{\rho \in \mathcal{G}} \{\rho(x, y)\}$, then $\mathcal{U}_{\mathcal{G}} \subsetneqq \mathcal{U}_{\hat{\rho}}$ on $X \times X$.

(b) Surely, $\mathbb{1}_\rho =. \mathbb{1} : (X, \mathcal{U}_{\mathcal{G}}) \to (X, \mathcal{U}_\rho)$ is uniformly continuous $\forall \rho \in \mathcal{G}$. Verify that $\mathcal{U}_{\mathcal{G}} = \mathcal{U}_{\mathcal{P}\{\mathbb{1}\}}$, where $\mathbb{1} = \{\mathbb{1}_\rho : \rho \in \mathcal{G}\}$.

(c) The evaluation map $\mathfrak{e} : (X, \mathcal{U}_{\mathcal{G}}) \to \prod_{\mathcal{G}} (X, \mathcal{U}_\rho)$, $x \rightsquigarrow \langle x \rangle_{\mathcal{G}}$ is an into unimorphism. In particular, every uniform space admits an embedding into a product of pseudometric spaces.

(d) Why does it follow that a separated uniform space admits an embedding into a product of metric spaces? [*Hint*: If d_ρ is the metric on $X/\mathfrak{R}_\rho$ (Example 3.17), then $\mathfrak{e} : (X, \mathcal{U}_{\mathcal{G}}) \to \prod_{\mathcal{G}} (X/\mathfrak{R}_\rho, \mathcal{U}_{d_\rho})$, $x \rightsquigarrow \langle \{x\} \rangle_{\mathcal{G}}$ is an into unimorphism.]

4C. The Gage of a Uniformity

(a) Given a collection $\mathcal{G}$ of pseudometrics on X, prove [notation as in Example 4.9(a)] that $\mathcal{G} = \tilde{\mathcal{G}}$ iff $\rho_1 + \rho_2 \in \mathcal{G}\ \forall \rho_1, \rho_2 \in \mathcal{G}$, and a pseudometric $d \in \mathcal{G}$ if $\forall \varepsilon > 0$, some $\rho(x, y) < \delta \Rightarrow d(x, y) < \varepsilon$ $(\rho \in \mathcal{G})$.

(b) If $(X, \mathfrak{T})$ is completely regular, the collection of all continuous pseudometrics on X is the gage for $\mathcal{U}_F$. (Compare this with Problem 1G(f).)

4D. Topological Preservation of Gage Spaces

Completely regular spaces are preserved under surjective homeomorphisms. Therefore (Corollary 4.3) such is also true for the property of being a gage space. Sharpen this as follows:

Let $h : (X, \mathfrak{T}(\mathcal{G})) \to (\tilde{X}, \tilde{\mathfrak{T}})$ be an onto homeomorphism, where $\mathcal{G}$ is a collection of pseudometrics on X. Then $\tilde{\mathcal{G}} = \mathcal{G} \circ (h^{-1} \times h^{-1}) = \{\rho \circ (h^{-1} \times h^{-1}) : (\tilde{x}_1, \tilde{x}_2) \rightsquigarrow \rho\{h^{-1}(\tilde{x}_1), h^{-1}(\tilde{x}_2)\} : \rho \in \mathcal{G}\}$ is a gage for $\tilde{\mathfrak{T}}$. [*Hint*: Confirm the fact that $S_{\rho_0(h^{-1} \times h^{-1}), \varepsilon}(\tilde{x}) = h\{S_{\varepsilon, \rho}(h^{-1}(\tilde{x}))\}\ \forall \varepsilon > 0$, $\tilde{x} \in \tilde{X}$ and $\rho \in \mathcal{G}$.]

4E. Completely Regular Spaces Are Gage Spaces

Use each of the approaches below to prove that a completely regular space $(X, \mathfrak{T})$ is a gage space.

(a) The mapping $\mathbf{e}:(X,\mathfrak{T})\longrightarrow \prod_{C(X;\mathbb{R})} I$ is an embedding for $I=[0,1]$ carrying the usual metric d_1. [*Hint*: The topological product space $\prod_{C(X;\mathbb{R})} I$ is a gage space (Why?). Consider subspaces and homeomorphs of gage spaces.] This proof may be easier if the results of Problem 4F are invoked.

(b) Let $\mathcal{G}=\{\rho_f: f\in C(X;\mathbb{R})\}$, where $\rho_f(x,y)=|f(x)-f(y)|$ is the pseudometric determined by $f\in C(X;\mathbb{R})$. Then $\mathcal{U}_{\mathcal{G}}\in\mathfrak{U}(X)$ is compatible with $\mathfrak{T}$.

4F. Projective Limits of Gages

(a) Let $\mathcal{F}=\{f_\alpha: X\to(Y_\alpha,\mathcal{U}_\alpha);\ \mathcal{U}_\alpha\in\mathfrak{U}(Y_\alpha)\,\forall\alpha\in\mathcal{Q}\}$ and let $\mathcal{G}_{\beta(\alpha)}=\{\rho_\beta:\beta\in B(\alpha)\}$ denote the gage of $\mathcal{U}_\alpha\ \forall\alpha\in\mathcal{Q}$. For each $\alpha\in\mathcal{Q}$ and $\beta\in B(\alpha)$ define $d_{\alpha,\beta}=\rho_\beta\circ(f_\alpha\times f_\alpha):(x_1,x_2)\rightsquigarrow\rho_\beta\{f_\alpha(x_1),f_\alpha(x_2)\}$. Show that $\mathcal{G}\subset\tilde{\mathcal{G}}$, where $\mathcal{G}=\{d_{\alpha,\beta}:\alpha\in\mathcal{Q},\beta\in B(\alpha)\}$ and $\tilde{\mathcal{G}}$ is the gage of $\mathcal{U}_{\mathcal{P}(\mathcal{F})}\in\mathfrak{U}(X)$. Use this to prove that $\mathcal{U}_{\mathcal{G}}=\mathcal{U}_{\tilde{\mathcal{G}}}=\mathcal{U}_{\mathcal{P}(\mathcal{F})}$ on $X\times X$. [Thus $\tilde{\mathcal{G}}$ is the gage (for $\mathcal{U}_{\mathcal{P}(\mathcal{F})}$) generated by $\mathcal{G}$.]

(b) Let $(X,\mathcal{U}_\pi)$ be the product uniform space of $\{(X_\alpha,\mathcal{U}_\alpha):\mathcal{U}_\alpha\in\mathfrak{U}(X_\alpha)\,\forall\alpha\in\mathcal{Q}\}$ and let $\mathcal{G}_{B(\alpha)}$ be the gage of $\mathcal{U}_\alpha\ \forall\alpha\in\mathcal{Q}$. Then $\mathcal{U}_\pi=\mathcal{U}_{\mathcal{G}}$, where $\mathcal{G}=\{d_{\alpha,\beta}=\rho_\beta\circ(\pi_\alpha\times\pi_\alpha):\alpha\in\mathcal{Q},\ \beta\in B(\alpha)$ and $\rho_\beta\in\mathcal{G}_{B(\alpha)}\}\subset\tilde{\mathcal{G}}$. Specifically, $\mathcal{G}$ generates the gage $\tilde{\mathcal{G}}$ for the product uniformity $\mathcal{U}_\Pi$.

(c) If E is a subset of a topological space $(X,\mathfrak{T})$ which admits a gage $\mathcal{G}$, then $\mathcal{G}_E=\{\rho:E\times E\to\mathbb{R}^+;\rho\in\mathcal{G}\}$ is a gage for $(E,\mathfrak{T})$.

4G. Pseudometrizibility Via Countable Entourages

The counterpart to Corollary 4.4 goes as follows: Let $(X,\mathcal{U})$ be a uniform space with base $\{U_n:n\in\mathbb{N}\}$, it being assumed that every U_n is symmetric and satisfies $U_{n+1}^3\subset U_n$.

(a) Define $\lambda(x,y)=0$ if $(x,y)\in\bigcap_{n=1}^{\infty}U_n$ and $\lambda(x,y)=1$ if $(x,y)\in U_n-U_{n+1}$. Then λ is a semipseudometric (Problem 1A) which does not necessarily satisfy the triangle inequality.

(b) This deficiency in triangle inequality is overcome by taking $\rho(x,y)=\inf\left\{\sum_{i=1}^{k}\lambda(x_{i-1},x_i)\text{: all finite sequences } x=x_0,x_1,\ldots,x_k=y\right\}$.

(c) Show that $U_n\subset U_{2^{-n},\rho}\subset U_{n-1}\ \forall n\in\mathbb{N}$ and use this to prove that $\mathcal{U}=\mathcal{U}_\rho$ on $X\times X$. [*Hint*: Prove inductively that $\lambda(x_0,x_n)\le 2\sum_{i=1}^{n}\lambda(x_{i-1},x_i)\ \forall$ sequences $\{x_0,x_1,\ldots,x_n\}\subset X$.]

4H. $\mathcal{F}$-Invariant Gages and Uniformities

Let $\mathcal{F}$ be a collection of bijections from a set X onto a set X. A pseudometric ρ on X is $\mathcal{F}$-invariant iff $\rho=\rho\circ(f\times f)\ \forall f\in\mathcal{F}$. A collection $\mathcal{G}$ of pseudometrics on X is

$\mathfrak{F}$-invariant whenever each $\rho \in \mathcal{G}$ is $\mathfrak{F}$-invariant. Analogously, an entourage $U \in \mathcal{U} \in \mathfrak{U}(X)$ is $\mathfrak{F}$-invariant if $U = (f \times f)^{-1} U \ \forall f \in \mathfrak{F}$, and $\mathcal{U} \in \mathfrak{U}(X)$ is termed $\mathfrak{F}$-invariant if it has an $\mathfrak{F}$-invariant base (i.e., a base of $\mathfrak{F}$-invariant entourages).

Let $\mathcal{U}_{\mathcal{G}} \in \mathfrak{U}(X)$ be determined by a collection $\mathcal{G}$ of pseudometrics on X, let $\tilde{\mathcal{G}}$ be the gage generated by $\mathcal{G}$, and denote the collection of all $\mathfrak{F}$-invariant $\mathcal{U}_{\mathcal{G}} \times \mathcal{U}_{\mathcal{G}}$-uniformly continuous pseudometrics on X by $\mathcal{G}_{\mathfrak{F}}$.

(a) If $\mathcal{G}$ is $\mathfrak{F}$-invariant, then so is $\mathcal{U}_{\mathcal{G}}$. [*Hint*: Show that each $U_{\varepsilon,\rho} (\rho \in \mathcal{G})$ satisfies $U_{\varepsilon,\rho} = U_{\varepsilon,\rho\circ(f\times f)} = (f \times f) U_{\varepsilon,\rho} \ \forall f \in \mathfrak{F}$.]

(b) By definition $\mathcal{G}_{\mathfrak{F}} \subset \tilde{\mathcal{G}}$ and $\mathcal{U}_{\mathcal{G}_{\mathfrak{F}}} \subset \mathfrak{U}_{\tilde{\mathcal{G}}} = \mathcal{U}_{\mathcal{G}}$. Prove that $\mathcal{U}_{\mathcal{G}} = \mathcal{U}_{\mathcal{G}_{\mathfrak{F}}}$ if $\mathcal{U}_{\mathcal{G}}$ is $\mathfrak{F}$-invariant. [*Hint*: $U \cap U^{-1}$ is $\mathfrak{F}$-invariant $\forall \mathfrak{F}$-invariant entourage $U \in \mathcal{U}_{\mathcal{G}}$ (Why?). Assume, therefore, that $\mathcal{B} = \{U_n : n \in \mathbb{N}\}$ is a base for $\mathcal{U}_{\mathcal{G}}$ with each symmetric entourage U_n satisfying $U_{n+1}^3 \subset U_n$. The pseudometric ρ on X (Problem 4G) is $\mathfrak{F}$-invariant, $\mathcal{U}_{\mathcal{G}} \times \mathcal{U}_{\mathcal{G}}$ uniformly continuous, and satisfies $U_{2^{-n-1},\rho} \subset U_n \ \forall n \in \mathbb{N}$.]

(c) Use (a), (b) to prove that a $\mathcal{U} \in \mathfrak{U}(X)$ is $\mathfrak{F}$-invariant iff the collection of all $\mathfrak{F}$-univariant $\mathcal{U} \times \mathcal{U}$-uniformly continuous pseudometrics on X is the gage for $\mathcal{U}$.

(d) If $\mathcal{U} \in \mathfrak{U}(X)$ has a countable base and is $\mathfrak{F}$-invariant, then $\mathcal{U} = \mathcal{U}_\rho$ for some $\mathfrak{F}$-invariant pseudometric ρ on X.

References for Further Study

On Uniformities and Gages: Dugundji [25]; Kelley [50].

5. Total Boundedness

A subset E of a pseudometric space (X, ρ) may be called "ε-sized" if its $\operatorname{diam} E = \sup_{E \times E} \rho(x, y) < \varepsilon$. If $\{x_n\}$ is a sequence of X, then $\{T_n = \{x_m : m \geq n\};\ n \in \mathbb{N}\}$ is the base for a filter $\mathfrak{F}$ in X (called the filter associated with $\{x_n\}$). If $\{x_n\}$ is ρ-Cauchy and $\varepsilon > 0$, then $\operatorname{diam} T_m < \varepsilon$ for large enough m and so $\mathfrak{F}$ contains and ε-sized set $\forall \varepsilon > 0$. The extension of these notions to uniform spaces begins with

DEFINITION 5.1. Let $(X, \mathcal{U})$ be a uniform space. A set $E \subset X$ is said to be U-sized ($U \in \mathcal{U}$) if $E \times E \subset U$. By a $\mathcal{U}$-Cauchy filter is meant a filter $\mathfrak{F}$ in X which contains a U-sized set $\forall U \in \mathcal{U}$.

Remark. Every $\{\mathcal{U}_E = \mathcal{U} \cap E \times E$-Cauchy$\}$ filter $\mathcal{G}$ in E is the base for a $\{\mathcal{U}$-Cauchy$\}$ filter $\mathcal{G}\langle X \rangle = \{A \subset X : A$ contains some $G \in \mathcal{G}\}$ in X. Note for instance, that $x \in \bar{E} \subset X$ implies $\{\mathcal{U}[x] \cap E\}\langle E \rangle$ is $\mathcal{U}_E$-Cauchy ($\forall U_E = U \cap (E \times E)$ and $V \in \mathcal{U}$ such that $V^{-1} \circ V \subset U$, the set $V[x] \cap E$ is U_E-sized).

THEOREM 5.2. (i) In a uniform space every convergent filter is Cauchy and every Cauchy filter converges to its accumulation points.

(ii) The uniformly continuous image of a Cauchy filter is Cauchy.

Proof. (i) Suppose the filter $\mathcal{F}$ converges to x_0 in $(X, \mathcal{U})$. Given any $U \in \mathcal{U}$, choose a symmetric $V \in \mathcal{U}$ contained in U. Since $F \subset V[x_0]$ for some $F \in \mathcal{F}$, one has $F \times F \subset V \subset U$ and confirmation that $\mathcal{F}$ is Cauchy. Assume next that $\mathcal{F}$ is Cauchy and accumulates at $x_0 \in X$. For any $U \in \mathcal{U}$ let W be a closed symmetric entourage contained in $\mathcal{U}$. Then $\bar{F}_0 \times \bar{F}_0 = \overline{F_0 \times F_0} \subset \bar{W} = W$ for some $F_0 \in \mathcal{F}$ and from $x_0 \in \bigcap_{\mathcal{F}} \bar{F} \subset \bar{F}_0 \subset W[x_0] \subset U[x_0]$ follows $\mathcal{F} \to x_0 \in X$.

(ii) Suppose $f: (X, \mathcal{U}) \to (Y, \mathcal{V})$ is a uniformly continuous surjection and $\mathcal{F}$ is Cauchy in X. For each $V \in \mathcal{V}$, there is a $U \in \mathcal{U}$ satisfying $(f \times f)U \subset V$. Since $F \times F \subset U$ for some $F \in \mathcal{F}$, it follows that $f(F) \times f(F) \subset (f \times f)U \subset V$. Thus $f(\mathcal{F}) = \{f(F) : F \in \mathcal{F}\}$ is $\mathcal{V}$-Cauchy. ■

A pseudometric space (X, ρ) is totally bounded if $\forall \varepsilon > 0$ there is a finite sequence of ε-sized subsets which cover X. This, of course, is equivalent to requiring that X have a finite cover $\{S_\varepsilon(x_i) : 1 \le i \le n(\varepsilon)\}$ $\forall \varepsilon > 0$. The corresponding generalization is

DEFINITION 5.3. A uniform space $(X, \mathcal{U})$ is totally bounded if X has a finite U-sized cover $\forall U \in \mathcal{U}$. The collection of all totally bounded uniformities for a set X is noted $\mathfrak{U}_{tb}(X)$.

One calls $E \subset (X, \mathcal{U})$ totally bounded in X whenever $\mathcal{U}_E \in \mathfrak{U}_{tb}(E)$; that is, $(E, \mathcal{U}_E)$ is totally bounded.

Total boundedness may be characterized and related to compactness by way of filters.

THEOREM 5.4. The following are equivalent for $(X, \mathcal{U})$:

(i) $\mathcal{U} \in \mathfrak{U}_{tb}(X)$.
(ii) Every maximal filter in X is Cauchy.
(iii) Every filter in X is contained in a Cauchy filter.

Proof. If $U \in \mathcal{U}$ in $\mathfrak{U}_{tb}(X)$, then X is covered by U-sized sets $\{E_1, \ldots, E_n\}$. Since every filter in X contains X, every maximal filter contains some $E_i \subset X$ and is therefore Cauchy. Thus (i)$\Rightarrow$(ii)$\Leftrightarrow$(iii) (the latter equivalence is clear since every filter is contained in a maximal filter) and it remains to prove (ii)$\Rightarrow$(i). Assume $\mathcal{U} \in \mathfrak{U}_{tb}(X)$ so that there exist $U, V \in \mathcal{U}$ with $V^{-1} \circ V \subset U$ such that no finite collection $\{V[x_i] : i = 1, 2, \ldots, n\}$ (of $V^{-1} \circ V$-sized sets!) covers X. Since the finite intersections of sets $A_x = X - V[x]$ are therefore nonempty, there is a maximal filter $\mathcal{M} \supset A = \{A_x : x \in X\}$. (Simply order the set of filters $\{\mathcal{F} \supset A : \bigcap_{i=1}^{n} F_{\alpha_i} \neq \varnothing$

for $F_{\alpha_i} \in \mathfrak{F}\}$ by inclusion and use Zorn's lemma.) Furthermore, every V-sized set $E \subset X$ is contained in some $V[x]$ so that $\mathfrak{M}$ cannot be Cauchy (clearly, $E \in \mathfrak{M}$ implies $E \cap X - V[x] \neq \varnothing$ which is impossible). ■

A compact uniform space is totally bounded [Theorems 5.4 and 5.1(i) with Appendix t.7]. In fact,

COROLLARY 5.5 A countably compact uniform space is totally bounded.

Proof. If $(X, \mathcal{U})$ is not totally bounded, X has no U-sized cover for some $U \in \mathcal{U}$. Given any $x_1 \in X$, there is an $x_2 \in X$ such that $(x_1, x_2) \notin U$ (otherwise $X \subset U[x_1]$). Similarly, there is an $x_3 \in X$ such that $(x_3, x_i) \notin U$ $(i = 1, 2)$. Proceed inductively to obtain a sequence $\{x_n\} \in X$ such that $(x_m, x_n) \notin U$ for $m \neq n$. Then $\{x_n\}$ has no accumulation point (otherwise $(x_m, x_n) \in U$ for large enough $m, n \in \mathbb{N}$). ■

A uniform space $(X, \mathcal{U})$ is totally bounded iff it has a base, each of whose entourages U affords a finite U-sized cover of X. The following characterization of total boundedness will also be useful. For a {finite} cover $\mathscr{C}$ of X, call $\bigcup_{\mathscr{C}} E \times E$ a {finite} square cover of $\Delta \in X \times X$.

THEOREM 5.6. The following are equivalent for $(X, \mathcal{U})$:

(i) $\mathcal{U} \in \mathfrak{U}_{tb}(X)$.

(ii) $\forall U \in \mathcal{U}$, there is a finite set $\{x_1, \ldots, x_n\} \subset X$ such that $X = \bigcup_{i=1}^{n} U[x_i]$.

(iii) $\mathcal{U}$ has a base of finite square covers of Δ.

Proof. (i)$\Rightarrow$(ii). For any $U \in \mathcal{U} \in \mathfrak{U}_{tb}(X)$ choose a symmetric entourage $V \subset U$ and let $X = \bigcup_{i=1}^{n} A_i$, where each A_i is V-sized. Taking $x_i \in A_i$ $(i = 1, 2, \ldots, n)$ yields $X \subset \bigcup_{i=1}^{n} V[x_i] \subset \bigcup_{i=1}^{n} U[x_i]$. Next (ii)$\Rightarrow$(iii). Given any $V \in \mathcal{U}$, choose a symmetric $U \in \mathcal{U}$ such that $U^4 \subset V$ and assume $X = \bigcup_{i=1}^{n} U[x_i]$. Then $W = \bigcup_{i=1}^{n} U^2[x_i] \times U^2[x_i] \subset V$ and we need only verify that $W \in \mathcal{U}$. Any fixed $y \in X$ belongs to some $U[x_i]$. Therefore $U[y] \subset U^2[x_i] \subset W[y]$, and it follows that $U^2 \subset W$ and $W \in \mathcal{U}$. Obviously (iii)$\Rightarrow$(i). ■

Total boundedness is a property of uniformities. One $\mathscr{U} \in \mathfrak{U}(X)$ may be totally bounded, whereas another uniformity compatible with $\mathfrak{T}_{\mathscr{U}}$ may not.

Example 5.7. The usual metric $d_1(x, y) = |x - y|$ is equivalent to the metric $\rho(x, y) = |x/(1+|x|) - y/(1+|y|)|$ on $\mathbb{R}$, since $\underset{x \rightsquigarrow x/(1+|x|)}{\mathbb{R} \to (-1, 1)}$ is a homeomorphism. Therefore $\mathscr{U}_{d_1} \notin \mathfrak{U}_{tb}(\mathbb{R})$ and $\mathscr{U}_\rho \in \mathfrak{U}_{tb}(\mathbb{R})$ are both compatible with $\mathfrak{T}(d_1) = \mathfrak{T}(\rho)$ on $\mathbb{R}$.

Our conspicuous lack of attention to covering uniformities is explainable by Theorem 2.3. Specifically, all concepts for $\mathfrak{c} \in \mathfrak{U}(X)$ can be defined in terms of $\mathscr{U}_\mathfrak{c} \in \mathfrak{U}(X)$.

Example 5.8. (a) the standard definition that $E \subset (X, \mathfrak{c})$ is $\mathscr{C}$-sized ($\mathscr{C} \in \mathfrak{c}$) if $E \subset G$ for some $G \in \mathscr{C}$ means precisely that E is $U_\mathscr{C}$-sized. Similarly, defining $(X, \mathfrak{c})$ to be totally bounded if every $\mathscr{C} \in \mathfrak{c}$ has a finite subcover is equivalent to X having a finite $U_\mathscr{C}$-sized cover $\forall \mathscr{C} \in \mathfrak{c}$.

(b) In terms of sequences, $\{x_n\} \in (X, \mathscr{U})$ is Cauchy iff $\forall U \in \mathscr{U}$ one has $(x_m, x_n) \in U$ for $m, n > N(U)$. Analogously, $\{x_n\} \in (X, \mathfrak{c})$ is Cauchy iff $\forall \mathscr{C} \in \mathfrak{c}$ there is an $N(\mathscr{C})$ such that $x_m, x_n \in G$ for some $G \in \mathscr{C}$.

Invariance properties of total boundedness include

THEOREM 5.9. (i) Every subset and finite union of totally bounded sets is totally bounded.

(ii) The closure of a totally bounded set is totally bounded.

(iii) Total boundedness is preserved by uniformly continuous surjections.

Proof. Since (i) is clear, we begin with (ii). If $E \subset (X, \mathscr{U})$ is totally bounded and V is a closed entourage in $\mathscr{U}$, then $E \subset \bigcup_{i=1}^{n} V[x_i]$ implies $\bar{E} \subset \overline{\bigcup_{i=1}^{n} V[x_i]} = \bigcup_{i=1}^{n} V[x_i]$ and Corollary 1.18 completes our proof. For (iii) let $f: (X, \mathscr{U}) \to (Y, \mathscr{V})$ be a uniformly continuous surjection and suppose $\mathscr{U} \in \mathfrak{U}_{tb}(X)$. If $V \in \mathscr{V}$, then $W = (f \times f)^{-1} V \in \mathscr{U}$ and $X = \bigcup_{i=1}^{n} W[x_i]$ implies $Y = f(X) \subset \bigcup_{i=1}^{n} f\{W[x_i]\} \subset \bigcup_{i=1}^{n} V[f(x_i)]$. ■

THEOREM 5.10. Let $\mathfrak{F} = \{f_\alpha : X \to (Y_\alpha, \mathscr{V}_\alpha); \mathscr{V}_\alpha \in \mathfrak{U}(Y_\alpha)\ \forall \alpha \in \mathscr{A}\}$. Then $\mathscr{I}_{\mathscr{P}(\mathfrak{F})} \in \mathfrak{U}_{tb}(X)$ iff every $f_\alpha(X) \subset (Y_\alpha, \mathscr{V}_\alpha)$ is totally bounded.

Proof. The necessary part of the proof follows from Theorem 5.9 (iii) by definition of $\mathscr{U}_{\mathscr{P}(\mathfrak{F})}$ and Definition 5.3. For the sufficiency proof assume (for convenience only!) that each $f_\alpha \in \mathfrak{F}$ is surjective and let $U = \bigcap_{i=1}^{n} (f_{\alpha_i} \times f_{\alpha_i})^{-1} V_{\alpha_i}$ $(V_{\alpha_i} \in \mathscr{V}_{\alpha_i}, \alpha_i \in \mathscr{A})$ be any $\mathscr{U}_{\mathscr{P}(\mathfrak{F})}$-base entourage. Since each $\mathscr{V}_{\alpha_i} \in \mathfrak{U}_{tb}(Y_{\alpha_i})$, one has $Y_{\alpha_i} = \bigcup_{j=1}^{n_{\alpha_i}} E_j^{(\alpha_i)}$ with $E_j^{(\alpha_i)} \times E_j^{(\alpha_i)} \subset V_{\alpha_i}$ and $X = f_{\alpha_i}^{-1}(Y_{\alpha_i}) = \bigcup_{j=1}^{n_{\alpha_i}} f_{\alpha_i}^{-1}(E_j^{(\alpha_i)})$ for each $i = 1, 2, \ldots, n$. In other words, X is covered by $\mathscr{C} = \bigwedge_{i=1}^{n} \mathscr{C}_{\alpha_i}$, where $\mathscr{C}_{\alpha_i} = \{f_{\alpha_i}^{-1}(E_j^{(\alpha_i)}) : j = 1, 2, \ldots, n_{\alpha_i}\}$. It remains only to note that each $\bigcap_{i=1}^{n} f_{\alpha_i}^{-1}(E_j^{(\alpha_i)}) \in \mathscr{C}$ is U-sized since $f_{\alpha_i}^{-1}(E_j^{(\alpha_i)}) \times f_{\alpha_i}^{-1}(E_j^{(\alpha_i)}) \subset (f_{\alpha_i} \times f_{\alpha_i}) V_{\alpha_i}$ $\forall i = 1, 2, \ldots, n$ and $j = 1, 2, \ldots, n_{\alpha_i}$. ■

Theorem 5.9(i) follows from Example 3.6(a). From Example 4.7(a) follows

COROLLARY 5.11. A product of uniform spaces is totally bounded iff each factor space is totally bounded.

Combining Theorem 5.10 with Theorem 3.16 also yields

COROLLARY 5.12. A uniform space is totally bounded iff its associated separated uniform space is totally bounded.

Example 5.13 (Examples 5.7 and 3.8). Since $(\mathbb{R}, \mathscr{U}_{d_1})$ is not totally bounded, neither is $(X, \mathscr{U}_{\mathscr{P}\{C(X;\mathbb{R})\}})$. Note however that $\mathscr{U}_{\mathscr{P}\{CB(X;\mathbb{R})\}} \in \mathfrak{U}_{tb}(X)$, since $(f(X), \mathscr{U}_{d_1})$ is totally bounded $\forall f \in CB(X; \mathbb{R})$.

Recall (Corollary 4.3, Theorem 1.19) that a topological space $(X, \mathfrak{T})$ has a compatible, separated uniformity iff $\mathfrak{T}$ is $T_{3\frac{1}{2}}$ (i.e., completely regular and T_2). We can now prove

THEOREM 5.14. A topological space $(X, \mathfrak{T})$ is $T_{3\frac{1}{2}}$ iff there is a separated $\mathscr{U} \in \mathfrak{U}_{tb}(X)$ compatible with $\mathfrak{T}$.

Proof. The necessary part only needs verification. Let $\mathbf{e} : (X, \mathfrak{T}) \to \mathbf{e}(X) \subset \prod_{C(X;\mathbb{R})} I$, $x \rightsquigarrow \{f(x)\}_f$, denote the standard embedding of $(X, \mathfrak{T})$ into the product of uniform spaces $I = [0, 1]$ with $\mathscr{U}_{d_1}$. Since $(I, \mathscr{U}_{d_1})$ is totally bounded and separated, the product uniformity $\mathcal{U}_\pi$ of $\{\mathcal{U}_{d_1} \in \mathfrak{U}(I_f = I) : f \in C(X; \mathbb{R})\}$ has these properties and so therefore has $\mathcal{U}_{\mathcal{P}(\mathrm{e})} \in \mathfrak{U}(X)$. The proof will be complete once we show that $\mathfrak{T}_{\mathcal{U}_{\mathcal{P}(\mathrm{e})}} = \mathfrak{T}$. First, $\mathcal{U}_{\mathcal{P}(\mathrm{e})} = \mathcal{U}_{\mathcal{P}\{C(X;I)\}}$ is compatible with $\mathfrak{T}_{\mathcal{P}(\mathrm{e})} = \mathfrak{T}_{\mathcal{P}\{C(X;I)\}} \subset \mathfrak{T}_{\mathcal{P}\{C(X;\mathbb{R})\}} = \mathfrak{T}$ [Problem 3C, Theorem 3.2, Example 3.8(a)]. On the other hand, $u \in \mathfrak{T}$ implies $\mathrm{e}(u) \in \mathfrak{T}_\pi$ in $\mathrm{e}(X)$ and so $u = \mathrm{e}^{-1}\{\mathrm{e}(u)\} \in \mathfrak{T}_{\mathcal{P}(\mathrm{e})}$. Thus $\mathfrak{T} \subset \mathfrak{T}_{\mathcal{P}(\mathrm{e})}$. ■

PROBLEMS

5A. The Cech Uniformity $\mathcal{U}_{\beta(X)}$

The finest totally bounded uniformity compatible with $(X, \mathfrak{T})$ is called the Cech uniformity, denoted $\mathcal{U}_{\beta(X)}$.

(a) $\mathcal{U}_{\beta(X)} = \mathcal{U}_{\mathcal{P}\{CB(X;\mathbb{R})\}} \subset \mathcal{U}_{\mathcal{P}\{C(X;\mathbb{R})\}}$.

(b) A mapping $f: (X, \mathcal{U}_{\beta(X)}) \to (\mathbb{R}, \mathcal{U}_{d_1})$ is uniformly continuous iff $f \in CB(X; \mathbb{R})$.

(c) Let $\mathfrak{F}$ denote the uniformly continuous functions from a uniform space $(X, \mathcal{U})$ into $(I, \mathcal{U}_{d_1})$. Then $\mathcal{U}_{\mathcal{P}\{\mathfrak{F}\}}$ is the finest totally bounded uniformity compatible with $\mathfrak{T}_\mathcal{U}$ which is coarser than $\mathcal{U}$. In particular, $\mathcal{U}_{\mathcal{P}\{C(X;I)\}} = \mathcal{U}_{\mathcal{P}\{CB(X;\mathbb{R})\}} = \mathcal{U}_{\beta(X)}$.

5B. Pseudocompactness and Total Boundedness

A topological space $(X, \mathfrak{T})$ is pseudocompact iff $C(X; \mathbb{R}) = CB(X; \mathbb{R})$; that is, every real-valued continuous function on X is bounded. This is equivalent to requiring that $f \in C(X; \mathbb{R})$ attain its maximum and minimum value on X.

(a) For a $T_{3\frac{1}{2}}$ space $(X, \mathfrak{T})$, the following are equivalent: (i) $(X, \mathfrak{T})$ is pseudocompact; (ii) $(X, \mathcal{U})$ is totally bounded $\forall \mathcal{U} \in \mathfrak{U}(X)$ compatible with $\mathfrak{T}$; (iii) $\mathcal{U}_F = \mathcal{U}_{\beta(X)}$ on $X \times X$ (i.e., $\mathcal{U}_F$ is totally bounded). [*Hint*: Note that $(X, \mathfrak{T})$ is pseudocompact iff each countable open cover of X has a finite subfamily whose closures cover X.]

(b) Every complete, pseudocompact uniform space is fine. [*Hint*: If $(X, \mathcal{U})$ is complete and $\mathfrak{T}_\mathcal{U}$ pseudocompact, then $\mathfrak{T}_\mathcal{U}$ is compact.]

6. Completeness

The pseudometric notions of completion and compactification are generalized to uniform spaces. It is then shown that there is a bijection between the T_2 compactifications and totally bounded separated uniformities compatible with a $T_{3\frac{1}{2}}$ space. What follows is information about the existence of $\mathscr{U}_c$ as well as a characterization of $T_{3\frac{1}{2}}$ spaces which have a unique compatible uniformity.

DEFINITION 6.1. A uniform space $(X, \mathscr{U})$ is complete {sequentially complete} if every Cauchy filter {sequence} in X converges. In such a situation $\mathscr{U}$ is termed a complete {sequentially complete} uniformity for X.

A subset $E \subset (X, \mathscr{U})$ is complete {sequentially complete} in $(X, \mathscr{U})$ whenever $(E, \mathscr{U}_E)$ is complete {sequentially complete}.

A few pertinent remarks are illustrated in

Example 6.2. (a) $\mathscr{U}_{\mathfrak{I}} \in \mathfrak{U}(X)$ is complete since $\{X\}$ is the only $\mathscr{U}_{\mathfrak{I}}$-Cauchy filter in X and $\{X\}$ converges to every $x \in X$. At the other extreme, $\mathscr{U}_{\mathfrak{D}} \in \mathfrak{U}(X)$ is complete, since $\Delta \in \mathscr{U}_{\mathfrak{D}}$ implies that $x = \Delta[x]$ is $\Delta^{-1} \circ \Delta = \Delta$-sized and every $\mathscr{U}_{\mathfrak{D}}$-Cauchy filter contains some one point set $\{x\} \in X$ to which it converges. Since $\mathscr{U}_{\mathfrak{I}} \subset \mathscr{V} \subset \mathscr{U}_{\mathfrak{D}}\ \forall \mathscr{V} \in \mathfrak{U}(X)$, completeness and order by inclusion are not necessarily related.

(b) Suppose $\mathscr{U} \subset \mathscr{V}$, where $\mathscr{U}, \mathscr{V} \in \mathfrak{U}(X)$ satisfy $\mathfrak{T}_{\mathscr{U}} = \mathfrak{T}_{\mathscr{V}}$ on X. Then $\mathscr{U}$-completeness implies $\mathscr{V}$-completeness, since a $\mathscr{V}$-Cauchy filter is $\mathscr{U}$-Cauchy, therefore $\mathfrak{T}_{\mathscr{U}} = \mathfrak{T}_{\mathscr{V}}$ convergent.

(c) Complete uniform spaces are sequentially complete and sequentially complete pseudometric spaces are complete. It need not be true, however, that a sequentially complete uniform space $(X, \mathscr{U})$ be complete—even if $\mathfrak{T}_{\mathscr{U}}$ is 1°! Reason: A sequentially compact uniform space is sequentially complete and totally bounded. Therefore (Theorem 6.7) a sequentially compact, noncompact uniform space is sequentially complete, noncomplete (totally bounded). The first ordinal space is an example of one such uniform space.

The preceding discussion serves as a good introduction to

THEOREM 6.3. A uniform space with a countable base (i.e., a pseudometrizible uniform space) is complete iff it is sequentially complete.

Proof. The sufficiency part only requires verification. Let $\mathscr{B} = \{V_n\colon V_{n+1}^2 \subset V_n;\ n \in \mathbb{N}\}$ be a base of symmetric entourages for a sequentially complete $\mathscr{U} \in \mathfrak{U}(X)$ and let $\mathfrak{F}$ be any $\mathscr{U}$-Cauchy filter in X. If $F_n \in \mathfrak{F}$ is V_n-sized and $x_n \in F_n\ \forall n \in \mathbb{N}$, then $\{x_n\}$ is $\mathscr{U}$-Cauchy and so converges to some $x_0 \in X$. Now for any $V_m \in \mathscr{B}$ one has $x_n \in V_m[x_0]$ and $F_n \cap V_m[x_0] \neq \varnothing\ \forall n \geq N = N(m)$. In particular, $F_{N(m)+m} \subset V_m[x_0]$ and $V_m[x_0] \in \mathfrak{F}$, which proves that $\mathfrak{F} \to x_0 \in X$. ■

Lest the reader cry bias, we reprove Theorem 6.3 for a uniform space $(X, \mathfrak{c})$. The arguments parallel those above and should be helpful in proving the covering counterparts of other results.

Proof. Let $\mathfrak{b}=\{\mathscr{C}_n : \mathscr{C}_{n+1} \overset{*}{<} \mathscr{C}_n \ \forall n \in \mathbb{N}\}$ be a base for a sequentially complete $\mathfrak{c} \in \mathfrak{U}(X)$ and consider any $\mathfrak{c}$-Cauchy filter $\mathfrak{F}$ in X. If $x_n \in F_n$, where $F_n \in \mathfrak{F}$ is $\mathscr{C}_n$-sized, then $\{x_n\}$ is $\mathfrak{c}$-Cauchy. (If $\mathscr{C}_k \in \mathfrak{b}$ and $n \geq k+1$, then $x_n \in F_n \subset G_n$ for some $G_n \in \mathscr{C}_n \overset{*}{<} \mathscr{C}_k$. Thus $m, n \geq k+1$ and $z \in F_m \cap F_n$ imply $x_m, x_n \in St(z, \mathscr{C}_{k+1}) \subset G_k$ for some $G_k \in \mathscr{C}_k$.) If $x_n \to x_0 \in X$ and $St(x_0, \mathscr{C}_m)$ is any $\mathfrak{T}_\mathfrak{c}$-nbd of x_0, there is a $N = N(m+2)$ such that $n > N$ implies $x_n \in St(x, \mathscr{C}_{m+2}) \subset G_{m+1}$ for some $G_{m+1} \in \mathscr{C}_{m+1}$. Since $x_n \in F_n \subset G_{m+1} \in \mathscr{C}_{m+1} \ \forall n > N+(m+1)$, we have $F_m \subset St(G_{m+1}, \mathscr{C}_{m+1}) \subset G_m$ for some $G_m \in \mathscr{C}_m$; but $x_0 \in G_m$, so that $F_m \subset St(x_0, G_m)$; that is, $St(x_0, G_m) \in \mathfrak{F}$ and $\mathfrak{F} \to x_0 \in X$. ■

COROLLARY 6.4. A pseudometric ρ on X is complete iff $(X, \mathscr{U}_\rho)$ is complete.

Proof. A sequence in X is ρ-Cauchy iff it is $\mathscr{U}_\rho$-Cauchy. Since $\mathfrak{T}(\rho) = \mathfrak{T}_{\mathscr{U}_\rho}$, the sequential completeness of p is equivalent to that of $\mathscr{U}_\rho$. ■

Completeness, like total boundedness, is a uniformity-dependent concept.

Example 6.5. Both $\mathscr{U}_{d_1}$ and $\mathscr{U}_\rho$ are compatible with $\mathfrak{T}(d_1)$ on $\mathbb{R}$ in Example 5.7. Although $\mathscr{U}_{d_1}$ is complete, $\mathscr{U}_\rho$ is not. (The ρ-Cauchy sequence $\{n : n \in \mathbb{N}\}$ does not converge.)

THEOREM 6.6. (i) A closed subset of a complete uniform space is complete, and a complete subset of a separated uniform space is closed.

(ii) A closed set E in a product-uniform space $\prod_{\mathfrak{a}} (X_\alpha, \mathscr{U}_\alpha)$ is $\mathscr{U}_\Pi$-complete iff each projection $\pi_\alpha(E) \subset (X_\alpha, \mathscr{U}_\alpha)$ is complete. In particular, a product of uniform spaces is complete iff each factor space is complete.

Proof. (i) If $E \subset (X, \mathscr{U})$ is closed and $\mathfrak{F}$ is a $\mathscr{U}_E = \mathscr{U} \cap (E \times E)$-Cauchy filter in E, the (Definition 5.1) $\mathscr{U}$-Cauchy filter $\mathfrak{F}\langle X \rangle$ converges to some $x_0 \in X$. Therefore $\mathfrak{F} \to x_0 \in \bar{E} = E$ and $(E, \mathscr{U}_E)$ is complete. Next, if $\mathscr{U}$ is separated and $E \subset X$ is complete, $x \in \bar{E}$ implies that $\mathfrak{F} \to x$ for some filter $\mathfrak{F}$ in E. Since $\mathfrak{F}$ is $\mathscr{U}_E$-Cauchy and $\mathscr{U}$ is separated, $\mathfrak{F}$ necessarily converges to some unique $x_0 \in E$ which is precisely x. Thus E is closed in $(X, \mathscr{U})$. For (ii), suppose that each $\pi_\alpha(E) \subset (X_\alpha, \mathscr{U}_\alpha)$ is complete and let $\mathscr{K}$ be a $\mathscr{U}_\pi$-Cauchy filter in $\prod_{\mathscr{Q}} X_\alpha$. Since each $\pi_\alpha\{\mathscr{K}\} = \{\pi_\alpha(K): K \in \mathscr{K}\}$ is $\mathscr{U}_\alpha \cap (\pi_\alpha(E) \times \pi_\alpha(E))$-Cauchy and convergent to some $x_\alpha \in \pi_\alpha(E)$, one has $\mathscr{K} \to \{x_\alpha\}_{\mathscr{A}} \in \bar{E} = E$ in $\left(\prod_{\mathscr{Q}} X_\alpha, \mathscr{U}_\pi\right)$. Assume, conversely, that $E \subset \prod_{\mathscr{Q}} X_\alpha$ is $\mathscr{U}_\pi$-complete and let $\mathscr{G}_\alpha$ be a $\mathscr{U}_\alpha \cap (\pi_\alpha(E) \times \pi_\alpha(E))$-Cauchy filter in $\pi_\alpha(E)\ \forall \alpha \in \mathscr{Q}$. Then [Example 3.7, (d)] $\mathscr{G}_\pi \subset E$ is $\mathscr{U}_\pi \cap (E \times E)$-Cauchy and converges to some $\{x_\alpha\}_{\mathscr{Q}} \in E$. Consequently $\pi_\beta\{G_\beta\} = \mathscr{G}_\beta \to x_\beta \in \pi_\beta(E)$ and the completeness of each $\pi_\beta(E) \subset (X_\beta, \mathscr{U}_\beta)$ $(\beta \in \mathscr{Q})$ is established. ∎

Remark. The conditions in Theorem 6.6 cannot be relaxed. Clearly $(0, 1]$ is incomplete in $([0, 1], \mathscr{U}_{d_1})$ and $\{1/n: n \in \mathbb{N}\}$ is a closed, incomplete subset of $((0, 1], \mathscr{U}_{d_1})$. Every proper subset E of the nonseparated space $(X, \mathscr{U}_{\mathscr{I}})$ is complete (a filter in E is $\mathscr{U}_{\mathscr{I}} \cap E \times E)$-convergent to every $x \in E$) but not $\mathscr{I}$-closed in X.

THEOREM 6.7. A uniform space is compact iff it is complete and totally bounded.

Proof. Theorem 5.4, Appendix t.7.

Remark. Each partial converse fails since $\mathscr{U}_{\mathscr{D}} \in \mathfrak{U}(X)$ and $\mathscr{U}_{d_1} \in \mathfrak{U}_{tb}(X)$ are noncompact for $X = [0, 1)$.

Some caution is in order here. Given a property enjoyed by total boundedness, one might conjecture that compactness and completeness do or do not simultaneously share this property. Such need not be true.

Example 6.8. (Theorem 5.9). Compactness, but not completeness, is preserved by uniformly continuous surjections. Assuredly, every uniform space $(X, \mathscr{U})$ is the uniformly continuous image of the complete space $(X, \mathscr{U}_{\mathscr{D}})$ under $\mathbb{1}: (X, \mathscr{U}_{\mathscr{D}}) \to (X, \mathscr{U})$.

(*a*) Suppose $f: X \to (Y, \mathscr{V})$ is surjective. Then $\mathscr{U}_{\mathscr{P}(f)}$ is complete iff $\mathscr{V}$ is complete. Verification: If $\mathscr{U}_{\mathscr{P}(f)}$ is complete and $\mathscr{K}$ is $\mathscr{V}$-Cauchy in Y, then $f^{-1}(\mathscr{K}) = \{f^{-1}(K): K \in \mathscr{K}\}$ is $\mathscr{U}_{\mathscr{P}(f)}$-Cauchy and convergent to some $x_0 \in X$. Therefore $\mathscr{K} = f(\mathscr{K}) \to f(x_0) \in Y$. If $\mathscr{V}$ is complete and $\mathscr{G}$ is $\mathscr{U}_{\mathscr{P}(f)}$-Cauchy, then $f(\mathscr{G})$ is $\mathscr{V}$-Cauchy and convergent to some $y_0 \in Y$. Since f is surjective, $\mathscr{G} \to x\ \forall x \in f^{-1}(y_0)$.

Remark. Since completeness is nonhereditary, surjectiveness is essential here.

(b) Suppose $f: (X, \mathscr{U}) \to (Y, \mathscr{V})$ is a unimorphism. Then $\mathscr{U}$ is complete iff $\mathscr{V}$ is complete. This follows from Example 3.13 and (a) above. It follows from (a) and Theorem 3.16 that a uniform space is complete iff its associated separated uniform space is complete. Similarly, (a) and Example 3.13 imply that $\mathcal{U}$ is complete iff $\mathcal{V}$ is complete for a unimorphism $f:(X, \mathcal{U})\to(Y, \mathcal{V})$.

Uniform Space Completions

Of fundamental importance is the fact that a (pseudo)metric space (X, ρ) can be isometrically embedded in a 1-1 manner onto a dense subspace of a complete (pseudo)metric space $(\check{X}, \check{\rho})$. The standard approach consists of defining a (pseudo)metric $\check{\rho}$ on $\check{X}=\{\text{all } \rho\text{-Cauchy sequences } \{x_n\}\in X\}$ by $\check{\rho}(\{x_n\}, \{y_n\})=\lim_n \rho(x_n, y_n)$. Accordingly, $\mu:(X, \rho)\to(\check{X}, \check{\rho})$, $x \rightsquigarrow \{x\}$, is an injection of (X, ρ) into a dense subspace of $(\check{X}, \check{\rho})$. The pair $(\mu; (\check{X}, \check{\rho}))$ is called a completion of (X, ρ). In similar fashion

DEFINITION 6.9. A {separated} completion of a uniform space $(X, \mathcal{U})$ is a pair $(\mu; (\check{X}, \check{\mathcal{U}}))$ consisting of a complete {separated} uniform space $(\check{X}, \check{\mathcal{U}})$ and a unimorphism μ of $(X, \mathcal{U})$ onto a dense subspace of $(\check{X}, \check{\mathcal{U}})$.

The most logical matter to consider next is

THEOREM 6.10. Every uniform space $(X, \mathcal{U})$ has a completion $(\mu; (\check{X}, \check{\mathcal{U}}))$. If $\mathcal{U}$ is separated, $(\check{X}, \check{\mathcal{U}})$ is separated and unique up to a unimorphism.

Proof. Let $\mathcal{G}$ be the gage of $\mathcal{U}$ so that $\mathcal{U}=\mathcal{U}_{\mathcal{G}}$ (Theorem 4.8). Then [Problem 4B(c)] $\mathfrak{e}:(X, \mathcal{U})\to\prod_{\mathcal{G}}(X, \mathcal{U}_\rho)$, $x \rightsquigarrow \{x\}_{\mathcal{G}}$, is an into unimorphism. For each $\rho\in\mathcal{G}$ let $(\mu_\rho; (\tilde{X}, \tilde{\rho}))$ denote the completion of (X, ρ). Since each μ_ρ is an injective isometry, each $\mu_\rho:(X, \mathcal{U}_\rho)\to(\tilde{X}, \mathcal{U}_{\tilde{\rho}})$ is an into unimorphism, and since $\tilde{\mathfrak{e}}:\prod_{\mathcal{G}}(X, \mathcal{U}_\rho)\to\prod_{\mathcal{G}}(\tilde{X}, \mathcal{U}_{\tilde{\rho}})$, $\{x\}_{\mathcal{G}} \rightsquigarrow \{(x)\}_{\mathcal{G}}$, is an into unimorphism so also is $\mu=\tilde{\mathfrak{e}}\mathfrak{e}:(X, \mathcal{U})\to\prod_{\mathcal{G}}(\tilde{X}, \mathcal{U}_{\tilde{\rho}})$. From the completeness of each $(\tilde{X}, \mathcal{U}_{\tilde{\rho}})$ (Corollary 6.4) follows that of $\prod_{\mathcal{G}}(\tilde{X}, \mathcal{U}_{\tilde{\rho}})$ (Theorem 6.6). Therefore $\check{X}=\overline{\mu(X)}$ is $\check{\mathcal{U}}=\mathcal{U}_\pi\cap(\check{X}\times\check{X})$-complete and $(\mu; (\check{X}, \check{\mathcal{U}}))$ is a completion of $(X, \mathcal{U})$. Suppose, next, that $\mathcal{U}$ is separated. Let $(X^*, \mathcal{U}^*)$ be the separated uniform space associated with a completion $(\mu; (\check{X}, \check{\mathcal{U}}))$ of $(X, \mathcal{U})$. Then $(X^*, \mathcal{U}^*)$ is complete (Example 6.8(b)) and $X^*=\nu(\check{X})=\nu\{\overline{\mu(X)}\}\subset\nu\{\overline{\mu(X)}\}\subset X^*$;

that is, $\nu\mu(X)$ is dense in $(X^*, \mathscr{U}^*)$. Since $\check{\mathscr{U}} = \mathscr{U}_{\mathscr{P}(\nu)}$ (Theorem 4.16) and $\check{\mathscr{U}} \cap (\mu(X)\times\mu(X)) = \mathscr{U}_{\mathscr{P}\{\nu/\mu(X)\}}$ is separated, $\nu:(\mu(X), \check{\mathscr{U}}) \to (X^*, \mathscr{U}^*)$ is an into unimorphism. This demonstrates that $(\nu\mu; (X^*, \mathscr{U}^*))$ is a separated completion of $(X, \mathscr{U})$. The uniqueness of $(X^*, \mathscr{U}^*)$, up to a unimorphism, will follow from Corollary 6.12 below. ■

Completeness can be characterized in terms of uniformly continuous extensions.

THEOREM 6.11. A {separated} uniform space $(X, \mathscr{V})$ is complete iff every uniformly continuous mapping $f:(E, \mathscr{U}_E) \to (Y, \mathscr{V})$ on a subset E of a uniform space $(X, \mathscr{U})$ has a {unique} uniformly continuous extension $\bar{f}:(\bar{E}, \mathscr{U}_{\bar{E}}) \to (Y, \mathscr{V})$.

Proof. Suppose $(Y, \mathscr{V})$ is complete and let $f:(E, \mathscr{U}_E) \to (Y, \mathscr{V})$ be uniformly continuous. For each $x \in \bar{E}$ the $\mathscr{V}$-Cauchy filter $f\{(\mathscr{U}[x] \cap E)\langle E\rangle\}$ converges in Y. Pick one such limit point and denote it y_x. Evidently $\bar{f}(x) = \begin{Bmatrix} f(x), & x \in E \\ y_x, & x \in \bar{E} \end{Bmatrix}$ extends f and it remains only to show that $\bar{f}$ is $\mathscr{U}_{\bar{E}}$-uniformly continuous. Given any $V \in \mathscr{V}$, let $W \in \mathscr{V}$ be a closed symmetric entourage such that $W^2 \subset V$ and (by f's $\mathscr{U}_E$-uniform continuity) pick an open symmetric $U \in \mathscr{U}$ satisfying $U_E \subset (f\times f)^{-1}W$. If $(x_1, x_2) \in U_{\bar{E}}$, the open set $U_{\bar{E}}[x_1] \cap U_{\bar{E}}[x_2]$ containing $x_2 \in \bar{E}$ necessarily contains some $z \in E$. Therefore $x_i \in U_{\bar{E}}[z] = U[z] \cap \bar{E}$ $(i = 1, 2)$, and $U_{\bar{E}}[z] \subset \overline{U_E[z]}$ (since $y \in U[z] \cap \bar{E} \Rightarrow \varnothing \neq W[y] \cap \{U[z] \cap E\} = W[y] \cap U_E[z]\ \forall W \in \mathscr{U}$). This means that $\bar{f}(x_i) \in \overline{f\{U_E[z]\}} \subset W[f(z)]$ $(i = 1, 2)$; therefore $(\bar{f}(x_1), \bar{f}(x_2)) \in W^2 \subset W$ and so $U_{\bar{E}} \subset (\bar{f}\times\bar{f})^{-1}V$ as required. For the converse suppose $\mathscr{V}$ is not complete and let $\mathscr{K}$ be a $\mathscr{V}$-Cauchy filter which does not converge in Y. If $(\mu; (\check{Y}, \check{\mathscr{V}}))$ is a completion of $(Y, \mathscr{V})$, then $\mu^{-1}:(\mu(X), \check{\mathscr{V}}) \to (X, \mathscr{V})$ has no uniformly continuous extension to $(\check{Y}, \check{\mathscr{V}})$. (If such an extension F exists, then $\mathscr{U}_{\mathscr{P}(F)} \subset \check{\mathscr{V}}$ on $\check{Y}\times\check{Y}$ and since $(F\times F)^{-1}\{(\mu\times\mu)^{-1}\check{V}\} \subset \check{V}\ \forall \check{V} \in \check{\mathscr{V}}$ one has $\check{\mathscr{V}} \subset \mathscr{U}_{\mathscr{P}(F)}$. The $\mathscr{U}_{\mathscr{P}(F)} = \check{\mathscr{V}}$-Cauchy filter $F^{-1}(\mathscr{K})$ therefore converges to some $\check{y} \in \check{Y}$ and (since F is a continuous surjection) $\mathscr{K} = F\{F^{-1}(\mathscr{K})\} \to F(\check{y}) \in Y$, which is impossible.)

To prove $\bar{F}$ unique when $\mathscr{V}$ is separated suppose that F_1, F_2 are two uniformly continuous extensions of $f:(E, \mathscr{U}_E) \to (Y, \mathscr{V})$ to $(\bar{E}, \mathscr{U}_{\bar{E}})$. The mapping $\xi: \underset{x \rightsquigarrow (F_1(x), F_2(x))}{\bar{E} \longrightarrow Y\times Y}$ is continuous. And since $E \subset \xi^{-1}(\Delta) = \{x \in X : F_1(x) = F_2(x)\}$ and $\xi^{-1}(\Delta)$ is closed, $\bar{E} = \xi^{-1}(\Delta)$. ■

COROLLARY 6.12. A unimorphism between dense subspaces $(A, \mathscr{U}_A)$ and $(B, \mathscr{V}_B)$ of complete separated uniform spaces $(X, \mathscr{U})$ and

$(Y, \mathcal{V})$ can be uniquely extended to a unimorphism between $(X, \mathcal{U})$ and $(Y, \mathcal{V})$.

Proof. Let $F:(X, \mathcal{U}) \rightarrow (Y, \mathcal{V})$ be the unique uniformly continuous extension of $f:(A, \mathcal{U}_A) \rightarrow (Y, \mathcal{V})$. For each $y \in Y = \bar{B}$ the $\mathcal{U}$-Cauchy filter $(f^{-1}\{V[y] \cap B\})\langle X \rangle$ converges to a unique $x_y \in X$. Since $F\{\mathcal{U}[x_y] \cap A\} = \mathcal{V}[y] \cap B$, it follows that $F(x_y) = y$ and F is bijective. Verification that F^{-1} is uniformly continuous is simple. If $\Phi:(Y, \mathcal{V}) \rightarrow (X, \mathcal{U})$ is the uniformly continuous extension of $f^{-1}:(B, \mathcal{V}_B) \rightarrow (X, \mathcal{U})$, then $F\{\Phi(y)\} = y \ \forall y \in Y$ and $F^{-1} = \Phi$. ∎

Remark. Note that F need be neither injective nor surjective if f is only a uniformly continuous bijection.

Completions and Compactifications

A compactification of a topological space $(X, \mathfrak{T})$ is a pair $(h; (\tilde{X}, \tilde{\mathfrak{T}}))$, where h is a homeomorphism of $(X, \mathfrak{T})$ onto a dense subset of a compact space $(\tilde{X}, \tilde{\mathfrak{T}})$. This can be refined for uniform spaces as

DEFINITION 6.13. A uniform compactification of a uniform space $(X, \mathcal{U})$ is a pair $(\mu; (\tilde{X}, \tilde{\mathcal{U}})$ that consists of a compact uniform space $(\tilde{X}, \tilde{\mathcal{U}})$ and a unimorphism μ of $(X, \mathcal{U})$ onto a dense subspace of $(X, \tilde{\mathcal{U}})$.

Remark. Every uniform compactification of $(X, \mathcal{U})$ is a compactification of $(X, \mathfrak{T}_{\mathcal{U}})$. If $\mathcal{U}$ is compatible with $(X, \mathfrak{T})$, then every uniform compactification of $(X, \mathcal{U})$ is a compactification of $(X, \mathfrak{T})$.

Uniform compactification is a uniformity dependent concept. A uniform space having a uniform compactification is necessarily totally bounded (Corollary 5.5, Theorem 5.9). The converse is also true by

THEOREM 6.14. The completion of a totally bounded uniform space is a uniform compactification.

Proof. If $(\mu; (\check{X}, \check{\mathcal{U}})$ is a completion of the totally bounded uniform space $(X, \mathcal{U})$, then (Theorem 5.9) $(\overline{\mu(x)}, \check{\mathcal{U}}) = (\check{X}, \check{\mathcal{U}})$ is totally bounded and Theorem 6.7 prevails. ∎

Since complete regularity is hereditary and topological, every topological space having a T_2 compactification is necessarily $T_{3\frac{1}{2}}$. Theorem 5.14 and Definition 6.13$^+$ yield the converse part of

COROLLARY 6.15. A topological space is $T_{3\frac{1}{2}}$ iff it has a T_2 compactification.

A topological space has more than one compactification. Two compactifications of $(X, \mathfrak{T})$ will be ordered $(h_1; (\tilde{X}_1, \tilde{\mathfrak{T}}_1)) < (h_2; (\tilde{X}_2, \tilde{\mathfrak{T}}_2))$ if there is a continuous extension $F: (\tilde{X}_2, \tilde{\mathfrak{T}}_2) \to (\tilde{X}_1, \tilde{\mathfrak{T}}_1)$ of $h_1 h_2^{-1}$. In this context $(h_i, (\tilde{X}_i, \tilde{\mathfrak{T}}_i))$ are equal iff $(h_1; (\tilde{X}_1, \tilde{\mathfrak{T}}_1)) < (h_2; (\tilde{X}_2, \tilde{\mathfrak{T}}_2))$ and $(h_2, (\tilde{X}_2, \tilde{\mathfrak{T}}_2)) < (h_1; (\tilde{X}_1, \tilde{\mathfrak{T}}_1))$. This, of course, is equivalent to F above being a homeomorphism. One more point: namely (Theorem 1.25), continuous mappings (homeomorphisms) between T_2 compactifications are uniformly continuous (unimorphisms).

Let $\mathfrak{U}_{t_2b}(X, \mathfrak{T})$ denote the collection of separated totally bounded uniformities compatible with a $T_{3\frac{1}{2}}$ space $(X, \mathfrak{T})$, and let $\mathcal{K}_2(X, \mathfrak{T})$ be the quotient set of equivalence classes, under "equality" relation, of all T_2 compactifications of $(X, \mathfrak{T})$. If $(h; (\tilde{X}, \tilde{\mathfrak{T}})) \in \mathcal{K}_2(X, \mathfrak{T})$, then $\tilde{X}$ carries the fine uniformity $\mathcal{U}_F$ (Corollary 1.26) and $(h; (\tilde{X}, \tilde{\mathcal{U}}_F))$ is the completion of $(X, \mathcal{U}_{\mathcal{P}(h)}) \in \mathfrak{U}_{t_2b}(X, \mathfrak{T})$. Verification is simple. The $\tilde{\mathcal{U}}_F - \mathcal{U}_{\mathcal{P}(h)}$ uniform continuity of h^{-1} follows from $(h^{-1} \times h^{-1})^{-1}\{(h \times h)^{-1} U\} = U \; \forall U \in \tilde{\mathcal{U}}_F$. Furthermore, $\tilde{\mathcal{U}}_F \cap (h(X) \times h(X))$ is compatible with $\tilde{\mathfrak{T}}$ on $h(X)$ and Theorems 3.2 and 5.10 apply.

We now have all the ingredients needed for

THEOREM 6.16. The mapping $\xi: \underset{(h;(\tilde{X},\tilde{\mathfrak{T}})) \rightsquigarrow \mathcal{U}_{\mathcal{P}(h)}}{\mathcal{K}_2(X, \mathfrak{T}) \to \mathfrak{U}_{t_2b}(X, \mathfrak{T})}$ is a biorder preserving bijection.

Proof. Given any $\mathcal{U} \in \mathfrak{U}_{t_2b}(X, \mathfrak{T})$, let $\mu_{\mathcal{U}}$ be a unimorphism of $(X, \mathcal{U})$ onto a dense subspace of the unique totally bounded, complete, separated uniform space $(\check{X}_{\mathcal{U}}, \check{\mathcal{U}})$. Surely $(\mu_{\mathcal{U}}; (\check{X}_{\mathcal{U}}, \mathfrak{T}_{\check{\mathcal{U}}})) \in \mathcal{K}_2(X, \mathfrak{T})$ (Theorem 6.14). Since $\mu_{\mathcal{U}}: (X, \mathcal{U}) \to (\mu_{\mathcal{U}}(X), \check{\mathcal{U}})$ is a unimorphism, $\mathcal{U} = \mathcal{U}_{\mathcal{P}(\mu_{\mathcal{U}})}$ (Example 3.13) and this proves ξ to be surjective. That ξ is injective is also clear, since $\mathcal{U}_{\mathcal{P}(h_1)} = \mathcal{U}_{\mathcal{P}(h_2)}$ implies $(h_1; (\tilde{X}_1, \tilde{\mathfrak{T}}_1)) = (h_2; (\tilde{X}_2, \tilde{\mathfrak{T}}_2))$ by Theorem 6.11 (both $h_1 h_2^{-1}, h_2 h_1^{-1}$ have uniformly continuous extensions). Now suppose $(h_1; (\tilde{X}_1, \tilde{\mathfrak{T}}_1)) < (h_2; (\tilde{X}_2, \tilde{\mathfrak{T}}_2))$. Take $\tilde{\mathcal{U}}_i$ as the fine uniformity on $\tilde{X}_i$ ($i = 1, 2$) and let $F: (\tilde{X}_2, \tilde{\mathcal{U}}_{2_F}) \to (X_1, \tilde{\mathcal{U}}_{1_F})$ be a uniformly continuous extension of $h_1 h_2^{-1}$. Then $(F \times F)^{-1} \tilde{\mathcal{U}}_{1_F} \subset \tilde{\mathcal{U}}_{2_F}$ implies $(h_2 \times h_2)\{(h_1 \times h_1)^{-1} \mathcal{U}_1\} = (h_1 h_2^{-1} \times h_1 h_2^{-1})^{-1} \tilde{\mathcal{U}}_{1_F} \subset \tilde{\mathcal{U}}_{2_F}$ and $\mathcal{U}_{\mathcal{P}(h_1)} = (h_1 \times h_1)^{-1} \mathcal{U}_1 \subset (h_2 \times h_2)^{-1} \mathcal{U}_2 = \mathcal{U}_{\mathcal{P}(h_2)}$. Thus ξ is order preserving. To show that ξ^{-1} also preserves order, consider any $\mathcal{U}_1, \mathcal{U}_2 \in \mathfrak{U}_{t_2b}(X, \mathfrak{T})$ satisfying $\mathcal{U}_1 \subset U_2$ and let $(\mu_i; (\check{X}_i, \check{\mathcal{U}}_i))$ be the completion of $(X, \mathcal{U}_i)$ $(i = 1, 2)$. Then $\mu_1 \mu_2^{-1}: (\mu_2(X), \check{\mathcal{U}}_2) \to (\mu_1(X), \check{\mathcal{U}}_1) \subset (\check{X}_1, \check{\mathcal{U}}_1)$ has a uniformly continuous extension to $(\check{X}_2, \check{\mathcal{U}}_2)$ and $\xi^{-1}(\mathcal{U}_1) = (\mu_1; (\check{X}_1, \mathfrak{T}_{\check{\mathcal{U}}_1})) < (\mu_2; (\check{X}_2, \mathfrak{T}_{\check{\mathcal{U}}_2})) = \xi^{-1}(\mathcal{U}_2)$ follows. ■

The Stone–Cech compactification $\beta(X)$ of $(X, \mathfrak{T})$ (Appendix t.7) is the largest (i.e., upper bound or last) element in the partially ordered set

$(\mathcal{K}_2(X, \mathfrak{T}), <)$. It is no accident therefore (Problems 3C, 5B) that $\mathcal{U}_{\beta(X)} = \mathcal{U}_{\mathcal{P}\{C(X;I)\}} = \mathcal{U}_{\mathcal{P}(e)} \in \mathfrak{U}_{t_2b}(X, \mathfrak{T})$ is the ξ-image of $\beta(X)$ in Theorem 6.16. In this vein $(\mu; (\check{X}, \check{\mathcal{U}}))$ is the completion of $(X, \mathcal{U}_{\beta(X)})$ iff $(\mu; (\check{X}, \mathfrak{T}_{\mathcal{U}})) = \beta(X)$.

Surely $(\mathbb{1}; (X, \mathfrak{T}))$ is the smallest element in $(\mathcal{K}_2(X, \mathfrak{T}), <)$ when $(X, \mathfrak{T})$ is compact, T_2 (Theorem 1.25, Theorem 6.11). This, in fact, is a special case of

THEOREM 6.17. The statements below are equivalent for a $T_{3\frac{1}{2}}$ space $(X, \mathfrak{T})$:

(i) $(X, \mathfrak{T})$ is locally compact.
(ii) $\mathcal{K}_2(X, \mathfrak{T})$ has a smallest element.
(iii) The coarse uniformity $\mathcal{U}_c$ exists for $(X, \mathfrak{T})$.

Proof. Statements (ii), (iii) are equivalent by Theorem 6.16. Our proof that (i), (ii) are equivalent uses the fact that dense, locally compact subsets of a T_2 space are open.

(i)⇒(ii). Suppose $(X, \mathfrak{T})$ is locally compact, noncompact and let $X_\infty \in \mathcal{K}_2(X, \mathfrak{T})$ denote its one point compactification (Appendix t.7). For any other $(h; (\tilde{X}, \tilde{\mathfrak{T}})) \in \mathcal{K}_2(X, \mathfrak{T})$, the surjection $F: (\tilde{X}, \tilde{\mathfrak{T}}) \to X_\infty$ taking $x \in h(X)$ into $(\mathbb{1}h^{-1})(x)$ and $x \in \tilde{X} - h(X)$ to ∞ extends $\mathbb{1}h^{-1}: (h(X), \tilde{\mathfrak{T}}) \to (X, \mathfrak{T}_\infty)$. We show that F is continuous, from which follows $X_\infty < (h; (\tilde{X}, \tilde{\mathfrak{T}}))$ and confirmation that (ii) holds. First, note that $F^{-1}(X) = h(X) \subset \tilde{X}$ is $\tilde{\mathfrak{T}}$-open (local compactness is preserved by homeomorphisms). Next $u \in \mathfrak{T} \subset \mathfrak{T}_\infty$ implies $F^{-1}(u) = h(u) \subset h(X)$ is $\tilde{\mathfrak{T}}$-open and $F^{-1}(u) \in \tilde{\mathfrak{T}}$ in $\tilde{X}$. Finally, $w \cup \infty = v \in \mathfrak{T}_\infty$ implies $X - w$ is $\mathfrak{T}$-compact and $h(X - w)$ is $\tilde{\mathfrak{T}}$-compact, therefore $\tilde{\mathfrak{T}}$-closed in $\tilde{X}$. Consequently, $F^{-1}(v) = F^{-1}(x_\infty) - F^{-1}(X_\infty - v) = \tilde{X} - F^{-1}(X - w) = \tilde{X} - h(X - w)$ is $\tilde{\mathfrak{T}}$-open in $\tilde{X}$. All cases considered, we conclude that F is continuous.

(ii)⇒(i). A compactification $(h; (\tilde{X}, \tilde{\mathfrak{T}}))$ having $\aleph(\tilde{X} - h(X)) > 1$ cannot be the smallest element in $\mathcal{K}_2(X, \mathfrak{T})$. To be sure, $X_\infty < (h; (\tilde{X}, \tilde{\mathfrak{T}}))$ by the continuity of $F: (\tilde{X}, \tilde{\mathfrak{T}}) \to X_\infty$ above, and $X_\infty \neq (h; (\tilde{X}, \tilde{\mathfrak{T}}))$, since no surjective extension of the bijection $\mathbb{1}h^{-1}$ can be injective. Now suppose $(h; (\tilde{X}, \tilde{\mathfrak{T}}))$ is the smallest element in $\mathcal{K}_2(X, \mathfrak{T})$. Assuredly, $\aleph(\tilde{X} - h(X)) \leq 1$. If $\aleph(\tilde{X} - h(X)) = 0$, then $(X, \mathfrak{T})$ is compact. Otherwise, $\aleph(\tilde{X} - h(X)) = 1$ so that $h(X) \subset \tilde{X}$ is $\tilde{\mathfrak{T}}$-open and $(h(X), \tilde{\mathfrak{T}})$, therefore $(X, \mathfrak{T})$ is locally compact. ■

Remark. If $\mathcal{U}_c$ exists, then $\mathcal{U}_c = \mathcal{U}_{\mathcal{P}(\mathbb{1})}$ for $\mathbb{1}: (X, \mathfrak{T}) \to X_\infty$. Note also that $\mathcal{U}_c \in \mathfrak{U}_{t_2b}(X)$ and $\mathcal{U}_c \subset \mathcal{U}_{\beta(X)}$. (Also see Problem 1H.)

COROLLARY 6.19. The following are equivalent for a locally compact T_2 space $(X, \mathfrak{T})$:

(i) $(X, \mathfrak{T})$ has a unique compatible uniformity.
(ii) $\aleph(\mathcal{K}_2(X, \mathfrak{T}))=1$.
(iii) $\aleph(\beta(X)-\mathfrak{e}(X))\leq 1$.

Proof. Clearly, (i)$\Leftrightarrow$(ii) by Theorem 6.16 and (ii)$\Rightarrow$(iii) since $\beta(X)$ is the smallest member of $\mathcal{K}_2(X, \mathfrak{T})$ when (ii) holds. Finally, suppose $X \lneqq \beta(X)$ and let $F:\beta(X)\to X_\infty$ be any continuous surjection extending $\mathbb{1}e^{-1}:\mathfrak{e}(X)\to X\in X_\infty$. Since $X_\infty\neq\beta(X)$, the closed mapping F is non-homeomorphic. In particular, F is not injective and $F^{-1}(\infty)=F^{-1}(X_\infty-X)=\beta(X)-\mathfrak{e}(X)$ contains more than one point. In other words, (iii)$\Rightarrow$(ii). ■

The Tietze extension theorem states that a topological space X is T_4 (i.e., both normal and T_2) iff every real-valued function on a closed subset of X has a continuous extension to X. This can be used to prove

THEOREM 6.19 (Katetov). A bounded, real-valued uniformly continuous function on a subspace $(A, \mathcal{U}_A)$ of $\mathcal{U}\in\mathfrak{U}_{t_2b}(X)$ has a bounded uniformly continuous extension to $(X, \mathcal{U})$.

Proof. Let $\bar{f}:(\bar{A}, \mathcal{U}_{\bar{A}})\to(\mathbb{R}, \mathcal{U}_{d_1})$ be the uniformly continuous extension of the bounded, uniformly continuous mapping $f:(A, \mathcal{U}_A)\to(\mathbb{R}, \mathcal{U}_{d_1})$ (Theorem 6.11). Assume, furthermore, that $(h;(\tilde{X}, \tilde{\mathfrak{T}}))$ is the T_2 compactification of $(X, \mathfrak{T}_{\mathcal{U}})$. Since $(\overline{h(A)}, \tilde{\mathfrak{T}})$ is compact, T_2 (therefore T_4), there is a continuous extension $F_1:(\overline{h(A)}, \tilde{\mathfrak{T}})\to(\mathbb{R}, d_1)$ of $\bar{f}h^{-1}:(h(A), \tilde{\mathfrak{T}})\to(\mathbb{R}, d_1)$. Let $F_2:(\tilde{X}, \tilde{\mathfrak{T}})\to(\mathbb{R}, d_1)$ be a continuous extension of F_1 and define $F=F_2h$. Clearly, F extends f to $(X, \mathcal{U})$, and F is bounded because F_2 is bounded on $(\tilde{X}, \tilde{\mathfrak{T}})$. Also, F is $\mathcal{U}-\mathcal{U}_{d_1}$ uniformly continuous (Theorem 3.4), since $(\tilde{X}, \tilde{\mathfrak{T}})$ carries $\mathcal{U}_{\mathcal{P}(F_2)}$ (Corollary 1.26) and $\mathcal{U}=\mathcal{U}_{\mathcal{P}(h)}$ (Theorem 6.16). ■

Remark. Since F_2 extends f, boundedness of f above is necessary. It is not sufficient, since $f(x)=x$ on a subspace $(A, \mathcal{U}_{d_1}\cap A\times A)$ has a uniformly continuous extension to $(\mathbb{R}, \mathcal{U}_{d_1})$.

PROBLEMS

6A. Additional Completeness Characterizations

A family of subsets ξ of a uniform space $(X, \mathcal{U})$ is said to have small sets if ξ contains a U-sized set $\forall U\in\mathcal{U}$. Prove the following properties of $(X, \mathcal{U})$ equivalent:

(i) $(X, \mathcal{U})$ is complete.

(ii) Every family of closed sets having small sets and the fip (Appendix t.7) has nonempty intersection.

(iii) $\mathcal{F}\langle X\rangle$ converges $\forall \mathcal{U}_E$-Cauchy filter $\mathcal{F}$ on a dense subset E of $(X, \mathcal{U})$. [*Hint*: (i)$\Rightarrow$(ii) since $\mathcal{G} = \{$finite intersections of sets in $\xi\}$ generates a $\mathcal{U}$-Cauchy filter $\mathcal{F}$ in X and $\varnothing \neq \bigcap_{\mathcal{F}} F \subset \bigcap_{E} E$. For (ii)$\Rightarrow$(iii) note that $\mathcal{F}\langle X\rangle \supset \overline{\mathcal{F}\langle X\rangle} = \{\bar{F} : F \in \mathcal{F}\langle X\rangle\}$ is $\mathcal{U}$-Cauchy and $x_0 \in \bigcap_{\mathcal{F}\langle X\rangle} \bar{F}$ implies $\mathcal{F}\langle X\rangle \to x_0$. For (iii)$\Rightarrow$(i), let $\mathcal{F}$ be $\mathcal{U}$-Cauchy in X and let $\mathcal{H} \subset \mathcal{F}$ be the $\mathcal{U}$-Cauchy filter determined by $\{U[F] : U \in \mathcal{U}, F \in \mathcal{F}\}$. Since E is dense, $H \cap E \neq \varnothing \forall H \in \mathcal{H}$ and $\mathcal{H} \cap E$ determines a $\mathcal{U}_E$-Cauchy filter $\mathcal{H}'$ in E such that $\mathcal{H}'\langle X\rangle \subset \mathcal{H}$ in X.]

6B. The T_2 Completion of a Uniform Space

Notation as in Definitions 3.14 and 6.9. Show that $(\check{X}/\mathfrak{R}, \mathcal{U}_{\check{X}/\mathfrak{R}})$ and $(\check{X}/\mathfrak{R}, \check{\mathcal{U}}_{X/\mathfrak{R}})$ are unimorphic; that is, the procedures involves below are commutative.

$$\begin{array}{ccccc} & & (X, \mathcal{U}) & & \\ & \swarrow & & \searrow^{\nu} & \\ (\check{X}, \check{\mathcal{U}}) & & & & (X/\mathfrak{R}, \mathcal{U}_{X/\mathfrak{R}}) \\ & \searrow^{\nu'} & & \swarrow^{\mu} & \\ & & (\check{X}/\mathfrak{R}, \mathcal{U}_{\check{X}/\mathfrak{R}}) = (\check{X}/\mathfrak{R}, \check{\mathcal{U}}_{X/\mathfrak{R}}) & & \end{array}$$

Remark. One calls $(\check{\mu} = \mu\nu; (\check{X}/\mathfrak{R}, \check{\mathcal{U}}_{X/\mathfrak{R}}))$ the T_2 completion of $(X, \mathcal{U})$.

6C. Commutativity via Induced Mappings

For every uniformly continuous function $f : (X_1, \mathcal{U}_1) \to (X_2, \mathcal{U}_2)$ there exist unique uniformly continuous functions f', $\check{f}$ such that the diagrams below are commutative.

$$\begin{array}{ccc} (X_1, \mathcal{U}_1) & \xrightarrow{f} & (X_2, \mathcal{U}_2) \\ \downarrow{\scriptstyle \nu_1} & & \downarrow{\scriptstyle \nu_2} \\ (X_1/\mathfrak{R}, \mathcal{U}_{X_1/\mathfrak{R}}) & \xrightarrow{f'} & (X_1/\mathfrak{R}, \mathcal{U}_{X_2/\mathfrak{R}}) \end{array} \qquad \begin{array}{ccc} (X_1, \mathcal{U}_1) & \xrightarrow{f} & (X_2, \mathcal{U}_2) \\ \downarrow{\scriptstyle \check{\mu}_1} & & \downarrow{\scriptstyle \check{\mu}_2} \\ (\check{X}_1/\mathfrak{R}, \check{\mathcal{U}}_{X_1}{}^{/\mathfrak{R}}) & \xrightarrow{\check{f}} & (\check{X}_2/\mathfrak{R}, \check{\mathcal{U}}_{X_2/\mathfrak{R}}) \end{array}$$

6D. Completions and Projective Limits

Let $(\check{\mu}; (\check{X}/\mathfrak{R}, \check{\mathcal{U}}_{X/\mathfrak{R}}))$ be the T_2 completion of $(X, \mathcal{U}_{\mathcal{P}(\mathcal{F})})$, where $\mathcal{F} = \{f_\alpha : X \to (Y_\alpha, \mathcal{U}_\alpha) : \mathcal{U}_\alpha \in \mathfrak{U}(Y_\alpha) \forall \alpha \in \mathcal{Q}\}$. Define (Problem 6C) $\check{\mathcal{F}} = \{\check{f}_\alpha : X/\mathfrak{R} \to (\check{Y}_\alpha/\mathfrak{R}, \check{\mathcal{U}}_{X_\alpha/\mathfrak{R}}) \forall \alpha \in \mathcal{Q}\}$. Then $\check{\mathcal{U}}_{X/\mathfrak{R}} = \mathcal{U}_{\mathcal{P}\{\check{\mathcal{F}}\}}$ on $\check{X}/\mathfrak{R} \times \check{X}/\mathfrak{R}$. In other words, the T_2 completion of a projective limit is the projective limit of its T_2 completions. [*Hint*: If $\mathcal{F}' = \{f'_\alpha : (X/\mathfrak{R}, \mathcal{U}_{X/\mathfrak{R}}) \to (Y_\alpha/\mathfrak{R}, \mathcal{U}_{Y_\alpha/\mathfrak{R}}) \forall \alpha \in \mathcal{Q}\}$ and $\mathcal{G} = \{\nu_\alpha : \alpha \in \mathcal{Q}\}$, then $\mathcal{U}_{\mathcal{P}\{\mathcal{F}\}} = \mathcal{U}_{\mathcal{P}\{\mathcal{G}\mathcal{F}\}} = \mathcal{U}_{\mathcal{P}\{\mathcal{F}'\nu\}}$ and $\mathcal{U}_{\mathcal{P}\{\mathcal{F}\}} = (\nu \times \nu)^{-1} \mathcal{U}_{P\{\mathcal{F}'\}}$ (Theorem 3.16, Corollary 3.5). Similarly, $\mathcal{U}_{\mathcal{P}\{\mathcal{F}'\}} = (\mu \times \mu)^{-1} \mathcal{U}_{\mathcal{P}\{\mathcal{F}\}}$ so that $\mathcal{U}_{\mathcal{P}\{\mathcal{F}\}} = (\check{\mu} \times \check{\mu})^{-1} \mathcal{U}_{\mathcal{P}\{\check{\mathcal{F}}\}}$. Since $\mathcal{U}_{\mathcal{P}\{\check{\mathcal{F}}\}}$ is separated, $\check{\mu} : (X/\mathfrak{R}, \mathcal{U}_{X/\mathfrak{R}}) \to (\check{\mu}(X), \mathcal{U}_{\mathcal{P}\{\check{\mathcal{F}}\}})$ is an onto unimorphism. Thus $(\check{\mu}(X), \check{\mathcal{U}}_{X/\mathfrak{R}})$ and $(\check{\mu}(X), \mathcal{U}_{\mathcal{P}\{\check{\mathcal{F}}\}})$ are unimorphic and Corollary 6.12 applies.]

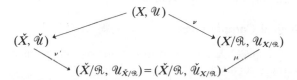

6E. Generalization of a Metric Completion

Let $\check{X}$ be the collection of all $\mathcal{U}$-Cauchy filters in a uniform space $(X, \mathcal{U})$. For each symmetric $U \in \mathcal{U}$, define $\check{U} = \{(\mathfrak{F}, \mathfrak{G}) \in \check{X} \times \check{X} : \mathfrak{F} \cap \mathfrak{G}$ contains a U-sized set$\}$. Then $\check{\mathcal{B}} = \{\check{U} : \text{symmetric } U \in \mathcal{U}\}$ is the base for a uniformity $\check{\mathcal{U}} \in \mathfrak{U}(X)$ such that

(i) $\mu : (X, \mathcal{U}) \xrightarrow[x \rightsquigarrow \mathcal{W}_x = \{E \supset X : x \in E\}]{} (\check{X}, \check{\mathcal{U}})$ is an into unimorphism.

(ii) $\mu(X)$ is dense in $\check{X}, \check{\mathcal{U}})$ and $(\check{X}, \check{\mathcal{U}})$ is complete. [*Hint*: Use Problem 6A(iii) to prove $(\check{X}, \check{\mathcal{U}})$ complete.]

6F. Metric Extension

Let E be a dense subset of a separated uniform space $(X, \mathcal{U})$. Assume further that $\mathcal{U}_E = \mathcal{U}_\rho$ for some metric ρ on E. Then ρ has a unique metric extension $\bar{\rho}$ to X such that $\mathcal{U}_{\bar{\rho}} = \mathcal{U}$. [*Hint*: (Theorem 4.6, Theorem 6.11) Let $\bar{\rho}$ be the uniformly continuous extension of $\rho : (E \times E, \mathcal{U}_\rho) \to (\mathbb{R}, \mathcal{U}_{d_1})$ to $(X \times X, \mathcal{U})$. Then $\mathcal{U} \subset \mathcal{U}_{\bar{\rho}}$ and the pseudometric $\bar{\rho}$ is a metric by Example 4.10.]

6G. Topological Completeness

A topological space $(X, \mathfrak{T})$ is said to be topologically complete if there is a complete uniformity compatible with $\mathfrak{T}$.

(a) A completely regular space is topologically complete iff $\mathcal{U}_F$ is complete.

(b) Every paracompact uniformizible space is topologically complete [*Hint*: Use Problems 1G(e) and 2F to prove $\mathcal{U}_F$ complete.]

(c) Why is every metrizible space $(X, \mathfrak{T})$ topologically complete? [*Caution*! A complete uniformity compatible with $\mathfrak{T}$ need not be metrizible (Problem 4A).]

6H. The Coarse Uniformity, $\mathcal{U}_c$

If $(X, \mathfrak{T})$ is locally compact and T_2, then (Appendix t.8).

(a) $\mathcal{U}_{\mathcal{P}\{C_0(X;\mathbb{R})\}} \in \mathfrak{U}_{tb}(X)$ is compatible with $\mathfrak{T}$ and $\aleph(X/\mathfrak{R} - \check{\mu}(X)) \le 1$. [*Hint*: If $(X, \mathfrak{T})$ is not compact, $\{X - K : \bar{K}$ is $\mathfrak{T}$-compact$\}$ is a $\mathcal{U}_{\mathcal{P}\{C_0(X;\mathbb{R})\}}$-Cauchy filter in X.]

(b) Use (a) to prove that $\mathcal{U}_{\mathcal{P}\{C_0(X;\mathbb{R})\}} = \mathcal{U}_c \in \mathfrak{U}_{t2b}(X, \mathfrak{T})$.

6I. Product Cech Uniformities

(a) Show that $\beta(X) \times \beta(Y) = \{\mathbf{e} \times \mathbf{e}; (\mathbf{e}(X) \times \mathbf{e}(Y), \mathfrak{T}_\pi \times \mathfrak{T}_\pi)\}$ is a T_2 compactification of $X \times Y$, where X and Y are $T_{3\frac{1}{2}}$ spaces.

(b) Why does $\beta(X \times Y) = \beta(X) \times \beta(y)$ iff $\mathcal{U}_{\beta(X \times Y)} = \mathcal{U}_{\beta(X)} \times \mathcal{U}_{\beta(Y)}$? [*Hint*: Use problem 6D, Theorem 6.16.]

(c) Can you give an example for which $\mathcal{U}_{\beta(X)} \times \mathcal{U}_{\beta(Y)} \subsetneq \mathcal{U}_{\beta(X \times Y)}$? [*Hint*: $G \times G$ cannot be pseudocompact when G is a non compact, locally compact T_2 group.]

Remark. The question when $\mathscr{U}(X\times Y)=\mathscr{U}(X)\times\mathscr{U}(Y)$ for $\mathscr{U}(X)$ denoting $\mathscr{U}_{\beta(X)}$, $\mathscr{U}_{\mathscr{P}\{C(X:\mathbb{R})\}}$ or $\mathscr{U}_F$ on $X\times X$ is investigated by Hager [37].

References for Further Study

On extensions of uniformly continuous mappings: Atsuji [5]; Corson and Isbell [22].

7. Function Spaces

The functions $\mathfrak{F}(X;\,Y)$ from a set X to a set Y is a vector space under pointwise operations. Accordingly, each uniformity for Y and collection of subsets of X determine a projective limit uniformity for $\mathfrak{F}(X;\,Y)$. Our main interest lies in the compactness, completeness, and continuity properties of such function spaces.

$\mathscr{S}$-Convergence

Definition 7.1. Let $\xi_X=\{\mathfrak{e}_x:\underset{f\rightsquigarrow f(x)}{\mathfrak{F}(X;\,Y)\to Y}:x\in X\}$.

(a) If $(Y,\mathfrak{T})$ is a topological space, $\mathfrak{T}_{\mathscr{P}\{\xi_X\}}$ (abbreviated $\mathfrak{T}_p$ or $\mathfrak{T}_\sigma$) is called the topology of pointwise convergence on $\mathfrak{F}(X;\,Y,\mathfrak{T})$.

(b) If $(Y,\mathscr{V})$ is a uniform space, $\mathscr{U}_{\mathscr{P}\{\xi_X\}}$ (abbreviated $\mathscr{U}_p$) is called the uniformity of pointwise convergence for $\mathfrak{F}(X;\,Y,\mathscr{V})$.

Notation. Since (Theorem 3.4) $\mathfrak{T}_{\mathscr{U}_p}=\mathfrak{T}_p$ relative to $(Y,\mathfrak{T}_{\mathscr{V}})$, we let $\mathfrak{F}_p(X;\,Y)$ denote both $(\mathfrak{F}(X;\,Y,\mathscr{V}),\mathscr{U}_p)$ and $(\mathfrak{F}(X;\,Y,\mathfrak{T}),\mathfrak{T}_p)$ when no confusion is likely.

Convergence in $\mathfrak{F}_p(X;\,Y)$ is termed "pointwise convergence" because $\mathfrak{F}_p(X;\,Y)$ is homeomorphic to $\left(\prod_X Y,\mathfrak{T}_\pi\right)$. In fact,

Theorem 7.2. The mapping $\mathfrak{e}:\underset{f\rightsquigarrow\{f(x)\}_f}{\mathfrak{F}_p(X;\,Y,\mathscr{V})\to\left(\prod_X Y,\mathscr{U}_\pi\right)}$ is a unimorphism. Therefore $\mathscr{U}_p=\mathscr{U}_{\mathscr{P}(\mathfrak{e})}$ and

(i) $\mathfrak{F}_p(X;\,Y,\mathscr{V})$ is separated iff $(Y,\mathscr{V})$ is separated.

(ii) Each $\mathfrak{e}_x\in\xi_X$ is a uniformly continuous open surjection.

(iii) A filter $\mathscr{G}\subset\mathfrak{F}_p(X;\,Y,\mathscr{V})$ converges to $f\in\mathfrak{F}(X;\,Y,\mathscr{V})$ iff $\mathfrak{e}_x\mathscr{G}=\{g(x):g\in\mathscr{G}\}\subset Y$ converges to $f(x)\in Y\ \forall x\in X$.

(iv) A filter $\mathscr{G}\subset\mathfrak{F}_p(X;\,Y,\mathscr{V})$ is Cauchy iff each $\mathfrak{e}_x\mathscr{G}$ is $\mathscr{V}$-Cauchy.

Proof. Clearly $\mathfrak{e}$ is a bijection. let $\mathcal{V}_x = \mathcal{V} \in \mathfrak{U}(Y_x = Y)\, \forall x \in X$. If $V = \bigcap_{i=1}^{n} (\pi_{x_i} \times \pi_{x_i})^{-1} V_{x_i}$ is any $\mathcal{U}_\pi$-base entourage, then $(\mathfrak{e} \times \mathfrak{e})^{-1} V = \bigcap_{i=1}^{n} (\mathfrak{e}_{x_i} \times \mathfrak{e}_{x_i})^{-1} V_{x_i} \in \mathcal{U}_p$ and so $\mathfrak{e}$ is $\mathcal{U}_p - \mathcal{U}_\pi$ uniformly continuous. The uniform continuity of $\mathfrak{e}^{-1}$ follows from Theorem 3.4 and the $\mathcal{U}_\pi$-uniform continuity of every $\pi_x = \mathfrak{e}_x \mathfrak{e}^{-1}$ $(x \in X)$. Finally, Example 3.13 yields $\mathcal{U}_p = \mathcal{U}_{\mathcal{P}(\mathfrak{e})}$, and (i)–(iv) follow from the corresponding properties for product spaces. ■

THEOREM 7.3. A subset $\mathcal{K} \subset \mathcal{F}_p(X; Y)$ is compact {complete} if $\mathcal{K}$ is closed and every $\overline{\{f(x): f \in \mathcal{K}\}} \subset Y$ is compact {complete}. The conditions are also necessary for $\mathcal{K}$'s compactness when Y is T_2.

Proof. The compactness {completeness} of each $\overline{\mathfrak{e}_x(\mathcal{K})} = \overline{\{f(x): f \in \mathcal{K}\}}$ in Y implies that of $\prod_X \overline{\mathfrak{e}_k(\mathcal{K})}$ in $\prod_X Y$. Since $\mathfrak{e}(\mathcal{K}) \subset \prod_X \overline{\mathfrak{e}_x(\mathcal{K})}$ and $\mathfrak{e}(\mathcal{K})$ is closed in $\prod_X Y$, the compactness {completeness} of $\mathfrak{e}(\mathcal{K}) \subset \prod_X Y$ and consequently that of $\mathcal{K} \subset \mathcal{F}_p(X; Y)$ is assured. If Y is T_2 and $\mathcal{K} \subset \mathcal{F}_p(X; Y)$ is compact, then $\mathcal{K}$ is $\mathcal{F}_p$-closed and $\overline{\mathfrak{e}_x(\mathcal{K})} = \mathfrak{e}_x(\mathcal{K}) \subset Y$ is compact $\forall x \in X$. ■

Remark. A complete subset $\mathcal{K}$ of a separated space $\mathcal{F}_p(X; Y, \mathcal{V})$ is closed, but each $\overline{\mathfrak{e}_x(\mathcal{K})} \subset Y$ need not be complete [Example 6.8(a)].

There are simply not enough $\mathcal{F}_p$-open sets for $\mathcal{F}_p$ to be considered a natural topology on $\mathcal{F}(X; Y)$.

Example 7.4. Let $\mathcal{Q}$ be the collection of finite subsets of an uncountable set X and let χ_A be the characteristic function of $A \in \mathcal{Q}$. The filterbase $\{\chi_A: A \in \mathcal{Q}\}$ converges to $\chi_X \in \mathcal{F}_p(X; I, d_1)$. Although each $\chi_{\{x\}}$ $(x \in X)$ is quite different from χ_X, the oddity $\chi_X \in \overline{\{\chi_{\{x\}}: x \in X\}}^{\mathcal{F}_p}$ prevails.

This seeming contradiction in Example 7.4 can be avoided if $\mathcal{F}(X; I, d_1)$ carries a topology $\mathcal{T} \supset \mathcal{T}_p$ under which all χ_A $(A \in \mathcal{Q})$ are not contained in any $\mathcal{T}$-nbd of χ_X. One such case is the following: Consider any $f \in \mathcal{F}(X; Y)$. For each $x \in X$ let w_x be a Y-nbd of $f(x) \in Y$ and take $W_{w_x}(f) = \{g \in \mathcal{F}(X; Y): g(x) \in w_x\ \forall x \in X\}$. Then $\{W_{w_x}(f): x \in X\}$ defines a nbd base at f for a topology $\mathcal{T}$ with the desired property. (Note, for instance, in Example 7.4 above no $\chi_A \in W_{s_\varepsilon}(\chi_X)$ unless $A = X$.) One more point: A mapping $g \in \mathcal{F}(X; Y)$ which is W_{w_x}-close to f above is uniformly close in the sense that $g(x) \in w_x\{f(x)\}\ \forall x \in X$. All this points to defining a uniform type topology on $\mathcal{F}(X; Y)$, which we proceed to introduce in

DEFINITION 7.5. Let $(Y, \mathcal{V})$ be a uniform space. For each $V \in \mathcal{V}$ define $W_V = \{(f, g) \in \mathcal{F}(X, Y; \mathcal{V}) \times \mathcal{F}(X; Y, \mathcal{V}): (f(x), g(x)) \in V\ \forall x \in X\}$. Then $\mathcal{B} = \{W_V: V \in \mathcal{V}\}$ is the base for a uniformity for $\mathcal{F}(X; Y, \mathcal{V})$ which is called the uniformity of uniform convergence, denoted $\mathcal{U}_u$. The corresponding uniform topology, written $\mathfrak{T}_u$ is called the topology of uniform convergence. $\mathcal{F}_u(X; Y, \mathcal{V})$ denotes $\mathcal{F}(X; Y, \mathcal{V})$ with $\mathcal{U}_u$ or $\mathfrak{T}_u$. Convergence in $\mathcal{F}_u(X; Y, \mathcal{V})$ is termed "uniform convergence."

Remark. Since $W_V = \bigcap_X (\mathbf{e}_x \times \mathbf{e}_x)^{-1} V$, the sets $\left\{W_V[f] = \mathbf{e}^{-1}\left(\prod_X V[f(x)]\right): V \in \mathcal{V} \text{ and } f \in \mathcal{F}(X; Y, \mathcal{V})\right\}$ constitute a $\mathfrak{T}_u$-nbd base and so $\mathcal{U}_p \subset \mathcal{U}_u$ on $\mathcal{F}(X; Y, \mathcal{V}) \times \mathcal{F}(X; Y, \mathcal{V})$.

Let $\mathcal{V}_{\prod_X}$ denote the uniformity on $\prod_X Y$ (which is finer than the product uniformity $\mathcal{V}_\pi$) having base entourages of the form $\bigcap_X (\pi_x \times \pi_x)^{-1} V$ $(V \in \mathcal{V})$. In this notation

THEOREM 7.6. (i) $\mathbf{e}: \mathcal{F}_u(X; Y, \mathcal{V}) \to \left(\prod_X Y, \mathcal{V}_{\prod_X}\right)$ is a unimorphism.

(ii) A filter $\mathcal{G} \subset \mathcal{F}_u(X; Y, \mathcal{V})$ converges to $f \in \mathcal{F}(X; Y)$ iff $\mathcal{G}$ is $\mathcal{U}_u$-Cauchy and pointwise convergent to f.

Proof. (i) The $\mathcal{U}_u$-uniform continuity of $\mathbf{e}$ is clear since

$$(\mathbf{e} \times \mathbf{e})^{-1}\left\{\bigcap_{i=1}^{n} \bigcap_X (\pi_x \times \pi_x)^{-1} V_i\right\} = \bigcap_{i=1}^{n} W_{V_i} \in \mathcal{U}_u.$$

On the other hand,

$$(\mathbf{e}^{-1} \times \mathbf{e}^{-1})^{-1} \bigcap_{i=1}^{n} W_{V_i} = \bigcap_X (\pi_x \times \pi_x)^{-1}\left\{\bigcap_{i=1}^{n} V_i\right\} \in \mathcal{V}_{\prod_X}$$

assures that $\mathbf{e}^{-1}$ is uniformly continuous.

(ii) Sufficiency only requires proof. Let W_V be any $\mathcal{U}_u$-base entourage. We assume (Corollary 1.17) that $V \in \mathcal{V}$ is closed and symmetric. Since $\mathcal{G}$ is $\mathcal{U}_u$-Cauchy, there is some $G \in \mathcal{G}$ such that $G \times G \subset W_V$. For any fixed $g' \in G$ note that $g(x) \in V[g'(x)]\ \forall g \in G$ and $x \in X$; that is, $G(x) \subset V[g'(x)]\ \forall x \in X$. Now use the hypothesis that $G(x) \to f(x)\ \forall x \in X$. Given any $U \in \mathcal{V}$, a member $\tilde{G} \in \mathcal{G}$ (depending on U) can be found such that $\tilde{G}(x) \subset U[f(x)]$. Thus $\tilde{G}(x) \cap G(x) \subset U[f(x)] \cap V[g'(x)] \neq \phi\ \forall U \in \mathcal{V}$ so that $f(x) \in \overline{V[g'(x)]} = V[g'(x)]\ \forall x \in X$. Since each $g' \in G$ belongs to $W_V[f]$, it follows that $G \subset W_V[f]$ and $\mathcal{G}$ is $\mathcal{U}_u$-convergent to f. ■

Our topologies on $\mathfrak{F}(X; Y, \mathscr{V})$ (the uniformities on $\prod_X Y$) lie at the extreme ends of the spectrum. Somewhere in between is

Definition 7.7. Let $\mathscr{S}$ be a nonempty collection of nonempty subsets of a set X. For each $S_\alpha \in \mathscr{S}$, let $\mathfrak{F}_{u_\alpha}(S_\alpha; Y, \mathscr{V})$ be as in Definition 7.5 with $X = S_\alpha$. The projective limit uniformity generated by $\mathscr{A} = \{a_\alpha : \mathfrak{F}(X; Y, \mathscr{V}) \to \mathfrak{F}_{u_\alpha}(S_\alpha; Y, \mathscr{V}) : S_\alpha \in \mathscr{S}\}$ ($f \rightsquigarrow f/S_\alpha$) is called the uniformity of $\mathscr{S}$-convergence for $\mathfrak{F}(X; Y, \mathscr{V})$, denoted $\mathscr{U}_{\mathscr{S}}$.

$\mathfrak{F}_{\mathscr{S}}(X; Y, \mathscr{V})$ denotes $\mathfrak{F}(X; Y, \mathscr{V})$ with $\mathscr{U}_{\mathscr{S}}$ or $\mathfrak{T}_{\mathscr{S}} = \mathfrak{T}_{\mathscr{U}_{\mathscr{S}}}$ (the topology of $\mathscr{S}$-convergence).

Example 7.8. (a) If $\mathscr{S}$ denotes the finite subsets of X, then $\mathfrak{F}_{\mathscr{S}}(X; Y, \mathscr{V}) = \mathfrak{F}_p(X; Y, \mathscr{V})$. If $\mathscr{S} = \{X\}$, then $\mathscr{U}_{\mathscr{S}}$ and $\mathscr{U}_u$ share the same base $\{W_V : V \in \mathscr{V}\}$ and so $\mathfrak{F}_{\mathscr{S}}(X; Y, \mathscr{V}) = \mathfrak{F}_u(X; Y, \mathscr{V})$.

(b) Suppose $\mathfrak{R}$ is another collection of nonempty subsets of X such that each $R \in \mathfrak{R}$ is contained in a finite union of members of $\mathscr{S}$. Then $\mathscr{U}_{\mathfrak{R}} \subset \mathscr{U}_{\mathscr{S}}$ on $\mathfrak{F}(X; Y, \mathscr{V}) \times \mathfrak{F}(X; Y, \mathscr{V})$. (Note that for $W_{S,V} = \{f, g \in \mathfrak{F}(S; Y, \mathscr{V}) : (f(x), g(x)) \in V \; \forall x \in \mathscr{S}\}$ and $R_\alpha \subset \bigcup_{i=1}^n S_{\beta_i}$, one obtains $\bigcap_{i=1}^n (a_{\beta_i} \times a_{\beta_i})^{-1} W_{S_{\beta_i}, V} \subset (a_\alpha \times a_\alpha)^{-1} W_{R_\alpha, V}$.) In particular, $\mathscr{U}_p \subset \mathscr{U}_{\mathscr{S}} \subset \mathscr{U}_u$ for every covering $\mathscr{S}$ of X.

(c) The uniformity {topology} of compact convergence is denoted $\mathscr{U}_\kappa \{\mathfrak{T}_\kappa\}$. This is $\mathscr{U}_{\mathscr{S}} \{\mathfrak{T}_{\mathscr{S}}\}$ for $\mathscr{S} = \mathfrak{C}$, where $\mathfrak{C}$ denotes the compact subsets of a topological space X. Clearly $\mathfrak{F}_\kappa(X; Y, \mathscr{V}) = \mathfrak{F}_u(X; Y, \mathscr{V})$ when X is compact and $\mathfrak{F}_\kappa(X; Y, \mathscr{V}) = \mathfrak{F}_p(X; Y, \mathscr{V})$ when X is discrete.

(d) The uniformity {topology} of precompact (i.e., totally bounded) convergence on a uniform space X is denoted $\mathscr{U}_\lambda \{\mathfrak{T}_\lambda\}$. Surely $\mathscr{U}_p \subset \mathscr{U}_\kappa \subset \mathscr{U}_\lambda$ on $\mathfrak{F}(X; Y, \mathscr{V}) \times \mathfrak{F}(X; Y, \mathscr{V})$.

(e) For each $y \in Y$ let $\phi_y \in \mathfrak{F}(X; Y, \mathscr{V})$ be the constant map $\phi_y(x) = y$. Then $\Phi : (Y, \mathscr{V}) \to \mathfrak{F}_{\mathscr{S}}(X; Y, \mathscr{V})$ ($y \rightsquigarrow \phi_y$) is an into unimorphism. Indeed, $(\Phi^{-1} \times \Phi^{-1})^{-1} V = (\Phi \times \Phi) V \in (\Phi(Y) \times \Phi(Y)) \cap \mathscr{U}_{\mathscr{S}}$), whereas $(\Phi \times \Phi)^{-1} \{(\Phi(Y) \times \Phi(Y)) \cap (a_\alpha \times a_\alpha)^{-1} W_{E_\alpha, V}\} = V \; \forall V \in \mathscr{V}$.

Returning to basics, let $\mathscr{V}_{\prod_{\mathscr{S}}}$ be the uniformity for $\prod_X Y$ having subbase entourages $\bigcap_{x \in S_\alpha} (\pi_\alpha \times \pi_\alpha)^{-1} V$ for $V \in \mathscr{V}$ and $S_\alpha \in \mathscr{S}$. Note, for example, that $\mathscr{V}_{\prod_{\mathscr{S}}} = \mathscr{V}_\pi$ if $\mathscr{S}$ consists of all finite subsets of X. Essentially, $\mathscr{V}_\pi \subset V_{\prod_{\mathscr{S}}} \subset V_{\prod_X}$ for every cover $\mathscr{S}$ of X.

THEOREM 7.9. (i) $\mathfrak{e}:\mathfrak{F}_{\mathscr{S}}(X;\ Y,\ \mathscr{V})\to\left(\prod_X Y,\ \mathscr{V}_{\prod_{\mathscr{S}}}\right)$ is a unimorphism.

(ii) Suppose $\aleph(Y)\geq 2$. Then $\mathfrak{F}_{\mathscr{S}}(X;\ Y,\ \mathscr{V})$ is separated iff $(Y,\ \mathscr{V})$ is separated and $\mathscr{S}$ covers X.

(iii) The properties below are equivalent for a filter $\mathcal{G}\subset\mathfrak{F}_{\mathscr{S}}(X;\ Y,\ \mathscr{V})$:

(a) $\mathcal{G}$ converges to $f\in\mathfrak{F}_{\mathscr{S}}(X;\ Y)$.

(b) $\mathcal{G}$ is a $\mathscr{U}_{\mathscr{S}}$-Cauchy and pointwise convergent to f on $\bigcup_{\mathscr{S}} S_\alpha$.

(c) $\mathcal{G}$ converges uniformly to f on each $S_\alpha\in\mathscr{S}$ (i.e., $\mathcal{G}/S_\alpha\to f/S_\alpha\in\mathfrak{F}_{u_\alpha}(S_\alpha;\ Y,\ \mathscr{V})\ \forall S_\alpha\in\mathscr{S}$).

Proof. (i) is straightforward; proceed as in Theorems 7.2 and 7.6. For (ii) suppose that $\mathfrak{F}_{\mathscr{S}}(X;\ Y,\ \mathscr{V})$ is separated. Then $(\Phi(Y),\ \mathscr{U}_{\mathscr{S}})$; therefore $(Y,\ \mathscr{V})$ is separated. Also, $\mathscr{S}$ covers X. Otherwise, some $x_0\in X-\bigcup_{\mathscr{S}} S_\alpha$ and $f(x_0)=y_1\neq y_2=f(X-x_0)$ yields the contradiction that $(f,\Phi_{y_2})\in\bigcap_{\mathscr{V},\mathscr{S}}(a_\alpha\times a_\alpha)^{-1}W_{S_\alpha,V}$. On the other hand, $\mathscr{S}$ covers X and $\mathscr{V}$ being separated implies that $\mathscr{U}_{\mathscr{S}}$ is separated [Theorem 7.2, Example 7.8(b)]. We begin (iii) with (a)$\Rightarrow$(b). If $\mathcal{G}\to f\in\mathfrak{F}_{\mathscr{S}}(X;\ Y,\ \mathscr{V})$, then $\mathcal{G}$ is $\mathscr{U}_{\mathscr{S}}$-Cauchy (Theorem 5.2). Since $\mathfrak{e}_x a_\alpha:\mathfrak{F}_{\mathscr{S}}(X;\ Y,\ \mathscr{V})\to(Y,\ \mathscr{V})$ is uniformly continuous $\forall x\in S_\alpha\in\mathscr{S}$, one has $\mathfrak{e}_x a_\alpha(\mathcal{G})=\mathcal{G}(x)\to f(x)=\mathfrak{e}_x a_\alpha(f)\ \forall x\in\bigcup_{\mathscr{S}} S_\alpha$. Next, (b)$\Rightarrow$(c), since $a_\alpha(\mathcal{G})=\mathcal{G}/S_\alpha$ is $\mathscr{U}_{u_\alpha}$-Cauchy $\forall S_\alpha\in\mathscr{S}$ and, since $\mathcal{G}/S_\alpha$ converges pointwise to $f/S_\alpha\ \forall S_\alpha\in\mathscr{S}$, it follows (Theorem 7.6) that $\mathcal{G}/S_\alpha\to f/S_\alpha\in\mathfrak{F}_u(S_\alpha;\ Y,\ \mathscr{V})\ \forall S_\alpha\in\mathscr{S}$. Finally, (c)$\Leftrightarrow$(a) is an obvious consequence of the fact that $\mathscr{U}_{\mathscr{S}}=U_{\mathscr{P}\{a_\alpha:S_\alpha\in\mathscr{S}\}}$. ■

If $\{f_n\}\in C(X;\ Y)$ converges pointwise to f, then $f\in\overline{C(X;\ Y)}\subset\mathfrak{F}_p(X;\ Y)$. If $\{f_n\}$ converges uniformly to f, then $f\in C(X;\ Y)$ (thereby generalizing the well-known fact that the limit of a uniformly convergent sequence of real-valued continuous functions is continuous). All this is encompassed in

THEOREM 7.10. If $\mathscr{S}$ is a collection of sets in X such that every $x\in X$ is an interior point of some member of $\mathscr{S}$, then $C(X;\ Y,\ \mathscr{V})$ is a closed subset of $\mathfrak{F}_{\mathscr{S}}(X;\ Y,\ \mathscr{V})$.

Proof. Suppose $f\in\overline{C(X;\ Y,\ \mathscr{V})}\subset\mathfrak{F}_{\mathscr{S}}(X;\ Y,\ \mathscr{V})$. To conclude that $f\in C(X;\ Y,\ \mathscr{V})$ we prove that $f(\mathfrak{N}_x)\to f(x)\ \forall x\in X$. (Here $\mathfrak{N}_x$ denotes the nbd filter at $x\in X$.) Fix any $x_0\in X$, let $S_{\alpha_0}\in\mathscr{S}\cap\mathfrak{N}_{x_0}$, and for any

$V \in \mathscr{V}$ pick a symmetric U such that $U^3 \subset V$. Then $g \in \{(a_{\alpha_0} \times a_{\alpha_0})^{-1} W_{S_{\alpha_0}, U}\}[f] \cap C(X; Y, \mathscr{V})$ implies $(f(x), g(x)) \in U\ \forall x \in S_{\alpha_0}$. Since $g(\mathfrak{N}_{x_0})$ is $\mathscr{V}$-Cauchy, some set $E \in \mathfrak{N}_{x_0}$ is contained in S_{α_0} and satisfies $g(E) \times g(E) \subset U$. In particular, $(g(x), g(x_0)) \in U$ and so $(f(x), f(x_0)) \subset U^3\ \forall x \in E$. Thus $f(E) \subset V[f(x_0)]$ and $f(\mathfrak{N}_{x_0}) \to f(x_0)$ as required. ■

THEOREM 7.11. $\mathfrak{F}_{\mathscr{S}}(X; Y, \mathscr{V})$ is complete iff $(Y, \mathscr{V})$ is complete.

Proof. If $\mathfrak{F}_{\mathscr{S}}(X; Y, \mathscr{V})$ is complete and $\mathcal{G} \subset (Y, \mathscr{V})$ is Cauchy, then [Example 7.8(e)] $\Phi(\mathcal{G})$ converges to some $f \in \mathfrak{F}_{\mathscr{S}}(X; Y, \mathscr{V})$. Therefore $\mathcal{G} = \mathfrak{e}_x \Phi(\mathcal{G}) \to \mathfrak{e}_x(f) = f(x) \in Y$ for $x \in \bigcup_{\mathscr{S}} S_\alpha$. Assume, conversely, that $(Y, \mathscr{V})$ is complete and let $\mathscr{H}$ be a Cauchy filter in $\mathfrak{F}_{\mathscr{S}}(X; Y, \mathscr{V})$. Then each $\mathscr{V}$-Cauchy filter $\mathfrak{e}_x(\mathscr{H})$ converges to some $y_x \in Y$ so that $\mathscr{H} \to \tilde{f} \in \mathfrak{F}_p(X; Y, \mathscr{V})$, where $\tilde{f}(x) = y_x\ \forall x \in X$. Consequently (Theorem 7.9) $\mathscr{H} \to f \in \mathfrak{F}_{\mathscr{S}}(X; Y, \mathscr{V})$. ■

Equicontinuity and Uniform Equicontinuity

Total boundedness and compactness in $\mathfrak{F}_{\mathscr{S}}(X; Y)$ are intimately tied to continuity notions in

DEFINITION 7.12. A subset $\Phi \subset \mathfrak{F}(X; Y, \mathscr{V})$, where X is a topological space {uniform space $(X, \mathscr{U})$} is said to be equicontinuous {uniformly equicontinuous} at $x \in X$ if $\forall V \in \mathscr{V}$, there is an X-nbd $u(x)$ {a $U \in \mathscr{U}$} such that $\phi(u(x)) \subset V[\phi(x)]$ {such that $\phi(U[x]) \subset V[\phi(x)]$} $\forall \phi \in \Phi$. Accordingly Φ is termed equicontinuous {uniformly equicontinuous} if it is equicontinuous {uniformly equicontinuous} at every $x \in X$.

Remark. Note that $\Phi \subset \mathfrak{F}(X, \mathscr{U}; Y, \mathscr{V})$ is uniformly equicontinuous iff $\bigcap_{\Phi} (\phi \times \phi)^{-1} V \in \mathscr{U}\ \forall V \in \mathscr{V}$.

The following general result is due to Bourbaki [13].

THEOREM 7.13. Let $\mathscr{S} = \{S_\alpha : \alpha \in \mathfrak{A}\}$ be a collection of subsets of a set T and consider a mapping $\psi: T \times X \to (Y, \mathscr{V})$, where X is a topological {uniform} space and $(Y, \mathscr{V})$ is a uniform space. For each $S_\alpha \in \mathscr{S}$ define $\Phi_{S_\alpha} = \{\phi_t : \underset{x \rightsquigarrow \psi(t, x)}{X \longrightarrow Y}\ ; t \in S_\alpha\}$. Then $j: \underset{x \rightsquigarrow \psi(\cdot, x)}{X \to \mathfrak{F}_{\mathscr{S}}(T; Y, \mathscr{V})}$ is continuous at x_0 {uniformly continuous} iff Φ_{S_α} is equicontinuous at x_0 {uniformly equicontinuous} $\forall S_\alpha \in \mathscr{S}$.

Proof. We give the equicontinuity proof only, since the uniform equicontinuity proof follows similarly. Assume that j is continuous at $x_0 \in X$. For any $S_\alpha \in \mathscr{S}$ and $V \in \mathscr{V}$ there is an X-nbd $u(x_0)$ such that $j\{u(x_0)\}\subset\{(a_\alpha\times a_\alpha)^{-1}W_{S_\alpha,V}\}[\psi(\,,x_0)]$; that is, $(\psi(\,,x_0),\psi(\,,x))\in(a_\alpha\times a_\alpha)^{-1}W_{S_\alpha,V}\ \forall x\in u(x_0)$. This means that $\psi(t,x)\in V[\psi(t,x_0)]\ \forall t\in S_\alpha$ and $x\in u(x_0)$. In other words, $\phi_t(x)\in V[\phi_t(x_0)]\ \forall x\in u(x_0)$ and $\phi_t\in\Phi_{S_\alpha}$, so that Φ_{S_α} is equicontinuous at x_0. Suppose, conversely, that $W(j(x_0))=\left\{\bigcap_{i=1}^{n}(a_{\alpha_i}\times a_{\alpha_i})^{-1}W_{S_{\alpha_i},V}\right\}[\psi(\,,x_0)]$ is any $\mathfrak{T}_{\mathcal{S}}$-base nbd of $j(x_0)$. Since each $\Phi_{S_{\alpha_i}}$ is equicontinuous at x_0, there exists X-nbds $u_{\alpha_i}(x_0)$ $(i=1,2,\ldots,n)$ each of which satisfies $\phi_t(x)\in V[\phi_t(x_0)]\ \forall x\in u_{\alpha_i}(x_0)$ and $\phi_t\in\Phi_{S_{\alpha_i}}$. Taking $u(x_0)=\bigcap_{i=1}^{n}u_{\alpha_i}$, one easily verifies that $j\{u(x_0)\}\subset W(j(x_0))$ and j is continuous at $x_0\in X$. ■

Corollary 7.14. A set $\Phi\subset\mathfrak{F}(X;\,Y,\mathcal{V})$ is equicontinuous at $x_0\in X$ {is uniformly equicontinuous} iff $j:\underset{x\rightsquigarrow e_x}{X\to\xi_X}\subset\mathfrak{F}_u(X;\,Y,\mathcal{V})$ is continuous at x_0 {is uniformly continuous}.

Proof. Take $T=\Phi=\mathcal{S}$ and let $\psi:\underset{(\phi,x)\rightsquigarrow\phi(x)}{\Phi\times X\to(Y,\mathcal{V})}$ in Theorem 7.13. ■

Equicontinuity is frequently preserved in taking closures.

Theorem 7.15. A set $\Phi\subset\mathfrak{F}(X;\,Y,\mathcal{V})$ is equicontinuous at $x_0\in X$ {is uniformly equicontinuous} iff $\bar{\Phi}^p$ is equicontinuous at $x_0\in X$ {is uniformly equicontinuous}. If $\mathcal{S}$ covers X, pointwise closure $\bar{\Phi}^p$ above can be replaced by $\mathcal{S}$-convergence closure $\bar{\Phi}^{\mathcal{S}}$.

Proof. Sufficiency is clear. Assume that $f\in\bar{\Phi}^p$, where $\Phi\subset\mathfrak{F}(X;\,Y,\mathcal{V})$ is equicontinuous at $x_0\in X$ and let $\mathcal{G}$ be a filter in Φ which converges pointwise to f. Given any closed $V\in\mathcal{V}$ and X-nbd $u(x_0)$ such that $(\phi(x),\phi(x_0))\in V\ \forall x\in u(x_0)$ and $\phi\in\Phi$, one has $(f(x),f(x_0))\in\bar{V}=V\ \forall x\in u(x_0)$. This then establishes that $\bar{\Phi}^p$ is equicontinuous at x_0. The proof for the case of uniform equicontinuity follows similarly. If $\mathcal{S}$ covers X, then $\Phi\subset\bar{\Phi}^{\mathcal{S}}\subset\bar{\Phi}^p$ and all three collections either are or are not simultaneously equicontinuous at x_0 (uniformly equicontinuous). ■

A few important equicontinuity properties are given below.

Example 7.16. (a) $\mathcal{U}_\kappa=\mathcal{U}_p$ for every equicontinuous set $\Phi\subset C(X;\,Y,\mathcal{V})$. Because of Example 7.8(b), only $\mathcal{U}_\kappa\subset\mathcal{U}_p$ on $\Phi\times\Phi$ needs verification. Given any $\mathcal{U}_\kappa$-subbase entourage $(a_K\times a_K)^{-1}W_{K,V}$ on $\Phi\times\Phi$, choose a symmetric $U\in\mathcal{V}$ satisfying $U^5\subset V$. For each $x\in X$ let u_x be an

X-open nbd of x such that $(\phi(x'), \phi(x)) \in U\ \forall x \in u_x$ and $\phi \in \Phi$. Then $K \subset \bigcup_{i=1}^{n} u_{x_i}$ and $\forall (x', x'') \in U_{x_i}$, one has $(\phi(x', \phi(x'')) \in U^2\ \forall \phi \in \Phi$. Taken together, these last two statements yield $\left\{\bigcap_{i=1}^{n} (\mathfrak{e}_{x_i} \times \mathfrak{e}_{x_i})^{-1} U\right\} \cap (\Phi \times \Phi) \subset \{(a_K \times a_K)^{-1} W_{K,V}\} \cap (\Phi \times \Phi)$.

(b) If $\Phi \subset C(X; Y, \mathscr{V})$ is equicontinuous and X is compact, then $\mathscr{U}_p = \mathscr{U}_u$ on $\Phi \times \Phi$. Verification follows from (a) above and the fact that $\mathfrak{F}_\kappa(X; Y, \mathscr{V}) = \mathfrak{F}_u(X; Y, \mathscr{V})$.

(c) $\bar{\Phi}^p = \bar{\Phi}^\kappa$ for every equicontinuous set $\Phi \subset C(X; Y, \mathscr{V})$. That $\bar{\Phi}^\kappa \subset \bar{\Phi}^p$ is clear by Example 7.8(b). Now suppose that $f \in \bar{\Phi}^p$, let $(a_K \times a_K)^{-1} W_{K,V}$ be any $\mathscr{U}_\kappa$-subbase entourage, and choose a symmetric $U \subset \mathscr{V}$ such that $U^3 \subset V$. Since $\bar{\Phi}^p$ is equicontinuous, there is an X-open nbd u_x of $x \in X$ such that $(\phi(x'), \phi(x)) \in U\ \forall x \in u_x$ and $\phi \in \bar{\Phi}^p$. Let $K \subset \bigcup_{i=1}^{n} u_{x_i}$. It follows from $f \in \bar{\Phi}^p$ that some $\phi_0 \in \left\{\bigcap_{i=1}^{n} (\mathfrak{e}_{x_i} \times \mathfrak{e}_{x_i})^{-1} U\right\}[f] \cap \Phi$, and a quick calculation reveals that $\phi_0 \in \{(a_K \times a_K)^{-1} W_{K,V}\}[f] \cap \Phi$. Since $K \in \mathscr{C}$ and $V \in \mathscr{V}$ were arbitrary, $f \in \bar{\Phi}^\kappa$ as required.

(d) Suppose $\Phi \subset \mathfrak{F}(X; Y, \mathscr{V})$ is equicontinuous on compacta.† It need not be true that $\Phi \subset C(X; Y, \mathscr{V})$ (although $\Phi/K \subset C(K; Y, \mathscr{V})\ \forall K \in \mathscr{C}$). If X is locally compact, then $\Phi \subset C(X; Y, \mathscr{V})$ is equicontinuous. Indeed, let $K(x)$ be an X-compact nbd of $x \in X$. Given any $V \in \mathscr{V}$, there is a $K(x)$-nbd $u(x)$ such that $(\phi(x'), \phi(x)) \in V\ \forall x' \in u(x)$ and $\phi \in \Phi/K(x)$. Since $u = K(x) \cap v(x)$ (where $v(x)$ is an X-nbd of x), it follows that $u(x)$ is an X-nbd and Φ is equicontinuous at $x \in X$.

(e) $\mathscr{U}_\kappa = \mathscr{U}_p$ on $\Phi \times \Phi$ (although Φ need not be equicontinuous) if Φ is equicontinuous on compacta. Verification is easy; simply let $\xi_K = \{\mathfrak{e}_x : (\Phi, \mathscr{U}_{u_K}) \to (Y, \mathscr{V}); x \in K\}$ and assume that Φ carries $\mathscr{U}_p$. Since $\mathfrak{e}_x a_K : (\Phi/K, \mathscr{U}_{\mathfrak{P}(\xi_K)}) \to (Y, \mathscr{V})$ is uniformly continuous $\forall x \in K$ and since $\mathscr{U}_{\mathfrak{P}\{\xi_K\}} = \mathscr{U}_{u_K}$ on $\Phi/K \times \Phi/K$ by (b) above, it follows (Theorem 3.4) that every a_K $(K \in \mathscr{C})$ is $\mathscr{U}_p$-uniformly continuous and $\mathscr{U}_\kappa \subset \mathscr{U}_p$.

THEOREM 7.17. Assume that $\Phi \subset \mathfrak{F}(X; Y, \mathscr{V})$ satisfies $\Phi/S_\alpha \subset C(S_\alpha; Y, \mathscr{V})\ \forall S_\alpha \in \mathscr{S}$. If Φ is $\mathscr{U}_{\mathscr{S}}$-totally bounded, then Φ is equicontinuous on members of $\mathscr{S}$ and $\{\phi(x) : \phi \in \Phi\} \subset (Y, \mathscr{V})$ is totally bounded $\forall x \in \bigcup_{\mathscr{S}} S_\alpha$.

The converse also holds when $\mathscr{S} = \mathscr{C}$.

† $\Phi \subset \mathfrak{F}(X; Y, \mathscr{V})$ is equicontinuous on members of $\mathscr{S}$ if $\Phi/S_\alpha \subset C(S_\alpha; Y, \mathscr{V})$ is equicontinuous $\forall S_\alpha \in \mathscr{S}$.

Proof. Suppose $\Phi \subset \mathfrak{F}_{\mathcal{S}}(X; Y, \mathcal{V})$ is totally bounded. Fix any $x_0 \in S_\alpha$ in $\mathcal{S}$ and for $V \in \mathcal{V}$ pick a symmetric $U \in \mathcal{V}$ such that $U^3 \subset V$. Since $\alpha_\alpha(\Phi) = \Phi/S_\alpha \subset \mathfrak{F}_{u_\alpha}(S_\alpha; Y, \mathcal{V})$ is totally bounded (Theorem 5.10), there exist $\{\phi_1, \ldots, \phi_n\} \in \Phi/S_\alpha$ such that $\Phi/S_\alpha \subset \bigcup_{i=1}^{n} W_{S_\alpha, U}[\phi_i]$ (Theorem 5.6). if $\phi \in \Phi/S_\alpha$, then $(\phi(x), \phi_i(x)) \in U\ \forall x \in S_\alpha$ and some $i = 1, 2, \ldots, n$; but each $\phi_i \in C(S_\alpha; Y, \mathcal{V})$ so that there are S_α-nbds $u_i(x_0)$ $(i = 1, 2, \ldots, n)$, each satisfying $\phi_i\{u_i(x_0)\} \subset U[\phi_i(x_0)]$. If $u(x_0) = \bigcap_{i=1}^{n} u_i(x_0)$, then $(\phi(x), \phi(x_0)) \in V\ \forall x \in u(x_0)$ and $\phi \in \Phi/S_\alpha$. Thus Φ/S_α is equicontinuous at $x_0 \in S_\alpha$. Since $\mathcal{U}_{\mathcal{P}\{e_x : x \in \bigcup_{\mathcal{S}} S_\alpha\}} \subset \mathcal{U}_{\mathcal{S}}$, surely $U_{\mathcal{P}\{e_x : x \in \bigcup S_\alpha\}}$ is totally bounded and our second assertion follows. Last Φ is $\mathcal{U}_\kappa = \mathcal{U}_p$ totally bounded when $S = \mathcal{C}$ and both conditions above hold [Theorem 5.10, Example 7.16(e)]. ■

THEOREM 7.18 (Ascoli). Suppose $(Y, \mathcal{V})$ is separated. Then a set $\Phi \subset C(X; Y, \mathcal{V})$ is $\mathfrak{T}_\kappa$-compact iff

(i) Φ is closed in $\mathfrak{F}_\kappa(X; Y, \mathcal{V})$,
(ii) $\overline{\{\phi(x) : \phi \in \Phi\}} \subset Y$ is compact $\forall x \in X$,
(iii) Φ is equicontinuous on compacta.

Proof. If Φ is $\mathfrak{T}_\kappa$-compact, it is $\mathfrak{T}_p$-compact and (i), (ii) hold (Theorem 7.3). Condition (iii) is a consequence of Theorem 7.17. In the other direction (iii) implies $\mathcal{U}_\kappa = \mathcal{U}_p$ on $\Phi \times \Phi$. With (i) this gives $\Phi = \bar{\Phi}^\kappa = \bar{\Phi}^p$. An application of (ii) now confirms the compactness of $(\Phi, \mathfrak{T}_p) = (\Phi, \mathfrak{T}_\kappa)$. ■

Remark. When X is locally compact, (Example 7.16) Φ is equicontinuous iff it is equicontinuous on compacta.

PROBLEMS

7A. $\mathfrak{T}_{\mathcal{S}}$-Convergence

$\mathfrak{T}_u$-convergence implies $\mathfrak{T}_\kappa$-convergence; therefore $\mathfrak{T}_p$-convergence in $\mathfrak{F}(X; Y, \mathcal{V})$. Show that the reverse implications are generally false. [*Hint*: Consider $\{x^n : n \in \mathbb{N}\}$ on $I = [0, 1]$ and $\{(1 - |x|/n)\chi_{[0,n)}(x) : n \in \mathbb{N}\}$ on $\mathbb{R}$.]

7B. Dini's Theorem

If a monotonically increasing net $\varphi \subset C(X; \mathbb{R})$ converges pointwise to $f \in C(X; \mathbb{R})$, then φ is $\mathfrak{T}_\kappa$-convergent to f. [*Note*: A net φ (Appendix t.6) is motononically increasing if $\beta > \alpha$ implies $\varphi_\beta(x) > \varphi_\alpha(x)\ \forall x \in X$. The members of φ need not be monotone.]

7C. Bounded Mappings

A function f from a set X to a metric space (Y, d) is said to be bounded if $f(X)$ has a finite diameter in Y.

(a) The collection $B(X; Y, d)$ of bounded functions in $\mathfrak{F}(X; Y, d)$ is both open and closed in $\mathfrak{F}_u(X; Y, d)$. If X is a topological space, then $CB(X; Y, d) = B(X; Y, d) \cap C(X; Y, d)$ is closed in $\mathfrak{F}_u(X; Y, d)$.

(b) Suppose X is a connected space and $\Phi \subset C(X; \mathbb{R}, d_1)$ is equicontinuous. If $\Phi(x_0) = \{\phi(x_0) : \phi \in \Phi\}$ is a bounded set in $(\mathbb{R}, d_1)$ for one point $x_0 \in X$, then $\Phi(x)$ is bounded $\forall x \in X$.

(c) Extend (b) by demonstrating that Φ is uniformly bounded on each compact subset of X.

7D. Additional $C_{\mathcal{S}}(X; Y, \mathcal{V})$ Properties

(a) If $\bar{\mathcal{S}} = \{\bar{S}_\alpha : S_\alpha \in \mathcal{U}\}$, then (Example 7.8) $\mathcal{U}_{\mathcal{S}} \subset \mathcal{U}_{\bar{\mathcal{S}}}$ for $\mathfrak{F}(X; Y, \mathcal{V})$. Show that $C_{\mathcal{S}}(X; Y, \mathcal{V}) = C_{\bar{\mathcal{S}}}(X; Y, \mathcal{V})$. [*Hint*: If $f, g \in C(X; Y, \mathcal{V})$ satisfy $(f(x), g(x)) \in V\ \forall x \in S_\alpha$, then $(f(x), g(x)) \in \bar{V}\ \forall x \in \bar{S}_\alpha$.]

(b) If $C_{\mathcal{S}}(X; Y, \mathcal{V})$ is separated, then $(Y, \mathcal{V})$ is separated (Theorem 7.9). Show conversely, that $C_{\mathcal{S}}(X; Y, \mathcal{V})$ is separated if $X = \overline{\bigcup_{\mathcal{S}} S_\alpha}$ and $(Y, \mathcal{V})$ is separated. [*Hint*: Verify for $f, g \in C(X; Y, \mathcal{V})$ that $f = g$ on $A \subset X$ implies $f = g$ on $\bar{A} \subset X$.]

(c) Let $\tilde{C}(X; Y, \mathcal{V})$ denote the collection of mappings from X to $(Y, \mathcal{V})$ whose restrictions to each $S_\alpha \in \mathcal{S}$ are continuous. Then $\tilde{C}(X; Y, \mathcal{V})$ is closed in $\mathfrak{F}_{\mathcal{S}}(X; Y, \mathcal{V})$. [*Hint*: Use Theorem 7.10. if $f \in \overline{\tilde{C}(X; Y, \mathcal{V})}^{\mathcal{S}}$, then $f/S_\alpha = a_\alpha(f) \in \overline{a_\alpha\{\tilde{C}(X; Y, \mathcal{V})\}} \subset \overline{\tilde{C}(X; Y, \mathcal{V})} = \tilde{C}(S_\alpha; Y, \mathcal{V}) \subset \mathfrak{F}_{u_\alpha}(S_\alpha; Y, \mathcal{V})$.]

(d) If X is locally compact or 1°, then $C(X; Y, \mathcal{V})$ is closed in $\mathfrak{F}_\kappa(X; Y, \mathcal{V})$. [*Hint*: Verify that $\tilde{C}(X; Y, \mathcal{V}) = C(X; Y, \mathcal{V})$.]

(e) The set $UC(X; Y)$ of uniformly continuous mappings from $(X, \mathcal{U})$ to $(Y, \mathcal{V})$ is closed in $\mathfrak{F}_u(X, \mathcal{U}; Y, \mathcal{V})$. [*Hint*: If $f \in \overline{UC(X; Y)}$ and $V \in \mathcal{V}$ is symmetric, there is a $\phi_0 \in UC(X; Y)$ and $U \in \mathcal{U}$ such that $(f(x), \phi_0(x)) \in V\ \forall x \in X$ and $(\phi_0 \times \phi_0)U \subset V$. Thus $(f \times f)U \subset V^3$.]

7E. Ascolli–Arzela Theorem

Let $\Phi \subset C(X; Y, d)$, where X is a compact T_2 space and (Y, d) is a compact metric space.

(a) The following are equivalent: (i) Φ is equicontinuous; (ii) $\bar{\Phi}^u$ is compact; (iii) Every sequence in $\bar{\Phi}^u$ contains a convergent subsequence. [*Hint*: $\mathcal{U}_u$ for $C(X; Y, d)$ is determined by the metric $\rho(f, g) = \sup_X d\{f(x), g(x)\}$.]

(b) The equivalence in (a) need not hold if X or Y are locally compact but not compact. [*Hint*: Both $\Phi = \{\text{constant functions}\}: I = [0, 1] \to \mathbb{R}$ and $\Psi = \{f_n : \underset{x \rightsquigarrow 1/1+(x+1)^n}{\mathbb{R} \longrightarrow I}\ ;\ n \in \mathbb{N}\}$ are equicontinuous.]

7F. Compact Open Topology

Assume that X, Y are topological spaces. let $\mathcal{C}$ denote the compact subsets of X and let $\mathfrak{O}$ denote the open sets in Y. The topology on $\mathfrak{F}(X; Y)$ having subbase elements $u_{K,\mathcal{O}} = \{f \in \mathfrak{F}(X; Y) : f(K) = \bigcup_K f(x) \subset \mathcal{O}\}$ for $K \in \mathcal{C}$ and $\mathcal{O} \in \mathfrak{O}$ is called the compact open topology, denoted $\mathfrak{T}_{co}$.

(a) Verify that $\mathfrak{T}_p \subset \mathfrak{T}_{co} \subset \mathfrak{T}_u$ on $\mathfrak{F}(X; Y, \mathcal{V})$ and $\mathfrak{T}_{co} = \mathfrak{T}_\kappa$ on every set $\Phi \subset X(X; Y, \mathcal{V})$.

(b) $\mathfrak{F}_{co}(X; Y)$ is T_i $(i = 0, 1, 2)$ {regular, completely regular} if Y is T_i {regular, completely regular}. The converse also holds for $C_{co}(X; Y)$.

(c) Give examples for which $f_n \to f \in \mathfrak{F}_p(X; Y)$, but $f_n \not\to f \in \mathfrak{F}_{co}(X; Y)$ and $g_n \to g \in \mathfrak{F}_{co}(X; Y, \mathcal{V})$, but $g_n \not\to g \in \mathfrak{F}_u(X; Y, \mathcal{V})$.

7G. Joint Continuity

A topology on $\Phi \subset \mathfrak{F}(X; Y, \mathcal{V})$ is called jointly continuous if the mapping $P : \underset{(\phi, x) \rightsquigarrow \phi(x)}{\Phi \times X \longrightarrow Y}$ is continuous.

(a) If $\mathfrak{T}$ is jointly continuous on Φ and $\mathfrak{T}' \supset \mathfrak{T}$, then $\mathfrak{T}'$ is jointly continuous. The discrete topology is the largest jointly continuous topology on Φ. There is no smallest jointly continuous topology on Φ in general.

(b) Suppose $\mathfrak{T}$ is the smallest jointly continuous topology on $\Psi \subset C(X; Y, \mathcal{V})$. Then a filter $\mathcal{G}$ on Ψ is $\mathfrak{T}$-convergent to $\phi_0 \in \Psi$ iff $P\{\mathcal{G} \times \mathfrak{N}_x\} \to \phi_0(x)\ \forall x \in X$. ($\mathfrak{N}_x$, as usual, denotes the X-nbd filter at $x \in X$.)

(c) Give examples confirming that neither $\mathfrak{T}_p$ nor $\mathfrak{T}_{co}$ are jointly continuous on Φ in general.

(d) If Φ is equicontinuous, then $\mathfrak{T}_p$ is the smallest jointly continuous topology on Φ [*Hint*: If $P : \Phi_{\mathfrak{T}} \times X \to Y$ is continuous, each $e_x : \Phi_{\mathfrak{T}} \to Y$ $(x \in X)$ is continuous.]

(e) If X is a T_2 or regular locally compact space, then $\mathfrak{T}_{co}$ is the smallest jointly continuous topology on any set $\Psi \subset C(X; Y)$. [*Hint*: If $\psi(x) \in \mathcal{O}$, there is an X-compact nbd $K(x)$ such that $\psi(K) \subset \mathcal{O}$. Then $P\{(K, \mathcal{O}) \times K\} \subset \mathcal{O}$ and it remains only to verify that $\mathfrak{T} \supset \mathfrak{T}_{co}$ for every jointly continuous topology $\mathfrak{T}$ on Ψ.]

(f) If each $x \in X$ is an interior point of some member of $\mathcal{S}$, then $\mathfrak{T}_{\mathcal{S}}$ is jointly continuous on $C(X; Y, \mathcal{V})$.

(g) $\Phi \subset \mathfrak{F}(X; Y, \mathcal{V})$ is equicontinuous if there exists a compact jointly continuous topology for Φ [*Hint*: (Kelley [50]) Fix any $x_0 \in X$ and symmetric $V \in \mathcal{V}$. For each $\phi \in \Phi_{\mathfrak{T}}$, there is a $\mathfrak{T}$-open nbd $G(\phi)$ and an X-nbd $w(x_0)$ such that $P\{(G \times w_0)(\phi, x_0)\} \subset V[\phi(x_0)]$. Then $\psi(w) \subset V^2[\psi(x_0)]\ \forall \psi \in G$ and by $\Phi_{\mathfrak{T}}$'s compactness there is a nbd $u(x_0) = \bigcap_{i=1}^n w_i(x_0)$ such that $\phi(u) \subset V^2[\phi(x_0)]\ \forall \phi \in \Phi$.]

(h) Let X be a topological space and let $\Phi \subset C(Y, \mathscr{V}; Z, \mathscr{W})$ be equicontinuous, where $(Y, \mathscr{V})$ and $(Z, \mathscr{W})$ are uniform spaces. Then $\underset{(\phi, f) \rightsquigarrow \phi \circ f}{P_0\colon \Phi_p \times C_p(X; Y, \mathscr{V}) \to C_p(X; Z, \mathscr{W})}$ is continuous. [*Hint*: Use Definition 7.1 and (d) above to prove that each $\underset{(\phi, f) \rightsquigarrow \phi\{f(x)\}}{e_x P_0\colon \Phi_p \times C_p(X; Y, \mathscr{V}) \to (Z, \mathscr{W})}$ $(x \in X)$ is continuous. Then apply Theorem 3.4.]

(i) If $X = (X, \mathscr{U})$ is a uniform space in (h) and Φ is uniformly equicontinuous, then $P_0\colon \Phi_u \times C_u(X, U; Y, \mathscr{V}) \to C_u(X, U; Z, \mathscr{W})$ is uniformly continuous.

References for Further Study

On Function Spaces: Bourbaki [12]; Kelley [50].

CHAPTER II

Topological Groups

Perspective. A topology on a group which renders the group operations continuous has compatible uniformities determined by its nbds of the identity e. This chapter is devoted to the study of such uniformizible spaces with enriched continuous group structure. Basic properties and notions are considered in sections 8–11. Section 12, which supplements this, forms the basis for our introduction to harmonic analysis in Chapter V. Specifically, it is shown that locally compact T_2 groups have a natural theory of invariant measure and integration—a particularization of which is Lebesgue measure and integration on the real line.

8. Topological Groups are Uniformizible

The subtitle above is verified shortly. We begin with

DEFINITION 8.1. A topological group is a group G with a topology $\mathfrak{T}$ on G such that both $\mathfrak{m}: \underset{(x,y) \rightsquigarrow xy}{G \times G \to G}$ and $\mathfrak{i}: \underset{x \rightsquigarrow x^{-1}}{G \to G}$ are continuous. In this context $\mathfrak{T}$ is said to be compatible with G's group structure. The collection of all such compatible topologies on G is denoted $T(G)$.

Remark. Note (Problem 8A) that continuity of both $\mathfrak{m}$ and $\mathfrak{i}$ is equivalent to the continuity of $\xi: \underset{(x,y) \rightsquigarrow xy^{-1}}{G \times G \to G}$.

Example 8.2. (a) A group with discrete or indiscrete topology is a topological group. Not every topology on a group G, however, is compatible with G's group structure, since $\mathfrak{T} \in T(G)$ iff $\mathfrak{T}$ is completely regular (Theorem 8.8, Corollary 4.3).

(b) Clearly, $\mathbb{R} = (\mathbb{R}, +; d_1)$ and $\mathbb{R}_\bullet = (\mathbb{R} - \{0\}, \cdot; d_1)$ are T_2 groups. For any $n \in \mathbb{N}$ define $d_n(x, y) = \sqrt[n]{\sum_{i=1}^{n} |x_i - y_i|^n}$ $\forall x = (x_1, \ldots, x_n)$ and

$y=(y_1,\ldots,y_n)$ in $\mathbb{R}^n$. Then $(\mathbb{R}^n,+;d_n)$ is a locally compact T_2 group. Continuity of ξ is a consequence of $d_n(x-y,x_0-y_0)\leq d_n(x,x_0)+d_n(y,y_0)\,\forall x_0,x,y_0,y\in\mathbb{R}^n$.

Example 8.3. (a) Let $M_n(\mathbb{F})$ denote the linear algebra of $n\times n$ matrices over $\mathbb{F}$ $(=\mathbb{R}$ or $\mathbb{C})$ and define $\|A\|_0=\max_{1\leq i,j\leq n}|a_{ij}|\,\forall A=(a_{ij})\in M_n(\mathbb{F})$.

Then $(M_n(\mathbb{F}),+;\|\,\|_0)$ is a Banach algebra (Definition 23.1), since $(\mathbb{F}^{n^2},\|\,\|_0)$ is complete and $\underset{A=(a_{ij})\rightsquigarrow(a_{11},\ldots,a_{1n};\ldots;a_{n1},\ldots,a_{nn})}{(M_n(\mathbb{F}),\|\,\|_0)\longrightarrow(\mathbb{F}^{n^2},\|\,\|_0)}$ is a congruence under $\|x_0\|_0=\max_{1\leq i\leq n^2}|x_i|$ for $x=(x_1,\ldots,x_{n^2})\in\mathbb{F}^{n^2}$.

(b) The multiplicative group $G_n(\mathbb{F})$ of nonsingular (i.e., invertible) members of $M_n(\mathbb{F})$ is open in $(M_n(\mathbb{F}),\|\,\|_0)$ by Theorem 23.6; and $(G_n(\mathbb{F}),\cdot;d)$ is a metrizible group under $d(A,B)=\|A-B\|_0$. Specifically, the continuity of each $\underset{A\rightsquigarrow a_{ij}}{\pi_{ij}:\mathbb{F}^{n^2}\to\mathbb{F}}$ $(i,j=1,2,\ldots,n)$ assures that $\underset{C=AB\rightsquigarrow c_{ij}}{\mathbb{F}^{n^2}\to\mathbb{F}}$; therefore $\underset{(A,B)\rightsquigarrow AB}{\mathfrak{m}:\mathbb{F}^{n^2}\times\mathbb{F}^{n^2}\to\mathbb{F}^{n^2}}$ is continuous. The continuity of each π_{ij} also implies the continuity of $\underset{A\rightsquigarrow|A|}{\mathbb{F}^{n^2}\to\mathbb{F}}$ and $\underset{A_{ij}\rightsquigarrow|A_{ij}|}{\mathbb{F}^{(n-1)^2}\to\mathbb{F}}$, where A_{ij} is the $(n-1)\times(n-1)$ matrix obtained by deleting the ith row and jth column of A. Thus $\mathfrak{i}:A\to A^{-1}=adj\,A/|A|$ is continuous.

THEOREM 8.4. If G is a topological group, then

(i) $\mathfrak{i}$ is a homeomorphism;

(ii) $\underset{x\rightsquigarrow gx}{{}_g\mathfrak{m}:G\to G}$ and $\underset{x\rightsquigarrow xg}{\mathfrak{m}_g:G\to G}$ are homeomorphisms $\forall g\in G$.

Furthermore, G is homogeneous (i.e., $x\neq y$ implies $h(x)=y$ for some homeomorphism $h:G\to G$).

Proof. Both $\mathfrak{i}$ and $\mathfrak{m}_g$ (the proof for ${}_g\mathfrak{m}$ follows similarly) are bijective and (i) is proved by observing that $\mathfrak{i}^{-1}(x)=\mathfrak{i}^{-1}\{(x^{-1})^{-1}\}=x^{-1}=\mathfrak{i}(x)\,\forall x\in G$. To verify that $\mathfrak{m}_g$ is bicontinuous write $\mathfrak{m}_g=\mathfrak{m}\xi_g$ for $\underset{x\rightsquigarrow(x,g)}{\xi_g:G\to G\times G}$. Since ξ_g $(g\in G)$ is continuous, so is $\mathfrak{m}_g$ and $\mathfrak{m}_g^{-1}=\mathfrak{m}\xi_{g^{-1}}$. The homogeneity of G is clear, since $\mathfrak{m}_{x^{-1}y}$ is a homeomorphism $\forall x,y\in G$. ∎

COROLLARY 8.5. For a topological group G, suppose $A\subset G$ is open, $B\subset G$ is closed, and $C,D\subset G$ are arbitrary. Then

(i) $\overline{C^{-1}}=\bar{C}^{-1}$,

(ii) $\overline{xCy}=x\bar{C}y\;\forall x,y\in G$,

(iii) CA and AC are open

(iv) Bx and xB are closed $\forall x\in G$,

(v) $\bar{C}\bar{D}\subset\overline{CD}$.

Proof. Since $\mathfrak{i}$ is a homeomorphism $\overline{C^{-1}} = \overline{\mathfrak{i}(C)} = \mathfrak{i}(\bar{C}) = \bar{C}^{-1}$. The proof of (ii) follows similarly, since ${}_x\mathfrak{m} \circ \mathfrak{m}_y : \underset{g \rightsquigarrow xgy}{G \to G}$ is a homeomorphism $\forall x, y \in G$. Both (iii) and (iv) follow from Theorem 8.4 (ii), since $xA = {}_x\mathfrak{m}(A)$ $\{xB = {}_x\mathfrak{m}(B)\}$ and $CA = \bigcup_{x \in C} {}_x\mathfrak{m}(A)$. Finally, the continuity of $\mathfrak{m}$ yields $\bar{C}\bar{D} = \mathfrak{m}(\bar{C} \times \bar{D}) = \mathfrak{m}(\overline{C \times D}) \subset \overline{\mathfrak{m}(C \times D)} = \overline{CD}$ as asserted in (v). ■

Remark. The product of two closed subsets of G need not be closed unless one of them is compact. (See Problem 8F.)

COROLLARY 8.6. If h is a homomorphism between topological groups G and $\tilde{G}$, then

(i) $h(AB) = h(A)\,h(B) \subset \overline{h(A)}\,\overline{h(B)} \subset \overline{h(AB)}$ $\forall A, B \subset G$.

(ii) $h^{-1}(\tilde{E})h^{-1}(\tilde{F}) \subset h^{-1}(\tilde{E}\tilde{F})$ and $\overline{h^{-1}(\tilde{E})}\,\overline{h^{-1}(\tilde{F})} \subset \overline{h^{-1}(\tilde{E}\tilde{F})}$ $\forall \tilde{E}, \tilde{F} \subset \tilde{G}$.

(iii) $h(A)$ and $\overline{h(A)}$ are symmetric† in $\tilde{G}$ for symmetric $A \subset G$. Dually, $h^{-1}(\tilde{A})$ and $\overline{h^{-1}(\tilde{A})}$ are symmetric in G for symmetric $\tilde{A} \subset G$.

Proof. Apply Theorem 8.4 to Corollary 8.5.

The preceding results also enable us to prove

THEOREM 8.7. Let H be a subgroup of a topological group G.

(i) H and $\bar{H}$ are topological groups.
(ii) If H is normal, so is $\bar{H}$.
(iii) $\bar{H}$ is abelian if H is abelian and G is T_2.
(iv) $\bar{H} = H$ if H is open in G.

Proof. That H is a topological group is clear, since the restriction of the continuous mapping $\xi: \underset{(x,y) \rightsquigarrow xy^{-1}}{G \times G \to G}$ to $H \times H$ is continuous. Furthermore, $H^2 = H = H^{-1}$ implies $\overline{\bar{H}\bar{H}^{-1}} \subset \overline{HH^{-1}} = \bar{H}$ (Corollary 8.5) so that $\bar{H}$ is a subgroup of G and (i) is proved. If H is normal, $xHx^{-1} = H$ and $x\bar{H}x^{-1} = \overline{xHx^{-1}} = \bar{H}$ $\forall x \in G$. Therefore $\bar{H}$ is normal and (ii) is verified. For (iii) let $h(x, y) = \mathfrak{m}\{\mathfrak{m}(x, y), \mathfrak{m}\{i(x), i(y)\}\} = xyx^{-1}y^{-1}$ $\forall x, y \in G$. Since h is continuous and G is T_2, both $\{e\}$ and $h^{-1}(e)$ are closed in G. From $H \times H \subset h^{-1}(e)$ (which holds because H is abelian) one obtains $\bar{H} \times \bar{H} = \overline{H \times H} \subset h^{-1}(e)$ and confirmation that $\bar{H}$ is abelian. Finally, (iv) follows from $G - H = (G - H)H = \bigcup_{x \in G - H} xH$, which is open if H is open. ■

† A subset E of a group G is said to be symmetric if $E = E^{-1}$. Thus $E \cap E^{-1}$ is symmetric $\forall E \subset G$.

Remark. *Caution*! $(H, \square; \mathfrak{T})$ may be both a topological group and topological subspace of $(G, \circ; \mathfrak{T})$, but not a topological subgroup of $(G, \circ; \mathfrak{T})$. That H be a subgroup of G in Theorem 8.7 cannot be omitted (since $G_n(\mathbb{F})$ is also dense in the nonmultiplicative group $(M_n(\mathbb{F}), \| \|_0)$ of Example 8.3).

Example 8.8. $\mathbb{R} = (\mathbb{R}, +; d_1)$ has no proper open subgroup $H \neq \{0\}$. A proper subgroup of $\mathbb{R}$ is closed iff it is countable. There are exactly $2^{\aleph_0} = c$ closed proper subgroups of $\mathbb{R}$. For verification let H be an open (therefore closed) subgroup of $\mathbb{R}$ which contains arbitrarily small positive numbers. The $H = \bar{H} = \mathbb{R}$. Every closed proper subgroup K of $\mathbb{R}$ therefore contains a smallest positive number $r_K \in \mathbb{R}$ and $K = \{nr_K : n \in \mathbb{I}\}$.

DEFINITION 8.9. Let G and $\tilde{G}$ be topological groups having nbd systems $\mathfrak{N}_e$ and $\mathfrak{N}_{\tilde{e}}$ at the identity $e \in G$ and $\tilde{e} \in \tilde{G}$, respectively. A homomorphism $h: G \to \tilde{G}$ is said to be

(a) open {nearly open} at $e \in G$ if $\forall u \in \mathfrak{N}_e$ there is a $\tilde{u} \in \mathfrak{N}_{\tilde{e}}$ such that $h(u) \supset \tilde{u}$ $\{\overline{h(u)} \supset \tilde{u}\}$. If $h(w)$ $\{\overline{h(w)}\}$ is a $\tilde{G}$-nbd $\forall G$-nbd w, then h is termed open {nearly open};

(b) nearly continuous at $e \in G$ if $\overline{h^{-1}(\tilde{u})} \in \mathfrak{N}_e$ $\forall \tilde{u} \in \mathfrak{N}_e$. If $\overline{h^{-1}(\tilde{w})}$ is a G-nbd $\forall G$-nbd $\tilde{w}$, then h is called nearly continuous.

THEOREM 8.10. A homomorphism $h: G \to \tilde{G}$ is continuous {nearly continuous, open, nearly open} iff it is continuous {nearly continuous, open, nearly open} at $e \in G$.

Proof. Sufficiency only requires verification. Suppose h is continuous at $e \in G$. For any $x \in G$ and $\tilde{G}$-nbd $\tilde{w}$ of $h(x) = \tilde{x}$ there is a $u \in \mathfrak{N}_e$ such that $h(u) \subset \tilde{w}\tilde{x}^{-1}$. Since $h(ux) \subset \tilde{w}$, the continuity of h follows. If h is nearly continuous at e, then $\overline{h^{-1}(\tilde{w}\tilde{x}^{-1})} \in \mathfrak{N}_e$ and for any $x \in h^{-1}\tilde{x}$, one obtains (Corollary 8.6) $\mathfrak{m}_x\{\overline{h^{-1}(\tilde{w}\tilde{x}^{-1})}\} \subset \overline{h^{-1}(\tilde{w}\tilde{x}^{-1})h^{-1}(\tilde{x})} \subset \overline{h^{-1}(\tilde{w})}$. Since $\mathfrak{m}_x\{\overline{h^{-1}(\tilde{w}\tilde{x}^{-1})}\}$ is a G-nbd, so also is $\overline{h^{-1}(\tilde{w})}$. Thus h is nearly continuous. Finally, h is open {nearly open) at e iff $\forall x \in G$ and G-nbd $u(x)$ there is a $\tilde{G}$-nbd $\tilde{v}$ of $\tilde{x} = h(x)$ such that $h(x) \supset \tilde{v}\{\overline{h(u)} \supset \tilde{v}\}$. In particular, $h(x)$ is an interior point of $h(u)$ {of $\overline{h(u)}$} $\forall x \in G$ and the openness {near openness} of h follows. ■

Let G be a topological group with $\mathfrak{N}_e$ denoting the collection of G-nbds of the identity $e \in G$. We already know (Theorem 8.4) that ug and gu are G-nbds of $g \in G$ $\forall u \in \mathfrak{N}_e$ and that, conversely, vg^{-1} and $g^{-1}v$ belong to $\mathfrak{N}_e$ $\forall G$-nbd v of $g \in G$. Thus knowledge about the G-nbds of e is equivalent to knowing the local nbd properties at each $g \in G$. This underlines the fundamental nature of

THEOREM 8.11. If G is a topological group, $\mathfrak{N}_e$ satisfies

(G_1) for every $u \in \mathfrak{N}_e$ there is a $v \in \mathfrak{N}_e$ such that $v^2 \subset u$;
(G_2) $u^{-1} \in \mathfrak{N}_e\ \forall u \in \mathfrak{N}_e$;
(G_3) for all $u \in \mathfrak{N}_e$ and $x \in G$ some $v \in \mathfrak{N}_e$ satisfies $xvx^{-1} \subset u$.

Conversely, every filter $\mathfrak{N}$ of subsets of a group G which satisfies $(G_1)-(G_3)$ determines a unique $\mathfrak{T} \in T(G)$ having $\mathfrak{N}$ as its nbd system at $e \in G$.

Proof. (G_1), (G_2) follow from the continuity of $\mathfrak{m}$ and $\mathfrak{i}$ at $e \in G$. Since ${}_{x^{-1}}\mathfrak{m}$ and $\mathfrak{m}_x$ are homeomorphisms, $\{{}_{x^{-1}}\mathfrak{m} \circ \mathfrak{m}_x\}(u) = x^{-1}ux \in \mathfrak{N}_e\ \forall u \in \mathfrak{N}_e$. Therefore $v \subset x^{-1}ux$ for $v \in \mathfrak{N}_e$ implies $xvx^{-1} \subset u$ and (G_3) is established. The converse proof uses the fact (Theorem 1.6$^+$, Definition 8.1$^+$) that (G_1), (G_2) together are equivalent to the existence, for each $u \in \mathfrak{N}$, of a $v \in \mathfrak{N}$ satisfying $vv^{-1} \subset u$. In particular, $e \in vv^{-1} \subset u$ implies $x \in ux\ \forall x \in \mathfrak{N}$ so that $\mathfrak{N}_x = \{ux : u \in \mathfrak{N}\}\ \{x \in G)$ satisfies (N_1) of Appendix t.1. That (N_2) also holds is clear, since $ux \cap vx = (u \cap v)x\ \forall u,\ v \in \mathfrak{N}$ and $x \in G$. Next, $ux \subset w$ implies $u \subset wx^{-1}$ and $wx^{-1} \in \mathfrak{N}$. Thus $w \in \mathfrak{N}_x$ and (N_3) holds. To prove (N_4) consider any $u \in \mathfrak{N}$ and take $v \in \mathfrak{N}$ such that $v^2 \subset u$. Then $vy \subset v^2x \subset ux\ \forall y \in vx$; that is, $ux \in \mathfrak{N}_y\ \forall y \in vx \in \mathfrak{N}_x$. We have thus established that $\{\mathfrak{N}_x : x \in G\}$ is the nbd system for a topology $\mathfrak{T}$ on G (as is also $\{{}_x\mathfrak{N} : x \in G\}$, where ${}_x\mathfrak{N} = \{xu : u \in \mathfrak{N}\}$). It remains only to demonstrate that $\mathfrak{T} \in T(G)$. For any $x, y \in G$ and $u \in \mathfrak{N}_e$ choose $v \in \mathfrak{N}_e$ satisfying $vv^{-1} \subset u$ so that $vx \times y(x^{-1}vx) \in \mathfrak{N}_x \times {}_y\mathfrak{N}$. Then $\xi\{vx \times y(x^{-1}vx)\} = (vx)\{x^{-1}v^{-1}x)y\} \subset vv^{-1}xy^{-1} \subset uxy^{-1}$, from which follows the continuity of ξ and the proof of the theorem. ■

Remark. If $\mathfrak{N}$ is only a filterbase in G satisfying (G_1)–(G_3), then $\mathfrak{N}_x$ and ${}_x\mathfrak{N}$ are $\mathfrak{T}$-nbd bases at $x \in G$.

Remark. Note that $\mathfrak{N}_e = \{xux^{-1} : u \in \mathfrak{N}_e\}$ for each fixed $x \in G$ since $v = x(x^{-1}vx)x^{-1}$ $\forall v \in \mathfrak{N}_e$.

Remark. The G-nbds are open sets if $\forall u \in \mathfrak{N}_e$ and $x \in u$ there is a $v \in \mathfrak{N}_e$ such that $vx \subset u$.

Example 8.12. (a) Let G be the composition group of affine transformations $\{f_{a,b} : \underset{x \rightsquigarrow ax+b}{\mathbb{R} \longrightarrow \mathbb{R}} : a, b \in \mathbb{R}$ and $a > 0\}$ which has identity $f_{1,0}$. Define $u_{\varepsilon,\delta}(f_{1,0}) = \{f_{a,b} \in G : |a-1| < \varepsilon$ and $|b| < \delta\}$. Then $\{u_{\varepsilon,\delta}(f_{1,0}) : 0 < \varepsilon, \delta \in \mathbb{R}\}$ is a filterbase in G satisfying (G_1)–(G_3) and the corresponding topology is compatible with G's group structure.

(b) Suppose $\mathfrak{M}$ denotes the composition group of Möbius transformations $\{f_{a,b,c,d} : \underset{z \rightsquigarrow (az+b)/(cz+d)}{\mathbb{C} \longrightarrow \mathbb{C}} : a, b, c, d \in \mathbb{C}$ and $ad - bc \neq 0\}$. If

$u_{\varepsilon,\delta}(f_{1,0,0,1})=\{f_{a,b,c,d}\in\mathfrak{M}: |a-1|, |d-1|<\varepsilon$ and $|b|, |c|<\delta\}$, the family $\{u_{\varepsilon,\delta}(f_{1,0,0,1}): 0<\varepsilon.\ \delta\in\mathbb{R}\}$ determines a $\mathfrak{T}\in T(\mathfrak{M})$. This is not too surprising since $(\mathfrak{M},\mathfrak{T})$ is essentially $(G_2(\mathbb{C}), d)$ in Example 8.3. Simply identify $f_{a,b,c,d}\in\mathfrak{M}$ with $\begin{pmatrix} a & b \\ c & d\end{pmatrix}\in G_2(\mathbb{C})$. Remark: $G_2(\mathbb{C})$, therefore $\mathfrak{M}$ is not abelian. (Two Möbius transforms commute iff they have the same fixed points.)

Every topological group is uniformizible!

THEOREM 8.13. Let $(G,\mathfrak{T})$ be a topological group with nbd system $\mathfrak{N}_e$ at $e\in G$. For each $w\in\mathfrak{N}_e$ define $U_w=\{(x, y)\in G\times G: yx^{-1}\in w\}$ and $\mathcal{C}_w=\{wx: x\in G\}$. Then

(i) $\mathfrak{B}=\{U_w: w\in\mathfrak{N}_e\}$ is the base for a uniformity $\mathcal{U}_G\in\mathfrak{U}(G)$,

(ii) $\mathfrak{b}=\{\mathcal{C}_w: w\in\mathfrak{N}_e\}$ is the base for a uniformity $\mathfrak{c}_G\in\mathfrak{U}(G)$,

(iii) $\{U[e]: U\in\mathcal{U}_G\}=\mathfrak{N}_e=\{St(e,\mathcal{C}):\mathcal{C}\in\mathfrak{c}_G\}$. Therefore $\mathfrak{T}_{\mathcal{U}_G}=\mathfrak{T}=\mathfrak{T}_{\mathfrak{c}_G}$ on G.

Proof. Clearly $U_{w_1}\cap U_{w_2}=U_{w_1\cap w_2}\in\mathfrak{B}\ \forall w_1, w_2\in\mathfrak{N}_e$. Every $U_w\in\mathfrak{B}$, has $\Delta\in U_w$ (since $x\in U_w[x]=wx\ \forall x\in G$) and satisfies $U_w^{-1}=\{(x, y)\in G\times G: xy^{-1}\in w\}=U_{w^{-1}}\in\mathfrak{B}$. Furthermore (by the continuity of ξ), there exist $w_1, w_2\in\mathfrak{N}_e$ satisfying $w_1w_2^{-1}\subset w$ so that $v=w_1\cap w_2^{-1}\in\mathfrak{N}_e$ and $U_v^2\subset U_{v^2}\subset U_w$. The proof of (i) is therefore complete. For $\mathcal{C}_{w_1}, \mathcal{C}_{w_2}\in\mathfrak{b}$, choose $w\in\mathfrak{N}_e$ such that $ww^{-1}\subset w_1\cap w_2$. Then $St(x,\mathcal{C}_w)\subset(w_1\cap w_2)x\ \forall x\in G$ (since $y\in St(x,\mathcal{C}_w)$ implies $yx^{-1}\in ww^{-1}\subset w_1\cap w_2$) and $\mathcal{C}_w\overset{b}{<}\mathcal{C}_{w_1}\wedge\mathcal{C}_{w_2}$. Thus (ii) follows from Definition 2.1$^+$. Next $U_w[e]=w\subset St(e,\mathcal{C}_w)\ \forall w\in\mathfrak{N}_e$ yields $\{U[e]: U\in U_G\}=\mathfrak{N}_e\supset\{St(e,\mathcal{C}):\mathcal{C}\in\mathfrak{c}_G\}$ for these three filters in G. Also, for any $w\in\mathfrak{N}_e$ and $v\in\mathfrak{N}_e$ chosen such that $vv^{-1}\subset w$ one has $St(e,\mathcal{C}_v)\subset w$. Therefore $\mathfrak{N}_e\subset\{St(e,\mathcal{C}):\mathcal{C}\in\mathfrak{c}_G\}$ and all of (iii) follows (see Definitions 1.12 and 2.6). ■

Remark. Note that $U_w\in\mathfrak{B}$ need not be symmetric despite the fact that $U_w^{-1}=U_{w^{-1}}$ (unless $xy^{-1}=yx^{-1}\ \forall x, y\in G$ or $u=u^{-1}\ \forall$ open $u\in\mathfrak{N}_e$).

Example 8.14. As above, define $U_w=\{(x, y)\in G\times G: yx^{-1}\in w\}$ and ${}_wU=\{(x, y)\in G\times G: x^{-1}y\in w\}\ \forall w\in\mathfrak{N}_e$. Then $\mathfrak{B}_G=\{U_w: w\in\mathfrak{N}_e\}$, ${}_G\mathfrak{B}=\{{}_wU: w\in\mathfrak{N}_e\}$ and $\mathfrak{B}={}_G\mathfrak{B}\cup\mathfrak{B}_G$ are bases for compatible uniformities $\mathcal{U}_G$, ${}_G\mathcal{U}$, and $\mathcal{U}$ on $G\times G$, respectively. It is easily verified [Problem 1F(a)] that $\mathcal{U}=\sup\{{}_G\mathcal{U}, \mathcal{U}_G\}$.

Although ${}_G\mathcal{U}=\mathcal{U}_G$ if G is abelian, if $yx^{-1}=xy^{-1}\ \forall x, y\in G$ or if $(G,\mathfrak{T})$ is compact (Corollary 1.26), this is not true in general. For instance, there is no ${}_{u_{\varepsilon',\delta'}}U\in{}_G\mathfrak{B}$ such that ${}_{u_{\varepsilon',\delta'}}U\subset U_{\varepsilon,\delta}\in\mathfrak{B}_G$ in Example 8.12(a).

Verification: If $g_{a,b} \in u_{\varepsilon',\delta'}$ with $a > 1$ and $b > 0$, then $g_{a,b} = f_{\alpha,\beta}^{-1}(f_{\alpha,\beta} g_{a,b}) \in u_{\varepsilon',\delta'}$ and $(f_{\alpha,\beta}, f_{\alpha,\beta} g_{a,b}) \in {}_{u_{\varepsilon',\delta'}}U \; \forall f_{\alpha,\beta} \in G$. For $\alpha = 3\delta/\delta'$ and $\beta = \delta/\varepsilon'$, however, $(f_{\alpha,\beta} g_{a,b}) f_{\alpha,\beta}^{-1} = f_{a,\alpha\beta - \beta(a-1)} \notin u_{\varepsilon,\delta}$ so that $(f_{\alpha,\beta}, f_{\alpha,\beta} g_{a,b}) \notin U_{u_{\varepsilon,\delta}}$.

Unless specified otherwise, $\mathcal{U}_G \in \mathfrak{U}(G)$ is assumed for a topological group G.

What immediately emerges from Theorem 8.13 is

THEOREM 8.15. A topological group with nbd system $\mathfrak{N}_e$ at $e \in G$ is completely regular. Furthermore,

(i) G has a nbd base of symmetric open (closed) sets.

(ii) $\text{Int}\, A = \{a \in A : ua \subset A \text{ for some } u \in \mathfrak{N}_e\}$ and $\bar{A} = \bigcap_{\mathfrak{N}_e} uA \; \forall A \subset G$.

(iii) G is T_k $(k = 0, 1, 2)$ iff $\{e\} = \bigcap_{\mathfrak{N}_e} u$ $(= \bigcap_{\mathfrak{N}} v$, where $\mathfrak{N}$ is any nbd base at $e \in G)$.

(iv) G is locally compact iff there is a compact $u \in \mathfrak{N}_e$; this being equivalent to G having a base of relatively compact {open} nbds at $e \in G$.

Proof. Complete regularity, (i), (ii) and (iii) follow, respectively, from Theorem 1.21, Corollary 1.17, Corollary 1.18, and Theorem 1. 19. The first part of (iv) holds by definition of local compactness and Theorem 8.4(ii). If $u \in \mathfrak{N}_e$ is compact and $v \in \mathfrak{N}_e$, then $u \cap v \in \mathfrak{N}_e$ and (by (i) above) there is a closed $w_1 \in \mathfrak{N}_e$ {and an open $w_2 \in \mathfrak{N}_e$ such that $w_2 \subset w_1$}, satisfying $w_1 \subset u \cap v \subset v$. Since $w_1 \subset u$ is compact {$w_2 \subset u$ is relatively compact}, all of (iv) is proved. ■

In terms of Definition 2.10 (and in comparison with Theorems 2.11 and 1.25), consider

THEOREM 8.16. A continuous homomorphism $h: G \to \tilde{G}$ is $\mathfrak{c}_G - \mathfrak{c}_{\tilde{G}}$ (equivalently, $\mathcal{U}_G - \mathcal{U}_{\tilde{G}}$) uniformly continuous.

Proof. For each $\tilde{u} \in \mathfrak{N}_{\tilde{e}}$ there is a $u \in \mathfrak{N}_e$ such that $h(u) \subset \tilde{u}$. This means that $h(ux) \subset \tilde{u}h(x)$ and $ux \subset h^{-1}\{\tilde{u}h(x)\} \; \forall x \in G$. Thus $\mathcal{C}_u < h^{-1}(\mathcal{C}_u)$ as required. ■

From Corollary 4.4 {Corollary 4.5} also follows.

THEOREM 8.17. A topological group G is pseudometrizible {metrizible} iff G is $1°$ {$1°$ and T_2}.

Proof. If $\mathfrak{N}_e = \{u_n : n \in \mathbb{N}\}$, then $\{U_{u_n} : n \in \mathbb{N}\}$ is a base for $\mathscr{U}_G \in \mathfrak{U}(G)$. Furthermore, $\mathfrak{T}(\rho) = \mathfrak{T}_{\mathscr{U}_\rho} = \mathfrak{T}_{\mathscr{U}_G}$ coincides with the topology on G for any pseudometric {metric} ρ on G such that $\mathscr{U}_G = \mathscr{U}_\rho$. ■

Completeness for Topological Groups

Completeness, as well as total boundedness, notions for a topological group G are defined relative to $\mathscr{U}_G$. Thus, for example, a 1° group G is complete iff it is sequentially complete (Theorems 8.17 and 6.3). Actually $\mathscr{U}_G$ can be replaced by ${}_G\mathscr{U}$ (and sometimes by $\mathscr{U}$) as demonstrated in

Example 8.18. Since $\mathfrak{i} : (G, \mathscr{U}_G) \to (G, {}_G\mathscr{U})$ is a unimorphism (Problem 8E), $\mathscr{U}_G$ is complete iff ${}_G\mathscr{U}$ is complete [Example 6.8, (c)]. Furthermore, $\mathscr{U}_G \subset \mathscr{U}$ and $\mathfrak{T}_{\mathscr{U}_G} = \mathfrak{T}_{\mathscr{U}}$ (is the topology on G) so that $\mathscr{U}_G$-completeness implies $\mathscr{U}$-completeness. A partial converse holds since $\mathscr{U} = \mathscr{U}_{\mathfrak{P}(\mathbb{1})}$, where $\mathbb{1} = \{\mathbb{1} : G \to (G, \mathscr{U}_G)$ and $\mathbb{1} : G \to (G, {}_G\mathscr{U})\}$. Essentially, a filter $\mathfrak{F}$ in G is $\mathscr{U}$-Cauchy iff $\mathfrak{F}$ is both $\mathscr{U}_G$ and ${}_G\mathscr{U}$-Cauchy. If $\mathscr{U}_G$ and ${}_G\mathscr{U}$ share the same Cauchy filters (i.e., $\mathfrak{F}$ is $\mathscr{U}_G$-Cauchy iff $\mathfrak{F}$ is ${}_G\mathscr{U}$-Cauchy), then $\mathscr{U}$-completeness implies $\mathscr{U}_G$-completeness. To be sure, every $\mathscr{U}_G$-Cauchy filter is ${}_G\mathscr{U}$-Cauchy; therefore $\mathscr{U}$-Cauchy and $\mathfrak{T}_{\mathscr{U}} = \mathfrak{T}_{\mathscr{U}_G}$ convergent.

THEOREM 8.19. A topological group G is complete iff there is a complete nbd $u \in \mathfrak{N}_e$.

Proof. If $\mathscr{U}_G$ is complete and $u \in \mathfrak{N}_e$ is closed, then u is complete (Theorem 6.6). On the other hand, suppose $u \in \mathfrak{N}_e$ is complete and let $\mathfrak{F}$ be a $\mathscr{U}_G$-Cauchy filter in G. Then $F_0 x_0^{-1} \subset u$ for every U_u-sized set $F_0 \in \mathfrak{F}$ and $x_0 \in F_0$. Fix any such pair (x_0, F_0) and observe that $F \cap u x_0 \neq \varnothing\ \forall F \in \mathfrak{F}$. Since $\mathfrak{m}_{x_0}$ is a unimorphism (Problem 8E), the $\mathscr{U}_G \cap (u \times u)$-Cauchy filter $\mathfrak{F} x_0^{-1} \cap u$ converges to some $y_0 \in u$ and so $\mathfrak{F} \to y_0 x_0 \in G$. Since $\mathfrak{F} \cap u x_0$ accumulates at y_0, one has $\varnothing \neq (F \cap u x_0) \cap v y_0 \subset F \cap v y_0\ \forall v \in \mathfrak{N}_e$ and $F \in \mathfrak{F}$, which means $\mathfrak{F} \to y_0$ (Theorem 5.2). ■

A pseudometric ρ on a group G is said to be right invariant if $\forall x \in G$, one has $\rho(yx, zx) = \rho(y, z)\ \forall y, z \in G$. Accordingly, a pseudometrizible group G is termed right complete if $\mathscr{U}_G = \mathscr{U}_\rho$ for a complete right invariant pseudometric ρ on G. Note [Problem 4A(a)] that ρ is compatible with the topology of G. Left invariance and completeness are defined similarly (cf. Problem 8M).

COROLLARY 8.20. A locally compact group is complete; a locally compact, 1° group is right complete.

Proof. The first assertion follows from Theorem 8.15(iv), Corollary 1.27, and Theorem 6.7. The second follows from the first in conjunction with Corollary 6.4, Problem 8M(a), and Problem 4H(d). ■

Locally compact T_2 spaces are $T_{3\frac{1}{2}}$ (Appendix t.7). A much stronger conclusion obtains for topological groups:

THEOREM 8.21. A locally compact, T_2 group is T_4.

Proof. Let $f: E \to (\mathbb{R}, d_1)$ be continuous, where E is a closed subset of a locally compact T_2 group G. Our proof consists in demonstrating (Tietze's characterization of normality) that f has a continuous extension to G. Let G_∞ and E_∞ be the one-point compactifications of G and E, respectively. Since $(E, \mathcal{U}_E)$ is complete and separated, $\mathbb{1}: E \to E$ has a unique uniformly continuous extension $\bar{\mathbb{1}}: \bar{E} = E_\infty \to E$ (Theorem 6.11). Therefore (Theorems 1.25, 6.7 and 6.20) $f\bar{\mathbb{1}}: E_\infty \to (\mathbb{R}, U_{d_1})$ has a unique uniformly continuous extension $\bar{F}$ to G_∞, which extends f to G. ■

Given {Cauchy} filters $\mathcal{A}, \mathcal{B}$ in a {topological} group G, one can easily verify that $\mathcal{A}\mathcal{B} = \{AB: (A, B) \in \mathcal{A} \times \mathcal{B}\}$ is a {Cauchy} filterbase in G. We use this for viewing Theorem 6.12, Corollary 6.12, in the context of topological groups.

THEOREM 8.22. Let H be a subgroup of a $\{T_2\}$ group G and let $\tilde{G}$ be a complete T_2 group. Then every continuous homomorphism $h: H \to \tilde{G}$ has a {unique} continuous homomorphic extension $\bar{h}$ to $\bar{H} \subset G$.

Proof. Since h is $\mathcal{U}_H - \mathcal{U}_{\tilde{G}}$ uniformly continuous (Theorems 8.7, 8.13 and Corollary 8.16), we need only verify that $\bar{h}$ is a homomorphism. For notation let $\mathcal{H}_x = \mathcal{U}_G[x] \cap H \ \forall x \in G$. Fix $x_1, x_2 \in \bar{H}$. For any $w \in \mathfrak{N}_e$ and $u, v \in \mathfrak{N}_e$ chosen such that $(ux_1)(vx_2) \subset wx_1x_2$, one has $(U_u[x_1] \cap H)(U_v[x_2] \cap H) \subset U_w[x_1x_2] \cap H$. Therefore $\{\mathcal{H}_{x_1}, \mathcal{H}_{x_2}\}\langle H\rangle \supset \mathcal{H}_{x_1x_2}\langle H\rangle$ for these $\mathcal{U}_H$-Cauchy filters in H. Since h is uniformly continuous and $\tilde{G}$ is complete $h\{\mathcal{H}_{x_1x_2}\langle H\rangle\}$ and so $h\{\mathcal{H}_{x_1}\langle H\rangle\}h\{\mathcal{H}_{x_2}\langle H\rangle\} = h(\{\mathcal{H}_{x_1}\mathcal{H}_{x_2}\}\langle H\rangle)$ converges to some $\tilde{y}_{x_1x_2} \in \tilde{G}$; but each $h\{\mathcal{H}_{x_i}\langle H\rangle\}$ converges to some $\tilde{y}_{x_i} \in \tilde{G}$ and, since $\tilde{G}$ is T_2, we have $\bar{h}(x_1x_2) = \tilde{y}_{x_1x_2} = \tilde{y}_{x_1}\tilde{y}_{x_2} = \bar{h}(x_1)\,\bar{h}(x_2)$ as required. ■

COROLLARY 8.23. Let $h: H \to \tilde{H}$ be a topological isomorphism, where H and $\tilde{H}$ are subgroups of complete T_2 groups G and $\tilde{G}$, respectively. Then $\bar{h}: \bar{H} \to \bar{\tilde{H}}$ is a topological isomorphism.

Connectedness Considerations

The strengthened version of Theorems 8.16 and 8.19 over arbitrary uniform spaces stems from the added group structure of the underlying set. The following two results further accentuate this difference between general uniform spaces and topological groups. First, note [Theorem 8.7, (iv)] that G is disconnected if it contains a proper open subgroup $H \neq \emptyset$.

THEOREM 8.24. For a connected topological group G

(i) $G = \bigcup_{n \geq 1} u^n \; \forall u \in \mathfrak{N}_e$.

(ii) If G is T_2, either $G = \{e\}$ or $\aleph(G) > \aleph_0$.

(iii) G is separable if some $u \in \mathfrak{N}_e$ is open and separable.

Proof. Assume for (i) that $u \in \mathfrak{N}_e$ is symmetric [Theorem 8.15, (i)]. Then $\bigcup_{n \geq 1} u^n$ is an open subgroup of G and so coincides with G. For (ii) let $f: G \to I = [0, 1]$ be continuous with $f(x_1) = 0$ and $f(x_2) = 1$ for some $x_1, x_2 \in G$ (remember that G is completely regular). Since the continuous image of a connected set is connected, all values $t \in [0, 1]$ are images under f of elements in G. To prove (iii) suppose that $\{x_n\} \in u$ satisfies $\overline{\{x_n\}}^{\mathfrak{T}_u} = u$ for the G-induced topology $\mathfrak{T}_u$ on u. From $\overline{\{x_n\}}^{\mathfrak{T}_u} = u \cap \overline{\{x_n\}}^{\mathfrak{T}_u}$ follows $\overline{\{x_n\}}^{\mathfrak{T}_u} \subset \overline{\{x_n\}}$. On the other hand, $u = u \cap \overline{\{x_n\}}$ and $\overline{\{x_n\}} \subset u = \overline{\{x_n\}}^{\mathfrak{T}_u}$. Therefore $\overline{\{x_n\}} = \overline{\{x_n\}}^{\mathfrak{T}_u} = u$ and the connectedness of G assures that $G = \overline{\{x_n\}}$. ■

Remark. The converse of (i) above is false. If $u \in \mathfrak{N}_0$ is symmetric in $(\mathbb{R}, +; d_1)$, then $u \cap Q \in \mathfrak{N}_e$ is symmetric in $(Q, +; d_1)$ (the nonconnected, additive group of rationals) and $Q = \bigcup_{n \geq 1} (u \cap Q)^n$.

Every 2° topological space is 1°. On the other hand,

COROLLARY 8.25. A locally compact, connected 1° group is 2°.

Proof. Let u be an open, relatively compact nbd of $e \in G$, where G is locally compact, connected, and 1°. Then $\bar{u}$ (being a compact, pseudometric space) is separable and so is u, therefore G. It remains only to note that a separable, pseudometric space is 2°. ■

PROBLEMS

8A. Quasi- Semi-, and Joint Topological Groups

A topology $\mathfrak{T}$ on a group G is quasi-topological {semitopological} if $\mathfrak{m}: \underset{(x,y) \rightsquigarrow xy}{G \times G \to G}$ {if $\mathfrak{i}: \underset{x \rightsquigarrow x^{-1}}{G \to G}$} is continuous. If $\mathfrak{m}$ is continuous in each argument separately, $\mathfrak{T}$ is jointly topological.

(a) $(G, \mathfrak{T})$ is quasi-topological iff $\mathfrak{m}$ is continuous at (e, e), and both ${}_g\mathfrak{m}$ and $\mathfrak{m}_g$ are continuous at e $\forall g \in G$.

(b) A complete pseudometric, joint topological group (G, d) is quasi-topological. Proceed as follows:

(i) For each open set $E \subset G$ and $\varepsilon > 0$ there is an open subset E_1 of E and $\delta > 0$ such that $d(x, e) < \delta$ implies $d(xy, y) \leq \varepsilon$ $\forall y \in E_1$. [*Hint*: $B_n = \{y \in G : d(x, e) < 1/n \Rightarrow d(xy, y) \leq \varepsilon\}$ is closed in G. Since $E \subset \bigcup_{n=1}^{\infty} B_n$ and $E \in \mathrm{Catg_{II}}(G)$ (cf Appendix t.3), $E \cap B_{n_0} \in \mathrm{Catg_{II}}(G)$ for some $n_0 \in \mathbb{N}$. Complete (i) by showing that $E_1 = B_{n_0}$ is an open subset of E.]

(ii) Obtain inductively a sequence $\{\delta_n > 0\}$ and a sequence $\{E_n\}$ of closed sets such that each E_n is an open subset of E_{n-1} and $d(x, e) < \delta_n$ implies $d(xy, y) \leq 1/n$ $\forall y \in E_n$. Assume diam $E_n < 1/n$ so that (Problem 6A) $y_0 \in \bigcap_{n=1}^{\infty} E_n$ for some $y_0 \in G$. Given $\varepsilon > 0$, there is a $\delta > 0$ such that $d(y, y_0) < \delta$ implies $d(yy_0^{-1}, e) < \varepsilon$. If $n > 1/\delta$, then $\mathfrak{m}\{[S_{\delta_{2n}}(e) \cap E_{2n}y_0^{-1}] \times [S_{\delta_{2n}}(e) \cap E_{2n}y_0^{-1}]\} \subset S_\varepsilon(e)$. Conclude from this and (a) above that (G, d) is quasi-topological.

(c) A 2°, Catg II, quasi-topological group G is semitopological. [*Hint*: If $F \subset G$ is closed, $F^{-1} = \pi_1\{G \times F \cap \mathfrak{m}^{-1}(e)\}$ is a Borel set (π_1 denotes projection onto the first factor) and (Cech [16], p. 391) $\mathfrak{i}: G_0 \to G$ is continuous for some $G - G_0 \in \mathrm{Catg_I}(G)$. Assume now that $x_n \to e$ for $\{x_n\} \in G$. Since $x_n^{-1}(G - G_0) = G - x_n^{-1}G_0 \in \mathrm{Catg_I}(G)$ $\forall n \in \mathbb{N}$ and $G - G_0 \cap \left(\bigcap_{n=1}^{\infty} x_n^{-1}G_0\right) \subset (G - G_0) \cup \left\{\bigcup_{n=1}^{\infty} (G - x_n^{-1}G_0\right\} \not\subseteq G$, one has $x_0 \in \bigcap_{n=1}^{\infty} x_n^{-1}G_0$ for some $x_0 \in G$. Therefore $x_n x_0 \in G_0$ and $x_0^{-1}x_n^{-1} = (x_n x_0)^{-1} \to x_0^{-1}$.]

Remark. (b), (c) are due to Montgomery [63].

(d) A complete metric, joint topological group (G, d) is semitopological. [*Hint*: (Waelbroeck [102] Pick any $x_{n_1} \in \{x_n\}$, where $x_n \to e$. Since $x_{n_1}x_n \to x_{n_1}$, there is an $n_2 \in \mathbb{N}$ such that $d(x_{n_1}, x_{n_1}x_{n_2}) < 2^{-2}$. Proceed inductively to obtain a sequence $p_k = x_{n_1}x_{n_2} \cdots x_{n_k}$ (with $x_{n_k} \in \{x_n\}$) such that $d(p_k, p_{k+1}) < 2^{-(k+1)}$ and $d(p_k x_{n_j}^{-1}, p_{k+1}x_{n_j}^{-1}) < 2^{-(k+1)}$ $\forall 1 \leq j \leq k+1$ and $k \in \mathbb{N}$. Define $p = \lim_k p_k$ and

$q_j = \lim_k p_k x_{n_j}^{-1}\ \forall j \in \mathbb{N}$. Then $d(p, q_j) < 2^{2-j}$ and $px_{n_j}^{-1} = q_j \to p$ as $j \to \infty$. Conclude that $x_{n_j}^{-1} \to e$ and $x_n^{-1} \to e$.]

(e) Let $G = (\mathbb{R}, +)$ carry the (1°, non 2°) topology $\mathfrak{T}_l$ having base $\{[a, b) : a, b \in \mathbb{R}$ and $a < b\}$. Then $(G, \mathfrak{T}_l)$ is quasi-topological (therefore joint topological) but not semitopological. [*Hint*: For any $(x_0, y_0) \in G \times G$ and $\mathfrak{T}_l$-nbd $[a, b)$ of $x_0 + y_0$, let $\varepsilon = b - (x_0 + y_0)$. Then $\mathfrak{m}\{[x_0, x_0 + \varepsilon/3) \times [y_0, y_0 + \varepsilon/3)\} \subset [x_0 + y_0, b) \subset [a, b)$. However, no $\mathfrak{T}_l$-nbd $[\alpha, \beta)$ of $0 \in \mathbb{R}$ satisfies $\mathfrak{i}\{[\alpha, \beta)\} \subset [0, b)$.]

Remark. Membership in $T(G)$ is independent of order, since $\mathfrak{I} \subset \mathfrak{T}_l \subset \mathfrak{D}$ and $\mathfrak{I}, \mathfrak{D} \in T(G)$, whereas $\mathfrak{T}_l \notin T(G)$.

(f) Every quasi-topological {semitopological} group is a quasi-uniformizible {t·semiuniformizible} space (Definition 1.6, Theorem 1.21).

8B. Compatible Group Topology via Subgroups

Let $\mathscr{H}$ be a collection of subgroups of a group G.

(a) $\mathscr{H}$ is a subbase for a filterbase $\mathfrak{B}$ in G which satisfies (G_1), (G_2) of Theorem 8.11. Furthermore, $\mathfrak{B}$ satisfies (G_3) if each subgroup in $\mathscr{H}$ is normal.

(b) Suppose $\mathfrak{B}$ satisfies $(G_1)-(G_3)$. The generated topology $\mathfrak{T} \in T(G)$ is T_2 iff $\{e\} \in \mathfrak{B}$.

(c) Show that $\mathscr{H}_0 = \{gHg^{-1} : H \in \mathscr{H}$ and $g \in G\}$ is the subbase for a $\mathfrak{T} \in T(G)$.

8C. Extremal Topologies on a Group

(a) Let $\mathfrak{T}_\alpha \in T(G)\ \forall \alpha \in \mathfrak{A}$. Then $\sup_{\mathfrak{A}} \mathfrak{T}_\alpha \in T(G)$ but not necessarily $\inf_{\mathfrak{A}} \mathfrak{T}_\alpha \in T(G)$. [Note that $\sup_{\mathfrak{A}} \mathfrak{T}_\alpha$ is the topology having subbase $\{\bigcup_{\mathfrak{A}} \mathfrak{T}_\alpha\}$ and $\inf_{\mathfrak{A}} \mathfrak{T}_\alpha = \bigcap_{\mathfrak{A}} \mathfrak{T}_\alpha$ on G.]

(b) Let $\mathfrak{T}_F$ denote the finest, nondiscrete T_2 topology that can be defined on a set X. Then a topology $\mathfrak{T}$ on X satisfies $\mathfrak{T} = \mathfrak{T}_F$ iff X has exactly one $\mathfrak{T}$-accumulation point x_0 and $\mathfrak{N}_{x_0} \cap X - x_0$ is a maximal filter in $X - x_0$ (where $\mathfrak{N}_{x_0}$ denotes the $\mathfrak{T}$-nbd filter at $x_0 \in X$). Prove that $\mathfrak{T}_F \notin T(G)$ for any group G.

8D. Topological Matrix Groups

(Notation as in Example 8.3). For $A = (a_{ij}) \in M_n(\mathbb{F})$ define $A^* = \overline{A^t}$, where $A^t = (a_{ji})$ is the transpose of A and $\bar{A} = (\bar{a}_{ij})$ is the complex conjugate of A.

(a) Each of the mappings $G_n(\mathbb{F}) \to G_n(\mathbb{F})$ is a homeomorphism:

(i) $A \rightsquigarrow A^{-1}$
(ii) $A \rightsquigarrow \bar{A}$
(iii) $A \rightsquigarrow A^t$
(iv) $A \rightsquigarrow A^*$
(v) $A \rightsquigarrow A^{-t} = (A^t)^{-1}$

(b) Prove that $(M_n, \| \|_2)$ is a Banach algebra relative to $\|A\|_2 = \sqrt{\sum_{i,j}^{n} |a_{ij}|^2}$. [*Hint*: Consider the mapping $\rho: (M_n, \| \|_2) \to (\mathbb{F}^{n^2}, \| \|)$.]

(c) $A \in G_n(\mathbb{F})$ is said to be orthogonal {Hermetian} iff $A^{-1} = A^t$ {iff $A = A^t$}. We call $A \in G_n(\mathbb{C})$ unitary iff $A^{-1} = A^+$. The multiplicative subgroups of $G_n(\mathbb{F})$ consisting of orthogonal {unitary} matrices are denoted $O_n(\mathbb{F})$ $\{U_n(\mathbb{C})\}$. Prove

(i) $O_n(\mathbb{F})$ and $U_n(\mathbb{C})$ are topological subgroups of $G_n(\mathbb{C})$.
(ii) $O_n(\mathbb{C}) \cap U_n(\mathbb{C}) = O_n(\mathbb{R}) = O_n(\mathbb{C}) \cap G_n(\mathbb{R})$.
(iii) $O_n(\mathbb{F})$ and $U_n(\mathbb{C})$ are closed subgroups of $G_n(\mathbb{C})$.
(iv) $O_n(\mathbb{R})$ and $U_n(\mathbb{C})$ are compact subgroups of $G_n(\mathbb{C})$.

[*Hint*: Use $A \in U_n(\mathbb{C})$ iff $\sum_{k=1}^{n} a_{ik}\bar{a}_{kj} = \delta_{ij}$ $(i, j = 1, 2, \ldots, n)$ to prove that $\|A\|_2 = 1$ and use (ii), (iii) for proving $O_n(\mathbb{R})$ compact.]

(v) What can you say about $\{A \in G_n(\mathbb{F}): A = A^*\}$? What if G is abelian?

8E. Uniform Continuity

(a) (Notation as in Example 8.14). A homomorphism $h: G \to \tilde{G}$ is continuous iff h is $\mathcal{U} - \tilde{\mathcal{U}}$ uniformly continuous.

(b) Both $\mathfrak{i}: (G, \mathcal{U}_G) \to (G, {}_G\mathcal{U})$ and $\mathfrak{i}: (G, \mathcal{U}) \to (G, \mathcal{U})$ are unimorphisms.

(c) Let $\mathcal{V}$ denote $\mathcal{U}_G$ $\{{}_G\mathcal{U}$ or $\mathcal{U}\}$ for G. Then ${}_a\mathfrak{m} \circ \mathfrak{m}_b: (G, \mathcal{V}) \to (G, \mathcal{V})$, $x \rightsquigarrow axb$, is a unimorphism $\forall a, b \in G$, whereas $\mathfrak{m}: (G \times G, \mathcal{V} \times \mathcal{V}) \to (G, \mathcal{V})$, $(x, y) \rightsquigarrow xy$, need not be uniformly continuous.

8F. Corollary 8.5 Revisited

(a) Suppose $C \subset G$ is compact and $B \cap C \neq \varnothing$. Then there is a $u \in \mathfrak{N}_e$ such that $Bu \cap Cu = \varnothing = uB \cap uC$. [*Hint*: [Theorem 9.15(ii)] $\overline{Bww^{-1}} = \bigcap_{v \in \mathfrak{N}_e} Bww^{-1}v = \bigcap_{u \in \mathfrak{N}_e} Bu = \bar{B} = B$ and $\overline{Bww^{-1}} \cap C = B \cap C = \varnothing$ implies $C \subset G - \overline{Bww^{-1}}$ $\forall w \in \mathfrak{N}_e$. Therefore $C \subset \bigcup_{i=1}^{n} (G - \overline{Bw_iw_i^{-1}})$ for some $\{w_i\} \in \mathfrak{N}_e$. If $u_1 = \bigcap_{i=1}^{n} w_i$, then $Bu_1u_1^{-1} \cap C \subset B \cap C = \varnothing$ and $Bu_1 \cap Cu_1 = \varnothing$. Take $u = u_1 \cap u_2$, where $u_2 \in \mathfrak{N}_e$ satisfies $u_2B \cap u_2C = \varnothing$.]

(b) Both BC and CB are closed if $C \subset G$ is compact. [*Hint*: If $x \notin BC$, then $B^{-1}x \cap C = \varnothing$ and $B^{-1}xu \cap Cu = \varnothing$ for some $u \in \mathfrak{N}_e$. Therefore $xuu^{-1} \cap BC = \varnothing$ and $x \notin \overline{BC}$.

(c) Give an example in which BC is not closed for closed $B \subset G$ and (noncompact) $C \subset G$. [*Hint*: What if B is the set of integers in $G = (\mathbb{R}, +; d_1)$ and $C = B\alpha$ with α irrational?]

8G. Banach–Kuratowski–Pettis Theorem

(Notation as in Appendix t.4.) Given a topological group G, define $\mathcal{B}_{ao_{II}}(G) = \mathcal{B}_{ao}(G) \cap \mathrm{Catg}_{II}(G)$, and let $\mathcal{Q}(G)$ $\{\mathcal{Q}_{II}(G)\}$ denote the collection of subsets of G, each of which contains some $B \in \mathfrak{B}_{a0}(G)$ {some $B \in \mathfrak{B}_{ao_{II}}(G)$}. Prove the following (Pettis [77]):

(a) $BB^{-1} \in \mathfrak{N}_e\ \forall B \in \mathfrak{B}_{ao_{II}}(G)$. [*Hint*: For any $x_0 \in \operatorname{Int} \mathbb{B}_{II}$ and open $u \in \mathfrak{N}_e$ such that $u^{-1}ux_0 \in \operatorname{Int} \mathbb{B}_{II}$ one has $u \subset BB^{-1}$. (If $E_x = [G-(\operatorname{Int} \mathbb{B}_{II} - B)] \cap x^{-1}[G-(\operatorname{Int} \mathbb{B}_{II} - B)]$ for $x \in u$, then $G - E_x \subset \{(\mathbb{B}_{II} - B) \cup x^{-1}(\mathbb{B}_{II} - B)\} \in \operatorname{Catg}_I(G)$. Since G is Gatg II, both E_x and ux_0 are $\operatorname{Catg}_{II}(G)$ so that $E_x \cap ux_0 \neq \varnothing$. If $y \in E_x \cap ux_0$, then $y \cup (x^{-1}y) \in \{G-(\operatorname{Int} \mathbb{B}_{II} - B)] \cap \operatorname{Int} \mathbb{B}_{II}\} \subset B$ and one obtains $y \in BB^{-1}$.]

(b) $BB^{-1} \in \mathfrak{N}_e\ \forall B \in \mathcal{Q}_{II}(G)$.

(c) Every subgroup $H \in \mathfrak{B}_{II}(G)$ is an open set.

8H. Open, Closed Subgroups

(a) A subgroup H of a topological group G is open {closed} iff $\operatorname{Int} H \neq \varnothing$ {$H \cap u$ is closed for some closed $u \in \mathfrak{N}_e$}.

(b) The center $\{x \in G: xy = yx\ \forall y \in G\}$ of a T_2 group G is a closed normal subgroup of G.

(c) Every discrete normal subgroup of a connected T_2 group G is a closed subgroup of G's center.

(d) The component $C(e)$ of a topological group G is a closed normal subgroup of G. (Note that $C(e)$ is defined as the largest connected subset of G that contains e.) [*Hint*: The continuous image ${}_{x^{-1}}\mathfrak{m}\{C(e)\}$ of $C(e)$ is connected and contains e. Therefore $\bigcup_{x \in C(e)} x^{-1}C(e) = \{C(e)\}^{-1}C(e) \subset C(e)$ and $C(e)$ is a subgroup of G. For normality, $e \in x^{-1}C(e) = ({}_{x^{-1}}\mathfrak{m} \circ \mathfrak{m}_x)\{C(e)\}$ implies $x^{-1}C(e)x \subset C(e)\ \forall x \in G$.]

(e) The following are equivalent: G contains no proper open subgroup; G contains no proper subgroup $H \in \mathfrak{B}_{ao_{II}}(G)$ (see Problem 8G); $G = \bigcup_{n \geq 1} u^n\ \forall u \in \mathfrak{N}_e$.

(f) Every compact open $u \in \mathfrak{N}_e$ contains a compact open and closed subgroup of G. [*Hint*: Given compact $F \subset u$, show that $Fv \subset u$ for some symmetric $v \in \mathfrak{N}_e$. Then $v^n \subset u\ \forall n \in \mathbb{N}$ and $\bigcup_{n \geq 1} v^n \subset u$.]

(g) If G is compact, then $u \in \mathfrak{N}_e$ in (f) above contains an open and closed normal subgroup of G. [*Hint*: Consider the subgroup $\bigcap_{x \in G} x\left(\bigcup_{n \geq 1} v^n\right)x^{-1}$ of G.]

(h) In a locally compact, totally disconnected group every $u \in \mathfrak{N}_e$ contains a compact open subgroup.

8I. Additional Connectedness Properties

(a) A topological group is disconnected iff it contains a proper open subgroup. Give an example of such a group.

(b) A topological group G is totally disconnected iff $C(e) = \{e\}$. Show also that G is totally disconnected if G is T_2 and every $u \in \mathfrak{N}_e$ contains an open subgroup of G.

(c) A locally compact group G is connected iff any of the conditions in Problem 8H(e) hold.

(d) A connected group G is Catg II iff $C(e) \in \text{Catg}_{\text{II}}(G)$.

(e) A connected group contains no proper subgroup $H \in \mathcal{C}_{\text{II}}(G)$ (see Problem 8G). [*Hint*: Use Theorem 8.24 and the fact that $HH^{-1} = H$.]

8J. Locally Compact Groups

(a) An LCT_2 group G is metrizible iff $e \in G$ is a G_δ set. [*Hint*: The necessary proof follows from Theorems 8.17 and 8.15(i), (ii). Suppose, conversely, that $e = \bigcap_{n=1} O_n$ for open sets $\{O_n : n \in \mathbb{N}\}$ in G. Take $v_1 = O_1$ and let v_2 be a relatively compact open set satisfying $e \in v_2 \subset \bar{v}_2 \subset v_1 \cap O_2 \subset v_1$. Obtain inductively a sequence of relatively compact open sets $\{v_n\}$ satisfying $e \in v_n \subset \bar{v}_n \subset v_{n-1} \cap O_n \subset v_{n-1}$ and show that $\{v_n : n \in \mathbb{N}\}$ is a G-nbd base at $e \in G$. (Otherwise, there is an open set E containing e such that $v_n \not\subset E$ and $\varnothing \neq \bar{v}_n \cap (G-E) \subset \bar{v}_1\ \forall n \geq 2$; the compactness of $\bar{v}_1$ and fip of $\{\bar{v}_n \cap (G-E)\}$ yields a contradiction.)]

(b) Every compact open $u \in \mathfrak{N}_e$ for a topological group G contains a compact open subgroup of G. [*Hint*: If $H \subset u$ is compact, $Hv \subset u$ for some symmetric $v \in \mathfrak{N}_e$ and $\bigcup_{n \geq 1} v^n \subset u$.]

(c) If G is compact, $u \in \eta_e$ in (b) above contains an open normal subgroup of G. $\left[\textit{Hint}: \text{Consider } \bigcap_{x \in G} x\left(\bigcup_{n \geq 1} v^n\right)x^{-1} \subset G\right]$

(d) In a locally compact, totally disconnected group every $u \in \mathfrak{N}_e$ contains a compact open subgroup.

(e) Let h be a continuous homomorphism from a subgroup $(H, +; d_1) \subset (\mathbb{R}, +; d_1)$ into a locally compact group G. Either h is an onto homeomorphism or $\overline{h(H)}$ is a compact abelian subgroup of G.

8K. Generating Subsets of a Topological Group

A subset X of a topological group G is said to generate G if no proper closed subgroup of G contains X.

(a) Every {normal} subset X of G generates a unique minimal {normal} subgroup $gp(X)$ of G. [*Hint*: Consider the intersection of all {normal} subgroups of G that contain X.]

(b) Suppose $h: G \to G$ is a continuous homomorphism such that $h/X = \mathbb{1}$ for $X \subset G$. If X generates G, then $h = \mathbb{1}$ on G. [*Hint*: $X \subset H = \{y \in G : h(y) = y\} \subset \bar{H}$.]

8L. Free Topological Groups

Begin with the fact that a topological space X is completely regular iff there exists a topological group $G(X)$ (called a free topological group of X) that satisfies

(F_1) X is a subspace of $G(X)$ which generates $G(X)$;

(F_2) every continuous mapping ψ of X into a topological group $\tilde{G}$ has a continuous homomorphism extension $\Psi\colon G(X) \to \tilde{G}$.

(a) Show that $G(X)$ is unique up to a topological isomorphism. [*Hint*: Suppose $\tilde{G}(X)$ satisfies (F_1), (F_2) and let $\Psi_G\colon G(X) \to \tilde{G}(X)$ and $\Psi_{\tilde{G}}\colon \tilde{G}(X) \to G(X)$ be continuous homomorphism extensions of $\mathbb{1}\colon X \to \tilde{G}(X)$ and $\mathbb{1}\colon X \subset \tilde{G}(X) \to X$, respectively. Use Problem 8K(b) to prove that Ψ_G is an onto topological isomorphism.]

(b) $G(X)$ is connected iff X is connected. [*Hint*: If X is connected, $X \subset C(e)$ [the component of $e \in G(X)$] and so $C(e) = \overline{C(e)} = G(X)$. If X is disconnected, some continuous surjection $\psi\colon X \to (\pm 1, \cdot\,; \mathfrak{D})$ has a surjective extension to $G(X)$.]

(c) Show that X is a closed subset of $G(X)$. Why does this confirm the existence of nonnormal topological groups (see Corollary 8.21)? [*Hint*: Suppose X is completely regular but nonnormal.]

Remarks. (Markov [60]) All statements above remain true if "group" is replaced by "abelian group" throughout. Subgroups of free topological groups are not necessarily free topological groups (Graev [34]. However, open subgroups of a free topological group are free (a recent, yet unpublished, result of R. Brown and L. Hardy).

8M. Left {Right} Invariance in Topological Groups

The terminology here follows that in Problem 4H. Thus a pseudometric ρ on a group G {a uniformity $\mathcal{U}$ for G} is left invariant if ρ {if $\mathcal{U}$} is ${}_GM$-invariant, where ${}_GM = \{{}_x\mathfrak{m} : x \in G\}$. Right invariance (with M_G replacing ${}_GM$) and invariance are defined similarly.

(a) Show that ${}_G\mathcal{U}$ $\{\mathcal{U}_G\}$ is the left {right} invariant for a topological group G. [*Remark* [Problem 4H(d)]. The left invariant, ${}_G\mathcal{U} \times {}_G\mathcal{U}$-uniformly continuous {right invariant, $\mathcal{U}_G \times \mathcal{U}_G$-uniformly continuous} pseudometrics on G constitute the gage for ${}_G\mathcal{U}$ {for $\mathcal{U}_G$}.

(b) Every 1° group has a compatible left invariant pseudometric ρ_l and a compatible right invariant pseudometric ρ_r.[*Hint*: Use Problems 4H(d) and 4B to obtain $\mathfrak{T}(\rho_l) = \mathfrak{T}_{\mathcal{U}_{\rho_l}} = \mathfrak{T}_{{}_G\mathcal{U}} = \mathfrak{T}_{\mathcal{U}_G} = \mathfrak{T}_{\mathcal{U}_{\rho_r}} = \mathfrak{T}(\rho_r)$.]

(c) Give an example for which the one-sided invariant pseudometric ρ_l and ρ_r in (b) above are not identical. In particular, show that G need not be defined by an invariant pseudometric. [*Hint*: Example 8.14.]

(d) A pseudometric ρ on G is invariant iff $\rho(ax, by) \le \rho(a, x) + \rho(b, y)$ $\forall a, b, x, y \in G$.

(e) For a left invariant pseudometric on G the following are equivalent: ρ is right invariant; ρ is inverse invariant (i.e., $\rho(x^{-1}y^{-1}) = \rho(x, y)\, \forall x, y \in G$); ρ is invariant under inner automorphisms (i.e., $\rho(gxg^{-1}, gyg^{-1}) = \rho(x, y)\, \forall$ fixed $g \in G$ and all pairs $(x, y) \in G \times G$).

(f) A topological group G has a compatible invariant pseudometric iff G has a countable nbd-base at e, each member of which is invariant under all inner automorphisms on G.

Remark. (e), (f) are due to Klee [53]. If G is T_2, the word "pseudometric" can be replaced by "metric."

8N. T_2 Group Completion

Every T_2 group $(G, \mathfrak{T})$ has a unique (up to a topological isomorphism) T_2 completion. Proceed as follows:

(a) Let $(\mu; \check{G}, \check{\mathcal{U}})$ be the separated completion of the uniform space $(G, \mathcal{U})$ (Example 8.14, Theorem 6.10). Since $\mathfrak{i}:(G, \mathcal{U}) \to (G, \mathcal{U})$, $x \rightsquigarrow x^{-1}$, is a unimorphism, $\mu \mathfrak{i} \mu^{-1} = \mathfrak{i}_0 : (\mu(G), \check{\mathcal{U}}) \to (\check{G}, \check{\mathcal{U}})$ has a unique continuous extension $\mathbb{1}_0 : (\check{G}, \check{\mathcal{U}}) \to (\check{G}, \check{\mathcal{U}})$.

(b) The continuity of multiplication implies the continuity of $\mu \mathfrak{m}(\mu^{-1} \times \mu^{-1}) = \mathfrak{m}_0 : (\mu(G) \times \mu(G), \check{\mathcal{U}} \times \check{\mathcal{U}}) \to (G, \check{\mathcal{U}})$, $(\check{x}_1, \check{x}_2) \rightsquigarrow x_1 x_2$. (Note that neither $\mathfrak{m}$ nor $\mathfrak{m}_0$ need be uniformly continuous by Problem 8E.) Show that there is a continuous extension M_0 of $\mathfrak{m}_0$ to $\check{G} \times \check{G}$. [*Hint*: If $\check{x}_1, \check{x}_2 \in \check{G}$, then $\mathcal{H}_i = \{\check{U}[\check{x}_i] \cap \mu(G)\}\langle\mu(G)\rangle$ is $\check{\mathcal{U}} \cap (\mu(G) \times \mu(G))$-Cauchy $(i = 1, 2)$ and $\mathcal{H}_1 \mathcal{H}_2$ is a Cauchy filterbase in $\mu(G)$. The generated Cauchy filter on $\check{G}$ converges to a unique element $\widetilde{x_1 x_2}$ (Problem 6A) and $M_0(\check{x}_1, \check{x}_2) = \widetilde{x_1 x_2}$ meets our needs.]

(c) Verify that $\mu(G)$ is a group, with identity $\tilde{e} = \mu(e)$ and inverse $\tilde{x}^{-1} = \mathfrak{i}_0(\check{x})\, \forall x \in \mu(G)$, under composition $\tilde{x}_1 \tilde{x}_2 = \mathfrak{m}_0(\tilde{x}_1, \tilde{x}_2)\, \forall \tilde{x} \in \mu(G)$. Additionally, $\mu : (G, \mathfrak{T}) \to (\mu(G), \mathfrak{T}_{\check{\mathcal{U}}})$ is a topological isomorphism.

(d) The set $\check{G}$ under composition $\check{x}_1 \check{x}_2 = M_0(\check{x}_1, \check{x}_2)$ is a group having identity $\check{e} = \tilde{e}$ and inverse of $\check{g}^{-1} = \mathbb{1}_0(\check{g})$. [*Hint*: If $\phi, \eta : X \to Y$ are continuous, Y is T_2 and $\phi/D = \eta/D$ on a dense subset $D \subset X$, then $\phi = \eta$ on X. Let $\phi(\check{x}_1, \check{x}_2, \check{x}_3) = \check{x}_1(\check{x}_2 \check{x}_3)$ and let $\eta(\check{x}_1, \check{x}_2, \check{x}_3) = (\check{x}_1 \check{x}_2)\check{x}_3$. Since ϕ, η agree on $\mu(G) \times \mu(G) \times \mu(G)$, they agree on $\check{G} \times \check{G} \times \check{G}$ and the associativity of multiplication follows. If $\phi(\check{x}) = \{\mathbb{1}_0(\check{x})\}\check{x}$ and $\eta(\check{x}) = \check{x}\{\mathbb{1}_0(\check{x})\}$, then $\phi/\mu(G) = \eta/\mu(G) = \check{e}$ implies $\phi = \eta = \check{e}$ on $\check{G}$.]

(e) Show that $(\check{G}, \mathfrak{T}_{\check{\mathcal{U}}})$ is a T_2 group which has $\check{\mathcal{U}}$ as its two-sided uniformity. [*Hint*: If $\check{\mathcal{V}}$ is the two-sided uniformity of $(\check{G}, \mathfrak{T}_{\check{\mathcal{U}}})$, then $\mathbb{1} : (\mu(G), \check{\mathcal{V}} \cap [\mu(G) \times \mu(G)]) \to (G, \mathcal{U})$ has a uniformly continuous extension to $(\check{G}, \check{\mathcal{W}})$. Thus $\check{\mathcal{W}} = \check{\mathcal{U}}$.]

(f) Why is $(\check{G}, \mathfrak{T}_{\check{\mathcal{U}}})$ unique up to a topological isomorphism?

(g) If G is abelian, so is $\check{G}$. [*Hint*: Argue as in (d) above, taking $\phi = \mathfrak{m}_0$ and $\eta(\tilde{x}_1, \tilde{x}_2) = \tilde{x}_2 \tilde{x}_1$.]

(h) Give an example of a T_2 group whose uniform space completion is not a topological group (Waelbroeck [102], p. 117).

8O. Right (Left) Group Completions

The pair $(\mu; \check{G})$ is a right (left) completion of a topological group G if μ is a topological isomorphism of G onto a dense subgroup of $\check{G}$ whose right uniformity $\mathcal{U}_{\check{G}}$ (left uniformity ${}_{\check{G}}\mathcal{U}$) is complete.

(a) A right (left) completion of G is a completion of G, but the converse fails. In fact, a T_2 group with a complete two-sided uniformity and incomplete right (left) uniformity has no right (left) completion. [*Hint*: Suppose $(\mu; \check{G})$ is the T_2 completion of G. Since $\mu:(G, \mathscr{U}_G)\to(\mu(G), \mathscr{U}_{\check{G}})$ and $\mu:(G, {}_G\mathscr{U})\to(\mu(G), {}_{\check{G}}\mathscr{U})$ are unimorphisms (Theorem 8.16, Problem 8E), $\mu(G)$ is $\check{G}$-closed and $\mathscr{U}_{\check{G}}$-incomplete.]

(b) A T_2 group G has a right (left) completion if $\mathscr{U}_G$ and ${}_G\mathscr{U}$ share the same Cauchy filters in G. [*Hint*: $\mathscr{U}_{\check{G}}$ and ${}_{\check{G}}\mathscr{U}$ will share the same Cauchy filters on $\check{G}$, which (Example 8.18) is right (left) complete.]

(c) Give an example of a T_2 group that has no right (left) completion. [*Hint*: (Dieudonne [23]): Let G be the composition group of all homeomorphisms of $I=[0, 1]$ onto itself with the right invariant metric topology given by $d(f, g)=\sup_I |f(x)-g(x)|$. Then G is $\mathscr{U}$-complete, since $\mathscr{U}=\mathscr{U}_\rho$ for $\rho(f, g)= d(f, g)+d(f^{-1}, g^{-1})$. Find a sequence $\{f_n\}\to f\in(G, \mathscr{U})$, where f is nonbijective. Then $\{f_n^{-1}\}$ is not ${}_G\mathscr{U}$-Cauchy.]

8P. Invariant Metric Completion

Every group with an invariant metric ρ is isomorphically isometric to a dense subgroup of a group with a complete invariant metric. Fill in the details to each proof below.

(a) Let $(\mu; \check{G})$ be the completion of $(G, \mathscr{U}=\mathscr{U}_\rho)$ (Problems 8M and 8N). Then $\tilde{\rho}(\tilde{x}, \tilde{y})=\rho(x, y)$ is an invariant metric on $\mu(G)$ such that $\mathscr{U}_{\tilde{\rho}}=\check{\mathscr{U}}$ on $\mu(G)\times\mu(G)$. Use Problem 6F to obtain an extension $\bar{\tilde{\rho}}$ of ρ to $\check{G}\times\check{G}$ and show that $\bar{\tilde{\rho}}$ meets our requirements.

(b) (Klee [53]) Let $\mu; (\check{G}, \check{\rho})$) be the metric space completion of (G, ρ) (cf p. 47) and define multiplication in $\check{G}$ termwise; that is, $\{x_n\}\{y_n\}=\{x_ny_n\}\ \forall\{x_n\}, \{y_n\}\in \check{G}$. Then $\check{G}$ is a group, $\mu: G\to\check{G}$ is an isomorphism, and $\check{\rho}$ is a complete invariant metric for $\check{G}$.

(c) Why is the theorem generally untrue for topological groups with only a right (left) invariant metric? [*Hint*: Consider Problems 6B and 6J.]

8Q. Topologies on Composition Groups of Homeomorphisms

(a) If $\mathscr{H}(X, X)$ is an equicontinuous family of homeomorphisms of a uniform space $(X, \mathscr{U})$ onto itself, then $\mathfrak{i}: \mathscr{H}_p^{-1}(X, X)\underset{h^{-1}\rightsquigarrow h}{\to}\mathscr{H}_p(X, X)$ is continuous. [*Hint*: It suffices (Theorem 3.4) to show that each $\mathfrak{e}_{x_0}\mathfrak{i}: \mathscr{H}_p^{-1}(X, X)\to(X, \mathscr{U})$ is continuous. For $h_0\in\mathscr{H}(X, X)$ with $y_0=h_0(x_0)$ and any symmetric $V\in\mathscr{U}$ there is a symmetric $U\in\mathscr{U}$ satisfying $h\{U[x_0]\}\subset V[h(x_0)]\ \forall h\in\mathscr{H}(X, X)$. Therefore $\mathfrak{e}_{x_0}\mathfrak{i}\{(e_{y_0}\times e_{y_0})^{-1}U[h_0^{-1}]\}\subset V[e_{x_0}i(h_0^{-1})]$.]

(b) $\mathscr{H}_p(X, X)$ is a topological group if $\mathscr{H}(X, X)$ is a composition group. In this case the left group uniformity ${}_{\mathscr{H}_p(X,X)}\mathscr{U}$ is finer than the pointwise uniformity $\mathscr{U}_p$; both uniformities coincide when $\mathscr{H}$ is uniformly equicontinuous.

(c) Suppose $(X, \mathscr{U})$ is a complete, separated uniform space and $\mathscr{H} \subset C(X; X)$ is uniformly equicontinuous. Let $\mathscr{V}$ be the two-sided (separated!) uniformity for $\mathscr{H}_p(X, X)$ and denote the set of limit points of all $\mathscr{V}$-Cauchy filters by $\mathscr{H}'(X, X)$. Then $\mathscr{H}'(X, X)$ is a uniformly equicontinuous composition group. If $\mathscr{V}'$ is the two-sided uniformity of $\mathscr{H}'_p(X, X)$, then $(\mathbb{1}: (\mathscr{H}'_p(X, X), \mathscr{V}'))$ is the T_2 completion of $(\mathscr{H}_p(X, X), \mathscr{V})$.

(d) If $\Gamma(X, X)$ denote the composition group of all homeomorphisms from X onto itself and $(X, \mathscr{U})$ is compact, then $\Gamma_u(X; X, \mathscr{U})$ is a topological group. [*Hint*: Use Corollary 1.26 to prove $\mathfrak{i}$ continuous.]

(e) $\Gamma_{co}(X, X)$ $\{\Gamma_\kappa(X, X)\}$ need not be a topological group even if X is a locally compact, T_2 {locally compact, separated uniform} space. [*Hint*: Consider $X = \{0, 2^n : n \in \mathbb{I}\} \subset (\mathbb{R}, d_1)$.]

(f) Suppose that $(X, \mathscr{U})$ is a separated, locally compact uniform space. For each $K \in \mathscr{C}$ and $U \in \mathscr{U}$ define $W_{K,U} = \{h, k \in \Gamma(X, X): (h(x), k(x)) \in U$ and $(h^{-1}(x), k^{-1}(x)) \in U \ \forall x \in K\}$. Then $\{W_{K,U}: K \in \mathscr{C}$ and $U \in \mathscr{U}\}$ is the subbase for a uniformity $\mathscr{U}_\beta$ finer than $\mathscr{U}_\kappa$ on $\Gamma(X, X) \times \Gamma(X, X)$. Prove also that $\Gamma_\beta(X, X) = (\Gamma(X, X), \mathfrak{T}_\beta)$ is a topological group. [Note that $\mathfrak{T}_\beta$ abbreviates $\mathfrak{T}_{\mathscr{U}_\beta}$.]

(g) $\Gamma_\beta(X, X)$ is complete relative to its two-sided uniformity. If $(X, \mathscr{U})$ is complete, so is $(\Gamma(X, X), \mathscr{U}_\beta)$.

(h) If $X = \{n + 2^{-m} : n \in \mathbb{I}$ and $m \in \mathbb{N}\}$ is the locally compact T_2 subspace of $(\mathbb{R}, d_1)$, neither $\mathscr{U}_\kappa$ nor $\mathscr{U}_\beta$ is comparable with a one- or two-sided uniformity of $(\Gamma(X, X), \mathfrak{T}_\beta)$.

(i) *Wanted*! A proof that $\mathfrak{T}_\kappa = \mathfrak{T}_\beta$ on $\Gamma(X, X)$ if $(X, \mathscr{U})$ in (f) above is also locally connected.

Remark. Bourbaki [12] offers an especially rich source of material concerning topologies on composition groups of homeomorphisms. The preceding problem is included as a small sample.

REFERENCES FOR FURTHER STUDY

On General Topological Groups: Bourbaki [12]; Husain [44]; Montgomery and Zippin [64]; Pontryagin [80].
On Free Topological Groups: Graev [34]; Markov [60].
On Homotopy Groups: Hilton [42]; McCarty [61]; Pontryagin [80].

9. Projective and Inductive Limits

If H is a subgroup of a topological group G, then $\mathfrak{T}_{\mathscr{P}(\mathbb{1})}$ is the induced topology of G on H (Example 3.6) and $(H, \mathfrak{T}_{\mathscr{P}(\mathbb{1})})$ is a topological group (Theorem 8.7). The generalization to projective limit of topological groups by homomorphisms is modeled along Theorems 3.2 and 2.4.

THEOREM 9.1. For any collection $\mathscr{H} = \{h_\alpha : G \to (G_\alpha, \mathfrak{T}_\alpha) : \mathfrak{T}_\alpha \in T(G_\alpha) \ \forall \alpha \in \mathfrak{A}\}$ of homomorphisms on a group G:

(i) $\mathfrak{T}_{\mathcal{P}(\mathcal{H})} \in T(G)$;

(ii) $\mathfrak{T}_{\mathcal{P}(\mathcal{H})}$ is T_2 if $\mathcal{H}$ is injective (i.e., $\bigcap_{\mathcal{A}} h_\alpha^{-1}(e_\alpha) = e \in G$) and each $\mathfrak{T}_\alpha$ ($\alpha \in \mathcal{A}$) is T_2. If $\mathfrak{T}_{\mathcal{P}(\mathcal{H})}$ is T_2, then $\mathcal{H}$ is injective.

(iii) $\mathfrak{T} = \mathfrak{T}_{\mathcal{P}(\mathcal{H})}$ for $\mathfrak{T} \in T(G)$ is equivalent to homomorphisms $h: G \to (G, \mathfrak{T})$ for any topological group G, being continuous iff every $h_\alpha h: G \to (G_\alpha, \mathfrak{T}_\alpha)$ is continuous.

Proof. (i) If $x, y \in G$ and $w = \bigcap_{i=1}^{n} h_{\alpha_i}^{-1}(w_{\alpha_i})$ is any $\mathfrak{T}_{\mathcal{P}(\mathcal{H})}$-nbd of $xy^{-1} \in G$, then $h_{\alpha_i}(xy^{-1}) = h_{\alpha_i}(x)h_{\alpha_i}(y^{-1}) \in w_{\alpha_i}$ $\forall 1 \le i \le n$. Since $\mathfrak{T}_{\alpha_i} \in T(G_{\alpha_i})$, there exist $\mathfrak{T}_{\alpha_i}$-nbds u_{α_i} of $h_{\alpha_i}(x)$ and v_{α_i} of $h_{\alpha_i}(y^{-1})$ such that $u_{\alpha_i}v_{\alpha_i}^{-1} \subset w_{\alpha_i}$. If $u = \bigcap_{i=1}^{n} h_{\alpha_i}^{-1}(u_{\alpha_i})$ and $v = \bigcap_{i=1}^{n} h_{\alpha_i}^{-1}(v_{\alpha_i})$, then $u \times v$ is a nbd of $(x, y) \in G \times G$ such that $\xi(u \times v) \subset w$. Thus $\mathfrak{T}_{\mathcal{P}(\mathcal{H})} \in T(G)$. For (ii) first suppose that every $\mathfrak{T}_\alpha$ is T_2 and let $\mathcal{H}$ be injective. If $\mathfrak{N}_{e_\alpha}$ denotes the $\mathfrak{T}_\alpha$-nbd system at $e_\alpha \in G_\alpha$, then $x \in \bigcap_{\mathcal{A}} \bigcap_{\mathfrak{N}_{e_\alpha}} h_\alpha^{-1}\{u(e_\alpha)\}$ implies $h_\alpha(x) \in \bigcap_{\mathfrak{N}_{e_\alpha}} h_\alpha^{-1}\{u(e_\alpha)\} = e_\alpha$ $\forall \alpha \in \mathcal{A}$ so that $x = e$. Thus $\bigcap_{\mathfrak{N}_e} u = \{e\}$ and $\mathfrak{T}_{\mathcal{P}(\mathcal{H})}$ is T_2. Conversely, if $\mathfrak{T}_{\mathcal{P}(\mathcal{H})}$ is T_2 and $x \neq e$ satisfies $x \notin \bigcap_{\mathfrak{N}_e} u$, then $x \notin h_\alpha^{-1}\{u(e_\alpha)\}$ and $h_\alpha(x) \neq e_\alpha$ for some $\alpha \in \mathcal{A}$. To prove (iii) simply invoke Theorem 3.4. ■

Example 9.2. (a) Given a collection $\{G_\alpha : \alpha \in \mathcal{A}\}$ of topological groups, let $G = \prod_{\mathcal{A}} G_\alpha$ be the product group under pointwise multiplication $(x_\alpha)_{\mathcal{A}}(y_\alpha)_{\mathcal{A}} = (x_\alpha y_\alpha)_{\mathcal{A}}$. Then each projection $\pi_{\alpha_0}: G \to G_{\alpha_0}$ ($\alpha_0 \in \mathcal{A}$) is a homomorphism and $\mathfrak{T}_{\mathcal{P}(\pi)} \in T(G)$ for $\pi = \{\pi_{\alpha_0} : \alpha_0 \in \mathcal{A}\}$. Since π is injective, $\mathfrak{T}_{\mathcal{P}(\pi)}$ is T_2 iff G_α is T_2.

The pair $(G, \mathfrak{T}_\pi)$, with $\mathfrak{T}_\pi$ denoting $\mathfrak{T}_{\mathcal{P}(\pi)}$, is called the topological product group of $\{G_\alpha : \alpha \in \mathcal{A}\}$.

(b) If $\mathcal{H} = \{h_\alpha : G \to (G_\alpha, \mathfrak{T}_\alpha) : \mathfrak{T}_\alpha \in T(G_\alpha)\, \forall \alpha \in \mathcal{A}\}$ is an injective collection of homomorphisms, then $(G, \mathfrak{T}_{\mathcal{P}(\mathcal{H})})$ is topologically isomorphic to a subgroup of $\prod_{\mathcal{A}} G_\alpha$. To be sure, the isomorphism $\mathfrak{e}(x) = \{h_\alpha(x)\}_{\mathcal{A}}$ from G into $\prod_{\mathcal{A}} G_\alpha$ is continuous, since $h_\alpha = \pi_\alpha \mathfrak{e}$ is $\mathfrak{T}_{\mathcal{P}(\mathcal{H})}$-continuous $\forall \alpha \in \mathcal{A}$. In like manner the continuity of each $\pi_\alpha = h_\alpha \mathfrak{e}^{-1}$ ($\alpha \in \mathcal{A}$) establishes the continuity of $\mathfrak{e}^{-1}$.

In comparison with Definition 7.7 (and Example 7.8) consider

Example 9.3. If X is a topological space and G is a topological group, the set $\mathfrak{T}(X; G)$ is a group with identity $F_e = e\chi_X$ and inverse

$(\mathfrak{i}\circ f)(x)=f(x)^{-1}$ under pointwise multiplication $(fg)(x)=f(x)g(x)\ \forall f, g\in \mathfrak{F}(X;G)$. Accordingly, $\mathfrak{F}_p(X;G)$ is a topological group with topological subgroup $C_p(X;G)$.

Suppose $\forall(S_\alpha, u, f)\in \mathcal{S}\times \mathfrak{N}_{F_e}\times \mathfrak{F}(X;G)$, there is a $w\in \mathfrak{N}_{F_e}$ such that $f(x)\,wf(x)^{-1}\subset u\ \forall x\in S_\alpha$. Then $\mathfrak{F}_{\mathcal{S}}(X;G)$ is a topological group. For verification it suffices to show that $\mathfrak{S}=\{(a_\alpha\times a_\alpha)^{-1}W_{S_\alpha,V_u}[F_e]:S_\alpha\in\mathcal{S}$ and $u\in\mathfrak{N}_{F_e}\}$ satisfies $(G_1)-(G_3)$ of Theorem 8.11. Begin by noting that $(a_\alpha\times a_\alpha)^{-1}W_{S_\alpha,V_u}[F_e]=\{f\in\mathfrak{F}(X;G):f(x)^{-1}\subset u\ \forall x\in S_\alpha\}$. Then G_1 holds, since $v\in\mathfrak{N}_{F_e}$ with $v^2\subset u$ yields $(a_\alpha\times a_\alpha)^{-1}W_{S_\alpha,V_{v^2}}[F_e]\subset(a_\alpha\times a_\alpha)^{-1}W_{S_\alpha,V_u}[F_e]$. Furthermore, G_2 holds because $\{(a_\alpha\times a_\alpha)^{-1}W_{S_\alpha,V_u}[F_e]\}^{-1}\subset(a_\alpha\times a_\alpha)^{-1}W_{S_\alpha,V_{u^{-1}}}[F_e]\in\mathfrak{S}$. By hypothesis $f(x)\,wf(x)^{-1}\subset u\ \forall x\in S_\alpha$ implies that $f\{(a_\alpha\times a_\alpha)^{-1}W_{S_\alpha,V_w}[F_e]\}f^{-1}\subset(a_\alpha\times a_\alpha)^{-1}W_{S_\alpha,V_u}[F_e]$ and so G_3 is fulfilled. The conclusion is that $\mathfrak{F}_{\mathcal{S}}(X;G)$ is a topological group whose right uniformity is precisely $U_{\mathcal{S}}$.

Inductive Limits

Certain uniform space deficiencies vanish in the enriched context of topological groups. For instance, compare Example 3.10(b), (c) with

THEOREM 9.4. Let $h:G\to\tilde{G}$ be an onto homomorphism, where G is a topological group and $\tilde{G}$ is a group. Then $h\mathfrak{N}_e=\{h(u):u\in\mathfrak{N}_e\}$ is a nbd base at $\tilde{e}\in\tilde{G}$ for $\mathfrak{T}_{\mathfrak{g}(h)}\in T(\tilde{G})$. Furthermore, $\mathfrak{T}_{\mathfrak{g}(h)}=\mathfrak{T}_{\mathcal{U}_{\mathfrak{g}(h)}}$ on G and $\mathcal{U}_{\tilde{G}}=\mathcal{U}_{\mathfrak{g}(h)}$ on $G\times G$.

Proof. Clearly, $h\mathfrak{N}_e$ satisfies (G_1)–(G_3) of Theorem 8.11. If $\tilde{\mathfrak{T}}\in T(\tilde{G})$ is the corresponding topology with nbd base $h\mathfrak{N}_e$ at $\tilde{e}\in\tilde{G}$, then $h:G\to(\tilde{G},\tilde{\mathfrak{T}})$ is continuous and open (Theorem 8.10). Continuity of h assures that $\tilde{\mathfrak{T}}\subset\mathfrak{T}_{\mathfrak{g}(h)}$. The reverse inclusion also holds since $E\in\mathfrak{T}_{\mathfrak{g}(h)}$ implies $h^{-1}(E)$ is G-open and $E=h\{h^{-1}(E)\}\in\tilde{\mathfrak{T}}$. Now $\mathcal{U}_{\tilde{G}}\subset\mathcal{U}_{\mathfrak{g}(h)}$ since $\mathfrak{T}_{\mathfrak{g}(h)}\in T(\tilde{G})$ and h is $\mathcal{U}_G-\mathcal{U}_{\tilde{G}}$ uniformly continuous (Theorem 8.16). In the other direction the continuity of $\mathbb{1}:(G,\tilde{\mathfrak{T}})\to(G,\tilde{\mathfrak{T}}_{\mathcal{U}_{\mathfrak{g}(h)}})$ implies $\mathcal{U}_{\tilde{G}}-\mathcal{U}_{\mathfrak{g}(h)}$ uniform continuity and $\mathcal{U}_{\mathfrak{g}(h)}\subset\mathcal{U}_{\tilde{G}}$. ∎

A normal subgroup H of a group G determines an equivalence relation $\mathfrak{R}$ on G by $(x,y)\in\mathfrak{R}$ iff $yx^{-1}\in H$ and the quotient set $G/H=\{Hx:x\in G\}$ is a group under multiplication $(Hx)(Hy)=Hxy\ \forall x,y\in G$.

DEFINITION 9.5. Let $\nu:\underset{x\rightsquigarrow Hx}{G\to G/H}$, where H is a normal subgroup of a topological group G. Then $(G/H,\mathfrak{T}_{\mathfrak{g}(\nu)})$ is termed the quotient topological group of G by H. One calls $(G/\{\bar{e}\},\mathfrak{T}_{\mathfrak{g}(\nu)})$ the T_2 group associated with G. (Compare with Definition 3.14.)

Remark. Note (Theorem 9.4) that $\mathcal{U}_{\mathcal{I}(\nu)}$ is precisely the right uniformity $\mathcal{U}_{G/H}$ of $(G/H, \mathfrak{T}_{\mathcal{I}(\nu)})$. Therefore $\mathcal{U}_{\mathcal{I}(\nu)}$ and $\mathfrak{T}_{\mathcal{I}(\nu)}$ are also written $\mathcal{U}_{G/H}$ and $\mathfrak{T}_{G/H}$ when convenient.

For future reference we include

THEOREM 9.6. The homomorphism $\nu: G \to (G/H, \mathfrak{T}_{\mathcal{I}(\nu)})$ is a continuous open surjection that is also closed if $H = \{\bar{e}\}$. Moreover, $(G/H, \mathfrak{T}_{\mathcal{I}(\nu)})$ is T_2 {discrete} iff H is closed {open}.

Proof. Everything but the closedness of ν has been established in Theorem 9.4. If $H = \{\bar{e}\}$ and $E \subset G$ is closed, then $\nu(x) \notin \nu(E)$ implies that $x \in E$ and $xu \cap E = \emptyset$ for some $u \in \mathfrak{N}_e$. For any $v \in \mathfrak{N}_e$ satisfying $v^2 \subset u$, it follows from $\{\bar{e}\} = \bigcap_{\mathfrak{N}_e} w \subset v$ that $x\{\bar{e}\}\, v \cap E = \emptyset$ and $\nu(x)\, \nu(v) \cap \nu(E) = \emptyset$. Therefore $\nu(x) \in \overline{\nu(E)}$ and ν is proved closed. The last assertion of the theorem follows from Theorem 8.15 and the fact that a topological space is T_1 {discrete} iff every point is closed {open}. ■

Example 9.7. Let $h: G \to \tilde{G}$ be an onto homomorphism between topological groups $G, \tilde{G}$ and let $H = h^{-1}(\tilde{e})$. Then there exists (Example 3.15) a unique group isomorphism $\mu_h: \underset{Hx \rightsquigarrow h(x)}{G/H \to \tilde{G}}$ such that $h = \mu_h \nu$ and $\mathcal{U}_{\mathcal{I}(h)} = \mathcal{U}_{\mathcal{I}(\mu_h)}$ is compatible with $\mathfrak{T}_{\mathcal{I}(h)} = \mathfrak{T}_{\mathcal{I}(\mu_h)} \in T(G)$. Also (Theorems 9.6 and 8.10), h is continuous {nearly continuous, open, nearly open} iff μ_h is continuous {nearly continuous, open, nearly open}.

The following illustration is important for later developments.

Example 9.8. Let $S'(\mathbb{C})$ denote the multiplicative group of complex numbers having modulus one, together with the two dim Euclidean topology. Since $h: \underset{x \rightsquigarrow \exp\{2\pi i x\}}{(\mathbb{R}, +; d_1) \to S'(\mathbb{C})}$ is continuous, open at $0 \in \mathbb{R}$ and $h^{-1}(1) = \mathbb{I}$, it follows that $\mu_h: \underset{\mathbb{I}+x \rightsquigarrow \exp\{2\pi i x\}}{(\mathbb{R}/\mathbb{I}, \mathfrak{T}_{\mathcal{I}(\nu)}) \to S'(\mathbb{C})}$ is an onto topological isomorphism. For this reason $\mathbb{R}/\mathbb{I}$ is called the circle group or the one dim torus group. It is not too difficult to verify that $\mathbb{R}/\mathbb{I}$ [therefore $S'(\mathbb{C})$] is a compact, T_2, 2° abelian group. (See Problem 9A.)

Topological Direct Products

A group G is said to be a direct product of normal subgroups H, K (denoted $G = H \otimes K$) if $G = HK$ and $H \cap K = \{e\}$. Each subgroup H, K is called the algebraic supplement of the other in G.

If $G = H \otimes K$, every $x \in G$ has a unique representation $x = kh$ $(h \in H, k \in K)$ and $\mu: \underset{(h,k) \rightsquigarrow hk}{H \times K \to HK}$ is an onto group isomorphism. For notation let $\pi_H : H \times K \to H$ and take $p_H = \pi_H \mu^{-1}$; the group homomorphisms π_K and p_K are defined similarly. A simple computation shows that $p_H^2 = p_H$, $p_K^2 = p_K$ and $p_H p_K = e = p_K p_H$; and $(p_H p_K) = \mathbb{1} = (p_K p_H)$ under pointwise multiplication, since $p_H(hk) p_K(hk) = hk = kh = p_K(hk) p_H(hk)$.

If $G = H \otimes K$ and $\mathfrak{T} \in T(G)$, we let $\mathfrak{T}_\pi \in T(H \times K)$ denote the product of the $\mathfrak{T}$-induced topologies $\mathfrak{T}_H$ on H and $\mathfrak{T}_K$ on K. Note that the continuity of $\mathfrak{m} : G \times G \to G$ implies continuity of $\mu : (H \times K, \mathfrak{T}_\pi) \to G$. Since it need not be true that μ^{-1} is continuous, we introduce

Definition 9.9. A topological group G is the topological direct product of normal subgroups H, K (denoted, $G \stackrel{t}{=} H \otimes K$) if $\mu : (H \times K, \mathfrak{T}_\pi) \to G = HK$ is a topological isomorphism. Each subgroup $(H, \mathfrak{T}_H)$, $(K, \mathfrak{T}_K)$ is then called the topological supplement of the other in G.

Remark. Note that $\mathfrak{T}_{\mathcal{P}\{p_H, p_K\}} \in T(G)$ is finer than the original topology $\mathfrak{T}$ on G, since $p_H^{-1}(u \cap H) \cap p_K^{-1}(u \cap K) \subset u^2 \; \forall u \in \mathfrak{N}_e$. Essentially, $\mu^{-1} : G \to (H \times K, \mathfrak{T}_\pi)$ is continuous iff both p_H, p_K are continuous (Theorem 3.4), which is equivalent to $\mathfrak{T} = \mathfrak{T}_{\mathcal{P}\{p_H, p_K\}}$ on G.

Our objective now is to extend the first law of group isomorphisms (the algebraic part of Theorem 9.10) to topological groups; the corresponding extension of the second law is considered in Problem 9D. If H is a subgroup of a topological group G and K is a normal subgroup of G, then $H \cap K$ is a normal subgroup of H and K is a normal subgroup of subgroup $HK \subset G$. Let $\mathfrak{T}_H$, $\mathfrak{T}_K$, and $\mathfrak{T}_{HK}$ denote the G-induced topology on H, K, and HK, respectively. Then $\underset{hk \rightsquigarrow Khk = Kh}{\nu_{HK} : (HK, \mathfrak{T}_{HK}) \to (HK/K, \mathfrak{T}_{\mathcal{G}(\nu_{HK})})}$ and $\underset{h \rightsquigarrow (H \cap K)k}{\nu_H : (H, \mathfrak{T}_H) \to (H/H \cap K, \mathfrak{T}_{\mathcal{G}(\nu_H)})}$ are continuous open surjective homomorphisms. Furthermore,

Theorem 9.10. $\underset{(H \cap K)h \rightsquigarrow Kh}{\eta : (H/H \cap K, \mathfrak{T}_{\mathcal{G}(\nu_H)}) \to (HK/K, \mathfrak{T}_{\mathcal{G}(\nu_{HK})})}$ is a continuous surjective isomorphism.

Proof. Since $H \subset HK$ and $H \cap K \subset K$, every $(H \cap K)h$ is contained in exactly one member $Kh \in HK/K$ and η is therefore well defined. That η is a surjective homomorphism is clear, since $\eta\{(H \cap K)h\} = Khk = Kh \; \forall Khk \in HK/K$, whereas $\eta\{(H \cap K)h_1 (H \cap K)h_2\} = \eta\{(H \cap K)h_1 h_2\} = Kh_1 Kh_2 = \eta\{(H \cap K)h_1\}\eta\{(H \cap K)h_2\} \; \forall h_1, h_2 \in H$. If $\eta\{(H \cap K)h_1\} = Kh_1 = Kh_2 = \eta\{(H \cap K)h_2\}$, then $(H \cap K)h_1 h_2^{-1} \subset H \cap K h_1 h_2^{-1} = H \cap K$ and $(H \cap K)h_1 \subset (H \cap K)h_2$. The same argument yields the reverse inclusion and establishes that η is injective. To prove η continuous let $\dot{E}$ be any $\mathfrak{T}_{\mathcal{G}(\nu_{HK})}$-open nbd of the identity element $K \in HK/K$. Then $E = \nu^{-1}(\dot{E})$ is a $\mathfrak{T}_{HK}$-open nbd of e and [Corollary 8.5(iii)] KE is a $\mathfrak{T}_{HK}$-open nbd of

$K \in HK$. Consequently, the $\mathfrak{T}_H$-open set $H \cap KE$ contains $H \cap K$ and $(H \cap K)(H \cap KE)$ is a $\mathfrak{T}_{\mathfrak{g}(\nu_H)}$-open nbd of the identity $H \cap K \in H/H \cap K$. It remains only to invoke Theorems 8.15(c) and 8.10 after noting that $\eta\{(H \cap K)(H \cap KE)\} \subset KE = \dot{E}$. ■

COROLLARY 9.11. Let $\nu_H: G \to G/H$ and $\nu_K: G \to G/K$ for a group $G = H \otimes K$. If $\mathfrak{T} \in T(G)$, then both $\eta_H: (H, \mathfrak{T}_H) \to (G/K, \mathfrak{T}_{\mathfrak{g}(\nu_K)})$, $h \rightsquigarrow Kh$, and $\eta_K: (K, \mathfrak{T}_K) \to (G/H, \mathfrak{T}_{\mathfrak{g}(\nu_H)})$, $k \rightsquigarrow Hk$, are continuous surjections.

It would be nice if η in Theorem 9.10 (therefore η_H and η_K in Corollary 9.11) were open. Unfortunately this need not be the case.

Example 9.12. (Pontryagin [80]). In $G = (\mathbb{R}^2, +; d_2)$ let $H = \{(x, \alpha x): x \in \mathbb{R}\}$ be the line with irrational slope α and let $K = \{(m, n) \in \mathbb{R}^2: m, n \in \mathbb{I}\}$. Note that $G_0 = H + K$ is a proper, dense subgroup of G. It follows from $H \cap K = \{(0, 0)\}$ that $(H/H \cap K, \mathfrak{T}_{\mathfrak{g}(\nu_H)})$ is discrete and, since $(H + K/K, \mathfrak{T}_{\mathfrak{g}(\nu_{HK})})$ is nondiscrete, η is not open. Observe also that $G_0 = H \otimes K$ and η_H is not open.

Suppose $\mathfrak{T} \in T(G)$, where $G = H \otimes K_1 = H \otimes K_2$. Although K_i ($i = 1, 2$) are group isomorphs (both are isomorphs of G/H), they need not be topological isomorphs because of the remarks above. Happily, the situation improves for topological direct products.

THEOREM 9.13. Let $\mathfrak{T} \in T(G)$, where $G = H \otimes K$. Then the following are equivalent:

(i) $\mathfrak{T} = \mathfrak{T}_{\mathfrak{P}\{p_H, p_K\}}$;
(ii) p_H is continuous;
(iii) p_K is continuous;
(iv) η_K is open;
(v) η_H is open.

Proof. Let $\mathfrak{i}_H: H \to H$, $h \rightsquigarrow h^{-1}$, and observe that

$$p_K = \mathfrak{m}(\mathbb{1} \times \mathfrak{i}_H p_H): G \xrightarrow{\mathbb{1} \times \mathfrak{i}_H p_H} G \times H \to K, \quad hk \rightsquigarrow (hk, h^{-1}) \rightsquigarrow k.$$

If p_H is continuous, then $\mathfrak{i}_H p_H$ and so p_K is continuous. In like manner p_K continuity implies p_H continuity and (i)⇔(ii)⇔(iii) follows. Furthermore, (ii)⇔(iv) and (iii)⇔(v), since $\eta_K = \overset{-1}{p_H}$ and $\eta_H = \overset{-1}{p_K}$. ■

Every homomorphism $p: G \to G$ satisfying $p^2 = p$ induces a direct product $G = p(G) \otimes p^{-1}(e)$.

Corollary 9.14. A normal subgroup H of a topological group $(G, \mathfrak{T})$ has a topological supplement in G iff there is a continuous surjective homomorphism $p: G \to H$ such that $p^2 = p$.

Proof. If p satisfies the above condition, $G = H \otimes p^{-1}(e)$ and $p = p_H$ so that $p^{-1}(e)$ is the topological supplement of $H \subset G$. ■

Problems

9A. Elementary Quotient Group Properties

Assume throughout that H is a normal subgroup of a topological group G and let $G/H = (G/H, \mathfrak{T}_{\mathfrak{g}(\nu)})$.

(a) G/H is compact {locally compact, connected, 1°, 2°} if G is compact {locally compact, connected, 1°, 2°}.

(b) If H is compact, $\nu: G \to G/H$ is a closed mapping. [*Hint* (Problem 8F): $HF \subset G$ is closed for closed $F \subset G$. If $Hx \in G/H - \nu(F)$, then $x \in u \subset G - HF$ for some open $u \subset G$ and so $Hx \in \nu(u) \subset G/H - \nu(F)$.]

(c) If H and G/H are compact {connected}, so is G.

(d) $G/C(e)$ is totally disconnected. [*Hint*: If $G/C(e)$ is not totally disconnected and $\dot{C}$ is the component of $C(e) \in G/C(e)$, then $\nu\{C(e)\} \subsetneq \dot{C}$ and $\nu^{-1}(\dot{C})$ is disconnected. This necessitates that $\dot{C} \subset G$ be disconnected.]

(e) G is complete iff $G/\{\bar{e}\}$ is complete. In general, G/H need not be complete even if G is complete and H is closed. [*Hint*: Example 6.9(b), Problem 20C.]

(f) If G has a left invariant metric d and H is closed, then $\check{d}(Hx, Hy) = d(\{Hx\}, \{Hy\})$ (the distance between sets $\{Hx\}$ and $\{Hy\}$ in G) is a left invariant metric for G/H.

9B. Topologically Isomorphic Quotient Groups

(a) Let $\mathbb{I}$ be the subgroup of integers in $\mathbb{R} = (\mathbb{R}, +; d_1)$. The torus $\mathbb{R}/\mathbb{I} \times \mathbb{R}/\mathbb{I}$ is topologically isomorphic to $\mathbb{R} \times \mathbb{R}/\mathbb{I} \times \mathbb{I}$.

(b) Is it always true that $G/H \times G/H$ is topologically isomorphic to $(G \times G)/(H \times H)$ for a normal subgroup H of a topological group G?

9C. Semidirect Products

Let H, K be groups with identity element e_H, e_K, respectively, and let

$$\begin{array}{c} H \to \mathcal{Q}(K) \\ x \rightsquigarrow \sigma_x \end{array}$$

be a homomorphism of H into the composition group $\mathcal{Q}(K)$ of isomorphisms from K onto K.

(a) The set $H \times K$ is a group under multiplication $(x, y)(x', y') = (xx', \sigma_{x'}(y)y')$. [This group is called the semidirect product of H by K, denoted $H \boxtimes K$.]

(b) Both $\begin{array}{c} j_H : H \to H' = H \times e_K \\ x \rightsquigarrow (x, e_K) \end{array}$ and $\begin{array}{c} j_K : K \to K' = e_H \times K \\ y \rightsquigarrow (e_H, y) \end{array}$ are isomorphisms. Furthermore, H' is a subgroup of $H \boxtimes K$ and K' is a normal subgroup of $H \boxtimes K$.

(c) Each coset of $H \boxtimes K/K'$ contains exactly one element of H.

(d) Let $f : H \to G$ and $g : K \to G$ be homomorphisms into a group G such that $g(\sigma_x(y)) = f(x^{-1})\, g(y) f(x)\, \forall (x, y) \in H \times K$. Then there is a unique homomorphism $h : H \boxtimes K \to G$ such that $f = hj_H$ and $g = hj_K$ [*Hint*: Consider $h(x, y) = f(x)\, g(y)$.]

(e) $\mathfrak{T}_H \times \mathfrak{T}_K \in T(H \boxtimes K)$ if $\mathfrak{T}_H \in T(H)$, $\mathfrak{T}_K \in T(K)$ and

$$\begin{array}{c} \eta : (H \times K, \mathfrak{T}_H \times \mathfrak{T}_K) \to (K, \mathfrak{T}_K) \\ (x, y) \rightsquigarrow \sigma_x(y) \end{array}$$

is a continuous surjection. [*Hint*: Use the continuity of multiplication in H and the continuity of η to show that $(a', z')(b', y') = (a'b', \sigma_{b'}(z')y')$ is near $(ab, \sigma_b(z)y) = (a, z)(b, y)$ if (a', z') and (b', y') are near (a, z) and (b, y), respectively.]

(f) If $(H, \mathfrak{T}_H)$ and $(K, \mathfrak{T}_K)$ are locally compact, so is $(H \boxtimes K, \mathfrak{T}_H \times \mathfrak{T}_K)$.

9D. Second Law of Isomorphisms

Let $H \subset K$, where H and K are normal subgroups of a topological group G. Define $\nu_H : G \to G/H$, $\nu_K : G \to G/K$ and $\nu : G/H \to (G/H)/(K/H)$. Then $(G/K, \mathfrak{T}_{\mathfrak{s}(\nu_K)})$ is topologically isomorphic to $[(G/H)/(K/H), \mathfrak{T}_{\mathfrak{s}(\nu)}]$. [*Hint*: $h = \nu\nu_H$: $G \to [(G/H)/(K/H), \mathfrak{T}_{\mathfrak{s}(\nu)}]$ is a continuous, open homomorphism and since $K = h^{-1}(K/H)$ the mapping $\mu_h : G/K \to (G/H)/(K/H)$ is an isomorphism (Example 9.7).]

9E. Locally Isomorphic Groups

Two topological groups G, $\tilde{G}$ are called locally isomorphic if there exist nbds $u \in \mathfrak{N}_e$ and $\tilde{u} \in \mathfrak{N}_{\tilde{e}}$ and a surjective homeomorphism $h : u \to \tilde{u}$ such that (i) $h(xy) = h(x)h(y)$ if $x, y \in u$ implies $xy \in u$, and (ii) $h^{-1}(\tilde{x}\tilde{y}) = h^{-1}(\tilde{x})h^{-1}(\tilde{y})$ if $\tilde{x}, \tilde{y} \in \tilde{u}$ implies $\tilde{x}\tilde{y} \in \tilde{u}$. One calls h a local isomorphism between G, $\tilde{G}$.

(a) Conditions (i), (ii) assure that $h(e) = \tilde{e}$ and $h(x^{-1}) = h(x)^{-1}\, \forall x \in u$ such that $x^{-1} \in u$.

(b) A homeomorphism satisfying (i) need not satisfy (ii). [*Hint*: Consider

$$h:(\mathbb{R},+;d_1)\to(\mathbb{R},+;d_1)\to(\mathbb{R}/\mathbb{I},+;\mathfrak{T}_{\mathfrak{g}(\nu)})$$
$$x \rightsquigarrow nx \rightsquigarrow \mathbb{I}+nx$$

on a sufficiently small nbd of $0\in\mathbb{R}$.]

(c) If there are nbds $u\in\mathfrak{N}_e$ and $\tilde{u}\in\mathfrak{N}_{\tilde{e}}$ satisfying (i) for a homeomorphism h, there are nbds $v\in\mathfrak{N}_e$ and $\tilde{v}\in\mathfrak{N}_{\tilde{e}}$ such that h satisfies (i), (ii).

(d) Produce a local isomorphism between $(\mathbb{R},+;d_1)$ and $S'(\mathbb{C})$.

(e) If H is a discrete normal subgroup of a topological group G, then G and G/H are locally isomorphic. [*Hint*: Pick $u\in\mathfrak{N}_e$ in G such that $u\cap H=\{e\}$ and let $v\in\mathfrak{N}_e$ satisfy $vv^{-1}\subset u$. Then $h=\kappa/v:v\to(h(v),\mathfrak{T}_{\mathfrak{g}(\nu)})$ is a homeomorphism that satisfies (i) and so (c) applies.]

(f) Every topological isomorphism is a local isomorphism, but the converse fails. [*Hint*: Show that $(\mathbb{R},+;d_1)$ and $S'(\mathbb{C})$ are not homeomorphic.]

(g) Reestablish the veracity of (d) via (e), (f), and the fact that $\underset{\mathbb{I}+x\rightsquigarrow\exp\{\pi i x\}}{\mathbb{R}/\mathbb{I}\longrightarrow S'(\mathbb{C})}$ is a topological isomorphism.

9F. Free Quotients

Every topological group is topologically isomorphic to a topological quotient of a free topological group (Problem 8L). [*Hint*: Suppose that $F(G)$ is the free topological group of a topological group G and let $\Psi:F(G)\to G$ be a continuous homomorphism that extends $\mathbb{1}:G\to G$. Then Ψ is open ($u(E)\cap G\in\mathfrak{N}_e$ $\forall F(G)$-nbd $u(E)$ of the identity $E\in F(G)$) and $\mu_\Psi:F(G)/\Psi^{-1}(0)\to G$ is a topological isomorphism.]

10. Open Mapping and Closed Graph Theorems

Given topological groups G and $\tilde{G}$, it is generally not true that

(oh) every continuous homomorphism of G onto $\tilde{G}$ is open;

(cg) every homomorphism with closed graph of $\tilde{G}$ into G is continuous.

Indeed (oh) fails for $\mathbb{1}:(G,\mathfrak{D})\to G$ and (cg) fails for $\mathbb{1}^{-1}$ when G is a nondiscrete T_2 group. Our goal here is to briefly introduce the conditions under which (oh) and (cg) hold. This theme is picked up and further developed in Section 21.

DEFINITION 10.1. A function $f\in\mathfrak{F}(X;Y)$, where X, Y are topological spaces, is said to have a closed graph if any of the following equivalent conditions hold:

(i) $g_r(f)=\{(x,f(x)):x\in X\}$ is a closed subset of $X\times Y$.

(ii) For each $x\in X$ and $y\neq f(x)\in Y$ there exist nbds $u(x)$ and $v(y)$ such that $f(u)\cap v=\varnothing$.

(iii) For each filter $\mathcal{G}$ in X with $\mathcal{G} \to x \in X$ and $f\mathcal{G} \to y \in Y$ one has $y = f(x)$.

Definition 10.1 is further illustrated in

Example 10.2 (a) The continuity of $f \in \mathcal{F}(X; Y)$ and closedness of $g_r(f) \subset X \times Y$ are generally unrelated. For instance, $g_r(\mathbb{1}) = \Delta \subset X \times X$ is closed and $\mathbb{1}: X \to (X, \mathfrak{D})$ is not continuous when X is a nondiscrete T_2 space. On the other hand, $\mathbb{1}: X \to X$ is continuous and $g_r(\mathbb{1}) \subset X \times X$ is not closed if X is non-T_2. (*Note*: We are recalling the fact that X is T_2 iff $\Delta \subset X \times X$ is closed.)

(b) If f is continuous and Y is T_2, then $g_r(f)$ is closed. This follows from (iii), since a topological space is T_2 iff every convergent filter has a unique limit.

(c) If $g_r(f)$ is closed, then $f(X)$ is T_1. Furthermore, $f(X)$ is T_2 if f is open and X is $T_1\{T_2\}$ if f is injective {continuous injection}. Verification: $f(x_1) \neq f(x_2)$ implies $f(u) \cap \{f(X) \cap v\} = \varnothing$ for nbds $u(x_1) \subset X$ and $\{f(X) \cap v\}(f(x_2)) \subset f(X)$. Therefore $f(x_2) \notin f(X) \cap v$ and $f(X)$ is T_1. If f is open, then $f(u)$ is also an $f(X)$-nbd and so $f(X)$ is T_2. Suppose f is injective and $x_1 \neq x_2$. Then $f(x_1) \neq f(x_2)$ and $u \cap f^{-1}\{f(X) \cap v\} \subset f^{-1}\{f(u)\} \cap f^{-1}(v) = \varnothing$ assures that $x_2 \notin u(x_1)$ and X is T_1. Finally, suppose f is a continuous injection. Then $f^{-1}: f(X) \to X$ is open and (since $g_r(f) = g_r(f^{-1})$ is closed in $f(X) \times X$) it follows from above that X is T_2.

For topological groups, Theorem 8.15, and the notions of Example 9.7 play a role in

THEOREM 10.3. Let $h: G \to \tilde{G}$ be a homomorphism between topological groups $G, \tilde{G}$. Then $g_r(h)$ is a subgroup of $G \times \tilde{G}$, and the statements

(i) $g_r(h)$ is closed,
(ii) $h^{-1}(\tilde{e})$ is closed,
(iii) $G/h^{-1}(\tilde{e})$ is T_2,
(iv) $g_r(\nu)$ is closed,
(v) $h(G)$ is T_2,
(vi) $g_r(\mu_h)$ is closed

are related as follows:

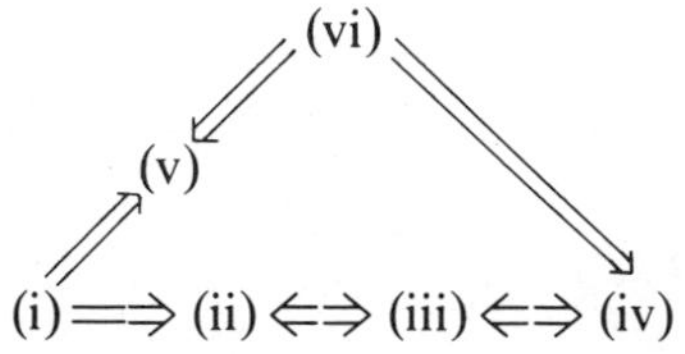

If h is continuous {open}, then (v) $\Rightarrow$ (i), (vi) {then (ii) $\Rightarrow$ (i)}.

Proof. Verification that $g_r(h)$ is a subgroup of $G\times h(G)\subset G\times\tilde{G}$ is straightforward. That (i) $\Rightarrow$ (ii) is also clear, since for any filter $\mathfrak{F}\subset h^{-1}(\tilde{e})$ converging to $x_0\in h^{-1}(\tilde{e})$ one has $h\mathfrak{F}\to\tilde{e}$ and $h(x_0)=\tilde{e}$. Proofs of the other implications in the diagram are simple applications of Theorem 9.6 in conjunction with Examples 9.7 and 10.2(b),(c). These, in turn, establish the remaining implications when h is continuous {open}. For instance, suppose h is open and let (ii) $\Leftrightarrow$ (iii) hold. If $\mathfrak{F}\to x_0\in G$ and $h\mathfrak{F}\to\tilde{x}_0\in\tilde{G}$ for some filter $\mathfrak{F}\subset G$, then $\nu\mathfrak{F}\to\nu x_0$ and $\nu\mathfrak{F}=\mu_h^{-1}\{h\mathfrak{F}\}\to\mu_h^{-1}(\tilde{x}_0)$. Therefore $\nu(x_0)=\mu_h^{-1}(\tilde{x}_0)$ and $\tilde{x}_0=h(x_0)$ so that (i) is proved true. ■

Conditions under which (oh) and (cg) hold can be given once we prove

LEMMA 10.4 (Kelley [50]). Let K be a closed subset of $(X,\rho)\times(Y,\mathscr{V})$, where (X,ρ) is a complete pseudometric space and $(Y,\mathscr{V})$ is a uniform space. Assume further that $\forall\alpha>0$, there is a $V_\alpha\in\mathscr{V}$ satisfying $V_\alpha[y]\subset\overline{K[S_{\alpha,\rho}(x)]}\ \forall(x,y)\in K$. Then $\forall\alpha,\varepsilon>0$, one has $\overline{K[S_{\alpha,\rho}(x)]}\subset K[S_{\alpha+\varepsilon,\rho}(x)]\ \forall(x,y)\in K$.

Proof. If $A\subset X$ and $y\in\overline{K[A]}$, then $y\in\overline{K[B]}$ for some $B\subset X$ with arbitrarily small diameter such that $A\cap B\neq\varnothing$. Indeed, assume that $V_\alpha\in\mathscr{V}$ is symmetric, take $y'\in K[A]\cap V_\alpha[y]$ and let $a\in A$ be chosen such that $(a,y')\in K$. Then $y\in V_\alpha[y']\subset\overline{K[S_{\alpha,\rho}(a)]}$, where $S_{\alpha,\rho}(a)=B$ has diameter 2α. Having established the above assertion, we show that $y\in\overline{K[S_{\alpha,\rho}(x)]}$ implies $y\in K[S_{\alpha+\varepsilon,\rho}(x)]\ \forall(x,y)\in K$. Let $A_0=S_{\alpha,\rho}(x)$ and inductively select sets $A_n\subset X$ having diameter less than $\varepsilon/2^{n+1}$ such that $y\in\overline{K[A_n]}$ and $A_n\cap A_{n+1}\neq\varnothing$. Choose an $x_n\in A_n\cap A_{n+1}\ \forall n\in\mathbb{N}$. The ρ-Cauchy sequence $\{x_n\}$ so obtained converges to some $x_0\in X$. Evidently, $\rho(x,x_0)<\alpha+\varepsilon$. For each pair of nbds $u(x_0)\subset X$ and $w(y)\subset Y$, one also has $K[u]\cap w\neq\varnothing$ (since $y\in\overline{K[A_n]}\subset\overline{K[u]}$ for some A_n). Thus $K\cap(u\times w)\neq\varnothing$ and it follows that $(x_0,y)\in\bar{K}=K$. In particular, $y\in K[x_0]\subset K[S_{\alpha+\varepsilon,\rho}(x)]$ as asserted. ■

THEOREM 10.5. Let $\tilde{G}$ be a topological group and let G be a right complete pseudometrizible group.†

(i) A nearly open homomorphism $h:G\to\tilde{G}$ having closed $g_r(h)\subset G\times\tilde{G}$ is open.

(ii) A nearly continuous homomorphism $\tilde{h}:\tilde{G}\to G$ having closed $g_r(\tilde{h})\subset\tilde{G}\times G$ is continuous.

† See Corollary 8.20^-.

Proof. Assume, for convenience, that h and $\tilde{h}$ are surjective. First, let ρ be as above and take $K = g_r(h)$ so that $K[A] = h(A)\ \forall A \subset X$. Given any $\alpha > 0$, there is (by the near openness of h at $e \in G$) a $\tilde{G}$-nbd $\tilde{u}_\alpha \in \mathfrak{N}_{\tilde{e}}$ such that $V_{\tilde{u}_\alpha}[\tilde{e}] \subset \overline{h\{S_{\alpha,\rho}(e)\}}$. Since ρ is right invariant, $\rho(x_1, x_2) = \rho(e, x_2 x_1^{-1})$ and $S_{\alpha,\rho}(x) = S_{\alpha,\rho}(e) \cdot x\ \forall x \in G$. Therefore $V_{\tilde{u}_\alpha}[\tilde{y}] = V_{\tilde{u}_\alpha}[\tilde{e}]\tilde{y} \subset \overline{h\{S_{\alpha,\rho}(e)\}} \cdot h(x) = \overline{h\{S_{\alpha,\rho}(x)\}}\ \forall \tilde{y} = h(x)$ and it follows by Lemma 10.4 that h is open. (If $\tilde{y} = h(x)$ for $x \in S_{\varepsilon,\rho}(e)$ and $\varepsilon' = \max\{\rho(e, x), \varepsilon - \rho(e, x)\}$, then $V_{\varepsilon'/3}[y] \subset \overline{h\{S_{2\varepsilon'/3,\rho}(x)\}} \subset h\{S_{\varepsilon,\rho}(e)\}$ and $\tilde{y} \in \operatorname{Int} h\{S_{\varepsilon,\rho}(x)\}$.) For the proof of (ii) let $K = \{(\tilde{h}(\tilde{x}), \tilde{x}) : \tilde{x} \in \tilde{G}\}$ so that $K[A] = \tilde{h}^{-1}(A)\ \forall A \subset X$. Since $\overline{\tilde{h}^{-1}\{S_{\alpha,\rho}(e)\}} \supset V_{\tilde{u}_\alpha}[\tilde{e}]$ for some $\tilde{u}_\alpha \in \mathfrak{N}_{\tilde{e}}$, the above also ensures that $\tilde{h}^{-1}$ maps G-open sets into $\tilde{G}$-open sets. ■

Compact sets behave as points in a regular space, and so

THEOREM 10.6. Properties (i), (ii) of Theorem 10.5 hold if G is a locally compact group.

Proof. (i) Assuming that $g_r(h)$ is closed, we first show that $h(u)$ is $\tilde{G}$-closed $\forall G$-compact $u \in \mathfrak{N}_e$. For $\tilde{y} \in G - \overline{h(u)}$, the $G \times \tilde{G}$-compact set $u \times \tilde{y}$ is contained in the open set $G \times \tilde{G} - g_r(h)$ and, since $G \times \tilde{G}$ is regular, there are closed nbds $W(u) \subset G$ and $\tilde{W}(\tilde{y}) \subset \tilde{G}$ such that $W \times \tilde{W} \subset G \times \tilde{G} - g_r(h)$. Thus $\tilde{W}(\tilde{y}) \subset \tilde{G} - h(u)$ and $h(u) \subset \tilde{G}$ is closed as asserted. Suppose now that h is nearly open. For any $v \in \mathfrak{N}_e$ and compact $u \in \mathfrak{N}_e$ contained in v one has $\overline{h(u)} = h(u) \subset h(v)$. In particular, $h(v) \in \mathfrak{N}_{\tilde{e}}$ and h is open (Theorem 8.10). The proof of (ii) follows similarly.

In juxtaposition with Theorem 10.5 should be

THEOREM 10.7. If G is separable or Lindelöf and $\tilde{G}$ is Catg II, then

(i) every surjective homomorphism $h: G \to \tilde{G}$ is nearly open;
(ii) every homomorphism $\tilde{h}: \tilde{G} \to G$ is nearly continuous.

Proof. (i) Suppose that G is separable with countably dense subset $\{x_n\}$. For any $u \in \mathfrak{N}_e$ let $v \in \mathfrak{N}_e$ be symmetric, open, and such that $v^2 \subset u$. Then $\{vx_n : n \in \mathbb{N}\}$ covers G and, since $\tilde{G} = \bigcup_{n=1}^{\infty} h(v)\,h(x_n)$ is Catg II, some $\overline{h(v)\,h(x_n)} = \overline{h(v)}\,h(x_n)$ has a nonempty interior. Because $\overline{h(v)}\,h(x_n)$ and $\overline{h(v)}$ are homeomorphic (Theorem 8.4), $\operatorname{Int} \overline{h(v)} \neq \varnothing$ as well. If $\tilde{y} \in \operatorname{Int} \overline{h(v)}$, then (Corollary 8.6) $\tilde{y}^{-1} \in \operatorname{Int} \overline{h(v)}$. Thus $\tilde{e} = \tilde{y}\tilde{y}^{-1}$ is an interior point of $\overline{h(v)}\,\overline{h(v)} \subset \overline{h(v^2)} \subset \overline{h(u)}$ and the near openness of h follows. If G

is Lindelöf, take $\{vx_n : n \in \mathbb{N}\}$ as the countable subcover of $\{vx : x \in G\}$ and proceed as above to prove h nearly open.

(ii) Let $\tilde{h} = \mu_{\tilde{h}} \nu : \tilde{G} \to \tilde{G}/\tilde{h}^{-1}(e) \to \tilde{h}(\tilde{G}) \subset G$ and suppose, as above, that $G = \bigcup_{n=1}^{\infty} vx_n$ for some symmetric open $v \in \mathfrak{N}_e$ and countable set $\{x_n\} \in G$. Since $\tilde{G}/\tilde{h}^{-1}(e) = \mu_{\tilde{h}}^{-1}\{\tilde{h}(\tilde{G})\} = \bigcup_{n=1}^{\infty} \mu_{\tilde{h}}^{-1}(vx_n)$ is Catg II (Appendix t.3), some $\overline{\mu_{\tilde{h}}^{-1}(vx_n)}$ and so $\overline{\mu_{\tilde{h}}^{-1}(v)}$ has a nonempty interior. This means that $\mu_{\tilde{h}}^{-1}$ is nearly open and $\mu_{\tilde{h}}$; therefore $\tilde{h}$ is nearly continuous. ■

THEOREM 10.8. Let G be a separable or Lindelöf group and let $\tilde{G}$ be Catg II $\{T_2$ and Catg II$\}$. Then (cg) $\{$(oh)$\}$ holds if G is right complete or locally compact.

Proof. Theorem 10.7 followed by Theorems 10.5 and 10.6. ■

COROLLARY 10.9. The (cg) $\{$(oh)$\}$ property holds whenever G is compact and $\tilde{G}$ is Catg II $\{T_2$, Catg II$\}$.

Proof. A compact space is both locally compact and Lindelöf. ■

PROBLEM

10A. Category, Continuity, and Openness

(Pettis [78]), Let h be a homomorphism between topological groups G, $\tilde{G}$. Using the results of Appendix t.4, prove that

(a) The following statements are equivalent:

(i) G is Catg II and $h : G \to \tilde{G}$ is continuous;
(ii) $h \in C(A; \tilde{G})$ for some set $A \in \mathcal{C}_{\mathrm{II}}(G)$;
(iii) for some $A \in \mathcal{C}_{\mathrm{II}}(G)$ there is a sequence $\{f_n\} \in C_p(A; \tilde{G})$ that converges to h/A;
(iv) $h^{-1}(\tilde{v}) \in \mathcal{C}_{\mathrm{II}}(G)$ $\forall$nbd $\tilde{v} \in \tilde{G}$.

[*Hint*: Show that (i) ⇒ (ii) ⇒ (iii). A proof that (iii) ⇒ (iv) is difficult but is provided in the above reference. For (iv) ⇒ (i) let $\tilde{v} \in \mathfrak{N}_{\tilde{e}}$ and pick $\tilde{u} \in \mathfrak{N}_{\tilde{e}}$ such that $\tilde{u}\tilde{u}^{-1} \subset \tilde{v}$. Then $\{h^{-1}(\tilde{u})\}\{h^{-1}(\tilde{u})\}^{-1} \subset h^{-1}(\tilde{v})$ and $h^{-1}(\tilde{v}) \in \mathfrak{N}_e$ (Problem 8G).]

(b) The properties below are also equivalent:

(i) $\tilde{G}$ is Catg II and h is open;

(ii) $h(v) \in \mathcal{C}_{\mathrm{II}}(\tilde{G})\ \forall \text{nbd } v \in G$;
(iii) there is a set $A \subset G$ such that $h(A) \in \mathcal{C}_{\mathrm{II}}\{h(G)\}$ and $h: A \to h(A)$ is open.

[*Hint*: For (i) $\Rightarrow$ (ii) take $A = G$ and use the fact that $\tilde{G}$ has a nbd base of open sets. If (ii) holds, G is Catg II. Moreover, for any $v \in \mathfrak{N}_e$ and $u \in \mathfrak{N}_e$ satisfying $uu^{-1} \subset v$, one has $\tilde{e} \in h(u)\{h(u)\}^{-1} \subset h(v)$ and (iii) holds. Assuming (iii), show that $\mu_h^{-1}/h(A): h(A) \to G/h^{-1}(\tilde{e})$ is continuous. By (a) above $\mu_h^{-1}: h(G) \to G/h^{-1}(\tilde{e})$ is continuous and h is open. Why does $h(A) \in \mathcal{C}_{\mathrm{II}}(h(G))$ ensure that $\tilde{G}$ is Catg II?]

(c) Using the results above, prove that a continuous homomorphism h from a Catg II group G into a topological group $\tilde{G}$ is open iff h is nearly open and $\forall v \in \mathfrak{N}_e$ there is a $u \in \mathfrak{N}_e$ contained in v such that $h(u) \subset \mathfrak{B}(\tilde{G})$.

11. Unitary Representations and Character Groups

This section, geometric in nature, extends some earlier notions and lays the foundation for the analytic considerations of Chapter V. The overall theme here, in one form or another, deals with continuous homomorphisms (called representations of degree equal dim Z) from a topological group onto subgroups of $L(Z, Z)$—the composition group of continuous linear operators on a Hilbert space Z. Our primary interest lies in finite dim representations, since they are easy to get a grip on and quickly yield a number of interesting results without leading too far astray. Another reason is that such representations constitute a traditional approach to abstract harmonic analysis on compact groups. Although our path is different, some cross pollination is provided by way of the material below and the problems that follow.

Unitary Representations

DEFINITION 11.1. A homomorphism $\underset{g \rightsquigarrow \mathcal{R}_g}{\mathcal{R}: G \to \mathfrak{S}(Z)}$ from a topological group G onto a composition group $\mathfrak{S}(Z) \subset L(Z, Z)$ is called a $\mathfrak{S}$-representation of G if it is strongly continuous (i.e., $\underset{g \rightsquigarrow \mathcal{R}_g(z)}{\mathfrak{e}_z \mathcal{R}: G \to (Z, \|\ \|)}$ is continuous $\forall z \in Z$).

Remark. Note that $\mathfrak{e}_z \mathcal{R}$ is continuous iff it is continuous at $e \in G$ since $\|\mathcal{R}_g(z) - \mathcal{R}_{g_0}(z)\| = \|\mathcal{R}_{g_0}\{\mathcal{R}_{g_0^{-1}g}(z) - z\}\| \le \|\mathcal{R}_{g_0}\|\,\|\mathcal{R}_{g_0^{-1}g}(z) - z\|$ for the $L(Z, Z)$-norm $\|\ \|$ defined in Corollary 13.25.

A mapping $\mathcal{R}: G \to L(Z, Z)$ is weakly continuous if $\underset{g \rightsquigarrow (\mathcal{R}_g(y), z)}{f_{y,z}\mathcal{R}: G \longrightarrow \mathbb{C}}$ is continuous $\forall y, z \in Z$. Strongly continuous mappings are weakly continuous.

THEOREM 11.2. A homomorphism $\mathscr{U}: G \to \mathfrak{E}(Z)$ onto a unitary subgroup $\mathfrak{E}(Z)$ of $L(Z, Z)$ is weakly continuous iff it is strongly continuous.

Proof. Strong continuity implies weak continuity by the preceding remarks. The converse utilizes $\|\mathscr{U}_g(z)-z\|^2=(\mathscr{U}_g(z)-z,\ \mathscr{U}_g(z)-z)=2\{\|z\|^2-R_e(\mathscr{U}_g(z), z)\}$. If $\mathscr{U}$ is weakly continuous and $g \to e$, then $\mathscr{U}_g \to \mathscr{U}_e = \mathbb{1}$ implies [Example 15.4(b)] that $R_e(\mathscr{U}_g(z), z)=R_e(f_{z,z}\mathscr{U}_g)\to R_e(f_{z,z}\mathbb{1})=\|z\|^2$ and so $\mathscr{U}$ is strongly continuous. ■

A subspace X of a vector space Y is $\mathscr{A}$-invariant for a collection $\mathscr{A}\subset L(Y, Y)$ if $\bigcup_{\mathscr{A}} A(X)\subset X$. Accordingly, $\mathscr{A}\subset \mathfrak{L}(Y, Y)$ is termed irreducible if $\{0\}$ and Y are the only $\mathscr{A}$-invariant subspaces of Y. Otherwise, $\mathscr{A}$ is said to be reducible.

DEFINITION 11.3. A representation $\mathscr{R}: G \to L(Z, Z)$ is irreducible (reducible) if $\{\mathscr{R}_g : g \in G\}\subset L(Z, Z)$ is irreducible (reducible).

If X is a closed, $\mathscr{R}$-invariant subspace of Z (i.e., $\bigcup_G \mathscr{R}_g(X)\subset X$), the representation $\mathscr{R}/X : \underbrace{G\to \mathfrak{E}(X)}_{g \rightsquigarrow \mathscr{R}_g/X}\subset L(X, X)$ is called the restriction of $\mathscr{R}$. Note that $\mathscr{R}/X$ may be irreducible—even though $\mathscr{R}$ is not.

Remark. $\overline{[\bigcup_{\mathfrak{A}} Y_\alpha]}$ is $\mathscr{R}$-invariant for $\mathscr{R}$-invariant subspaces $\{Y_\alpha : \alpha \in \mathfrak{A}\}\subset Z$. To be sure, $y\in\overline{[\bigcup_{\mathfrak{A}} Y_\alpha]}$ implies $y=\lim_n \sum_{i=1}^n \beta_i y_{\alpha_i}$ $(y_{\alpha_i}\in Y_{\alpha_i})$ and $\mathfrak{E}(Z)\subset L(Z, Z)$ ensures that $\mathscr{R}_g(y)=\lim_n \mathscr{R}_g\left(\sum_{i=1}^n \beta_i y_{\alpha_i}\right)\in\overline{[\bigcup_{\mathfrak{A}} Y_{\alpha_i}]}\ \forall g\in G$.

The following result is also known as Schur's lemma.

THEOREM 11.4. Let $\mathscr{R}\in\mathfrak{L}(X, X)$ and $\mathfrak{S}\in\mathfrak{L}(Y, Y)$ be irreducible operators on vector spaces X and Y, respectively. If $T\in\mathfrak{L}(X, Y)$ satisfies $\mathscr{R}T = T\mathfrak{S}$, then $T=0$ or T is bijective.

Proof. The subspace $T(X)\subset Y$ is $\mathfrak{S}$-invariant and so $T(X)=\{0\}$ or $T(X)=Y$. If $T\neq 0$, the $\mathscr{R}$-invariance of $T^{-1}(0)\subset X$ yields $T^{-1}(0)=\{0\}$. ■

COROLLARY 11.5. Let $\mathscr{A}\subset\mathfrak{L}(X, X)$ be irreducible for a finite dim vector space X. Every $A_0\in\mathfrak{L}(X, X)$ satisfying $A_0\mathscr{A}=\mathscr{A}A_0$ has the form $A_0=\lambda_{A_0}\mathbb{1}$ for some $\lambda_{A_0}\in\mathbb{C}$. In particular, $\dim X=1$ if $A_1A_2=A_2A_1\ \forall A_1, A_2\in\mathscr{A}$.

Proof. Since X is finite dim, A_0 has an eigenvalue λ_{A_0} and $(A_0-\lambda_{A_0}\mathbb{1})A = A(A_0-\lambda_{A_0}\mathbb{1})\ \forall A\in\mathscr{A}$ implies $A-\lambda_{A_0}\mathbb{1}=0$. The second assertion follows from the fact that $\mathscr{A}=\{\lambda_A\mathbb{1}: A\in\mathscr{A}\}$ is irreducible and every subspace of X is $\mathscr{A}$-invariant. ■

Example 11.6 (a) One-dim representations of a group G are obviously irreducible. If G is abelian, then irreducible finite dim representations and irreducible unitary representations are also one-dim. Both assertions hold since $\{\mathscr{R}_g: g\in G\}\subset[\mathbb{1}]$ in $L(Z, Z)$—the first case by Corollary 11.5, the second by (e.g., Sugiura [94]). A unitary representation $\mathscr{U}: G\to L(Z, Z)$ is irreducible iff $[\mathbb{1}]=\{A\in L(Z, Z): \mathscr{U}_g A = A\mathscr{U}_g\ \forall g\in G\}$.

Remark. Irreducible unitary representations on a compact T_2 group are finite dim (Naimark [69]).

(b) Each $e^{i\theta}\in S'(\mathbb{C})$ in Example 9.8 may be viewed as a unitary operator $\underset{z\rightsquigarrow ze^{i\theta}}{e^{i\theta}:\mathbb{C}\to\mathbb{C}}$. Accordingly, every irreducible unitary representation of $S'(\mathbb{C})$ is of the form $\underset{e^{i\theta}\rightsquigarrow e^{in\theta}}{\lambda_n: S'(\mathbb{C})\to S'(\mathbb{C})}$ $(n\in\mathbb{N})$. Reason: a unitary representation $\underset{e^{i\theta}\rightsquigarrow e^{i\phi(\theta)}}{\mathscr{U}: S'(\mathbb{C})\to S'(\mathbb{C})}$ satisfies $\phi(q\theta)=q\phi(\theta)$ for every rational number q. Since $\phi(\theta+2\pi)=\phi(\theta)+2n_0\pi$ for some $n_0\in\mathbb{I}$, one has $\phi(2\pi q)=2\pi n_0 q$ and the continuity of $\phi:\mathbb{R}\to\mathbb{R}$ yields $\phi(2\pi\theta)=2n_0\pi\theta\ \forall\theta\in\mathbb{R}$. In other words, $\mathscr{U}=\lambda_{n_0}$.

(c) The irreducible unitary representations of

$$\mathbb{R}=(\mathbb{R}, +; d_1) \text{ is } \{\underset{t\rightsquigarrow e^{i\theta t}}{\chi_\theta:\mathbb{R}\to S'(\mathbb{C})}: \theta\in\mathbb{R}\}.$$

The reader can verify that $\underset{t\rightsquigarrow\mathscr{U}(t)}{\mathscr{U}:\mathbb{R}\to S'(\mathbb{C})}$ is differentiable with $\mathscr{U}'(t)=\mathscr{U}'(0)\mathscr{U}(t)$. It then follows that $\alpha=\mathscr{U}'(0)$ is purely imaginary and $\mathscr{U}=\chi_{i^{-1}\alpha}$.

Remark. $\chi_{\theta_1}=\chi_{\theta_2}$ iff $\theta_1=\theta_2$ is a consequence of $\chi'_{\theta_1}(0)=\chi'_{\theta_2}(0)$.

A Hilbert space Z is said to be an (internal) direct sum of closed subspaces $\{Z_\alpha: \alpha\in\mathscr{Q}\}$ (written $Z=\sum_{\mathscr{Q}}\oplus Z_\alpha$) if $\{Z_\alpha:\alpha\in\mathscr{Q}\}$ are pairwise orthogonal and $\overline{[\bigcup_{\mathscr{Q}} Z_\alpha]}=Z$. This ensures that every $z\in Z$ has a unique representation $z=\sum_{\mathscr{Q}} z_\alpha$ $(z_\alpha\in Y_\alpha)$ and norm $\|z\|=\sqrt{\sum_{\mathscr{Q}}\|z_\alpha\|^2}$ (uniqueness of representation is equivalent to $\sum_{\mathscr{Q}} z_\alpha=0\Rightarrow z_\alpha=0\ \forall\alpha\in\mathscr{Q}$, which is certainly the case here, since $\|\sum_{\mathscr{Q}} z_\alpha\|^2=\sum_{\mathscr{Q}}\|z_\alpha\|^2$).

DEFINITION 11.7. A representation $\mathscr{R}: G\to L(Z, Z)$ is a direct sum of representations $\mathscr{R}_\alpha: G\to L(Z_\alpha, Z_\alpha)$ (written, $\mathscr{R}=\sum_{\mathscr{Q}}\oplus\mathscr{R}_\alpha$) if $Z=\sum_{\mathscr{Q}}\oplus Z_\alpha$ and $\mathscr{R}_g(z)=\sum_{\mathscr{Q}}\mathscr{R}_{\alpha_g}(z_\alpha)\ \forall g\in G$.

Remark. If Z is finite dim, each subspace $Z_\alpha \subset Z$ is closed (Corollary 13.13) and $Z = [\bigcup_{\mathfrak{A}} Z_\alpha]$.

Example 11.8. (a) If $\mathcal{U}: G \to L(Z, Z)$ is unitary and $Y \subset Z$ is $\mathcal{U}$-invariant, then $Y^\perp = \{z \in Z: (y, z) = 0\}$ and $\bar{Y}^\perp$ are $\mathcal{U}$-invariant. [By Definition 11.3⁺, since $y \in Y$ and $z \in Y^\perp$ yield $0 = (\mathcal{U}_g(y), z) = (y, \mathcal{U}_{g^{-1}}(z))$ and $\mathcal{U}_{g^{-1}}(z) \in Y^\perp \; \forall g \in G$.] Since $Z = \bar{Y} \oplus \bar{Y}^\perp$, one has $\mathcal{U} = \mathcal{U}/\bar{Y} \oplus \mathcal{U}/\bar{Y}^\perp$. Note that the unitary representations $\mathcal{U}/\bar{Y}$ and $\mathcal{U}/\bar{Y}^\perp$ may be reducible and $\bar{Y}$ $\{\bar{Y}^\perp\}$ may be further decomposed into a subdirect sum.

(b) If $\mathcal{U}$ is non unitary, $\bar{Y}^\perp$ need not be $\mathcal{U}$-invariant and $\mathcal{U} = \mathcal{U}/\bar{Y} \oplus \mathcal{U}/\bar{Y}^\perp$ need not follow. For instance, consider $\mathcal{R}: G = \mathbb{C} \to \mathfrak{E}(Z) \subset L(Z, Z)$,

$$z \rightsquigarrow \mathcal{R}_z = \begin{pmatrix} 1 & z \\ 0 & 1 \end{pmatrix}$$

where $Z = \mathbb{C}^2$ carries the standard inner product $((\alpha_1, \beta_1), (\alpha_2, \beta_2)) = \alpha_1 \bar{\alpha}_2 + \beta_1 \bar{\beta}_2$. Although $Y = (\mathbb{C}, 0)$ is $\mathcal{R}$-invariant, $Y^\perp = (0, \mathbb{C})$ is not. In fact, Y is the only one dim, $\mathcal{R}$-invariant subspace of Z and so $\mathcal{R}$ has no direct sum decomposition.

THEOREM 11.9. Every finite dim unitary representation $\mathcal{U}$ of G is a (finite) direct sum of irreducible unitary representations of G.

Proof. Use induction on dim Z. Assume the theorem true for n-dim representations of G and let Z be $n+1$ dim. If $\mathcal{U}$ is reducible, some m-dim $(1 \leq m \leq n)$ subspace $Y \subset Z$ is $\mathcal{U}$-invariant and $\mathcal{U} = \mathcal{U}/Y \oplus \mathcal{U}/Y^\perp$. Since $1 \leq \dim Y^\perp \leq n$, it follows (by induction hypothesis) that $\mathcal{U}/Y$ and $\mathcal{U}/Y^\perp$; therefore $\mathcal{U}$ is a direct sum of irreducible unitary representations of G. ∎

Since infinite dim Hilbert spaces have uncountable dim (Problem 13G), inductive arguments frequently give way to some form of ordering principle. This is easily illustrated once we introduce

DEFINITION 11.10. A representation $\mathcal{R}: G \to L(Z, Z)$ is said to be cyclic if $\overline{[\mathcal{R}_g(z_0): g \in G]} = Z$ for some $z_0 \in Z$ (one calls z_0 a cyclic vector for $\mathcal{R}$).

THEOREM 11.11. Every unitary representation of G is a direct sum of cyclic unitary representations of G.

Proof. Fix any $z_1 \neq 0$ in Z and let $Z_1 = [\mathcal{U}_g(z_1): g \in G]$. Then $\bar{Z}_1$ is $\mathcal{U}$-invariant in Z and $\mathcal{U}/\bar{Z}_1$ is cyclic. If $\bar{Z}_1 = Z$, then $\mathcal{U}$ is cyclic and our proof is complete. Otherwise, $\bar{Z}_1 \neq Z$ and there is some $z_2 \neq 0$ in $\bar{Z}_1^\perp$. Since $\bar{Z}_1^\perp$ is $\mathcal{U}$-invariant, $Z_2 = [\mathcal{U}_g(z_2): g \in G] \subset \bar{Z}_1^\perp$ and $\mathcal{U}/\bar{Z}_2$ is cyclic. As above, $\bar{Z}_2 = \bar{Z}_1^{\perp}$ implies $\mathcal{U} = \mathcal{U}/\bar{Z}_1 \oplus \mathcal{U}/\bar{Z}_1^{\perp}$ is a cyclic decomposition and our proof is complete. Assume that the proof is not complete and let $\mathcal{S}$ be the collection of all systems $S_{\mathcal{A}} = \{Z_\alpha : \alpha \in \mathcal{A}\}$, where each system $S_{\mathcal{A}}$ consists of pairwise orthogonal, closed, $\mathcal{U}$-invariant subspaces of Z, on each of which the restriction of $\mathcal{U}$ is cyclic. Clearly, $\mathcal{S}$ contains $\{\bar{Z}_1, \bar{Z}_2\}$ as one such system. Partially order $\mathcal{S}$ by inclusion. Then every totally ordered subset $\{S_{\mathcal{A}} : \mathcal{A} \in a\}$ has an upper bound $\bigcup_{\mathcal{A}} S_{\mathcal{A}} \in \mathcal{S}$ and (by Zorn's lemma) $\mathcal{S}$ has a maximal member S_0. This, in fact, requires that $Z = \sum_{S_0} \oplus Z_\alpha$ and $\mathcal{U} = \sum_{S_0} \oplus \mathcal{U}/Z_\alpha$. Otherwise, there is a $z_0 \neq 0$ in $\left\{\sum_{S_0} \oplus Z_\alpha\right\}^{\perp}$ and $\mathcal{U}/\bar{Z}_0$ is cyclic for $Z_0 = [\mathcal{U}_g(z_0) : g \in G]$. Accordingly, $S_0 \cup \{\bar{Z}_0\} \in \mathcal{S}$ contains S_0 and this contradicts the maximality of $S_0 \in \mathcal{S}$. ■

Example 11.12. Let $Z = \mathbb{C}^2$ in Example 11.8(b) carry the orthonormal basis $\{e_1 = (1, 0), e_2 = (0, 1)\}$. The two-dim unitary representation $\mathcal{U} : \underset{e^{i\theta} \rightsquigarrow \mathcal{U}_\theta}{S'(\mathbb{C}) \to \mathfrak{E}(Z)}$ with $\mathcal{U}_\theta(e_1) = (\alpha_\theta, 0)$ and $\mathcal{U}_\theta(e_2) = (0, \beta_\theta)$ is reducible by Example 11.6(a). To illustrate Theorem 11.11 let $Z_1 = [\mathcal{U}_\theta(e_1) : \theta \in [0, 2\pi)] = [(e_\theta, 0) : \theta \in [0, 2\pi)] = (\mathbb{C}, 0)$ so that $Z_1^{\perp} = (0, \mathbb{C})$. Then $\mathcal{U} = \mathcal{U}/Z_1 \oplus \mathcal{U}/Z_1^{\perp}$ for $\mathcal{U}/Z_1$ $(e^{i\theta}) \rightsquigarrow \mathcal{U}_\theta(z_1, 0) = (z_1\alpha_\theta, 0)$ and $\mathcal{U}/Z_1^{\perp}$ $(e^{i\theta}) \rightsquigarrow \mathcal{U}_\theta(0, z_2) = (0, z_2\beta_\theta)$. The unitary representations $\mathcal{U}/Z_1$ and $\mathcal{U}/Z_1^{\perp}$ have respective cyclic vectors e_1 and e_2 in Z.

Character Groups

The focus now is on one-dim unitary representations. Specifically, G is a topological group and $S'(\mathbb{C})$ is the compact metrizible group of Example 9.8.

DEFINITION 11.13. A continuous homomorphism $\chi : G \to S'(\mathbb{C})$ is called a character of G. The abelian group G' of all characters of G under pointwise multiplication $(\chi_1\chi_2)(x) = \chi_1(x)\chi_2(x)\ \forall x \in G$ is called the character group (or dual) of G'. Note that χ_G is the identity element in G' and $(1/\chi)(x) = 1/\chi(x)$ is the inverse of $\chi \in G'$. (We use $1/\chi$ to avoid possible confusion with the inverse mapping χ^{-1} if it exists.)

Notation: It will be useful to let $\langle x, \chi \rangle$ denote $\chi(x)$ for $(x, \chi) \in G \times G'$. In this context one can easily verify that $\langle x^{-1}, \chi \rangle = \langle x, \chi \rangle^{-1} = \langle x, 1/\chi \rangle = \overline{\langle x, \chi \rangle}\ \forall (x, \chi) \in G \times G'$.

A variety of topologies (uniformities) can be defined on $G' \subset C(G; S'(\mathbb{C}))$. Since our main interest lies with G'_κ (Example 7.8), we let $S_\varepsilon = \{z \in S'(\mathbb{C}) : |z-1| < \varepsilon\}$ and define $W_{K,S_\varepsilon} = \{\chi \in G' : \chi(x) \in S_\varepsilon \ \forall x \in K\}$ in order to first establish

THEOREM 11.14. G'_κ is a topological group,† which is T_2 if G is T_2.

Proof. $\mathfrak{N}_{\chi_G} = \{W_{K,S_\varepsilon} : \text{compact } K \subset G \text{ and } \varepsilon > 0\}$ is a filter base in G', since $W_{K_1 \cup K_2, S_{\varepsilon_1+\varepsilon_2}} \subset W_{K_1,S_{\varepsilon_1}} \cap W_{K_2,S_{\varepsilon_2}}$ $\forall$ compact $K_1, K_2 \subset G$ and $\varepsilon_1, \varepsilon_2 > 0$. If $W_{K,S_\varepsilon} \in \mathfrak{N}_{\chi_G}$ and $\delta < \varepsilon/2$, then $\chi_1, \chi_2 \in W_{K,S_\delta}$ implies $|\chi_1(x)\chi_2(x) - 1| \leqq |\chi_1(x) - 1|\,|\chi_2(x)| + |\chi_1(x)|\,|\chi_2(x) - 1| \leqq 2\delta < \varepsilon \ \forall x \in K$. Thus $\{W_{K,S_\delta}\}^2 \subset W_{K,S_\varepsilon}$ and property (G_1) of Theorem 8.11 holds. From $|\chi(x) - 1| = |1 - 1/\chi(x)|$ also follows $\{W_{K,S_\varepsilon}\}^{-1} = W_{K,S_\varepsilon}$ and property (G_2); and (G_3) holds because $\chi W_{K,S_\varepsilon} 1/\chi = W_{K,S_\varepsilon} \ \forall \chi \in G'$ and $W_{K,S_\varepsilon} \in \mathfrak{N}_{\chi_G}$. Therefore $\mathfrak{N}_{\chi_G}$ is a $\mathfrak{T}_\kappa = \mathfrak{T}_{co} \in T(G')$ nbd base at $\chi_G \in G'$. If G is T_2, then $\mathfrak{T}_p \subset \mathfrak{T}_\kappa$ and so $\mathfrak{T}_\kappa$ is T_2. ■

Remark. We mention here (Corollaries 27.12 and 29.9) that each $W_{K,S_\varepsilon} \in \mathfrak{N}_{\chi_G}$ is $\mathfrak{T}_{co} = \mathfrak{T}_\kappa$ open in G', and $W_{K',S_\varepsilon} = \{x \in G : \langle x, \chi\rangle \in S_\varepsilon \ \forall \chi \in K'\}$ is G-open $\forall$ compact set $K' \subset G'_\kappa$.

Our emphasis on $\mathfrak{T}_\kappa \in T(G')$ is not whimsical. It is strongly tied (by Ascoli's theorem) to locally compact T_2 properties of groups, the manner in which is made clear once we define $\mathscr{Q}(S_\varepsilon) = \sup\{\arg|z| : z \in \bar{S}_\varepsilon\}$ and prove

THEOREM 11.15. For a locally compact, T_2 group G:

(i) G' is locally compact and T_2.
(ii) G' is 2° if G is 2°.
(iii) G' is compact {discrete} if G is discrete {compact}.

Proof. The proof of (i) is based on Theorem 8.15(iv) once we show that $W_{\bar{u},S_\varepsilon} \in \mathfrak{N}_{\chi_G}$ is $\mathfrak{T}_\kappa$-relatively compact for relatively compact $u \in \mathfrak{N}_e$. Begin by observing that $W_{\bar{u},S_\varepsilon}$ is equicontinuous (otherwise, for any $\varepsilon_0 > 0$ choose $n \in \mathbb{N}$ such that $\mathscr{Q}(S_\varepsilon) < n\mathscr{Q}(S_{\varepsilon_0})$ and let $v \in \mathfrak{N}$ satisfy $v^n \subset u$. Then $|\arg \chi(x)| \gneqq \mathscr{Q}(S_{\varepsilon_0})$ for some $\chi \in W_{\bar{u},S_\varepsilon}$ and $x \in v$, and the contradiction $|\arg \chi(x^n)| = |\arg\{\chi(x)\}^n| \geq n\mathscr{Q}(S_{\varepsilon_0}) > \mathscr{Q}(S_\varepsilon)$ for $x^n \in u$ follows). Now $\overline{W_{\bar{u},S_\varepsilon}}^{\mathfrak{T}_\kappa}$ is equicontinuous (Theorem 7.15) and [since $S'(\mathbb{C})$ is compact] $\overline{W_{\bar{u},S_\varepsilon}(x)}^{\mathfrak{T}_\kappa} \subset S'(\mathbb{C})$ is relatively compact $\forall x \in G$. Therefore (Theorem 7.18) $\overline{W_{\bar{u},S_\varepsilon}}^{\mathfrak{T}_\kappa}$ is compact and the proof of (i) is complete.

† It is assumed henceforth that G' means G'_κ.

(ii) Let $\mathfrak{N}=\{u_n : n\in\mathbb{N}\}$ be a G-nbd system of relatively compact open sets. Then $\{W_{\bigcup_{i=1}^m u_i, S_{1/n}} : m, n\in\mathbb{N}\}$ is a G'-nbd base at χ_G (clearly, $W_{\bigcup_{i=1}^m u_i, S_{1/n}} \subset W_{K,S_\varepsilon}$ for $1/n<\varepsilon$ and cover $\{u_1\cdots u_m\}$ of K) and G' is metrizible. It suffices therefore to prove G' separable. Let $\mathfrak{N}'=\{v_n : n\in\mathbb{N}\}$ be an $S'(\mathbb{C})$-open nbd base and let Λ denote the collection of all finite index sets $\gamma=\{n_1, n_2, \ldots, n_k\}\subset\mathbb{N}$. From each nonempty $V_\gamma=\bigcap_\gamma W_{\bar{u}_n, v_n}\in\mathfrak{N}_{\chi_G}$ choose a representative χ_γ and let $M=\{\chi_\gamma : \gamma\in\Lambda\}$. Surely M is countable. We show it to be dense in G' by demonstrating that $\chi W_{K,v}\cap M\neq\varnothing$ $\forall$ compact $K\subset G$, open set $v(1)$ in $S'(\mathbb{C})$ and $\chi\in G'$. For each $x\in K$ there is a nbd $v_{n(x)}\in\mathfrak{N}'$ of $\chi(x)$ such that $v_{n(x)}\subset\chi(x)v$ and (since χ is continuous) some $u_{n(x)}\in\mathfrak{N}$ such that $\chi\{u_{n(x)}\}\subset v_{n(x)}$. Since $\{u_{n(x)} : x\in K\}$ covers K and K is compact, $K\subset\bigcup_{i=1}^m u_{n(x_i)}$ for some $\gamma_K=\{n(x_1), \ldots, n(x_m)\}\in\Lambda$. Now $\chi\in V_{\gamma_K}\neq\phi$ implies that some $\chi_{\gamma_K}\in V_{\gamma_K}\cap M$, and since $\chi_{\gamma_K}(x)\in v_{n(x)}\subset\chi(x)v$ $\forall x\in K$ we have $\chi_{\gamma_K}\in\chi W_{K,v}\cap M$ as asserted.

(iii) If G is compact, then $\chi_G=W_{G,S_\varepsilon}\in\mathfrak{N}_{\chi_G}$ and so G' is discrete. (If $\chi\neq\chi_G$ belongs to W_{G,S_ε}, then $\chi(x)\neq 1$ and $|\arg\chi(x^n)|>\mathfrak{a}(S_\varepsilon)$ for large enough $n\in\mathbb{N}$.) If G is discrete, then $W_{\{e\},S_\varepsilon}=G'$ and the compactness of G' follows from that of $\overline{W_{\{e\},S_\varepsilon}}^{\mathfrak{T}_\kappa}=G'$. ■

THEOREM 11.16. Let G_1, G_2 be topological groups. Then $\eta: G_1'\times G_2'\to(G_1\times G_2)'$, $(\chi_1,\chi_2)\rightsquigarrow\chi_1\times\chi_2$ with $(\chi_1\times\chi_2)(x_1, x_2)=\chi_1(x_1)\ \chi_2(x_2)$ $\forall(x_1, x_2)\in G_1\times G_2$ is an onto topological isomorphism.

Proof. Clearly, η is a homomorphism since $\eta\{(\chi_1,\chi_2)(\chi_1',\chi_2')\}=\eta(\chi_1\chi_1',\chi_2\chi_2')=\chi_1\chi_1'\times\chi_2\chi_2'=\chi_1\chi_2\times\chi_1'\chi_2'=\eta(\chi_1,\chi_2)\ \eta(\chi_1',\chi_2')$. If $\chi_1\times\chi_2=\chi_1'\times\chi_2'$, then $\chi_1(x_1)=(\chi_1\times\chi_2)(x_1, e_2)=(\chi_1'\times\chi_2')(x_1, e_2)=\chi_1'(x_1)$ $\forall x_1\in G$ and $\chi_1=\chi_1'$. Similarly, $\chi_2=\chi_2'$ and so η is injective. It is also surjective, since every $\chi\in(G_1\times G_2)'$ is the image of (χ_1,χ_2), where $\chi_1=\chi(\ , e_2)$ and $\chi_2=\chi(e_1,\ )$. Finally, η is continuous and open because $W_{K_1\cup e_1\times K_2\cup e_2, S_\varepsilon}\subset\eta\{W_{K_1,S_\varepsilon}\times W_{K_2,S_\varepsilon}\}\subset W_{K_1\times K_2, S_{2\varepsilon}}$ $\forall\varepsilon>0$ and compact $K_i\subset G_i$ $(i=1,2)$. ■

Moving on to the dual of a quotient group requires

DEFINITION 11.17. Let H be a subset of a topological group G and let Φ be a subset of G'. Then $H^\circ=\{\chi\in G' : \langle x,\chi\rangle=1\ \forall x\in H\}$ is called the annihilator of H in G' and $\Phi^{\bigcirc}=\{x \in G:\langle x, \chi\rangle=1\ \forall \chi \in \Phi\}$ is the annihilator of Φ in G.

Remark. $H^{\bigcirc}=\bigcap_H \langle x, \chi^{-1}\rangle(1)=\bigcap_H \mathrm{e}_x^{-1}(1)$ is a closed subgroup of G' (by Theorem 7.2(ii), since $\mathfrak{T}_p \subset \mathfrak{T}_\kappa$) and $\Phi^{\bigcirc}=\bigcap_\Phi \chi^{-1}(1)$ is a closed subgroup of G.

Let $\nu: G \to G/H$, where H is a normal subgroup of a topological group G. If $\chi \in H^{\bigcirc}$, then $\chi(x_1)=\chi(x_2)$ for $Hx_1=Hx_2$ and (Example 9.7) $\mu_\chi \in (G/H)'$. This sets the stage for

THEOREM 11.18. Let H be a closed normal subgroup of a locally compact T_2 group G. Then $\zeta: H^{\bigcirc} \to (G/H)'$ is an onto topological isomorphism.

Proof. The isomorphism ζ is clearly surjective, since $\zeta(\mu\nu)=\mu$ with $\mu\nu \in H^{\bigcirc}$ for every $\mu \in (G/H)'$. Since ν is a continuous open surjection (Theorem 9.6), $\dot{K} \subset G/H$ is compact iff $\dot{K}=\nu(K)$ for some compact set $K \subset G$. Reason: $\dot{K} \subset \bigcup_{i=1}^n \nu(u_i)$ for relatively compact open sets $\{u_1, \ldots, u_n\} \subset G$ (Theorem 8.15(iv)). Furthermore, G/H is T_2 so that $\dot{K} \subset G/H$ is closed and $K=\nu^{-1}(\dot{K}) \cap \bigcup_{i=1}^n \bar{u}_i \subset G$ meets our needs. The continuity {openness} of ζ is now a simple consequence of the fact that $\zeta(W_{K,S_\varepsilon} \cap H^{\bigcirc})=W_{\dot{K},S_\varepsilon}\ \forall \varepsilon>0$ and compact $K \subset G$ {compact $\dot{K} \subset G/H$}. ■

Example 11.19. (a) Let $G=\{g, g^2, \ldots, g^{n-1}, g^n=e\}$ be a (discrete) nth order cyclic group. If $\chi \in G'$, then $\chi(g)=e^{i\theta(g)}$ and $\chi(g^k)=e^{ik\theta(g)}$. Also, $1=\chi(g^n)=e^{in\theta(g)}$ implies $\theta(g)=2\pi m/n$ for some $m \in \mathbb{I}$. Therefore $G'=\{\chi_m: g^k \rightsquigarrow e^{(2mk\pi i)/n}; m \in \mathbb{I} \cap [1, n]\}$. In fact, $G'=G$, since $\chi_m \rightsquigarrow g^m$ is a topological isomorphism.

(b) If $G=\{e=g^0, g, g^2, \ldots\}$ is a (discrete) infinite cyclic group, every $\chi \in G'$ is determined by its value $\chi(g)=e^{i\theta} \in S'(\mathbb{C})$. It is easily verified that $G'=\{\chi_\theta(g^n)=e^{in\theta}: \theta \in [0, 2\pi)\}$; therefore $\mathbb{I}'=\{\chi_\theta: n \rightsquigarrow e^{in\theta}; \theta \in [0, 2\pi)\}$ is compact. The compactness of $\mathbb{I}'$ also follows from $\mathbb{I}'=S'(\mathbb{C})$, since $\eta(\chi_\theta)=e^{i\theta}$ is a topological isomorphism between $\mathbb{I}'$ and $S'(\mathbb{C})$. (To be sure, η is continuous, since $\eta\{W_{\mathbb{I} \cap [-2,2], S_\alpha}\} \subset S_\varepsilon\ \forall S_\varepsilon$ with $\alpha=\mathcal{Q}(S_\varepsilon)$, and η is open by virtue of Theorems 11.15, 8.17 and 10.8.)

(c) It follows from (b) that $\mathbb{I}^{\bigcirc}=\mathbb{I}$, since $\exp\{in\theta\}=1$ iff $\theta=2\pi m$ for some $m \in \mathbb{I}$, and $\chi_{2\pi m} \rightsquigarrow m$ is a topological isomorphism. From Theorem 11.18 also follows $\mathbb{I}=\mathbb{I}^{\bigcirc}=(\mathbb{R}/\mathbb{I})'=\{S'(\mathbb{C})\}'$.

(d) $\{S'(\mathbb{C})\}' = \{\chi_n : e^{i\theta} \rightsquigarrow e^{in\theta}; n \in \mathbb{I}\}$ can be established by (c) above or by (e) below [since $\chi_\theta(e^{i\alpha}) = \chi_\theta[e^{i(\alpha+2\pi)}]$ implies $\theta \in \mathbb{I}$]. Combining (b), (c) also yields $\mathbb{I}'' = \{S'(\mathbb{C})\}' = \mathbb{I}$ and $\{S'(\mathbb{C})\}'' = \mathbb{I}' = S'(\mathbb{C})$, where $G'' = ((G')', \mathfrak{T}_\kappa)$ is called the second dual of G.

(e) $\mathbb{R}' = \{\chi_\theta : t \rightsquigarrow e^{it}; \theta \in \mathbb{R}\}$ and $\mathbb{R} = \mathbb{R}'$, since $\theta \rightsquigarrow \chi_\theta$ is a topological isomorphism. Confirmation that $\chi \in \mathbb{R}'$ of the above form uses the fact that $\int_0^\varepsilon \chi(t)\,dt = \delta > 0$ for some $\varepsilon > 0$. Specifically, $\chi(t_0) = (1/\delta)\chi(t_0) \int_0^\varepsilon \chi(t)\,dt = 1/\delta \int_{t_0}^{t_0+\varepsilon} \chi(t)\,dt$ has a continuous derivative and $\chi'(t+t_0) = \chi'(t)\,\chi(t_0)$ (with $\chi'(t_0) = \chi'(0)\,\chi(t_0)$ and $|\chi(t_0)| = 1$) yields $\chi = \chi_{-i\chi'(0)}$.

(f) An application of Theorem 11.16 to (e) above reveals that $\{\mathbb{R}^n\}' = \{\chi_\theta : (t_1, \ldots, t_n) \rightsquigarrow e^{i(t_1\theta_1 + \cdots + t_n\theta_n)}; \theta = (\theta_1 \ldots \theta_n) \in \mathbb{R}^n\}$ and $\{\mathbb{R}^n\}' = \prod_n \mathbb{R}'$.

Example 11.19(c) raises the question: Does $G'' = G$ for every locally compact T_2 abelian (LCT$_2$A) group G? Our answer is affirmative. Since every $\xi_x : G' \to S'(\mathbb{C})$, $\chi \rightsquigarrow \chi(x)$ $(x \in G)$ belongs to G'', the continuous homomorphism $\delta : G \dot{\to} G''$, $x \rightsquigarrow \xi_x$ emerges as a logical candidate for investigation; but this, in turn, brings up another issue. Specifically, there exist [Problem 11A(c)] T_2 abelian groups G having $G' = \{\chi_G\}$. One therefore wonders which topological groups have injective or nontrivial duals. The answer, to be proved later (Corollary 29.10) is

THEOREM 11.20. (Gelfand–Raikov). The dual G' of an LCT$_2$A group G is injective (i.e., $\chi(x) = 1\ \forall \chi \in G'$ implies $x = e$).

Theorem 11.20 ensures that δ is injective and takes us one step closer to proving (cf Theorem 29.17)

THEOREM 11.21. (Pontryagin duality theorem). If G is an LCT$_2$A group, then $\delta : G \to G''$ is an onto topological isomorphism.

The most immediate consequences of Theorems 11.20 and 11.21 are included below; others are given in Section 29.

COROLLARY 11.22. An LCT$_2$A group G is compact {discrete} iff G' is discrete {compact}.

Proof. Write $G = G'' = (G')'$ and apply Theorem 11.21 to G'. ■

COROLLARY 11.23. If H is a closed subgroup of an LCT$_2$A group G, then $H = H^{\circ\circ}$ and $H' = G'/H^{\circ}$.

Proof. If $x_0 \in H$, then $\xi_{x_0}(\chi) = 1\ \forall \chi \in H^{\circ}$ so that $\xi_{x_0} \in H^{\circ\circ} = (H^{\circ})^{\circ}$ and $H \subset H^{\circ\circ}$. The reverse inclusion utilizes Theorem 12.6. If $x_0 \notin H$, then $Hx_0 \neq H$ and some $\mu_\chi \in (G/H)'$ satisfies $1 \neq \mu_\chi(Hx_0) = \chi(x_0) = \xi_{x_0}(\chi)$. Thus $\xi_{x_0} \notin H^{\circ\circ}$ and $H^{\circ\circ} \subset H$. The second assertion follows from $(G'/H^{\circ})' = H^{\circ\circ} = H$ and Theorem 11.21. ■

COROLLARY 11.24. Every $\chi \in H'$ has an extension $\tilde{\chi} \in G'$. If $x \notin H$, then $\tilde{\chi}$ may be chosen such that $\tilde{\chi}(x) \neq 1$.

Proof. If $\chi \in H' = G/H^{\circ}$, then $\chi = H^{\circ}\tilde{\chi}$ for some $\tilde{\chi} \in G'$ and $\chi(x) = \tilde{\chi}(x)\ \forall x \in H$. If $\tilde{\chi}(x) = 1$ for $x \in G - H$, then $\chi_0(x) \neq 1$ for some $\chi_0 \in H^{\circ}$ and $\tilde{\tilde{\chi}} = \tilde{\chi}\chi_0 \in G'$ is an extension of χ such that $\tilde{\tilde{\chi}}(x) \neq 1$. ■

Concluding Remarks

1. Duality theory is primarily concerned with LCT$_2$A groups, despite the fact that commutativity has been ignored almost everywhere in this section. This is easily explained. Commutativity plays no role until we arrive at Theorems 11.20 and 11.21. Since G', therefore G'', is abelian, we see that Theorem 11.21 fails for nonabelian groups. Another argument in favor of commutativity can be advanced. Specifically, the dual of every locally compact T_2 group is the dual of an LCT$_2$A group (Problem 11F) so that nothing is lost in restricting one's attention to LCT$_2$A groups. Such is the case in Chapter V.

2. Why has $S'(\mathbb{C})$ been assigned the special role it plays in defining dual groups? The answer again relates to Theorem 11.21. Suppose G is an LCT$_2$A group: let $\tilde{S}'(\mathbb{C})$ denote the group of continuous homomorphisms from $S'(\mathbb{C})$ to G and let $\tilde{\tilde{S}}'(\mathbb{C})$ be the group of continuous homomorphisms from $\tilde{S}'(\mathbb{C})$ to G. Then (Pontryagin [80], Hewitt and Ross [41]) $\tilde{\tilde{S}}'(\mathbb{C}) = S'(\mathbb{C})$ iff $S'(\mathbb{C}) = G$ (equality here means topological isomorphism). Thus $S'(\mathbb{C})$ is the unique (up to a topological isomorphism) LCTA$_2$ group for which dual groups can be defined, subject to the condition that Theorem 11.21 holds.

PROBLEMS

11A. Groups with No Finite Dim, Irreducible, Unitary Representations

(Hewitt and Ross [41]). Let G be a topological group and let $G_0 = \{g \in G : \mathcal{R}_g = \mathcal{R}_e\ \forall$ finite dim, irreducible, unitary representation $\mathcal{R}$ of $G\}$.

(a) Why can the word "irreducible" be omitted in the above definition of G_0?

(b) Prove that G_0 is a closed invariant subgroup of G. [*Hint*: $G_0 = \bigcap \{g \in G : \mathcal{R}_g = \mathcal{R}_e \ \forall$ finite dim unitary respresentation $\mathcal{R}$ of $G\}$.]

(c) Suppose H is an infinite, invariant subgroup of G for which there are finitely many conjugacy classes $\{e\}, H_1, \ldots, H_m$ of H relative to inner automorphisms (i.e., $H_j = g_j H g_j^{-1}$ for some $g_j \in G$). Assume, moreover (Problem 8K), that $G = gp(H)$ and $H = gp(H_j)\ \forall j = 1, 2, \ldots, m$. Then the trivial homomorphism is the only homomorphism of G into a compact T_2 group. In particular, $G = G_0$. [*Hint*: Assume that φ is a homomorphism of G into a compact T_2 group $\tilde{G}$. If $\varphi(H_j) = \{\tilde{e}\}$ for some j, then $\varphi(G) = \{\tilde{e}\}$. Now prove that $\varphi(H_j) \cap \tilde{W} \neq \varnothing\ \forall \tilde{W} \in \mathfrak{N}_{\tilde{e}} \Rightarrow \varphi(H_j) = \{\tilde{e}\}$. Use this to show that $\varphi(H_j) \neq \{\tilde{e}\}\ \forall j = 1, 2, \ldots, n$ implies that $\varphi(H)$ is an infinite discrete subgroup of $\tilde{G}$, which is impossible.]

11B. Characters of a finite Dim Representation

Let $\{e_i : 1 \leq i \leq n\}$ be a basis for a vector space Z and let $\{f_j : 1 \leq j \leq n\}$ be the dual basis satisfying $f_j(r_i) = \delta_{i,j}$. For $T \in L(Z, Z)$ define $\operatorname{tr}(T) = \sum_{i=1}^{n} f_i\{T(e_i)\}$.

(a) Show that $\operatorname{tr}: L(Z, Z) \to \mathbb{C}$ is linear and satisfies

(i) $\operatorname{tr}(T_1 T_2) = \operatorname{tr}(T_2 T_1)$;

(ii) $\operatorname{tr}(STS^{-1}) = \operatorname{tr}(T)\ \forall$ isomorphism $S \in L(Z, Z)$;

(iii) $\operatorname{tr}(T) = \sum_{i=1}^{n} (T(e_i), e_i)$ and $\operatorname{tr}(T^*) = \overline{\operatorname{tr}(T)}$ if Z is a Hilbert space.

(b) The character of a representation $\mathcal{R}: G \to L(Z, Z)$ is defined as $\chi_{\mathcal{R}}: G \underset{g \rightsquigarrow \operatorname{tr}(\mathcal{R}_g)}{\longrightarrow} \mathbb{C}$. Why is $\chi_{\mathcal{R}}$ continuous? If $\mathcal{R}$ is unitary, $\chi_{\mathcal{R}}(x^{-1}) = \overline{\chi_{\mathcal{R}}(x)}\ \forall x \in G$.

(c) Every $\chi \in G'$ is the character of a one-dim representation; namely, $\mathcal{R} = \chi$. The character of a representation of a compact T_2 group G belongs to G' iff $\dim \mathcal{R} = 1$.

(d) Assume that G is compact T_2. Then $\chi_{\mathcal{R}_1} = \chi_{\mathcal{R}_2}$ iff $\mathcal{R}_1$ and $\mathcal{R}_2$ are equivalent representations of G. [*Hint*: Use (a) and Problem 11A(b).]

11C. Unitary Representations and Positive Definite Functions

A complex-valued function φ on a group G is positive definite if $\sum_{j,k=1}^{n} \alpha_j \bar{\alpha}_k \varphi(g_j g_k^{-1}) \geq 0\ \forall g_1, \ldots, g_n \in G$ and $\{\alpha_1 \cdots \alpha_n\} \in \mathbb{C}$. The continuous positive definite functions on a T_2 group G are denoted $P(G)$.

(a) If $\mathcal{U}: G \to L(Z, Z)$ is a unitary representation of G, then $\varphi: G \underset{g \rightsquigarrow (\mathcal{U}_g(z), z)}{\longrightarrow} \mathbb{C}$ belong to $P(G)\ \forall z \neq 0$ in Z. [*Hint*: $|\varphi(g) - \varphi(e)| = |(\{\mathcal{U}_g - \mathbb{1}\}(z), z)| \leq \|\mathcal{U}_g(z) - z\| \|z\|$ and the strong continuity of $\mathcal{U}$ guarantees the continuity of φ at $e \in G$.]

(b) Every $\varphi \not\equiv 0$ in $P(G)$ satisfies $\varphi(g) = (\mathscr{U}_g(z), z)$ for some unitary representation $\mathscr{U}$ and $z \neq 0$ in Z. [*Hint*: Let $\mathscr{Y} = [\varphi_g : g \in G] \subset \mathfrak{F}(G; \mathbb{C})$ and define $\langle f, F \rangle = \sum_{j,k} \alpha_j \bar{\beta}_k \varphi(g_i g_k^{-1})$ for $f = \sum_{j=1}^{n} \alpha_j \varphi_{g_j}$ and $F = \sum_{k=1}^{m} \beta_k \varphi_k$ in $\mathscr{Y}$. Then (Theorem 28.2) $\langle\, ,\rangle$ is a sesquilinear functional on $\mathscr{Y}$, and (Appendix A.3) $\mathscr{X} = \{f \in \mathscr{Y} : \langle f, f \rangle = 0\} = \{f \in \mathscr{Y} : \langle f, F \rangle = 0\ \forall F \in \mathscr{Y}\}$ is a vector subspace of $\mathscr{Y}$ whose quotient space $\mathscr{Z} = \mathscr{Y}/\mathscr{X}$ carries the inner product $(\mathscr{X} + f, \mathscr{X} + F) = \langle f, F \rangle$. Let Z be the Hilbert space completion of $\mathscr{Z}$. Then each $g \in G$ determines a linear operator $\underset{f \rightsquigarrow f_g}{t_g : X \to X}$ with $t_g(\mathscr{X}) \subset \mathscr{X}$, and each t_g determines an isometry $\underset{\mathscr{X}+f \rightsquigarrow \mathscr{X}+f_g}{T_g : \mathscr{Z} \longrightarrow \mathscr{Z}}$ which can be uniquely extended to an isometry $\mathscr{U}_g \in L(Z, Z)$. Show that $\underset{g \rightsquigarrow \mathscr{U}_g}{\mathscr{U} : G \to L(Z, Z)}$ and $z = \mathscr{X} + \varphi_e$ meet our requirements.]

11D. Annihilator Groups

Prove each of the following assertions:

(i) $H^{\circ} = \{\chi \in G' : {}_x\chi = \chi \text{ on } G\ \forall x \in H\}$. [*Hint*: Use Theorem 11.18.]

(ii) (Notation as in Problem 8K) $\overline{g(H_1 \cup H_2)}^{\circ} = (H_1 \cup H_2)^{\circ} = H_1^{\circ} \cap H_2^{\circ} = \overline{g(H_1)}^{\circ} \cap \overline{g(H_2)}^{\circ}\ \forall$ pair of subsets $H_1, H_2 \subset G$.

(iii) A group $H \subset G$ is open iff $H^{\circ} \subset G'$ is compact. [*Hint*: Use Theorems 9.6 and 11.18.]

11E Dual Pairs

A dual pair $\langle H, K \rangle$ is a system consisting of two topological groups H, K and a multiplication $\underset{(h,k) \rightsquigarrow hk = kh}{H \times K \to S'(\mathbb{C})}$ satisfying:

1. $(h_1 h_2)k = (h_1 k)(h_2 k)$ and $h(k_1 k_2) = (hk_1)(hk_2)\ \forall h, h_i \in H$ and $k, k_i \in K$.
2. $h_n \to h \in H$ implies $h_n k \to hk\ \forall k \in K$, and $k_n \to k$ implies $hk_n \to hk\ \forall h \in H$.

(a) If $\langle H, K \rangle$ is a dual pair, $H \subset K'$ and $K \subset H'$. [*Hint*: Interpret $\underset{h \rightsquigarrow hk}{k : H \to S'(\mathbb{C})}$ and $\underset{k \rightsquigarrow hk}{h : K \to S'(\mathbb{C})}$.]

(b) If $H_1 \subset H$ and $K_1 \subset K$, then $H_1^{\circ} = \{k \in K : hk = 1\ \forall h \in H_1\}$ and $K_1^{\circ} = \{h \in H : hk = 1\ \forall k \in K_1\}$ are closed subgroups of K and H, respectively.

(c) A dual pair $\langle H, K \rangle$ is H-injective $\{K$-injective$\}$ if $H^{\circ} = \{e_K\}$ $\{$if $K^{\circ} = \{e_H\}\}$. The pair $\langle H, K \rangle$ is injective if it is both H- and K-injective. Taking "evaluation" as multiplication, show that $\langle G, G' \rangle$ is an injective dual pair for every locally compact T_2 group G. If G is also abelian, $\langle G', G \rangle$ is injective.

(d) If H, K are LCT_2A groups and $\langle H, K \rangle$ is injective, then $H = K'$ and $K = H'$.

(e) Show that (d) fails for nonabelian groups. [*Hint* (Pontryagin [80]): Let H be a discrete group with two independent generators x_1, x_2 (i.e., every $x \in H$ has a unique representation $x = x_1^{n_1} x_2^{n_2}$ for $n_1, n_2 \in \mathbb{I}$) and take $K = (\mathbb{R}, +; d_1)$. Pick any $\alpha, \beta \in \mathbb{R}$ such that α/β is irrational. Then $\langle H, K \rangle$ is injective under $\underset{(x_1^{n_1} x_2^{n_2}, t) \rightsquigarrow \exp\{2\pi i y\}}{H \times K \longrightarrow S'(\mathbb{C})}$, where $(n_1 \alpha t)(n_2 \beta t) \equiv y \pmod 1$. However, $H' \neq K$ and $K' \neq H$.]

11F. Duality and Commutativity

Define $(G')^{\circ} = \{x \in G : \chi(x) = 1 \ \forall \chi \in G'\}$ for a locally compact, T_2 group G.

(a) $(G')^{\circ}$ is a closed, normal subgroup of G. Prove this directly and also by demonstrating that $(G')^{\circ} = \bar{C}_G$, where $C_G = \left\{x \in G : x = \prod_{i=1}^{n} a_i b_i a_i^{-1} b_i^{-1} \ (a_i, b_i \in G) \text{ and } n \in \mathbb{N}\right\}$.

(b) For each $\chi \in G'$, let $\mu_\chi(\bar{C}_G x) = \chi(x)$. Then $\mu_\chi \in (G/\bar{C}_G)'$ and $\underset{\chi \rightsquigarrow \mu_\chi}{G' \to (G/\bar{C}_G)'}$ is a topological isomorphism.

(c) $G/\bar{C}_G$ is abelian. [*Hint*: For a normal subgroup $A \subset G$ show that G/A is abelian iff $C_G \subset A$.]

(d) The preceding results suggest that many dual group statements do not require commutativity. Show, for instance, that Corollary 11.22 still holds for all locally compact T_2 groups.

(e) Problem 11E(e) confirms that duality theory for LCT_2A groups is not equivalent to that for locally compact T_2 groups. If $\langle H, K \rangle$ is an injective dual pair for locally compact T_2 groups H and K, then $\langle H/\bar{C}_H, K/\bar{C}_K \rangle$ is an injective dual pair. Therefore $(H/\bar{C}_H)' = K/\bar{C}_K$ and $(K/\bar{C}_K)' = H/\bar{C}_H$.

11G. Self-Dual Groups

An LCT_2A group G is said to be self-dual if $G = G'$ (see Example 11.19):

(a) Why is $G \times G'$ self-dual?

(b) $\prod_{i=1}^{n} G_i$ is self-dual iff each G_i is self-dual.

(c) For $1 < n \in \mathbb{N}$ let $\mathbb{I}/n = \{[0], [1], \ldots, [n-1]\}$ be the T_2 cyclic group under addition mod n [where $[k] = \{k' \in \mathbb{I} : k' = k \pmod n\}$]. Then $\mathbb{I}/n$ is self-dual. [*Hint*: $(\mathbb{I}/n)' = \{\chi_m([k]) = e^{2\pi i k m/n} : m \in \mathbb{N} \cap [1, n]\}$.]

(d) Using (c) above, prove that every finite T_2 abelian group is self-dual. [*Hint*: Such a group $G = \prod_{k=1}^{s} \mathbb{I}/k$, where each $k = p_k^{n_k} > 1$ (p_k, prime).]

(e) If H is a compact, open subgroup of a self-dual group G, then H and G/H are self-dual.

11H. Discrete Abelian Groups and Connectedness of Duals

Let G be a discrete abelian group. Then G' is connected iff G has no element of finite order. [*Hint*: If $x^n = e$ for $x \in G$, then $H_n = \{x, x^1, \ldots, x^{n-1}, x^n = e\}$ and so

G' is disconnected. (Consider $\nu: G' \to G'/H_n^{\circ} = H_n' = H_n$ in Problem 11G.) If $C(\chi_G)$ is the component of $\chi_G \in G'$, then $G'/C(\chi_G)$ is compact T_2 and totally disconnected (Problem 9A(d)). Let (problem 8I(a)) $\dot{H}$ be a proper, open subgroup of $G'/C(\chi_G)$ and take $H = \nu^{-1}(\dot{H})$. Then $1 < \aleph(G'/C(\chi_G) < \infty$ and $G/H = H^{\circ} = G'' = G$ ensure that H°, therefore G, has an element of finite order.]

11I. The Bohr Compactification

Given a LCT$_2$A group G, let $G'_\kappa = (G', \mathfrak{T}_\kappa)$ and $G''_\kappa = (G'_\kappa)'_\kappa$ with $G'_{\mathfrak{D}} = (G', \mathfrak{D})$ and $\check{G} = (G'_{\mathfrak{D}})'_\kappa$ for the discrete topology $\mathfrak{D}$ on G'.

(a) Show that $\underset{x \rightsquigarrow e_x}{\delta: G \to G''_\kappa} \subset \check{G}$ is a continuous isomorphism. [*Hint*: A subset of G is compact iff it is finite.]

(b) $(G'_\kappa)'$ is a dense subgroup of the compact T_2 group $\check{G}$. [*Hint*: If $H = \overline{(G'_\kappa)'}^{\mathfrak{T}_\kappa} \neq \check{G}$, then a nontrivial $\phi \in (\check{G}/H)'$ and a nontrivial character $\underset{x \rightsquigarrow \phi(Hx)}{\phi\nu: \check{G} \to S'(\mathbb{C})}$ can be found such that $\phi\nu \in H^{\circ}$. Show therefore that some $\chi_0 \neq \chi_G$ belongs to G' and $\chi_0(x) = \xi_x(\chi_0) = 1\ \forall x \in G$ yields the contradiction that $\chi_0 = \chi_G$.]

(c) The mapping $\delta: G \to \check{G}$ need not be open. Why? [*Hint*: If δ is open, G''_κ is locally compact and $G''_\kappa = \check{G}$ (Corollary 8.20).]

(d) *Wanted*! A short proof that G is compact iff $G''_\kappa = \check{G}$ iff δ is open.

(e) For every $\{\chi_1 \cdots \chi_n\} \in G'_\kappa$, every $\varepsilon > 0$ and every homomorphism $h: G'_\kappa \to S'(\mathbb{C})$ there is an $x \in G$ such that $|h(\chi_i) - \chi_i(x)| < \varepsilon\ \forall i = 1, 2, \ldots, n$. [*Hint*: If $h: G'_\kappa \to S'(\mathbb{C})$ is a homomorphism, then $h \in G'_{\mathfrak{D}}$ and h is $\mathfrak{T}_\kappa$-continuous (i.e., $h \in \check{G}$).]

Remark. Although δ is not a homeomorphism, $(\delta, \check{G})$ is called the Bohr compactification of G.

References for further Study

On characters and dual groups: Hewitt and Ross [41]; Pontryagin [80].

On the Bohr compactification and almost periodic functions: Loomis [58]; Hewitt and Ross [41]; Dunkl and Ramirez [27]; Rudin [85]; Katznelsson [49].

12. Haar Measure and Integration†

Unless specified otherwise, G denotes a locally compact T_2 group with the notation of Definition 8.1, Theorem 8.4. Here $C_0(G) = C_0(G; \mathbb{R})$. For $f: G \to \mathbb{R}$ define $f_x = f \circ \mathfrak{m}_x$ $\{{}_xf = f \circ {}_x\mathfrak{m}\}$ and $\hat{f} = f \circ \mathfrak{i}$ (where "$\circ$" denotes composition of mappings. We use $f_1 f_2$ to denote the product map

† A review of Appendix M explains much of the notation employed here.

$(f_1f_2)(x)=f_1(x)f_2(x)\,\forall x\in G.$) It follows from ${}_xf\{G-{}_{x^{-1}}\mathfrak{m}(C)\}=f(G-C)=\hat{f}\{G-\mathfrak{i}^{-1}(C)\}$ and Theorem 8.4 that ${}_xf$ (similarly, f_x) and f belong to $C_0(G)\,\forall f\in C_0(G)$ and $x\in G$.

Definition 12.1. A left Haar integral for G is a positive linear functional I such that $I({}_xf)=I(f)\,\forall f\in C_0(G)$ and $x\in G$.

A left Haar measure for G is a regular Borel measure μ satisfying $\mu(xE)=\mu(E)\,\forall E\in R_\sigma(\mathcal{C})$ and $x\in G$. Right Haar integral and measure for G are defined similarly.

The objective of this section is to prove that G has a left Haar integral $I_G\neq 0$ and a left Haar measure $\mu_G\neq 0$, both of which are unique up to a multiplication constant. (A special case of this is the Lebesque integral and measure on $\mathbb{R}=(\mathbb{R},+;d_1)$.) The existence of either left Haar entity I_G, μ_G ensures the existence of the corresponding right Haar entity, and conversely. For instance, let $\Psi(f)=I_G(\hat{f})\,\forall f\in C_0(G)$ and let $\nu(E)=\mu_G(E^{-1})\,\forall E\in R_\sigma(\mathcal{C})$. Then $\Psi(f_x)=I_G(\hat{f_x})=I_G({}_{x^{-1}}\hat{f})=I(\hat{f})=\Psi(f)$ and $\nu(Ex)=\mu_G(x^{-1}E^{-1})=\mu_G(E^{-1})=\nu(E)$. Thus every "left invariance" result implies and is implied by the corresponding "right invariance" result. Our approach is to produce I_G first and then show that μ_G is the associated regular Borel measure in the Riesz Markoff Theorem (Appendix m.8).

A word of caution is in order before proceeding. Some texts (e.g., Rudin [85]; Naimark [69]; Hewitt and Ross [41]) define Borel sets as members of $\mathfrak{R}_\sigma(\mathfrak{C})=\mathfrak{R}_\sigma(\mathfrak{O})$, where $\mathfrak{C}$ and $\mathfrak{O}$ denote the closed and open subsets of G, respectively. In general, $\mathfrak{R}_\sigma(\mathcal{C})\subset\mathfrak{R}_\sigma(\mathfrak{O})$ so that μ_G need not be defined on open (closed) subsets of G. This seeming restriction, however, is not serious. (See the discussion on pp. 121–122.)

Using the notation of Example 8.14, we first prove

Theorem 12.2. Every $f\in C_0(G)$ is ${}_G\mathcal{U}$ and $\mathcal{U}_G$ uniformly continuous.

Proof. Given any $\varepsilon>0$, assume that $f(G-C)=0$ for some $C\in\mathcal{C}$ and (Theorem 8.15) let $u\in\mathfrak{N}_e$ be compact and symmetric. Then $w=\{y\in G:|f(xy)-f(x)|<\varepsilon\;\forall x\in Cu\}\in\mathfrak{N}_e$ is open. For verification first observe that $F:G\times G\longrightarrow\mathbb{R}$, $(y,x)\rightsquigarrow f(xy)-f(x)$ is continuous and $w=\{y\in G: F(y,x)\in S_\varepsilon(0)\,\forall x\in Cu\}$. Furthermore, $\forall x\in Cu$ there exist open sets $u_x(x)$ and $v_x(x)$ such that $F(u_x\times v_x)\subset S_\varepsilon$, which means (since $Cu=\mathfrak{m}(C\times u)$ is compact) that Cu is covered by some $\{u_{x_1}\cdots u_{x_n}\}$. If $y_0\in w$ and one defines $v_0(y_0)=\bigcap_{i=1}^n v_{x_i}$, then $F(v_0\times Cu)\subset S_\varepsilon$ implies $v_0(y_0)\subset w$ and

$y_0 \in \operatorname{Int} w$ so that w is open as asserted. Now $z \in v = u \cap w$ satisfies $|f(xz) - f(x)| < \varepsilon \ \forall x \in Cu$ and $|f(xz) - f(x)| = 0 \ \forall x \in G - Cu$ (since $C \subset Cu$ implies $f(G - Cu) = 0$ and $f(xz) \neq 0$ implies $x \in Cu$). Thus $|f(xz) - f(x)| < \varepsilon \ \forall x \in G$ and fixed $z \in v(e)$. For $z_0 = xz$ this yields $|f(z_0) - f(x)| < \varepsilon \ \forall x \in G$ and $x^{-1}z_0 \in v(e)$; that is, ${}_vU \subset (f \times f)^{-1} V \in {}_G\mathcal{U}$. The proof that $(f \times f)^{-1} V \in \mathcal{U}_G \ \forall \varepsilon > 0$ follows similarly. ■

COROLLARY 12.3. Both mappings $\underset{x \rightsquigarrow {}_x f}{G \to C_0(G)}$ and $\underset{x \rightsquigarrow f_x}{G \to C_0(G)}$ are continuous in x for each $f \in C_0(G)$.

Proof. $\|{}_x f - {}_y f\|_0 \leq \|{}_x f - f\|_0 + \|f - {}_y f\|_0 < \varepsilon \ \forall x, y^{-1} \in v_1 v_2^{-1}$ in $\mathfrak{N}_e$ follows from $\|{}_{x_i} f - f\|_0 < \varepsilon/2$ for $x_i \in V_i(e)$ $(i = 1, 2)$. ■

Let $f \in C_0^+(G)$ with $f(G - C) = 0$ and suppose that $\varphi \not\equiv 0$ in $C_0^+(G)$ satisfies $\inf_u \varphi(x) \geq \eta > 0$ for some G-open set u. Assuredly, the compact set $C \subset G$ is covered by some $\{x_1^{-1}u, \ldots, x_n^{-1}u\}$. If $\|f\|_0 = \sup_C f(x)$ and $\alpha_i \geq (\|f\|_0/\eta) \ \forall i = 1, 2, \ldots, n$, then $f \leq \sum_{i=1}^n \alpha_i ({}_{x_i}\varphi)$ in the sense that $f(x) \leq \sum_{i=1}^n \alpha_i \varphi(x_i x) \ \forall x \in G$. It pays to introduce

DEFINITION 12.4 The collection $\{\alpha_1 \cdots \alpha_n\}$ above is called a covering of f by φ. We denote by $(f; \varphi)$ the $\inf \left\{ \sum_i \alpha_i : \{\alpha_1 \cdots a_n\} \text{ is a cover of } f \text{ by } \varphi \right\}$.

THEOREM 12.5. The number $(f; \varphi)$ is finite, nonnegative, and satisfies:

(i) $({}_x f; \varphi) = (f; \varphi) \ \forall x \in G$;
(ii) $(f_1 + f_2; \varphi) = (f_1; \varphi) + (f_2; \varphi)$;
(iii) $(\alpha f; \varphi) = \alpha (f; \varphi) \ \forall \alpha > 0$;
(iv) $f_1 \leq f_2 \Rightarrow (f_1; \varphi) \leq (f_2; \varphi)$;
(v) $(f; \psi) \leq (f; \varphi)(\varphi; \psi)$;
(vi) $\dfrac{1}{(\psi; f)} \leq \dfrac{(f; \varphi)}{(\psi; \varphi)} \leq (f; \psi)$.

Proof. First note that $f \leq \sum_{i=1}^n \alpha_i ({}_{x_i}\varphi)$ implies $\|f\|_0 / \|\varphi\|_0 \leq \sum_{i=1}^n \alpha_i \leq n\|f\|_0/\eta$ and so $\|f\|_0/\|\varphi\|_0 \leq (f; \varphi) \leq n\|f\|_0/\eta$. It follows that $(f; \varphi) = 0$ iff $f \equiv 0$. To prove (i) begin with ${}_x f \leq \sum_{i=1}^n \alpha_i ({}_{x_i x}\varphi)$ and $({}_x f; \varphi) \leq \sum_{i=1}^n \alpha_i$ and obtain

$({}_xf;\varphi)\le(f;\varphi)$. A symmetrical argument yields $(f;\varphi)=({}_{x^{-1}}\{{}_xf\};\varphi)\le(f;\varphi)$ as required. For (ii) suppose that $f_1\le\sum_{i=1}^{n}\alpha_i({}_{x_i}\varphi)$ and $f_2\le\sum_{j=1}^{m}\beta_j({}_{y_j}\varphi)$. Then $(f_1+f_2;\varphi)\le\sum_{i=1}^{n}\alpha_i+\sum_{j=1}^{m}\beta_j$ and $(f_1+f_2;\varphi)\le(f_1;\varphi)+(f_2;\varphi)$ follows by taking the inf over all covers of f_k by φ $(k=1,2)$. If $f\le\sum_{i=1}^{n}\alpha_i({}_{x_i}\varphi)$, then $(\alpha f;\varphi)\le\alpha\sum_{i=1}^{n}\alpha_i$ and $(\alpha f;\varphi)\le\alpha(f;\varphi)$. The same reasoning yields $(f;\varphi)=(1/\alpha\{\alpha f\};\varphi)\le 1/\alpha(\alpha f;\varphi)$ and establishes (iii). The proof of (iv) is trivial. (v) If $f\le\sum_{i=1}^{m}\alpha_i({}_{x_i}\varphi)$ and $\varphi\le\sum_{j=1}^{n}\beta_j({}_{y_j}\psi)$, then $f\le\sum_{i=1}^{m}\alpha_i\left\{\sum_{j=1}^{n}\beta_j({}_{y_j}\psi)\right\}=\sum_{i,j}\alpha_i\beta_j({}_{y_jx_i}\psi)$ ensures that $(f;\psi)\le\sum_{i,j}\alpha_i\beta_j=\left(\sum_{i=1}^{m}\alpha_i\right)\left(\sum_{j=1}^{n}\beta_j\right)$, and $(f;\psi)\le(f;\varphi)(\varphi;\psi)$ follows by taking infs over all covers of f by φ and φ by ψ. Clearly, (vi) follows from (v), since $(\psi,\varphi)\le(\psi;f)(f;\varphi)$ and $(f;\varphi)\le(f;\psi)(\psi;\varphi)$. ■

Fix $0\ne f_0\in C_0^+(G)$ and define $I_\varphi(f)=(f;\varphi)/(f_0;\varphi)$ for every $0\ne\varphi$, $f\in C_0^+(G)$. It follows from Theorem 12.5 that $I_\varphi\ge0$ satisfies $1/(f_0;f)\le\Phi_\varphi(f)\le(f;f_0)$ and $I_\varphi({}_xf)=I_\varphi(f)\,\forall x\in G$. Also, $I_\varphi(\alpha f)=\alpha I_\varphi(f)\,\forall\alpha>0$ and $I_\varphi(f_1+f_2)\le I_\varphi(f_1)+I_\varphi(f_2)$. If this last inequality were an equality and the preceding equality held $\forall\alpha\in\mathbb{R}$, then I_φ would be a left invariant integral for G. Overcoming these deficiencies constitutes the heart of the proof that G possesses a left Haar integral.

For notation let $C_{0_E}^+(G)=\{f\in C_0^+(G):f(G-E)=0\}$. Since G is T_4 (Corollary 8.21), $C_{0_v}^+(G)\ne\varnothing\;\forall v\in\mathfrak{N}_e$ and linearity of I_φ can be approximated in the sense of

THEOREM 12.6. For every $f_1,f_2\in C_0^+(G)$ and $\varepsilon>0$ there is a $v\in\mathfrak{N}_e$ such that $I_\varphi(f_1)+I_\varphi(f_2)\le I_\varphi(f_1+f_2)+\varepsilon\;\forall\varphi\ne0$ in $C_{0_v}^+(G)$.

Proof. Assume that $(f_1+f_2)(G-C)=0$ for $C\in\mathcal{C}$ and let $f\in C_0^+(G)$ satisfy $f(C)=1$. For any $\delta,\varepsilon>0$ and corresponding $F=f_1+f_2+\delta f$ define $h_i(x)=f_i(x)/F(x)$ for $F(x)\ne0$ and $h_i(x)=0$ otherwise. Clearly, $F,h_1,h_2\in C_0^+(G)$ and $h_1+h_2\le1$. Furthermore (Theorem 12.2), for any $\varepsilon'>0$ there is a $v\in\mathfrak{N}_e$ such that $|h_i(z_0)-h_i(x)|<\varepsilon'\;\forall x^{-1}z_0\in v$ $(i=1,2)$. Now consider any $\varphi\in C_{0_v}^+(G)$ and let $F\le\sum_{j=1}^{n}\alpha_j({}_{x_j^{-1}}\varphi)$. Since each ${}_{x_j^{-1}}\Phi(x)=0$ implies $x_j^{-1}x\in v$ and $|h_i(x)-h_i(x_j)|<\varepsilon'$ $(i=1,2)$, one has $f_i(x)=(Fh_i)(x)\le$

$\left[\sum_{j=1}^{n} \alpha_j\varphi(x_j^{-1}x)\right]h_i(x)=\sum_{j=1}^{n} \alpha_i\varphi(x_j^{-1}x)h_i(x)\leq \sum_{j=1}^{n} \alpha_j\varphi(x_j^{-1}x)[h_i(x_j)+\varepsilon']$ and $(f_i;\varphi)\leq \sum_{i=1}^{n} \alpha_j[h_i(x_j)+\varepsilon']$. Therefore $(f_1;\varphi)+(f_2;\varphi)\leq \sum_{j=1}^{n} \alpha_j(1+2\varepsilon')$ and $(f_1;\varphi)+(f_2;\varphi)\leq(F;\varphi)(1+2\varepsilon')$ follows by taking the inf over all covers of f by φ. Division by $(f_0;\varphi)$ further yields $I_\varphi(f_1)+I_\varphi(f_2)\leq I_\varphi(f_1+f_2)+2\varepsilon' I_\varphi(f_1+f_2)+\delta(1+2\varepsilon')I_\varphi(F)$; but $I_\varphi(F)<\infty$ so that $2\varepsilon' I_\varphi(f_1+f_2)+\delta(1+2\varepsilon')I_\varphi(F)<\varepsilon$ for sufficiently small δ, ε' and our proof is therefore complete. ■

We are now ready for the principal result of this section; namely,

THEOREM 12.7. Every locally compact T_2 group G has a nonzero, left invariant Haar integral that is unique up to a multiplicative constant.

Proof. Existence: For each $f\not\equiv 0$ in $C_0^+(G)$ let E_f be the interval $[(f_0;f)^{-1},(f;f_0)]$ and take $E_f=\{0\}$ if $f\equiv 0$. Then each $\{I_\varphi(f)\}_f$ is a "point" in the compact T_2 product space $X=\prod_{C_0^+(G)} E_f$. For each G-nbd $v\in\mathfrak{N}_e$ let $F_v=\overline{\{I_\varphi:\varphi\in C_{0_v}^+(G)\}}\subset X$. The collection $\{F_v:v\in\mathfrak{N}_e\}$ has the fip (Appendix t.7), since for any $\{F_{v_1}\cdots F_{v_n}\}$,there is a $\tilde{\varphi}\subset C_0^+(G)$ such that $\tilde{\varphi}\left(G-\bigcap_{i=1}^{n} v_i\right)=0$, and so $\Phi_{\tilde{\varphi}}\in\bigcap_{i=1}^{n} F_{v_i}$. Let $J\in\bigcap_{\mathfrak{N}_e} F_v=X$ be interpreted as the mapping $C_0^+(G)\underset{f\rightsquigarrow\pi_f(J)}{\longrightarrow}\mathbb{R}$, where π_f denotes the projection of X onto the fth factor space E_f. Now, given any X-nbd $u(J)=\{\xi:|\xi(f_i)-J(f_i)|<\varepsilon\ \forall f_i\in C_0^+(G)\ (i=1,2,\ldots,n)\}$ and any $v\in\mathfrak{N}_e$, there is some $I_{\varphi_0}\not\equiv 0$ in $u(J)\cap\{I_\varphi:\varphi\in C_{0_v}^+(G)\}$ and J (for this fixed Φ_{φ_0}) satisfies $\forall f\in C_{0_v}^+(G)$:

(i) $J(f)=0$ iff $f\equiv 0$;
(ii) $J({}_xf)=J(f)\ \forall x\in G$;
(iii) $J(\alpha f)=\alpha J(f)\ \forall\alpha>0$;
(iv) $J(f_1+f_2)=J(f_1)+J(f_2)$.

Indeed, $J(0)=0$ by definition. If $J(f)=0$, then $|I_\varphi(f)|<\varepsilon\ \forall\varepsilon>0$ so that $I_\varphi(f)=0$ and $f\equiv 0$. Clearly, (ii) follows from $|J({}_xf)-J(f)|\leq|J({}_xf)-I_\varphi({}_xf)|+|I_\varphi({}_xf)-J(f)|$ and $I_\varphi({}_xf)=I_\varphi(f)$. For (iii) use $|J(\alpha f)-\alpha J(f)|\leq|J(\alpha f)-I_\varphi(\alpha f)|+|I_\varphi(\alpha f)-\alpha J(f)|=|J(\alpha f)-I_\varphi(\alpha f)|+\alpha|I_\varphi(f)-J(f)|$. To verify (iv) let $f_3=f_1+f_2$ and choose any $\varepsilon>0$. Then (Theorem 12.6) there is a $v\in\mathfrak{N}_e$ such that $I_\varphi(f_1)+I_\varphi(f_2)\leq I_\varphi(f_1+f_2)+\varepsilon/4\ \forall\varphi\not\equiv 0$ in $C_{0_v}^+(G)$. By the approximation there is a $\psi \in C_{0_v}^{+}(G)$ such that $|I_\psi(f_i) - J(f_i)| < \varepsilon/4$ for each $i = 1, 2, 3$. Accordingly, $|J(f_1) + J(f_2) - J(f_1 + f_2)| \leq |J(f_1) - I_\psi(f_1)| + |J(f_2) - I_\psi(f_2)| + |I_\psi(f_1) + I_\psi(f_2) - I_\psi(f_1 + f_2)| + |I_\psi(f_1 + f_2) - J(f_1 + f_2)| < \varepsilon$ and (iv) follows, since $\varepsilon > 0$ is arbitrary. Having thus established (i)–(iv), we now show that J can be extended to a left invariant linear functional on $C_0(G)$. Since every $f \in C_0(G)$ can be written $f = f^+ - f^-$ with $f^{\pm} \in C_0^+(G)$ (Appendix m.2), define $I_G(f) = J(f^+) - J(f^-)$. Note that I_G is well defined, since $f^+ - f^- = \phi - \psi$ for $\phi, \psi \in C_0^+(G)$ implies $f^+ + \psi = \phi + f^-$ and $J(f^+) + J(\psi) = J(f^+ + \psi) = J(\phi + f^-) = J(\phi) + J(f^-)$. Now $I_G(f_1 + f_2) = J\{(f_1 + f_2)^+\} - J\{(f_1 + f_2)^-\} = \{J(f_1^+) - J(f_1^-)\} - \{J(f_2^+) - J(f_2^-)\} = I_G(f_1) + I(f_2)$. For any $\beta = -\alpha$ $(\alpha > 0)$, one also has $I_G(\beta f) = \alpha I_G(-f) = -\alpha[J\{(-f)^+\} - J\{(-f)^-\}] = \beta[J(f^+) - J(f^-)] = \beta I_G(f)$. Finally, ${}_xf = {}_xf^+ - {}_xf^-$ yields $I_G({}_xf) = J({}_xf^+) - J({}_xf^-) = J(f^+) - J(f^-) = I_G(f)$. This completes the existence proof of the theorem.

Uniqueness: Suppose I_i $(i = 1, 2)$ are left Haar integrals for G and (Riesz–Markoff theorem. Appendix m.8) let μ_i $(i = 1, 2)$ be the corresponding regular Borel measures satisfying $I_i(f) = \int_G f\, d\mu_i$ $\forall f \in C_0(G)$. Fix $f_0 \in C_0^+(G)$ satisfying $\int_G f_0\, d\mu_1 = 1$ and define $\lambda = \int_G f_0\, d\mu_2$. We claim that $I_2 = \lambda I_1$. Verification requires the use of Fubini's theorem (Appendix m.4). For any $f \in C_0(G)$, one has

$$I_2(f) = \int_G f(x)\, d\mu_2(x) = \int_G \left[\int_G f_0(y)\, d\mu_1(y)\right] f(x)\, d\mu_2(x)$$

$$= \int_G \left[\int_G f_0(x^{-1}y)\, d\mu_1(y)\right] f(x)\, d\mu_2(x)$$

$$= \int_G \left[\int_G f_0(x^{-1}y) f(x)\, d\mu_1(y)\right] d\mu_2(x)$$

$$= \int_G \left[\int_G f_0(x^{-1}y) f(x)\, d\mu_2(x)\right] d\mu_1(y)$$

$$= \int_G \left[\int_G {}_{y^{-1}}\{\hat{f}_0(x) f(yx)\}\, d\mu_2(x)\right] d\mu_1(y)$$

$$= \int_G \left[\int_G \hat{f}_0(x) f(yx)\, d\mu_2(x)\right] d\mu_1(y)$$

$$= \int_G \left[\int_G \hat{f}_0(x) f(yx)\, d\mu_1(y) \right] d\mu_2(x)$$

$$= \int_G \left[\int_G f(yx)\, d\mu_1(y) \right] \hat{f}_0(x)\, d\mu_2(x)$$

$$= \int_G \left[\int_G \hat{f}(x^{-1}y^{-1})\, d\mu_1(y) \right] \hat{f}_0(x)\, d\mu_2(x)$$

$$= \int_G \left[\int_G {}_{x^{-1}}\hat{f}(y^{-1})\, d\mu_1(y) \right] \hat{f}_0(x)\, d\mu_2(x)$$

$$= \int_G \left[\int_G f(y)\, d\mu_1(y) \right] \hat{f}_0(x)\, d\mu_2(x)$$

$$= \left(\int_G \hat{f}_0(x)\, d\mu_2(x) \right) \left(\int_G f(y)\, d\mu_1(y) \right) = \lambda I_1(f). \quad \blacksquare$$

On a LCT$_2$A group G every left Haar integral I_G is both right invariant and inversion invariant (i.e., $I_G(\hat{f}) = I_G(f)\ \forall f \in C_0(G)$). In general, however, a left Haar integral need be neither right nor inversion invariant (cf Example 12.9). The nexus between left, right, and inversion invariance is not too difficult to unravel. Fix any $x_0 \in G$ and define $J(f) = I_G(f_{x_0})\ \forall f \in C_0(G)$. Then $J({}_xf) = I_G\{{}_x(f_{x_0})\} = I_G(f_{x_0}) = J(f)\ \forall x \in G$ and $f \in C_0(G)$; that is, J, too, is a left Haar integral and there is a number $\Delta(x_0) > 0$ such that $I_G(f_{x_0}) = \Delta(x_0) I_G(f)\ \forall f \in C_0(G)$.

DEFINITION 12.8. The nonnegative function $\Delta: G \to \mathbb{R}$ above is called the modular function of G, and G is termed unimodular if $\Delta(x) = 1\ \forall x \in G$.

Example 12.9. Every compact T_2 group and every discrete group is unimodular. if G is compact T_2 and $f = \chi_G$, then $I_G(f) = \mu_G(G) < \infty$. Therefore $I_G(f) = I_G(f_x) = \Delta(x) I_G(f)$ implies $\Delta(x) = 1\ \forall x \in G$. Suppose G is discrete so that $f \in C_0(G)$ iff $f(G-F) = 0$ for some finite set F. Then $I_G(\chi) = I_G(\zeta_{xe}) = I_G({}_{x^{-1}}\{\chi_e\}) = I_G(\chi_e) \neq 0$ and $\Delta(x) I_G(\chi_e) = I_G(\{\chi_e\}_x) = I_G(\chi_{x^{-1}}) = I_G(\chi_e)\ \forall x \in G$.

Evidently $\Delta: G \to \mathbb{R}_\bullet$ is a homomorphism since $\Delta(x_1x_2) I_G(f) = I_G(f_{x_1x_2}) = \Delta(x_2) I_G(f_{x_1}) = \Delta(x_1)\Delta(x_2) I_G(f)\ \forall f \in C_0(G)$ and $x_1, x_2 \in G$. With a little more effort one can also prove

THEOREM 12.10. The modular function is continuous. Furthermore, $I_G(f)=I_G(\Delta\hat{f})$ and $I_G(\hat{f})=I_G(\Delta f)\ \forall f\in C_0(G)$.

Proof. Consider any $f\in C_0(G)$ with $I_G(f)=1$ and let $u\in\mathfrak{N}_e$ be symmetric and compact. Since Nu is compact [where $N=N(f)$ is the support of f], $G-Nu\subset G-N$ and $(f_x-f)(G-Nu)=f_x(G-Nu)=0\ \forall x\in Nu$. This yields $|\Delta(x)-1|=|\Delta(x)I_G(f)-I_G(f)|=|I_G(f_x-f)|\le\sup_{Nu}|f_x-f|\mu_G(Nu)\le\|f_x-f_e\|_0\mu_G(Nu)\ \forall x\in u$. Now, given any $\varepsilon>0$, there is (Corollary 12.3) a $v\in\mathfrak{N}_e$ such that $\|f_x-f_e\|_0<\varepsilon/\mu_G(Nu)\ \forall x\in v$ and the continuity of Δ follows (Theorem 8.10), since $|\Delta(x)-1|<\varepsilon\ \forall x\in u\cap v$ in $\mathfrak{N}_e$. ■

If $f\in C_0(G)$, then $\tilde{f}=\Delta\hat{f}\in C_0(G)$ and $\widetilde{{}_xf}=\Delta(x)\tilde{f}_{x^{-1}}\ \forall x\in G$. Therefore $J(F)=I_G(\tilde{f})$ is a left Haar integral for G and $J=\alpha I_G$ for some $\alpha>0$. Accordingly, $I_G(f)=(1/\alpha)J(f)=(1/\alpha)I_G(\tilde{f})=(1/\alpha^2)I_G(\tilde{\tilde{f}})=(1/\alpha^2)I_G(f)$ and $\alpha=1$, so that $I_G(f)=J(f)=I_G(\tilde{f})=I_G(\Delta f)$ and $I_G(\hat{f})=I_G(\Delta\hat{\hat{f}})=I_G(\Delta f)\ \forall f\in C_0(G)$.

COROLLARY 12.11. G is unimodular iff I_G is right {inverse} invariant.

Proof. I_G is right invariant iff G is unimodular, in which case I_G is inverse invariant. If I_G is inverse invariant, it is also right invariant, since $I_G(f_x)=I_G(\widehat{f_x})=I_G({}_{x^{-1}}\hat{f})=I_G(\hat{f})=I_G(f)\ \forall x\in G$ and $f\in C_0(G)$. ■

If $I_G(f)=\int_G f\,d\mu_G\ \forall f\in C_0(G)$, intuition suggests that μ_G and I_G share the same invariance properties. This is indeed the case and verification begins with

THEOREM 12.12. Let T be a homeomorphism of G onto itself and let μ be a regular Borel measure for G. Define $f^T(x)=f(Tx)\ \forall x\in G$ and $\nu(E)=\mu\{T(E)\}\ \forall E\in\mathcal{R}_\sigma(\mathcal{C})$. Then

(i) ν is a regular Borel measure for G;

(ii) if $\tilde{\mu}$ is a regular Borel measure for G such that $\int_G f^T d\mu=\int_G f\,d\tilde{\mu}\ \forall f\in C_0(G)$, then $\mu\{T^{-1}(E)\}=\tilde{\mu}(E)\ \forall E\in\mathcal{R}_\sigma(\mathcal{C})$.

Proof. (i) Since T preserves compact (therefore, Borel) sets, ν is a Borel measure for G. If $E\subset G$ is Borel, then $T(E)$ is Borel and (Appendix m.8) there exist sequences $\{D_n\} \in \mathcal{C}$ with $D_n \uparrow T(E)$ and $\{\mathcal{O}_n\} \in \mathfrak{O}$ with $\mathcal{O}_n \downarrow T(E)$. Taking $D_n = T(C_n)$ $(C_n \in \mathcal{C})$ and $\mathcal{O}_n = T(U_n)$ $(U_n \in \mathfrak{O})$ yields $C_n \subset E \subset U_n \; \forall n \in \mathbb{N}$, from which follows $\nu(C_n) = \mu(D_n) \uparrow \mu\{T(E)\} = \nu(E)$ and $\nu(U_n) = \mu(\mathcal{O}_n) \downarrow \mu\{T(E)\}$. Thus $\sup\{\nu(C): C \subset E$ and $C \in \mathcal{C}\} = \inf\{\nu(u): E \subset u$ and $u \in \mathfrak{O}\}$ and the regularity of ν is proved. (ii) Let $C \in \mathcal{C}_\delta$ and suppose that (Appendix m.8) $\{f_n\} \in C_0(G)$ satisfies $f_n \downarrow \chi_C$. Then $f_n^T \downarrow \{\chi_C\}^T = \chi_{T^{-1}(C)}$ and $\mu\{T^{-1}(C)\} = \int_G \chi_{T^{-1}(C)}\, d\mu \downarrow \int_G f_n^T\, d\mu = \int_G f\, d\tilde{\mu} \downarrow \int_G \chi\, d\tilde{\mu} = \tilde{\mu}(C)$, which confirms that $\mu\{T^{-1}(E)\} = \tilde{\mu}(E) \; \forall E \in \mathfrak{R}_\sigma(\mathcal{C})$. ■

THEOREM 12.13. Let μ_G and ν_G be the unique, regular Borel measures associated with left and right, Haar integrals for G, respectively. For each $x \in G$ and $E \in \mathfrak{R}_\sigma(\mathcal{C})$

(i) $\mu_G(xE) = \mu_G(E)$ (iv) $\nu_G(Ex) = \nu_G(E)$

(ii) $\mu_G(Ex) = \Delta(x^{-1})\mu_G(E)$ (v) $\nu_G(xE) = \Delta(x)\, \nu_G(E)$

(iii) $\mu_G(E^{-1}) = \int_E \Delta\, d\mu_G$ (vi) $\nu_G(E^{-1}) = \int_E \hat{\Delta}\, d\nu_G$

Proof. For (i) take $T = {}_x\mathfrak{m}$ (therefore, $T^{-1} = {}_{x^{-1}}\mathfrak{m}$) and let $\tilde{\mu} = \mu = \mu_G$ in Theorem 12.12. Since $I_G(f^T) = I_G(f) \; \forall f \in C_0(G)$, we have $\mu_G(xE) = \mu_G(E) \; \forall E \in \mathfrak{R}_\sigma(\mathcal{C})$. For (ii) take $T = \mathfrak{m}_{x^{-1}}$ and use $\int_G f^T\, d\mu_G = \Delta(x^{-1}) \int_G f\, d\mu_G = \int_G f\, d\{\Delta(x^{-1})\mu_G\} \; \forall f \in C_0(G)$ to conclude that $\mu_G(Ex) = \mu_G\{\mathfrak{m}_x(E)\} = \Delta(x^{-1})\mu_G(E) \; \forall E \subset \mathfrak{R}_\sigma(\mathcal{C})$. To prove (iii) assume that I_G has been extended to $L_1(G, \mu_G)$ (see Note 3 on p. 122) and use Theorem 12.10 to obtain $\mu_G(E^{-1}) = \int_G \chi_{E^{-1}}\, d\mu_G = \int \hat{\chi}_E\, d\mu_G = \int_G \Delta \chi_E\, d\mu_G = \int_E \Delta\, d\mu_G$. Similar reasoning establishes (iv)–(vi). Specifically, $J(f) = I_G(\hat{f})$ is right invariant and $J_G(f) = \alpha I_G(\hat{f})$ for some $\alpha > 0$. Take $T = \mathfrak{i}$ in Theorem 12.12 and use $\int_G \hat{f}\, d\nu_G = \int_G \hat{f}\, d\{\alpha\mu_G\} \forall f \in C_0(G)$ to obtain $\nu_G(E) = \alpha\mu_G(E^{-1}) \forall E \in \mathfrak{R}_\sigma(\mathcal{C})$. Surely (iv), (v), and (vi) follow immediately from (i), (ii), and Theorem 12.10, respectively. ■

Reasoning similar to that in Corollary 12.11 also yields

COROLLARY 12.14. G is a unimodular iff μ_G is right {inverse} invariant.

Since μ_G deserves equal time with I_G, we include the counterpart to "$I_G(f) \geq 0 \; \forall f \in C_0^+(G)$" in

THEOREM 12.15. $\mu_G(u) > 0$ for every open Borel set $u \neq \varnothing$ in G.

Proof. Since $\mu_G \not\equiv 0$, one has $\mu_G(C) > 0$ for some $C \in \mathcal{C}$. Therefore $C \subset \bigcup_{i=1}^{n} x_i u$ for some $\{x_1 \cdots x_n\} \subset G$ and $0 < \mu_G(C) < \sum_{i=1}^{n} \mu_G(x_i u) = n\mu_G(u)$ follows. ■

A few important consequences of Theorem 12.15 are given below.

COROLLARY 12.16. (i) G is compact iff $\mu_G(G) < \infty$, and G is discrete iff $\mu_G(e) > 0$.

(ii) Suppose G is nondiscrete and $E \in \mathcal{R}_\sigma(\mathcal{C})$ has $\mu_G(E) = \beta < \infty$. Then for any $\alpha \in [0, \beta]$ there is a compact set $C_\alpha \subset E$ such that $\alpha < \mu_G(C_\alpha)$.

Proof. (i) The sufficiency part only needs verification. Suppose G is noncompact and let $v \in \mathfrak{N}_e$ be compact. Since G cannot be covered by a finite collection $\{x_i v\}$, one can inductively obtain a sequence $\{x_n : n \in \mathbb{N}\} \subset G$ such that $x_n \notin \bigcup_{i=1}^{n-1} x_i v \; \forall n \in \mathbb{N}$. Let $u \in \mathfrak{N}_e$ be symmetric, open, and satisfying $u^2 \subset v$. Then $\{x_n u : n \in \mathbb{N}\}$ are disjoint and (since u is bounded) $\mu_G(G) \geq \sum_{n=1}^{\infty} \mu_G(x_n u) = \sum_{n=1}^{\infty} \mu_G(u) = \infty$. Suppose next that $\mu_G(e) > 0$. Then every relatively compact set $u \subset G$ has a finite number of elements (otherwise $\mu_G(x) > 0 \; \forall x \in G$ implies the contradiction $\mu_G(\bar{u}) = \infty$). Therefore for any relatively compact open set $u = \{x_1, \ldots, x_{n-1}, e\}$ the set $\{e\} = \bigcap_{i=1}^{n-1} u \cap (G - x_i)$ is open. (ii) Since μ_G is regular, $\mu_G(C) > \alpha$ for some compact set $C \subset E$ and (since $\mu_G(x) = 0 \; \forall x \in G$) there exist open sets $\mathcal{O}_x$ such that $\mu_G(\bar{\mathcal{O}}_x) < \beta - \alpha \; \forall x \in G$. Now $C = \bigcup_{k=1}^{n} \bar{\mathcal{O}}_{x_k}$ for some $\{x_1 \cdots x_k\}$ and a smallest integer $n_0 \geq 2$ can be found such that $\alpha \leq \mu_G\left\{C \cap \bigcup_{k=1}^{n_0} \bar{\mathcal{O}}_{x_k}\right\}$. The compact set $C_\alpha = C \cap \left\{\bigcup_{k=1}^{n_0 - 1} \bar{\mathcal{O}}_{x_k}\right\} \subset E$ has $\mu_G(C_\alpha) > \alpha$. ■

Remark. It is a nice little exercise to establish further the existence of a σ-compact set $C_\alpha \subset E$ such that $\alpha = \mu_G(C_\alpha)$.

COROLLARY 12.17. $F_1 = F_2$ in $C(G)$ iff $\int_G fF_1\, d\mu_G = \int_G fF_2\, d\mu_G$ $\forall f \in C_0(G)$.

Proof. Suppose $F(x_0) = F_1(x_0) - F_2(x_0) > 0$ for some $x_0 \in G$. Since $F = F_1 - F_2 \in C(G)$, there is a relatively compact, open nbd $u(x_0)$ such that $F(x) > 0\ \forall x \in u$. Let $f \in C_0^+(G)$ satisfy $f(x_0) = 1$ and $f(G-u) = 0$ and assume that $v(x_0)$ is a relatively compact, open nbd such that $f(x) > 0\ \forall x \in v$. Then $(fF)(x) = f(x)F(x) \geq 0$ on $u(x_0)$ and $(fF)(G-u) = 0$. In fact, $(fF)(x) > 0$ for $x \in w(x_0) = u(x_0) \cap v(x_0)$. Since $w(x_0)$ is a Borel open set, $\int_G fF_1\, d\mu_G - \int_G fF_2\, d\mu_G = \int_G fF\, d\mu_G \geq \int_w fF\, d\mu_G > 0$. Therefore $\int_G fF_2\, d\mu_G \leq \int_G fF_1\, d\mu_G$ implies $F_2 \leq F_1$ and the reverse inequalities follow similarly. ■

Clearly, $(f, g) = \int_G f\bar{g}\, d\mu_G$ defines an inner product on $C_0(G)$. If G is discrete, $\{[\mu_G(e)]^{-1}\delta_{x_0,x} : x_0 \in G\}$ is a complete orthonormal system in $C_0(G)$. The other extreme is considered in

Example 12.18. Let G be compact and T_2. Then $\int_G \chi\, d\mu_G = 0$ for $\chi \neq \chi_G$, and $(\chi_1, \chi_2) = \int_G \chi_1\chi_2^{-1}\, d\mu_G = 0$ for $\chi_1 \neq \chi_2$. [The second assertion follows from the first, which holds, since $\chi(x_0) \neq 1$ yields $\int_G \chi(x)\, d\mu_G(x) = \chi(x_0)\int_G \chi(x_0^{-1}x)\, d\mu_G(x) = \chi(x_0)\int_G \chi(x)\, d\mu_G(x)$.] Accordingly, G' is a complete orthonormal system in $L_1(G)$ for normalized μ_G (i.e., $\mu_G(G) = 1$); completeness holding by Corollary 29.11 (Appendix a.2).

Restricted and Extended Haar Concepts

Note 1. If ν_G on $\mathfrak{R}_\sigma(\mathfrak{C}_\delta)$ is the Baire restriction of μ_G on $\mathfrak{R}_\sigma(\mathfrak{C})$, then ν_G is regular and satisfies Theorem 12.13 $\forall E \in \mathfrak{R}_\sigma(\mathfrak{C}_\delta)$ and $x \in G$. Furthermore, $\nu_G(u) > 0$ for every nonempty open (Baire) set $u \in \mathfrak{R}_\sigma(\mathfrak{C}_\delta)$. In this sense ν_G is the restricted left Haar measure for G.

Note 2. Let $\mu_{\mathcal{T}G}$ be the weakly Borel extension of μ_G to $\mathfrak{R}_\sigma(= \mathfrak{R}_\sigma(\mathfrak{D}) = \mathfrak{R}_\sigma(\mathfrak{F}))$ given by $\mu_{\mathcal{T}G}(E) = \sup\{\mu_G(C) : C \subset E$ and $C \in \mathfrak{C}\}\ \forall E \in \mathfrak{R}_\sigma$. Then $\mu_{\mathcal{T}G}(xE) = \mu_{\mathcal{T}G}(E)$ and $\mu_{\mathcal{T}G}(Ex) = \Delta(x)^{-1}\mu_{\mathcal{T}G}(E)\ \forall E \in \mathfrak{R}_\sigma$ and $x \in G$. One also has $\mu_{\mathcal{T}G}(u) > 0$ for every nonempty open set $u \in \mathfrak{R}_\sigma$. In this regard $\mu_{\mathcal{T}G}$ is viewed as the extended left Haar measure for G (such is the case in Chapter V).

Remark. Although a Borel measure is inner regular iff it is outer regular, this need not be true for weakly Borel measures unless they are finite (Berberian [10]). Since $\mu r_G(C) = \mu_G(C)\ \forall C \in \mathcal{C}$, the weakly Borel measure μr_G is inner regular. It need not be outer regular. In fact, (Berberian [9]) μr_G is regular (i.e., outer regular) iff G is discrete or σ-compact.

Note 3. Suppose $(X, \Sigma; \nu)$ is an arbitrary measure space and ν_0 is the restriction of ν to a σ-ring $\Sigma_0 \subset \Sigma$. If $f \in L_1(X; \nu_0)$, then $f \in L_1(X; \nu)$ and $\int_X f\,d\nu_0 = \int_X f\,d\nu$. (Indeed, if $f \geq 0$ and $0 \leq \varphi_n \uparrow f$ for ν_0-simple functions $\{\varphi_n\}$, then $\int_X \varphi_n\,d\nu_0 = \int_X \varphi_n\,d\nu$ and $\int_X f\,d\nu_0 = \lim_n \int_X \varphi_n\,d\nu_0 = \lim_n \int_X \varphi_n\,d\nu = \int_X f\,d\nu$.)

The preceding remark clearly demonstrates that $C_0(G) \subset L_1(G, \nu_G) \subset L_1(G, \mu_G) \subset L_1(G, \mu r_G)$. Let $(\tilde{\mathfrak{R}}_\sigma, \tilde{\mu}_G)$ represent the pair $(\mathfrak{R}_\sigma(\mathcal{C}_\delta), \nu_G)$, $(\mathfrak{R}_\sigma(\mathcal{C}), \mu_G)$ or $(\mathfrak{R}_\sigma, \mu r_G)$. Since $\int_G {}_x\{\chi_E\}\,d\tilde{\mu}_G = \int_G \chi_{x^{-1}E}\,d\tilde{\mu}_G = \tilde{\mu}_G(x^{-1}E) = \tilde{\mu}_G(E) = \int_G \chi_E\,d\tilde{\mu}_G$ and $\int_G \{\chi_E\}_x\,d\tilde{\mu}_G = \int_G \chi_{Ex^{-1}}\,d\tilde{\mu}_G = \tilde{\mu}_G(Ex^{-1}) = \Delta(x) \int_G \chi_E\,d\tilde{\mu}_G\ \forall E \in \tilde{\mathfrak{R}}_\sigma$, one has $\xi_G\, {}_x\{\varphi_n\}\,d\tilde{\mu}_G = \int_G \varphi_n\,d\tilde{\mu}_G$ and $\int_G \{\varphi_n\}_x\,d\tilde{\mu}_G = \Delta(x) \int_G \varphi_n\,d\tilde{\mu}_G$ for every $\tilde{\mu}_G$-integrable simple function φ_n. In particular, $\int_G {}_xF\,d\tilde{\mu}_G = \int_G F\,d\tilde{\mu}_G$ and $\int_G F_x\,d\tilde{\mu}_G = \Delta(x) \int_G F\,d\tilde{\mu}_G\ \forall F \in L_1(G, \tilde{\mu}_G)$ and $x \in G$. Since $|\int_G F\,d\tilde{\mu}_G| \leq \|F\|_1$, the positive linear functional $I_{\tilde{\mu}_G}(F) = \int_G F\,d\tilde{\mu}_G$ is continuous on $L_1(G, \tilde{\mu}_G)$. If I_{μ_G} on $L_1(G, \mu_G)$ is interpreted as the left Haar integral for G, then I_{ν_G} on $L_1(G, \nu_G)$ and $I_{\mu r_G}$ on $L_1(G, \mu r_G)$ are the restricted and extended left Haar integrals for G, respectively.

Note 4. Since μ_G is regular, $C_0(g)$ is dense in $L_1(G, \mu_G)$ (Appendix m.8) and I_G may be extended by continuity to $L_1(G, \mu_G)$. This extension (also denoted I_G) preserves the properties of I_G on $C_0(G)$. For instance (Corollary 12.3), $\underset{x \rightsquigarrow {}_xF}{G \to (L_1(G, \mu_G), \|\ \|_1)}$ is continuous in x for each $F \in L_1(G, \mu_G)$. Simply choose $f \in C_0(G)$ such that $\|{}_xf - {}_xF\|_1 = \|f - F\|_1 < \varepsilon/3$ and let $\|{}_xf - f\|_1 < \varepsilon/3\ \forall x \in v_1(e)$. Then $\|{}_xF - F\|_1 \leq \|{}_xF - {}_xf\|_1 + \|{}_xf - f\|_1 + \|f - F\|_1 < \varepsilon\ \forall x \in v_1$.

Remark. The reader can easily verify that the same results hold for $L_p(G, \mu_G)$ $(1 \leq p < \infty)$.

Computation and Examples of Haar Concepts

We begin with the simplest applications to unimodular groups.

Example 12.19. (a) If $G=\{x_1\cdots x_n\}$ is discrete, then (Example 12.9) $I_G(f)=\sum_{i=1}^{n} f(x_i)$ and $\mu_G(E)=\sum_E \mu_G(x)=\mu_G(e)\cdot\aleph(E)$. If G is an infinite discrete group, $I_G(f)=\sum_{i=1}^{n} I_G(f(x_i)\chi_{x_i})=\left\{\sum_{i=1}^{n} f(x_i)\right\}I_G(\chi_e)$ and $\mu_G(E)=\mu_G(e)\cdot\aleph(E)$.

(b) If $G=(\mathbb{R}_0, +; d_1)$, every $f\in C_0(G)$ vanishes outside some finite interval and is therefore Riemann integrable. Thus $I_G(f)=\int_{-\infty}^{\infty} f(x)\,dx$ is a Haar integral and the corresponding μ_G is ordinary Lebesgue measure. The analog for $G=(\mathbb{C}, +; d_2)$ gives $I_g(f)=\int_{\mathbb{C}} f(z)\,dz = \int_{-\infty}^{\infty}\int_{-\infty}^{\infty} f(x+iy)\,dx\,dy$ and $\mu_G=dx\,dy$. Extension of I_G and μ_G to $\mathbb{R}^n$ and $\mathbb{C}^n=\mathbb{R}^{2n}$ follow similarly.

(c) If $G=S'(\mathbb{C})$ in Example 9.8, then $I_G(f)=\int_0^{2\pi} f\{e^{it}\}\,dt$ and μ_G is Lebesgue measure on $[0, 2\pi]\subset\mathbb{R}$.

(d) Let $G=\mathbb{R}\bullet$ and consider $I(f)=\int_{-\infty}^{\infty} f(t)\,F(t)\,dt$, where $F\in C(G)$ is to be found subject to the condition that $I(f)=I({}_xf)\ \forall f\in C_0(G)$ and $x\in G$. Then $\int_{-\infty}^{\infty} f(t)\,F(t)\,dt=\int_{-\infty}^{\infty} f(xt)\,F(t)\,dt=(1/|x|)\int_{-\infty}^{\infty} f(t')\,F(t'/t)\,dt'\ \forall f\in C_0(G)$ and (Corollary 12.17) $F(t)=(1/|x|)F(t/|x|)\ \forall x\in G$. Taking $t=x$ gives $F(x)=F(x/|x|)\cdot 1/|x|)$. Therefore I and $I_G(f)=\int_{-\infty}^{\infty} [f(t)/|t|]\,dt$ is a Haar integral for G, and the corresponding Haar measure is $\mu_G(E)=\int_{E^{-1}} dt/t$.

Example 12.20. The method in (d) above may be applied to nonunimodular groups. Let $G=\{(x, y)\in\mathbb{R}\times\mathbb{R}: x\neq 0\}$ with $(x_1, y_1)(x_2, y_2)=(x_1x_2, x_1y_2+y_1)$ be topologized as a subspace of $\mathbb{R}^2$ and consider integration relative to Lebesgue measure on $\mathbb{R}^2$. Assume the existence of an $F\in C(G)$ such that $I(f)=\iint_G f(x, y)F(x, y)\,dx\,dy\ \forall f\in C_0(G)$ and use the left invariance of I to determine F. Thus $\iint_G f(x, y)\,F(x, y)\,dx\,dy=I(f)=I({}_{(x_0,y_0)}f)=\iint_G f(x_0x, x_0y+y_0)F(x, y)\,dx\,dy=(1/x_0^2)\iint_G f(x', y')F[x'/x_0', (y'-y_0)/x_0]\,dx'\,dy'$ (via $x'=x_0x$ and $y'=x_0y+y_0$). Consequently, $1/x_0^2\,F[x/x_0, (y-y_0)/x_0]=F(x, y)\ \forall(x, y)\in G$ so that $1/x_0^2\,F(1, 0)=F(x_0, y_0)$; that is, $F(x, y)=1/x^2\,F(1, 0)$ and $I_G(f)=\iint_G [f(x, y)/x^2]\,dx\,dy$ is a left Haar integral for G. Accordingly, $\mu_G(E) = \iint_{E^{-1}} (1/|x|)\,dx\,dy$, where $E^{-1} = \{(1/x, -y/x) \in G : (x, y) \in E\}\ \forall E \in \mathfrak{R}_\sigma(\mathcal{C})$.

A simple computation confirms that $J_G(f) = \iint_G [f(x, y)/|x|]\,dx\,dy$ is a right Haar integral for G. Therefore $J_G(f) = \alpha I_G(\hat{f}) = \alpha I_G(\Delta f)$ for some $\alpha > 0$. Thus $\iint_G [f(x, y)/|x|]\,dx\,dy = \alpha \iint_G [f(x, y)/x^2]\,\Delta(x, y)\,dx\,dy\ \forall f \in C_0(G)$ yields $\Delta(x, y) = (|x|/\alpha)\ \forall (x, y) \in G$. Since $\Delta\{(x, y)^{-1}\} = \Delta(1/x, -y/x) = 1/|x|\alpha$, one has $1 = \Delta(x, y)\,\Delta\{(x, y)^{-1}\} = 1/\alpha^2$ and $\alpha = 1$. Obviously G is not unimodular.

The substitutions in Examples 12.19(d) and 12.20 suggest another frequently used technique for computing Haar entities and modular functions. It is assumed here that the reader is familiar with the elementary theory of functions of several variables, in particular, Riemann integration and Jacobians $\mathfrak{J}(T)$ of transformations in $\mathbb{R}^n$. Consider the following three conditions:

($*_1$) G is an open subset of $\mathbb{R}^n$ and, with the induced topology, is a topological group relative to some defined composition. (Observe that

$$\mathfrak{m}: \underset{(x,\,y)\rightsquigarrow xy}{G \times G \subset \mathbb{R}^{2n} \to G \subset \mathbb{R}^n}$$

may be viewed as the function $\mathfrak{m}(x_1 \cdots x_n, y_1 \cdots y_n)$, where $x = (x_1 \cdots x_n)$ and $y = (y_1 \cdots y_n)$ are points in G.)

($*_2$) The partial derivatives $\partial F_i/\partial x_j$ and $\partial F_i/\partial y_j$ $(i, j = 1, 2, \ldots, n)$ of $F_i = \pi_i \circ \mathfrak{m}$ exist and are continuous on $G \times G$ (here π_i denotes the projection of $\mathbb{R}^n$ onto the ith factor $\mathbb{R}$).

($*_3$) $\mathfrak{J}({}_x\mathfrak{m})$ and $\mathfrak{J}(\mathfrak{m}_x)$ are constant; that is, the Jacobians depend only on x for each fixed $x \in G$.

Conditions ($*_2$) ($*_3$) are satisfied if (as in our considerations) each $\pi_i(xy) = F_i(x_1 \cdots x_n, y_1 \cdots y_n) = \sum_{j,k} c_{j,k}^{(i)} x_j x_k + d_i$ $(i = 1, 2, \ldots, n)$, where the real numbers $c_{j,k}^{(i)}$ do not depend on x, y. One can also easily verify that $|\mathfrak{J}({}_e\mathfrak{m})| = 1$, that $\mathfrak{J}({}_{x_1x_2}\mathfrak{m}) = \mathfrak{J}({}_{x_1}\mathfrak{m})\mathfrak{J}({}_{x_2}\mathfrak{m})$, and that $l: x \rightsquigarrow |\mathfrak{J}({}_x\mathfrak{m})|$ is a continuous homomorphism from G to $\mathbb{R}_\bullet$. The dual statement also applies to $r: x \rightsquigarrow |\mathfrak{J}(\mathfrak{m}_x)|$. If $f \in C_0(G)$ and $x = (x_1 \cdots x_n) \in G$, we write $\int_G f(x)\,dx$ as an abbreviation of the ordinary n-dim Riemann integral $\iint \cdots \int_G f(x_1 \cdots x_n)\,dx_1 \cdots dx_n$.

The stage has now been set for proving

THEOREM 12.21. If G satisfies ($*_1$)–($*_3$), then

(i) $I_G(f) = \int_G f(x)[|\mathfrak{J}({}_x\mathfrak{m})|]^{-1}\,dx$ is a left Haar integral.

(ii) $J_G(f) = \int_G f(x)[|\mathfrak{J}(\mathfrak{m}_x)|]^{-1}\, dx$ is a right Haar integral.

(iii) The modular function is $\Delta(x) = |\mathfrak{J}({}_x\mathfrak{m})|/|\mathfrak{J}(\mathfrak{m}_x)|$.

Proof. Since $I_G(f)$ is linear and positive (the product measure in $\mathbb{R}^n$ is Haar, $G \subset \mathbb{R}^n$ is open, and Theorem 12.15 applies), it remains to prove that I_G is left invariant. Given a surjection $T: X = G \underset{x \rightsquigarrow y = Tx}{\rightarrow} Y = G$, the well-known formula for transformations of multiple integrals yields $\int_Y \varphi(y)\, dy = \int_{TX} \varphi(Tx)|\mathfrak{J}(T)|\, dx$. Now take $T = {}_{x_0}\mathfrak{m}: X = G \underset{x \rightsquigarrow x_0 x}{\rightarrow} Y = G$. Then ${}_{x^{-1}}f$, therefore ${}_{x_0^{-1}}f/|\mathfrak{J}({}_y\mathfrak{m})|$ belongs to $C_0(Y)\ \forall f \in C_0(X)$ and

$$I_G({}_{x_0^{-1}}f) = \int_{Y=G} \frac{({}_{x^{-1}}f)(y)}{|\mathfrak{J}({}_y\mathfrak{m})|}\, dy = \int_{{}_{x_0}\mathfrak{m}(X)=G} \frac{\{{}_{x_0^{-1}}f \circ {}_{x_0}\mathfrak{m}\}(x)|\mathfrak{J}({}_{x_0}\mathfrak{m})|}{|\mathfrak{J}({}_{x_0 x}\mathfrak{m})|}\, dx$$

$$= \int_G \frac{f(x)}{J({}_x m)}\, dx = I_G(f)$$

$\forall f \in C_0(X = G)$ and $x \in G$. The proof of (ii) follows similarly and is therefore omitted. If J_G is a right Haar integral for G, then $J_G(f) = \alpha I_G(\Delta f)$ for some $\alpha > 0$. Note that from $r, l \in C(G)$ and $\int_G f(x)[|\mathfrak{J}(\mathfrak{m}_x)|]^{-1}\, dx = \int_G f(x)\,\Delta(x)[\alpha/|\mathfrak{J}({}_x\mathfrak{m})|]\, dx\ \forall f \in C_0(G)$, there follows (Corollary 12.17) that $\Delta(x) = (1/\alpha)[|\mathfrak{J}({}_x\mathfrak{m})|/|\mathfrak{J}(\mathfrak{m}_x)|]$; but $1 = \Delta(e) = (1/\alpha)[|\mathfrak{J}({}_e\mathfrak{m})|/|\mathfrak{J}(\mathfrak{m}_e)|] = 1/\alpha$ $\alpha = 1$, as required for the proof of (iii). ■

Example 12.22. (a) In $G = \mathbb{R}_\bullet$ one has $\mathfrak{J}({}_t\mathfrak{m}) = t$ and $I_G(f) = \int_{-\infty}^{\infty} [f(t)/|t|]\, dt = J_G(f)$ [Example 12.19(c)]. In $G = (\mathbb{C} - \{0\}, \cdot\,; d_2)$, we obtain $|\mathfrak{J}({}_{(x+iy)}\mathfrak{m}| = |x + iy|^2$. Therefore

$$I_G(f) = \int_{\mathbb{C}} \frac{f(z)}{|z|^2}\, dz = \int_{-\infty}^{\infty} \int_{-\infty}^{\infty} \frac{f(x+iy)}{x^2 + y^2}\, dx\, dy = J_G(f).$$

(b) Return to Example 12.20, where ${}_{(x_0, y_0)}\mathfrak{m}: G \underset{(x,y) \rightsquigarrow (x_0 x, x_0 y + y_0)}{\longrightarrow} G$. Here $\mathfrak{J}({}_{(x_0, y_0)}\mathfrak{m}) = x_0^2$. Since $\mathfrak{m}_{(x_0, y_0)}(x, y) = (xx_0, xy_0 + y)$ and $\mathfrak{J}(\mathfrak{m}_{(x_0, y_0)}) = x_0$, we have proved anew that $\Delta(x, y) = |x|$.

Example 12.23. (a) Let $G_2(\mathbb{R})$ be as in Example 8.3. For any $A \in G_2(\mathbb{R})$ one has $|\mathfrak{J}({}_A\mathfrak{m})| = |A|^2 = |\mathfrak{J}(\mathfrak{m}_A)|$. Therefore $G_2(\mathbb{R})$ is a nonabelian, unimodular group. Identifying $G_2(\mathbb{R})$ with $\mathbb{R}^4$ also establishes that

$$I_G(f)=\int_{-\infty}^{\infty}\int_{-\infty}^{\infty}\int_{-\infty}^{\infty}\int_{-\infty}^{\infty}\frac{f(x_1,x_2,x_3,x_4)}{(x_1x_4-x_2x_3)^2}\,dx_1\,dx_2\,dx_3\,dx_4$$

is a Haar integral for $G_2(\mathbb{R})$.

(b) If $G_n(\mathbb{R})$ carries the $\mathbb{R}^{n^2}$-induced topology, then $|\mathcal{J}({}_A\mathfrak{m})|=|A|^n=|\mathcal{J}(\mathfrak{m}_A)|\ \forall A\in G_n(\mathbb{R})$. In this context

$$I_G(f)=\int_{-\infty}^{\infty}\cdots\int_{-\infty}^{\infty}\frac{f(x_1\cdots x_n,x_{n+1}\cdots x_{2n},\ldots,x_{n(n-1)}\cdots x_{n^2})}{|A|^n}\,dx_1\cdots dx_{n^2}$$

is a Haar integral for $G_n(\mathbb{R})$.

(c) If G is the subgroup $\left\{A=\begin{pmatrix}x & y\\ 0 & 1\end{pmatrix}: 0\neq x,\ y\in\mathbb{R}\right\}$ of $G_2(\mathbb{R})$, then G identifies with $\mathbb{R}^2$-(x axis) under $A \rightsquigarrow (x,y,0,1)\rightsquigarrow(x,y)$. We have already met this group in Examples 12.20 and 12.22(b). Although $I_G(f)=\int_{-\infty}^{\infty}\int_{-\infty}^{\infty}[f(x,y)]/x^2\,dx\,dy$ follows from (a) above, one cannot conclude that I_G is also right invariant, since G is not unimodular. This demonstrates that a subgroup of a unimodular group need not be unimodular (see problem 12C).

(d) The subgroup G of upper triangular matrices in $G_n(\mathbb{R})$ may be regarded as an open subset of $\mathbb{R}^{n(n+1)/2}$. In this context,

$$I_G(f)=\int_{-\infty}^{\infty}\cdots\int_{-\infty}^{\infty}\frac{f(x_1\cdots x_n,x_{n+2}\cdots x_{2n},x_{2n+3}\cdots x_{3n},\ldots,x_{n^2})}{|x_1^n x_{n+2}^{n-1}x_{n+3}^{n-2}\cdots x_{n^2}|}\,dx_1\cdots dx_{n^2}$$

is a left Haar integral for G. A right Haar integral has the form

$$J_G(f)=\int_{-\infty}^{\infty}\cdots\int_{-\infty}^{\infty}\frac{f(x_1\cdots x_n,x_{n+2}\cdots x_{2n},x_{2n+3}\cdots x_{3n},\ldots,x_{n^2})}{|x_1 x_{n+2}^{2}x_{2n+3}^{3}\cdots x_{n^2}^{n}|}\,dx_1\cdots dx_{n^2}.$$

Example 12.24. Let $(H,\mathcal{H})$ and $(K,\mathcal{K})$ be locally compact T_2 measurable groups (i.e., $\mathcal{H}$ and $\mathcal{K}$ are the Borel σ-rings of H and K, respectively).

(a) If μ_H and μ_K are left Haar measures for H and K, respectively, then $\mu_H\times\mu_K$ is a left Haar measure for $(H\times K,\mathcal{R}_\sigma(\mathcal{H}\times\mathcal{K}))$. Indeed, for any $(h_0, k_0) \in H \times K$ and Borel set $E \in \mathfrak{R}_\sigma(\mathcal{H} \times \mathcal{K})$, there is $(h_0, k_0)E = \{(h_0h, k_0k) : (h, k) \in E\}$. Since μ_K is left invariant and the "h_0h-section of $(h_0, k_0)E$" in H can be written $\{(h_0, k_0)E\}_{h_0h} = \{k_0k \in K : (h_0h, k_0k) \in E\} = \{k_0k : (h, k) \in E\} = k_0E_h$, one has

$$(\mu_H \times \mu_K)\{(h_0, k_0)E\} = \int_H \mu_K\{(h_0, k_0)E\}_{h_0h}\, d\mu_H = \int_H \mu_K(k_0E_h)\, d\mu_H =$$

$\int_H \mu_K(E_h)\, d\mu_H = (\mu_H \times \mu_K)(E)$ as required.

(b) If I and J are left Haar integrals for H and K, respectively, then $I \times J$ is a left Haar integral for $H \times K$.

For verification make use of Fubini's theorem to get $(I \times J)\{_{(h_0,k_0)}f\} = J_k[I_h\{f(h_0h, k_0k)\}] = J_k[I_h\{f(h, k_0k)\}] = I_h[J_k\{f(h, k_0k)\}] = I_h[J_k\{f(h, k)\}] = (I \times J)(f)\ \forall f \in C_0(H \times K)$ and $(h_0, k_0) \in H \times K$.

(c) Let Δ_H, Δ_K, and $\Delta_{H\times K}$ denote the modular functions of H, K and H×K, respectively. Then $\Delta_{H\times K}(h, k) = \Delta_H(h)\, \Delta_K(k)\ \forall (h, k) \in H \times K$. In particular, $H \times K$ is unimodular iff both H and K are unimodular. This follows from

$$\begin{aligned}
\Delta_{H\times K}(h_0, k_0)(I \times J)(f) &= (I \times J)\{f_{(h_0,k_0)}\} \\
&= J_k[I_h\{f(hh_0, kk_0)\}] \\
&= \Delta_H(h_0) J_k[I_h\{f(h, kk_0)\}] \\
&= \Delta_H(h_0) I_h[J_k\{f(h, kk_0)\}] \\
&= \Delta_H(h_0)\, \Delta_K(k_0) I_h[J_k\{f(h, k)\}] \\
&= \Delta_H(h_0)\, \Delta_K(k_0)(I \times J)(f).
\end{aligned}$$

Obviously $\Delta_{H\times K} \equiv 1$ if $\Delta_H = \Delta_K \equiv 1$. The converse also holds, since

$$\begin{aligned}
\Delta_H(h) = \Delta_H(h)\, \Delta_K(e_K) = \Delta_{H\times K}(h, e_K) = 1 &= \Delta_{H\times K}(e_H, k) \\
&= \Delta_H(e_H)\, \Delta_K(k) = \Delta_K(k)\ \forall (h, k) \in H \times K.
\end{aligned}$$

PROBLEMS

12A. Haar Integrals and Quotient Groups

Let H be a closed, normal subgroup of a locally compact T_2 group G. Prove that for each left Haar integral I_H and left Haar integral $I_{G/H}$ there is a left Haar integral I_G duch that $I_G(f) = I_{G/H}(\xi_f)$, where $\xi_f(Hx) = I_H(_xf)\ \forall x \in G$. Proceed as follows:

(a) Suppose that $f \in C_0(G)$ with $f(G - K) = 0$ for compact $K \subset G$; let $x \in G$ and define $_xf'$ as the restriction of $_xf$ to H. Then $_xf' \in C_0(H)$ with $_xf'(H - x^{-1}K \cap H) = 0$.

(b) Show that $\xi_f(Hx) = I_H({}_xf')$ is well defined and $\xi_f \in C_0(G/H)$ with $\xi_f(G/H - HK) = 0$. Furthermore, $\xi_f \geq 0$ {and $\xi_f \neq 0$} if $f \geq 0$ {and $f \neq 0$}.

(c) Use (b), (c) to verify that $C_0(G) \to C_0(G/H)$, $f \rightsquigarrow \xi_f$ is an onto congruence. [*Hint*: For $\varphi \in C_0(G/H)$ with $\varphi(G/H - HC) = 0$ for compact $C \subset G$ choose $\zeta \in C_0^+(G)$ such that $\zeta(C) = 1$ and define $\psi(x) = [\zeta(x)\varphi(Hx)]/\xi_\zeta(Hx)\chi_{G-\zeta^{-1}(0)}(x)$. Then $\varphi = \xi_{\psi'}$.]

(d) Why is $I_G(f) = I_{G/H}(\xi_f)$ let invariant? [*Hint*: First prove that ${}_{x_0H}\xi_f = \xi_{x_0f}$ $\forall x_0 \in G$ and $f \in C_0(G)$.]

(e) Let $H = \mathbb{I}$ carry $\mu_H(n) = 1$ $\forall n \in \mathbb{I}$ and let $G/H = S'(\mathbb{C})$ carry $dx/\sqrt{2\pi}$, where dx is Lebesgue measure on $\mathbb{R}$. The corresponding Haar integrals in (d) require that $I_G(F) = I_{G/H}(\xi_F) = 1/\sqrt{2\pi} \int_{-\infty}^{\infty} F(x)\,dx$ $\forall F \in C_0(G)$. [*Hint*: Justify that $\sum_{-\infty}^{\infty} \int_0^1 f(x+n)\,dx = \sum_{-\infty}^{\infty} \int_n^{n+1} F(x)\,dx = \int_{-\infty}^{\infty} F(x)\,dx$.]

12B. Haar Measure and Quotient Groups

Assume that H in Problem 12A is compact and normal.

(a) If μ_G is a left Haar measure for G, then $\mu_{G/H}(E) = \mu_G\{\nu^{-1}(E)\}$ is a left Haar measure for G/H. [*Hint*: $\nu^{-1}(E) \subset G$ is compact {Borel} $\forall$ compact {Borel} set $E \subset G/H$, and $\nu^{-1}(H \times E) = x\nu^{-1}(E)$ $\forall x \in G$ and $E \subset G/H$.]

(b) Let Δ_G and $\Delta_{G/H}$ denote the modular functions on G and G/H, respectively. Then $\Delta_{G/H}(Hx) = \Delta_G(x)$ $\forall x \in G$. In particular, G/H is unimodular iff G is unimodular. [*Hint*: Show that $\Delta_{G/H}(Hx)\mu_{G/H}(E) = \mu_{G/H}(EHx) = \mu_G\{\nu^{-1}(EHx)\} = \Delta_G(x)\mu_{G/H}(E)$.]

(c) In general, the compactness of H in (b) above cannot be removed. (Consider noncompact $H = G$.) Show, however, that G is unimodular iff G/Z is unimodular (where Z denotes the center of G).

(d) How would you state and prove the analog of (a) and (b), using left Haar integrals? [*Hint*: If I_G is a left Haar integral for G, define $I_{G/H}(F) = I_G(F \circ \nu^{-1})$ $\forall F \in C_0(G/h)$.]

12C. Unimodular {Sub}Groups

(a) If G is compact, $\Delta(G)$ is a compact subgroup of $\mathbb{R}$. Use this to confirm (Example 12.9) that $\Delta \equiv 1$. [*Hint*: What is the only compact subgroup of $\mathbb{R}_\bullet$?)

(b) If H is an open subgroup of G, then $\Delta_G(h) = \Delta_H(h)$ $\forall h \in H$. Prove this and conclude that an open subgroup of a unimodular group is unimodular.

(c) A closed subgroup of a unimodular group need not be unimodular. Why? [*Hint*: Consider Example 12.23(c).]

(d) If H is a closed, normal subgroup of G, then $\Delta_G(h) = \Delta_H(h)$ $\forall h \in H$. Therefore H is unimodular if G is. [*Hint*: Assume that I_G is the left Haar integral defined in Problem 12A. Show that $\xi_{f_h} = \Delta_H(h)\xi_f$ $\forall h \in H$ and obtain $\Delta_G(h)\,I_G(f) = \Delta_H(h)\,I_G(f)$ $\forall f \in C_0(G)$.]

(f) Ker $(\Delta)=\{x\in G:\Delta(x)=1\}$ is the largest closed, normal unimodular subgroup of G. Show further that Ker (Δ) contains every open unimodular subgroup of G but not necessarily every closed unimodular subgroup of G.

(g) If Ker (Δ) contains all closed unimodular subgroups of G, then G is unimodular. [*Hint*: $H_x=\{x, x^2, \dots, x^n, \dots\}$ is abelian $\forall x\in G$. Each $H_x\subset$ Ker (Δ) ensures that $G\subset$ Ker (Δ).]

12D. Nonunimodular Groups and Inverse Sets

Let μ_G be a left Haar measure for a nonunimodular group G. Produce an open set $E\subset G$ such that $\mu_G(E)<\infty$ and $\mu_G(E^{-1})=\infty$. [*Hint*: If $\Delta(x_0^{-1})\geq 2$ and $u^2\subset \Delta^{-1}\{(\frac{1}{2}, 2)\}$ for a symmetric $u\in\mathfrak{N}_e$, then $\{x_0^n u: n\in\mathbb{N}\}$ are pairwise disjoint. There exists a sequence of symmetric open nbds $E_k(e)\subset u$ such that $\mu_G(E_k)\leq 2^{-k}$ $\forall k\in\mathbb{N}$ and an increasing sequence $\{n_k: k\in\mathbb{N}\}$ of natural numbers such that $2^{-n_k+1}\leq\mu_G(E_k)$ $\forall k\in\mathbb{N}$. Take $E=\bigcup_{k=1}^{\infty} x_0^{n_k}E_k$ and use Theorem 12.13.]

12E. Nonunimodular Groups Via Semidirect Products

Let $H\boxtimes K$ be the semidirect product (Problem 9C) of locally compact T_2 groups H and K.

(a) If I_H and I_K are left Haar integrals for H and K, respectively, then $I_H\times I_K$ is a left Haar integral for $H\boxtimes K$. [Note that ${}_{(h_0,k_0)}f(h,k)=f(h_0h, \xi_h(k_0)k)$, where $\underset{h\rightsquigarrow\xi_h}{\xi: H\to\mathfrak{A}(K)}$ is the homomorphism that defines $H\boxtimes K$.]

(b) For each fixed $h_0\in H$ there is a $\delta(h_0)>0$ such that $I_K\{f(\xi_{h_0}(k))\}=\delta(h_0)\,I_K\{f(k)\}$ $\forall f\in C_0(K)$. Furthermore, $\delta: H\to\mathbb{R}_\bullet$ is a continuous homomorphism.

(a)′ Let μ_H and μ_K be left Haar measures for H and K, respectively. Then $\mu_H\times\mu_K$ is a left Haar measure for $(H\boxtimes K, \mathfrak{R}_\sigma(\mathcal{S}\times\mathfrak{T}))$, where $\mathcal{S}$ and $\mathfrak{T}$ are the Borel σ-rings of H and K. [*Hint*: It suffices to consider compact sets $C_s\in\mathcal{S}$ and $C_t\in\mathfrak{T}$. Define $(h_0, k_0)(C_s\times C_t)=\{(h_0h, \xi_h(k_0)k): (h,k)\in C_s\times C_t\}$ and make use of $(\mu_H\times\mu_K)(E)=\int_K \mu_H(E_h)\,d\mu_K$ $\forall E\in\mathfrak{R}_\sigma(\mathcal{S}\times\mathfrak{T})$. Do not confuse "the section $E_k\subset\mathfrak{T}$" with Eh which has no meaning here!]

(b)′ The analog of (b) for measures requires that $\mu_K\{\xi_{h_0}(B)\}=\delta(h_0)\mu_K(B)$ $\forall B\in\mathfrak{T}$. Prove that this is so.

(c) Use (a), (b) or (a)′, (b)′ to obtain $\Delta_{H\boxtimes K}(h,k)=\Delta_H(h)\,\Delta_K(k)\,\delta(h)$ $\forall(h,k)\in H\times K$. (Compare this with Example 12.24(c).)

(d) Semidirect products can be used to generate nonunimodular groups. Let H and K be unimodular. Then $H\boxtimes K$ is unimodular only if $\delta\equiv 1$; that is, only if ξ is the trivial homomorphism. In particular, the semidirect product of unimodular groups by a nontrivial homomorphism is nonunimodular. Prove all assertions above.

(e) Show that G in Example 12.23(c) is the semidirect product of $H=\mathbb{R}_\bullet$ and $K=\mathbb{R}$. Furthermore, $\underset{h\rightsquigarrow hm}{\xi: H\to\mathfrak{A}(K)}$ is a nontrivial homomorphism. Use this to that $\Delta_{H\boxtimes K}(h,k)=|h|$ $\forall(h,k)\in G$.

(f) Suppose H is not unimodular and let $K=\mathbb{R}$. If $\xi: H \to \mathcal{A}(K)$ with $\xi_h(k)=$ $\Delta_H(h^{-1})k$, then $\delta(h)=\Delta_H(h^{-1})$.
$h \rightsquigarrow \xi_h$

(g) Prove that every locally compact T_2 group G may be viewed as a subgroup of the unimodular group $G \boxtimes \mathbb{R}$.

12F. Measure and Topology in Groups

(a) All $\mathfrak{T}$-properties of a locally compact T_2 group $(G, \mathfrak{T})$ can be described via μ_G. Show that $\rho(E, F)=\mu_G(E \,\Delta F)$ is a pseudometric on $\mathfrak{R}_\sigma(\mathfrak{C})$ and prove the following:

(i) If $\mu_G(E)<\infty$ and $f(x)=\rho(xE, E)\, \forall x \in G$, then f is continuous. Conclude therefore that $S_\varepsilon(E)=\{x \in G: \rho(xE, E)<\varepsilon\}$ is $\mathfrak{T}$-open $\forall E \in \mathfrak{R}_\sigma(\mathfrak{C})$ with $\mu_G(E)<\infty$. [*Hint*: Use the regularity of μ_G.]

(ii) $\mathcal{U}_e=\{S_\varepsilon(E): \mu_G(E)<\infty$ and $\varepsilon>0\}$ is a $\mathfrak{T}$-nbd base at $e \in G$. [*Hint*: For each $u \in \mathfrak{N}_e$ there exists an $S_\varepsilon(E) \subset u$.]

(iii) Show (in comparison with Problem 8G) that $EE^{-1} \in \mathfrak{N}_e\, \forall E \in \mathfrak{R}_\sigma(\mathfrak{C})$ with $\mu_G(E)>0$.

(iv) A set $B \subset G$ is bounded (i.e., $B \subset C$ for some compact set $C \subset G$) iff $B \subset \{x \in G: \rho(xE, E) \leq \varepsilon\}$ for some Baire set E and $\varepsilon>0$ such $\varepsilon<2\mu(E)<\infty$.

(b) The measure theoretic analog of a T_2 group can be formulated as follows: A measure group is a σ-finite measure space $(G, \mathfrak{R}; \mu)$, where G is a group and both $\mathfrak{s}: G \times G \to G \times G$ and $\mathfrak{s}^{-1}$ preserve measurable sets. $(G, \mathfrak{R}; \mu)$ if left invariant if both
$(x,y) \rightsquigarrow (x,xy)$
μ and $\mathfrak{R}$ are left invariant ($\mathfrak{R}$ is left invariant if $xE \in \mathfrak{R}\, \forall W \in \mathfrak{R}$ and $x \in G$). One terms $(G, \mathfrak{R}, \mu)$ separated whenever $\{e\}=\bigcap_{\varepsilon>0} \{S_\varepsilon(E): 0<\mu(e)<\infty\}$.

(i) If ν is the restriction of μ_G [in (a) above] to the Baire subsets $\mathfrak{R}_\delta$ of $(G, \mathfrak{T})$, then $(G, \mathfrak{R}_\delta; \nu)$ is a separated left measure group. [*Hint*: Note that $\mathfrak{R}_\delta \times \mathfrak{R}_\delta$ is the σ-ring of Baire sets in $G \times G$ and $\mathfrak{s}$ is a homeomorphism.]

(ii) Every separated left measure group $(G, \mathfrak{R}; \mu)$ determines a locally compact T_2 topology on G. This is the Weil topology having $\{S_\varepsilon(E): 0<\varepsilon/2<\mu(e)<\infty$ and $E \in \mathfrak{R}\}$.

12G. Equivalent Unitary Representations

Two representations $\mathscr{R}: G \to L(Z, Z)$ and $\mathscr{R}': G \to L(Z', Z')$ are said to be equivalent if there is an onto congruence $T: Z \to Z'$ such that $\mathscr{R}_g=T^{-1}\mathscr{R}'_g T\; \forall g \in G$. Assume that G is compact and T_2.

(a) Let the representation $\mathscr{R}: G \to L(Z, Z)$ satisfy $\sup_G |||\mathscr{R}_g|||<\infty$. Then there is another inner product on Z whose norm is equivalent to the initial norm and with respect to which $\mathcal{R}$ is unitary. [*Hint*: If $(\,,\,)$ is the original inner product with norm $\|\;\|$, then $\langle y, z\rangle = \int_G (\mathcal{R}_g(y), \mathcal{R}_g(z))\, d\mu_G(g)$ is also an inner product on Z. Moreover, $\mathcal{R}_{g_0}$ is $\langle\,,\,\rangle$-unitary $\forall g_0 \in G$. The $\langle\,,\,\rangle$-induced norm satisfies $\langle\!\langle z\rangle\!\rangle^2 = \langle z, z\rangle = \int_G \|\mathcal{R}_g(z)\|^2\, d\mu_G(g) \le |||\mathcal{R}_g|||^2 \|z\|^2 = M^2\|z\|^2$ for $M = \sup_G |||\mathcal{R}_g|||$. On the other hand, $\|z\|^2 = \|\mathcal{R}_{g^{-1}}\{\mathcal{R}_g(z)\}\|^2 \le M\|\mathcal{R}_g(z)\|^2$ yields $\|z\|^2 = \int_G \|z\|^2\, d\mu_G(g) \le M \int_G \|\mathcal{R}_g(z)\|^2\, d\mu_G(g) = M\langle\!\langle z\rangle\!\rangle^2$.]

(b) Every finite dim representation $\mathcal{R}: G \to L(Z, Z)$ is equivalent to a unitary representation of G. [*Hint*: Assume that $Z = \mathbb{C}^n$. Since $\sup_G |||\mathcal{R}_g||| < \infty$ (Corollary 12.13), $\mathbb{C}^n$ carries $\langle\,,\,\rangle$ under which $\mathcal{R}'$ [defined as $\mathcal{R}$ on $(\mathbb{C}^n, \langle\,,\,\rangle)$] is unitary. Pick any $\langle\,,\,\rangle$-orthonormal basis $\mathcal{B}'$ for $\mathbb{C}^n$ and let T be the matrix expressing $\mathcal{B}'$ in terms of $\{e_i = (\delta_{i,n}): i = 1, 2, \ldots, n\}$. Then $\mathcal{R}_g = T^{-1}\mathcal{R}'_g T\ \forall g \in G$.]

References for Further Study

On Haar integration and measure: Hewitt and Ross [41]; Nachbin [67].

On measure and topology in groups: Halmos [38].

On representations and Haar measure: Hewitt and Ross [41]; Naimark [69].

CHAPTER III

Topological Vector Spaces

Perspective. This chapter is concerned with topological vector spaces (TVSps). Such spaces specialize topological groups and encompass the important class of locally convex TVSps which generalize normed linear spaces. This places TVSps in perspective to the theme of topologies compatible with increasingly enriched algebraic structure. Applications to vector-valued measure spaces provide further embellishment. The remainder of this chapter is devoted to the geometry of locally convex TVSps both for its own intrinsic beauty and its importance in classical mathematics.

13. TVSps and Topological Groups

TVSps are first viewed as topological additive (abelian) groups. As such, nbd properties and other related concepts can be localized at the origin. In the opposite direction sharp contrasts between TVSps and the theory of topological groups are given—these differences being due essentially to the scalar multiplication property of the former. We begin with some notation and a few preliminaries.

A vector space X over the field $\mathbb{F}$ is called real or complex depending on whether $\mathbb{F}=\mathbb{R}$ or $\mathbb{F}=\mathbb{C}$, respectively. We let $\mathfrak{L}(X, Y)$ denote the vector space of linear mappings from X to another vector space Y and take $L(X, Y)=\mathfrak{L}(X, Y)\cap C(X, Y)$ when X, Y also carry topologies. In this vein $X^*=\mathfrak{L}(X;\mathbb{F})$ and $X'=L(X;\mathbb{F})$ are termed the algebraic and continuous duals of X, respectively.

DEFINITION 13.1. A family $\mathcal{F}=\{f_\alpha\in\mathfrak{L}(X, Y_\alpha):\alpha\in\mathcal{Q}\}$ is called surjective if $\bigcap_{\mathcal{Q}} f_\alpha^{-1}(y_\alpha)\neq\varnothing\ \forall y=(y_\alpha)_{\mathcal{Q}}\in\prod_{\mathcal{Q}} Y_\alpha$.

If $Y_\alpha = Y\ \forall\alpha\in\mathcal{Q}$, then $\mathcal{F}\subset\mathfrak{L}(X, Y)$ is said to be

(i) injective if $\bigcap_{\mathcal{Q}} f_\alpha^{-1}(0)=\{0\}$;

(ii) concentrating if $\forall\ \{f_{\alpha_0}, f_{\alpha_1}, \ldots, f_{\alpha_n}\} \subset \mathfrak{F}$, either $f_{\alpha_0} \in [f_{\alpha_1}, \ldots, f_{\alpha_n}]$ (the vector subspace of $\mathfrak{L}(X, Y)$ spanned by $\{f_{\alpha_k} : 1 \leq k \leq n\}$) or for each $y_0 \neq 0$ in Y there is an $x_0 \in X$ such that $f_{\alpha_k}(x_0) = y_0\delta_{k,0}$.

Remark. Every $f_\alpha \in \mathfrak{F}$ is surjective when $\mathfrak{F}$ is surjective (concentrating). Essentially, $\mathfrak{F}$ is surjective (injective) iff $\mathfrak{e} : X \to \prod_{\mathfrak{A}} Y$, $x \rightsquigarrow \{f_\alpha(x)\}$ is surjective (injective).

Example 13.2. X^* is both injective and concentrating for a vector space X. Every $x_0 \neq 0$ can be extended to a basis $\mathfrak{B}$ for X and $f(x_0) = 1$ for $f \in X^*$ defined by $f : x = \sum_{i=1}^{n} \alpha_i x_i\ (x_i \in \mathfrak{B}) \rightsquigarrow \sum_{i=1}^{n} \alpha_i$. To verify that X^* is concentrating it suffices to consider any $f_0 \neq 0$ and linearly independent set $\{f_1, \ldots, f_n\} \in X^*$. Evidently (ii) holds for $n = 0$. If (ii) holds for $n \in \mathbb{N}$ and $\{f_0, f_1, \ldots, f_{n+1}) \in X^*$, there exist $x_k \in X$ such that $f_k(x_j) = \delta_{j,k}\ \forall j, k = 0, 1, \ldots, n+1$. Thus $x - \sum_{k=1}^{n+1} f_0(x)\, x_k \in \bigcap_{k=1}^{n+1} f_k^{-1}(0)\ \forall x \in X$. Therefore $f_0(\tilde{x}) = 0\ \forall \tilde{x} \in \bigcap_{k=1}^{n+1} f_k^{-1}(0)$ so that $f_0 = \sum_{k=1}^{n+1} f_0(x_k) f_k$, or $f_0(x_0) \neq 0$ for $\tilde{x}_0 \in \bigcap_{k=1}^{n+1} f_k^{-1}(0)$, in which case every scalar $y_0 \neq 0$ is the f_0 image of $x_0 = y_0/[f_0(\tilde{x}_0)]\tilde{x}_0 \in \bigcap_{k=1}^{n+1} f_k^{-1}(0)$.

The above also demonstrates that every linearly independent set $\{f_1, \ldots, f_n\} \in X^*$ determines a linearly independent set $\{x_1, \ldots, x_n\} \in X$ such that $f_i(x_j) = \delta_{i,j}\ \forall 1 \leq i, j \leq n$.

DEFINITION 13.3. A topological vector space (TVS) is a vector space X (over the field $\mathbb{F} = \mathbb{R}$ or $\mathbb{C}$) with a topology $\mathfrak{T}$ on X such that both

$$\underset{(x,\,y) \rightsquigarrow x+y}{\mathfrak{m} : X \times X \to X} \quad \text{and} \quad \underset{(\alpha,x) \rightsquigarrow \alpha x}{\mathfrak{s} : \mathbb{F} \times X \to X}$$

are continuous. In this context $\mathfrak{T}$ is said to be compatible with X's linear structure. The collection of all such compatible topologies on X is denoted $T(X)$.†

Remark. Note that $\underset{((\alpha,x),y) \rightsquigarrow \alpha x + y}{\xi : (\mathbb{F} \times X) \times X \to X}$ is continuous. In fact, the identification of $(\mathbb{F} \times X) \times 0$ with $\mathbb{F} \times X$ and $1 \times X$ with X assures that ξ is continuous iff both $\mathfrak{m}$ and $\mathfrak{s}$ are simultaneously continuous. (See also Problem 13A.)

† Since every TVS is a topological group, $T(X)$ above subsumes this notation in Definition 8.1.

Although every normed linear space (NLS) is a T_2VS (by virtue of $\|(x+y)-(x_0+y_0)\| \le \|x-x_0\| + \|y-y_0\|$ and $\|\alpha x - \alpha_0 x_0\| \le |\alpha|\,\|x-x_0\| + |\alpha-\alpha_0|\,\|x_0\|$ and $\mathcal{G} \in T(X)$ for a vector space X, not every topology on X is compactible with X's linear structure [Example 8.2(a)]. This also follows (Theorem 1.22) from the fact that a TVS X carries uniformities $\mathcal{U}_X$ and $\mathfrak{c}_X$ having respective bases $\{U_u : u \in \mathfrak{N}_0\}$ and $\{\mathcal{C}_u : u \in \mathfrak{N}_0\}$, where $U_u = \{(x, y) \in X \times X : y \in u + x\}$ and $\mathcal{C}_u = \{u + x : x \in X\}$. In this context (Theorem 8.16) every $f \in L(X, Y)$ is $\mathcal{U}_X - \mathcal{U}_Y$ uniformly continuous.

Many elementary TVS properties are particularizations of already familiar results for topological groups.

THEOREM 13.4. For a TVS X

(i) $\underset{x \rightsquigarrow x+y}{\mathfrak{m}_y : X \to X}$ and $\underset{x \rightsquigarrow \alpha x}{\mathfrak{s}_\alpha : X \to X}$, therefore $\underset{x \rightsquigarrow \alpha x + y}{\xi_{\alpha,y} : X \to X}$ are homeomorphisms $\forall y \in X$ and $0 \ne \alpha \in \mathbb{F}$. In particular, $u(0)$ is an X-nbd of 0 iff $\alpha u + x$ is a nbd of x $\forall x \in X$ and $\alpha \ne 0$.

(ii) Suppose $A \subset X$ is open, $B \subset X$ is closed, and $C, D \subset X$ are arbitrary. Then

(a) $\overline{\alpha C + x} = \alpha \bar{C} + x$ $\forall x \in X$ and $\alpha \in \mathbb{F}$.
(b) $A + C$ is open.
(c) $B + C$ is closed if C is compact.
(d) $\overline{C + D} \subset \bar{C} + \bar{D}$.

(iii) If $\mathfrak{N}_0$ is an X-nbd base at $0 \in X$, then $\operatorname{Int} E = \{x \in E : u + x \subset E$ for some $u \in \mathfrak{N}_0\}$ and $\bar{E} = \bigcap_{\mathfrak{N}_0} \{E + u(0)\}$. Furthermore, X is T_k $(k = 0, 1, 2)$ iff $\bigcap_{\mathfrak{N}_0} u = \{0\}$.

(iv) X is pseudometrizible {metrizible} iff it is 1° at $0 \in X$ {and T_2}.

Proof. Theorem 8.4, Corollary 8.5, and Problem 8F; Theorem 8.15 and Theorem 8.17. ■

Scalar multiplication now enables us to introduce

DEFINITION 13.5. A subset K of a TVS X is said to be

(i) absorbent if $\forall x \in X$, there is an $\varepsilon > 0$ such that $x \in \lambda K$ $\forall |\lambda| \ge \varepsilon$; (this is equivalent to requiring that there be a $\delta > 0$ such that $\lambda x \in K$ $\forall |\lambda| \le \delta$);

(ii) balanced if $\alpha K \subset K$ $\forall |\alpha| \le 1$;

(iii) convex if $\forall x, y \in K$ one has $\alpha x + (1-\alpha)y \in K$ $\forall \alpha \in [0, 1]$;

(iv) absolutely convex (abbreviated, abs. convex) if it is both balanced and convex.

The intersection of all balanced {convex, abs convex} sets which contain K is called the balanced hull K_b {convex hull K_c, balanced convex hull K_{bc}} of K. Note (Problem 14A) that $K_{bc} = (K_b)_c \neq (K_c)_b$ in general.

Some useful, frequently referred to properties of balanced convex sets are given below.

THEOREM 13.6. Let K be a subset of a vector space X. Then

(i) $K_b = \{\alpha x : x \in K \text{ and } |\alpha| \leq 1\}$.

(ii) $K_c = \left\{ \sum_{i=1}^{n} \alpha_i x_i : x_i \in K \text{ and } \alpha_i > 0 \text{ with } \sum_{i=1}^{n} \alpha_i = 1 \right\}$.

(iii) $K_{bc} = \left\{ \sum_{i=1}^{n} \alpha_i x_i : x_i \in K \text{ and } \sum_{i=1}^{n} |\alpha_i| \leq 1 \right\}$.

(iv) If K is balanced, $\alpha K \subset \beta K$ for $|\alpha| \leq |\beta|$.

(v) If K is convex, then $\sum_{i=1}^{n} \alpha_i K = \left(\sum_{i=1}^{n} \alpha_i \right) K$ for positive $\{\alpha_i\}$.

(vi) $\sum_{i=1}^{n} \alpha_i K = \left(\sum_{i=1}^{n} |\alpha_i| \right) K$ if K is abs convex.

(vii) An abs convex set K is absorbent iff $X = [K] = \bigcup_{n=1}^{\infty} nK$, where $[K]$ denotes the vector subspace of X spanned by K.

Proof. (i) Since $B = \{\alpha x : x \in X \text{ and } |\alpha| \leq 1\}$ is balanced and contains K, it contains K_b. On the other hand, $B \subset K_b$, since $B \subset C$ for each balanced set $C \supset K$. For (ii) first assume that C is a convex set that contains K. Then $\sum_{i=1}^{n} \alpha_i x_i \in C$ $\forall \{x_1, \ldots, x_n\} \in K$ and positive $\{\alpha_1 \cdots \alpha_n\}$ such that $\sum_{i=1}^{n} \alpha_i = 1$. (Suppose inductively that the claim is true for $1 \leq k \leq n-1$ and let $\sum_{i=1}^{n} \alpha_i = 1$. Then $\sum_{i=1}^{n-1} \alpha_i = 0$ implies $\sum_{i=1}^{n} \alpha_i x_i + \alpha_n x_n = x_n \in K \subset C$, and $\sum_{i=1}^{n-1} \alpha_i = \delta > 0$ implies $\sum_{i=1}^{n} \alpha_i x_i = \delta \sum_{i=1}^{n-1} (\alpha_i/\delta) x_i + (1-\delta) x_n \in C$.) The proof of (ii) now proceeds as in (i). Next, suppose $x = \sum_{i=1}^{n} \alpha_i x_i$ for $x_i \in K$ and $\sum_{i=1}^{n} |\alpha_i| \leq 1$. If

$\sum_{i=1}^{n} |\alpha_i| = 0$, then $x = 0 \in K_{bc}$. If $\sum_{i=1}^{n} \alpha_i x_i = \delta < 1$, take $\beta_i = |\alpha_i|/\delta$ and write $x = \sum_{i=1}^{n} \beta_i(\alpha_i/\beta_i)x_i$. From $|\alpha_i/\beta_i| = \delta < 1$ follows $(\alpha_i/\beta_i)x_i \in K_b$ and $\left(\text{since } \sum_{i=1}^{n} |\beta_i| = 1\right)$ we also have $x \in (K_b)_c = K_{bc}$. In the other direction, every $x = \sum_{i=1}^{n} \beta_i(\alpha_i x_i) \in K_{bc}$ (where $x_i \in K$, $|\alpha_i| \le 1$ and $\beta_i > 0$ satisfy $\sum_{i=1}^{n} \beta_i = 1$) has $\sum_{i=1}^{n} |\beta_i \alpha_i| \le \sum_{i=1}^{n} |\beta_i|$ and therefore belongs to the right-hand side of the equality in (iii). Turning to (iv), it becomes clear that $\beta = 0$ implies $\alpha = 0$ and $\alpha K = \beta K = \{0\}$. If $\beta \ne 0$, then $|\alpha/\beta| \le 1$ and $(\alpha/\beta)x \in K\ \forall x \in K$. In particular, $\alpha K \subset \beta K$. One part of (v) always holds, since $\left(\sum_{i=1}^{n} \alpha_i\right)K \subset \sum_{i=1}^{n} \alpha_i K\ \forall K \subset X$. If $y = \alpha_1 x_1 + \alpha_2 x_2$ for $x_1, x_2 \in K$, then $\alpha_1/(\alpha_1+\alpha_2)x_1 + \alpha_2/(\alpha_1+\alpha_2)x_2 \in K$ yields $y \in (\alpha_1+\alpha_2)K$. Proceed by induction to obtain $\sum_{i=1}^{n} \alpha_i K \subset \left(\sum_{i=1}^{n} \alpha_i\right)K$. Clearly (vi) holds if $\alpha_1 = \alpha_2 = 0$. If $|\alpha_1| + |\alpha_2| \ne 0$, then $\alpha_1/(|\alpha_1|+|\alpha_2|)K + \alpha_2/(|\alpha_1|+|\alpha_2|)K \subset K_{bc} = K$ and so $\alpha_1 K + \alpha_2 K \subset (|\alpha_1|+|\alpha_2|)K = |\alpha_1|K + |\alpha_2|K \subset \alpha_1 K + \alpha_2 K$. This proves (vi) for $n = 2$ and the complete proof of (vi) follows by induction. Clearly, $X \subset \bigcup_{n=1}^{\infty} nK \subset [K]$ for any absorbent set K. If K is abs convex and $x = \sum_{i=1}^{n} \alpha_i x_i \in [K] = X$, then $x \in \lambda K\ \forall |\lambda| \ge \varepsilon = \sum_{i=1}^{n} |\alpha_i|$ establishes that K is absorbent. ∎

In practice, one usually begins with a filterbase of subsets containing the origin which determines a unique topology compatible with the vector space's linear structure.

Theorem 13.7. A TVS X has a nbd base $\mathfrak{N}_0$ at $0 \in X$ which satisfies

(i) $u(0)$ is absorbent and balanced $\forall u \in \mathfrak{N}_0$
(ii) for each $u \in \mathfrak{N}_0$ there is a $v \in \mathfrak{N}_0$ such that $v + v \subset u$.

Conversely, every filterbase $\mathfrak{N}$ of subsets of a vector space X that satisfies (i), (ii) determines a unique topology $\mathfrak{T} \in T(X)$ having $\mathfrak{N}$ as a nbd base at $0 \in K$.

Proof. Every X-nbd $v(0)$ is absorbent. Since $\mathfrak{s}$ is continuous at $(0, x)\in \mathbb{F}\times X$, one has $\mathfrak{s}\{S_\delta(0)\times(w+x)\}\subset v$ for some $\delta>0$ and X-nbd $w(0)$. Thus $\alpha x\in v\ \forall|\alpha|<\delta$ and repeating this argument with $x=0$ shows that $u(0)=\bigcup_{|\alpha|\le\delta}\alpha w\subset v$. Since u is a balanced nbd of 0 (Theorem 13.4), the collection $\mathfrak{N}_0$ of a balanced nbds is a nbd base at $0\in X$ which satisfies (i). The continuity of $\mathfrak{m}$ at $(0, 0)\in X\times X$ also assures that $\mathfrak{N}_0$ satisfies (ii).

For the converse situation define $\mathfrak{N}_x=\mathfrak{N}+x\ \forall x\in X$ and let M_x be the filter generated by each $\mathfrak{N}_x$. Since $x\in x+u(0)\ \forall u\in\mathfrak{N}$, property N_1 of Appendix t.1 holds. Given any $x+u\in\mathfrak{N}_x$, let $v\in\mathfrak{N}$ satisfy $v+v\subset u$. If $y\in x+v(0)$, then $y+v\subset x+u$ and (since $y+v\in\mathfrak{N}_y$) $x+u\in M_y$. Therefore N_4 holds and the unique topology $\mathfrak{T}$ on X having M_x as its nbd system at x also has $\mathfrak{N}_x$ as a nbd base at $x\in X$. It remains only to verify that $\mathfrak{T}\in T(X)$. Fix x_0, $y_0\in X$ and let $u(0)\in\mathfrak{N}$. Then $u(0)+x_0+y_0$ is a $\mathfrak{T}$-nbd of x_0+y_0 and $\mathfrak{m}\{(v(0)+x_0)\times(v(0)+y_0)\}\subset u(0)+x_0+y_0$ for $v\in\mathfrak{N}$ satisfying $v+v\subset u$. This proves $\mathfrak{m}$ continuous. Since v is absorbent, there is a $\delta>0$ such that $\lambda x_0\in v\ \forall|\lambda|\le\delta$. Let $\alpha_0\in\mathbb{F}$ be fixed, take $\alpha\in\mathbb{F}$ such that $|\alpha-\alpha_0|<\varepsilon=\min\{1,\delta\}$ and let $x\in v+x_0$. Then $\alpha x-\alpha_0x_0=\alpha(x-x_0)+(\alpha-\alpha_0)x_0\in v+v\subset u$ and $\mathfrak{s}\{S_\varepsilon(\alpha_0)\times(v+x_0)\}\subset u+x_0$ establishes the continuity of $\mathfrak{s}$. ■

Remark. It is important to note that $\forall v\in\mathfrak{N}_0$ and $\lambda\in\mathbb{F}$ there is a $w\in\mathfrak{N}_0$ such that $w\subset v$. Use (ii) and induction to obtain $2^n w\subset v\ \forall n\in\mathbb{N}$. If n is large enough to ensure that $|\lambda|\le 2^n$, then $\lambda 2^{-n}w\subset w$ and $\lambda w\subset 2^n\, w\subset v$.

COROLLARY 13.8. A TVS has a nbd base of balanced open (closed) sets.

Proof. If $w(0)\in\mathfrak{N}_0$ is chosen open in Theorem 13.7, then $u(0)$ is open (Theorem 13.4) and the first part of the corollary follows. For the second half use the fact that a TVS (being a regular space) has a nbd base of closed sets at $0\in X$. Given any balanced $v\in\mathfrak{N}_0$, let $u(0)$ be a balanced nbd such that $\bar{u}\subset v$. It remains only to prove that $\bar{u}$ is balanced, which is precisely our next result. ■

THEOREM 13.9. Suppose that X is a TVS. Then the closure of a balanced {convex} set is balanced {convex} and the interior of a convex set is convex.

Proof. Suppose $x\in\bar{B}$, where $B\subset X$ is balanced. For any balanced nbd $u\in\mathfrak{N}_0$ there is an $x_0\in B\cap(u+x)$. In particular, $\lambda x_0\in\lambda\{B\cap(u+x)\}\subset B\cap(u+\lambda x)$ ensures that $\lambda x\in\bar{B}\ \forall|\lambda|\le 1$. Now suppose that B is convex and let $x, y\in\bar{B}$. Given any $u\in\mathfrak{N}_0$ and balanced $v\in\mathfrak{N}_0$ with $v+v\subset u$, there are elements $x_0\in B\cap(v+x)$ and $y_0\in B\cap(v+y)$. For each $\alpha \in [0, 1]$ one therefore has $\alpha x_0 + (1-\alpha)y_0 \in \alpha B \cap (v + \alpha X) + (1-\alpha)B \cap \{v + (1-\alpha)y\} \subset B \cap \{u + \alpha x + (1-\alpha)y\}$ and $\alpha x + (1-\alpha)y \in \bar{B}$. Finally, let $x, y \in \text{Int } B$ for a convex set $B \subset X$. Fix $\alpha \in [0, 1]$, let $z = \alpha x + (1-\alpha)y \in B$ and take $E = \alpha u + (1-\alpha)v$, where u and v are nbds in B of x and y, respectively. Since $E = \bigcup_{x_u \in u} \{\alpha x_u + (1-\alpha)v\} \subset B$ is a nbd of z, one has $z \in \text{Int } B$ as required. ■

Remark. Note that the interior of the balanced set $\{(x, y) \in \mathbb{R}^2 : x, y \geq 0\}$ does not contain the origin and is therefore not balanced.

The following result will be useful later on:

THEOREM 13.10. Let $(X, \mathfrak{T})$ be a complete TVS. If $\mathfrak{T}' \in T(X)$ is finer than $\mathfrak{T}$ and has a nbd base of $\mathfrak{T}$-closed sets at 0, then $(X. \mathfrak{T}')$ is complete.

Proof. We may assume (Theorems 13.7 and 13.9) that the $\mathfrak{T}$-closed, $\mathfrak{T}'$-base nbds are all balanced. If $\mathfrak{F}$ is any $\mathfrak{T}'$-Cauchy filter in X, then $\mathfrak{F} \xrightarrow{\mathfrak{T}} x_0 \in X$. Given any $\mathfrak{T}'$-nbd $u(0)$, let $v(0)$ be a $\mathfrak{T}$-closed balanced $\mathfrak{T}'$-nbd satisfying $v + v \subset u$ and choose any $x_v \in F_v$, where $F_v \in \mathfrak{F}$ is v-sized. Then $F_v \subset x_v + v$ and $x_0 \in \overline{F_v}^{\mathfrak{T}} \subset \overline{x_v + v}^{\mathfrak{T}} = x_v + v$ so that $F_v \subset x_0 + v + v \subset x_0 + u$ and $\mathfrak{F} \xrightarrow{\mathfrak{T}'} x_0$ follows. ■

In comparison with Theorem 8.10, consider

THEOREM 13.11. Let $f \in \mathfrak{L}(X, Y)$, where X and Y are TVSps.

(i) f is continuous {nearly continuous, open, nearly open} iff f is continuous {nearly continuous, open, nearly open} at $0 \in X$.

(ii) If $Y = \mathbb{F}$, then $f \not\equiv 0$ is an open mapping and the following are equivalent:

(a) f is continuous;
(b) $f^{-1}(0)$ is closed in X;
(c) f is bounded on some $u \in \mathfrak{N}_0$.

Proof. Only (ii) needs verification. Suppose that $E \subset X$ is open and $x_0 \in E$. Then $E - x_0$ is absorbent and for $f(x) = 1$ there is a $\delta > 0$ such that $\lambda x \in E - x_0 \ \forall |\lambda| \leq \delta$. Moreover, $S_\delta\{f(x_0)\} = f(x_0) + S_\delta(0) \subset f(E)$. Clearly, $(a) \Rightarrow (b)$. Since $f \not\equiv 0$, one has $f(x_0) = 1$ for some $x_0 \in X$. If $f^{-1}(0)$ is closed, there is a balanced $u \in \mathfrak{N}_0$ such that $\{x_0 + u\} \cap f^{-1}(0) = \varnothing$ and $u \subset$

$\{x \in X : |f(x)| < 1\}$ establishes that $(b) \Rightarrow (c)$. Surely (c) $\Rightarrow$ (a), since $\sup_u |f(x)| < M$ implies $f[(\varepsilon/M)u] \subset S_\varepsilon(0)\ \forall \varepsilon > 0$. ■

We settle now, once and for always, the case of finite dim T_2VSps.

THEOREM 13.22. Every n-dim T_2VS is topologically isomorphic to $\mathbb{F}^n$.

Proof. It is assumed that $\mathbb{F}^n$ carries the norm $\|\alpha\|_2 = \sqrt{\sum_{i=1}^{n} |\alpha_i|^2}$ for $\alpha = (\alpha_1 \cdots \alpha_n) \in \mathbb{F}^n$. If $\{x_1, \ldots, x_n\}$ is a basis for a T_2VS $(X, \mathfrak{T})$, the linear bijection $t : \alpha \leadsto x = \sum_{i=1}^{n} \alpha_i x_i$ determines a norm $\|x\| = \|\alpha\|_2$ on X under which t is a congruence. We claim that $\mathfrak{T}$ coincides with this norm topology $\|\ \|$ on X. Given any $u \in \mathfrak{N}_0$, let $v(0)$ be a balanced nbd satisfying $\sum_n v \subset u$ and take $\delta = \min_{1 \le i \le n} \delta_i$, where $\delta_i > 0$ are chosen such that $\lambda x_i \subset v\ \forall |\lambda| \le \delta_i$. Then $t\{S_{\delta, \|\,\|_2}(0)\} \subset S_{\delta, \|\,\|}(0) \subset u$ and $\mathfrak{T} \subset \|\ \|$ follows, as does the continuity of t. Since $\{\alpha \in \mathbb{F}^n : \|\alpha\|_2 = \varepsilon\}$ is compact, $B_\varepsilon = \{x \in X : \|x\| = \varepsilon\}$ is $\mathfrak{T}$-compact and closed in X. If $u \cap B_\varepsilon = \varnothing$ for balanced $u \in \mathfrak{N}_0$, then $u \subset S_{m, \|\,\|}(0)$ for some $m \in \mathbb{N}$ [otherwise some $x_0 \in u \cap \{X - S_{n, \|\,\|}(0)\}\ \forall n \in \mathbb{N}$ and $\|x\| \ge n > \varepsilon$ implies that $\varepsilon x_0/\|x_0\| \in u \cap B_\varepsilon$] and it follows that $(\varepsilon/m)u \subset S_{\varepsilon, \|\,\|}(0)$. Consequently, $\|\ \| \subset \mathfrak{T}$ on X. ■

Remark. The T_2 condition above cannot be omitted, since an n-dim indiscrete TVS is not homeomorphic to $\mathbb{F}^n$.

COROLLARY 13.13. (i) Every finite dim subspace of a T_2VS is closed.

(ii) $\mathfrak{L}(X, Y) = L(X, Y)$ if Y is a TVS and X is a finite dim T_2VS.

Proof. View an n-dim subspace M of a T_2VS X as a uniform subspace of the separated uniform space X. Since M and $\mathbb{F}^n$ are unimorphic (Theorem 8.16), M is complete [Example 6.9(c)] and closed in X (Theorem 6.6). The proof of (i) is therefore complete. (ii) If $\{x_1 \cdots x_n\}$ is a basis for the T_2VS X, then X carries the norm $\left\|\sum_{i=1}^{n} \alpha_i x_i\right\| = \sqrt{\sum_{i=1}^{n} |\alpha_i|^2}$ and the proof that $f \in \mathfrak{L}(X, Y)$ is continuous is similar (substitute $y_i = f(x_i)$ for x_i) to the continuity proof of t in Theorem 13.12. ■

Recall (Theorem 8.15) that a TVS is locally compact iff it has a relatively compact $u \in \mathfrak{N}_0$, in which case (Theorem 5.9) it has a totally bounded nbd of 0. A good measure of the effect of scalar multiplication can be seen in

THEOREM 13.14. The following are equivalent for a T_2VS X:

(i) dim X is finite;
(ii) X is locally compact;
(iii) X has a totally bounded nbd of 0.

Proof. Obviously (i)$\Rightarrow$(ii) (Theorem 13.11) and (ii)$\Rightarrow$(iii). If $u \in \mathfrak{N}_0$ is balanced and totally bounded, $u \subset \bigcup_{i=1}^{n} (x_i + \frac{1}{2}u)$ for some finite set $\{x_1 \cdots x_n\} \subset X$. Then $H = \left[\bigcup_{i=1}^{n} x_i\right]$ is finite dim and so closed in X. We claim that $H = X$, thereby establishing that (iii)$\Rightarrow$(i). Suppose to the contrary that $x_0 \in X - H$ and define $S = \{\alpha \neq 0 \text{ in } \mathbb{F} : H \cap (x_0 + \alpha u) \neq \varnothing\}$. Since u is balanced, $S \neq \varnothing$. Now $x_0 + w \subset X - H$ for $w \in \mathfrak{N}_0$ implies $\alpha_0 w \subset u$ for some $\alpha_0 \neq 0$. (Indeed, let $u \subset \bigcup_{i=1}^{m} (x_i + w_1)$ for a balanced $w_1 \in \mathfrak{N}_0$ satisfying $w_1 + w_1 \subset w$, assume $\alpha_i x_i \in w_1$ for $\alpha_i \neq 0$, and take $|\alpha_0| = \min_{1 \leq i \leq m} \{1, |\alpha_i|\}$. Then $\alpha_0 u \subset \bigcup_{i=1}^{m} (\alpha_i x_i + \alpha_0 w_1) \subset w$.) Since $x_0 + \alpha u \subset X - H$ for $|\alpha| \leq |\alpha_0|$, we have $S \subset \{\alpha \in \mathbb{F} : |\alpha| \geq |a_0|$ and $\delta = \inf_S |\alpha| > 0$. This, however, leads to the following contradiction: For any $\alpha \in S$ satisfying $|\alpha| > 2\delta$ and any $y_0 \in H \cap (x_0 + \alpha u)$ the inclusion $(1/\alpha)(y_0 - x_0) \in u \subset H + \frac{1}{2}u$ yields $y_0 - x_0 \in \alpha x_h + (\alpha/2)u$ for some $x_h \in H$, and this means that $\alpha/2 \in S$ satisfies $|\alpha| > 2\delta$. ■

This is a good place to emphasize that all statements for topological groups do not filter down to TVSps; there is a price to pay for this enriched TVS structure in that scalar multiplication must also be satisfied. For example, a vector space X with the discrete topology $\mathfrak{D}$ is a (locally compact) T_2 group but not necessarily a TVS. In fact, $(X, \mathfrak{D})$ is a TVS iff $X = \{0\}$.

A few comparisons of TVSps with earlier uniform spaces are given below.

Example 13.15. Let $\mathfrak{F}(X; \mathbb{F})$ be the vector space of mappings from a set X to the field $\mathbb{F}$. For each $\varepsilon > 0$ and subset $E \subset X$, define $u_{E,\varepsilon}(0) = \{f \in \mathfrak{F}(X; \mathbb{F}) : \sup_E |f(x)| < \varepsilon\}$.

(a) $\mathfrak{F}_p(X;\mathbb{F})$ is a complete T_2VS having $\{u_{E,\varepsilon}:\varepsilon>0$ and finite sets $E\subset X\}$ as a nbd base at 0. That $\mathfrak{T}_p\in T(\mathfrak{F}(X;\mathbb{F}))$ is left as an easy exercise. Completeness follows from Theorem 7.3. It is worth noting here that $\mathfrak{T}_p$ is metrizible iff $\aleph(X)\leq\aleph_0$. To be sure, $\aleph(X)\leq\aleph_0$ implies $\{u_{\bigcup_{i=1}^{n} x_i, 1/n}:n\in\mathbb{N}\}$ is a nbd base at 0, whereas $\aleph(X)>\aleph_0$ ensures that no countable collection $\{u_{E_n,\varepsilon_n}:n\in\mathbb{N}\}$ can be a nbd base at 0 (since $x\notin\bigcup_n E_n$ yields $u_{E_n,\varepsilon_n}\not\subset u_{\{x\},1}\ \forall n\in\mathbb{N}$).

(b) If X is a topological space $C_\kappa(X;\mathbb{F})$ is a T_2VS since $\mathfrak{N}_0=\{u_{K,\varepsilon}:\varepsilon>0$ and compact $K\subset X\}$ satisfies Theorem 13.7 and $\bigcap_{\mathfrak{N}_0} u_{K,\varepsilon}=\{0\}$; for instance, $u_{K,\varepsilon}$ is absorbent, since for any $f\in C(X;\mathbb{F})$ one has $\sup_K|f|=M_f<\infty$ and $\lambda f\in u_{K,\varepsilon}\ \forall|\lambda|<\varepsilon/M_f$. Note that all $u_{K,\varepsilon}\in\mathfrak{N}_0$ are open, since $F\in u_{K,\varepsilon}$ implies $\sup_K|F|=M_F<\varepsilon$ and $F+u_{K,\varepsilon-M_F}\subset u_{K,\cdot}$. An Application of Problem 7D also reveals that $C_\kappa(X;\mathbb{F})$ is complete if X is either locally compact or 1°.

(c) $\mathfrak{F}_{\mathcal{S}}(X;\mathbb{F})$ cannot be a TVS if some $f\in\mathfrak{F}(X;\mathbb{F})$ is unbounded on a member $E\in\mathcal{S}$ since $u_{E,\varepsilon}\in\mathfrak{N}_0$ will not be absorbent. Nevertheless, $u_{E,\varepsilon}$ is balanced and satisfies $u_{E,\varepsilon/2}+u_{E,\varepsilon/2}\subset u_{E,\varepsilon}$ so that $\mathfrak{F}_{\mathcal{S}}(X;\mathbb{F})$ is still a topological group.

A TVS is pseudometrizible iff it is 1°, in which case the pseudometric may be assumed translation invariant† (see Problem 13L). Two concrete illustrations of TVSps with a translation invariant pseudometric are given below.

Example 13.16. Let $\mathcal{S}$ be the vector space of all sequences $\{x=(x_1,x_2,\ldots):x_i\in\mathbb{F}\}$.

(a) If $\{\lambda_n\}$ is any fixed sequence of positive numbers satisfying $\sum_{n=1}^{\infty}\lambda_n<\infty$, then $d(x,y)=\sum_{n=1}^{\infty}\lambda_n|x_n-y_n|/(1+|x_n-y_n|)$ is a translation invariant metric whose topology $\mathfrak{T}(d)$ is compatible with $\mathcal{S}$'s linear structure. Verification that $\mathfrak{T}(d)\in T(\mathcal{S})$: Clearly, $|a+b|/(1+|a+b|)\leq|a|/(1+|a|)+|b|/(1+|b|)\ \forall a,b\in\mathbb{F}$ implies $d(\tilde{x}+\tilde{y},x+y)\leq d(\tilde{x},x)+d(\tilde{y},y)$ and the continuity of $\mathfrak{m}$ follows. If $-1<a\leq b$, then $a/(1+a)\leq b/(1+b)$ (since $x/(1+x)$ is increasing for $x>-1$) so that $\gamma|x_n|/(1+\gamma|x_n|)\leq|x_n|/(1+|x_n|)$ and $d(\gamma x,0)\leq d(x,0)$ for $|\gamma|\leq 1$. To prove $\mathfrak{s}:\mathbb{F}\times X\to X$ continuous let $S_\varepsilon(\alpha X)=\{z\in X:d(z,\alpha x)<\varepsilon\}$ be any $\mathfrak{T}(d)$-nbd

† A pseudometric ρ on a vector space X is translation invariant if $\rho(x+z,y+z)=\rho(x,y)\ \forall x,y\in X$. This, of course, is equivalent to requiring that $\rho(x,y)=\rho(x-y,0)\ \forall x,y\in X$.

of αx. Then $d(\beta y, \alpha x) = d\{\beta y - \alpha x, 0\} \leq d\{(\beta - \alpha)x, 0\} + d\{\alpha(y - x), 0\} + d\{(\beta - \alpha)(y - x), 0\} \leq d\{(\beta - \alpha)x, 0\} + 2\varepsilon/3$ for $|\beta - \alpha| < 1$ and $d(y - x, 0) < \varepsilon/3$. (We have assumed that $|\alpha| \leq 1$. If $|\alpha| > 1$, then $d\{\alpha(y - x), 0\} \leq |\alpha|\, d(y - x, 0)$ and ε is replaced by $\varepsilon/|\alpha|$.) Since $\sum_{n=1}^{\infty} \lambda_n < \infty$, some $k \in \mathbb{N}$ exists such that $\sum_{k+1}^{\infty} \lambda_n < \varepsilon/6$. Let $\Gamma = \max_{1<n\leq k} \lambda_n$ and $M = \max_{1<n\leq k} |x_n|$ and take $\varepsilon' = \varepsilon/6k$. Then $\sum_{n=1}^{k} \lambda_n |\beta - \alpha||x_n|/(1 + |\beta - \alpha||x_n|) < \varepsilon/6$ and $d\{(\beta - a)x, 0\} < \varepsilon/3$ for $|\beta - \alpha| < \varepsilon'' = \min\{\varepsilon', 1\}$. In particular, $\mathfrak{s}$ maps $S_{\varepsilon''}(a) \times S_{\varepsilon/3}(x)$ into $S_\varepsilon(\alpha x)$ and is therefore continuous.

(b) A sequence $\{x_n = (x_n^{(1)}, x_n^{(2)}, \cdots) : n \in \mathbb{N}\}$ is d-convergent to $y = (y_1, y_2, \ldots) \in \mathcal{S}$ iff $\lim_n x_n^{(k)} = y_k\ \forall k \in \mathbb{N}$. Thus $\mathcal{S}$ is complete. We leave as an exercise confirmation that $\mathcal{S}$ is separable and noncompact.

Example 13.17. Let $\mathbb{S} = \mathbb{S}(X, \mu)$ be the space of measurable functions on a finite measure space $(X, \mathfrak{R}_\sigma; \mu)$. For each $\varepsilon, \delta > 0$ define $u_{\varepsilon,\delta}(0) = \{f \in \mathbb{S} : \mu\{x \in X : |f(x)| > \varepsilon\} < \delta\}$. Then $\mathfrak{N}_0 = \{u_{\varepsilon,\delta} : \varepsilon, \delta > 0\}$ satisfies Theorem 13.7. To see that $u_{\varepsilon,\delta}$ is absorbent consider any $g \in \mathbb{S}$. Then (e.g., see Natanson [73]) there is a bounded function $h \in \mathbb{S}$ with $\sup_X |h| < M$ such that $\mu\{x \in X : g(x) \neq h(x)\} < \delta$. If $M = \gamma\varepsilon$ for $\gamma > 0$, then $g \in u_{M,\delta} = \gamma u_{\varepsilon,\delta}$ and (since $u_{\beta,\delta} \subset u_{\alpha,\delta}$ for $\beta > \alpha > 0$) it follows that $g \in \lambda u_{\varepsilon,\delta}\ \forall |\lambda| \geq M/\varepsilon$.

(a) If $\mathfrak{T} \in T(\mathbb{S})$ has $\mathfrak{N}_0$ as its nbd base at 0, then $\chi_{\{x\}} \in \bigcap_{\mathfrak{N}_0} u_{\varepsilon,\delta}\ \forall x \in X$ and so $\mathfrak{T}$ is not T_2. However, $\mathfrak{T}$ is pseudometrizible, since it has $\{u_{(1/n),(1/m)} : n, m \in \mathbb{N}\}$ as a nbd base at $0 \in \mathbb{S}$. In fact, $\rho(f, g) = \int_X |f(x) - g(x)|/(1 + |f(x) - g(x)|)\, d\mu$ is a translation invariant pseudometric satisfying $\mathfrak{T}(\rho) = \mathfrak{T}$ on $\mathbb{S}$. Simply define $E_\varepsilon(f) = \{x \in X : |f| > \varepsilon\}$ and use

$$\frac{\varepsilon}{1+\varepsilon}\mu\{E_\varepsilon(f)\} \leq \int_{E_\varepsilon(f)} \frac{|f|}{1+|f|}\, d\mu \leq \rho(f, 0) \leq \mu\{E_\varepsilon(f)\} + \varepsilon\mu\{X - E_\varepsilon(f)\}$$

to show that $S_{\delta\varepsilon}(0) \subset u_{\varepsilon,\delta}(0)$ and $u_{\varepsilon',\varepsilon/2}(0) \subset S_\varepsilon(0)\ \forall \varepsilon, \delta > 0$ and $\varepsilon' < \min\{\varepsilon/2, (\varepsilon/2\mu(X))\}$. We note in passing that $\mathfrak{T}$ is T_2 and ρ is a metric if $\mathbb{S}$ is viewed as the quotient space under the equivalence relation $f = g$ iff $f = g$ a.e. on X.

(b) The above inequalities also demonstrate that $\mathfrak{T}(\rho)$-convergence is equivalent to convergence in measure. Since $f_m - f_n \xrightarrow{\text{meas}} 0$ implies

$f_n \xrightarrow{\text{meas}} f$ for some $f \in \mathcal{S}$ (e.g., see Halmos [38]), it follows that ρ is a complete pseudometric for $\mathcal{S}$.

Boundedness in a TVS

Example 13.15 may be generalized, and this serves as motivation for briefly focusing on another key notion associated with a TVS.

DEFINITION 13.18. A subset B of a TVS X is termed bounded if $\forall u(0) \in \mathfrak{N}_0$ there is an $\alpha > 0$ such that $B \subset \alpha u$. Furthermore, B is said to be absorbed by $E \subset X$ if there is an $\alpha > 0$ such that $B \subset \lambda E \; \forall |\lambda| > \alpha$. Note, therefore, that $B \subset X$ is bounded iff it is absorbed by every $u \in \mathfrak{N}_0$.

Of considerable importance later on is

DEFINITION 13.19. A TVS X is said to be quasi-complete if every closed bounded subset of X is complete.

Remark. Every product of quasi-complete spaces is quasi-complete, since $\overline{\prod_{\mathcal{a}} \pi_\alpha(B)} = \prod_{\mathcal{a}} \overline{\pi_\alpha(B)} \; \forall B \subset \prod_{\mathcal{a}} X_\alpha$.

Remark. Completeness implies quasi-completeness, which in turn implies sequential completeness (by Theorem 13.20$^+$). The reverse implications are generally not true. Every semireflexive space (Definition 20.1) is quasi-complete, and $(l_1\sigma(l_1, l_\infty))$ is a sequentially complete, nonquasi-complete TVS (Problem 19E).

A TVS X is a uniform space. In this context (Theorem 5.6) $B \subset X$ is totally bounded iff $\forall u \in \mathfrak{N}_0$ there are $\{x_1, \ldots, x_n) \subset B$ such that $B \subset \bigcup_{i=1}^{n} \{u + x_i\} = u + \bigcup_{i=1}^{n} x_i$. The simplest (total) boundedness properties of a TVS are summarized in

THEOREM 13.20. (i) Every totally bounded set is bounded.
(ii) If $A, B \subset X$ are (totally) bounded, so are

(a) subsets of B;
(b) $\bar{B}$ and B_b;
(c) $A \cup B$, $A + B$, and $\alpha B \; \forall \alpha \in F$;
(d) $f(B) \subset Y$ for $f \in L(X, Y)$.

Proof. If $B \subset X$ is totally bounded and $u \in \mathfrak{N}_0$, pick $v \in \mathfrak{N}_0$ such that $v + v \subset u$. Then $B \subset v + \bigcup_{i=1}^{n} x_i$ for $\{x_1 \cdots x_n\} \subset B$ and there is a $\lambda \geq 1$ such that $x_i \in \lambda v\ \forall 1 \le i \le n$. The boundedness of B now follows from $B \subset v + \lambda v \subset \lambda u$. Most of (ii) is obvious or a consequence of Theorem 5.9. We verify only (b). Given any $u \in \mathfrak{N}_0$, pick (Corollary 13.8) a closed balanced nbd $w(0) \subset u(0)$ and an $\alpha > 0$ such that $B \subset \alpha w$. From the boundedness of B and $(\bar{B})_b \subset \overline{(\alpha w)}_b = \alpha w \subset \alpha u$ follows the boundedness of $(\bar{B})_b$ as well as that of $\bar{B}$ and B_b. Suppose, next, that B is totally bounded and let $(\mu; \check{X}, \check{\mathcal{U}}_X)$ be the completion of $(X, \mathcal{U}_X)$. Then $\overline{\mu(B)}$ is compact and $\{\overline{\mu(B)}\}_b$ is totally bounded in $(\check{X}, \check{\mathcal{U}}_X)$ (Theorem 6.8, Problem 13C). It follows that $\{\mu(B)\}_b$, therefore $\mu^{-1}\{\mu(B)\}_b$ and $B_b \subset \mu^{-1}\{\mu(B)\}_b$ are totally bounded. ■

Remark. Note that Cauchy sequences are totally bounded in a TVS.

In a TVS X every bounded nbd $B(0)$ gives rise to a nbd base of bounded sets $\{(1/n)B : n \in \mathbb{N}\}$. The collection of closed, balanced, bounded subsets of X is fundamental, meaning that every bounded subset of X is contained in some such closed, balanced, bounded set. These assertions are easily verified. Perhaps not so obvious is a counterexample to Theorem 13.20(i).

Example 13.21. Surely, $B = \{e_n = (\delta_{m,n})_{m \in \mathbb{N}} : n \in \mathbb{N}\} \subset l_p$ $(2 \le p < \infty)$ is bounded, since $B \subset \{x \in l_p : \|x\|_p = 1\} \subset (2/\varepsilon) S_\varepsilon(0)\ \forall \varepsilon > 0$. However, $\|e_i - e_j\|_p = \sqrt[p]{2}$ and $S_\varepsilon(e_i) \cap S_\varepsilon(e_j) = \varnothing$ for $i \ne j$ and $\varepsilon < \sqrt[p]{2}$. Therefore no finite collection $\{e_1, \ldots, e_k\} \subset B$ satisfies $B \subset \bigcup_{i=1}^{k} \{S_\varepsilon(0) + e_i\} = \bigcup_{i=1}^{k} S_\varepsilon(e_i)$ for $\varepsilon < \sqrt[p]{2}$, and B is not totally bounded.

The following useful characterization of boundedness is employed frequently:

THEOREM 13.22. A subset B of a TVS X is bounded iff $\lambda_n x_n \to 0$ for all sequences $\{x_n\} \in B$ and $\{\lambda_n\} \in \mathbb{F}$ such that $\lambda_n \to 0$.

Proof. Suppose $B \subset X$ is bounded and let $u \in \mathfrak{N}_0$ be balanced. Then $\lambda B \subset u$ for some $\lambda \ne 0$. If $\lambda_n \to 0$ and $\{x_n\} \in B$, then $|\lambda_n| < |\lambda|$ and $\lambda_n x_n = (\lambda_n/\lambda)\lambda x_n \subset u\ \forall n > N(\lambda)$. Thus $\lambda_n x_n \to 0$. On the other hand, if B is not bounded, there is a sequence $\{y_n\} \in B$ such that $y_n \notin nu\ \forall n \in \mathbb{N}$. This, of course, is impossible, since $(1/n)y_n \to 0$. ■

DEFINITION 13.23. A mapping $f \in \mathfrak{L}(X, Y)$ between TVSps X, Y is said to be bounded if it preserves bounded sets. $B\mathfrak{L}(X, Y)$ denotes the subspace of bounded $f \in \mathfrak{L}(X, Y)$.

A partial converse to $L(X, Y) \subset B\mathfrak{L}(X, Y)$ [Theorem 13.18(a)] is

THEOREM 13.24. $L(X, Y) = B\mathfrak{L}(X, Y)$ whenever X is 1°.

Proof. Suppose $f \in B\mathfrak{L}(X, Y) - L(X, Y)$ so that $f^{-1}(v)$ is not an X-nbd for some Y-nbd $v(0)$. If $\mathfrak{N}_0 = \{u_1 \supset u_2 \supset \cdots u_n \supset u_{n+1} \supset \cdots\}$ is an X-nbd base at 0 satisfying Theorem 13.7, then $u_n \not\subset f^{-1}(v)$ and some $x_n \in u_n$ satisfies $f(x_n) \notin nv$ for each $n \in \mathbb{N}$. This, however, contradicts Theorem 13.20(d) in that $B = \{x_n : n \in \mathbb{N}\}$ is bounded, but $f(B)$ is not. ■

Remark. Spaces for which $L(X, Y) = B\mathfrak{L}(X, Y)$ are considered in Section 19.

COROLLARY 13.25. Let (X, p) be seminormed and suppose (Y, q) is {semi} normed.

(i) $f \in L(X, Y)$ iff there is a $\lambda > 0$ such that $q\{f(x)\} \leq \lambda p(x)\ \forall x \in X$. Thus

$$|||f||| = \sup_{p(x)\neq 0} \frac{q\{f(x)\}}{p(x)} = \sup_{p(x)\leq 1} q\{f(x)\} = \sup_{p(x)=1} q\{f(x)\}$$

is a {semi} norm on $L(X, Y)$.

(ii) $L(X, Y)$, $|||\ |||$ is Banach if (Y, q) is complete.

Proof. The existence of $\lambda > 0$ ensures that $f \in L(X, Y)$. If no such $\lambda > 0$ exists, $q\{f(x_n)\} > np(x)$ for $x_n \in X$ and so $q\{f(x_n/p(x_n))\} > n$ for $x_n/p(x_n) \in S_{1,p(0)}$ implies $f \notin B\mathfrak{L}(X, Y)$. This also establishes that $\{q\{f(x)\}/p(x) : p(x) \neq 0\}$ has a lub $|||f||| \leq \lambda$ for $f \in L(X, Y)$. Therefore $|||\ |||$ is well defined and easily shown to be a {semi} norm such that $q\{f(x)\} \leq |||f|||\, p(x)\ \forall x \in X$. Accordingly, $\sup_{p(x)} q\{f(x)\} \leq |||\ |||$. Also, by definition, every $\varepsilon > 0$ determines an x_ε with $p(x_\varepsilon) \neq 0$ such that

$$|||f||| - \varepsilon \leq \frac{q\{f(x_\varepsilon)\}}{p(x_\varepsilon)} = q\left\{f\left(\frac{x_\varepsilon}{p(x_\varepsilon)}\right)\right\} \leq \sup_{p(x)=1} q\{f(x)\}.$$

This (being true $\forall \varepsilon > 0$) means that $|||f||| = \sup_{p(x)=1} q\{f(x)\} \leq \sup_{p(x)\leq 1} q\{f(x)\}$. Next, suppose (Y, q) is complete and let $\{f_n\} \subset L(X, Y)$ be $|||\ |||$-Cauchy. Then $\{f_n(x)\}$ is q-Cauchy and convergent in Y for each fixed $x \in X$. Pick one such limit $y_x \in Y$ and define $F(x) = y_x\ \forall x \in X$. Clearly, $F \in \mathfrak{L}(X, Y)$. Since $\{f_n\}$ is $|||\ |||$-Cauchy, one has $\sup_n |||f_n||| \leq M < \infty$ for some $M > 0$. Thus $q\{f_n(x)\} \leq Mp(x)\ \forall n \in \mathbb{N}$ implies $\lim_n q\{f_n(x)\} \leq Mp(x)$. Since p is continuous and $q\{F(x)\}=\lim_n q\{f_n(x)\}$, we also have $q\{F(x)\}\le M\,p(x)\;\forall x\in X$. In other words, $F\in L(X, Y)$. Finally, $q\{f_n(x)-f_m(x)\}\le \varepsilon p(x)\;\forall m,\ n\ge N(\varepsilon)$ implies $q\{F(x)-f_m(x)\}=\lim_n q\{f_n(x)-f_m(x)\}\le \varepsilon\, p(x)$ for $m\ge N(\varepsilon)$, and there follows $f_n \xrightarrow{|||\,|||} F$ as required. ■

$\mathscr{S}$-*Topologies on* $B\mathfrak{L}(X, Y)$

For the promised extensions of Example 13.15 it will be useful to recall (Definition 7.7) that $a_\alpha^{-1}(W_V[f])=\{g\in\mathfrak{F}(X;\,Y,\mathscr{V}):g(x)\in V[f(x)]\;\forall x\in E_\alpha\}$ is a $\mathfrak{T}_{\mathscr{S}}$-subbase nbd $\forall E_\alpha\in\mathscr{S}$ and $V\in\mathscr{V}$. Now assume that X, Y are TVSps and let $\mathscr{S}$ be a collection of X-bounded sets such that $\forall B_1, B_2\in\mathscr{S}$ some $B_3\in\mathscr{S}$ satisfies $B_1\cup B_2\subset B_3$. (For example, $\mathscr{S}$ may be taken to be $\mathscr{S}_{\mathfrak{b}}=\{$all bounded subsets of $X\}$.) For each $B\in\mathscr{S}$ and Y-nbd $u(0)$ we define $w_{B,u}=\{f\in B\mathfrak{L}(X, Y):f(B)\subset u\}$ and prove

THEOREM 13.26. If X, Y, and $\mathscr{S}$ are as above and $\mathfrak{N}_0$ is a Y-nbd base of balanced sets, then $\mathscr{W}_0=\{w_{B,u}: B\in\mathscr{S}$ and $u\in\mathfrak{N}_0\}$ is a nbd base at 0 for $\mathfrak{T}_{\mathscr{S}}\in T(B\mathfrak{L}(X, Y))$. Furthermore, $\mathfrak{T}_{\mathscr{S}}$ is T_2 if Y is T_2 and $X=[\bigcup_{\mathscr{S}} B]$.

Proof. $\mathscr{W}_0$ satisfies Theorem 13.7 and is therefore the nbd base at 0 for a unique topology $\mathfrak{T}\in T(B\mathfrak{L}(X, Y))$. Since $U_u\in\mathscr{U}_Y$ and $a_B^{-1}(W_{U_u}[0])=w_{B,u}$, both topologies $\mathfrak{T}$ and $\mathfrak{T}_{\mathscr{S}}$ share the same subbase sets and so $\mathfrak{T}=\mathfrak{T}_{\mathscr{S}}$. Finally, suppose that $X=[\bigcup_{\mathscr{S}} B]$ and Y is T_2. If $f\not\equiv 0$ in $B\mathfrak{L}(X, Y)$, then $f(x_0)\neq 0$ for some $x_0\in B\in\mathscr{S}$ and $f(x_0)\notin u(0)$ for some $u\in\mathfrak{N}_0$. Therefore $f\notin\bigcap_{\mathscr{W}_0} w_{B,u}$ and so $\mathfrak{T}_{\mathscr{S}}$ is T_2 (Theorem 8.15). ■

Remark. Note that $w_{B,u}$ is convex if $u\in\mathfrak{N}_0$ is convex.

Every TVS $B\mathfrak{L}_{\mathscr{S}}(X, Y)$ is a topological subgroup of the topological additive group $\mathfrak{L}_{\mathscr{S}}(X, Y)\subset\mathfrak{F}_{\mathscr{S}}(X, Y)$ [Example 13.15(c)]. A few other statements worth noting are included below.

Example 13.27. (a) $\mathfrak{L}(X, Y)$ is closed in $\mathfrak{F}_{\mathscr{S}}(X, Y)$ if $\mathscr{S}$ covers X. This follows (Example 7.8) from $\mathfrak{T}_{\mathscr{S}_\sigma}\subset\mathfrak{T}_{\mathscr{S}}$ on $\mathfrak{F}(X, Y)$ and the fact that $\mathfrak{L}(X, Y)$ is closed in $\mathfrak{F}_\sigma(X, Y)$. Specifically, $f\in\overline{\mathfrak{L}(X, Y)}^\sigma\subset\mathfrak{F}(X, Y)$ implies that some filter $\mathfrak{G}\in\mathfrak{L}(X, Y)$ satisfies $\mathfrak{G}(x)\to f(x)\;\forall x\in X$ (Theorem 7.2) and $\lim\mathfrak{G}(\alpha x_1+x_2)=\alpha\lim\mathfrak{G}(x_1)+\lim\mathfrak{G}(x_2)$ ensures that $f\in\mathfrak{L}(X, Y)$.

(b) $B\mathfrak{L}(X, Y) = L(X, Y)$ is closed in $\mathfrak{F}_\kappa(X, Y)$ when X is 1° (Theorem 13.24, Problem 7D). If every $x \in X$ is an interior point of some $B \in \mathcal{S}$, then $L(X, Y) = C(X, Y) \cap \mathfrak{L}(X, Y)$ is closed in $\mathfrak{F}_{\mathcal{S}}(X, Y)$ [Theorem 7.10, Example 7.8(b)].

(c) Observe (Theorem 7.11) that $\mathfrak{L}_{\mathcal{S}}(X, Y)$ in (a) $\{\mathfrak{L}_{\mathcal{S}}(X, Y)$ in (b)$\}$ is complete if Y [equivalently, $\mathfrak{F}_{\mathcal{S}}(X, Y)$] is complete. In general, however, $\mathfrak{L}_{\mathcal{S}}(X, Y)$ is not complete (cf. Theorem 18.33).

(d) (See Example 7.16.) $\Phi_\sigma = \Phi_\lambda$ for every equicontinuous set $\Phi \subset L(X, Y)$. Reason: If $B \in \mathcal{S}_\lambda$ and $u \in \mathfrak{N}_0$, there exist balanced nbds $w(0) \in \mathfrak{N}_0$ and $v(0) \subset X$ such that $w + w + w \subset u$ and $\Phi(v) \subset w$. Therefore $B \subset \bigcup_{i=1}^{n} (v + x_i)$ for $\{x_1 \cdots x_n\} \subset B$ implies $\phi_0 + W_{\sum_{i=1}^{n} x_i, w} \subset \phi_0 + W_{B,u}\ \forall \phi_0 \in \Phi$ and $\mathfrak{T}_{\mathcal{S}} \subset \mathfrak{T}_{\mathcal{S}_\sigma}$ on Φ follows.

(e) Every equicontinuous set $\Phi \subset L(X, Y)$ is $\mathfrak{T}_{\mathcal{S}_b}$-bounded, therefore $\mathfrak{T}_{\mathcal{S}}$-bounded [Example 7.8(b)]. Why? For any $\mathfrak{T}_{\mathcal{S}_b}$-nbd $W_{B,u} \subset L(X, Y)$ one has $\Phi(v) \subset u$ for some X-nbd $v(0)$ and $\Phi \subset \lambda W_{B,u}$ is a consequence of $B \subset \lambda v$ for some $\lambda > 0$.

(f) If X and Y are NLSps, then $f_n \to 0 \in L_{\mathcal{S}}(X, Y)$ implies $|||f_n||| \to 0$. Indeed, $f_n \xrightarrow{\mathfrak{T}_{\mathcal{S}}} 0$ requires that $f_n \in W_{S_1(0), S_{1/m}(0)}$ for $n > N(m)$. If $|||f_n||| \not\to 0$, then $1/m > \sup_{S_1(0)} |||f_n(x)||| = |||f_n||| > \varepsilon$ for some $\varepsilon > 0$ and $n > N(m)$, which is impossible for large enough $m \in \mathbb{N}$.

The following result is of fundamental importance:

THEOREM 13.28 (Banach Steinhaus). Suppose X is a Catg II TVS and Y is a TVS. Then $\Phi \subset L(X, Y)$ is equicontinuous iff $\Phi(x) \subset Y$ is bounded $\forall x \in X$.

Proof. The necessary part of the proof follows from Example 13.27(e) and Definition 7.1. For sufficiency consider any Y-nbd $u(0)$ and let $v(0)$ be a closed balanced nbd satisfying $v + v \subset u$. Clearly, $E = \bigcap_{\Phi} \phi^{-1}(v)$ is closed, balanced, and absorbent so that $X = \bigcup_{n=1}^{\infty} nE$ and $\operatorname{Int} E \neq \varnothing$ (Theorems 13.3 and 13.4). In particular, $(E - E)(0)$ is a nbd of $0 \in X$ for which $\Phi(E - E) \subset \Phi(E) - \Phi(E) \subset v - v \subset u$. ■

Remark. This theorem is extended to a larger class of TVSps in Theorem 19.17.

COROLLARY 13.29. Suppose $(X, \|\ \|)$ is a Banach space and $(Y, \|\ \|)$ is a NLS. If $\sup_{\Phi} \|\phi(x)\| < \infty\ \forall x \in X$, then $\sup_{\Phi} |||\phi||| < \infty$.

Proof. Since Φ is equicontinuous, $\Phi(\overline{S_\lambda(0)}) \subset \overline{S_1(0)}$ for some $\lambda > 0$ and $(\lambda/\|x\|)\|\phi(x)\| = \|\phi(x/\|x\|)\| \leq 1\ \forall x \in X$ ensures that $|||\phi||| \leq (1/\lambda)\ \forall \phi \in \Phi$. ■

PROBLEMS

13A. Continuity Considerations

(a) Show (Definition 13.3) that $\mathfrak{m}$ is continuous iff each $\mathfrak{m}_x$ is continuous and $\mathfrak{m}$ is continuous at $(0, 0)$. Similarly, $\mathfrak{s}$ is continuous iff $\mathfrak{s}$ is continuous at $(0, 0) \in \mathbb{F} \times X$ and continuous in each argument separately. Exactly where does the continuity of $\mathfrak{s}$ break down for $(X, \mathfrak{D})$ when $X \neq \{0\}$?

(b) Let X, Y, Z be TVSps such that X is Catg II and Y is 1°. Then a bilinear mapping $m: X \times Y \to Z$ is continuous at $(0, 0)$ if it is continuous in each argument separately. [*Hint*: Assume $\{v_n(0): n \in \mathbb{N}\}$ is a Y-nbd base at 0. Given any Z-nbd $u(0)$, let $w(0)$ be a closed balanced Z-nbd such that $w - w \subset u$ and define $E_n = m^{-1}/[x \cap v_n](w)$. Then $H(0) \subset E_{n_0} - E_{n_1}$ for some x-nbd $H(0)$ and $n_0, n_1 \in \mathbb{N}$, and $m(H(0) \times v_{n_0} \cap v_{n_1}) \subset u(0)$.]

(c) Every nonzero (complex-valued) linear functional $f = f_1 + if_2$ on a complex TVS X has a decomposition $f = f_{\text{Re}} + if_{\text{Im}}$, where $f_{\text{Re}}, f_{\text{Im}} \in L(X; \mathbb{R})$. Furthermore, $f \in X'$ iff $f_{\text{Re}}, f_{\text{Im}} \in L(X; \mathbb{R})$. [*Hint*: Show that $f_{\text{Im}}(x) = -if_{\text{Re}}(ix)\ \forall x \in X$.]

(d) Suppose $B \subset X$ is abs convex and $f \in \mathfrak{L}(X, Y)$, where Y, Y are TVSps with X having a nbd base of abs convex sets. Then $f \in L(X, Y)$ iff f is continuous at $0 \in B$. [*Hint*: $(x_0 + u) \cap B \subset 2u \cap (B - B) \subset 2(u \cap B)$ for every abs convex nbd $u(0) \subset X$.]

13B. Balanced and Convex Hulls

(a) Every subset K of a vector space X has $K_{bc} = (K_b)_c$.

(b) The balanced hull of a convex set need not be convex. [*Hint*: For $K = \{(x, 1-x) \in \mathbb{R}^2 : x \in [0, 1]\}$ the "segment" $\{(x, x-1) : x \in [0, 1]\}$ joining $(0, -1)$ and $(1, 0)$ in K_b is not contained in K_b.] Proceed to give an example for which $K_{bc} \neq (K_c)_b$.

For (c)–(f) below, X is assumed to be a TVS.

(c) Show that K_b, K_c, and K_{bc} are open if $K \subset X$ is open. [*Hint*: Let $x = \sum_{k=1}^{n} \alpha_k x_k$ for $x_k \in K$ and $\alpha_k > 0$ satisfying $\sum_{k=1}^{n} \alpha_k = 1$. If $v_k \subset K$ is a nbd of $x_k\ \forall k = 1, 2, \ldots, n$, then $\sum_{k=1}^{n} \alpha_k v_k \subset K_c$ is a nbd of x.]

(d) Neither K_b nor K_c need be closed if $K \subset X$ is closed. [*Hint*: If $K = \{(n, 1/n): n \in \mathbb{N}\} \subset \mathbb{R}^2$, then $(0, 0) \in \overline{K_b} - K_b$. Since K_c is the area between $y < 1$ and (including) the polygonal path joining points $(n, 1/n) \in K$, one has $(x, 1) \in \overline{K_b} - K_c$ for $x > 1$.]

(e) Verify that $\overline{K_c} = \overline{(\bar{K})_c}$ $\{\overline{K_{bc}} = \overline{(\bar{K})_{bc}}\}$ is the intersection of all closed convex {abs convex} sets containing $K \subset X$.

13C. Compactness and Completeness Considerations

(a) In a TVS X compactness {completeness} is preserved under scalar multiplication and finite unions. Show further that a finite sum of compact sets is compact and that X is compact iff $X=\{0\}$.

(b) K_b is compact if $K\subset X$ above is compact. [*Hint*: Verify that $\underset{(\alpha,x)\rightsquigarrow \alpha x}{\overline{S_1(0)}\times K\to K_b}$ is continuous.] Show why K_c need not be compact.

(c) If $\{K_1,\ldots,K_n\}$ are compact convex sets in X, then $\left(\bigcup_{i=1}^{n} K_i\right)_c$ is compact.

[*Hint*: $B_1=\left\{(\alpha_1\cdots\alpha_n)\in\mathbb{F}^n:\alpha_i\geq 0 \text{ and } \sum_{i=1}^{n}\alpha_1=1\right\}$ is compact and

$$\underset{(\alpha_1\cdots\alpha_n),x_1,\cdots,x_n\rightsquigarrow\sum_{i=1}^{n}\alpha_i x_i}{B_1\times K_1\times\cdots\times K_n\to\left(\bigcup_{i=1}^{n}K_i\right)_c}$$ is continuous.]

(d) Suppose X is a real {complex} TVS. If $K\subset X$ is compact and convex, then K_{bc} $\{\overline{K_{bc}}\}$ is compact. [*Hint*: If $\mathbb{F}=\mathbb{R}$ and $A=\{(\alpha_1,\alpha_2)\in\mathbb{R}\times\mathbb{R}:\alpha_1+\alpha_2\leq 1 \text{ for } \alpha_1,\alpha_2\geq 0\}$, then $K_{bc}=\{\alpha_1x_1-\alpha_2x_2:x_i\in K \text{ and } (\alpha_1,\alpha_2)\in A\}$ and $\underset{((\alpha_1,\alpha_2),x_1,x_2)\rightsquigarrow\alpha_1x_1-\alpha_2x_2}{A\times K\times K\longrightarrow K_{bc}}$ is continuous. If $\mathbb{F}=\mathbb{C}$, then $\tilde{K}=K+iK\subset X$ is convex and compact and $e^{i\theta}K\subset\tilde{K}$ implies $K_{bc}\subset\tilde{K}$.]

(e) If X in (c) is real {complex}, then $\left(\bigcup_{i=1}^{n}K_i\right)_{bc}$ $\left\{\overline{\left(\bigcup_{i=1}^{n}K_i\right)_{bc}}\right\}$ is compact. [*Hint*: For $\mathbb{F}=\mathbb{R}$, each $(K_i)_{bc}$ is compact and $\left\{\bigcup_{i=1}^{n}(K_i)_{bc}\right\}_c=\left(\bigcup_{i=1}^{n}K_i\right)_{bc}$ is compact. For $\mathbb{F}=\mathbb{C}$ use the fact that each $\left\{\bigcup_{i=1}^{n}\overline{(K_i)_{bc}}\right\}_c$ is compact.]

(f) Let $K\subset X$ be compact and suppose X has a nbd base of abs convex sets. Then K_c is compact iff K_c is complete. In particular, $\overline{K_c}$ $\{K_c\}$ is compact for every compact subset K of a quasi-complete LCTVS {LCT$_2$VS} (cf Definition 14.1).

13D. TVS Complexifications

(a) If X is a real TVS, then $X^+=X\times X$ with the product topology is a complex TVS under $(x_1,y_1)+(x_2,y_2)=(x_1+x_2,y_1+y_2)$ and $(\alpha+i\beta)(x_1,y_1)=(\alpha x_1-\beta y_1,\beta x_1+\alpha y_1)\,\forall x_1,x_2,y_1,y_2\in X$ and $\alpha,\beta\in\mathbb{R}$. Furthermore, $\underset{x\rightsquigarrow(x,0)}{X\to X^+}$ is a (real) topological isomorphism.

(b) For each $f\in\mathfrak{L}(X,\mathbb{R})$ define $f^+(x,y)=f(x)+i\,f(y)$. Then $f^+\in\mathfrak{L}(X^+,\mathbb{C})$ and $\underset{f\rightsquigarrow f^+}{L(X,\mathbb{R})\to L(X^+,\mathbb{C})}$ is a real isomorphism.

(c) Suppose $(X,\|\ \|)$ is a real NLS and define $|(x,y)|=\|x\|+\|y\|\ \forall x,y\in X$. Then $(X^+,|\ |)$ is a real NLS and the topological isomorphism in (a) is now a (real) congruence.

(d) If $\|(x,y)\|^+ = 1/\sqrt{2}\sup_\theta |e^{i\theta}(x,y)|\ \forall x, y \in X$, then $(X^+, \|\ \|^+)$ is a complex NLS under which $(X, \|\ \|) \to (X^+, \|\ \|^+)$ is a (real) congruence. Thus $(X^+, \|\ \|^+)$ is complete iff $(X, \|\ \|)$ is complete. [*Hint*: Show that $(X, \|\ \|)$ is complete iff $(X^+, |\ |)$ is complete and use the fact that $|\ |$ and $\|\ \|^+$ are equivalent. Specifically, $(1/\sqrt{2})|(x,y)| \le \|(x,y)\|^+ \le |(x,y)|\ \forall x, y \in X$.].

(e) For each $f \in L(X, X)$ define $f^+(x,y) = (f(x), f(y))$. Then

$$(L(X,X), |||\ |||) \underset{f \rightsquigarrow f^+}{\to} (L(X^+, X^+), |||\ |||);$$

therefore $(X, \|\ \|) \underset{x \rightsquigarrow m_x^+}{\to} (L(X^+, X^+), |||\ |||)$ is a (real) congruence. [*Hint*: From $|||f||| = \sup_{x \ne 0} \|f(x)\|/\|x\| \le \sup_{(x,y)\ne(0,0)} |f^+(x,y)|/|(x,y)| \le |||f|||$ follows $|||f||| = |f^+|$. Therefore $\|f^+(x,y)\|^+ = 1/\sqrt{2}\sup_\theta |f^+\{e^{i\theta}(x,y)\}| \le 1/\sqrt{2}\,|f^+| \sup_\theta |e^{i\theta}(x,y)| = |||f|||\,\|(x,y)\|^+$ yields $|||f^+||| \le |||f|||$. Proceed!]

(f) If $(X, (\ ,\ ))$ is a real inner product space, $(X^+, (\ ,\ )^+)$ is a complex inner product space under $((x_1, y_1), (x_2, y_2))^+ = (x_1, x_2) + (y_1, y_2) - i[(x_1, y_2) - (y_1, x_2)]$. Furthermore, $(\ ,\ )^+$ extends $(\ ,\ )$ and $X \underset{x \rightsquigarrow (x,0)}{\to} X^+$ is a (real) congruence relative to the inner product norms.

13E. TVS Completions

A completion of a TVS X is a pair $(h, \check{X})$, where $\check{X}$ is a complete TVS and h is a topological isomorphism of X onto a dense vector subspace of $\check{X}$.

(a) Every TVS X has a completion, that is unique and separated if X is T_2. [*Hint*: Assume (Problem 8P) that $(h, \check{X})$ is the completion of the topological additive abelian group X and use Theorem 8.22 to show that $\check{X}$ is a vector space and $\hat{\mathfrak{s}}: \mathbb{F} \times \check{X} \underset{(\alpha,\check{x}) \rightsquigarrow \alpha\check{x}}{\to} \check{X}$ is continuous. For uniqueness use Corollary 8.23.]

(b) Let $\bar{E}^{\vee}$ denote the closure in $\check{X}$ of $E \subset X$. Then $\overline{\mathfrak{N}_0}^{\vee} = \{\bar{u}^{\vee} \subset \check{X} : u \in \mathfrak{N}_0\}$ is an $\check{X}$-nbd base at 0 for each X-nbd base $\mathfrak{N}_0$. [*Hint*: In a topological space $A \cap \bar{B} \subset \overline{A \cap B}\ \forall$ open set A and arbitrary set B.]

(c) Why is the completion of a metrizible TVS metrizible?

(d) Why is $C_0(\Omega)$ not $\mathfrak{T}_\kappa$-complete? [*Hint*: Conclude from (c) that $C_0(\Omega)$ is dense in $C_\kappa(\Omega)$.]

13F. Connectedness of TVSps

A TVS is both connected and locally connected. [*Hint*: $\{0\} \subset \bigcap_X [x] \subset \bigcup_X [x] = X$ and (verify!) each one-dim subspace $[x \ne 0] \subset X$ is connected.]

13G. Category and Dimension

(a) A vector subspace M of a TVS X is either dense in X or else $\text{Catg}_{\text{I}}(X)$.

(b) An infinite dim Catg II T_2VS has uncountable dim. [*Hint*: If $\{x_1, x_2, \ldots,\}$ is a basis for X, then $X = \bigcup_{n=1}^{\infty} [x_1, \ldots, x_n]$. Use Corollary 13.13 and (a) above.]

(c) Does (b) hold if the space is not T_2? [*Hint*: Consider $(X, \mathcal{I})$ with dim $X = \aleph_0$.]

13H. The Absorption Theorem

Let B be a convex bounded subset of a TVS X and let $A \subset X$ be a closed convex set that absorbs each $x \in B \cup (-B)$.

(a) Show that A absorbs B if B is a Catg II topological space. [*Hint*: (Kelley and Namioka [52]): $B = \bigcup_{m=1}^{\infty} \left\{ \bigcap_{n=m}^{\infty} (nA \cap B) \right\}$ so that $y_0 \in \text{Int}\left\{ \bigcap_{n=N_0}^{\infty} (nA \cap B) \right\}$ in B for some $N_0 \in \mathbb{N}$. If $u \in \mathfrak{N}_0$ satisfies $(y_0 + u) \cap B \subset nA \ \forall n \geq N_0$, then [Theorem 13.20(ii)] $\lambda(B - B) \subset u$ for some $\lambda \in (0, 1)$. In particular, $y_0(1-\lambda) + \lambda B \subset y_0 + \lambda(B-B) \cap B \subset nA \ \forall n \geq N_0$. Pick $M_0 > N_0$ such that $-y_0(-\lambda) \in \alpha A \ \forall \alpha \geq M_0$ and obtain $\frac{1}{2}\lambda B = \frac{1}{2}\{y_0(1-\lambda) + \lambda B\} + \frac{1}{2}\{-y_0(1-\lambda)\} \subset \alpha A \ \forall \alpha \geq M_0$.]

(b) Why does A absorb B if B is compact? [*Hint*: Show that a compact subset of a T_3 space is a Catg II space.]

13I. Completely Continuous Mappings

A mapping $f \in \mathfrak{L}(X, Y)$, where X and Y are TVSps, is completely continuous (also called compact) if $f(u) \subset K$ for some X-nbd $u(0)$ and compact set $K \subset Y$.

(a) $L_{cc}(X, Y) = \{\text{completely continuous } f \in \mathfrak{L}(X, Y)\}$ is a vector subspace of $L(X, Y)$.

(b) $f \in L_{cc}(X, Y)$ takes bounded sets into relatively compact sets. The converse also holds if X is normable.

(c) Suppose X is normable and Y is T_2. Then $f \in L(X, Y)$ is completely continuous if $\dim f(X) \subset Y$ is finite. [*Hint*: Bounded sets are relatively compact in $\mathbb{F}^n$. Use Theorem 13.12, Corollary 13.14.]

(d) $L_{cc}(X, Y) = \mathfrak{L}(X, Y)$ if X, Y are T_2 and $\dim X < \aleph_0$. In general, $L_{cc}(X, Y) \neq L(X, Y)$. [*Hint*: No congruence $f \in L(X, X)$ is completely continuous on an infinite dim NLS X (Theorem 13.14).]

(e) fg and gf belong to $L_{cc}(X, X) \ \forall (f, g) \in L_{cc}(X \times X) \times L(X, X)$. Therefore $\mathcal{P}(f) \in L_{cc}(X, X)$ for every polynomial $p(f)$ in f without constant term. Why is it possible that $f^{-1} \notin L(X, X)$ if f^{-1} exists? [*Hint*: Consider $ff^{-1} = \mathbb{1}$ in (d).]

(f) If Y is metrizible and $f \in L_{cc}(X, Y)$, then $f(X)$ is a separable subspace of Y. [*Hint*: If $f(u) \subset K$, then $f(X) = \bigcup_{n=1}^{\infty} nf(u)$ and $f(u)$ is separable.]

(g) $L_{cc}(X, Y)$ is $|||\,|||$-closed in $L(X, Y)$ if X is a NLS and Y is Banach. [*Hint*: Suppose $f_n \xrightarrow{|||\,|||} f \in \mathfrak{L}(X, Y)$ for $\{f_n\} \in L_{cc}(X, Y)$ and let $\sup_B \|x\| < M_B$ for $B \subset X$. Given $\varepsilon > 0$, there is an $n_0 \in \mathbb{N}$ such that $n > n_0$ implies $|||f_n - f||| < \varepsilon/3M_B$ and $f_n(B) \subset \bigcup_{i=1}^{m_n} S_{\varepsilon/3}(f_n(x_i))$ for $\{f_n(x_i): 1 \le i \le m_n\} \subset f_b(B)$. Show that $f(B) \subset Y$ is totally bounded.]

13J. Uniformity for a Metrizible TVS

(a) Every metrizible TVS X has a translation invariant metric ρ such that $\mathcal{U}_\rho = \mathcal{U}_X$ on $X \times X$. [*Hint*: Use $(x, y) \in U_u$ iff $(x - y, 0) \in U_u$ $\forall u \in \mathfrak{N}_0$ to prove that $\rho(x, y) = \rho(x - y, 0)$ $\forall x, y \in X$ in Problem 4G.]

(b) If $\mathfrak{T}(d) \in T(X)$ for a metric d on a vector space X, does it necessarily follow that $\mathcal{U}_d = \mathcal{U}_X$? [*Hint*: $\mathcal{U}_d \neq \mathcal{U}_X$ for $d(x, y) = \tan^{-1}|x - y|$ on $X = \mathbb{R}$.]

(c) $\mathcal{U}_d = \mathcal{U}_X$ if $(X, \|\,\|)$ is a NLS and $d(x, y) = \|x - y\|$.

(d) If $0 < p < 1$, then $\|x\|_p = \sqrt[p]{\int_X |x|^p \, d\mu}$ is not a norm on $L_p(X, \mathfrak{R}_\sigma, \mu)$ (verify!). Nevertheless, $d_p(x, y) = \|x - y\|_p^p$ is a complete translation invariant metric such that $\mathfrak{T}(d_p) \in T(L_p(X, \mathfrak{R}_\sigma, \mu))$. Compare $\mathcal{U}_{d_p}$ with $\mathcal{U}_{L_p}(X, \mathfrak{R}_\sigma, \mu)$.

(e) Prove the analogous statements for l_p $(0 < p < 1)$.

References for Further Study

On TVSps: Bourbaki [13]; Horvath [43]; Kelley and Namioka [52]; Kothe [55].

On ordered TVSps: Peressini [77]; Schaeffer [88].

On the finest types of TVSps: Waelbroeck [102].

On category and TVSps: Saxon [85], [87].

On Riesz theory of compact mappings: Robertson and Robertson [84].

14. Locally Convex TVSps

The most important TVSps in analysis are those with convex nbd bases at the origin. Such spaces encompass the class of NLSps and are shown to be characterized by their continuous seminorms. Our emphasis on scalar multiplication marks the juncture from which TVS theory veers from that of topological groups.

DEFINITION 14.1. A TVS is called locally convex (LCTVS) if it has a nbd base at 0 consisting of convex sets.

THEOREM 14.2. Every LCTVS has a nbd base $\mathfrak{N}_0$ at 0 such that

(i) $u(0)$ is abs convex and absorbent $\forall u(0) \in \mathfrak{N}_0$;
(ii) for each $u \in \mathfrak{N}_0$ there is a $v(0) \in \mathfrak{N}_0$ satisfying $v+v \subset u$.

Conversely, every filterbase $\mathfrak{N}$ of subsets of a vector space X that satisfies (i), (ii) determines a unique locally convex topology $\mathfrak{T} \in T(X)$ having $\mathfrak{N}$ as a nbd base at 0.

Proof. If $w(0)$ is any nbd of 0 and $v(0)$ is a balanced nbd of 0 contained in $w(0)$, then $v \subset (v)_{bc} = v_c \subset w$ and $\mathfrak{N}_0 = \{$abs convex, absorbent sets$\}$ is a nbd base at 0. If $u \in \mathfrak{N}_0$ and $\tilde{v}+\tilde{v} \subset u$ for a balanced nbd $\tilde{v}(0)$, then $v+v \subset \tilde{v}$ for any abs convex nbd $v \in \mathfrak{N}_0$. The converse part of the theorem is clear by virtue of Theorem 13.7. ■

Remark. If $\mathfrak{N}$ only satisfies (i) above, $\tilde{\mathfrak{N}} = \{\alpha u : u \in \mathfrak{N}, \alpha \in \mathbb{F}\}$ satisfies (i), (ii) (since $(\alpha/2)u + (\alpha/2)u \subset u \ \forall \alpha u \in \tilde{\mathfrak{N}}$).

COROLLARY 14.3. Every LCTVS has a nbd base at 0 of closed abs convex sets.

Proof. Given any nbd $u(0)$, let $v(0)$ be a closed balanced nbd contained in $u(0)$ and let $w(0)$ be an abs convex nbd contained in $v(0)$. Then (Theorem 13.9) $\overline{w(0)}$ is an abs convex nbd contained in $u(0)$. ■

Remark. If B is a {totally} bounded subset of an LCTVS X, so is B_c and B_{bc}. Reason: If B is bounded, proceed as in Theorem 13.20 to get $B \subset \alpha w$ for a closed abs convex nbd $w \subset u$. Then $B_c \subset (\alpha w)_c = \alpha w \subset u$. Since B_c is bounded, so also is $(B_b)_c = B_{bc}$. If B is totally bounded and $u, v \in \mathfrak{N}_0$ satisfy $v+v \subset u$, then $B \subset v + \bigcup_{i=1}^{n} x_i$ $(x_i \in B)$ and $B_c \subset v + K_c$ for $K = \{x_1, \ldots, x_n\}$. Since K_c is compact (Problem 13C). $K_c \subset v + \bigcup_{j=1}^{m} x'_j$ for $\{x'_1 \cdots x'_m\} \subset K_c \subset B_c$ and so $B_c \subset u + \bigcup_{j=1}^{m} x'_j$.

COROLLARY 14.4. Every collection $\mathfrak{N}$ of abs convex absorbent sets of a vector space X generates a coarsest locally convex topology $\mathfrak{T} \in T(X)$ under which every $v \in \mathfrak{N}$ is a $\mathfrak{T}$-nbd of 0.

Proof. Let $\mathfrak{T}$ be the topology having nbd base $\mathfrak{N}_0 = \left\{ \bigcap_{i=1}^{n} \varepsilon_i v_{\alpha_i} : v_{\alpha_i} \in \mathfrak{N} \right.$ and $\varepsilon_i > 0\}$ at 0. If $\tilde{\mathfrak{T}} \in T(X)$ is locally convex and every $v \in \mathfrak{N}$ is a $\tilde{\mathfrak{T}}$-nbd of 0, then every $u \in \mathfrak{N}_0$ is a $\tilde{\mathfrak{T}}$-nbd and $\mathfrak{T} \subset \tilde{\mathfrak{T}}$. ■

Seminorms offer perhaps the most useful technique for defining compatible, locally convex topologies on a vector space.

DEFINITION 14.5. A seminorm is a nonnegative, finite, real-valued mapping p on a vector space X which satisfies $p(\alpha x)=|\alpha|p(x)$ and $p(x+y)\leq p(x)+p(y)\ \forall x, y\in X$ and $\alpha\in\mathbb{F}$.

Remark. Together, $p(x)\leq p(y)+p(x-y)$ and $p(y)\leq p(x)+p(y-x)$ yield $|p(x)-p(y)|\leq p(x-y)\ \forall x, y\in X$.

Remark. Note that $S_{\varepsilon,p}(0)=\{x\in X: p(x)<\varepsilon\}$ and $\bar{S}_{\varepsilon,p}(0)=\{x\in X: p(x)\leq\varepsilon\}$ are both abs convex and absorbent. If q is another seminorm on X, then $S_{1,p}(0)\subset\bar{S}_{1,q}(0)$ implies $q\leq p$ (otherwise, $0\leq p(x_0)<\alpha<q(x_0)$ and $(1/\alpha)x_0\in S_{1,p}(0)-\bar{S}_{1,q}(0)$ for some $x_0\in X$).

A seminorm p determines a pseudometric $d_p(x, y)=p(x-y)$ and it is this pseudometric topology that is assumed in

THEOREM 14.6. The following are equivalent for p on X:

(i) p is continuous;
(ii) p is continuous at $0\in X$;
(iii) $S_{\varepsilon,p}(0)$ is a nbd of $0\in X$ for some (therefore all!) $\varepsilon>0$.

Proof. Clearly (i)$\Rightarrow$(ii), and (ii)$\Rightarrow$(iii) since $u(0)=p^{-1}\{(-\varepsilon,\varepsilon)\}=S_{\varepsilon,p}(0)$ for some nbd $u(0)$. The inequality in the above remark also ensures that (iii)$\Rightarrow$(i). ■

Remark. If X is a TVS and $f(x)\leq p(x)$ on X for $f\in X^*$, then the continuity of p implies that of f, since (Theorem 13.11) f is bounded on $S_\varepsilon(0)$. Essentially, $f\in X^*$ is continuous iff $|f|(x)=|f(x)|$ is continuous.

THEOREM 14.7. Given any collection $\mathcal{P}$ of seminorms on a vector space X, there is a coarsest, locally convex topology $\mathcal{T}(\mathcal{P})\in T(X)$ under which every $p\in\mathcal{P}$ is continuous.

Proof. Take $\mathcal{N}_0=\left\{\bigcap_{i=1}^{n} S_{\varepsilon_i,p_{\alpha_i}}(0): \varepsilon_i>0 \text{ and } p_{\alpha_i}\in\mathcal{P}\right\}$ in Corollary 14.4 and apply Theorem 14.6. ■

Some useful properties of $\mathcal{T}(\mathcal{P})$ are collected below. The straightforward proofs are omitted.

THEOREM 14.8. (i) $\mathcal{T}(\mathcal{P})$ is T_2 iff $\forall x\neq 0$, there is a $p\in\mathcal{P}$ such that $p(x)>0$.

(ii) A filter $\mathcal{F}$ converges to $x_0\in X$ iff $\forall p\in\mathcal{P}$ and $\varepsilon>0$ there is a set $F\in\mathcal{F}$ such that $p(x_0-x)<\varepsilon\ \forall x\in F$. In particular, a sequence $x_n\to x_0$ iff

$\forall p \in \mathcal{P}$ and $\varepsilon > 0$ there is an $N(\varepsilon, p) \in \mathbb{N}$ such that $p(x_0 - x_n) < \varepsilon \ \forall n > N(\varepsilon, p)$.

(iii) If f is a linear mapping from a TVS Y into $(X, \mathfrak{T}(\mathcal{P}))$, the following are equivalent:

(a) f is continuous;

(b) pf is continuous $\forall p \in \mathcal{P}$;

(c) for each $p \in \mathcal{P}$ there is a continuous seminorm q on Y such that $p\{f(y)\} \le q(y) \ \forall y \in Y$.

Theorem 14.7 has a converse in that every compatible, locally convex topology on a vector space X is determined by its collection of continuous seminorms. This is made clear once we lay bare the nexus between seminorms on X and abs convex absorbing subsets of X.

DEFINITION 14.9. The Minkowski functional of an absorbent set $K \subset X$ is defined as $p_K(x) = \inf\{\lambda > 0 : x \in \lambda K\}$. Clearly, $K \subset \bar{S}_{1,p_K}(0) = \{x \in X : p_K(x) \le 1\}$.

Remark. If K is convex, $p_K(x+y) \le p_K(x) + p_K(y) \ \forall x, y \in X$ since $(x, y) \in \alpha K \times \beta K$ implies $x + y \in \alpha K + \beta K = (\alpha + \beta)K$ and $p_K(x+y) \le \alpha + \beta \ \forall \alpha, \beta > 0$. If K is convex and contains 0, then $S_{1,p_K}(0) = \{x \in X : p_K(x) < 1\} \subset K$ since $p_K(x) = \alpha \in (0, 1)$ yields $x = (1-\alpha)0 + \alpha(1/\alpha)x \in K$.

Remark. Note that $p_K(\alpha x) = \alpha p_K(x) \ \forall x \in X$ and $\alpha > 0$ (therefore $p_K(0) = 0$), since $x \in \lambda K$ for $\lambda > 0$ yields $\alpha x \in \alpha\lambda K$ and $p_K(\alpha x) \le \alpha\lambda$. Thus $p_K(\alpha x) \le \alpha p_K(x)$ and $p_K(x) = p_K\{(1/\alpha)(\alpha x)\} \le (1/\alpha) p_K(\alpha x)$ follows similarly. If K is balanced, $p_K(\alpha x) = |\alpha| p_K(x) \ \forall x \in X$ and $\alpha \in \mathbb{F}$ since $x \in \lambda K$ implies $(\alpha/|\alpha|)x \in \lambda K$ and $(1/|\alpha|) p_K(\alpha x) = p_K((\alpha/|\alpha|)x) \le \lambda$. Thus $p_K(\alpha x) \le |\alpha| p_K(x)$ and $p_K(x) = p_K\{(1/\alpha)(\alpha x)\} \le (1/|\alpha|) p_K(\alpha x)$.

The preceding remarks show that every abs convex, absorbent set $K \subset X$ determines a seminorm p_K on X. It is a simple matter to further establish

THEOREM 14.10. Suppose K is a convex absorbent set containing 0 in a TVS X. Then

(i) $\operatorname{Int} K \subset S_{1,p_K}(0) \subset K \subset \bar{S}_{1,p_K}(0) \subset \overline{S_{1,p_K}(0)} \subset \bar{K}$;

(ii) $\operatorname{Int} K = S_{1,p_K}(0)$ and $\bar{K} = \bar{S}_{1,p_K}(0)$ if p_K is continuous;

(iii) p_K is continuous iff $0 \in \operatorname{Int} K$.

Proof. (i) We have already established the second, third (therefore the last) inclusion. The first and fourth inclusions use the continuity of $\mathfrak{m}$ at $(1, x) \in \mathbb{F} \times X$. If $x_0 \in \operatorname{Int} K$, there is an $\varepsilon > 0$ such that $\lambda x_0 \in \operatorname{Int} K \ \forall |\lambda - 1| < \varepsilon$ and for $\lambda_0 \in (1, 1+\varepsilon)$ there follows $p_K(x_0) = (1/\lambda_0) p_K(\lambda_0 x_0) \le 1/\lambda_0 < 1$.

Similarly, $x_0 \in \bar{S}_{1,p_K}(0)$ implies $\lambda x_0 \in S_{1,p_K}(0)\ \forall |\lambda| < 1$ so that $x_0 = \lim_{\lambda \uparrow 1} \lambda x_0 \in \overline{S_{1,p_K}(0)}$. If p_K in (ii) is continuous, then $S_{1,p_K}(0) = p_K^{-1}\{(-1, 1)\}$ is open and $\bar{S}_{1,p_K}(0) = p_K^{-1}\{[-1, 1]\}$ is closed in X. Clearly (iii) follows from (i), (ii), and Theorem 14.6. ■

The promised converse of Theorem 14.7 is

THEOREM 14.11. If $\mathcal{P}$ denotes the collection of continuous seminorms on an LCTVS $(X, \mathfrak{T})$, then $\mathfrak{T} = \mathfrak{T}(\mathcal{P})$.

Proof. Let $\mathfrak{N}_0$ be a $\mathfrak{T}$-nbd base of closed abs convex sets. Then $\underset{u \rightsquigarrow p_u}{\mathfrak{N}_0 \to \mathcal{P}}$ is a bijection since $u = \bar{S}_{1,p_u}(0) = \bar{S}_{1,p_v}(0) = v$ for $p_u = p_v$, and $p = p_{\bar{S}_{1,p}(0)}\ \forall p \in \mathcal{P}$ (by Definition 14.5$^+$, after noting that $p(x) < 1 \Rightarrow p_{\bar{S}_{1,p(0)}}(x) \le 1$ and $p_{\bar{S}_{1,p(0)}}(x) < 1 \Rightarrow p(x) \le 1$). Accordingly, $\mathfrak{T} = \mathfrak{T}(\mathcal{P})$, since $S_{1,p_u}(0) \subset u = \bar{S}_{1,p_u}(0) \subset S_{2,p_u}(0)\ \forall u \in \mathfrak{N}_0$. ■

COROLLARY 14.12. A set $B \subset (X, \mathfrak{T})$ is bounded iff $p(B) \subset \mathbb{R}$ is bounded $\forall p \in \mathcal{P}$.

Proof. If $B \subset X$ is $\mathfrak{T} = \mathfrak{T}(\mathcal{P})$-bounded, there is a $\lambda_p > 0$ such that $B \subset \lambda_p S_{1,p}(0) = S_{\lambda_p,p}(0)\ \forall p \in \mathcal{P}$. In the other direction $\sup_B p(x) \le M_p < \infty$ implies $B \subset (2M_p/\varepsilon) S_{\varepsilon,p}(0)\ \forall \varepsilon > 0$. ■

Example 14.13. $B\mathfrak{L}_{\mathcal{S}}(X:\mathbb{F})$ (therefore $\mathfrak{F}_p(X:\mathbb{F})$ and $C_\kappa(X:\mathbb{F})$) in Example 13.15 is an LCT$_2$VS, since each $E \in \mathcal{S}$ determines a seminorm $\underset{f \rightsquigarrow \sup_E |f(x)|}{q_E: B\mathfrak{L}(X:\mathbb{F}) \longrightarrow \mathbb{R}}$ and $\mathfrak{T}_{\mathcal{S}} = \mathfrak{T}(\mathcal{P})$ for $\mathcal{P} = \{q_E : E \in \mathcal{S}\}$.

Less familiar and certainly more extreme are the notions in

Example 14.14. (a) Every vector space X has a finest compatible, locally convex topology $\mathfrak{T}_{F_c}$. Simply take $\mathfrak{T}_{F_c}$ as the topology having all abs convex absorbing subsets as its nbd base at 0. This obviously is the topology determined by the collection of all seminorms on X.

(b) $\mathfrak{T}_{F_c}$ is T_2, since $x_0 \ne 0$ may be included in an algebraic basis $\mathfrak{B}$ of X and $\left\{ \sum_{i=1}^{n} \alpha_i x_i : x_i \in \mathfrak{B} \text{ and } |\alpha_i| < 1 \right\}$ is an abs convex absorbent set which does not contain x_0. In terms of seminorms define $p_0(x) = |\alpha_0|$ for each $x = \alpha_0 x_0 + \sum_{i=1}^{n} \alpha_i x_i$ $(x_i \in \mathfrak{B})$ and obtain $p_0(x_0) = 1$ for $p_0 \in \mathcal{P}$.

(c) $L(X, Y) = \mathfrak{L}(X, Y)\ \forall$ LCTVS Y (in particular, $X^* = X'$) since $f^{-1}(w) \subset X$ is abs convex and absorbent for every such $w \subset Y$.

(d) If $B \subset X$ is $\mathfrak{T}_{F_c}$-bounded, then $\dim[B] < \aleph_0$. To be sure, every linearly independent set $\{x_n : n \in \mathbb{N}\}$ can be extended to a basis $\mathfrak{B}$ for X, and $p(x) = n\delta_{x,x_n}$ for $x \in \mathfrak{B}$, with p defined linearly elsewhere, is a seminorm on X such that $\{p(x_n) : n \in \mathbb{N}\}$ is unbounded.

Example 14.15. Not every TVS is an LCTVS. The space $\mathfrak{S} = \mathfrak{S}(X, \mu)$ in Example 13.17 is not locally convex if $\forall n \in \mathbb{N}$ there is a finite collection of disjoint sets of measure $1/n$ whose union covers X. Reason: No convex set $K \subset \mathfrak{S}$ satisfies $u_{\varepsilon_1,\delta_1}(0) \subset K \subset u_{\varepsilon_2,\delta_2}(0)$. For verification assume that $\varepsilon_1 < \varepsilon_2$ and $1/n = \delta_1 < \delta_2 < \mu(X)/2$. Then $X = \bigcup_{k=1}^{m} E_k$, where $\{E_k\}$ are disjoint and $\mu(E_k) = (1/2n)\ \forall k = 1, 2, \ldots, m$. Since $(m+1)\varepsilon_2 \chi_{E_k} \in u_{\varepsilon_1,\delta_1}(0) \subset K$ for each $1 \le k \le m$, the convexity of K implies (Theorem 13.6) that $\sum_{k=1}^{m} 1/m\{(m+1)\varepsilon_2\chi_{E_k}\} \in K_c = K$—and this is clearly impossible, since $\sum_{k=1}^{m} [(m+1)/m]\varepsilon_2\chi_{E_k} \notin u_{\varepsilon_2,\delta_2}(0)$.

Remark. Lebesgue measure on $X = [0, 1)$ fits the above framework.

Example 14.16. The metric TVS (L_p, d_p) in Problem 13K(d) is not locally convex if $\mu(E_n) > 0$ for a countable, disjoint collection $\{E_n\} \in \mathfrak{R}_\sigma$. Otherwise, there is an abs convex nbd $u(0)$ such that $S_\varepsilon(0) \subset u(0) \subset S_1(0)$ for some $\varepsilon > 0$ (here $S_\varepsilon(0) = \{f \in L_p : d_p(f, 0) < \varepsilon\}$. Let $f_n = \{\mu(E_n)\}^{-1/p}\chi_{E_n} \in L_p$. Since $(\varepsilon/2)^{1/p} f_n \in S_\varepsilon(0) \subset u$, one has $g_n = \sum_{k=1}^{n} 1/n\{(\varepsilon/2)^{1/p} f_k\} \subset u\ \forall n \in \mathbb{N}$. This cannot be true, since $d_p(g_n, 0) = (\varepsilon/2)n^{1-p} > 1$ for large $n \in \mathbb{N}$.

The metrizible TVS (l_p, d_p) is not locally convex (take $E_n - \{n\}$ above). Although $B = \{e_n : n \in \mathbb{N}\} \subset l_{p>1}$ is bounded (Example 13.21), B_c is not, since $B_c \not\subset \lambda S_1(0) = S_\lambda(0)$ for any $\lambda \in (0, \infty)$. This is clear, since for any $\lambda \in (0, \infty)$ a sufficiently large $m \in \mathbb{N}$ can be found such that $d_p(y, 0) = m^{1-p} > \lambda$ and $y = \sum_{k=1}^{m} (1/m)e_k \in B_c - S_\lambda(0)$. Local convexity is therefore essential in Corollary 14.3^+.

An abs convex subset B of a TVS is also an absorbent subset of $E_B = [\bigcup_{\mathbb{N}} nB] \subset X$. As such, the Minkowski functional p_B is a seminorm on E_B and there follows

THEOREM 14.17. Let B be an abs convex, bounded subset of a TVS $(X, \mathfrak{T})$. Then $\mathfrak{T}/E_B = \mathfrak{T}(p_B)$ on E_B. Furthermore, p_B is a norm if $\mathfrak{T}$ is T_2 and (E_B, p_B) is complete if B is $\mathfrak{T}$-sequentially closed.

Proof. Given any $\mathfrak{T}$-nbd $u(0)$, one has $B \subset \lambda u$ and $(1/\lambda)S_{1,p_B}(0) \cap E_B \subset (1/\lambda)B \cap E_B \subset u \cap E_B$ for some $\lambda > 0$. Therefore (Theorem 14.10) $\mathfrak{T}/E_B \subset \mathfrak{T}(p_B)$. Assume now that $\mathfrak{T}$ is T_2. For any $x_0 \in E_B - \{0\}$ there is a $\mathfrak{T}$-nbd $u(0) \not\ni x_0$ and a $\lambda_B > 0$ such that $B \subset \lambda_B u$. Since $x_0 \notin (1/\lambda_B)B$, it follows that $p_B(x_0) = \lambda_B p_B[(1/\lambda_B)x_0] > 0$ and so p_B is a norm on E_B. Finally, suppose B is $\mathfrak{T}$-sequentially complete and sequentially closed. If $\{x_n\} \in E_B$ is p_B-Cauchy, then $\{x_n\}$ is $\mathfrak{T}$-convergent to some $x_0 \in \overline{E_B}$; and $x_m - x_n \in \varepsilon B$ for $m, n > N(\varepsilon)$ by Definition 14.9$^+$. Thus $\lim_n (x_m - x_n) = x_m - x_0 \in \overline{\varepsilon B} = \varepsilon B$ for $m > N(\varepsilon)$ and $x_n \xrightarrow{\mathfrak{T}(p_B)} x_0 \in E_B$ follows. ■

COROLLARY 14.18. Let A be a closed, convex subset of a TVS X and let B be an abs convex, bounded, sequentially complete, and sequentially closed subset of X. If A absorbs every $x \in B$, then A absorbs B.

Proof. B is an abs convex, bounded closed subset of the complete seminormed space (E_B, p_B). As a complete pseudometric space, (B, p_B) is Catg II, and (by hypothesis) $A \cap E_B$ is abs convex, p_B-closed and absorbs every $x \in B$. Therefore (Problem 13H) B is absorbed by $A \cap E_B$ and by A also. ■

The following result due to Kolmogorov is basic.

THEOREM 14.19. A TVS {LCTVS} is normable iff it is T_2 and has a bounded, convex {bounded} nbd of 0.

Proof. The conditions are necessary, since $S_1(0)$ is a bounded, convex nbd of 0 in a NLS. Suppose, therefore, that E is a convex, bounded nbd of 0 in a T_2VS X and let $w(0)$ be a balanced nbd contained in E. Then (Theorem 13.9, Corollary 14.3) p_B is a seminorm on X for the abs convex, bounded set $B = \overline{(w)_c} \subset X$. If $x_0 \neq 0$, there is a balanced nbd $u(0) \not\ni x_0$ and a $\lambda_B > 0$ such that $B \subset \lambda_B u$. Since B is absorbent, one also has $x_0 \in \alpha B$ for some $\alpha > 0$. Together, the above statements ensure that $(1/\lambda_B \alpha)x_0 \in u$ and $\lambda_B \alpha > 1$; therefore $p_B(x_0) \geq \alpha > 0$. It remains only to show that $\mathfrak{T} = \mathfrak{T}(p_B)$ for the norm p_B on X. This, however, is clear, since $B = \bar{S}_{1,p_B}(0)$ and the $\mathfrak{T}(p_B)$-nbd base $\{\bar{S}_{\varepsilon,p_B}(0) = \varepsilon B : \varepsilon > 0\}$ is also a $\mathfrak{T}$-nbd base at $0 \in X$.

If X is an LCTVS and $\tilde{E}$ is an abs convex nbd of 0 contained in E, then $\tilde{E}$ is bounded and the above proof applies. ■

Remark. A TVS {LCTVS} X is normable iff there is a bounded, convex {bounded} nbd B of 0 such that $\bigcap_{\lambda>0} \lambda B = \{0\}$. This is true, since $\{\lambda B : \lambda > 0\}$ is a nbd base at $0 \in X$.

Example 14.20. Let $\mathfrak{D}^\infty(I)$ denote the vector space of $\mathbb{F}$-valued, infinitely differentiable functions on $I=[0, 1]\subset\mathbb{R}$. It is easily verified that $\mathfrak{T}_\kappa=\|\,\|_0=\mathfrak{T}_u\in T(\mathfrak{D}^\infty(I))$, where $\|\,\|_0$ is the sup norm on I. For each $n\in\mathbb{N}_0$, let $p_n(f)=\sup_I|f^{(n)}(x)|$ and take $\mathfrak{P}=\{p_n: n\in\mathbb{N}_0\}$. Then $\mathfrak{T}(\mathfrak{P})\in T(\mathfrak{D}^\infty(I))$ is a nonnormable, complete, metrizible LCT_2VS such that $\mathfrak{T}_\kappa\subset\mathfrak{T}(\mathfrak{P})$. Verification follows below:

$\mathfrak{T}(\mathfrak{P})$ is nonnormable since no nbd $S_{\varepsilon,p_n}(0)$ is bounded. Indeed, $g_k(x)=(\varepsilon/2)\,e^{-kk}\sin e^k x\in S_{\varepsilon,p_n}(0)\ \forall k\in\mathbb{N}_0$ and $\sup_k\sup_I|p_{n+1}(g_k)|=\sup_k(\varepsilon/2)\,e^k=\infty$. Therefore $p_{n+1}(S_{\varepsilon/2,p_n}(0))$ is unbounded in $\mathbb{R}$ and Corollary 14.12 applies. Next, metrizibility of $\mathfrak{T}(\mathfrak{P})$ follows from Theorem 13.4. To establish $\mathfrak{T}(\mathfrak{P})$ completeness consider (Theorem 6.3) any $\mathfrak{T}(\mathfrak{P})$-Cauchy sequence $\{f_m\}\in\mathfrak{D}^\infty(I)$. Then $\{f_m\}$ is $\mathfrak{T}(p_n)$-Cauchy $\forall n\in\mathbb{N}_0$. In particular, $\{f_m\}$ is $\mathfrak{T}(p_0)=\|\,\|_0$-Cauchy and (since $\|\,\|_0$ is a complete norm on $C(I;\mathbb{F})$) convergent to some $F\in C(I;\mathbb{F})$. Since $\{f_m^{(1)}\}$ is $\mathfrak{T}(p_1)=\|\,\|_0$-Cauchy (therefore uniformly convergent on I), one also has $f_m^{(1)}\to F^{(1)}\in C(I;\mathbb{F})$. Proceeding inductively, we obtain $f_m^{(n)}\xrightarrow{\mathfrak{T}(p_n)}F^{(n)}\in C(I;\mathbb{F})\ \forall n\in\mathbb{N}_0$ and this proves that $\{f_m\}\xrightarrow{\mathfrak{T}(\mathfrak{P})}F\in\mathfrak{D}^\infty(I)$.

Example 14.20 offers a good illustration of

DEFINITION 14.21. A Frechet space is a complete metrizible LCTVS (i.e., the locally convex topology is determined by a complete metric).

Example 14.22. (a) $\mathfrak{F}_p(X;\mathbb{F})$ in Examples 13.15 and 14.13 is Frechet iff $\aleph(X)\le\aleph_0$. Note, however, that $\mathfrak{F}_p(X;\mathbb{F})$ is nonnormable if $\aleph(X)=\aleph_0$, since no $\mathfrak{T}_p$-nbd $u_{x,\varepsilon}(0)$ is bounded ($y\ne x$ implies $2n\lambda\chi_{\{y\}}\in u_{x,\varepsilon}(0)-nu_{y,\lambda}(0)\ \forall n\in\mathbb{N}$).

(b) $C_\kappa(X;\mathbb{F})$ is Frechet if X is locally compact or 1°, and $X=\bigcup_{n=1}^\infty K_n$ for a strictly increasing sequence of compact sets $\{K_n\}\subset X$. (For each $u_{K,\varepsilon}(0)$ there is an $m\in\mathbb{N}$ such that $1/m<\varepsilon$ and $K\subset K_m$. Therefore, $u_{K_m,1/m}(0)\subset u_{K,\varepsilon}(0)$ and $\mathfrak{T}_\kappa$ is seen to be metrizible.) If X is also $T_{3\frac{1}{2}}$, then $C_\kappa(X;\mathbb{F})$ is nonnormable, since no nbd $u_{K,(1/n)}(0)$ is bounded. To be sure, for any $x\in K_{n+1}-K_n$ and $f\in C(X, I)$ satisfying $f(K_n)=\{0\}$ and $f(x)=1$ one has $2mf\in u_{K_n,1/n}(0)-mu_{K_{n+1},1/(n+1)}(0)\ \forall n\in\mathbb{N}$. Note that this nonnormability of $C_\kappa(X;\mathbb{F})$ may fail if $\{K_n\}$ is not an infinite, strictly increasing sequence. For example, $\mathfrak{T}_\kappa=\|\,\|_0$ on $C(X;\mathbb{F})$ when X is compact.

Remark. The conditions on X are surely fulfilled for any open set $X \subset \mathbb{F}^n$. If $X = \mathbb{F}^n$, then $X = \bigcup_{n=1}^{\infty} \bar{S}_n(0)$. Otherwise, take $K_n = \bar{S}_n(0) \cap \{x \in \mathbb{F}^n : \inf_{x_0 \in X} \|x - x_0\| \geq 1/n\}$.

(c) There exist complete metrizible TVSps that are not locally convex (i.e., not Frechet). One such illustration is (l_p, d_p) $(0 < p < 1)$ in Problem 13K, Example 14.16. Another is (L_p, d_p) in certain instances. Note also (Examples 13.17 and 14.15) that $\mathbb{S}(X, \mu)$ is a complete pseudometrizible TVS that need not be locally convex.

PROBLEMS

14A. LCTVS Completion

(a) A completion $\check{X}$ of an LCTVS X is an LCTVS. [*Hint*: See Problem 13E(b).]

(b) Suppose $\mathcal{P}$ is a family of seminorms that generates the topology of an LCT$_2$VS X. Then each $p \in \mathcal{P}$ determines a seminorm $\check{p}$ on $\check{X}$ and $\check{\mathcal{P}} = \{\check{p} : p \in \mathcal{P}\}$ generates the topology on $\check{X}$. [*Hint* (Theorem 6.12): For $\check{x} \in \check{X}$ and net $\varphi \in X$ converging to $\check{x}$ define $\check{p}(\check{x}) = \lim p(\varphi)$.]

14B. LCTVSps with a Countable Base

(a) If E is an abs convex, absorbent subset of a vector space X, then $\{(1/n)E : n \in \mathbb{N}\}$ is a nbd base for a pseudometrizible, locally convex topology $\mathcal{T} \in T(X)$.

(b) Show that each d below is a translation invariant pseudometric on an LCTVS X having nbd base $\mathfrak{N}_0 = \{u_n(0) : n \in \mathbb{N}\}$ at 0:

(i) Let $f(x) = \sum_{n=1}^{\infty} 2^{-n} \inf\{1, p_{u_n}(x)\}$ and define $d(x, y) = f(x - y)$.

(ii) Let $q_m(x) = \max_{1 \leq n \leq m} p_{u_n}(x)\ \forall m \in \mathbb{N}$ and suppose that $\sum_{n=1}^{\infty} \lambda_n < \infty$ for $\{\lambda_n > 0\}$.

Then define $d(x, y) = \sum_{n=1}^{\infty} \lambda_n [q_n(x - y)]/[1 + q_n(x - y)]$.

14C. Mackey's Countability Condition

Let $\{B_n : n \in \mathbb{N}\}$ be a sequences of bounded subsets of a TVS X having $\mathfrak{N}_0 = \{u_n(0) : n \in \mathbb{N}\}$ as a nbd base at 0. Then there is a sequence $\{\varepsilon_n > 0\}$ such that $\bigcup_{n=1}^{\infty} \varepsilon_n B_n$ is bounded. [*Hint*: $u_m(0) \subset \bar{S}_{1, p_{u_m}}(0)$ and each $n \in \mathbb{N}$ determines a sequence

$\{\alpha_{mn}>0: m\in\mathbb{N}\}$ such that $\sup_{B_n} p_{u_m}(x)\leq\alpha_{mn}\ \forall m\in\mathbb{N}$. Let $\alpha_m=\max_{1\leq n\leq m}\alpha_{mn}$ for each $m\in\mathbb{N}$. Since $\alpha_{mn}\leq\alpha_m$ for $m\geq n$, there is an $\varepsilon_n>0$ for each $n\in\mathbb{N}$ such that $\alpha_{mn}\leq(1/\varepsilon_n)\alpha_m\ \forall m\in\mathbb{N}$. Show that $B=\bigcap_{m=1}^{\infty}\bar{S}_{\alpha_m,p_{u_m}}(0)$ is a bounded set containing $\bigcup_{n=1}^{\infty}\varepsilon_n B_n$.]

14D. Spaces of Power Series

Let $R_0\in(0,\infty)$ and let X be the vector space of all functions $x(t)$ representable as a power series $x(t)=\sum_{n=1}^{\infty}\alpha_n t^n$ in a complex variable t for $|t|<R_0$. For each $r\in(0,R_0)$ the mapping $p_r(x)=\sum_{n=0}^{\infty}|\alpha_n|r^n$ defines a norm on X (verify!). If $\mathcal{P}=\{p_r: r\in(0,R_0)\}$, then $(X,\mathfrak{T}(\mathcal{P}))$ is an $\mathrm{LCT_2VS}$ and $\mathfrak{T}(\mathcal{P})$-convergence is equivalent to uniform convergence on each compact subset of $(0,R_0)$. [*Hint*: If $s\in(0,R_0)$ and $M_s=\sup_{|t|=s}|x(t)|$, then $|\alpha_n|<(M_s/s)\ \forall n\in\mathbb{N}$.]

14E. Spaces of Differentiable Functions

Let $\mathfrak{D}^k(\mathbb{R},\mathbb{F})$ $(k=0,1,\ldots,\infty)$ denote the vector space of $\mathbb{F}$-valued, kth differentiable functions on $\mathbb{R}$. Thus $\mathfrak{D}^\infty(\mathbb{R};\mathbb{F})\subset\mathfrak{D}^k(\mathbb{R};\mathbb{F})\subset\mathfrak{D}^0(\mathbb{R};\mathbb{F})=C(\mathbb{R};\mathbb{F})$.

(a) The collections of seminorms $\mathcal{P}=\{p_{m,n}(f)=\sup_{[-n,n]}|f^{(m)}(x)|: m,n\in\mathbb{N}_0\}$ and $\tilde{\mathcal{P}}=\{p_{m,n}(f)=\max_{1\leq k\leq n}\sup_{[-n,n]}|f^{(k)}(x)|: m,n\in\mathbb{N}_0\}$ generate the same topology $\xi\in T(\mathfrak{D}^k(\mathbb{R},\mathbb{F}))$ which is finer than $\mathfrak{T}_\kappa$.

(b) Convince yourself that ξ on $\mathfrak{D}^k(\mathbb{R};\mathbb{F})$ is the ξ-induced topology of $\mathfrak{D}^{k-1}(\mathbb{R};\mathbb{F})$. In particular, ξ on $\mathfrak{D}^\infty(\mathbb{R};\mathbb{F})$ coincides with the restriction of ξ on $\mathfrak{D}^k(\mathbb{R};\mathbb{F})\ \forall k\in\mathbb{N}$.

(c) $(\mathfrak{D}^k(\mathbb{R};\mathbb{F}),\xi)$ is a Frechet space $\forall k=0,1,\ldots,\infty$. [*Hint*: If $\{f_j: j\in\mathbb{N}\}\in\mathfrak{D}^k(\mathbb{R};\mathbb{F})$ is ξ-Cauchy, then $\{f_j\}$ is $p_{m,n}$-Cauchy $\forall m,n\in\mathbb{N}_0$. Therefore $\{f_j\}$ is $p_{0,n}=\|\ \|_0$-Cauchy $\forall n\in\mathbb{N}$ and $f_j\xrightarrow{\mathfrak{T}(\mathcal{P}_{0,n})}F\in C(\mathbb{R};\mathbb{F})$, where $\mathcal{P}_{0,n}=\{p_{0,n}: n\in\mathbb{N}_0\}$. Show that $f_j^{(m)}\xrightarrow{\mathfrak{T}(\mathcal{P}_{m,n})}F^{(m)}\in C(\mathbb{R};\mathbb{F})\ \forall m=1,2,\ldots,k$ and conclude that $\{f_j\}\xrightarrow{\xi}F\in\mathfrak{D}^k(\mathbb{R};\mathbb{F})$.]

14F. Rapidly Decreasing Functions

Let $\mathfrak{D}_r^\infty(\mathbb{R};\mathbb{F})$ be the vector subspace of $\mathfrak{D}^\infty(\mathbb{R};\mathbb{F})$, each of whose members f satisfies $\lim_{|x|\to\infty}|x|^n|f^{(m)}(x)|=0\ \forall m,n\in\mathbb{N}_0$. The members of $\mathfrak{D}_r^\infty(\mathbb{R};\mathbb{F})$ (e.g., $e^{g(x)}$ if $\lim_{|x|\to\infty}g(x)=0$) are called functions of rapid decrease.

(a) If $\zeta \in T(\mathfrak{D}_r^\infty(\mathbb{R};\mathbb{F}))$ is determined by the collection of seminorms $\{q_{m,n}(f) = \sup_{\mathbb{R}} |(1+|x|^n)f^{(m)}(x)| : m, n \in \mathbb{N}_0\}$, then $\mathfrak{T}_\kappa \subset \xi/\mathfrak{D}_r^\infty(\mathbb{R};\mathbb{F}) \subset \zeta$ on $\mathfrak{D}_r^\infty(\mathbb{R};\mathbb{F})$.

(b) $(\mathfrak{D}_r^\infty(\mathbb{R};\mathbb{F}), \zeta)$ is a Frechet space. [*Hint*: If $\{f_j : j \in \mathbb{N}\} \in \mathfrak{D}_r^\infty(\mathbb{R};\mathbb{F})$ is ζ-Cauchy, $f_j \xrightarrow{q_{0,0}} F \in C(\mathbb{R};\mathbb{F})$ and $f_j^{(m)} \xrightarrow{q_{m,0}} F^{(m)} \in C(\mathbb{R};\mathbb{F})\ \forall m \in \mathbb{N}_0$. For any fixed $m, n \in \mathbb{N}_0$ there is a constant $M_{m,n+1}$ such that $\sup_{\mathbb{R}} |(1+|x|^{n+1})f_j^{(m)}(x)| < M_{m,n+1}\ \forall j \in \mathbb{N}$. (Use the fact that $\{f_j\}$ is $q_{m,n+1}$-Cauchy.) Therefore $\sup_{\mathbb{R}} |x|^n |F^{(m)}(x)| < M_{m,n+1}[|x|^n/(1+|x|^{n+1})]$ and $F \in \mathfrak{D}_r^\infty(\mathbb{R};\mathbb{F})$. Go on to prove that $f_j \xrightarrow{\zeta} F$.]

14G. Slowly Increasing Functions

Let $\mathfrak{D}_i^\infty(\mathbb{R};\mathbb{F})$ be the vector subspace of $\mathfrak{D}^\infty(\mathbb{R};\mathbb{F})$, each of whose members f satisfies $\sup_{\mathbb{R}} |g(x)f^{(n)}(x)| < \infty\ \forall g \in \mathfrak{D}_r^\infty(\mathbb{R};\mathbb{F})$ and $n \in \mathbb{N}_0$. Elements in $\mathfrak{D}_i^\infty(\mathbb{R};\mathbb{F})$ are called functions of slow increase.

Prove that the collection of seminorms $\{p_{g,n}(f) = \sup_{\mathbb{R}} |g(x)f^{(n)}(x)| : g \in \mathfrak{D}_r^\infty(\mathbb{R};\mathbb{F})$ and $n \in \mathbb{N}_0\}$ generates a complete LCT$_2$ topology $\eta \in T(\mathfrak{D}_i^\infty(\mathbb{R};\mathbb{F}))$. [*Hint*: Suppose $\mathscr{H} \in \mathfrak{D}_i^\infty(\mathbb{R};\mathbb{F})$ is an η-Cauchy filter. For each $m \in \mathbb{N}$ and $\varepsilon > 0$ there is a $g_m \in \mathfrak{D}_r^\infty(\mathbb{R};\mathbb{F})$ satisfying $\sup_{[-m,m]} |g(x)| \geq 1$ and a set $H_m \in \mathscr{H}$ such that $\sup_{[-m,m]} |h^{(n)}(x) - \tilde{h}^{(n)}(x)| < \varepsilon$ $\forall h, \tilde{h} \in H_m$. Essentially, $\mathscr{H}^{(n)}$ is $\|\ \|_0$-Cauchy on $[-m, m]\ \forall n \in \mathbb{N}_0$: Since this is true $\forall m \in \mathbb{N}$ there is an $F \in \mathfrak{D}^\infty(\mathbb{R};\mathbb{F})$ such that $\forall n \in \mathbb{N}_0$, we have $\mathscr{H}^{(n)} \to F^{(n)}$ uniformly on each $[-m, m]$ (i.e., there is an $H_m \in \mathscr{H}$ such that $\sup_{[-m,m]} |h^{(n)}(x) - F^{(n)}(x)| < \varepsilon\ \forall h \in H_m$). Since $\mathscr{H}$ is $p_{g,n}$-Cauchy $\forall g \in \mathfrak{D}_r^\infty(\mathbb{R};\mathbb{F})$ and $n \in \mathbb{N}_0$, there is an $\hat{H} = \hat{H}(g, n) \in \mathscr{H}$ satisfying $\sup_{\mathbb{R}} |g(x)\{\hat{h}^{(n)}(x) - \hat{h}^{(n)}(x)\}| < \varepsilon\ \forall \hat{h}, \hat{h} \in \hat{H}$. For any $\bar{h} \in H_m \cap \hat{H}$ and every $h \in \hat{H}$ use $|g(x)\{F^{(n)}(x) - h^{(n)}(x)\}| \leq |g(x)\{F^{(n)}(x) - \bar{h}^{(n)}(x)\}| + |g(x)\{\bar{h}^{(n)}(x) - h^{(n)}(x)\}|$ to obtain $\sup_{\mathbb{R}} |g(x)\, F^{(n)}(x)| < \infty$ and conclude that $H \xrightarrow{\eta} F \in \mathfrak{D}_i^\infty(\mathbb{R};\mathbb{F})$.]

14H. Holomorphic Functions

Let $\mathscr{H}(\Omega)$ denote the vector space of holomorphic functions on a domain Ω of the complex plane. Thus $\mathscr{H}(\Omega) = \mathfrak{D}^\infty(\Omega;\mathbb{C})$ (the space of infinitely differentiable complex-valued functions on Ω).

(a) $\mathfrak{T}_\kappa \in T(\mathscr{H}(\Omega))$ is determined by both collections of seminorms $\mathfrak{P} = \{p_{m,n}(f) = \sup_{|z|\leq n} |f^{(m)}(z)| : m, n \in \mathbb{N}_0\}$ and $\mathfrak{Q} = \{q_n(f) = \sup_{K_n} |f(z)|$, where $K_n = \{|z| \leq n : \inf_{\Omega} |z - w| \geq 1/n\}\ \forall n \in \mathbb{N}\}$. [*Hint*: $p_{m,n}(f) \leq m!/n^m \sup_{|z|\leq n} |f(z)|$ and $\mathfrak{T}(\mathfrak{P}) \subset \mathfrak{T}_\kappa$.

If K is compact, then $\inf_K |z-\Omega|>0$ and $K\subset K_n$ for some $n\in\mathbb{N}$. If $z_0\in K_n$ and Γ is the circle $|z-z_0|=1/2n$, then

$$|f^{(m)}(z_0)|=\left|\frac{m!}{2\pi i}\int_\Gamma \frac{f(z)}{(z-z_0)^{m+1}}\,dz\right|\leq m!(2n)^m q_{2n}(f).$$

These assertions yield $\mathfrak{T}_\kappa\subset\mathfrak{T}(\mathcal{Q})$. For $\mathfrak{T}(\mathcal{Q})\subset\mathfrak{T}(\mathcal{P})$, show that $q_n(f)\leq p_{0,n}(f)$ $\forall n\in\mathbb{N}$.]

(b) $\mathcal{H}_\kappa(\Omega)$ is a Frechet space. [*Hint*: A $\mathfrak{T}_\kappa$-Cauchy sequence $\{f_n\}\in\mathcal{H}(\Omega)$ converges to some $F\in C(\Omega;\mathbb{C})$. Since the first partial derivatives of each f_n satisfy the Cauchy–Rieman equations, so do the first partials of F. *Alternate*: Show that $\mathcal{H}(\Omega)$ is a closed subset of the Frechet space $C_\kappa(\Omega;\mathbb{C})$.]

The spaces considered in Problems 14E to H arise in the theory of distributions. In this context $\mathbb{R}^n$ usually replaces $\mathbb{R}$ as the domain considered.

References for Further Study

On LCTVSps: See Section 13 references "On TVSps."
On the theory of distributions: Dunford and Schwartz [26]; Horvath [43]; Treves [98].

15. Projective and Inductive Limits

A projective limit of TVSps (LCTVSps) by linear transformations is a TVS (LCTVS), thereby continuing the trend set by Theorems 3.2 and 9.1. Subspaces and product spaces are viewed from this perspective. In the other direction every LCT_2VS is shown to be topologically isomorphic to a subspace of a product of NLSps. This illuminates the nature by which LCT_2VS properties are intertwined with those of NLSps.

Since a general inductive limit of TVSps (LCTVSps) need not be a TVS, we introduce a weaker type locally convex inductive limit topology. Quotient spaces and convex direct sums fit this framework.

Projective Limits

THEOREM 15.1. Let $\mathcal{F}=\{f_\alpha\in\mathfrak{L}(X,Y_\alpha):\alpha\in\mathcal{Q}\}$, where X is a vector space and each $Y_\alpha(\alpha\in\mathcal{Q})$ is a TVS. Then $\mathfrak{T}_{\mathcal{P}(\mathcal{F})}\in T(X)$ and

(i) $\mathfrak{T}_{\mathcal{P}(\mathcal{F})}$ is locally convex if every Y_α is locally convex.
(ii) $\mathfrak{T}_{\mathcal{P}(\mathcal{F})}$ is T_2 if $\mathcal{F}$ is injective and each Y_α is T_2. If $\mathfrak{T}_{\mathcal{P}(\mathcal{F})}$ is T_2, then $\mathcal{F}$ is injective.

(iii) For $\mathfrak{T} \in T(X)$ with $X_{\mathfrak{T}} = (X, \mathfrak{T})$ the following are equivalent:

(a) $\mathfrak{T} = \mathfrak{T}_{\mathcal{P}(\mathfrak{F})}$,
(b) $f \in L(Z, X_{\mathfrak{T}})$ iff $f_\alpha f \in L(Z, Y_\alpha)\ \forall \alpha \in \mathfrak{A}$ and TVS Z.

Proof. $(X, \mathfrak{T}_{\mathcal{P}(\mathfrak{F})})$ is a topological group (Theorem 9.1) and so $\mathfrak{m}$ is continuous. For confirmation that $\mathfrak{s}$ is continuous let $\mathfrak{N}_\alpha$ be a Y_α-nbd base at 0 and consider any $\mathfrak{T}_{\mathcal{P}(\mathfrak{F})}$-nbd $\beta_0 x_0 + \bigcap_{i=1}^{n} f_{\alpha_i}^{-1}(u_{\alpha_i})$ of $\beta_0 x_0$. For each $1 \le i \le n$ there is an $\varepsilon_{\alpha_i} > 0$ and $v_{\alpha_i}(0) \in \mathfrak{N}_{\alpha_i}$ such that $\beta y_{\alpha_i} \in u_{\alpha_i} + \beta_0 f_{\alpha_i}(x_0)$ for $|\beta - \beta_0| < \varepsilon_{\alpha_i}$ and $y_{\alpha_i} \in v_{\alpha_i} + f_{\alpha_i}(x_0)$. Taking $\varepsilon = \min_{1 \le i \le n} \varepsilon_{\alpha_i}$ thus ensures that $\mathfrak{s}\left\{S_\varepsilon(\beta_0) \times \bigcap_{i=1}^{n} f_{\alpha_i}^{-1}(v_{\alpha_i})\right\} \subset \beta_0 x_0 + \bigcap_{i=1}^{n} f_{\alpha_i}^{-1}(u_{\alpha_i})$ and that $\mathfrak{s}$ is therefore continuous. The proof of (i) is straightforward, whereas (ii) and (iii) follow directly from Theorem 9.1(iii) and Theorem 3.4, respectively. ■

COROLLARY 15.2. Suppose for each $\alpha \in \mathfrak{A}$ above there is a nonempty index set $\beta(\alpha)$ and corresponding collection $\mathcal{G}_{\beta(\alpha)} = \{g_\beta \in \mathfrak{L}(Y_\alpha, Z_\beta): Z_\beta \text{ is a TVS } \forall \beta \in \beta(\alpha)\}$. If $\mathfrak{T}_{\mathcal{P}\{\mathcal{G}_{\beta(\alpha)}\}} \subset \mathfrak{T}_\alpha$ (the topology on Y_α) $\forall \alpha \in \mathfrak{A}$, then $\mathfrak{T}_{\mathcal{P}\{\mathcal{G}\mathfrak{F}\}} \subset \mathfrak{T}_{\mathcal{P}(\mathfrak{F})}$ for $\mathcal{G}\mathfrak{F} = \{g_\beta f_\alpha : \alpha \in \mathfrak{A} \text{ and } \beta \in \beta(\alpha)\}$. In particular, $\mathfrak{T}_{\mathcal{P}(\mathfrak{F})} = \mathfrak{T}_{\mathcal{P}(\mathcal{G}\mathfrak{F})}$ if $\mathfrak{T}_\alpha = \mathfrak{T}_{\mathcal{P}(\mathcal{G}_{\beta(\alpha)})}\ \forall \alpha \in \mathfrak{A}$

Proof. Simply modify the arguments in Corollary 3.5. ■

Immediate consequences of Theorem 15.1 are

Example 15.3. (a) Every vector subspace M of a TVS {LCTVS} $(X, \mathfrak{T})$ is a TVS {LCTVS} with the induced topology, since $\mathfrak{T}/M = \mathfrak{T}_{\mathcal{P}(\mathbb{1})}$ for $\mathbb{1}: M \to X$. Furthermore, $\mathfrak{T}_{\mathcal{P}(\mathfrak{F})}/M = \mathfrak{T}_{\mathcal{P}\{\mathfrak{F}/M\}}$ for $\mathfrak{F}$ as in Theorem 15.1.

(b) If each X_α $(\alpha \in \mathfrak{A})$ is a TVS {LCTVS}, then $X = \prod_{\mathfrak{A}} X_\alpha$ with the product topology $\mathfrak{T}_\pi$ is called the product TVS {LCTVS}. Since $\pi = \{\pi_\alpha : X \to X_\alpha : \alpha \in \mathfrak{A}\}$ is injective, $\mathfrak{T}_\pi = \mathfrak{T}_{\mathcal{P}(\pi)}$ is T_2 iff each X_α is T_2. Note also that each $j_\alpha : X_\alpha \xrightarrow[x_\alpha \rightsquigarrow (x_\alpha \delta_{\alpha,\beta})_{\beta \in \mathfrak{A}}]{} X$ $(\alpha \in \mathfrak{A})$ is an into topological isomorphism (j_α is continuous, since every $\pi_\beta j_\alpha = \delta_{\alpha,\beta}\mathbb{1}$ $(\beta \in \mathfrak{A})$ is continuous, and j_α^{-1} is continuous, since $j_\alpha^{-1} = \pi_\alpha$). If $\mathfrak{T}_\pi$ is T_2, then $j_\alpha(X_\alpha) = \bigcap_{\mathfrak{A}-\alpha} \pi_\beta^{-1}(0)$ is closed in X.

(c) Given $\{\mathfrak{T}_\alpha : \alpha \in \mathfrak{A}\} \subset T(X)$ for a vector space X, write $X_\alpha = (X, \mathfrak{T}_\alpha)$ and take $\mathfrak{F} = \{\mathbb{1}_\alpha = \mathbb{1} : X \to X_\alpha : \alpha \in \mathfrak{A}\}$. Then $\mathfrak{T}_{\mathcal{P}(\mathfrak{F})} = \sup_{\mathfrak{A}} \mathfrak{T}_\alpha \in T(X)$ (Problem 3E).

This is a good place to accentuate the nature of $\mathfrak{T}_{\mathcal{P}(\mathfrak{F})}$ in terms of individual components $f \in \mathfrak{F}$.

Example 15.4. Let $\mathfrak{F} = \{f_\alpha \in \mathfrak{L}(X, Y_\alpha): Y_\alpha \text{ is a TVS } \forall \alpha \in \mathfrak{A}\}$.

(a) A subset $B \subset X$ is $\mathfrak{T}_{\mathcal{P}(\mathfrak{F})}$-bounded {totally bounded} iff $f_\alpha(B) \subset Y_\alpha$ is bounded {totally bounded} $\forall \alpha \in \mathfrak{A}$. Use Theorems 5.9 and 13.20 and the fact that if every $f_\alpha(B) \subset Y_\alpha$ is bounded and $v(0) = \bigcap_{i=1}^{n} f_{\alpha_i}^{-1}(u_{\alpha_i})$ is any $\mathfrak{T}_{\mathcal{P}(\mathfrak{F})}$-nbd there follows $B \subset (\max_{1 \le i \le n} \lambda_{\alpha_i})v$ for $f_{\alpha_i}(B) \subset \lambda_{\alpha_i} u_{\alpha_i}$ $(i = 1, 2, \ldots, n)$.

(b) If $Y_\alpha = Y \; \forall \alpha \in \mathfrak{A}$ and Y is a NLS, every $f_\alpha \in \mathfrak{F}$ is $\mathfrak{T}_{\mathcal{P}\{Y'\mathfrak{F}\}}$-bounded (therefore $\mathfrak{T}_{\mathcal{P}(\mathfrak{F})}$-bounded since $\mathfrak{T}_{\mathcal{P}\{Y'\mathfrak{F}\}} \subset \mathfrak{T}_{\mathcal{P}\{\mathfrak{F}\}}$). Specifically, $g\{f_\alpha(B)\}$ is $\mathbb{F}$-bounded $\forall (g, f_\alpha) \in Y' \times \mathfrak{F}$ and $\mathfrak{T}_{\mathcal{P}\{Y'\mathfrak{F}\}}$-bounded set $B \subset X$. It remains therefore to invoke Problem 17B(c).

(c) Suppose $Y_\alpha = Y \; \forall \alpha \in \mathfrak{A}$ and let $\mathbb{F} = \mathbb{C}$. Then the following are equivalent for a net φ in X:

(i) $\varphi \xrightarrow{\mathfrak{T}_{\mathcal{P}\{\mathfrak{F}\}}} x_0$,
(ii) $f_\alpha(\varphi) \to f_\alpha(x_0) \; \forall \alpha \in \mathfrak{A}$,
(iii) $R_e f_\alpha(\varphi) \to R_e f_\alpha(x_0) \; \forall \alpha \in \mathfrak{A}$,
(iv) $I_m f_\alpha(\varphi) \to I_m f_\alpha(x_0) \; \forall \alpha \in \mathfrak{A}$,

where $R_e(z)$ and $I_m(z)$ denote the real and imaginary parts of $z \in \mathbb{C}$, respectively. Clearly (i)$\Leftrightarrow$(ii)$\Rightarrow$(iii), (iv) follows from $R_e(z)$, $I_m(z) \le |z| \; \forall z \in \mathbb{C}$, and since we have $|f_\alpha(\varphi) - f_\alpha(x_0)| \le |R_e f_\alpha(\varphi) - R_e f_\alpha(x_0)| + |I_m f_\alpha(\varphi) - I_m f_\alpha(x_0)|$ only (iii)$\Leftrightarrow$(iv) needs verification; but (iv) follows from (iii) {(iii) follows from (iv)} and the observation that $I_m f_\alpha(x) = R_e\{if_\alpha(x)\} \; \forall x \in X$ $\{R_e f_\alpha(x) = I_m\{if_\alpha(x)\} \; \forall x \in X\}$.

(d) If $Y_\alpha = Y \; \forall \alpha \in \mathfrak{A}$ then $\mathfrak{T}_{\mathcal{P}\{Y'\mathfrak{F}\}} \subset \mathfrak{T}_{\mathcal{P}\{\mathfrak{F}\}}$ and $\varphi \xrightarrow{\mathfrak{T}_{\mathcal{P}\{Y'\mathfrak{F}\}}} x_0$ whenever $\varphi \xrightarrow{\mathfrak{T}_{\mathcal{P}\{\mathfrak{F}\}}} x_0$. The converse, of course, is not true in general. Let $\{e_n = (\delta_{m,n})_{m \in \mathbb{N}} : n \in \mathbb{N}\}$ be a complete orthonormal subset of a Hilbert space $(X, (\;,\;))$ and take $\mathfrak{F} = \{\mathbb{1}: X \to X\}$. Since $\|e_n\| = 1 \; \forall n \in \mathbb{N}$, one has $e_n \not\to 0$ relative to $\|\;\| = \mathfrak{T}_{\mathcal{P}(\mathbb{1})}$ on X. Now (Riesz representation theorem) each $f \in X'$ determines a unique $x \in X$ satisfying $f(y) = (y, x) \; \forall y \in X$. In particular, $f(e_n) = (e_n, x) \; \forall n \in \mathbb{N}$ and (Appendix a.2) $\sum_{n=1}^{\infty} |f(e_n)|^2 = \|x\|^2$ yields $\lim_n f(e_n) = 0$. Thus $e_n \to 0$ relative to $\mathfrak{T}_{\mathcal{P}\{X'\mathbb{1}\}}$ on X.

THEOREM 15.5. Let $\mathfrak{F} = \{f_\alpha \in \mathfrak{L}(X, Y): \alpha \in \mathfrak{A}\}$, where X is a vector space and Y is a TVS.

(i) $\mathfrak{T}_{\mathcal{P}(\mathfrak{F})}$ is 1° (i.e., pseudometrizible) if Y is 1° and $\dim[\mathfrak{F}] \leq \aleph_0$.

(ii) $\mathfrak{T}_{\mathcal{P}\{\mathfrak{F}\}}$ is seminormable if Y is seminormable and $\dim[\mathfrak{F}] < \aleph_0$. Suppose $\mathfrak{F}$ is concentrating and Y is not indiscrete. Then $\dim[\mathfrak{F}] \leq \aleph_0$ $\{\dim[\mathfrak{F}] < \aleph_0\}$ if $\mathfrak{T}_{\mathcal{P}\{\mathfrak{F}\}}$ is 1° {seminormable}.

Proof. If $\mathfrak{N}_0$ is a countable nbd base at $0 \in Y$ and $\mathfrak{F}_{\mathfrak{B}}$ is a countable basis for $[\mathfrak{F}] \subset \mathfrak{L}(X, Y)$, then $\left\{\bigcap_{i=1}^{n} f_{\alpha_i}^{-1}(u_{\alpha_i}) : f_{\alpha_i} \in \mathfrak{F}_{\mathfrak{B}} \text{ and } u_{\alpha_i} \in \mathfrak{N}_0\right\}$ is a countable $\mathfrak{T}_{\mathcal{P}\{\mathfrak{F}\}}$-nbd base at $0 \in X$ and (i) is proved. If $\mathfrak{F}_{\mathfrak{B}} = \{f_{\alpha_1} \cdots f_{\alpha_n}\}$ is a basis for $[\mathfrak{F}]$ in (ii) and B is a bounded convex nbd of $0 \in Y$ (Theorem 14.19), then $\bigcap_{i=1}^{n} f_{\alpha_i}^{-1}(B)$ is a $\mathfrak{T}_{\mathcal{P}\{\mathfrak{F}\}}$-bounded convex nbd of $0 \in X$. Indeed, for any other $\mathfrak{T}_{\mathcal{P}\{\mathfrak{F}\}}$-nbd $\bigcap_{j=1}^{m} f_j^{-1}(u)$, write $f_{\gamma_j} \in \mathfrak{F}$ as $f_{\gamma_j} = \sum_{t=1}^{n} \beta_{t_j} f_{\alpha_t}$ $(j = 1, 2, \ldots, m)$ and let $B \subset \lambda u(0)$ for $\lambda > 0$. Then $\bigcap_{i=1}^{n} f_{\alpha_i}^{-1}(B) \subset \lambda \max_{1 \leq j \leq m} \left\{\sum_{t=1}^{n} \beta_{t_j}\right\} \bigcap_{j=1}^{m} f_{\gamma_j}^{-1}(u)$.

Suppose $\mathfrak{F}$ is concentrating and $v(0)$ is a proper nbd of Y. Given any countable $\mathfrak{T}_{\mathcal{P}(\mathfrak{F})}$-nbd base $\{w_n(0) : n \in \mathbb{N}\}$, there is a $\mathfrak{T}_{\mathcal{P}(\mathfrak{F})}$-nbd $\bigcap_{i=1}^{m_n} f_{\alpha_{i(n)}}^{-1}(u_n) \subset w_n$ for each $n \in \mathbb{N}$. Then every $f \in \mathfrak{F}$ satisfies $f \in \left[\bigcup_{n=1}^{\infty} \bigcup_{i=1}^{m_n} f_{\alpha_{i(n)}}\right]$ and $\dim[\mathfrak{F}] \leq \aleph_0$ follows. Otherwise $\bigcap_{i=1}^{m_n} f_{\alpha_i}^{-1}(u_n) \not\subset f^{-1}(v)\ \forall n \in \mathbb{N}$ (since $x_0 \in \bigcap_{i=1}^{m_n} f_{\alpha_{i(n)}}^{-1}(u_n) - f^{-1}(v)$ for $f(x_0) = y_0 \in Y - v$ and $f_{\alpha_{i(n)}}(x_0) = 0\ \forall 1 \leq i \leq m_n$), thereby contradicting the fact that some $w_n(0) \subset f^{-1}(v)$. Given any $\mathfrak{T}_{\mathcal{P}(\mathfrak{F})}$-bounded convex nbd E of $0 \in X$, there is a $\mathfrak{T}_{\mathcal{P}(\mathfrak{F})}$-bounded nbd $\bigcap_{i=1}^{n} f_{\alpha_i}^{-1}(u) \subset E$ and $\mathfrak{F} \subset [f_{\alpha_i} \cdots f_{\alpha_n}]$ ensures that $\dim[\mathfrak{F}] \leq n$. This last inclusion is clear, since $f \notin [f_{\alpha_1} \cdots f_{\alpha_n}]$ implies $\bigcap_{i=1}^{n} f_{\alpha_i}^{-1}(u) \not\subset \lambda f^{-1}(v) = f^{-1}(\lambda v)\ \forall \lambda > 0$ (use λy_0 instead of y_0 above), contradicting the boundedness of $\bigcap_{i=1}^{n} f_{\alpha_i}^{-1}(u)$. ∎

In the next two corollaries $\mathfrak{F} = \{f_\alpha \in \mathfrak{L}(X, Y)\ \forall \alpha \in \mathfrak{A}\}$ and Y is a TVS with $Y^{\blacksquare}$ denoting either Y' or Y^*.

COROLLARY 15.5. $\mathfrak{T}_{\mathcal{P}\{Y^{\blacksquare}\mathfrak{F}\}}$ is metrizible {normable} iff $Y^{\blacksquare}\mathfrak{F}$ is injective and $\dim[Y^{\blacksquare}\mathfrak{F}] \leq \aleph_0$ $\{\dim[Y^{\blacksquare}\mathfrak{F}] < \aleph_0\}$.

Proof. $Y^{\blacksquare}\mathfrak{F} \subset X^*$, which is concentrating (Example 13.4). ■

COROLLARY 15.7. If $\mathfrak{T}_{\mathcal{P}\{Y^{\blacksquare}\mathfrak{F}\}}$ is normable, $\dim X$ is finite (and $\mathfrak{T} = \mathfrak{T}_{\mathcal{P}\{Y^{\blacksquare}\mathfrak{F}\}}$ $\forall \mathfrak{T} \in T(X)$ finer than $\mathfrak{T}_{\mathcal{P}\{Y^{\blacksquare}\mathfrak{F}\}}$). The converse also holds if $\mathfrak{T}_{\mathcal{P}\{Y^{\blacksquare}\mathfrak{F}\}}$ is T_2.

Proof. If $\mathfrak{T}_{\mathcal{P}\{Y^{\blacksquare}\mathfrak{F}\}}$ is normable, $\underset{x \rightsquigarrow \xi_x}{X \to [Y^{\blacksquare}\mathfrak{F}]^*}$ with $\xi_x(gf_\alpha) = g\{f_\alpha(x)\}$ is a vector space isomorphism and $\dim X \le \dim [Y^{\blacksquare}\mathfrak{F}]^* < \aleph_0$. The second statement and the converse both follow from Theorem 13.12. ■

Every subspace and every product of TVSps {LCTVSps} is a projective limit TVS {LCTVS}. In a sense these are the only examples of projective limit spaces.

THEOREM 15.8. If $\mathfrak{F} = \{f_\alpha \in \mathfrak{L}(X, Y_\alpha) : \alpha \in \mathfrak{A}\}$ is injective, then $(X, \mathfrak{T}_{\mathcal{P}\{\mathfrak{F}\}})$ is topologically isomorphic to a subspace of $\prod_{\mathfrak{A}} Y_\alpha$. Therefore every LCT$_2$VS X is topologically isomorphic to a subspace of a product of normed linear (Banach) spaces.

Proof. Since $\mathfrak{F}$ is injective, $\underset{x \rightsquigarrow (f_\alpha(x))}{\mathfrak{e} : X \to \prod Y_\alpha}$ is a vector space isomorphism. Furthermore, $\mathfrak{e}$ is continuous, since every $f_\alpha \in \mathfrak{F}$ is continuous, and $\mathfrak{e}^{-1}$ is continuous (Theorem 15.1(iii)), since $\pi_\alpha = f_\alpha \mathfrak{e}^{-1}$ is continuous $\forall \alpha \in \mathfrak{A}$.

Suppose $\mathcal{P}$ is a collection of continuous seminorms that determines the topology of an LCT$_2$VS X (Theorem 14.11). For each $p \in \mathcal{P}$ let $X_p = X/p^{-1}(0)$ carry the norm $\|p^{-1}(0) + x\|_p = p(x)$. Then $\mathfrak{T}(\mathcal{P}) = \mathfrak{T}_{\mathcal{P}\{\mathfrak{F}\}}$ for $\mathfrak{F} = \{\nu_p : X \to X_p : p \in \mathcal{P}\}$. Specifically, $\mathfrak{T}_{\mathcal{P}\{\mathfrak{F}\}} \subset \mathfrak{T}(\mathcal{P})$, since every $\nu_p \in \mathfrak{F}$ is $\mathfrak{T}(\mathcal{P})$-continuous, and $\mathfrak{T}(\mathcal{P}) \subset \mathfrak{T}_{\mathcal{P}\{\mathfrak{F}\}}$, since $S_{\varepsilon,p}(0) = \nu_p^{-1} S_{\varepsilon,\|\,\|_p}\{p^{-1}(0)\}$ $\forall p \in \mathcal{P}$. The T_2 property of $\mathfrak{T}(\mathcal{P})$ also ensures that $\mathfrak{F}$ is injective and X is topologically isomorphic to $\mathfrak{e}(X) \subset \prod_{\mathcal{P}} X_p$. ■

Inductive Limits

Suppose $\mathfrak{F} = \{f_\alpha \in \mathfrak{L}(X_\alpha, Y) : \alpha \in \mathfrak{A}\}$, where Y is a vector space and each X_α is a TVS. Unlike projective limits, the general inductive limit topology $\mathfrak{T}_{\mathcal{I}\{\mathfrak{F}\}}$ need not be compatible with Y's linear structure ($\mathfrak{T}_{\mathcal{I}\{\mathfrak{F}\}}$-nbds of 0 need not be absorbent and continuity of $\mathfrak{s}$ may fail). This necessitates a little retrenching to a weaker modified inductive limit topology $\mathfrak{T}_{\mathcal{I}'\{\mathfrak{F}\}} \in T(Y)$.

THEOREM 15.9. Let $\mathcal{F}=\{f_\alpha \in \mathcal{L}(X_\alpha, Y): \alpha \in \mathcal{A}\}$ be as above.

(i) $\mathcal{N}_0=\{$abs convex, absorbent sets $u \subset Y: f_\alpha^{-1}(u)$ is an X_α-nbd $\forall \alpha \in \mathcal{A}\}$ is a nbd base at $0 \in Y$ for a finest locally convex topology $\mathcal{T}_{\mathcal{I}'\{\mathcal{F}\}} \in T(Y)$ under which every $f_\alpha \in \mathcal{F}$ is continuous.

(ii) Suppose $\mathcal{T} \in T(Y)$ is locally convex and let $Y_{\mathcal{T}}=(Y, \mathcal{T})$. Then $\mathcal{T}=\mathcal{T}_{\mathcal{I}'\{\mathcal{F}\}}$ is equivalent to $f \in L(Y_{\mathcal{T}}, Z)$ iff $ff_\alpha \in L(X_\alpha, Z)\ \forall \alpha \in \mathcal{A}$ and TVS Z.

Proof. $\mathcal{N}_0$ satisfies Theorem 14.2 (since $\frac{1}{2}u(0)+\frac{1}{2}u(0)=u(0)\ \forall u \in \mathcal{N}_0$) and so $\mathcal{T}_{\mathcal{I}'\{\mathcal{F}\}}$ is well defined. If $\mathcal{T} \in T(Y)$ is locally convex and $f_\alpha \in L(X_\alpha, Y_{\mathcal{T}})\ \forall \alpha \in \mathcal{A}$, then every abs convex $\mathcal{T}$-nbd belongs to $\mathcal{N}_0$ and $\mathcal{T} \subset \mathcal{T}_{\mathcal{I}'\{\mathcal{F}\}}$. The proof of (ii) follows as in Theorem 3.11 by taking $f=\mathbb{1}$ for $Z=Y_{\mathcal{T}}$ and $Z=Y_{\mathcal{T}_{\mathcal{I}'(\mathcal{F})}}$. ■

The requirements on $\mathcal{N}_0$ can be weakened if those on $\mathcal{F}$ are strengthened.

COROLLARY 15.10. If $Y=[\bigcup_{\mathcal{A}} f_\alpha(X_\alpha)]$ in Theorem 15.9, then each of the following defines a $\mathcal{T}_{\mathcal{I}'\{\mathcal{F}\}}$-nbd base at $0 \in Y$:

(i) $\mathcal{N}_0=\{$abs convex $u \subset Y: f_\alpha^{-1}(u)$ is an X_α-nbd $\forall \alpha \in \mathcal{A}\}$.

(ii) $\mathcal{N}=\{\{\bigcup_{\mathcal{A}} f_\alpha(u_\alpha)\}_{bc}: u_\alpha \in \mathcal{N}_\alpha\}$, where $\mathcal{N}_\alpha$ is an X_α-nbd base at 0 satisfying Theorem 13.7.

One calls $(Y', \mathcal{T}_{\mathcal{I}'\{\mathcal{F}\}})$ an internal inductive limit of $\{X_\alpha\}$ by $\mathcal{F}$.

Proof. (i) If $u(0) \in \mathcal{N}_0$ and $y=\sum_{i=1}^n a_i f_{\alpha_i}(x_{\alpha_i})=\sum_{i=1}^n f_{\alpha_i}(\alpha_i x_{\alpha_i}) \in Y=[\bigcup_{\mathcal{A}} f_\alpha(X_\alpha)]$, there exist $\varepsilon_i>0$ such that $a_i x_{\alpha_i} \in \lambda f_{\alpha_i}^{-1}(u)\ \forall |\lambda| \geq \varepsilon_i$. Therefore $y \in n\lambda u\ \forall |\lambda| \geq \sum_{i=1}^n \varepsilon_i$ and $u(0)$ is seen to be absorbent. (ii) If $E=\{\bigcup_{\mathcal{A}} f_\alpha(u_\alpha)\}_{bc} \in \mathcal{N}$, then $u_\alpha \subset f_\alpha^{-1}(E)$ and $f_\alpha^{-1}(E) \in \mathcal{N}_\alpha\ \forall \alpha \in \mathcal{A}$. Given any abs convex $\mathcal{T}_{\mathcal{I}'\{\mathcal{F}\}}$-nbd $u(0) \subset Y$, pick $v_\alpha \in \mathcal{N}_\alpha$ satisfying $v_\alpha \subset f_\alpha^{-1}(u)\ \forall \alpha \in \mathcal{A}$ and obtain $\{\bigcup_{\mathcal{A}} f_\alpha(v_\alpha)\}_{bc} \subset u(0)$. ■

The TVS version of Corollary 3.12 is

COROLLARY 15.11. Suppose for each $\alpha \in \mathcal{A}$ there is a nonempty index set $\beta(\alpha)$ and a corresponding collection $\mathcal{G}_{\beta(\alpha)}=\{g_\beta \in \mathcal{L}(Z_\beta, X_\alpha): Z_\beta$

is a TVS $\forall \beta \in \beta(\alpha)\}$. If $\mathfrak{T}_\alpha \subset \mathfrak{T}_{\mathfrak{g}'\{\mathfrak{G}_{\beta(\alpha)}\}}$ $\forall \alpha \in \mathfrak{A}$, then $\mathfrak{T}_{\mathfrak{g}'\{\mathfrak{F}\}} \subset \mathfrak{T}_{\mathfrak{g}'\{\mathfrak{F}\mathfrak{G}\}}$ for $\mathfrak{F}\mathfrak{G} = \{f_\alpha g_\beta : \alpha \in \mathfrak{A}$ and $\beta \in \beta(\alpha)\}$. In particular, $\mathfrak{T}_{\mathfrak{g}'\{\mathfrak{F}\}} = \mathfrak{T}_{\mathfrak{g}'\{\mathfrak{F}\mathfrak{G}\}}$ if $\mathfrak{T}_\alpha = \mathfrak{T}_{\mathfrak{g}'\{\mathfrak{G}_{\beta(\alpha)}\}}$ $\forall \alpha \in \mathfrak{A}$.

Undoubtedly, the simplest example of an inductive limit occurs when $\mathfrak{F}$ consists of a single surjection. Quotient spaces fit this framework and the easily modified details in Theorems 9.4 to 9.6 serve as a basis for listing

THEOREM 15.12. Let h be a linear surjection from a TVS {LCTVS} X to a vector space Y. If $\mathfrak{N}_0$ is an X-nbd base at 0 that satisfies Theorem 13.7 (Theorem 14.2), then $h\mathfrak{N}_0 = \{h(u) : u \in \mathfrak{N}_0\}$ is a $\mathfrak{T}_{\mathfrak{g}(h)} \in T(Y)$-nbd base at 0 that also satisfies Theorem 13.7 (Theorem 14.2).

DEFINITION 15.13. Let M be a vector subspace of a TVS {LCTVS} X. Then $(X/M, \mathfrak{T}_{\mathfrak{g}(\nu)})$ is called the quotient TVS {LCTVS} of X by M. Note that $(X/M, \mathfrak{T}_{\mathfrak{g}(\nu)})$ is T_2 iff M is closed in X.

$(X/\{\bar{0}\}, \mathfrak{T}_{\mathfrak{g}(\nu)})$ is termed the T_2VS {LCT_2VS} associated with X.

Remark. $\nu : X \to (X/M, \mathfrak{T}_{\mathfrak{g}(\nu)})$ is a continuous, open, linear surjection which is also closed when $M = \{\bar{0}\}$.

Remark. $\mathcal{U}_{X/\{\bar{0}\}} = \mathcal{U}_{\mathfrak{g}(\nu)} \in \mathcal{U}(X/\{\bar{0}\})$ so that [Example 6.8, (c)] X is $\mathcal{U}_X$-complete iff $X/\{\bar{0}\}$ is $\mathcal{U}_{\mathfrak{g}(\nu)}$-complete.

Every open subgroup of a topological group is closed (Theorem 8.7), but an open vector subspace of TVS is necessarily the whole vector space (Theorems 9.6 and 13.14$^+$). Another distinction between TVSps and topological groups concerns norms.

Example 15.14. Let M be a vector subspace of a NLS $(X, \|\ \|)$. Then $|M + x_0| = \inf_{x \in M + x_0} \|x\|$ is a seminorm on X/M such that $\nu\{S_{\varepsilon,\|\ \|}(0)\} \subset S_{\varepsilon|\ |}(0)\ \forall \varepsilon > 0$. Consequently, $|\ | \subset \mathfrak{T}_{\mathfrak{g}(\nu)} \in T(X/M)$. If M is closed, then $|\ |$ is a norm and $|\ | = \mathfrak{T}_{\mathfrak{g}(\nu)}$ by Theorem 15.12, since $\nu\{S_{\varepsilon,\|\ \|}(0)\} = S_{\varepsilon,|\ |}(0)\ \forall \varepsilon > 0$ (specifically, $|M + x_0| < \varepsilon \Rightarrow \|x\| < \varepsilon$ for some $x \in M + x_0$ and $M + x_0 \in \nu(x) \in \nu\{S_{\varepsilon,|\ |}(0)\}$ follows).

Example 15.15. Let $\mathfrak{L}_M(X, Y) = \{f \in \mathfrak{L}(X, Y) : M \subset f^{-1}(0)\}$, where X, Y are vector spaces and M is a subspace of X. In essence $f \in \mathfrak{L}_M(X, Y)$ iff f is constant on $M + x\ \forall x \in X$. Therefore each $f \in \mathfrak{L}_M(X, Y)$ determines a unique $\mu_f \in \mathfrak{L}(X/M, Y)$ such that $f = \mu_f \nu$ (simply define $\mu_f(M + x) = f(x)\ \forall x \in X$). It is easily verified that μ_f is surjective {injective} iff f is surjective $\{M = f^{-1}(0)\}$ and that $\underset{f \rightsquigarrow \mu_f}{\mathfrak{L}_M(X, Y) \to \mathfrak{L}(X/M, Y)}$ is an onto isomorphism.

If X and Y are TVSps, $L_M(X, Y) \to L(X/M, Y)$ is an onto vector space isomorphism (Theorem 15.8). Furthermore, $L_M(X, Y) = L_{\bar{M}}(X, Y)$ so that $L(X, Y) = L_{\{\bar{0}\}}(X, Y)$ and $L(X/\{\bar{0}\}, Y)$, and, in particular, X' and $(X/\{\bar{0}\})$, are vector-space isomorphs.

Topological direct sums constitute another important application of inductive limits. By way of introduction, consider

Example 15.16. If $\{M_\alpha : \alpha \in \mathcal{Q}\}$ are subspaces of a vector space X and $\mathcal{F} = \{\mathbb{1}_\alpha = \mathbb{1} : (M_\alpha, \mathfrak{T}_\alpha) \to X\,; \mathfrak{T}_\alpha \in T(M_\alpha) \forall \alpha \in \mathcal{Q}\}$, then $\mathfrak{T}_{\mathcal{I}'\{\mathcal{F}\}} \in T(X)$ is the finest locally convex topology on X under which $\mathfrak{T}_{\mathcal{I}'\{\mathcal{F}\}/M_\alpha} \subset \mathfrak{T}_\alpha \ \forall \alpha \in \mathcal{Q}$. To be sure, every $\mathfrak{T}_{\mathcal{I}'\{\mathcal{F}\}/M_\alpha}$-nbd $u(0) \cap M_\alpha = \mathbb{1}_\alpha^{-1}(u)$ is a $\mathfrak{T}_\alpha$-nbd. If $\mathfrak{T} \in T(X)$ satisfies $\mathfrak{T}/M_\alpha \subset \mathfrak{T}_\alpha \ \forall \alpha \in \mathcal{Q}$, then every $\mathbb{1}_\alpha$ is $\mathfrak{T}_\alpha - \mathfrak{T}$ continuous. Local convexity of $\mathfrak{T}$ then ensures, that $\mathfrak{T} \subset \mathfrak{T}_{\mathcal{I}'\{\mathcal{F}\}}$.

Every inductive limit can be reduced to the above framework. For $\mathcal{F} = \{f_\alpha \in \mathcal{L}(X_\alpha, Y) : X_\alpha$ has nbd base $\mathfrak{N}_\alpha$ at $0 \ \forall \alpha \in \mathcal{Q}\}$ take $M_\alpha = f_\alpha(X_\alpha) \subset Y$ and let $\mathfrak{T}_\alpha \in T(M_\alpha)$ have nbd base $f_\alpha \mathfrak{N}_\alpha$ at $0 \in M_\alpha$.

If $X = \sum_{i=1}^{n} \oplus M_i$ above, then $\underset{(x_1,\ldots,x_n) \rightsquigarrow \sum_{i=1}^n x_i}{\prod_{i=1}^{n} M_i \to X}$ is an onto vector space isomorphism. This is easily generalized. Given any collection of TVSps $\{X_\alpha : \alpha \in \mathcal{Q}\}$, let $\sum_{\mathcal{Q}} \oplus X_\alpha$ denote the subspace of $\prod_{\mathcal{Q}} X_\alpha$ consisting of tuples with at most finitely many nonzero coordinates. If $e_\alpha = (\delta_{\alpha,\beta})_{\beta \in \mathcal{Q}} \ \forall \alpha \in \mathcal{Q}$, then every $x = (x_\alpha) \in \sum_{\mathcal{Q}} \oplus X_\alpha$ has a unique representation $x = \sum_{i=1}^{n} x_{\alpha_i} e_{\alpha_i}$, which we abbreviate $x = \sum_{i=1}^{n} x_{\alpha_i}$. Turning to topological considerations, we find that each $j_\beta : X_\beta \to j_\beta(X_\beta) \subset (\sum_{\mathcal{Q}} \oplus X_\alpha, \mathfrak{T}_\pi)$ $(\beta \in \mathcal{Q})$ is a topological isomorphism [Example 15.3(b)]. Therefore $\mathfrak{T}_\pi \subset \mathfrak{T}_{\mathcal{I}'(J)} \in T(\sum_{\mathcal{Q}} \oplus X_\alpha)$ for $J = \{j_\alpha : \alpha \in \mathcal{Q}\}$. Note that $\mathfrak{T}_{\mathcal{I}'(J)}$ is locally convex but that $\mathfrak{T}_\pi$ need not be. If $\mathfrak{T}_\pi$ is locally convex, then (Example 15.16) $\mathfrak{T}_\pi$ and $\mathfrak{T}_{\mathcal{I}'\{J\}}$ both induce the original topology on every X_α [more precisely on every $j_\alpha(X_\alpha)$]. It need not be true that $\mathfrak{T}_\pi = \mathfrak{T}_{\mathcal{I}'\{J\}}$.

THEOREM 15.17. If $\mathcal{Q} = \{\alpha_1, \ldots, \alpha_n\}$ is finite and each X_{α_i} $(\alpha_i \in \mathcal{Q})$ is an LCTVS, then $\mathfrak{T}_\pi = \mathfrak{T}_{\mathcal{I}'\{J\}}$ on $\sum_{\mathcal{Q}} \oplus X_\alpha = \prod_{\mathcal{Q}} X_\alpha$.

Proof. If $u(0)\subset \sum_{i=1}^{n} \oplus X_{\alpha_i} = \prod_{i=1}^{n} X_{\alpha_i}$ is an abs convex $\mathfrak{T}_{\mathcal{J}'\{J\}}$-nbd, then $j_{\alpha_i}^{-1}(u)\cap X_{\alpha_i}$ is an X_{α_i}-nbd $\forall 1\le i\le n$ and $E=1/n \prod_{i=1}^{n} j_{\alpha_i}^{-1}(u)\cap X_{\alpha_i}$ is a $\mathfrak{T}_\pi$-nbd contained in $u(0)$. This is clear, since $x=\sum_{i=1}^{n} j_{\alpha_i}(x)\in\sum_{n}(1/n)u = u\ \forall x=(x_{\alpha_i}\cdots x_{\alpha_n})\in E$. ■

Two types of topological direct sum can now be defined.

DEFINITION 15.18. (a) A TVS $(X,\mathfrak{T})$ is the topological direct sum of TVSps $\{X_\alpha:\alpha\in\mathcal{Q}\}$ if $\mathfrak{T}=\mathfrak{T}_\pi$ on $X=\sum_{\mathcal{Q}}\oplus X_\alpha$. We denote this $X\overset{t}{=}\sum_{\mathcal{Q}}\oplus X_\alpha$.

(b) An LCTVS $(X,\mathfrak{T})$ is a convex direct sum of LCTVSps $\{X_\alpha:\alpha\in\mathcal{Q}\}$ if $\mathfrak{T}=\mathfrak{T}_{\mathcal{J}'\{J\}}$ on $X=\sum_{\mathcal{Q}}\oplus X_\alpha$. This is denoted $X\overset{tc}{=}\sum_{\mathcal{Q}}\oplus X_\alpha$.

THEOREM 15.19. Let $X\overset{tc}{=}\sum_{\mathcal{Q}}\oplus X_\alpha$. Then X is T_2 iff each X_α is T_2, in which case each X_α is closed in X. The following properties also hold whenever X is T_2:

(i) $B\subset X$ is {totally} bounded iff there is a finite subset $\mathcal{Q}_0\subset\mathcal{Q}$ such that $\pi_\alpha(B)=\{0\}\ \forall\alpha\notin\mathcal{Q}_0$ and $\pi_\alpha(B)\subset X_\alpha$ is {totally} bounded $\forall\alpha\in\mathcal{Q}_0$.

(ii) A closed set $B\subset X$ is compact iff $\pi_\alpha(B)=0\ \forall\alpha\notin\mathcal{Q}_0$ and $\pi_\alpha(B)\subset X_\alpha$ is compact $\forall\alpha\in\mathcal{Q}_0$.

(iii) X is complete iff each X_α $(\alpha\in\mathcal{Q})$ is complete.

Proof. If each X_α is T_2, then $\mathfrak{T}_\pi$ is T_2 and each X_α (identified with $j_\alpha(X_\alpha)=\bigcap_{\mathcal{Q}-\alpha}\pi_\beta^{-1}(0)$) is $\mathfrak{T}_\pi$-closed in X. Since $\mathfrak{T}_\pi\subset\mathfrak{T}_{\mathcal{J}'\{J\}}=\mathfrak{T}$, our first assertion is clear. The sufficiency of (i) and (ii) and the neccssity of (iii) are given by Example 15.3(a) and Theorem 6.6, respectively. Let us verify the necessary part of (i) and the sufficiency of (iii). First, the {total} boundedness of $B\subset X$ implies that of $\pi_\alpha(B)\subset X_\alpha\ \forall\alpha\in\mathcal{Q}$. If there is a subset $\tilde{\mathcal{Q}}=\{\alpha_n:n\in\mathbb{N}\}$ of $\mathcal{Q}$ such that $x_{\alpha_n}\in\pi_{\alpha_n}(B)-\{0\}\ \forall\alpha_n\in\tilde{\mathcal{Q}}$, pick an X_{α_n}-balanced nbd $u_{\alpha_n}(0)\not\ni(1/n)x_{\alpha_n}\ \forall\alpha_n\in\tilde{\mathcal{Q}}$ and take $u_\alpha(0)=X_\alpha\ \forall\alpha\in\mathcal{Q}-\tilde{\mathcal{Q}}$. Then (Corollary 15.10) $u(0)=\{\bigcup_{\mathcal{Q}} j_\alpha(u_\alpha)\}_{bc}$ is a $\mathfrak{T}_{\mathcal{J}'\{J\}}=\mathfrak{T}$-nbd of $0\in\sum_{\mathcal{Q}}\oplus X_\alpha=[\bigcup_{\mathcal{Q}} j_\alpha(X_\alpha)]$. However, $(1/n)x_{\alpha_n}\notin u(0)\ \forall n\in\mathbb{N}$. This contradicts Theorem 13.22 and therefore confirms that $\{\alpha\in\mathcal{Q}:\pi_\alpha(B)\neq\{0\}\}$ is finite.

Assume for (iii) that $\mathfrak{N}$ is a $\mathfrak{T}=\mathfrak{T}_{\mathcal{I}'\{J\}}$ nbd base of abs convex closed sets. Evidently, $v=\bar{v}^{\mathfrak{T}}\subset\bar{v}^{\mathfrak{T}_\pi}\ \forall v(0)\in\mathfrak{N}$. If $x\in\bar{v}^{\mathfrak{T}_\pi}$, there is a finite subset $\mathcal{Q}_0\subset\mathcal{Q}$ such that $x_\alpha=0\ \forall\alpha\notin\mathcal{Q}_0$ and it follows (by Theorem 15.17) that $x\in\bar{v}^{\mathfrak{T}_\pi}\cap\sum_{\mathcal{Q}_0}\oplus X_\alpha=\overline{v\cap\sum_{\mathcal{Q}_0}\oplus X_\alpha}^{\mathfrak{T}_\pi}=\overline{v\cap\sum_{\mathcal{Q}_0}\oplus X_\alpha}^{\mathfrak{T}_{\mathcal{I}'\{J\}}}=\bar{v}^{\mathfrak{T}_{\mathcal{I}'\{J\}}}\cap\sum_{\mathcal{Q}_0}\oplus X_\alpha\subset\bar{v}^{\mathfrak{T}_{\mathcal{I}'\{J\}}}$. This establishes that every $v(0)\in\mathfrak{N}$ is $\mathfrak{T}_\pi$-closed. Now suppose that every X_α is complete and let $\mathcal{G}$ be a $\mathfrak{T}_{\mathcal{I}'\{J\}}$-Cauchy filter in $\sum_{\mathcal{Q}}\oplus X_\alpha$. Clearly, $\pi_\alpha\mathcal{G}$ has a unique limit $x_\alpha\in X_\alpha$ ($\alpha\in\mathcal{Q}$) and $\mathcal{G}\xrightarrow{\mathfrak{T}_\pi}x=(x_\alpha)\in\prod_{\mathcal{Q}}X_\alpha$. We claim that $x\in\sum_{\mathcal{Q}}\oplus X_\alpha$. Suppose to the contrary that there is an infinite set $\tilde{\mathcal{Q}}=\{\alpha_n: x_{\alpha_n}\neq 0\}\subset\mathcal{Q}$. For $u(0)$ above pick $v(0)\in\mathfrak{N}$ such that $v\subset u$ and let $y\in G$, where $G\in\mathcal{G}$ is $\frac{1}{2}v$-sized. Then $G\subset y+\frac{1}{2}v$ and $y_{\alpha_{n_0}}=0$, whereas $x_{\alpha_{n_0}}\neq 0$ for some $\alpha_{n_0}\in\tilde{\mathcal{Q}}$. Accordingly, $(1/n_0)\mathcal{G}\xrightarrow{\mathfrak{T}_\pi}(1/n_0)x$ implies that $(1/n_0)x\in(1/n_0)\bar{G}^{\mathfrak{T}_\pi}\subset(1/n_0)v+(1/2n_0)\bar{v}^{\mathfrak{T}_\pi}\subset(1/n_0)y+(1/2n_0)u$ and $(1/n_0)x\in u_{\alpha_{n_0}}$ contradicts the definition of $u_{\alpha_{n_0}}$. This proves our claim and we now show that $\mathcal{G}\xrightarrow{\mathfrak{T}_{\mathcal{I}'\{J\}}}x\in\sum_{\mathcal{Q}}\oplus X_\alpha$. For any $w(0)\in\mathfrak{N}$ and $\frac{1}{2}w$-sized set $\tilde{G}\in\mathcal{G}$, one has $x\in\bar{\tilde{G}}^{\mathfrak{T}_\pi}\subset x_g+\frac{1}{2}w$ and $x_g\in x+\frac{1}{2}w$ for $x_g\in\tilde{G}$. Therefore $\tilde{G}\subset x+w$ and the asserted convergence holds. ■

Remark. The reader can easily verify that a convex direct sum of quasi-complete spaces is quasi-complete.

Equality denotes vector space isomorphism in

THEOREM 15.20. If $\{X_\alpha:\alpha\in\mathcal{Q}\}$ is a collection of LCTVSps, then $\sum_{\mathcal{Q}}\oplus X'_\alpha=(\prod_{\mathcal{Q}}X_\alpha)'$ and $\prod_{\mathcal{Q}}X'_\alpha=(\sum_{\mathcal{Q}}\oplus X_\alpha,\mathfrak{T}_{\mathcal{I}'\{J\}})'$.

Proof. If $f=\sum_{\mathcal{Q}}f_\alpha\in\sum_{\mathcal{Q}}\oplus X'_\alpha$, then $f_{\alpha_i}\neq 0$ for some finite index set $\mathcal{Q}_0=\{\alpha_1,\ldots,\alpha_n\}\subset\mathcal{Q}$ and $F_f\{(x_\alpha)\}=\sum_{\mathcal{Q}}f_\alpha(x_\alpha)$ is defined $\forall(x_\alpha)\in\prod_{\mathcal{Q}}X_\alpha$. In fact, $F_f\in(\prod_{\mathcal{Q}}X_\alpha)'$, since $F_f\left\{\bigcap_{i=1}^{n}\pi_{\alpha_i}^{-1}(S_{\varepsilon/2n}(0))\right\}\subset S_\varepsilon(0)\ \forall\varepsilon>0$. It remains only to confirm that $\underset{f\rightsquigarrow F_f}{\phi:\sum_{\mathcal{Q}}\oplus X'_\alpha\to(\prod_{\mathcal{Q}}X_\alpha)'}$ is surjective; but every $F\in(\prod_{\mathcal{Q}}X_\alpha)'$ satisfies $\sup_{x\in u}|F(x)|<1$ for some $\mathfrak{T}_\pi$-nbd $u(0)=\bigcap_{j=1}^{m}\pi_{\alpha_j}^{-1}(u_{\alpha_j})$ and ensures further that $F_{j_\alpha}\in X'_\alpha$ satisfies $F_{j_\alpha}=0$ if $\alpha\neq\alpha_j$ $(j=1,2,\ldots,m)$. Thus

$\phi(\sum_{\mathcal{A}} F_{j_\alpha}) = F$ and our first equality is established. The proof of the second equality follows similarly. Specifically, $\psi: \prod_{\mathcal{A}} X'_\alpha \to (\sum_{\mathcal{A}} \oplus X_\alpha)'$ with $G_f(\sum_{\mathcal{A}} x_\alpha) =$
$f = (f_\alpha) \rightsquigarrow G_f$

$\sum_{\mathcal{A}} f_\alpha(x_\alpha)$ is a linear injection. If $G \in (\sum_{\mathcal{A}} \oplus X_\alpha)'$, then (Theorem 15.1) $Gj_\alpha \in X'_\alpha$ and $\psi\{(Gj_\alpha)\} = G$. ■

COROLLARY 15.21. Let $\mathfrak{F} = \{f_\alpha \in \mathfrak{L}(X, Y_\alpha) : \alpha \in \mathcal{A}\}$ be an injective collection of surjections from a vector space X to LCTVSps Y_α. Then $(X, \mathfrak{T}_{\mathcal{P}\{\mathfrak{F}\}})' = [\bigcup_{\mathcal{A}} Y'_\alpha f_\alpha]$.

Proof. Since every $f_\alpha \in \mathfrak{F}$ is surjective, each $Y'_\alpha \to Y'_\alpha f_\alpha$ is an iso-
$f \rightsquigarrow ff_\alpha$
morphism and $\xi: \sum_{\mathcal{A}} \oplus Y'_\alpha \to \sum_{\mathcal{A}} \oplus Y'_\alpha f_\alpha$ is an onto vector space isomorphism.
$\sum_{i=1}^{n} F_{\alpha_i} \rightsquigarrow \sum_{i=1}^{n} F_{\alpha_i} f_{\alpha_i}$

Since $\mathfrak{F}$ is injective, we identify $(X, \mathfrak{F}_{\mathcal{P}\{\mathfrak{F}\}})$ with $\mathfrak{e}(X) \subset \prod_{\mathcal{A}} Y_\alpha$ (Theorem 15.7). Actually, $\overline{\mathfrak{e}(X)} = \prod_{\mathcal{A}} Y_\alpha$, since for any $y = (y_\alpha) \in \prod_{\mathcal{A}} Y_\alpha$ and $\mathfrak{T}_\pi$-nbd $u(0) = \bigcap_{i=1}^{n} \pi_{\alpha_i}^{-1}(v_{\alpha_i})$ there follows $\mathfrak{e}(x) \in \{y + u(0)\} \cap \mathfrak{e}(X)$ for $x \in \bigcap_{i=1}^{n} f_{\alpha_i}^{-1}(y_{\alpha_i} + v_{\alpha_i})$. Piecing everything together yields $(X, \mathfrak{T}_{\mathcal{P}\{\mathfrak{F}\}})' = (\mathfrak{e}(X))' = (\prod_{\mathcal{A}} Y_\alpha)' = \sum_{\mathcal{A}} \oplus Y'_\alpha = \sum_{\mathcal{A}} \oplus Y'_\alpha f_\alpha = [\bigcup_{\mathcal{A}} Y'_\alpha f_\alpha]$, where equality means vector space isomorph. ■

An internal inductive limit is either a direct sum or a quotient as demonstrated in

THEOREM 15.22. If $\mathfrak{F} = \{f_\alpha \in \mathfrak{L}(X_\alpha, Y) : \alpha \in \mathcal{A})$ and $Y = [\bigcup_{\mathcal{A}} f_\alpha(X_\alpha)]$, then $(Y, \mathfrak{T}_{\mathcal{I}\{\mathfrak{F}\}})$ is topologically isomorphic to a quotient of $\sum_{\mathcal{A}} \oplus X_\alpha$.

Proof. The linear surjection $\zeta: X \overset{t}{=} \sum_{\mathcal{A}} \oplus X_\alpha \to (Y, \mathfrak{T}_{\mathcal{I}'\{\mathfrak{F}\}})$ is continuous
$\sum_{i=1}^{n} x_{\alpha_i} \rightsquigarrow \sum_{i=1}^{n} f_{\alpha_i}(x_{\alpha_i})$

by Theorem 15.1 and the fact that $\zeta j_\alpha = f_\alpha \;\forall \alpha \in \mathcal{A}$. Write $\zeta = \nu \mu_\zeta$ and take $M = \zeta^{-1}(0)$ in Example 15.15. Since $\mu_\zeta^{-1} f_\alpha = \nu j_\alpha$ is continuous $\forall \alpha \in \mathcal{A}$, the continuous linear bijection $\mu_\zeta: (X/M, \mathfrak{T}_{\mathcal{I}'(\nu)}) \to (Y, \mathfrak{T}_{\mathcal{I}'\{\mathfrak{F}\}})$ is also open. ■

COROLLARY 15.23. $(Y, \mathfrak{T}_{\mathscr{G}'\{\mathfrak{F}\}})' = \left\{F = (F_\alpha) \in \prod_{\mathscr{Q}} X'_\alpha : \sum_{i=1}^{n} F_{\alpha_i}(x_{\alpha_i}) = 0 \text{ if } \sum_{i=1}^{n} f_{\alpha_i}(x_{\alpha_i}) = 0 \text{ for } f_{\alpha_i} \in \mathfrak{F}\right\}$.

Proof. $(Y, \mathfrak{T}_{\mathscr{G}'\{\mathfrak{F}\}})' = (X/M, \mathfrak{T}_{\mathscr{G}(\nu)})' = L_M(X:\mathbb{C})$ for $M = \zeta^{-1}(0)$. ■

We round out this section by briefly considering a TVS topological direct sum of two subspaces. Notation parallels that in 9.9–9.14 (with addition replacing multiplication). Thus subspaces H, K of a TVS X are assumed to carry the induced topologies $\mathfrak{T}_H$ and $\mathfrak{T}_K$, respectively. If $\mathfrak{T} \in T(X)$ and $X = H \oplus K$, then $\underset{k \rightsquigarrow H+k}{\eta_K:(K, \mathfrak{T}_K) \to (X/H, \mathfrak{T}_{\mathscr{G}(\nu_H)})}$ and $\underset{h \rightsquigarrow K+h}{\eta_H:(H, \mathfrak{T}_H) \to (X/K, \mathfrak{T}_{\mathscr{G}(\nu_K)})}$ are continuous vector space isomorphs (e.g., η_K is injective, since $X = H \oplus K$, and continuous, since $\eta_K = \nu_H \mathbb{1}_K$ for $\underset{k \rightsquigarrow k}{\mathbb{1}_K: K \to K}$).

DEFINITION 15.24. If $X \overset{t}{=} H \oplus K$, each TVS $(H, \mathfrak{T}_H)$ and $(K, \mathfrak{T}_K)$ is called the topological supplement of the other.

Every vector subspace H of a TVS X has an algebraic supplement (since $X = H \oplus [\mathfrak{B}_K]$, where $\mathfrak{B} = \mathfrak{B}_H \cup \mathfrak{B}_K$ is the extension of any basis $\mathfrak{B}_H$ for H to a basis for X) and all such supplements of H are isomorphic, since they are all isomorphs of X/H. It need not be true, however, that H has a topological supplement. Such is the case, for example (Theorem 15.19), when H is nonclosed and X is T_2. It is necessary but not sufficient that H be closed in order to have a topological supplement in a T_2VS X.

Example 15.25. If X is a T_2VS with $X' = \{0\}$, no one dim subspace H has a topological supplement. reason: $X = H \oplus K$ implies that $\dim H = \dim X/K = 1$ and (Theorem 17.2) $K = f^{-1}(0)$ for some $f \neq 0$ in X^*. Since $f \notin X'$, it follows that K is not closed and $X \overset{t}{\neq} H \oplus K$. Note that H above is closed (Corollary 13.13). See also Theorem 17.14, Example 17.15.

THEOREM 15.26. The following are equivalent for $\mathfrak{T} \in T(X)$ and $X = H \oplus K$:

(i) $X \overset{t}{=} H \oplus K$;
(ii) $\mathfrak{T} = \mathfrak{T}_{\mathscr{P}\{p_H, p_K\}}$;

(iii) p_H $\{p_K\}$ is continuous;
(iv) η_H $\{\eta_K\}$ is open.

Proof. Modify the statements in the proof of Theorem 9.13. ■

Corollary 9.14 applied to Theorem 15.26 also yields

COROLLARY 15.27. A vector subspace H of a TVS X has a topological supplement iff there is a continuous linear surjection $p:X\to H$ satisfying $p^2=p$. All topological supplements of $(H, \mathfrak{T}_H)$ are topological isomorphs of $(X/H, \mathfrak{T}_{\mathfrak{I}(\nu_H)})$.

COROLLARY 15.28. A vector subspace H of a T_2VS X has a topological supplement if any of the following conditions hold:

(i) H is closed and dim $X/H<\infty$.
(ii) H is finite dim and has a closed algebraic supplement.
(iii) Dim $H<\infty$ and X' is injective.

Proof. Assume for convenience that $X=H\oplus K$. If (i) holds, then $\overset{-1}{\eta_K}$ is open (Corollary 13.13) and $X\overset{t}{=}H\oplus K$. Taking $H=X/K$ shows that the conclusion also holds for (ii). For (iii) suppose that $\{e_1, \ldots, e_h\}$ is a basis for H and let $f_j\in X'$ satisfy $f_j(e_i)=\delta_{i,j}$ $\forall 1\le i,\ j\le k$. Then $p:X\to H$ with $x \rightsquigarrow \sum_{i=1}^{h} f_i(x)e_i$ is a continuous linear surjection such that $p^2(x)= p\left\{\sum_{i=1}^{h} f_i(x)e_i\right\}=\sum_{i=1}^{h} f_i(x)p(e_i)=p(x)\ \forall x\in X$. The continuity of p follows from the continuity of the composite mappings

$$X\to \prod^{h}\mathbb{F}\longrightarrow \prod^{h} H\longrightarrow H$$
$$x \rightsquigarrow (f_i(x)) \rightsquigarrow (f_i(x)e_i) \rightsquigarrow \sum_{i=1}^{h} f_i(x)e_i$$

Taking $p_H=p$ therefore yields $X\overset{t}{=}H\oplus K$ for $K=\overset{-1}{p}(0)$. ■

Example 15.29. (a) Every algebraic supplement of $H\subset(X, \mathfrak{I})$ is a topological supplement. If $X=H\oplus K$, then η_H is open, since $\eta_H(H)=X\in \mathfrak{T}_{\mathfrak{I}(\nu_K)}$.

(b) Suppose $X=M\oplus N$, where M and N are closed vector subspaces of a Banach space $(X, \|\ \|)$. Then $X\overset{t}{=}M\oplus N$. Simply observe that $(M, \|\ \|)$ and $(X/N, \|\!|\ \|\!|)$ are Banach and use Theorem 10.8 to verify that η_M is open.

(c) Let H be a closed vector subspace of a Hilbert space $(X, (\ ,\ ))$. Then $H^{\perp} = \{x \in X : (h, x) = 0\ \forall h \in H\}$ is closed and $X = H \oplus H^{\perp}$. Thus $(X, \|\ \|) \overset{t}{=} H \oplus H$ for $\|x\| = \sqrt{(x, x)}$.

PROBLEMS

15A. $\mathfrak{T}_{\mathcal{P}\{Y'\mathfrak{F}\}}$-Convergence

Example 15.4(c) can be sharpened when $Y = (Y, \|\ \|)$ is a NLS. Prove that $\varphi \xrightarrow{\mathfrak{T}_{\mathcal{P}\{Y'\mathfrak{F}\}}} x_0 \in X$ iff $\forall f \in \mathcal{Q}$, there is an $M_f > 0$ satisfying $\sup \|f(\varphi)\| < M_f$ and $gf(\varphi) \to gf(x_0)\ \forall gf \in G\mathfrak{F}$ where $\overline{[G]} = Y'$.

Remark. See Example 20.20 for an illustration, where $gx_n \to 0\ \forall g \in [G]$ with $\overline{[G]} = X$, but $x_n \xrightarrow{\mathfrak{T}_{\mathcal{P}\{X'\mathbb{1}\}}} 0$ fails.

15B. Strict Inductive Limits

Let $\{E_n\}$ be an increasing sequence of subspaces of a vector space X such that $X = \bigcup_{n=1}^{\infty} E_n$ and take $\P = \{\mathbb{1}_n = \mathbb{1} : E_n \to X;\ n \in \mathbb{N}\}$. If $\mathfrak{T}_n \in T(X)$ is locally convex and $\mathfrak{T}_n / E_{n-1} = \mathfrak{T}_{n-1}\ \forall n \in \mathbb{N}$, then $(X, \mathfrak{T}_{\mathcal{I}'\{\P\}})$ is called the strict inductive limit of $\{E_n\}$.

(a) Show that $\mathfrak{T}_{\mathcal{I}'(\P)/E_n} = \mathfrak{T}_n\ \forall n \in \mathbb{N}$. [*Hint*: For $\mathfrak{T}_n \subset \mathfrak{T}_{\mathcal{I}'\{\P\}/E_n}$ let $v_n(0) \subset E_n$ be an abs convex $\mathfrak{T}_n$-nbd and obtain inductively a sequence of abs convex $\mathfrak{T}_{n+m}$-nbds $v_{n+m}(0) \subset E_{n+m}$ such that $v_{n+m} \subset v_{n+m+1}$ and $v_{n+m} \cap E_n = v_n\ \forall m \in \mathbb{N}$. Then $u(0) = \left\{\bigcup_{m=0}^{\infty} v_{n+m}\right\}_{bc}$ is a $\mathfrak{T}_{\mathcal{I}'\{\P\}}$-nbd such that $u(0) \cap E_n = v_n$. (If $z = \alpha x + \beta y \in E_n$ for $(x, y) \in \left(\bigcup_{m=1}^{\infty} v_{n+m}\right) \times v_n$ and $|\beta| < |\alpha| + |\beta| \leq 1$, then $x = (1/\alpha)(z - \beta y) \in \left\{\bigcup_{m=1}^{\infty} v_{n+m}\right\} \cap E_n \subset v_n$ and $z \in \{v_n\}_{bc} = v_n$.]

(b) $\mathfrak{T}_{\mathcal{I}'(\P)}$ is T_2 iff each $\mathfrak{T}_n$ is T_2. [*Hint*: If each $\mathfrak{T}_n$ is T_2 and $x_0 \neq 0$ in X, then $x_0 \notin v_{n+m}(0)$ for some $\mathfrak{T}_{n+m}$-nbd $v_{n+m} \subset E_{n+m}$. Thus $x_0 \notin u(0)$ for some $\mathfrak{T}_{\mathcal{I}'\{\P\}}$-nbd $u(0) \subset X$.]

(c) Show that $(X, \mathfrak{T}_{\mathcal{I}'\{\P\}})$ is complete iff each $(E_n, \mathfrak{T}_n)$ complete. [*Hint*: Assume that each $(E_n, \mathfrak{T}_n)$ is complete, let $\mathfrak{N}$ denote the $\mathfrak{T}_{\mathcal{I}'\{\P\}}$-nbd system at $0 \in X$, and suppose $\mathcal{G} \subset X$ is a $\mathfrak{T}_{\mathcal{I}'\{\P\}}$-Cauchy filter. Then $\mathcal{G} + \mathfrak{N}$ is a $\mathfrak{T}_{\mathcal{I}'\{\P\}}$-Cauchy filterbase which converges iff $\mathcal{G}$ converges. For the $\mathfrak{T}_{\mathcal{I}'\{\P\}}$-convergence of $\mathcal{G} + \mathfrak{N}$ show that $(\mathcal{G} + \mathfrak{N}) \cap E_{n_0}$ is a filterbase for some $n_0 \in \mathbb{N}$. (Suppose that $(G_n + w_n) \cap E_n = \varnothing\ \forall n \in \mathbb{N}$, where $\{G_n\} \in \mathcal{G}$ and $\{w_n(0)\}$ is a decreasing sequence of abs convex nbds in $\mathfrak{N}$. Then $u(0) = \left\{\bigcup_{n=1}^{\infty} (w_n \cap E_n)\right\}_{bc} \in \mathfrak{N}$ satisfies $(G_n + u) \cap E_n = \varnothing\ \forall n \in \mathbb{N}$. If $x \in G \cap E_{n_0}$ and $y \in G \cap G_{n_0}$ for $u(0)$-sized $G \in \mathcal{G}$, then $x = y + (x - y) \in (G_{n_0} + u) \cap E_{n_0}$, which is impossible.)]

Assume now that each E_n is $\mathfrak{T}_{n+1}$-closed in E_{n+1}.

(d) Each E_n is $\mathfrak{T}_{\mathscr{I}'\{\P\}}$-closed in X. In particular, each E_n is $\mathfrak{T}_n$-complete if $(X, \mathfrak{T}_{\mathscr{I}'\{\P\}})$ is complete. [*Hint*: E_n is $\mathfrak{T}_{n+m}$-closed in E_{n+m} $\forall m \in \mathbb{N}$. If $x_0 \in X - E_n$, then $x_0 \in E_{n+m}$ and $\{x_0 + v_{n+m}(0)\} \cap E_n = \varnothing$ for some $\mathfrak{T}_{n+m}$-nbd $v_{n+m}(0)$. In particular, $(x_0 + u(0)) \cap E_n = \varnothing$ for a $\mathfrak{T}_{\mathscr{I}'\{\P\}}$-nbd $u(0)$ such that $v_{n+m} = u \cap E_{n+m}$.]

(e) A subset $B \subset X$ is $\mathfrak{T}_{\mathscr{I}'\{\P\}}$-(totally) bounded iff $B \subset E_{n_0}$ is $\mathfrak{T}_{n_0}$-(totally) bounded for some $n_0 \in \mathbb{N}$. [*Hint*: Sufficiency is clear. If $B \subset X$ is $\mathfrak{T}_{\mathscr{I}'\{\P\}}$-bounded and $B \not\subset E_n$ $\forall n \in \mathbb{N}$, there is a sequence $\{x_n\} \in B$ and an increasing sequence $k(n) \in \mathbb{N}$ such that $x_n \in E_{k(n+1)} - E_{k(n)}$ $\forall n \in \mathbb{N}$. As in (a) above, there is a sequence $\{v_{k(n)}(0) \subset E_{k(n)} : n \in \mathbb{N}\}$ of abs convex $\mathfrak{T}_{k(n)}$-nbds satisfying $\{(1/n)x_n + v_{k(n+1)}\} \cap E_{k(n)} = \varnothing$ and $v_{k(n+1)} \cap E_{k(n)} = v_{k(n)}$ $\forall n \in \mathbb{N}$. Obtain the contradiction (Theorem 13.22) that $(1/n)x_n \notin \left\{ \bigcup_{n=1}^{\infty} v_{k(n)} \right\}_{bc}$ for large $n \in \mathbb{N}$.]

(f) $\mathfrak{T}_{\mathscr{I}'\{\P\}}$ is not 1°. [*Hint*: If $\{u_n(0) : n \in \mathbb{N}\}$ is a decreasing $\mathfrak{T}_{\mathscr{I}'\{\P\}}$ nbd base at $0 \in X$, then $B = \{x_n \in u_n - E_n : n \in \mathbb{N}\}$ is not $\mathfrak{T}_{\mathscr{I}'\{\P\}}$-bounded [by (e) above], which contradicts the fact that B is absorbed by every $u_n(0)$.]

15C. Examples of Strict Inductive Limits

(a) Let $\mathcal{C}$ denote the compact subsets of a locally compact T_2 space X. For each $K \in \mathcal{C}$ let $C_K(X; \mathbb{F})$ be the vector subspace of $C_0(X; \mathbb{F})$ whose members have their support in K and take $\P = \{\mathbb{1}_K = \mathbb{1} : (C_K(X; \mathbb{F}), \|\ \|_0) \to C_0(X; \mathbb{F})\ \forall K \in \mathcal{C}\}$. Verify that $\|\ \|_0 \subset \mathfrak{T}_{\mathscr{I}'\{\P\}} \in T(C_0(X; \mathbb{F})$. If $X = \bigcup_{n=1}^{\infty} K_n$ for $\{K_1 \subset K_2 \subset \cdots\} \subset \mathcal{C}$, then $(C_0(X; \mathbb{F}), \mathfrak{T}_{\mathscr{I}'\{\P\}})$ is the strict inductive limit of $(C_{K_n}(X; \mathbb{F}), \|\ \|_0)$. Show further that $(C_0(X; \mathbb{F}), \mathfrak{T}_{\mathscr{I}'\{\P\}})$ is a complete, nonmetrizible LCT_2VS. [*Hint*: $C_{K_n}(X; \mathbb{F}) \subset C_{K_{n+1}}(X; \mathbb{F})$ and $(C_{K_n}(X; \mathbb{F}), \|\ \|_0)$ is Banach $\forall n \in \mathbb{N}$.]

Remark. Note that $\|\ \|_0 \neq \mathfrak{T}_{\mathscr{I}'\{\P\}}$ in (ii). The results of (a) holds for any open set $X = \Omega$ in $\mathbb{F}^n$ (Example 14.22⁺).

(b) Let $\mathfrak{D}_0^\infty(\mathbb{R}; \mathbb{F})$ be the vector subspace of $\mathfrak{D}_r^\infty(\mathbb{R}; \mathbb{F})$ (cf problem 14F) whose members have compact support. For each $n \in \mathbb{N}$ let $\mathfrak{D}_{K_n}^\infty(\mathbb{R}; \mathbb{F})$ be the subspace of $\mathfrak{D}_0^\infty(\mathbb{R}; \mathbb{F})$ whose members have their support in $K_n = [-n, n]$. Now, taking $\xi_n \in T(D_{K_n}^\infty(\mathbb{R}; \mathbb{F}))$ as the topology determined by the seminorms $\{p_m(f) = \sup_{K_n} |f^{(m)}(x)| : m \in \mathbb{N}_0\}$, prove that $(\mathfrak{D}_0^\infty(\mathbb{R}; \mathbb{F}), \mathfrak{T}_{\mathscr{I}'\{\P\}})$ is the strict inductive limit of $\{(D_{K_n}^\infty(\mathbb{R}; \mathbb{F}), \xi_n) : n \in \mathbb{N}\}$. Furthermore show that

(i) $(\mathfrak{D}_{K_n}^\infty(\mathbb{R}; \mathbb{F}), \xi_n)$ is Frechet $\forall n \in \mathbb{N}$.
(ii) $(\mathfrak{D}_0^\infty(\mathbb{R}; \mathbb{F}), \mathfrak{T}_{\mathscr{I}'\{\P\}})$ is a complete nonmetrizible LCT_2VS.
(iii) $\mathfrak{T}_{\mathscr{I}'\{\P\}}$ is finer that ζ and ξ on $\mathfrak{D}_0^\infty(\mathbb{R}; \mathbb{F})$. [*Hint*: Use Example 15.16 after confirming that $\zeta / \mathfrak{D}_{K_n}^\infty(\mathbb{R}; \mathbb{F}) \subseteq \xi_n$ $\forall n \in \mathbb{N}$.]

15D. The Finest Convex Topology Revisited

(a) Show (Example 14.14) that $(X, \mathfrak{T}_{F_c}) \stackrel{tc}{=} \sum_{\mathfrak{B}} \oplus [x_\alpha]$, where $\mathfrak{B} = \{x_\alpha : \alpha \in \mathfrak{A}\}$ is an algebraic basis for X. Conclude from this that $(X, \mathfrak{T}_{F_c})$ is a complete LCT_2VS. [*Hint*: $\mathfrak{T}_{\mathcal{I}\{J\}} \in T(X)$ is locally convex and (Corollary 13.13) $j_\alpha : [x_\alpha] \to (X, \mathfrak{T}_{F_c})$ is continuous $\forall \alpha \in \mathfrak{A}$.]

(b) $(X, \mathfrak{T}_{F_c})$ is nonmetrizible if $\mathfrak{B}$ is countable. [*Hint*: $\mathfrak{T}_{F_c}$ is the strict inductive limit of $E_n = [x_1, \ldots, x_n]$ (with the Euclidean n-space norm) $\forall n \in \mathbb{N}$. Apply Problem 15B(f).]

(c) $(X, \mathfrak{T}_{F_c})$ is normable iff dim X is finite.

15E. Kernel and Range Topological Supplements

Let $f \in L(X, Y)$, where X and Y are TVSps. prove that

(a) $f^{-1}(0)$ has a topological supplement in X iff there is a $F \in L(Y, X)$ such that $fF = \mathbb{1}_Y : Y \to Y$;

(b) $f(X)$ has a topological supplement in Y iff there is a $F \in L(Y, X)$ such that $Ff = \mathbb{1}_X : X \to X$.

References for Further Study

On initial and final systems: Bourbaki [13]; Braconnier [14]; Horvath [43]; Kothe [55]; Sebastiaõ e Silva [90]; Takenouchi [97].

On strict inductive limits: Robertson and Robertson [84].

On topological direct sums: Edwards [28]; Köthe [55].

16. Vector-Valued Measure TVSps

This section serves as the vehicle for a brief, but relevant, side trip into vector-valued measure spaces with compatible linear topologies. Our results are due in part (Theorems 16.3 and 16.10, and Lemma 16.13, Problem 16A) to Suffel [95]. The intent here is to illustrate some earlier notions and form the basis for further applications in the sections that follow.

DEFINITION 16.1. The collection $\mathfrak{A}(\Omega, \Sigma; Y)$ of additive set functions from a measurable space (Ω, Σ) to a vector space $Y \neq \{0\}$ is a vector space under pointwise operations $(\mu + \nu)(E) = \mu(E) + \nu(E)$ and $(\alpha\mu)(E) = \alpha\mu(E)\ \forall E \in \Sigma$. If Y is a T_2VS, the subspace of countably additive set functions in $\mathfrak{A}(\Omega, \Sigma; Y)$ is denoted $\mathcal{C}\mathfrak{A}(\Omega, \Sigma; Y)$. Members of $\mathcal{C}\mathfrak{A}(\Omega, \Sigma; Y)$ are also called vector measures.

For each $\omega \in \Omega$ and $y \in Y$ define $\mu_{\omega,y} = y\chi_\omega$. Then $\mathcal{D}_\omega(\Omega, \Sigma; Y) = \{\mu_{\omega,y} : y \in Y\}$ and $\mathcal{D}(\Omega, \Sigma; Y) = [\bigcup_\Omega \mathcal{D}_\omega(\Omega, \Sigma; Y)]$ are vector subspaces of $\mathcal{C}\mathcal{a}(\Omega, \Sigma; Y)$.

Some additional illustrations of Definition 13.3 are given below.

Example 16.2. (a) $\mathcal{A} = \{A_E : X \to Y \ (E \in \Sigma)\} \subset \mathfrak{L}(X, Y)$ is injective for every vector space X satisfying $\mathcal{D}(\Omega, \Sigma; Y) \subseteq X \subseteq \mathcal{a}(\Omega, \Sigma; Y)$. If $\Sigma = \Sigma(\xi)$ is the σ-ring generated by a family of subsets $\xi \subset \Omega$, then $\mathcal{A}$ is injective iff $\mathcal{A}_\xi = \{A_E \in \mathcal{A} : E \in \xi\}$ is injective.
$$\mu \rightsquigarrow \mu(E)$$

(b) $\mathcal{A}_{\Sigma_0}$ is concentrating for a disjoint subcollection $\Sigma_0 \subset \Sigma$. (If $\{A_{E_0}, A_{E_1}, \ldots, A_{E_n}\} \subset \Sigma_0$ and $y_0 \neq 0$ in Y, pick $\omega \in E$ and obtain $A_{E_k}(\mu_\omega, y_0) = y_0 \delta_{k,0} \ \forall 0 \leq k \leq n$.)

(c) $\mathcal{A}$ is surjective if Σ is finite. Decompose Σ (using finite unions, intersections, and differences of these finite unions and intersections) into a collection Σ_0 of disjoint subsets. Then Σ_0 is finite and $\mathcal{A}_{\Sigma_0}$ is surjective. (Verification is by induction. Clearly, $A_{E_1}(\mu_\omega, y_1) = y_1$ for $\omega \in E_1 \in \Sigma_0$ and $y_1 \in Y$. Suppose there is a $\mu \in \mathcal{D}(\Omega, \Sigma; Y)$ for each $\{E_1, \ldots, E_n\} \subset \Sigma_0$ and $\{y_1, \ldots, y_n\} \in Y$ such that $A_{E_i}(\mu) = y_i \ \forall 1 \leq i \leq n$. Given any $y_{n+1} \in Y$ and $E_{n+1} \in \Sigma$, pick $\omega \in E_{n+1}$ and let $y = \mu(E_{n+1})$. Then $\nu = \mu + \mu_{\omega, y_{n+1} - y} \in \mathcal{D}(\Omega, \Sigma; Y)$ and $A_{E_i}(\nu) = y_i \ \forall 1 \leq i \leq n+1$.) Now $\aleph(\Sigma) < \aleph(\Sigma_0)$ and $\forall E_{ij} \subset E_i \in \Sigma$, one then has $A_{E_{ij}} = A_{E_i}$ on $\mathcal{D}_\omega(\Omega, \Sigma; Y)$ $(\omega \in E_{ij})$. Thus $\mathcal{A}$ is surjective.

(d) If Y is an $\mathrm{LCT_2VS}$, then $\{\mu(E) : E \in \Sigma\} \subset Y$ is bounded $\forall \mu \in \mathcal{C}\mathcal{a}(\Omega, \Sigma; Y)$ (Corollary 18.8) and so Σ is finite if (i.e., iff) $\mathcal{A}$ is surjective. Specifically, for any countably infinite subcollection $\{E_n : n \in \mathbb{N}\} \subset \Sigma$ and $y_0 \neq 0$ in Y surjectivity of $\mathcal{A}$ requires that some unbounded $\mu \in \mathcal{C}\mathcal{a}(\Omega, \Sigma; Y)$ satisfy $\mu(E_n) = ny_0 \ \forall n \in \mathbb{N}$.

The exact relationship between $\dim \mathcal{a}(\Omega, \Sigma; Y)$, $\dim Y$, and $\aleph(\Sigma)$ is based on

THEOREM 16.3. A σ-ring Σ is finite iff each countable collection of disjoint sets in Σ is finite. In particular, every countable σ-ring is finite.

Proof. Obviously every collection of disjoint sets in a finite σ-ring is finite. Suppose that Σ is infinite and let $\{E_n \neq \varnothing : n \in \mathbb{N}\}$ be any collection of disjoint sets in Σ. For each index subset $\Lambda \subset \mathbb{N}$ define $B_\Lambda = \bigcap_{n \in \Lambda} E_n - \bigcup_{\mathbb{N} - \Lambda} E_n$. If $m_0 \in \Lambda_\alpha - \Lambda_\beta$, then $B_{\Lambda_\alpha} \cap B_{\Lambda_\beta} = \varnothing$, since $\omega \in B_{\Lambda_\alpha}$ implies $\omega \in E_{m_0} \subset \bigcup_{m_0 \notin \Lambda_\beta} E_n$ and $\omega \notin B_{\Lambda_\beta}$. Thus distinct B_Λ are disjoint and we now show that $E_m = \bigcup_{m\in\Lambda} B_\Lambda \ \forall m \in \mathbb{N}$. Clearly $\bigcup_{m\in\Lambda} B_\Lambda \subset E_m$ since $B_\Lambda \subset E_m \ \forall m \in \Lambda$. For the reverse inclusion suppose that $\omega \in E_m$ and let $\Lambda_\omega = \{n \in \mathbb{N} : \omega \in E_n\}$. Then $m \in \Lambda_\omega$ and $\omega \notin E_n$ for $n \notin \Lambda_\omega$. Thus $\omega \in B_{\Lambda_\omega}$ and $E_m \subset B_{\Lambda_\omega} \subset \bigcup_{m\in\Lambda} B_\Lambda$ follows. Our first assertion is now complete, since $\{E_m = \bigcup_{m\in\Lambda} B_\Lambda : m \in \mathbb{N}\}$ are distinct and $\{B_\Lambda : \text{all indexes } \Lambda \subset \mathbb{N}\}$ contains a countably infinite number of disjoint sets in Σ.

If Σ is infinite, it has a collection of disjoint sets $\{F_n = \varnothing : n \in \mathbb{N}\}$ and $\{G_\Lambda = \bigcup_{m\in\Lambda} F_m : \Lambda \subset \mathbb{N}\}$ is an uncountable collection of distinct sets in Σ. (There are $2^{\aleph_0}$ index sets $\Lambda \subset \mathbb{N}$, each of which contains some $m \in \mathbb{N}$. Furthermore, $\Lambda_\alpha \neq \Lambda_\beta$ implies some $m_0 \in (\Lambda_\alpha - \Lambda_\beta) \cup (\Lambda_\beta - \Lambda_\alpha)$ and $G_{\Lambda_\alpha} \neq G_{\Lambda_\beta}$.) Thus Σ is uncountable. ■

COROLLARY 16.4. Let X be a vector space satisfying $\mathfrak{D}(\Omega, \Sigma; Y) \subseteq X \subseteq \mathcal{Q}(\Omega, \Sigma; Y)$. Then dim X is finite iff both Σ and dim Y are finite.

Proof. If Σ and dim Y are finite, so is $\dim \prod_\Sigma Y$. Since $\mathbf{e} : \mathcal{Q}(\Omega, \Sigma; Y) \to \prod Y$, $\mu \rightsquigarrow \{\mu(E)\}_\Sigma$ is an isomorphism, $\mathcal{Q}(\Omega, \Sigma; Y)$ and so X are finite dim. If dim X is finite, then $\mathfrak{D}_\omega(\Omega, \Sigma; Y)$ and so Y are finite dim. Furthermore, Σ is finite. Otherwise, Σ contains a countable collection of distinct sets $\{E_n : n \in \mathbb{N}\}$ and $\{\mu_{\omega_n, y\neq 0} : \omega_n \in E_n \ \forall n \in \mathbb{N}\}$ is an infinite linearly independent subset of $\mathfrak{D}(\Omega, \Sigma; Y)$. ■

Since $\mathscr{A} = \{A_E : E \in \Sigma\} \subset \mathfrak{L}(X, Y)$ above, we further establish

COROLLARY 16.5. Dim $[\mathscr{A}] \leq \aleph_0$ iff Σ is finite.

Proof. If Σ is finite, then $\mathscr{A}$ and dim $[\mathscr{A}]$ are finite. If Σ is infinite, it contains an uncountable collection Σ_0 of disjoint subsets and $\mathscr{A}_{\Sigma_0} = \{A_E : E \in \Sigma_0\}$ is linearly independent in $\mathfrak{L}(X, Y)$. (If $\sum_{i=1}^n \lambda_i A_E = 0$ for $E_i \in \Sigma_0$, pick $\omega_i \in E_i$ and obtain $0 = \left(\sum_{i=1}^n \lambda_i A_{E_i}\right)\mu_{\omega_k, y\neq 0} = \lambda_k y \ \forall 1 \leq k \leq n$.) Therefore dim $[\mathscr{A}] \geq$ dim $[\mathscr{A}_{\Sigma_0}] > \aleph_0$. ■

Remark. In general, $\mathscr{A}$ is not linearly independent. For distinct nonempty sets $E_1, E_2 \in \Sigma$, take $E_2' = E_2 - E_1$ and let $E_3 = E_1 \cup E_2'$. Then $A_{E_3} = A_{E_1} + A_{E_2}$.

Vector spaces X satisfying $\mathfrak{D}(\Omega, \Sigma; Y) \subseteq X \subseteq \mathcal{Q}(\Omega, \Sigma; Y)$ are numerous and easily illustrated. For a T_2VS Y, define $wk\mathcal{C}\mathcal{Q}(\Omega, \Sigma; Y) = \{\mu \in \mathcal{Q}(\Omega, \Sigma; Y) : f\mu \in \mathcal{C}\mathcal{Q}(\Omega, \Sigma; \mathbb{F}) \ \forall f \in Y'\}$. Evidently $\mathfrak{D}(\Omega, \Sigma; Y) \subset$

$\mathcal{CA}(\Omega, \Sigma; Y) \subset wk\mathcal{CA}(\Omega, \Sigma; Y)$, and a little probing (Corollary 17.9) reveals that $wk\mathcal{CA}(\Omega, \Sigma; Y) \subset \mathcal{A}(\Omega, \Sigma; Y)$ if Y is also locally convex. Actually, a good deal more can be said by following the lengthy, but not too difficult, details (Suffel [96]) establishing

THEOREM 16.6. If Y is an LCT_2VS and $\Omega \in \Sigma$, then $\mathcal{CA}(\Omega, \Sigma; Y) = wk\mathcal{CA}(\Omega, \Sigma; Y)$. Furthermore, $\mu \in wk\mathcal{CA}(\Omega, \Sigma; Y)$ whenever there is a sequence $\{\mu_n\} \in \mathcal{CA}(\Omega, \Sigma; Y)$ such that $\mu(E) = \lim_n \mu_n(E) \in Y \; \forall E \in \Sigma$.

Example 16.7. (a) A norm $\| \;\|$ on Y determines a norm $\|\mu\|_0 = \sup_\Sigma \|\mu(E)\|$ on $B\mathcal{A}(\Omega, \Sigma; Y, \| \;\|) = \{\mu \in \mathcal{A}(\Omega, \Sigma; Y, \| \;\|) : \sup_\Sigma \|\mu(E)\| < \infty\}$. Furthermore, $\mathcal{A}(\Omega, \Sigma; Y, \| \;\|)$ is $\| \;\|_0$-complete (i.e., Banach) if $(Y, \| \;\|)$ is Banach and $\Omega \in \Sigma$. Reason as follows: If $\{\mu_n\} \in B\mathcal{A}(\Omega, \Sigma; Y, \| \;\|)$ is $\| \;\|_0$-Cauchy, then $\{\mu_n(E)\} \subset Y$ is $\| \;\|$-Cauchy and convergent $\forall E \in \Sigma$. Taking $\mu(E) = \lim_n \mu_n(E) \; \forall E \in \Sigma$, we obtain $\mu_n \xrightarrow{\| \;\|_0} \mu \in B\mathcal{A}(\Omega, \Sigma; Y, \| \;\|)$. Assuredly μ is additive, since $\mu\left(\bigcup_{i=1}^k E_i\right) = \lim_m \mu_n\left(\bigcup_{i=1}^k E_i\right) = \lim_n \sum_{i=1}^k \mu_n(E_i) = \sum_{i=1}^k \mu(E_i)$ for disjoint $\{E_1, \ldots, E_k\} \subset \Sigma$. If $\|\mu_n - \mu_m\|_0 < \varepsilon$ for $n, m \geq N(\varepsilon) \in \mathbb{N}$, then $\|\mu_n\|_0 \leq \|\mu_{N(\varepsilon)}\|_0 + \varepsilon \; \forall n \geq N(\varepsilon)$ and $\|\mu_n\|_0 \leq \sum_{j=1}^{N(\varepsilon)} \|\mu_j\|_0 + \varepsilon \; \forall n \in \mathbb{N}$. Accordingly for any $E \in \Sigma$ we have $\|\mu(E)\| \leq \|\mu(E) - \mu_n(E)\| + \|\mu_n(E)\| \leq 2\varepsilon + \sum_{j=1}^{N(\varepsilon)} \|\mu_j\|_0 \; \forall n \geq N(\varepsilon)$ and $\sup_\Sigma \|\mu(E)\| < \infty$ follows. That $\mu_n \xrightarrow{\| \;\|_0} \mu$ is also clear, since we have $\|\mu_n(E) - \mu(E)\| = \|\mu_n(E) - \lim_m \mu_m(E)\| = \lim_m \|\mu_n(E) - \mu_m(E)\| \leq \varepsilon \; \forall n \geq N(\varepsilon)$.

(b) Let $\mathcal{D}(\Omega, \Sigma; Y, \| \;\|) \subseteq X \subseteq B\mathcal{A}(\Omega, \Sigma; Y, \| \;\|)$ and suppose X carries $\| \;\|_0$. Then $A_E \in L(X, Y)$ with $|||A_E||| = 1 \; \forall E \neq \varnothing$ in Σ, (since $\|A_E(\mu)\| \leq \|\mu\|_0 \; \forall \mu \in X$ and $\|A_E(\mu_{\omega,y})\| = 1$ for $\omega \in E$ and $\|y\| = 1$). Therefore $\mathfrak{T}_{\mathcal{P}\{\mathcal{A}\}} \subseteq \| \;\|_0$ on X. Equality holds if $\Sigma = \{E_1, \ldots, E_n\}$ is finite, since one then has $\bigcap_{i=1}^n A_{E_i}^{-1}\{S_{\varepsilon, \| \;\|}(0)\} \subset S_{\varepsilon, \| \;\|_0}(0) \; \forall \varepsilon > 0$.

Remark. Note that $|||A_E - A_F||| \geq 1$ for distinct $E, F \in \Sigma$ (since $\omega \in E - F$ implies $\|\mu_{\omega,y}\|_0 = \|y\| = \|\mu_{\omega,y}(E)\| = \|(A_E - A_F)\mu_{\omega,y}\| \leq |||A_E - A_F||| \; \|\mu_{\omega,y}\|_0$).

(c) We already know [Example 16.2(d)] that $\mathcal{CA}(\Omega, \Sigma; Y, \| \;\|) \subset B\mathcal{A}(\Omega, \Sigma; Y, \| \;\|)$. If $(Y, \| \;\|)$ is Banach and $\Omega \in \Sigma$, then $\mathcal{CA}(\Omega, \Sigma; Y, \| \;\|)$ is $\| \;\|_0$-closed in $B\mathcal{A}(\Omega, \Sigma; Y, \| \;\|)$ (therefore Banach under $\| \;\|_0$). Simply consider $\mu_n \xrightarrow{\| \|_0} \mu \in B\mathcal{A}(\Omega, \Sigma; Y, \| \|)$ for $\{\mu_n\} \in \mathcal{CA}(\Omega, \Sigma; Y, \| \|)$ and invoke Theorem 16.6 to obtain $\mu \in \mathcal{CA}(\Omega, \Sigma; Y, \| \|)$.

Remark. $\underset{Y \rightsquigarrow \mu_{\omega, y}}{(Y, \| \|) \to (\mathcal{D}_\omega(\Omega, \Sigma; Y, \| \|), \| \|_0)}$ is a congruence $\forall \omega \in \bigcup_{\Sigma} E$.

Example 16.8. The total variation of $\mu \in \mathcal{CA}(\Omega, \Sigma; Y, \| \|)$ is defined as

$$\|\mu\|_v = \sup \left\{ \sum_{i=1}^{n} \|\mu(E_i)\| : \text{disjoint } \{E_1, \ldots, E_n\} \subset \Sigma \right\}.$$

It is easily verified that $\| \|_v$ is a norm on $\mathcal{CA}_{FV}(\Omega, \Sigma; Y, \| \|) = \{\mu \in \mathcal{CA}(\Omega, \Sigma; Y, \| \|) : \|\mu\|_v < \infty\}$. Furthermore, $\Omega \in \Sigma$ implies $\|\mu(E)\| \le \|\mu(E)\| + \|\mu(\Omega - E)\| \le \|\mu\|_v$ and $\|\mu\|_0 \le \|\mu\|_v$ on $\mathcal{CA}_{FV}(\Omega, \Sigma; Y, \| \|)$ follows. If $(Y, \| \|)$ is also complete, so is $\| \|_v$. Verification proceeds as in Example 16.7(a). To see that $\mu_n \xrightarrow{\| \|_v} \mu \in \mathcal{CA}_{FV}(\Omega, \Sigma; Y, \| \|)$, assume the sets $\{E_1, \ldots, E_k\} \subset \Sigma$ are disjoint and let $\|\mu_m - \mu_n\|_v < \varepsilon/3$ for $m, n \ge N_0(\varepsilon) \in \mathbb{N}$. Now pick $N_i(\varepsilon) \in \mathbb{N}$ such that $\|\mu_m(E_i) - \mu(E_i)\| < \varepsilon/3k \; \forall m \ge N_i(\varepsilon)$ and observe that $\sum_{i=1}^{k} \|\mu(E_i)\| \le \sum_{i=1}^{k} \|\mu(E_i) - \mu_m(E_i)\| + \sum_{i=1}^{k} \|\mu_m(E_i) - \mu_n(E_i)\| + \sum_{i=1}^{k} \|\mu_n(E_i)\| \le \varepsilon/3 + \varepsilon/3 + (\varepsilon/3 + \|\mu_{N_0(\varepsilon)}\|_v)$ for $m, n \ge \sum_{j=0}^{k} N_j(\varepsilon)$. Thus $\|\mu\|_v < \infty$ and $\mu \in \mathcal{CA}_{FV}(\Omega, \Sigma; Y, \| \|)$ follows by taking the sup over all finite disjoint sequences in Σ. That $\mu_n \xrightarrow{\| \|_v} \mu$ is also clear since $\sum_{i=1}^{k} \|\mu_n(E_i) - \mu(E_i)\| = \sum_{i=1}^{k} \|\mu_n(E_i) - \lim_m \mu_m(E_i)\| = \lim_m \sum_{i=1}^{k} \|\mu_n(E_i) - \mu_m(E_i)\| \le \lim_m \|\mu_n - \mu_m\|_v \le \varepsilon$ for $n \ge N_0(\varepsilon)$.

Remark. Observe that $\mathcal{D}(\Omega, \Sigma; Y, \| \|) \subset \mathcal{CA}_{FV}(\Omega, \Sigma; Y, \| \|)$ and $\| \|_0 = \| \|_v$ on $\mathcal{D}_\omega(\Omega, \Sigma; Y, \| \|) \; \forall \omega \in \bigcup_{\Sigma} E$.

Example 16.9. Suppose that Ω is a topological space and Y is a T_2VS. Then $\mu \in \mathcal{A}(\Omega, \Sigma; Y)$ is termed regular if

(*) For each $E \in \Sigma$ and Y-nbd $u(0)$ there is an Ω-compact set $C \in \Sigma$ and an Ω-open set $\mathcal{O} \in \Sigma$ such that $C \subset E \subset \mathcal{O}$ and $\mu(A) - \mu(E) \in u(0) \; \forall A \in \Sigma$ satisfying $C \subset A \subset \mathcal{O}$.

(a) The collection $\mathcal{CA}_{\mathfrak{R}}(\Omega, \Sigma; Y)$ of regular measures in $\mathcal{CA}(\Omega, \Sigma; Y)$ is a subspace of $\mathcal{CA}(\Omega, \Sigma; Y)$. [For any fixed μ_1, μ_2 in $\mathcal{CA}_{\mathfrak{R}}(\Omega, \Sigma; Y)$ and $\alpha \in \mathbb{F}$ choose $\{C_i, \mathcal{O}_i\}$ satisfying (*) for μ_i $(i=1, 2)$. Then $C = C_1 \cap C_2$ and $\mathcal{O} = \mathcal{O}_1 \cup \mathcal{O}_2$ satisfies (*) for $\alpha\mu_1 + \mu_2$.] If Ω is T_1, then $\mathfrak{D}_\omega(\Omega, \Sigma; Y) \subset \mathcal{CA}_{\mathfrak{R}}(\Omega, \Sigma; Y)\ \forall \omega \in \bigcup_{\Sigma} E$. Fix $\mu_{\omega,y} \in \mathfrak{D}_\omega(\Omega, \Sigma; Y)$ and consider any $E \in \Sigma$ and $u(0) \subset Y$. If $\omega \in E$, take $C = \{\omega\}$ and $\mathcal{O} = \Omega$. Otherwise, take $\mathcal{O} = \Omega - \omega$ and let $C = \{\omega_0\}$ for $\omega_0 \in E \neq \emptyset$. If $E = \emptyset$, take $C = \emptyset = \mathcal{O}$.

(b) $\mathcal{CA}_{\mathfrak{R}}(\Omega, \Sigma; Y, \| \|)$ is closed in $(\mathcal{CA}(\Omega, \Sigma; Y, \| \|), \| \|_0)$ and therefore $\| \|_0$-complete if $(Y, \| \|)$ is Banach and $\Omega \in \Sigma$. Indeed, suppose $\mu_n \xrightarrow{\| \|_0} \mu \in \mathcal{CA}(\Omega, \Sigma; Y, \| \|)$ for $\{\mu_n\} \in \mathcal{CA}_{\mathfrak{R}}(\Omega, \Sigma; Y, \| \|)$. Given any $E \in \Sigma$ and $\varepsilon > 0$, choose $N \in \mathbb{N}$ such that $\|\mu - \mu_N\|_0 < \varepsilon/2$ and let $\{C, \mathcal{O}\} \in \Sigma$ satisfy (*) for μ_N relative to $u = S_{\varepsilon/2}(0) \subset Y$. Then $\{C, \mathcal{O}\}$ satisfies (*) for μ relative to $S_\varepsilon(0) \subset Y$ (by virtue of $\|\mu(A) - \mu(E)\| \leq \|\mu - \mu_N\|_0 + \|\mu_N(A) - \mu(E)\|$).

The question that naturally arises (or at least should!) is how "close" the above subspaces are to one another and to $\mathcal{A}(\Omega, \Sigma; Y)$.

Theorem 16.10. For each $\nu \in \mathcal{A}(\Omega, \Sigma; Y)$ and $\{E_1, \ldots, E_n\} \in \Sigma$ there is a $\mu \in \mathfrak{D}(\Omega, \Sigma; Y)$ satisfying $\mu(E_j) = \nu(E_j)\ \forall 1 \leq j \leq n$. Therefore $\mathfrak{D}(\Omega, \Sigma; Y) = \mathcal{A}(\Omega, \Sigma; Y)$ if Σ is finite.

Proof. For each $1 \leq k \leq n$ let $\left\{\Lambda_i^k : 1 \leq i \leq \binom{n}{k}\right\}$ denote the collection of subsets of $\{1, 2, \ldots, n\}$, each of which contains k natural numbers. Define $B_{\Lambda_i^k}$ as in Theorem 16.3 and take $\Lambda = \{\Lambda_i^k : B_{\Lambda_i^k} \neq \emptyset\}$. For each $\Lambda_i^k \in \Lambda$ pick one $\omega_{ik} \in B_{\Lambda_i^k}$ and let $y_{ik} = \nu(B_{\Lambda_i^k})$. Then $\mu = \sum_\Lambda \mu_{\omega_{ik}, y_{ik}} \in \mathfrak{D}(\Omega, \Sigma; Y)$ meets our requirements. Specifically, $\mu(E_j) = \sum_\Lambda \mu_{\omega_{ik}, y_{ik}}(E_j) = \sum_{j \in \Lambda_i^k} \mu_{\omega_{ik}, y_{ik}}(E_j) + \sum_{j \notin \Lambda_i^k} \mu_{\omega_{ik}, y_{ik}}(E_j)$ (this sum being taken over all $\Lambda_i^k \in \Lambda$). Since $j \notin \Lambda_i^k$ implies $E_j \cap B_{\Lambda_i^k} = \emptyset$ and $\omega_{ik} \in B_{\Lambda_i^k} - E_j$, the second summand equals zero and $\mu(E_j) = \sum_{j \in \Lambda_i^k} y_{ik} = \sum_{j \in \Lambda_i^k} \nu(B_{\Lambda_i^k}) = \nu\left(\bigcup_{j \in \Lambda_i^k} B_{\Lambda_i^k}\right) = \nu(E_j)$ for each $j = 1, 2, \ldots, n$. ■

Corollary 16.11. $\overline{\mathfrak{D}(\Omega, \Sigma; Y)}^{\mathcal{T}_{\mathcal{P}\{\mathcal{A}\}}} = \mathcal{A}(\Omega, \Sigma; Y)$ if Y is a TVS.

Proof. Every $\mathcal{T}_{\mathcal{P}\{\mathcal{A}\}}$-nbd $\nu + \bigcap_{i=1}^n A_{E_i}^{-1}(u)$ of $\nu \in \mathcal{A}(\Omega, \Sigma; Y)$ contains an element $\mu \in \mathfrak{D}(\Omega. \Sigma; Y)$. ■

Some earlier projective and inductive limit notions are illustrated below.

Example 16.12. Let $\mathfrak{D}(\Omega, \Sigma; Y) \subseteq X \subseteq \mathfrak{A}(\Omega, \Sigma; Y)$, where Y is a TVS.

(a) $(X, \mathfrak{T}_{\mathcal{P}\{\mathcal{A}\}})$ is locally convex $\{T_2\}$ iff Y is locally convex $\{T_2\}$. The condition on Y is sufficient by Theorem 15.11, Example 16.2(a). It is also necessary, since $\psi_\omega: Y \to (\mathfrak{D}_\omega(\Omega, \Sigma; Y), \mathfrak{T}_{\mathcal{P}\{\mathcal{A}\}})$ is an onto topological isomorphism $\forall \omega \in \bigcup_\Sigma E$. The isomorphism ψ_ω is continuous, since for any
$y \rightsquigarrow \mu_{\omega,y}$
$\mathfrak{T}_{\mathcal{P}\{\mathcal{A}\}}$-nbd $v(0) = \bigcap_{i=1}^n A_{E_i}^{-1}(u) \cap \mathfrak{D}_\omega(\Omega, \Sigma; Y)$ one has $\psi_\omega^{-1}(v) = Y$ if $\omega \notin \bigcup_{i=1}^n E_i$ and $\psi_\omega^{-1}(v) = u$ if $\omega \in \bigcup_{i=1}^n E_i$. Furthermore, ψ_ω is open, since $\psi_\omega(u) = A_E^{-1}(u) \cap \mathfrak{D}_\omega(\Omega, \Sigma; Y)$ for $E \ni \omega$.

(b) Suppose Y is a T_2VS and let $\omega_0 \in \bigcup_\Sigma E$. Then $\mathfrak{D}_{\omega_0}(\Omega, \Sigma; Y)$ is a closed subspace of $(X, \mathfrak{T}_{\mathcal{P}\{\mathcal{A}\}})$ and $A_E^{-1}(0)$ is the topological supplement of $\mathfrak{D}_{\omega_0}(\Omega, \Sigma; Y)$ in $(X, \mathfrak{T}_{\mathcal{P}\{\mathcal{A}\}})$ for each $E \ni \omega_0$. The first assertion holds, since $\psi_{\omega_0} A_E: X \to \mathfrak{D}_{\omega_0}(\Omega, \Sigma; Y)$ is a retraction† for $E \ni \omega_0$. The second assertion follows (Corollary 15.27) by noting that $\psi_{\omega_0} A_E: (X, \mathfrak{T}_{\mathcal{P}\{\mathcal{A}\}}) \to (\mathfrak{D}_{\omega_0}(\Omega, \Sigma; Y), \mathfrak{T}_{\mathcal{P}\{\mathcal{A}\}})$ is a continuous linear surjection such that $(\psi_{\omega_0} A_E)^2 = \psi_{\omega_0} A_E$ and $(\psi_{\omega_0} A_E)^{-1}(0) = A_E^{-1}(0)$.

(c) If Σ is finite, $(X, \mathfrak{T}_{\mathcal{P}\{\mathcal{A}\}})$ is the internal inductive limit of $\P = \{\mathbb{1}_\omega = \mathbb{1}: \mathfrak{D}_\omega(\Omega, \Sigma; Y) \to X; \omega \in \Omega\}$. Clearly $\mathfrak{T}_{\mathcal{P}\{\mathcal{A}\}} \subset \mathfrak{T}_{\mathcal{I}'\{\P\}}$ on $X = \mathfrak{D}(\Omega, \Sigma; Y) = \left[\bigcup_\Omega \mathbb{1}_\omega\{\mathfrak{D}_\omega(\Omega, \Sigma; Y)\}\right]$. If $u(0) \subset X$ is a $\mathfrak{T}_{\mathcal{I}'\{\P\}}$-nbd, then $\psi_\omega^{-1}\mathbb{1}_\omega^{-1}(u)$ is a Y-nbd and $j_\omega\{\psi_\omega^{-1}\mathbb{1}_\omega^{-1}(u)\}$ is a $\prod_{\mathcal{A}} Y$-nbd. Therefore $u = \mathfrak{e}^{-1}\{j_\omega \psi_\omega^{-1}\mathbb{1}_\omega^{-1}(u)\}$ is a $\mathfrak{T}_{\mathcal{P}\{\mathcal{A}\}}$-nbd and $\mathfrak{T}_{\mathcal{I}'\{\P\}} \subset \mathfrak{T}_{\mathcal{P}\{\mathcal{A}\}}$ follows.

Our next theorem makes use of the following:

LEMMA 16.13. Let X satisfy $\mathfrak{D}(\Omega, \Sigma; Y) \subseteq X \subseteq \mathfrak{A}(\Omega, \Sigma; Y)$ and suppose Y is a T_2VS. For each $\{E_1, \ldots, E_n\} \subset \Sigma$ and Y-nbd $w(0)$:

(i) There is a Y-nbd $u(0)$ and disjoint sets $\{F_m: 1 \le m \le 2^n - 1\} \subset \Sigma$ such that $\bigcap_{m=1}^{2^n-1} A_{F_m}^{-1}(u) \subset \bigcap_{j=1}^n A_{E_j}^{-1}(w)$.

† A continuous mapping $r: X \to A \subset X$ is called a retraction if $r/A = \mathbb{1}$. If X is T_2 and r is a retraction, then A is closed ($x \in \bar{A} - A$ implies $r(x) \neq x$ and $r(u) \subset v$ for disjoint nbds $u(x)$ and $v\{r(x)\}$. This is impossible, since $x \in \bar{A}$ also yields $r(a) = a \in u(x) \cap A$ and $a \in u \cap v$ for some $a \in A$.)

(ii) If $w \neq Y$ and $\Lambda_0 = \{1 \le j \le n : E \cap E_j \neq \phi\}$ for $E \in \Sigma$, then $\bigcap_{j=1}^{n} A_{E_j}^{-1}(u) \subset A_E^{-1}(w)$ implies $E \subset \bigcup_{\Lambda_0} E_j$. Equality holds whenever $\{E_1, \ldots, E_n\}$ are disjoint.

Proof. (i) Begin by observing that each E_j $(1 \le j \le n)$ can be written as a disjoint union $E_j = \bigcup_{k=1}^{n} \bigcup_{j \in \Lambda_i^k} B_{\Lambda_i^k}$. (Specifically, $\omega \in E_j$ implies $\Lambda_\omega = \Lambda_{i_0}^{k_0}$ for some $1 \le k_0 \le i_0$ and $1 \le i_0 \le \binom{n}{k_0}$ and there follows $\omega \in B_{\Lambda_{i_0}^{k_0}} \subset \bigcup_{j \in \Lambda_{i_0}^{k_0}} B_{\Lambda_i^{k_0}}$. Furthermore, $\alpha \in \bigcup_{j \in \Lambda_{i_1}^{k_1}} B_{\Lambda_{i_1}^{k_1}} \cap \bigcup_{j \in \Lambda_{i_2}^{k_2}} B_{\Lambda_{i_2}^{k_2}}$ with $k_1 \neq k_2$ contradicts $\alpha \in B_{\Lambda_{i_1}^{k_1}} \cap B_{\Lambda_{i_2}^{k_2}}$ for $\Lambda_{i_1}^{k_1} \neq \Lambda_{i_2}^{k_2}$ in Theorem 16.3). Altogether, there are $\sum_{k=1}^{n} \binom{n}{k} = 2^n - 1$ distinct sets of the form $B_{\Lambda_i^k}$ and for any balanced Y-nbd $u(0)$ satisfying $\sum_{2^n-1} u \subset w$ we have $\bigcap^{2^n-1} A_{B_{\Lambda_i^k}}^{-1}(u) \subset \bigcap_{j=1}^{n} A_{E_j}^{-1}(w)$. The first assertion in (ii) is clear, since $\omega \in E - \bigcup_{\Lambda_0} E_j$ contradicts $\mu_{\omega,y} \in \bigcap_{j=1}^{n} A_{E_j}^{-1}(u) - A_E^{-1}(w)$ for $y \notin w(0)$. In the same vein, if $\{E_1, \ldots, E_n\}$ are disjoint and $\omega \in \bigcup_{\Lambda_0} E_j - E$, then $\omega \in E_{j_0}$ for exactly one $j_0 \in \Lambda_0$ and $\mu_{\omega',y} - \mu_{\omega,y} \in \bigcap_{j=1}^{n} A_{E_j}^{-1}(u) - A_E^{-1}(w)$ for $\omega' \in E \cap E_{j_0}$. ■

THEOREM 16.14. $(X, \mathfrak{T}_{\mathcal{P}\{\mathcal{A}\}})$ is metrizible {normable} iff Σ is finite and Y is metrizible {normable}.

Proof. The conditions on Σ and Y are sufficient by Theorem 15.5. If $(X, \mathfrak{T}_{\mathcal{P}\{\mathcal{A}\}})$ is metrizible {normable}, so is $(\mathfrak{D}_\omega(\Omega, \Sigma; Y), \mathfrak{T}_{\mathcal{P}\{\mathcal{A}\}})$ $(\omega \in \bigcup_{\Sigma} E)$ and Y. It remains therefore to show that Σ is finite whenever $\mathfrak{T}_{\mathcal{P}\{\mathcal{A}\}}$ has a countable nbd base $\{v_n(0): n \in \mathbb{N}\}$ at $0 \in X$. For each $n \in \mathbb{N}$ there is a $\mathfrak{T}_{\mathcal{P}\{\mathcal{A}\}}$-nbd $\bigcap_{i=1}^{m_n} A_{E_{\alpha(i)}}(u_n) \subset v_n$. Assume [Lemma 16.13(i)] that the sets $\{E_{\alpha(i)}: 1 \le i \le m_n\}$ are disjoint and let $\mathscr{E} = \{E_{\alpha(1)}, \ldots, E_{\alpha(m_n)}: n \in \mathbb{N}\}$. If $E \in \Sigma$ and $w(0) \neq Y$, some $\bigcap_{i=1}^{m_n} A_{E_{\alpha(i)}}^{-1}(u_n) \subset A_E^{-1}(w)$ and so [Lemma 16.13(ii)] $E = \bigcup_{i=1}^{m_n} E_{\alpha(i)} \in \mathfrak{R}(\mathscr{E})$. Thus $\Sigma \subset \mathfrak{R}(\mathscr{E})$ and (since $\mathscr{E}$, therefore $\mathfrak{R}(\mathscr{E})$, is countable) Σ is finite by Theorem 16.3. ■

COROLLARY 16.15. Suppose $\mathfrak{D}(\Omega, \Sigma; Y) \subseteq X \subseteq B\mathfrak{A}(\Omega, \Sigma; Y)$ for a T_2VS Y.

(i) $(X, \mathfrak{T}_{\mathcal{P}\{\mathcal{A}\}})$ is normable iff Y is normable and $\mathfrak{T}_{\mathcal{P}\{\mathcal{A}\}} = \| \|_0$.

(ii) $(X, \mathfrak{T}_{\mathcal{P}\{Y'\mathcal{A}\}})$ is 1° {seminormable} iff Σ is finite and dim $Y' \leq \aleph_0$ {dim $Y' < \aleph_0$}.

(iii) Assume that Y is a LCT$_2$VS. Then $(X, \mathfrak{T}_{\mathcal{P}\{Y'\mathcal{A}\}})$ is normable iff dim $X < \aleph_0$.

Proof. (i) follows via Theorem 16.14 and Example 16.7(b). The proof of (ii) follows from Theorem 15.5 after observing that $\mathfrak{T}_{\mathcal{P}\{Y'\mathcal{A}\}}$ on X is the same as $\mathfrak{T}_{\mathcal{P}\{\mathcal{A}\}}$ relative to $\mathfrak{T}_{\mathcal{P}\{Y'\}} \in T(Y)$. Clearly, (iii) is a consequence of (ii) and Corollary 15.7. ■

Theorem 16.16. For a T_2VS Y:

(i) $\mathfrak{A}(\Omega, \Sigma; Y)$ is $\mathfrak{T}_{\mathcal{P}\{\mathcal{A}\}}$-{sequentially} complete iff Y is {sequentially} complete.

(ii) $\mathcal{C}\mathfrak{A}(\Omega, \Sigma; Y)$ is complete iff $\mathcal{C}\mathfrak{A}(\Omega, \Sigma; Y) = \mathfrak{A}(\Omega, \Sigma; Y)$ and Y is complete.

(iii) Y is sequentially complete if $\mathcal{C}\mathfrak{A}(\Omega, \Sigma; Y)$ is sequentially complete. The converse also holds if Y is an LCT$_2$VS and $\Omega \in \Sigma$.

Proof. The necessary part of (i) and the first assertion in (iii) are clear by virtue of Example 16.12(a), (b) and Theorem 6.6. If Y is complete and $\{\mu_\varphi\}$ is a $\mathfrak{T}_{\mathcal{P}\{\mathcal{A}\}}$-Cauchy net in $\mathfrak{A}(\Omega, \Sigma; Y)$, then $\{\mu_\varphi(E)\}$ is Y-Cauchy $\forall E \in \Sigma$ and we define $\mu(E) = \lim \mu_\varphi(E) \, \forall E \in \Sigma$. For disjoint sets $\{E_1, \ldots, E_n\} \in \Sigma$ and any Y-nbd $u(0)$ pick a balanced Y-nbd $v(0)$ such that $\sum_{n+1} v \subset u$. Then there is a μ_{φ_0} such that $\mu_\varphi(E_i) - \mu(E_i) \subset v \; \forall 1 \leq i \leq n$ and $\mu_\varphi\left(\bigcup_{i=1}^n E_i\right) - \mu\left(\bigcup_{i=1}^n E_i\right) \subset v \; \forall \mu_\varphi \geq \mu_{\varphi_0}$. This yields $\mu\left(\bigcup_{i=1}^n E_i\right) - \sum_{i=1}^n \mu(E_i) = \left\{\mu\left(\bigcup_{i=1}^n E_i\right) - \mu_\varphi\left(\bigcup_{i=1}^n E_i\right)\right\} + \left\{\mu_\varphi\left(\bigcup_{i=1}^n E_i\right) - \sum_{i=1}^n \mu_\varphi(E_i)\right\} + \sum_{i=1}^n \{\mu_\varphi(E_i) - \mu(E_i)\} \subset \sum_{n+1} v \subset u$. It follows therefore that $\left\{\mu\left(\bigcup_{i=1}^n E_i\right) - \sum_{i=1}^n \mu(E_i)\right\} \in \bigcap_{\mathfrak{N}_0} u = \{0\}$ and $\mu_\varphi \xrightarrow{\mathfrak{T}_{\mathcal{P}\{\mathcal{A}\}}} \mu \in \mathfrak{A}(\Omega, \Sigma; Y)$ [Example 15.4(c)]. The sufficiency proof for sequential completeness in (i) follows similarly, as does the second assertion in (iii) (using Theorem 16.6). Evidently (i) ensures that $\mathcal{C}\mathfrak{A}(\Omega, \Sigma; Y)$ is complete if Y is complete and $\mathcal{CA}(\Omega, \Sigma; Y) = \mathcal{A}(\Omega, \Sigma; Y)$. If $\mathcal{CA}(\Omega, \Sigma; Y)$ is complete, then $\mathcal{CA}(\Omega, \Sigma; Y)$ is $\mathfrak{T}_{\mathcal{P}\{\mathcal{A}\}}$-closed in $\mathcal{A}(\Omega, \Sigma; Y)$by Theorem 6.6 and $\overline{\mathfrak{D}(\Omega, \Sigma; Y)}^{\mathfrak{T}_{\mathcal{P}\{\mathcal{A}\}}} = \mathcal{CA}(\Omega, \Sigma; Y) = \mathcal{A}(\Omega, \Sigma; Y)$ by Corollary 16.11. ■

COROLLARY 16.17. $(\mathcal{CA}(\Omega, \Sigma; Y), \mathfrak{T}_{\mathcal{P}\{\mathcal{A}\}})$ is a Frechet {Banach} space iff Σ is finite and Y is Frechet {Banach}.

Proof. Theorems 16.14, 16.16, and 16.10. ■

Our last result here is concerned with compactness.

THEOREM 16.18. Let Y be a T_2VS. Then

(i) $\mathcal{K} \subset \mathcal{A}(\Omega, \Sigma; Y)$ is $\mathfrak{T}_{\mathcal{P}\{\mathcal{A}\}}$-relatively compact iff $\mathcal{K}(E) \subset Y$ is relatively compact $\forall E \in \Sigma$.

(ii) $\mathcal{K} \subset \mathcal{CA}(\Omega, \Sigma; Y)$ is $\mathfrak{T}_{\mathcal{P}\{\mathcal{A}\}}$-relatively compact iff $\mathcal{K}(E) \subset Y$ is relatively compact $\forall E \in \Sigma$ and $\bar{\mathcal{K}}^{\mathfrak{T}_{\mathcal{P}\{\mathcal{A}\}}}$ in $\mathcal{CA}(\Omega, \Sigma; Y)$ equals $\bar{\mathcal{K}}^{\mathfrak{T}_{\mathcal{P}\{\mathcal{A}\}}}$ in $\mathcal{A}(\Omega, \Sigma; Y)$.

Proof. (i) If $\bar{\mathcal{K}}^{\mathfrak{T}_{\mathcal{P}\{\mathcal{A}\}}} \subset \mathcal{A}(\Omega, \Sigma; Y)$ is compact, $A_E(\bar{\mathcal{K}}^{\mathfrak{T}_{\mathcal{P}\{\mathcal{A}\}}})$ is Y-compact and closed $\forall E \in \Sigma$. Therefore $A_E(\mathcal{K}) \subset A_E(\bar{\mathcal{K}}^{\mathfrak{T}_{\mathcal{P}\{\mathcal{A}\}}}) \subset \overline{A_E(\mathcal{K})}$ ensures that $\overline{\mathcal{K}(E)} = \overline{A_E(\mathcal{K})} - A_E(\bar{\mathcal{K}}^{\mathfrak{T}_{\mathcal{P}\{\mathcal{A}\}}})$ is Y-compact. Suppose, conversely, that $\overline{A_E(\mathcal{K})}$ is Y-compact $\forall E \in \Sigma$. Then $\prod_{\Sigma} \overline{A_E(\mathcal{K})}$ is $\prod_{\Sigma} Y$-compact and we claim (see Theorem 15.8) that the closure of $\mathfrak{e}(\mathcal{K})$ in $\mathfrak{e}\{\mathcal{A}(\Omega, \Sigma; Y)\}$ coincides with the closure of $\mathfrak{e}(\mathcal{K})$ in $\prod_{\Sigma} \overline{A_E(\mathcal{K})}$. This means that $\mathfrak{e}(\bar{\mathcal{K}}^{\mathfrak{T}_{\mathcal{P}\{\mathcal{A}\}}}) = \overline{\mathfrak{e}(\mathcal{K})} \subset \mathfrak{e}\{\mathcal{A}(\Omega, \Sigma; Y)\}$ and $\bar{\mathcal{K}}^{\mathfrak{T}_{\mathcal{P}\{\mathcal{A}\}}}$ is compact in $(\mathcal{A}(\Omega, \Sigma; Y), \mathfrak{T}_{\mathcal{P}\{\mathcal{A}\}})$. *Proof of Claim*: Let $\lambda \in \prod_{\Sigma} \overline{A_E(\mathcal{K})} \cap \overline{\mathfrak{e}(\mathcal{K})}$ (the closure of $\mathfrak{e}(\mathcal{K})$ in $\prod_{\Sigma} \overline{A_E(\mathcal{K})}$) and let $E_3 = E_1 \cup E_2$ for disjoint $E_1, E_2 \in \Sigma$. For any Y-nbd $u(0)$ and balanced Y-nbd $v(0)$ satisfying $v + v + v \subset u$ there is a $\mu \in \{\lambda + \bigcap_{i=1}^{3} \pi_{E_i}^{-1}(v)\} \cap \mathfrak{e}(\mathcal{K})$. Since $\pi_{E_i}(\lambda) - \mu(E_i) \in v$ $(i = 1, 2, 3)$, one has $\pi_{E_3}(\lambda) = \pi_{E_1}(\lambda) + \pi_{E_2}(\lambda)$. In essence, $\lambda \in \mathfrak{e}(\lambda_0) \in \mathfrak{e}\{\mathcal{A}(\Omega, \Sigma; Y)\} \cap \overline{\mathfrak{e}(\mathcal{K})}$ (the closure of $\mathfrak{e}(\mathcal{K})$ in $\mathfrak{e}\{\mathcal{A}(\Omega, \Sigma; Y)\}$) for $\lambda_0 \in \mathcal{A}(\Omega, \Sigma; Y)$ defined by $\lambda_0(E) = \pi_E(\lambda)$. Now suppose $\lambda \in \overline{\mathfrak{e}(\mathcal{K})} \cap \mathfrak{e}\{\mathcal{A}(\Omega, \Sigma; Y)\}$ and fix $E_0 \in \Sigma$. For any balanced Y-nbd $u(0)$ there is an $\mathfrak{e}(\mu) \in \{\lambda + \pi_{E_0}^{-1}(u)\} \cap \mathfrak{e}(\mathcal{K})$. Hence $\mu(E_0) \in \{\lambda(E_0) + u(0)\} \cap A_{E_0}(\mathcal{K})$ and [since $u(0)$ is arbitrary] $\lambda(E_0) \in \overline{A_{E_0}(\mathcal{K})}$. This establishes that $\lambda \in \prod_{\Sigma} \overline{A_E(\mathcal{K})} \cap \overline{\mathfrak{e}(\mathcal{K})}$ and proves our claim.

(ii) The conditions are necessary for $\mathcal{K}$'s relative compactness in $(\mathcal{CA}(\Omega, \Sigma; Y), \mathfrak{T}_{\mathcal{P}\{\mathcal{A}\}})$. To be sure, $\bar{\mathcal{K}}^{\mathfrak{T}_{\mathcal{P}\{\mathcal{A}\}}} \cap \mathcal{CA}(\Omega, \Sigma; Y)$ (the closure of $\mathcal{K}$ in

$\mathcal{CA}(\Omega, \Sigma; Y)$) is contained in $\bar{\mathcal{K}}^{\mathfrak{T}_{\mathcal{P}\{\mathcal{A}\}}}$ (the closure of $\mathcal{K}$ in $\mathcal{A}(\Omega, \Sigma; Y)$). If $\lambda \in \bar{\mathcal{K}}^{\mathfrak{T}_{\mathcal{P}\{\mathcal{A}\}}}$, then $\mu_\varphi \xrightarrow{\mathfrak{T}_{\mathcal{P}\{\mathcal{A}\}}} \lambda$ for some net $\{\mu_\varphi\} \subset \mathcal{K}$. Since $\{\mu_\varphi\}$ is $\mathfrak{T}_{\mathcal{P}\{\mathcal{A}\}}$-Cauchy and $\bar{\mathcal{K}}^{\mathfrak{T}_{\mathcal{P}\{\mathcal{A}\}}} \cap \mathcal{CA}(\Omega, \Sigma; Y)$ is complete, we have $\mu_\varphi \xrightarrow{\mathfrak{T}_{\mathcal{P}\{\mathcal{A}\}}} \lambda_0 \in \bar{\mathcal{K}}^{\mathfrak{T}_{\mathcal{P}\{\mathcal{A}\}}} \cap \mathcal{CA}(\Omega, \Sigma; Y)$. It follows, from $(\mathcal{A}(\Omega, \Sigma; Y), \mathfrak{T}_{\mathcal{P}\{\mathcal{A}\}})$ being T_2, that $\lambda = \lambda_0$ and $\bar{\mathcal{K}}^{\mathfrak{T}_{\mathcal{P}\{\mathcal{A}\}}} \subset \bar{\mathcal{K}}^{\mathfrak{T}_{\mathcal{P}\{\mathcal{A}\}}} \cap \mathcal{CA}(\Omega, \Sigma; Y)$. Therefore $\mathcal{K}$ is $\mathfrak{T}_{\mathcal{P}\{\mathcal{A}\}}$ relatively compact in $\mathcal{A}(\Omega, \Sigma; Y)$ and (i) applies. The conditions are also sufficient, since they imply that $\bar{\mathcal{K}}^{\mathfrak{T}_{\mathcal{P}\{\mathcal{A}\}}} \cap \mathcal{CA}(\Omega, \Sigma; Y) = \bar{\mathcal{K}}^{\mathfrak{T}_{\mathcal{P}\{\mathcal{A}\}}}$ is $\mathfrak{T}_{\mathcal{P}\{\mathcal{A}\}}$-compact in $\mathcal{A}(\Omega, \Sigma; Y)$. Therefore $\bar{\mathcal{K}}^{\mathfrak{T}_{\mathcal{P}\{\mathcal{A}\}}} \cap \mathcal{CA}(\Omega, \Sigma; Y)$ is $\mathfrak{T}_{\mathcal{P}\{\mathcal{A}\}}$-compact in $\mathcal{CA}(\Omega, \Sigma; Y)$. ■

PROBLEMS

16A. Atomic Measures

An element $\mu \in \mathcal{A}(\Omega, \Sigma; Y)$ is said to be atomic if it has an atom E [i.e., a set $E \in \Sigma$ such that $\mu(E) \neq 0$ and either $\mu(A) = \mu(E)$ or $\mu(A) = 0 \; \forall A \in \Sigma$ satisfying $A \subset E$]. Let $\mathcal{CA}_t(\Omega, \Sigma; Y)$ denote the collection of atomic measures in $\mathcal{CA}(\Omega, \Sigma; Y)$ for a T_2VS $Y \neq \{0\}$.

(a) $\mathcal{D}(\Omega, \Sigma; Y) - \{0\} \subset \mathcal{CA}_t(\Omega, \Sigma; Y)$. [*Hint*: Suppose $\mu = \sum_{i=1}^{n} \mu_{\omega_i, y_i} \neq 0$ for distinct $\omega_i \in \Omega$ and $y_i \neq 0$ $(i = 1, 2, \ldots, n)$. Then $\omega_{i_0} \in E_{i_0}$ in Σ for some $1 \leq i_0 \leq n$. Assume that $i_0 = 1$ and let $w_1 = \{\omega_1, \ldots, \omega_{k_1}\}$ be the renumbered subset of $\{\omega_i : 1 \leq i \leq n\}$ which satisfies $\omega_j \in E_1 \; \forall 1 \leq j \leq k_1$. If $k_1 = 1$, take $E_1 = E$. Otherwise define $\xi_1 = \{F \in \Sigma : \omega_j \in F \subset E_1$ for some $1 \leq j \leq k_1\}$. Either $w_1 \subset F \; \forall F \in \xi_1$ (in which case, take $E_1 = E$) or there is an $F \in \xi_1$ such that $\omega_j \in F$ for some $1 \leq j \leq k_1$. Let $w_2 = \{\omega_1, \ldots, \omega_{k_2}\}$ be the renumbered subset of w_1 such that $\omega_j \in F \; \forall 1 \leq j \leq k_2$. Clearly, $k_1 > k_2$. If $k_2 = 1$, take $F = E_2$. Otherwise define $\xi_2 = \{G \in \Sigma : \omega_j \in G \subset F$ for some $1 \leq j \leq k_2\}$. Either $w_2 \subset G \; \forall G \in \xi_2$ (in which case, take $F = E_2$) or there is an $H \in \xi_2$ such that $\omega_j \in H$ for some $1 \leq j \leq k_2$. Proceed to obtain a sequence $n \geq k_1 > k_2 > \cdots > k_p \geq 1$ such that $k_p = 1$ or $w_p \subset K \; \forall K \in \xi_p$. Convince yourself that $E_p = E$ is a μ-atom.]

(b) $\overline{\mathcal{CA}_t(\Omega, \Sigma; Y)}^{\mathfrak{T}_{\mathcal{P}\{\mathcal{A}\}}} = \mathcal{A}(\Omega, \Sigma; Y)$. [*Hint*: By assumption $\Sigma \neq \phi$ so that $(\mathcal{D}(\Omega, \Sigma; Y), \mathfrak{T}_{\mathcal{P}\{\mathcal{A}\}})$ is nondiscrete (Theorem 13.14⁺) and $0 \in \overline{\mathcal{D}(\Omega, \Sigma; Y) - \{0\}}^{\mathfrak{T}_{\mathcal{P}\{\mathcal{A}\}}} \subset \mathcal{D}(\Omega, \Sigma; Y)$. Therefore $\mathcal{D}(\Omega, \Sigma; Y) \subset \overline{\mathcal{D}(\Omega, \Sigma; Y) - \{0\}}^{\mathfrak{T}_{\mathcal{P}\{\mathcal{A}\}}}$ and we need only invoke Corollary 16.11.]

16B. $\mathcal{A}_{FV}(\Omega, \Sigma; Y)$ as a Topological Direct Sum

Assume that Σ is the σ-algebra of all subsets of Ω and define $l_1(\Omega) = \{f : \Omega \to \mathbb{F}; \|f\|_1 = \sum_{\Omega} |f(\omega)| < \infty\}$.

(a) For each $f \in l_1(\Omega)$, define $\mu_f(E) = \sum_{\omega \in E} f(\omega)\ \forall E \in \Sigma$. Then

$$\zeta: (l_1(\Omega), \|\ \|_1) \to (\mathcal{A}_{FV}(\Omega, \Sigma; \mathbb{F}), \|\ \|_v)$$
$$f \rightsquigarrow \mu_f$$

is an into congruence.

(b) Show that $\zeta\{l_1(\Omega)\} = \{\mu \in \mathcal{C}\mathcal{A}_{FV}(\Omega, \Sigma; \mathbb{F})$: there is a countable set $E_0 \in \Sigma$ such that $\mu(E) = 0\ \forall E \subset \Omega - E_0\}$.

(c) Define $\mathcal{A}_0(\Omega, \Sigma; \mathbb{F})$ as the vector subspace of $\mathcal{A}_{FV}(\Omega, \Sigma; \mathbb{F})$ each of whose members satisfies $\mu(E) = 0$ for every finite set $E \in \Sigma$. Then $(\mathcal{A}_{FV}(\Omega, \Sigma; \mathbb{F}), \|\ \|_v) \stackrel{t}{=} \zeta\{l_1(\Omega)\} \oplus \mathcal{A}_0(\Omega, \Sigma; \mathbb{F})$.

Remark. A more detailed development of this and related problems may be found in Edwards [28].

16C. Vague Topologies on $\mathcal{C}\mathcal{A}(\Omega, \Sigma; \mathbb{C})$

Let Σ be the σ-algebra of Borel subsets of a compact T_2 space Ω and denote the space of bounded measurable functions on Ω by $BM(\Omega, \mathbb{C})$. For each $\{f_{\alpha_1} \cdots f_{\alpha_n}\} \in C(\Omega, \mathbb{C})$ $\{BM(\Omega, \mathbb{C})\}$ and each $\varepsilon > 0$, define $v(0; f_{\alpha_1} \cdots f_{\alpha_n}; \varepsilon) = \{\mu \in \mathcal{C}\mathcal{A}(\Omega, \Sigma; \mathbb{C}): |\int_\Omega f_{\alpha_i}\, d\mu| < \varepsilon\ \forall 1 \le i \le n\}$.

(a) Show that $\mathfrak{N} = \{v(0); f_{\alpha_1} \cdots f_{\alpha_n}; \varepsilon): f_{\alpha_i} \in C(\Omega, \mathbb{C})$ and $\varepsilon > 0\}$ and $\mathfrak{N}_{B\mathcal{M}} = \{v(0; f_{\alpha_1} \cdots f_{\alpha_n}; \varepsilon): f_{\alpha_i} \in B\mathcal{M}(\Omega, \mathbb{C})$ and $\varepsilon > 0\}$ are nbd bases at 0 for locally convex topologies $\mathcal{O}_v, \mathcal{O}_{v_{B\mathcal{M}}} \in T\{\mathcal{C}\mathcal{A}(\Omega, \Sigma; \mathbb{C})\}$, respectively. Furthermore, $\mathcal{O}_{v_{B\mathcal{M}}}$ is T_2. [*Hint*: For $v(0) = v(0; f_{\alpha_1} \cdots f_{\alpha_n}; \varepsilon)$ and any $\mu_0 \in \mathcal{C}\mathcal{A}(\Omega, \Sigma; \mathbb{C})$, either $\|\mu_0\|_0 = 0$ and $\mu_0 \in v(0)$ or $\|\mu_0\|_0 \neq 0$ and $\lambda\mu_0 \in v$ for $\lambda = \varepsilon/(\max_{1 \le i \le n} \sup_\Omega |f_{\alpha_i}|)\|\mu_0\|_0$.]

(b) Let $M_0(\Omega, \Sigma; \mathbb{C}) = \{\mu \in \mathcal{C}\mathcal{A}(\Omega, \Sigma; \mathbb{C}): |\mu(E)| \le |\mu(F)|$ whenever $E \subset F$ for $E, F \in \Sigma\}$. Then $\mathcal{O}_v = \mathcal{O}_{v_{B\mathcal{M}}} = \mathfrak{T}_{\mathcal{P}\{\mathcal{A}\}}$ on $M_0(\Omega, \Sigma; \mathbb{C})$. [*Hint*: $\mathcal{O}_{B\mathcal{M}} \subset \mathfrak{T}_{\mathcal{P}\{\mathcal{A}\}}$, since $v(0; \chi_\Omega; \varepsilon) \subset v(0; A_{E_1} \cdots A_{E_n}; \varepsilon)\ \forall \{E_1 \cdots E_n\} \in \Sigma$. Show that $|\mu(\Omega)| < \|\mu\|_0 < 2|\mu(\Omega)|$ and obtain $v(0; A_\Omega; \max_{1 \le i \le n} \sup_\Omega |f_{\alpha_i}|) \subset v(0; f_{\alpha_1} \cdots f_{\alpha_n}; \varepsilon)$.]

(c) Why is $(M_0(\Omega, \Sigma; \mathbb{C}), \mathfrak{T}_{\mathcal{P}\{\mathcal{A}\}})$ not a TVS?

References for Further Study

Suffel [95], [96]; Varadarajan [101].

17. Hahn Banach Theorems

This section introduces a number of analytic and geometric results that play a central role in functional analysis. One immediate by-product is the characterization of those TVSps X having $X' = \{0\}$. Two other results of paramount importance, which distinguish LCTVSps from general TVSps, are also given; namely, the continuous dual of an LCT_2VS is injective and every continuous linear functional on a vector subspace of an LCTVS X has an extension to X'. We conclude in this spirit by proving the powerful Krein Milman theorem which characterizes each compact convex set in an LCT_2VS as the closed convex hull of a particular subset of itself.

DEFINITION 17.1. A maximal proper vector subspace of a vector space X is called a hyperplane in X.

Hyperplanes are intimately connected to linear functionals as is easily demonstrated in

THEOREM 17.2. (i) A subspace H of a vector space X is a hyperplane in X iff $\dim X/H = 1$ iff $H = f^{-1}(0)$ for some $f \neq 0$ in X^*.

(ii) In a TVS X a hyperplane H is either closed or dense. In fact, $h = f^{-1}(0)$ is closed iff $f \in X'$.

Proof. If H is a hyperplane in X and $x_0 \in X - H$, every $x \in X$ has a unique representation $x = x_h + \lambda x_0$ $(x_h \in H)$. Therefore $H = f^{-1}(0)$ for $f \in X^*$ defined by $f(x_h + \lambda x_0) = \lambda$. Since $\underset{H+x \rightsquigarrow f(x)}{\mu_f : X/H \to \mathbb{F}}$ is a vector space isomorphism, $\dim X/H = 1$. Now suppose $\dim X/H = 1$ so that $X/H = [H + x_0]$ and $X = H + [x_0]$ for some $x_0 \notin H$. If M is any vector subspace of X that contains H, then $x_0 \notin M$ implies $M = H$ and $x_0 \in M$ implies $M = X$. Therefore H is a hyperplane in X and the proof of (i) is complete. Suppose for (ii) that H is a hyperplane in the TVS X. Since $\bar{H}$ is a vector subspace of X, either $H = \bar{H}$ or $H = X$. Evidently $H = f^{-1}(0)$ is closed for $f \in X'$. If H is closed, X/H is a one dim T_2VS (Definition 15.13) and μ_f, therefore $f = \mu_f \nu$, is continuous (Corollary 13.13). ■

THEOREM 17.3 (H.B. Separation Theorem). If Ω is an open convex subset of a TVS X and M is a vector subspace of X such that $M \cap \Omega = \varnothing$, then $H \cap \Omega = \varnothing$ for some closed hyperplane $H \supset M$.

Proof. We use Zorn's lemma. Let $\mathscr{H}$ denote the collection of proper vector subspaces of X, each of which contains M and is disjoint with Ω. Clearly, $M \in \mathscr{H} \neq \varnothing$. Partially order $\mathscr{H}$ by inclusion. Then $\bigcup_{\mathscr{H}} H$ is an upper bound for every subset of $\mathscr{H}$ and therefore there is a maximal element $H = \bar{H}$ in $\mathscr{H}$ (otherwise, $\bar{H} = X$ implies $H \cap \Omega \neq \varnothing$). ■

COROLLARY 17.4. Every closed vector subspace M of an LCTVS X is the intersection of all closed hyperplanes in X which contain M.

Proof. If $x_0 \notin M$, then $u(x_0) \cap M = \varnothing$ for a convex open nbd $u(x_0)$ and $u(x_0) \cap H = \varnothing$ for some closed hyperplane $H \supset M$. ■

COROLLARY 17.5. Let A, B be nonempty, disjoint, convex subsets of a TVS X. If A is open, there is an $f \in X'$ such that $f(A) \cap f(B) = \varnothing$.

Proof. The set $A - B = \{x_a - x_b : (x_a, x_b) \in A \times B\}$ is open, convex and does not contain $M = \{0\}$. Therefore $f^{-1}(0) \cap (A - B) = \varnothing$ for some $f \in X'$ and $f(A) \cap f(B) = \varnothing$ follows (since $f(x_a) = f(x_b)$ for $(x_a, x_b) \in A \times B$ implies $x_a - x_b \in f^{-1}(0) \cap (A - B)$). ■

COROLLARY 17.6. If K is a convex subset of an LCTVS X and $x_0 \notin \operatorname{Int} K$ $\{x_0 \notin \bar{K}\}$, there is an $f \neq 0$ in X' such that $f(x_0) \notin \overline{f(\operatorname{Int} K)}$ $\{f(x_0) \notin \overline{f(K)}\}$.

Proof. Recall (Theorem 13.11(ii)) that every $f \in X^*$ is open. For the first assertion take $A = \operatorname{Int} K$ and $B = \{x_0\}$ in Corollary 17.5. If $x_0 \notin \bar{K}$, then $(ux_0) \cap K = \varnothing$ for some convex open nbd $u(0)$ and the proof follows by taking $A = u + x_0$ and $B = K$. ■

The problem of extending linear functionals essentially reduces to that of separating open convex sets by hyperplanes. We see this in

THEOREM 17.7 (H.B. Extension Theorem). Let p be a seminorm on a vector space X and let M be a vector subspace of X. If $f \in M^*$ satisfies $|f(x)| \leq p(x)$ on M, there is an $F \in X^*$ extending f such that $|F(x)| \leq p(x)$ on X.

Both f and F are continuous if p is continuous on a TVS X.

Proof. The theorem is trivially true for $f = 0$ Assume, therefore, that X carries $\mathfrak{T}(p)$ and consider any $x_0 \in M$ such that $f(x_0) = 1$. Since $\Omega = S_{1,p}(0) + x_0$ is open, convex, and disjoint from $f^{-1}(0)$, there is a closed hyperplane $H \supset f^{-1}(0)$ such that $H \cap \Omega = \varnothing$. As in Theorem 17.2, it follows from $x_0 \in \Omega - H$ that every $x \in X$ $\{$every $y \in M\}$ has a unique representation $x = x_h + \lambda x_0$ $(x_h \in H)$ $\{y = y_k + \delta x_0$, where $y_k \in K = f^{-1}(0) \subset M\}$. Therefore $H = F^{-1}(0)$ and $F(x_0) = 1$ for $F \in X^*$ defined by $f(x_h + \lambda x_0) = \lambda$. Clearly $F(x) = f(x)\ \forall x \in M$. If $p(x) < 1$ and $|F(x)| \geq 1$, then $x_0 - x/F(x) \in H \cap \Omega$. Consequently, $S_{1,p}(0) \subset S_{1,|F|}(0)$ and $|F| \leq p$ on X (by

Definition 14.5$^+$). The last assertion of the theorem follows from Theorem 14.6$^+$. ■

COROLLARY 17.8. Let p be a seminorm on a vector space X. For each $x_0 \neq 0$ in X there is an $F \in X^*$ such that $F(x_0) = p(x_0)$ and $|F(x)| \leq p(x)$ on X. If p is continuous on a TVS X, then $F \in X'$.

Proof. Take $M = [x_0]$ and define $f(\lambda x_0) = \lambda p(x_0)$. Then $f \in M^*$ and $f(x_0) = p(x_0)$. ■

The algebraic dual X^* of a vector space X is injective (Definition 13.3). Of far greater importance is

COROLLARY 17.9. X' is injective for an LCT_2VS X.

Proof. For any $x_0 \neq 0$ in X there exists [Theorems 14.11 and 14.8(i)] a continuous seminorm p such that $p(x_0) > 0$ and Corollary 17.8 applies. ■

One advantage LCTVSps have over general TVSps is

THEOREM 17.10. If M is a vector subspace of an LCTVS X, then every $f \in M'$ has a continuous extension to X'.

Proof. Since $f^{-1}\{S_1(0)\}$ is an abs convex M-nbd, one has $u(0) \cap M \subset f^{-1}\{S_1(0)\}$ for some abs convex X-nbd $u(0)$. Therefore (Theorem 14.10) p_u is a continuous seminorm satisfying $|f(x)| \leq p_u(x)\ \forall x \in M$ and it remains only to invoke Theorem 17.7. ■

THEOREM 17.11 (H.B. Approximation Theorem). Let $M_0 \subset M$, where M_0 and M are vector subspaces of an LCTVS X. Then $\bar{M}_0 = M$ iff $f/M_0 = 0$ implies $f/M = 0\ \forall f \in X'$.

Proof. The condition is clearly necessary, since $0 \in M_0'$ has a (unique) extension by continuity to $0 \in \bar{M}_0'$. For the sufficiency proof suppose that $x_0 \in M - \bar{M}_0$ and let $\nu: X \to (X/\bar{M}_0, \mathfrak{T}_{\mathfrak{g}(\nu)})$. Then $[\bar{M}_0 + x_0]$ is a one dim subspace of the T_2VS $(X/\bar{M}_0, \mathfrak{T}_{\mathfrak{g}(\nu)})$ and [Example 15.3(a), Corollary 13.13] the continuous mapping $f: [\bar{M}_0 + x_0] \to \mathbb{F}$, $\bar{M}_0 + \lambda x_0 \rightsquigarrow \lambda$ has an extension $F \in (X/\bar{M}_0, \mathfrak{T}_{\mathfrak{g}(\nu)})'$. This in turn produces an $\tilde{F} = F\nu \in X'$ satisfying $\tilde{F}(x_0) = 1$ and $\tilde{F} = 0$ on M_0. ■

COROLLARY 17.12. If $x_0 \notin \bar{M}_0$, then $\bar{M}_0 \subset f^{-1}(0)$ and $f(x_0) = 1$ for some $f \in X'$.

Proof. Take $M = X$ above. ■

We can now say when the extensions in Theorem 15.10 are unique

COROLLARY 17.13. The following are equivalent for a subset E of an LCTVS X:

(i) $f/[E] = 0$ implies $f = 0$ on X $\forall f \in X'$.
(ii) $\overline{[E]} = X$.
(iii) X is the smallest closed vector subspace of X containing E.
(iv) Every $f \in [E]'$ has a unique extension to X'.

Proof. For (i)⇔(ii) take $M_0 = [E]$ and $M = X$ in Theorem 17.11. Surely (ii)⇒(iii), since every closed subspace M of X which contains $[E]$ satisfies $X = \overline{[E]} \subset M$. Obviously (iii)⇒(ii) and (i)⇒(iv). If (i) does not hold, $f/[E] = 0$ for some $f \neq 0$ in X' and so $f = 0 \in [E]'$ has more than one extension. Therefore (iv) also fails. ■

It is also a simple matter to characterize TVSps X having $X' = \{0\}$. Such spaces, incidently, contain only dense hyperplanes (Theorem 15.2).

THEOREM 17.14. (i) A TVS X has $X' = \{0\}$ iff X contains no proper convex nbd.

(ii) An LCTVS $(X, \mathfrak{T})$ has $X' = \{0\}$ iff $\mathfrak{T} = \mathfrak{I}$.
(iii) An LCT$_2$VS X has $X' = \{0\}$ iff $X = \{0\}$

PROOF. One part of (i) is clear, since $f^{-1}\{S_1(0)\}$ is a proper convex nbd of $0 \in X$ $\forall f \neq 0$ in X'. On the other hand, $x_0 \in X - E$ for a convex nbd E implies that $A = \text{Int}\, E \neq \varnothing$ and $B = \{x_0\}$ satisfy the hypotheses of Corollary 17.5. Thus $X' \neq \{0\}$. The sufficiency part of (ii) is obvious. If $X' = \{0\}$ for an LCTVS $(X, \mathfrak{T})$ and $v(0)$ is a $\mathfrak{T}$-nbd of 0, then $u(0) \subset v(0)$ for some convex $\mathfrak{T}$-nbd. Since $u(0) = X$ [by (i) above], $v(0) = X$ and $\mathfrak{T} = \mathfrak{I}$. The proof of (iii) is an immediate consequence of Corollary 17.9. ■

Remark. Note [Theorem 13.9 and (i) above] that a Catg II TVS X has $X' = \{0\}$ iff $\bar{E} = X$ for every convex absorbent set $E \subset X$.

Theorem 17.14 considerably limits the class of TVSps having zero continuous dual. Two such spaces are provided in

Example 17.15. (a) If $I = [0, 1]$ carries Lebesgue measure in Example 14.16(a), then $L_p'(I) = \{0\}$ for $0 < p < 1$. Verification: Given any $f \in L_p^*(I)$, there is an $x = x(t)$ in $L_p(I)$ such that $d_p(x, 0) = 1$ and $f(x) = \varepsilon > 0$.

Since $\eta(s)=\int_0^s |x(t)|^p\,d\mu$ is uniformly continuous, there is a partition $0=s_0<s_1<\cdots<s_{n-1}<s_n=1$ of I such that $\eta(s_k)-\eta(s_{k-1})\le 1/n\ \forall 1\le k\le n$. From $x(t)=\sum_{k=1}^n x(t)\chi_{[s_{k-1},s_k]}(t)$ and $f(x)=\varepsilon$ there follows $|f\{x(t)\chi_{[s_{k_0-1},s_{k_0}]}(t)\}|\ge\varepsilon/n$ for some $1\le k_0\le n$. The sequence $x_n(t)=nx(t)\chi_{[s_{k_0-1},s_{k_0}]}(t)$ in $L_p(I)$ satisfies $d_p(x_n,0)\le n^{p-1}$. Therefore $x_n\to 0$ and $|f(x_n)|\ge\varepsilon\ \forall n\in\mathbb{N}$ precludes that $f\notin L_p'(I)$.

(b) $\mathcal{S}'([0,1))=\{0\}$ in Example 14.15, since $\{u_{(1/m),(1/n)}(0)\}_c=\mathcal{S}$ $\forall m,n\in\mathbb{N}$. Reason: If $f\in\mathcal{S}$, then $mf\chi_{[(k-1)/m,k/m)}\in u_{(1/m),(1/n)}(0)\ \forall 1\le k\le m$ and $\sum_{k=1}^m (1/m)\{mf\chi_{[(k-1)/m,k/m)}\}\in\{u_{(1/m),(1/n)}(0)\}_c$ by Theorem 13.6.

Krein Milman Theorem

A point x_0 in a convex subset E of a vector space X is said to be an extreme point of E if x_0 is an endpoint of every closed segment $[x,y]=\{\alpha x+(1-\alpha)y:\alpha\in[0,1]$ and $x,y\in E\}$ that contains it. Equivalently, x_0 belongs to no open segment $(x,y)=[x,y]-\{x,y\}$ joined by two points in E. The set of extreme points of E is denoted $\mathcal{E}(E)$. Thus, for example, $\mathcal{E}(E)=\{(x,y)\in E^2:x^2+y^2=1\}$ if E is the closed unit circle in E^2. Clearly, $\mathcal{E}(E)=\varnothing$ if E is open in a TVS X. It is also possible that $\mathcal{E}(E)=\varnothing$ when E is closed and bounded. The gist of our next theorem is that $\mathcal{E}(E)\ne\varnothing$ and $E=\overline{\{\mathcal{E}(E)\}_c}$ for every convex compact subset E of an $\mathrm{LCT_2VS}$ X.

A convex set $B\subset E$ is called a support of E if $[x,y]\subset B$ for every open segment $(x,y)\subset E$ satisfying $(x,y)\cap B\ne\varnothing$. We now denote the collection of all closed supports of $E\subset X$ by $\mathcal{S}(E)$ and go on to prove

THEOREM 17.16 (Krein Milman). If E is a convex compact subset of an $\mathrm{LCT_2VS}$ X, then $\mathcal{E}(B)\ne\varnothing\ \forall B\in\mathcal{S}(S)$ and $E=\overline{\{\mathcal{E}(E)\}_c}$.

PROOF. Fix any $B\in\mathcal{S}(E)$. Then $B\in\mathcal{S}(B)\ne\varnothing$ and $\mathcal{S}(B)$ is partially ordered by reverse inclusion $A_1<A_2$ iff $A_2\subset A_1$. Since B is compact, it follows (Appendix t.7) that every totally ordered set $\{A_\alpha:\alpha\in\mathcal{Q}\}\subset\mathcal{S}(B)$ has an upper bound $A=\bigcap_{\mathcal{Q}} A_\alpha\in\mathcal{S}(B)$. Therefore (Zorn's lemma) $\mathcal{S}(B)$ possesses a maximal element B_0 that we claim is an extreme point of B. If $a,b\in B_0$ are distinct, then (Corollary 15.9) $f(a)\ne f(b)$ for some $f\in X'$. We can, of course, write $f=f_{\mathrm{R_e}}+if_{\mathrm{I_m}}$ for $f_{\mathrm{R_e}},f_{\mathrm{I_m}}\in L(X;\mathbb{R})$ and use the compactness of B_0 to define $\alpha_0=\inf_{B_0} f_{\mathrm{R_e}}(x)$. A little computation shows that $\alpha_0\in\mathcal{E}\{f_{\mathrm{R_e}}(B_0)\}\subset\mathbb{R}$ or, what is equivalent, $\alpha_0\in\mathcal{S}\{f_{\mathrm{R_e}}(B_0)\}$. This necessitates that $B_0=B_0\cap f_{\mathrm{R_e}}^{-1}(\alpha_0)\in\mathcal{S}(B)$. Specifically, $x_0\in[x,y]\cap\tilde{B}_0$ for $[x,y]\subset B$ implies

$[x, y] \subset B_0$ and $f_{R_e}(z) \geq \alpha_0 = f_{R_e}(x_0)\ \forall z \in [x, y]$; therefore $f_{R_e}(z) = \alpha_0\ \forall z \in [x, y]$ and $[x, y] \subset \tilde{B}_0$. The ensuing contradiction $B_0 \subsetneqq \tilde{B}_0 \in \mathscr{S}(B)$ [by virtue of $f_{R_e}(a) \neq f_{R_e}(b)$] therefore proves that $\aleph(B_0) = 1$ and accordingly establishes that $B_0 \in \mathfrak{S}(B)$. Since $\mathfrak{S}(B) \neq \varnothing\ \forall B \in \mathscr{S}(E)$ and $\mathfrak{S} \in \mathscr{S}(E)$, there follows $\varnothing \neq \mathfrak{S}(E) \subset E$. The properties of E also ensure (Theorem 13.9) that $\overline{\{\mathfrak{S}(E)\}_c} \subset E$. For the reverse inclusion let $K = \overline{\{\mathfrak{S}(E)\}_c}$ and suppose that $x_0 \in E - K$. Viewing X as a real vector space ensures (Corollary 17.6) that $f(K) \cap f(x_0) = \varnothing$ for some $f \in L(X; \mathbb{R})$. If $f(x_0) > \sup_K f(x)$, define $B = \{y \in E : f(y) = \sup_E f(x)\}$ (for $f(x_0) < \inf_K f(x)$ take $B = \{y \in E : f(y) = \inf_E f(x)\}$). Then $B \in \mathscr{S}(E)$, which is impossible, since $B \subset E - K$ and $\mathfrak{S}(B) \neq \varnothing$. ■

THEOREM 17.17. If E is a compact subset of an LCT$_2$VS X such that $\overline{E_c}$ is compact, then $\mathfrak{S}(\overline{E_c}) \subset E$.

Proof. Given any closed, abs convex nbd $u(0)$, one has $E \subset \bigcup_{n=1}^{n} \{u + x_i\}$ for some $\{x_1 \cdots x_n\} \subset E$. The sets $A_i = \overline{(E \cap \{u + x_i\})_c}$ ($i = 1, 2, \ldots, n$) are convex, compact, and [Problem 13C(c)] satisfy $\overline{E_c} = \left\{\bigcup_{i=1}^{n} A_i\right\}_c$. Now every $x \in E_c$ can be written $x = \sum_{i=1}^{n} \alpha_i y_i$, where $y_i \in A_i \subset u + x_i$ and $\sum_{i=1}^{n} \alpha_i = 1$ for $\alpha_i \geq 0$. Therefore $x \in \mathfrak{S}(\overline{E_c})$ is some y_{i_0} ($1 \leq i_0 \leq n$) and $x \in E + u$. Since E is closed and $u(0)$ is arbitrary, $x \in E$ as required. ■

PROBLEMS

17A. $B\mathfrak{L}_{\mathscr{S}}(X, Y)$ Revisited

It is assumed here that X, Y are LVTVSps.

(a) Show that $\overline{L(X, Y)} = \mathfrak{L}_p(X, Y)$ if X is T_2. [*Hint*: Let $f_0 + w_{B,u(0)}(0)$ be any $\mathfrak{T}_p$-nbd of $f_0 \in \mathfrak{L}(X, Y)$, where $\mathfrak{B} = \{x_1 \cdots x_n\}$ is linearly independent in X. Then each $f_i : [x_1 \cdots x_n] \to \mathbb{F}$ has an extension $F_i \in X'$ such that $F_i(x_j) = \delta_{i,j}$. Let $y_i = f_0(x_i)$

$$\sum_{j=1}^{n} \alpha_j x_j \rightsquigarrow \alpha_i$$

and define $g_i : X \longrightarrow Y$, $x \rightsquigarrow F_i(x) y_i$ $\forall 1 \leq i \leq n$. Then $g = \sum_{i=1}^{n} g_i \in \{f_0 + w_{B,u}\} \cap L(X, Y)$.]

(b) Suppose $X = \bigcup_{\mathscr{S}} B$ is T_2 and Y is complete, T_2. Then $L_{\mathscr{S}}(X, Y)$ is complete if $\mathfrak{L}(X, Y) \bigcap_{\mathscr{S}} C(B, Y) \subset L(X, Y)$. [*Hint*: Every $L_{\mathscr{S}}(X, Y)$-Cauchy filter $\mathcal{G}$ is

$\mathfrak{T}_p$-Cauchy and convergent to some $f \in \mathfrak{L}_p(X, Y)$ (Example 13.27, Problem 6A). Use Theorem 7.9 to get $f \in \bigcap_{\mathcal{S}} C(B, Y)$.]

17B. Continuous Duals of {Semi-} Normed Spaces

Notation as in Corollary 13.25 for the seminormed space (X, p).

(a) Every $f \in X'_p$ satisfies $|||f||| \leq 1$ iff $|f(x)| \leq p(x)$ on X. Therefore $\sup_{|||f||| \leq 1} |f(x)| \leq p(x)\ \forall x \in X$. [*Hint*: $|f(x)| \leq 1$ if $p(x) \leq 1$ (since $\sup_{p(x) \leq 1} |f(x)| \leq 1$ for $|||f||| \leq 1$) and $|||f||| < 1$ if $|f(x)| \leq p(x)$ (otherwise $1 < \sup_{p(x) \leq 1} |f(x)| \leq \sup_{p(x) \leq 1} p(x)$).]

(b) If p is a norm on X, then $(X, p) \underset{x \rightsquigarrow J_x(f) = f(x)}{\rightarrow} (X'', |||\ |||)$ is a (usually into) congruence. *Hint*: $|||J_x||| = \sup_{|||f||| \leq 1} |f(x)| \leq p(x)\ \forall x \in X$. If $x \neq 0$, some $F \in X'$ with $|||F||| = 1$ satisfies $|F(x)| = p(x) \neq 0$. Therefore $|||J_x||| \geq |F(x)| = p(x)$.]

(c) A subset B of a NLS $(X, \|\ \|)$ is bounded iff $f(B)$ is bounded $\forall f \in X'$. [*Hint*: By Corollary 13.29 and (b) above $\sup_{x \in B} |J_x(f)| = \sup_{x \in B} |f(x)|\ \forall x \in X'$ implies $\sup_B \|x\| = \sup_B |||J_x||| < \infty$.]

17C. Generalized H.B. Extension Theorem

(a) Prove Theorem 17.7, assuming only that $p(x+y) \leq p(x) + p(y)$ and $p(\alpha x) = \alpha p(x)\ \forall x, y \in X$ and $\alpha \geq 0$. [*Hint*: Show that p determines a locally convex topology $\mathfrak{T}(p) \in T(X)$.]

(b) Let g be a scalar-valued function on a subset E of a vector space X. Then there is an $f \in X^*$ with $f/E = g$ iff there exists a seminorm p on X satisfying $\left|\sum_{i=1}^{n} \alpha_i g(x_i)\right| \leq p\left(\sum_{i=1}^{n} \alpha_i x_i\right)$ for every pair of sequences $\{x_1, \ldots, x_n\} \in X$ and $\{\alpha_1, \ldots, \alpha_n\} \in \mathbb{F}$. [*Hint*: If $f \in X^*$ and $f/E = g$, then $p(x) \leq |f(x)|$ meets the above requirements. Show that $f(x) = \sum_{i=1}^{n} \alpha_i g(x_i)\ \forall x = \sum_{i=1}^{n} \alpha_i x_i \in [E]$ is a well-defined member of $[E]^*$ satisfying $|f(x)| \leq p(x)$ on $[E]$. Proceed!]

17D. H.B. Theorems for Real TVSps

Assume here that X is a real TVS (i.e., a TVS over $\mathbb{R}$).

(a) Under the hypothesis of Theorem 17.3 prove that some $F \in X'$ satisfies $M \subset F^{-1}(0)$ and $F(\omega) \neq 0\ \forall \omega \in \Omega$. [*Hint*: $M \subset f^{-1}(0)$ for some $f \in X'$. Since Ω is open and convex, f does not change sign on Ω. Consider $F = f$ or $F = -f$.]

(b) Under the hypothesis of Corollary 17.5, prove that there is an $F \in X'$ and $\alpha > 0$ such that $F > \alpha$ on A and $F \leq \alpha$ on B. [*Hint*: $F(A) - F(B) = F(A - B) > 0$ for some $F \in X'$ and (since $B \neq \varnothing$) $\alpha = \inf_A F > -\infty$.]

(c) Show that $F < \alpha$ on B if B in (b) is open.

(d) Suppose X is locally convex and let A, B be disjoint convex sets in X. If A is closed and B is compact, there is an $F \in X'$ and $\alpha > 0$ such that $F < \alpha$ on A and $F > \alpha$ on B. [*Hint*: For each closed abs convex nbd $u(0)$ let $\mathcal{O}_u$ be the complement in X of $\overline{A + u}$. If $B \subset \bigcup_{i=1}^{n} \mathcal{O}_{u_i}$ and $v(0)$ is an open abs convex nbd satisfying $v + v \subset \bigcap_{i=1}^{n} \mathcal{O}_{u_i}$, then $\{A + v\} \cap \{B + v\} = \varnothing$.]

17E. An Exercise in Separation

(a) If A is a closed convex subset of a real LCTVS X and $x_0 \notin A$, then $f(x_0) > \sup_A f(x_0)$ for some $f \in X'$. [*Hint*: Use $B = \{x_0\}$ in Problem 17D(d).]

(b) If $\mathbb{F} = \mathbb{R}$ or $\mathbb{C}$ and A in (a) is also balanced, then $|F(x_0)| > \sup_A |F(a)|$ for some $F \in X'$. [*Hint*: Take $F(x) = f(x) - if(ix)$ for f in (a) and use $|F(x_0)| \geq \mathrm{R_e}\, F(x_0)| = |f(x_0)| > \alpha$.]

(c) Use (a) and (b) to prove that $\overline{E_c} = \{x_0 \in X : f(x_0) \geq \sup_E f(x) \ \forall f \in L(X; \mathbb{F})\}$ and $\overline{E_{bc}} = \{x_0 \in X : |f(x_0)| \geq \sup_E |f(x)| \ \forall f \in X'\}$ for every subset E of an LCTVS X.

17F. Convex Cones in Real-Ordered Spaces

A subset K of a real vector space X is called a cone if $\lambda K \subset K \ \forall \lambda > 0$. The cone K is pointed {convex} if $0 \in K$ {K is convex}. A pointed cone K is said to be salient if $\{0\}$ is the only vector subspace of X contained in K.

(a) A cone K is convex iff $K + K \subset K$, in which case $[K] = K - K$.

(b) Show that $K \cap (-K)$ is the largest subspace of X contained in a pointed convex cone K.

(c) A pointed cone is salient iff it contains no one dim subspace of X.

A real vector space X with a partial ordering $<$ is called ordered (OVS) if $x < y$ implies $x + z < y + z$ and $\lambda x < \lambda y \ \forall x, y, z \in X$ and $\lambda > 0$.

(d) $\mathfrak{P}(X) = \{x \in X : x > 0\}$ is a salient pointed convex cone in an OVS $(X, <)$.

(e) If K is a salient pointed convex cone in a vector space X, then $(X, <)$ is an OVS under $x < y$ iff $y - x \in K$. Show further that $<$ is the only ordering for which $K = \mathfrak{P}(X)$.

(f) An $f \in X^*$ for an OVS $(X, <)$ is called positive (denoted, $f \geq 0$) if $f(x) \geq 0\ \forall x \in \mathfrak{P}(X)$. Show that $\mathfrak{P}(X^*) = \{f \in X^* : f \geq 0\}$ is a salient pointed convex cone in X^*. Moreover, $X^* = \mathfrak{P}(X^*) \cup \{-\mathfrak{P}(X^*)\}$ iff $\dim X = 1$.

17G. Krein's Theorem

Assume throughout that M is a subspace of an OVS $(X, <)$.

(a) Suppose that $\forall x \in X$ there are $y_1, y_2 \in M$ such that $y_1 < x < y_2$. Then every $f \in \mathfrak{P}(M^*)$ has an extension to $\mathfrak{P}(X^*)$. [*Hint*: Show that $p(x) = \inf_M \{f(y) : y > x\}$ is finite-valued, has $p(x) \geq x\ \forall x > 0$, and satisfies the hypothesis of Problem 17C.]

(b) Why does the conclusion in (a) remain true if $\forall x \in X$ there is a $y \in M$ satisfying $x < y$?

(c) Let $X = \mathfrak{F}(\Omega; \mathbb{R})$ for a set Ω and suppose that M_0 is a vector subspace of M such that $\forall f \in M$ there exist $g_1, g_2 \in M_0$ satisfying $g_1 < f < g_2$. Then every $\psi \in \mathfrak{P}(M_0^*)$ has an extension $\Psi \in \mathfrak{P}(M^*)$.

(d) Suppose T is compact and T_2, and let $M = C(\Omega; \mathbb{R})$ in (c) carry $\| \|_0$. Then every $\psi \in \mathfrak{P}(M_0^*)$ has an extension $\Psi \in \mathfrak{P}(M') = M' \cap \mathfrak{P}(M^*)$. [*Hint*: Every $\Psi \in \mathfrak{P}(M)$ is $\| \|_0$-bounded (prove!).]

(e) (Krein) If X is a TVS and $\operatorname{Int} \mathfrak{P}(M) \neq \varnothing$ in X, then every $f \in \mathfrak{P}(M^*)$ has an extension $F \in \mathfrak{P}(M') = X' \cap \mathfrak{P}(X)$. [*Hint*: If $x_0 + u \subset \mathfrak{P}(X)$ for an X-balanced nbd u(u), then $x < \lambda x \in \mathfrak{P}(X)\ \forall x \in X$ and $\lambda > 0$ such that $x \in \lambda u$.]

17H. Analytic Vector-Valued Functions

Liouville's theorem, Cauchy's integral theorem, and Cauchy's integral formula all carry over to analytic vector-valued functions. A function $F : \Omega \to (X, \| \|)$, where Ω is a region in the complex plane and $(X, \| \|)$ is a NLS is said to be analytic {weakly analytic} in Ω if $F'(\lambda_0) = \lim_{\lambda \to \lambda_0} [F(\lambda) - F(\lambda_0)]/(\lambda - \lambda_0)$ exists {if $(fF)'(\lambda_0) = \lim_{\lambda \to \lambda_0} [fF(\lambda) - fF(\lambda_0)]/(\lambda - \lambda_0)$ exists $\forall f \in X'$} on each component of $\lambda_0 \in \Omega$.

(a) Every analytic function in Ω is weakly analytic with $(fF)'(\lambda_0) = f\{F'(\lambda_0)\}\ \forall f \in X'$ and $\lambda_0 \in \Omega$.

(b) If $F \in \mathfrak{F}(\Omega; X)$ is bounded (i.e., $\sup_{\mathbb{C}} \|F(\lambda)\| < \infty$) and entire (i.e., analytic throughout $\mathbb{C}$), then F is constant. [*Hint*: fF is constant and $f\{F(\lambda_1) - F(\lambda_2)\} = 0\ \forall f \in X'$. Use Corollary 17.9.]

(c) Let $F \in C(\Omega; X)$ for a Banach Space $(X, \| \|)$ and let $\Gamma \subset \Omega$ be a rectifiable curve joining $\alpha, \beta \in \Omega$. Subdivide Γ by points $\{\alpha = \lambda_1, \ldots, \lambda_n = \beta\} \in \Gamma$, let λ_i^* lie in the arc of Γ joining λ_i, λ_{i+1} and take $s = \max_{1 \leq i \leq n-1} |\lambda_{i+1} - \lambda_i|$. Show that it makes sense to define $\int_\Gamma F(\lambda)\, d\lambda \overset{\| \|}{=} \lim_{\substack{n \to \infty \\ \Delta \to 0}} \sum_{i=1}^{n-1} F(\lambda_i^*)(\lambda_{i+1} - \lambda_i)$.

(d) If $F \in \mathfrak{F}(\mathbb{C}; X)$ in (c) is analytic inside and continuous on a closed rectifiable curve Γ, then $\int_\Gamma F(\lambda)\, d\lambda = 0$. [*Hint*: $f(\int_\Gamma F(\lambda)\, d\lambda) = \int_\Gamma fF(\lambda)\, d\lambda = 0\ \forall f \in X'$.]

(e) Proceed as in analytic function theory to obtain $F^{(n)}(\lambda_0) = n!/2\pi i \int_\Gamma [F(\lambda)\, d\lambda]/[(\lambda - \lambda_0)^{n+1}]\ \forall n \in \mathbb{N}$ and λ_0 interior to Γ in (d).

Remark. A different definition of line integral is given in Example 20.6.

17I. More on $\mathbb{S}(X, \mu)$

Notation as in Example 13.17 and Problem 16A.

(a) If A is an atom, there is a unique complex number $\lambda = \lambda(A, f)$ for each $f \in \mathbb{S}(X, \mu)$ such that $\mu\{x \in A : f(x) \neq \lambda\} = 0$. [*Hint*: Fix any $f \in \mathbb{S}(X, \mu)$ and let $A(k, n) = A \cap f^{-1}\{\bar{S}_{1/n}(z_k)\}$, where $\{z_k : k \in \mathbb{N}\}$ is dense in $\mathbb{C}$ (with the Euclidean topology) and $\bar{S}_{1/n}(z_k) = \{z \in \mathbb{C} : |z - z_k| \leq 1/n\}$. Since $A = \bigcup_{k=1}^{\infty} A(k, n)$, each $n \in \mathbb{N}$ determines a k_n such that $\mu(A) = \mu\{A(k_n, n)\}$. If $B = \bigcap_{n=1}^{\infty} A(k_n, n)$, then $\mu(B) = \mu(A)$ and $\left[\text{since } f(B) \subset \bigcap_{n=1}^{\infty} \bar{S}_{1/n}(z_k)\right]$ $\aleph\{f(B)\} = 1.$]

(b) Let $\mathscr{A}(X, \mu)$ denote the collection of atoms in X and $\forall A \in \mathscr{A}(X, \mu)$, define $\Phi_A(f) = \int_A f\, d\mu\ \forall f \in \mathbb{S}(X, \mu)$. Then $\Phi_A \in \mathbb{S}'(X, \mu)$. [*Hint*: $|\Phi_A(f)| = |\lambda(A, f)\mu(A)| < \varepsilon$ for $d(f, 0) < \delta = \min\{\mu(A)/2, \varepsilon/2\}$.]

(c) $\Phi_A = \Phi_B$ for $A, B \in \mathscr{A}(X, \mu)$ iff $\mu(A \triangle B) = 0$. If $\mathscr{A}(X, \mu) \neq \varnothing$, there is a pairwise disjoint collection $\mathscr{A}' = \{A_n : n \in \mathbb{N}\} \subset \mathscr{A}(X, \mu)$ such that each $B \in \mathscr{A}(X, \mu)$ satisfies $\Phi_B = \Phi_{A_n}$ for some $A_n \in \mathscr{A}'$. [*Hint*: $(A, B) \in \mathfrak{R}$ iff $\Phi_A = \Phi_B$ defines an equivalence relation on $\mathscr{A}(X, \mu)$. If $\mathfrak{R}_A = \{B \in \mathscr{A}(X, \mu) : \Phi_A = \Phi_B\}$, then $\{\mathfrak{R}_A : \mu(A) \geq 1/n\}$ is finite and $\{\mathfrak{R}_A : A \in \mathscr{A}(X; \mu)\}$ is countable. Let $\mathscr{A}'$ contain a representative from each such $\mathfrak{R}_A$ $(A \in \mathscr{A}(X, \mu))$.]

(d) A subset $B \subset X$ is locally measurable if $B \cap E \in \mathfrak{R}_\sigma\ \forall E \in \mathfrak{R}_\sigma$. In this case $f\chi_B$ is measurable $\forall f \in \mathbb{S}(X, \mu)$ and $\psi_B(f) = \psi(f\chi_B) \in \mathbb{S}^*(X, \mu)\ \forall \psi \in \mathbb{S}^*(X, \mu)$. Prove that $\psi_B = 0\ \forall \psi \in \mathbb{S}'(X, \mu)$ if $\mathscr{A}(B, \mu) = \varnothing$. [*Hint*: If $\psi_B \neq 0$, then $\psi_B(f) = 1$ for some $f \in \mathbb{S}(X, \mu)$ and so $\chi \in \mathbb{S}'(X, \mu)$. Since $A = \{x \in X : |f\chi_B(x)| \neq 0\} \in \mathfrak{R}_\sigma$ contains no atoms, there is a sequence $\{A_n \in \mathfrak{R}_\sigma : n \in \mathbb{N}\} \subset A$ such that $\mu(A_n) = 2^{-n}\mu(A)\ \forall n \in \mathbb{N}$ (e.g., see Kelley and Namioka [52], p. 56). Let $g_n = 2^n f\chi_B\chi_{A_n}\ \forall n \in \mathbb{N}$. Then $d(g_n, 0) \leq 2^{-n}\mu(A)$ and $g_n \to 0$, but $|\psi(g_n)| \geq 1\ \forall n \in \mathbb{N}$ so that $\psi \notin \mathbb{S}'(X, \mu)$.]

(e) Use (b) and (d) to prove that $\mathbb{S}'(X, \mu) \neq 0$ iff $\mathscr{A}(X, \mu) \neq \varnothing$.

(f) If $\mathscr{A}(X, \nu) \neq \varnothing$, then $\mathbb{S}'(X, \mu) = [\Phi_A : A \in \mathscr{A}']$. [*Hint*: $B = X - \bigcup_{\mathscr{A}'} U_n$ is locally measurable and contains no atoms. Suppose $\psi \in \mathbb{S}'(X, \mu)$ and $f \in \mathbb{S}(X, \mu)$. Then $\mathscr{A}' = \{A_1, \ldots, A_n\}$ implies $\psi(f) = \sum_{k=1}^{n} \lambda(A_k, f)\, \psi(\chi_{A_k})$. If $\aleph(\mathscr{A}') = \aleph_0$, take $f_n = \sum_{k=1}^{n} f\chi_{A_k}$. From $\sum_{n=1}^{\infty} \mu(A_n)$, follows $f_n \xrightarrow{\text{measure}} f - f\chi_B$ and $\psi(f) = \sum_{n=1}^{\infty} \lambda(A_n, f)\psi(\chi_{A_n})$. Finally, take $g_n = [\psi(\chi_{A_n})]^{-1}\chi_{A_n}$ if $\psi(\chi_{A_n}) \neq 0$ and $g_n = 0$ otherwise. Then $g_n \xrightarrow{\text{measure}} 0$ and $\{n \in \mathbb{N} : \psi(\chi_{A_n}) \neq 0\}$ is finite.]

(g) $\mathfrak{S}'(X, \mu)$ is injective if $\mu(E)=0\ \forall E \in \mathfrak{R}_\sigma$ having $\mathscr{A}(E, \mu)=\varnothing$.

Remark. This problem is due to Murkerjee and Summers [66].

References for Future Study

On H.B. theorems and applications: Edwards [28]; Köthe [55].
On H.B. theorems and the problem of measure: Bachman and Narici [6].
On unique H.B. extensions: Phelps [19].

18. Duality Theory

If G is an LCT_2A group, G' is injective and $G''=G$ follows (Theorems 11.20 and 11.21). This was generalized to dual pairs of LCT_2A groups in Problem 11E. Since X' is injective for an LCT_2VS X, we consider LCT_2 topologies $\mathfrak{T} \in T(X')$ such that $(X', \mathfrak{T})'=X$. This too will be generalized to dual pairs $\langle X, Y\rangle$ of LCT_2VSps. Every LC topology $\mathfrak{T} \in T(Y)$ satisfying $(Y, \mathfrak{T})'=X$ lies between a weakest and strongest such topology. We next define $\mathfrak{T}_{\mathscr{S}}$-type (polar) topologies on X via collections $\mathscr{S}$ of bounded sets in Y. Every polar topology lies between a weakest and strongest polar topology on X. Furthermore, $\mathfrak{T} \in T(X)$ is locally convex iff $\mathfrak{T}$ is a polar topology of $\langle X, X'\rangle$. Virtually everything in this section serves to define, develop, and illustrate the properties of the above dual pair topologies.

DEFINITION 18.1. A system $\langle X, Y\rangle$ consisting of two vector spaces over the same field $\mathbb{F}$ and a bilinear mapping $\langle\ ,\ \rangle: X \times Y \to \mathbb{F}$ is called a dual pair. The dual pair $\langle X, Y\rangle$ is X-injective $\{Y$-injective$\}$ if $\langle X,\ \rangle = \{\langle x,\ \rangle : x \in X\} \subset Y^*$ is injective $\{\langle\ , Y\rangle = \{\langle\ , y\rangle : y \in Y\} \subset X^*$ is injective$\}$. Accordingly, $\langle X, Y\rangle$ is termed injective if it is both X and Y injective.

Remark. One calls $\sigma(X, Y)=\mathfrak{T}_{\mathscr{P}\{\langle\ , Y\rangle\}}$ the weak topology on X and $\sigma(Y, X)=\mathfrak{T}_{\mathscr{P}\{\langle X,\ \rangle\}}$ the weak topology on Y. Note (Theorem 15.1) that $\sigma(X, Y)$ $\{\sigma(Y, X)\}$ is T_2 iff $\langle X, Y\rangle$ is Y-injective $\{X$-injective$\}$.

The linear surjection $\underset{y \rightsquigarrow \langle\ , y\rangle}{\mu_X: Y \to \langle\ , Y\rangle \subset X^*}$ is an isomorphism iff $\langle X, Y\rangle$ is X-injective. Calling vector space isomorphs equal, there follows

THEOREM 18.2. If $\langle X, Y\rangle$ is a dual pair, then $(X, \sigma(X, Y))' = Y/\mu_X^{-1}(0)$. Furthermore, $(X, \sigma(X, Y))' = Y$ iff $\langle X, Y\rangle$ is X-injective.

Proof. If $f \in X' = (X, \sigma(X, Y))'$, then (Theorem 14.6) $S_{1,|f|}(0) = \{x \in X : |f(x)| < 1\}$ is a $\sigma(X, Y)$-nbd and so contains some nbd

$v(0; y_1, \ldots, y_n; \varepsilon)$. Since X^*, therefore X' is concentrating (Example 13.2), one has $f \in [\langle\ , y_1\rangle, \ldots, \langle\ , y_n\rangle] \subset \langle\ , Y\rangle = \mu_X(Y)$. Thus μ_X is surjective and $Y/\overset{-1}{\mu}_X(0)$ and X' are isomorphs. Assuredly, $Y/\overset{-1}{\mu}_X(0) = Y$ iff $\langle X, Y\rangle$ is X-injective. ■

Remark. By symmetry $(X, \sigma(Y, X))' = X/\overset{-1}{\mu}_Y(0)$ and $(Y, \sigma(Y, X))' = X$ iff $\langle X, Y\rangle$ is Y-injective.

COROLLARY 18.3. Let $E \subset Y$ for a Y-injective dual pair $\langle X, Y\rangle$. Then

(i) $\overline{[E]}^{\sigma(X,Y)} = Y$ iff $\langle X, Y\rangle$ is $[E]$-injective.
(ii) If $[E] \subsetneqq \overline{[E]}^{\sigma(Y,X)} = Y$, then $\sigma(X, [E]) \subsetneqq \sigma(X, Y)$.

Proof. (Corollary 17.13) $\overline{[E]}^{\sigma(Y,X)} = Y$ iff $f/\overline{[E]}^{\sigma(Y,X)} = 0$ implies $f = 0\ \forall f \in (Y, \sigma(Y, X))' = X$, which means $\langle x, \overline{[E]}^{\sigma(Y,X)}\rangle = 0$ implies $x = 0$ $\forall x \in X$. This is equivalent to $[E]$-injectivity. That (ii) holds is clear since $(X, \sigma(X, [E]))' = [E] \subsetneqq Y = (X, \sigma(X, Y))'$. ■

DEFINITION 18.4. A locally convex topology $\mathfrak{T} \in T(X)$ $\{\mathfrak{T} \in T(Y)\}$ is called a topology of the X-injective $\{Y$-injective$\}$ dual pair $\langle X, Y\rangle$ if $(X, \mathfrak{T})' = Y$ $\{$if $(Y, \mathfrak{T})' = X\}$.

Some simple but important observations should be made before proceeding.

Example 18.5. (a) $\sigma(X, Y)$ is the smallest topology of an X-injective dual pair $\langle X, Y\rangle$. In particular, $\sigma(X, X^*)$ and $\sigma(X, X')$ are the smallest topologies of $\langle X, X^*\rangle$ and $\langle X, X'\rangle$, respectively.

(b) One always has $(X^*, \sigma(X^*, X)) = \mathfrak{L}_p(X; \mathbb{F})$ for a vector space X (Definition 7.1). Since $\mathbb{F}$ with the usual Euclidean topology is complete, $(X^*, \sigma(X^*, X))$ is complete (Example 13.27). If $\langle X, Y\rangle$ is injective, then $Y = (X, \sigma(X, Y))' \subset X^*$ and $\sigma(X^*, X)/Y = \sigma(Y, X)$ showing that $(\bar{Y}^{\sigma(X^*,X)}, \sigma(X^*, X))$ is the completion of $(Y, \sigma(Y, X))$.

(c) If $(X, \|\ \|)$ is a NLS, then $\|\ \|$ is a topology of $\langle X, X'\rangle$ and (Corollary 13.25) $\|\|\ \|\|$ is a topology of $\langle X', X''\rangle$. It need not be true, however, that $\|\|\ \|\|$ is a topology of $\langle X', X\rangle$. Indeed, $X \subsetneqq (X', \|\|\ \|\|)'$ if $(X, \|\ \|)$ is incomplete [Problem 17B(c), (e)].

(d) $\sigma(X, Y)$-boundedness is equivalent to $\sigma(X, Y)$-total boundedness for a dual pair $\langle X, Y\rangle$. If $B \subset X$ is $\sigma(X, Y)$-bounded, then [Example 15.4(a)] $\langle B, y\rangle \subset \mathbb{F}$ is relatively compact and so totally bounded $\forall y \in Y$. Thus B is $\sigma(X, Y)$-totally bounded. The converse is clear by Theorem 13.20(i).

(e) For an injective dual pair $\langle X, Y\rangle$ $\dim X<\aleph_0$ iff $\dim Y<\aleph_0$ (Theorem 18.2). Therefore, $\sigma(X, Y)$ is normable iff $\dim X<\aleph_0$ (Theorems 13.12 and 15.5 and the fact that $\langle\ , Y\rangle=(X, \sigma(X, Y))'\subset X^*$ is concentrating). If Y above is also a Catg II TVS (e.g., $Y=X'$ for a NLS X), then $\sigma(X, Y)$-metrizibility is equivalent to $\dim X<\aleph_0$ (Problems 13G, 17B).

Certain topological notions are dual-pair dependent in that only the dual pair, and no particular topology of this pair, dominates. For example, $\mathcal{CA}(\Omega, \Sigma; Y)$ is $\langle Y, Y'\rangle$-dependent when Y is an LCT_2VS and $\Omega\in\Sigma$. This follows from $\mathcal{CA}(\Omega, \Sigma; Y, \mathfrak{T})=\text{wk}\ \mathcal{CA}(\Omega, \Sigma; Y, \mathfrak{T})=\mathcal{CA}(\Omega, \Sigma; Y, \sigma(Y, Y'))$ $\forall$ topology $\mathfrak{T}$ of $\langle Y, Y'\rangle$ [Theorem 16.6, Example 15.4(c)]. Two particularly important dual-pair dependent properties are closure and boundedness .

THEOREM 18.6. For any topology $\mathfrak{T}$ of an X-injective dual pair $\langle X, Y\rangle$

(i) $\bar{E}^{\mathfrak{T}}=\bar{E}^{\sigma(X,Y)}$ if $E\subset X$ is convex.

(ii) $B\subset X$ is $\mathfrak{T}$-bounded iff B is $\sigma(X, Y)$-bounded.

Proof. Evidently $\bar{E}^{\mathfrak{T}}\subset\bar{E}^{\sigma(X,Y)}$, and $\mathfrak{T}$-boundedness implies $\sigma(X, Y)$-boundedness by Example 18.5(a). If $x_0\notin\bar{E}^{\mathfrak{T}}$, then (Corollary 17.6) $\langle x_0, y_0\rangle\notin\overline{\langle E, y_0\rangle}$ for some $y_0\in Y=(X, \sigma(X, Y))'$ and $\{x_0+v(0; y_0; \varepsilon)\}\cap E=\varnothing$ for $\varepsilon=\inf_{x\in E}|\langle x_0, y_0\rangle-\langle x, y_0\rangle|>0$ ensures that $x_0\notin\bar{E}^{\sigma(X,Y)}$. To complete the proof of (ii), suppose B is $\sigma(X, Y)$-bounded and let $u(0)$ be any abs convex $\mathfrak{T}$-closed nbd. If p_u is the Minkowski functional of u and $X/\overset{-1}{p_u}(0)$ carries $\|x+\overset{-1}{p_u}(0)\|_{p_u}=p_u(x)$, then each $f\in(X/\overset{-1}{p_u}(0))'$ satisfies $f\nu=\langle\ ; y\rangle\in\langle\ , Y\rangle=X'$ for some $y\in Y$. Thus $\infty>\sup_{x\in B}|\langle x, y\rangle|=\sup_B|f\{\nu(x)\}|$ $\forall f\in(X/\overset{-1}{p_u}(0))'$ ensures [Problem 17B(c)] that $\nu(B)$ is $\|\ \|_{p_u}$-bounded and B is p_u-bounded. In particular, B is $\mathfrak{T}$-bounded, since $B\subset\lambda S_{1,p_u}(0)=\lambda u(0)$ for some $\lambda>0$. ■

COROLLARY 18.7. If $\mathfrak{T}$ is a complete topology of an X-injective dual pair $\langle X, Y\rangle$, then $(X, \mathfrak{T}')$ is complete for every topology $\mathfrak{T}'$ of $\langle X, Y\rangle$ finer than $\mathfrak{T}$.

Proof. $\mathfrak{T}'$ has a nbd base of abs convex closed sets. Each such set is also $\mathfrak{T}$-closed and Theorem 13.10 applies. ■

COROLLARY 18.8. If Y is a T_2VS, then $\mu\in\mathcal{CA}(\Omega, \Sigma; Y)$ is a bounded range for every topology of the dual pair $\langle Y, Y'\rangle$.

Proof. If $f \in Y'$, then (Appendix m.5) $f\mu \in \mathcal{CA}(\Omega, \Sigma; \mathbb{C})$ has a Jordan decomposition $f\mu = \nu_1 - \nu_2 + i(\nu_3 - \nu_4)$ by positive measures $\{\nu_k : 1 \le k \le 4\}$ on (Ω, Σ) and, since these ν_k are all bounded (otherwise, $\nu_k(E_n)\uparrow\infty$ for $\{E_n\} \in \Sigma$ contradicts $\infty > \nu_k(\bigcup_{n=1} E_n) \ge \nu_k(E_n)\ \forall n \in \mathbb{N}$), $\sup_{\Sigma} f\{\mu(E)\} < \infty$. Thus $f\{\mu(E) : E \in \Sigma\}$ is bounded $\forall f \in Y'$ and $\{\mu(E) : E \in \Sigma\}$ is $\sigma(Y, Y')$-bounded. ■

Polar Sets

If $\langle X, Y\rangle$ is a dual pair, each $y \in Y$ determines an X-seminorm $p_y(x) = |\langle x, y\rangle|$ whose closed sphere $\bar{S}_{1,p_y}(0) = \{x \in X : |\langle x, y\rangle| \le 1\}$ we may denote y°. In more general terms

DEFINITION 18.9. Let $A \subset X$ and $B \subset Y$ for a dual pair $\langle X, Y\rangle$. Then $A^{\circ} = \{y \in Y : \sup_{x \in A} |\langle x, y\rangle| \le 1\}$ is called the polar of A in Y and $B^{\circ} = \{x \in X : \sup_{y \in B} |\langle x, y\rangle| \le 1\}$ is called the polar of B in X.

If $\langle Y, Z\rangle$ is also a dual pair, one calls $A^{\circ\circ} = (A^{\circ})^{\circ} \subset Z$ the bipolar of A in Z (abbreviated, "bipolar of A" when $Z = X$). Note that the bipolar $A^{\circ\circ}$ of A always contains A.

The basic properties of polar sets are summarized in

THEOREM 18.10. For a dual pair $\langle X, Y\rangle$

(i) $A^{\circ} \subset Y$ is abs convex and $\sigma(X, Y)$-closed $\forall A \subset X$;
(ii) $B^{\circ} \subset A^{\circ}$ for $A \subset B$ in X. Thus $A^{\circ} = A^{\circ\circ\circ}\ \forall A \subset X$;
(iii) $\{\bigcup_{\mathcal{A}} A_\alpha\}^{\circ} = \bigcap_{\mathcal{A}} A_\alpha^{\circ}$ for $\{A_\alpha : \alpha \in \mathcal{A}\} \subset X$;
(iv) $(\lambda A)^{\circ} = 1/|\lambda| A^{\circ}\ \forall \lambda \ne 0$ and $A \subset X$. In particular, $A \subset X$ is $\sigma(X, Y)$-bounded iff $A^{\circ} \subset Y$ is absorbent.

Proof. (i) $A^{\circ} = \bigcap_{x \in A} \bar{S}_{1,p_x}(0)$ is clearly abs convex. Since every p_x $(x \in X)$ is $\sigma(Y, X)$-continuous, $\bar{S}_{1,p_x}(0) = \overline{S_{1,p_x}(0)}^{\sigma(Y,X)}$ (Theorem 14.10) and A° is seen to be $\sigma(Y, X)$-closed in Y. Assertion (ii) follows from (i) and the fact that $A \subset A^{\circ\circ}\ \forall A \subset X$. The proof of (iii) is straightforward, as in the first part of (iv) (since $|\langle\lambda x, y\rangle| = |\lambda||\langle x, y\rangle| = |\langle x, \lambda y\rangle|\ \forall (x, y) \in X \times Y$). Now $u(0)$ is a $\sigma(X, Y)$-nbd of 0 iff $u(0) = \bigcap_{i=1}^{n} y_i^{\circ}$ for some $\{y_1, \ldots, y_n\} \in Y$. If $A^{\circ} \subset Y$

is absorbent, $\bigcup_{i=1}^{n} y_i \in \lambda A^{\bigcirc}$ for some $\lambda > 0$. Therefore $(1/\lambda)A \subset (1/\lambda)A^{\bigcirc\bigcirc} = (\lambda A^{\bigcirc})^{\bigcirc} \subset \bigcap_{i=1}^{n} y_i^{\bigcirc} = u$ and A is $\sigma(X, Y)$-bounded. Conversely, if A is $\sigma(X, Y)$-bounded and $y \in Y$, then $A \subset \lambda y^{\bigcirc}$ and $y \in \lambda A^{\bigcirc}$ $\forall |\lambda| \geq \varepsilon$ for some $\varepsilon > 0$. ■

THEOREM 18.11. Let $\langle X, Y\rangle$ be an X-injective dual pair and suppose that $X \subset Z$ for some vector subspace Z of Y^*. Then $A^{\bigcirc\bigcirc} = \overline{A_{bc}}^{\sigma(Z,Y)}$ is the bipolar in Z of $A \subset X$.

Proof. Evidently $\overline{A_{bc}}^{\sigma(Z,Y)} \subset A^{\bigcirc\bigcirc}$ in Z by Theorem 18.10(i). If $z_0 \notin \overline{A_{bc}}^{\sigma(Z,Y)}$, then (Corollary 17.6) $\langle z_0, y_0\rangle \notin \overline{\langle A_{bc}, y_0\rangle}$ for some $y_0 \in Y = (Z, \sigma(Z, Y))'$ and $\sup_{z \in A_{bc}} |\langle z, y_0\rangle| < |\langle z_0, y_0\rangle|$ follows from $\langle A_{bc}, y_0\rangle$ being abs convex. If

$$y_1 = \frac{2y_0}{|\langle z_0, y_0\rangle| + \sup_{A_{bc}} |\langle z, y_0\rangle|}$$

then $\sup_{A_{bc}} |\langle z, y_1\rangle| < 1 < |\langle z_0, y_0\rangle|$ ensures that $y_1 \in A^{\bigcirc}$ in Y and $z_0 \notin A^{\bigcirc\bigcirc}$ in Z. ■

COROLLARY 18.12. If $\{A_\alpha : \alpha \in \mathscr{Q}\}$ is a collection of abs convex $\sigma(X, Y)$-closed subsets of X for an X-injective dual pair $\langle X, Y\rangle$, then $\{\bigcap_{\mathscr{Q}} A_\alpha\}^{\bigcirc} = \overline{\{\bigcup_{\mathscr{Q}} A_\alpha^{\bigcirc}\}_{bc}}^{\sigma(Y,X)}$.

Proof. $\{\bigcup_{\mathscr{Q}} A_\alpha^{\bigcirc}\}^{\bigcirc} = \bigcap_{\mathscr{Q}} A_\alpha^{\bigcirc\bigcirc} = \bigcap_{\mathscr{Q}} A_\alpha$ so that $\overline{\{\bigcup_{\mathscr{Q}} A_\alpha^{\bigcirc}\}_{bc}}^{\sigma(Y,X)} = \{\bigcup_{\mathscr{Q}} A_\alpha^{\bigcirc}\}^{\bigcirc\bigcirc} = \{\bigcap_{\mathscr{Q}} A_\alpha\}^{\bigcirc}$. ■

Example 18.13. Given a set $M \subset X$ for a dual pair $\langle X, Y\rangle$, one calls $M^\perp = \{y \in Y : \langle x, y\rangle = 0\ \forall x \in M\}$ the vector subspace of Y orthogonal to M. It is easily verified that $[M]^{\bigcirc} = [M]^\perp = M^\perp \subset M^{\bigcirc}$ in Y. If X is a TVS, one also has $[M]^{\bigcirc\bigcirc} = \overline{[M]} = M^{\perp\perp}$ in X.

(a) If X above is an LCTVS and M is a vector subspace of X, then M' and $X'/M^{\bigcirc}$ are vector-space isomorphs. Simply note (Corollary 17.10) that $\eta : X' \to M'$, $f \rightsquigarrow f/M$, is a linear surjection having $\eta^{-1}(0) = M^{\bigcirc}$.

(b) (Example 15.15) $M^{\bigcirc} = M^\perp = \mathfrak{L}_M(X, Y)$ for the dual pair $\langle X, \mathfrak{L}(X, Y)\rangle$ under $\langle x, f\rangle = f(x)$. It X, Y are TVSps, then $M^{\bigcirc} = L_M(X, Y)$ is the vector space isomorph of $L(X/M, Y)$. In particular, $(X/M, \mathfrak{T}_{\mathfrak{s}(\nu)})' = M^{\circ} \subset X'$ for $\langle X, X' \rangle$ and $\langle X/\overline{[M]}, ||| \; |||)' = \overline{[M]}^{\circ} \subset X'$ when $(X, \| \; \|)$ is a NLS.

Our next two theorems are basic and are referred to frequently.

THEOREM 18.14. For a TVS X

(i) u° is the same set in X' and X^* $\forall X$-nbd $u(0)$. If $\mathfrak{N}_0$ is an X-nbd base at 0, then $X' = \bigcup_{\mathfrak{N}_0} u^{\circ}$ for $u^{\circ} \subset X'$ (or for $u^{\circ} \subset X^*$).

(ii) $\Phi \subset X'$ is equicontinuous iff $\Phi \subset u^{\circ}$ in X' for some X-nbd $u(0)$.

Proof. Surely $u^{\circ} \cap X' \subset u^{\circ} \cap X^*$, whereas the reverse inclusion holds by virtue of Theorem 13.11. Also, $\Phi \subset X'$ is equicontinuous iff $\forall \varepsilon > 0$ there is an X-nbd $v(0)$ such that $\sup_{v(0)} |\phi(x)| < \varepsilon \; \forall \phi \in \Phi$ which, of course, means that $\Phi \subseteq [(1/\varepsilon)v]^{\circ}$. ■

THEOREM 18.15. (Alaoglu–Bourbaki) Let X be a TVS. Then $u^{\circ} \subset X'$ is $\sigma(X', X)$-compact $\forall$-nbd $u(0) \subset X$, and every equicontinuous set $\Phi \subset X'$ is $\sigma(X', X)$-relatively compact.

Proof. We know from $u \subset u^{\circ\circ}$ in X that $u^{\circ\circ}$ is absorbing. Therefore $u^{\circ} \subset X^*$ is $\sigma(X^*, X)$-bounded and $\sigma(X^*, X)$-compact [Theorem 10.10(iv), (i) and Example 18.5(d)]. It remains only to invoke Theorem 18.14 after observing that $\sigma(X', X) = \sigma(X^*, X)/X'$. ■

Remark. If $(X, \| \; \|)$ is a NLS, then $\bar{S}_{1, ||| \; |||}(0) = \{\bar{S}_{1, \| \; \|}(0)\}^{\circ} \subset X'$ is $\sigma(X', X)$-compact.

Transpose of a Linear Mapping

Our discussion here is to some extent motivated by the following simple assertions:

THEOREM 18.16. Every $F \in \mathfrak{L}(X, Y)$ is $\sigma(X, X^*) - \sigma(Y, Y^*)$ continuous. If X and Y are TVSps, then $F \in L(X, Y)$ is $\sigma(X, X') - \sigma(Y, Y')$ continuous.

Proof. For any $\sigma(Y, Y^*)$-nbd $v(0) = v(f_1, \ldots, f_n : \varepsilon)$ $(f_i \in Y^*)$, one has $f_i F \in X^*$ with $F\{v(0; f_1F, \ldots, f_nF; \varepsilon\} \subset v(0)$. Obviously $f_iF \in X'$ if $f_i \in Y'$ and $F \in L(X, Y)$. ■

Remark. The converse, of course, fails, since the identity operator on a TVS $(X, \mathfrak{T})$ is $\sigma(X, X') - \sigma(X, X')$ continuous but not $\sigma(X, X') - \mathfrak{T}$ continuous for $\mathfrak{T} \supsetneq \sigma(X, X')$.

Despite the above remark, weak continuity has its effect on linear mappings. This is made clear shortly. For now, note that each $F \in \mathcal{L}(X, Y)$ determines a unique mapping $F^t \in \mathcal{L}(Y^*, X^*)$ defined for each $y^* \in Y^*$ by $\{F^t(y^*)\}(x) = y^*\{F(x)\} \forall x \in X$. Actually, each $F \in \mathcal{L}(X, Y)$ turns $\langle X, Y^*\rangle$ into a dual pair under $\langle x, y^*\rangle = \langle F(x), y^*\rangle = \langle x, F^t y^*\rangle$.

DEFINITION 18.17. Let $F \in \mathcal{L}(X_1, X_2)$, where $\langle X_1, Y_1\rangle$ and $\langle X_2, Y_2\rangle$ are dual pairs. The unique mapping $F^t \in \mathcal{L}(Y_2, X_1^*)$ defined for each $y_2 \in Y_2$ by $\langle x, F^t(y_2)\rangle = \langle F(x), y_2\rangle \; \forall x \in X$ is called the transpose of F.

Remark. $F^t \in \mathcal{L}(Y_2, X_1^*)$ is $\sigma(Y_2, X_2) - \sigma(X_1^*, X_1)$ continuous, since (Theorem 15.1) $xF^t : (Y_2, \sigma(Y_2, X_2)) \longrightarrow \mathbb{F}$, $y \rightsquigarrow \{F^t(y)\}(x)$ is continuous $\forall x \in X_1$ (simply note that $v(0; F(x); \varepsilon/2)$ is a $\sigma(Y_2, X_2)$-nbd such that $(xF^t)\{v(0; F(x); \varepsilon/2)\} \subset S_\varepsilon(0) \; \forall \varepsilon > 0$).

If $\langle X_i, Y_i\rangle$ $(i = 1, 2)$ are X_i-injective, then Y_i is a subspace of X_i^* and $F^t \in \mathcal{L}(X_2^*, X_1^*)$ has a restriction to $\mathcal{L}(Y_2, X_1^*)$. Suppose $F \in L(X_1, X_2)$, where X_i carries $\sigma(X_i, Y_i)$. Then $F^t(y) = \langle F(\ ), y\rangle \in X_1' = Y_1 \; \forall y \in Y_2 = X_2'$ and $F^t \in L(X_2', X_1') = L(Y_2, Y_1)$. The converse is also true. If $F^t \in \mathcal{L}(Y_2, Y_1)$ and $v(0) = \{x \in X_2 : \sup_{1 \le i \le n} |\langle x, y_i\rangle| < \varepsilon\}$ for $\{y_1, \ldots, y_n\} \in Y_2$, then $u(0) = \{x \in X_1 : \sup_{1 \le i \le n} |\langle x, F^t(y_i)\rangle| < \varepsilon\}$ is a $\sigma(X_1, Y_1)$-nbd such that $F\{u\} \subset v$. All this is summarized in the first part of .

THEOREM 18.18. Let $F \in \mathcal{L}(X_1, X_2)$, where $\langle X_i, Y_i\rangle$ are X_i-injective dual pairs. Then

(i) $F^t \in \mathcal{L}(Y_2, Y_1)$ iff F is weakly continuous (i.e., $\sigma(X_1, Y_1) - \sigma(X_2, Y_2)$ continuous).

(ii) If $\langle X_i, Y_i\rangle$ are injective and F is weakly continuous, then $F^{tt} = F$ and $F^t \in \mathcal{L}(Y_2, Y_1)$ is weakly continuous.

Proof. (ii) follows from (i) since $F^t \in \mathcal{L}(Y_2, Y_1)$ implies $F^{tt} \in \mathcal{L}(X_1, Y_2^*)$ and $F^{tt}(x) = \langle x, F^t(\ )\rangle = \langle F(x),\ \rangle \in Y_2' = X_2 \; \forall x \in X_1$ ensures that F^t is weakly continuous and $F^{tt} = F$. ■

THEOREM 18.19. If $F \in \mathcal{L}(X_1, X_2)$ for dual pairs $\langle X_1, Y_1\rangle$ and $\langle X_2, Y_2\rangle$, then

(i) $\{F(E)\}^\circ = (F^t)^{-1}(E^\circ)$ in $Y_2 \; \forall E \subset X_1$ and E° in X_1^*.

(ii) $(F^t)^{-1}(0) = \{F(X_1)\}^\circ$ in Y_2. Thus F^t is injective iff $F(X_1)$ is $\sigma(X_2, Y_2)$-dense in X_2.

Proof. (i) is clear since $\{F(E)\}^{\circ}=\{y\in Y_2:\sup_{x\in E}|\langle F(x),y\rangle|\le 1\}=\{y\in Y_2:\sup_{x\in E}|\langle x,F^t(y)\rangle|\le 1\}=\{y\in Y_2:F^t(y)\in E^{\circ}$ in $X_1^*\}=(F^t)^{-1}(E^{\circ})$. Taking $E=X_1$ in (i) gives $(F^t)^{-1}(0)=\{F(X_1)\}^{\circ}$ in Y_2 as asserted in (ii). Accordingly, $0=(F^t)^{-1}(0)=\{F(X_1)\}^{\circ}$ iff $X_2=\{F(X_1)\}^{\circ\circ}=\overline{F(X_1)}^{\sigma(X_2,Y_2)}$. ■

Remark. If F is weakly continuous, $F^t\in\mathfrak{L}(Y_2,Y_1)$ so that the above holds for polars in Y_1.

Polar Topologies

If $\mathscr{S}$ is a collection of $\sigma(X,Y)$-bounded sets in X for a dual pair $\langle X,Y\rangle$, then $\mathscr{S}^{\circ}=\{B^{\circ}:B\in\mathscr{S}\}$ consists of abs convex sets in Y and (Corollary 14.4) there is a coarsest locally convex topology $\mathfrak{T}_{\mathscr{S}}\in T(Y)$ under which every $B^{\circ}\in\mathscr{S}^{\circ}$ is a $\mathfrak{T}_{\mathscr{S}}$-nbd of $0\in Y$.

In most cases it happens that $\mathscr{S}$ satisfies

(*) $\forall B_1,B_2\in\mathscr{S}$ one has $\lambda B_1\in\mathscr{S}\ \forall\lambda>0$ and $B_1\cup B_2\subset B_3$ for some $B_3\in\mathscr{S}$.

Condition (*) ensures that $[\bigcup_{\mathscr{S}}B]=\{\bigcup_{\mathscr{S}}B\}_{bc}$ and that $\mathscr{S}^{\circ}$ itself is a $\mathfrak{T}_{\mathscr{S}}$-nbd base at $0\in Y$ (letting $\bigcup_{i=1}^{n}B_{\alpha_i}\subset B_{\alpha_{n+1}}\in\mathscr{S}$ and taking $\varepsilon<\min_{1\le i\le n}\varepsilon_i$ gives $B^{\circ}\subset\bigcap_{i=1}^{n}\varepsilon_iB^{\circ}_{\alpha_i}$ for $B=(1/\varepsilon)B_{\alpha_{n+1}}\in\mathscr{S}$). Two other arguments can be made for assuming a priori that $\mathscr{S}$ satisfies (*). First $\{\mathscr{S}\}_{\xi}=\left\{\bigcup_{i=1}^{n}\lambda_iB_{\alpha_i}:\lambda_i\ne 0\text{ and }B_{\alpha_i}\in\mathscr{S}\right\}$ satisfies (*) and $\mathfrak{T}_{\{\mathscr{S}\}_{\xi}}=\mathfrak{T}_{\mathscr{S}}$ on Y (cf Theorem 18.22). Second, identifying Y with $(X,\sigma(X,Y))'=L(X;\mathbb{F})$ for an X-injective dual pair $\langle X,Y\rangle$ returns us to the framework of Theorem 13.26, in which case we view $Y_{\mathscr{S}}$ as the LCTVS $L_{\mathscr{S}}(X;\mathbb{F})$ having base nbds $w_{B,\varepsilon}(0)=\{y\in Y:\sup_{B}|\langle x,y\rangle|<\varepsilon\}=\varepsilon B^{\circ}$.

DEFINITION 18.20. Let $\mathscr{S}$ bc a collection of $\sigma(X,Y)$-bounded sets that satisfies (*). The locally convex topology $\mathfrak{T}_{\mathscr{S}}\in T(Y)$ having nbd base $\mathscr{S}^{\circ}$ is called the topology of $\mathscr{S}$-convergence on Y. Any $\mathfrak{T}_{\mathscr{S}}$ topology defined in this manner is called a polar topology on Y.

THEOREM 18.21. For $\langle X,Y\rangle$ and $\mathscr{S}$ above

(i) $\mathfrak{T}_{\mathscr{S}}$ is T_2 iff $X=\overline{[\bigcup_{\mathscr{S}}B]}^{\sigma(X,Y]}$ and $\langle X,Y\rangle$ is X-injective.

(ii) $(Y,\mathfrak{T}_{\mathscr{S}})'=\bigcup_{\mathscr{S}}B^{\circ\circ}$ for bipolars in Y^*.

Proof. First observe that $\mathfrak{T}_{\mathcal{S}}$ is determined by the collection of seminorms $\{q_B(y) = \sup_B |\langle x, y\rangle| : B \in \mathcal{S}\}$. If $\mathfrak{T}_{\mathcal{S}}$ is T_2, then $\{0\} = \bigcap_{\mathcal{S}} B^{\circ}$ yields $X = (\bigcap_{\mathcal{S}} B^{\circ})^{\circ} = (\bigcup_{\mathcal{S}} B)^{\circ\circ} = \overline{(\bigcup_{\mathcal{S}} B)}_{bc}^{\sigma(X,Y)} = \overline{[\bigcup_{\mathcal{S}} B]}^{\sigma(X,Y)}$ and $\langle X, Y\rangle = 0$ implies $q_B(y) = 0\ \forall\ B \in \mathcal{S}$; therefore $y = 0$ (Theorem 14.8). In the other direction $0 = q_B(y)\ \forall B \in \mathcal{S}$ implies $0 = \langle \bigcup_{\mathcal{S}} B, y\rangle$ and $0 = \langle \overline{[\bigcup_{\mathcal{S}} B]}^{\sigma(X,Y)}, y\rangle$ by Example 15.4(c). Therefore $y = 0$ and $\mathfrak{T}_{\mathcal{S}}$ is T_2 if $X = \overline{[\bigcup_{\mathcal{S}} B]}^{\sigma(X,Y)}$ and $\langle X, Y\rangle$ is X-injective. The proof of (ii) follows by Theorem 18.14. ∎

Of considerable utility is the malleability of $\mathcal{S}$ in the manner prescribed below.

THEOREM 18.22. Given a family $\mathcal{S}$ of $\sigma(X, Y)$-bounded sets in X,

(i) for any other collection $\mathfrak{R}$ of $\sigma(X, Y)$-bounded sets $\mathfrak{T}_{\mathfrak{R}} \subset \mathfrak{T}_{\mathcal{S}}$ iff each $R \in \mathfrak{R}$ is contained in some $B^{\circ\circ} \in \mathcal{S}^{\circ\circ}$.

(ii) $\mathfrak{T}_{\mathcal{S}}$ remains unchanged if $\mathcal{S}$ is replaced by any of the following:

(a) $$\mathcal{S}_0 = \{\text{subsets of members of } \mathcal{S}\}$$

(b) $$\mathcal{S}_{\xi} = \left\{ \bigcup_{i=1}^{n} B_{\alpha_i} : B_{\alpha_i} \in \mathcal{S} \right\}$$

(c) $$\mathcal{S}_{\bullet} = \{\lambda B : B \in \mathcal{S} \text{ and } \lambda \in \mathbb{F}\}$$

(d) $$\mathcal{S}_b = \{B_b : B \in \mathcal{S}\}$$

(e) $$\bar{\mathcal{S}}^{\sigma(X,Y)} = \{\bar{B}^{\sigma(X,Y)} : B \in \mathcal{S}\}$$

(f) $$\mathcal{S}^{\circ\circ} = \{B^{\circ\circ} \subset X : B \in \mathcal{S}\} = \bar{\mathcal{S}}_{bc}^{\sigma(X,Y)}$$

Proof. Both assertions are simple consequences of Theorems 18.10 and 18.11. In particular, $B^{\circ} = (B^{\circ\circ})^{\circ} = (\bar{B}_{bc}^{\sigma(X,Y)})^{\circ} \subset (\bar{B}^{\sigma(X,Y)})^{\circ} \subset B^{\circ} = (B_b)^{\circ}\ \forall b \in \mathcal{S}$. ∎

We now zero in on specific polar topologies.

DEFINITION 18.23. Let $\langle X, Y\rangle$ be a dual pair. On Y define and denote

(i) the weak topology $\sigma(Y, X) = \mathfrak{T}_{\mathcal{S}_{\sigma}}$, where $\mathcal{S}_{\sigma} = \{\text{finite sets in } X\}$;

(ii) the strong topology $\beta(Y, X) = \mathfrak{T}_{\mathcal{S}_{\beta}}$, where $\mathcal{S}_{\beta} = \{\sigma(X, Y)\text{-bounded subsets of } X\}$;

(iii) the Mackey topology $\tau(Y, X) = \mathfrak{T}_{\mathcal{S}_\tau}$, where $\mathcal{S}_\tau = \{$abs convex $\sigma(X, Y)$-compact subsets of $X\}$.

A few observations worth noting are collected below:

Example 18.24. (a) If $X = \bigcup_{\mathcal{S}} B^{\circ\circ}$ for a family $\mathcal{S}$ of $\sigma(X, Y)$-bounded sets, then each $x \in X$ satisfies $\sup_{y \in B^\circ} |\langle x, y\rangle| \leq 1$ for some $B \in \mathcal{S}$. Therefore each $\langle x, \ \rangle \in Y^*$ is $\mathfrak{T}_{\mathcal{S}}$-continuous and $\sigma(Y, X) \subset \mathfrak{T}_{\mathcal{S}}$. In other words, $\sigma(Y, X)$ is the weakest polar topology on Y subject to the condition that $X = \bigcup_{\mathcal{S}} B^{\circ\circ}$.

(b) If X is a separable LCT_2VS, then $(\Phi, \sigma(X', X))$ is metrizible for every equicontinuous set $\Phi \subset X'$. Verification: If $X = \overline{[x_n : n \in \mathbb{N}]}$ and $\mathcal{S}$ consists of all finite subsets of $\{x_n : n \in \mathbb{N}\}$, then $(X', \mathfrak{T}_{\mathcal{S}})$ is metrizible and $\mathfrak{T}_{\mathcal{S}} \subset \sigma(X', X)$. If $\Phi \subset X'$ is equicontinuous, $\Phi \subset u^\circ$ for some X-nbd $u(0)$. However, u° is $\sigma(X', X)$ compact and T_2 so that $\mathfrak{T}_{\mathcal{S}} = \sigma(X', X)$ on u°.

(c) Since $X = \bigcup_{\mathcal{S}_\sigma} B = \bigcup_{\mathcal{S}_\beta} B$, both $\sigma(Y, X)$ and $\beta(Y, X)$ are T_2 iff $\langle X, Y\rangle$ is X-injective. In this case $(Y, \mathfrak{T}_{\mathcal{S}})$ is simply the relativized subspace of $B\mathfrak{L}_{\mathcal{S}}(X; \mathbb{F})$ for $\mathcal{S} = \mathcal{S}_\sigma, \mathcal{S}_\beta$.

(d) Clearly, $\tau\langle Y, X\rangle \subset \beta(Y, X)$, since $\beta(Y, X)$ is the strongest polar topology on Y. In the other direction $X = \bigcup_{\mathcal{S}_\tau} B$ and $\sigma(Y, X) \subset \tau(Y, X)$ if $\langle X, Y\rangle$ is Y-injective. To be sure, $\mathcal{S}_\sigma^\circ = \{B^\circ : B \in \mathcal{S}_\sigma\}$ is a $\sigma(Y, X)$-nbd base and $X = (Y, \sigma(Y, X))' = \bigcup_{\mathcal{S}_\sigma} B^{\circ\circ}$ for polars in $Y' = X$. It remains only to note (Theorem 18.15) that the abs convex sets $B^{\circ\circ} \subset X$ are $\sigma(Y', Y) = \sigma(X, Y)$-compact.

(e) If $(X, \|\ \|)$ is a NLS, then $|||\ ||| = \beta(X', X)$ on X' and $\beta(X'', X')/X = \|\ \| \subset \beta(X, X')$ on X. Reason: $\bar{S}_{1,|||\ |||}(0) = \{\bar{S}_{1,\|\ \|}(0)\}^\circ$ implies $|||\ ||| \subset \beta(X', X)$, and [Theorem 18.6(ii))] $E \subset \lambda \bar{S}_{1,\|\ \|}(0)$ implies $\{\bar{S}_{1/\lambda,|||\ |||}(0)\}^\circ \subset E^\circ$ for $E \in \mathcal{S}_\beta$ so that $\beta(X', X) \subset |||\ |||$ follows. For the second assertion note (Problem 17B) that $\underset{x \rightsquigarrow \langle x, \ \rangle}{(X, \|\ \|) \to (X'', ||||\ ||||)}$ is an into conguence and $\bar{S}_{\varepsilon,\|\ \|}(0) = \{\bar{S}_{\varepsilon,\|\ \|}(0)\}^{\circ\circ} = \{\bar{S}_{1/\varepsilon,|||\ |||}(0)\}^\circ \ \forall \varepsilon > 0$.

Remark. Note [in conjunction with Example 18.5(c)] that $\beta(Y, X)$ need not be a topology of the dual pair $\langle X, Y\rangle$.

For a dual pair $\langle X, Y\rangle$, every $\beta(Y, X)$-bounded set is $\sigma(Y, X)$-bounded. Partial converses include

THEOREM 18.25. A set $B \subset Y$ is $\beta(Y, X)$-bounded if B is

(i) convex and $\sigma(Y, X)$-compact or
(ii) $\sigma(Y, X)$-bounded, closed, and sequentially complete.

Proof. Suppose B is convex and $\sigma(Y, X)$-compact. If $u(0)$ is any $\beta(Y, X)$-nbd, the $A^{\circ} \subset u$ for some $\sigma(X, Y)$-bounded set $A \subset X$. It follows [Theorem 18.10, Problem 13H(b)] that A°, therefore u, absorbs B and so B is $\beta(Y, X)$-bounded. If B satisfies (ii) and $A^{\circ} \subset u$ is as above, (E_B, p_B) is a Banach space (Theorem 14.17) and $\sup_A |\langle a, y\rangle| < \infty \ \forall y \in E_B$ [Example 15.4(a)]. Therefore $\langle A, \ \rangle : (E_B, p_B) \to \mathbb{F}$ is equicontinuous (Theorem 13.28) and $\langle A, \varepsilon B\rangle \subset \bar{S}_1(0)$ in $\mathbb{F}$ yields $B \subset (1/\varepsilon)A^{\circ} \subset (1/\varepsilon)u$ for some $\varepsilon > 0$. ∎

Every polar topology $\mathfrak{T}_{\mathscr{S}}$ is determined by its seminorms $\{q_B(y) = \sup_{x \in B} |\langle x, y\rangle| : B \in \mathscr{S}\}$, and every topology determined by seminorms is locally convex (Theorem 14.7). This chain is completed and equivalence is established by proving

THEOREM 18.26. If $(X, \mathfrak{T})$ is an LCTVS and $\mathscr{S}_e = \{$equicontinuous subsets of $X'\}$, then $\mathfrak{T} = \mathfrak{T}_{\mathscr{S}_e}$.

Proof. Let $\mathfrak{N}_0$ be a $\mathfrak{T}$-nbd base of abs convex closed sets so that (Theorems 18.6 and 18.11) $u = u^{\circ\circ} \ \forall u \in \mathfrak{N}_0$ and $\mathfrak{T} = \mathfrak{T}_{\mathfrak{N}_0^{\circ}}$ for $\mathfrak{N}_0^{\circ} = \{u^{\circ} : u \in \mathfrak{N}_0\}$. Then [Theorem 18.22(i)] $\mathfrak{T}_{\mathfrak{N}_0^{\circ}} \subset \mathfrak{T}_{\mathscr{S}_e}$ since $u^{\circ} = (u^{\circ})^{\circ\circ}$ and $u^{\circ} \in \mathscr{S}_e \ \forall u \in \mathfrak{N}_0$ (Theorem 18.14). On the other hand, $\mathfrak{T}_{\mathscr{S}_e} \subset \mathfrak{T}$ since $\mathscr{S}_e$ satisfies (*) (so that $\mathscr{S}_e^{\circ}$ is a $\mathfrak{T}_{\mathscr{S}_e}$-nbd base at $0 \in X$) and each $E \in \mathscr{S}_e$ satisfies $E \subset u^{\circ}$ for some $u \in \mathfrak{N}_0$. ∎

Theorem 18.26 and Example 18.24(e)⁺ lead one to ask which polar topologies are topologies of the dual pair.

THEOREM 18.27 (Mackey–Arens). A locally convex topology $\mathfrak{T} \in T(Y)$ is a topology of the Y-injective dual pair $\langle X, Y\rangle$ iff $\mathfrak{T} = \mathfrak{T}_{\mathscr{S}}$, where $\mathscr{S}$ is a cover of X by abs convex $\sigma(X, Y)$-compact sets.

Proof. If $(Y, \mathfrak{T})' = X$ for some locally convex $\mathfrak{T} = T(Y)$ and $\mathfrak{N}_0$ is any $\mathfrak{T}$-nbd base at $0 \in Y$, then $\mathfrak{N}_0^{\circ} \subset \mathscr{S}_e$ in Y' and $\mathfrak{T}_{\mathfrak{N}_0^{\circ}} = \mathfrak{T}_{\mathscr{S}_e} = \mathfrak{T}$ follows as in Theorem 18.26. Furthermore, each $u^{\circ} \in \mathfrak{N}_0^{\circ}$ is $\sigma(Y', Y) = \sigma(X, Y)$-compact (Theorem 18.15) and $X = Y' = \bigcup_{\mathfrak{N}_0^{\circ}} u^{\circ}$ by Theorem 18.14.

Suppose, conversely, that $\mathfrak{T}_{\mathscr{S}}$ is defined for $\mathscr{S}$ above [it being understood that $\mathscr{S}$ satisfies (*)]. Then $\mathscr{S}^{\circ}$ is a $\mathfrak{T}_{\mathscr{S}}$-nbd base at $0 \in Y$ and $(Y, \mathfrak{T}_{\mathscr{S}})' = \bigcup_{\mathscr{S}} B^{\circ\circ}$ for polars in $Y^* \supset X$; but $B^{\circ\circ} = \bar{B}^{\sigma(Y^*, Y)}$ for the injective dual pair $\langle Y^*, Y\rangle$ and (since B is $\sigma(X, Y) = \sigma(Y^*, Y)/X$ compact and $\sigma(X, Y)$ is T_2) one has $B = \bar{B}^{\sigma(Y^*, Y)}$ $\forall B \in \mathscr{S}$. Thus $(Y, \mathfrak{T}_{\mathscr{S}})' = \bigcup_{\mathscr{S}} B = X$ and $\mathfrak{T}_{\mathscr{S}}$ is a topology of $\langle X, Y\rangle$. ■

Corollary 19.28. $\tau(Y, X)$ is the finest topology of a Y-injective dual pair $\langle X, Y\rangle$. Therefore a locally convex $\mathfrak{T} \in T(Y)$ is a topology of $\langle X, Y\rangle$ above iff $\sigma(Y, X) \subset \mathfrak{T} \subset \tau(Y, X)$.

We have already met the Mackey topology in disguise.

Example 18.29 (Example 14.14). $\mathfrak{T}_{F_c} = \tau(X, X^*)$ on a vector space X, since $(X, \mathfrak{T}_{F_c})' = X^*$ and $\tau(X, X^*) \in T(X)$ is locally convex. Accordingly, the collection of abs convex absorbent subsets of X constitutes a $\tau(X, X^*)$-nbd base at 0.

(a) Every subspace M of a vector space X is $\sigma(X, X^*)$-closed. Indeed, $\mathfrak{T}_{F_c}$ is complete (Problem 15E) and $\tau(M, M^*) = \tau(X, X^*)/M$ so that M is $\tau(X, X^*)$-complete and closed in X.

(b) A locally convex topology $\mathfrak{T} \in T(X)$ coincides with $\tau(X, X')$ iff $\mathscr{S}_\tau \subset \mathscr{S}_e$ in X' [$\mathscr{S}_\tau$ denotes the $\tau(X, X')$ equicontinuous sets in X']. This follows from Theorem 18.14, Corollary 18.28, and Definition 18.23.

Example 18.30. Let $\mathfrak{F} = \{f_\alpha \in \mathfrak{L}(X, Y) : \alpha \in \mathfrak{A}\}$, where X is a vector space and Y is an LCTVS. A nice application of earlier notions proves anew (Corollary 15.21) that $(X, \mathfrak{T}_{\mathfrak{P}\{\mathfrak{F}\}})' = [Y'\mathfrak{F}] \subset X^*$.

By definition $[Y'\mathfrak{F}] \subset (X, \mathfrak{F}_{\mathfrak{P}\{\mathfrak{F}\}})'$. For the reverse inclusion write $Z = [Y'\mathfrak{F}]$ and observe that $\langle X, Z\rangle$ is an X-injective dual pair under $\langle x, z\rangle = z(x)$. Assume that $\mathfrak{N}_0$ is a Y-nbd base of abs convex closed sets. Then $u^{\circ} \subset Y'$ is $\sigma(Y', Y)$-compact $\forall u \in \mathfrak{N}_0$ (Theorem 18.15) and the continuity of each linear surjection $\eta_\alpha : (Y', \sigma(Y', Y)) \to (Z, \sigma(Y, X))$, $g \rightsquigarrow gf_\alpha$, ensures that $\eta_\alpha(u^{\circ}) \subset Z$ is $\sigma(Z, X)$-compact. Thus $\{\eta_\alpha(u^{\circ})\}^{\circ} \subset X$ is a $\tau(X, Z)$-base nbd contained in the $\mathfrak{T}_{\mathfrak{P}\{\mathfrak{F}\}}$-subbase nbd $\overset{-1}{f_\alpha}(u)$. It follows, therefore, that $\mathfrak{T}_{\mathfrak{P}\{\mathfrak{F}\}} \subset \tau(X, Z)$ and $(X, \mathfrak{T}_{\mathfrak{P}\{\mathfrak{F}\}})' \subset (X, \tau(X, Z))' = [Y'\mathfrak{F}]$.

Theorem 18.16 can be brought into sharper focus.

THEOREM 18.31. Let X, Y be LCT_2VSps. Then $F \in \mathfrak{L}(X, Y)$ is weakly continuous iff F is $\tau(X, X') - \tau(Y, Y')$ continuous. Either type continuity is equivalent to F being $\tau(X, X')$-$\mathfrak{T}$ continuous for every topology $\mathfrak{T}$ of $\langle Y, Y' \rangle$.

Proof. Mackey continuity implies weak continuity. Suppose, therefore, that F is weakly continuous and let $v(0)$ be any $\tau(Y, Y')$-nbd in Y. Then $E^\circ \subset v$ for some abs convex $\sigma(Y', Y)$-compact set $E \subset Y'$ and [Theorem 18.18(ii)] the weak continuity of $F^t: Y' = (Y, \sigma(Y, Y'))' \to X' = (X, \sigma(X, X'))'$ ensures that $F^t(E) \subset X'$ is $\sigma(X', X)$-compact. In particular (Theorem 18.19) $\{F^t(E)\}^\circ = F^{-1}(E^\circ)$ is a $\tau(X, X')$-nbd contained in $F^{-1}(v)$ and F is seen to be $\tau(X, X') - \tau(Y, Y')$ continuous. ■

COROLLARY 18.32. Let $\mathfrak{F} = \{f_\alpha \in \mathfrak{L}(X, X_\alpha) : \alpha \in \mathfrak{A}\}$ and $\mathfrak{G} = \{g_\alpha \in \mathfrak{L}(X_\alpha, X) : \alpha \in \mathfrak{A}\}$ where X is a vector space and $X_\alpha = (X_\alpha, \mathfrak{T}_\alpha)$ are all LCT_2VSps.

(i) $\mathfrak{T}_{\mathcal{P}\{\mathfrak{F}\}} = \sigma(X, X')$ if $\mathfrak{F}$ is injective and each X_α carries $\sigma(X_\alpha, X'_\alpha)$.
(ii) $\mathfrak{T}_{\mathcal{I}'\{\mathfrak{G}\}} = \tau(X, X')$ if $\mathfrak{T}_{\mathcal{I}'\{\mathfrak{G}\}}$ is T_2 and each X_α carries $\tau(X_\alpha, X'_\alpha)$.

Proof. By definition $\sigma(X, X') \subset \mathfrak{T}_{\mathcal{P}\{\mathfrak{F}\}}$ and $\mathfrak{T}_{\mathcal{I}'\{\mathfrak{G}\}} \subset \tau(X, X')$. The reverse inclusions also hold, since every $f_\alpha \in \mathfrak{F}$ is $\mathfrak{T}_{\mathcal{P}\{\mathfrak{F}\}}$-continuous (therefore weakly continuous) and every $g_\alpha \in \mathfrak{G}$ is $\mathfrak{T}_{\mathcal{I}'\{\mathfrak{G}\}}$-continuous (therefore Mackey continuous). ■

Grothendieck's Completion Theorem

Since every LCTVS carries a polar topology (Theorem 18.21), polar space completions provide an alternative approach to LCTVS completions (Problem 14A). Let X be an LCTVS covered by a collection $\mathcal{S}$ of bounded subsets [which, a fortiori satisfies (*)]. Then the T_2VS $(X^*, \mathfrak{T}_\mathcal{S})$, but not necessarily $(X', \mathfrak{T}_\mathcal{S})$, is complete by Theorem 18.21, Example 13.27(c). One logical candidate for the completion of $(X', \mathfrak{T}_\mathcal{S})$ is $\check{X} = \{f \in X^* : f/B \in C(B; \mathbb{F})\ \forall B \in \mathcal{S}\}$.

THEOREM 18.33. If X is an LCT_2VS and $\mathcal{S}$ is a collection of abs convex, closed bounded subsets that covers X, then $(\check{X}, \mathfrak{T}_\mathcal{S})$ is the T_2 completion of $(X', \mathfrak{T}_\mathcal{S})$. Thus $(X', \mathfrak{T}_\mathcal{S})$ is complete iff $X' = \check{X}$.

Proof. $(\check{X}, \mathfrak{T}_\mathcal{S})$ is complete, since $\check{X}$ is $\mathfrak{T}_\mathcal{S}$-closed in X^*. Indeed, if $\mathfrak{G} \xrightarrow{\mathfrak{T}_\mathcal{S}} f \in X^*$ for a filter $\mathfrak{G} \subset X$, then $\mathfrak{G}/B \xrightarrow{\text{uniformly}} f/B \in \mathfrak{F}(B; \mathbb{F})$ and (since

$\mathfrak{G}/B \in C(B; \mathbb{F})$ and B is bounded) $f/B \in C(B; \mathbb{F})\, \forall B \in \mathscr{S}$. Moving on, observe that $\langle X, X' \rangle$ and $\langle X, \check{X} \rangle$ are injective dual pairs. Let E° and $E^{\circlearrowright}$ denote polars of $E \subset X$ in X' and $\check{X}$, respectively. To prove $\overline{X'}^{\mathfrak{T}_{\mathscr{S}}} = \check{X}$ we show that $\{f + B^{\circlearrowright}\} \cap X' \neq \varnothing\ \forall f \in \check{X}$ and $B \in \mathscr{S}$. Since $f/B \in C(B; \mathbb{F})$, there is (Theorem 13.11) and abs convex $\sigma(X, X')$-closed nbd $u(0)$ such that $f \in (B \cap u)^{\circlearrowright}$. We claim that $(B \cap u)^{\circlearrowright} \subset B^{\circlearrowright} + u^{\circlearrowright}$. When verified, this ensures that $f \in B^{\circlearrowright} + g$ and $g \in f + B^{\circlearrowright}$ for some $g \in u^{\circlearrowright} = u^{\circ} \in X'$ (Theorem 18.14).

Proof of Claim. [Theorem 13.4(iii)]. $B^{\circlearrowright} + u^{\circlearrowright}$ is $\sigma(\check{X}, X)$-closed, since $u^{\circlearrowright}$ is $\sigma(\check{X}, X)$-compact (Theorem 18.15) and $B^{\circlearrowright}$ is $\sigma(\check{X}, X)$-closed. Furthermore, $B^{\circlearrowright} + u^{\circlearrowright}$ is abs convex and contains $B^{\circlearrowright} \cup u^{\circlearrowright}$. Therefore (Corollary 18.12) $B^{\circlearrowright} + u^{\circlearrowright} \supset \overline{(B^{\circlearrowright} \cup u^{\circlearrowright})}_{bc}^{\sigma(\check{X}, X)} = (B \cap u)$. ■

COROLLARY 18.34. Suppose $\mathfrak{T} \in T(X)$ is a topology of the X-injective dual pair $\langle X, Y \rangle$ and $\mathscr{S}$ covers X by abs convex, closed bounded subsets. Then $(\check{Y}, \mathfrak{T}_{\mathscr{S}})$ is the completion of $(Y, \mathfrak{T}_{\mathscr{S}})$, where $\check{Y} = \{f \in Y^* : f/B \in C(B; \mathbb{F})\, \forall B \in \mathscr{S}\}$.

Proof. $Y = (X, \mathfrak{T})'$ and $(Y, \mathfrak{T}_{\mathscr{S}})$ is T_2. ■

COROLLARY 18.35. Suppose $\langle X, Y \rangle$ is injective and $\mathfrak{T}_{\mathfrak{R}} \subset \mathfrak{T}_{\mathscr{S}}$ on Y relative to covers $\mathfrak{R}$ and $\mathscr{S}$ of X by $\sigma(X, Y)$-bounded sets. Then $(Y, \mathfrak{T}_{\mathfrak{R}})$-completeness implies $(Y, \mathfrak{T}_{\mathscr{S}})$-completeness.

Proof. The bounded sets in $\mathfrak{R}$ and $\mathscr{S}$ may be assumed abs convex. Then (Theorem 18.22) each $R \in \mathfrak{R}$ is contained in some $B \in \mathscr{S}$, and $Y \subset \check{Y}(\mathscr{S}) \subset \check{Y}(\mathfrak{R})$ for $\check{Y}(\mathscr{S}) = \{f \in Y^* : f/B$ is $\sigma(Y, X)$-continuous $\forall B \in \mathscr{S}\}$ and $\check{Y}(\mathfrak{R}) = \{f \in Y^* : f/R$ is $\sigma(Y, X)$-continuous $\forall R \in \mathfrak{R}\}$. Clearly, $Y = \check{Y}(\mathscr{S})$ if $Y = \check{Y}(\mathfrak{R})$. ■

COROLLARY 18.36. The completion of an LCT_2VS X is $(\check{X}, \mathfrak{T}_{\mathscr{S}_e})$, where $\check{X} = \{\psi \in X^* : \psi/E$ is $\sigma(X', X)$-continuous $\forall E \in \mathscr{S}_e$ in $X'\}$. In particular, X is complete iff $X = \check{X}$.

Proof. If $Y = (X', \sigma(X', X))$ in Corollary 18.34, then $X = (Y', \mathfrak{T}_{\mathscr{S}_e})$ by Theorem 18.26. ■

In general, $\sigma(X, X')$ is not 1° (Corollary 15.6, Example 19.7(b)] and so weak countable compactness (i.e., every sequence has a $\sigma(X, X')$-accumulation point) does not imply weak compactness. The whole picture changes in the presence of completeness.

THEOREM 18.37 (Eberlein). If X is an LCT$_2$VS that is $\tau(X, X')$-complete and $A \subset X$ is $\sigma(X, X')$-countably compact, then $\bar{A}^{\sigma(X,X')}$ is $\sigma(X, X')$-compact. In particular, $\bar{A}$ is compact if A is countably compact.

Proof. If $A \subset X$ is $\sigma(X, X')$-countably compact, then $\bar{A}^{\sigma(X'^*,X')}$ is $\sigma(X'^*, X')$-compact (Corollary 5.5, Example 18.5(b)] and (since $\sigma(X, X') = \sigma(X'^*, X')/A$) we need only verify that $\bar{A}^{\sigma(X'^*,X')} \subset X$. This will follow by Theorem 18.33 once it is established that $\psi \in \bar{A}^{\sigma(X'^*,X')}$ is continuous on every abs convex $\sigma(X', X)$-compact set $E \subset X'$. If this is not true, there is an $\varepsilon > 0$ such that every $\sigma(X', X)$-nbd $u(0)$ determines an $f_u \in E \cap u$ satisfying $|\psi(f_u)| > \varepsilon$ and (since $\psi \in \bar{A}^{\sigma(X'^*,X')}$) an $x_u \in A$ such that $|\psi(f_u) - f_u(x_u)| < \varepsilon/3$. Proceed inductively to obtain two sequences $\{f_n\} \in E$ and $\{x_n\} \in A$ such that $|\psi(f_k) - f_k(x_n)| < \varepsilon/3 \; \forall k \leq n$, $|f_n(x_k)| \leq \varepsilon/3 \; \forall k < n$ and $|\psi(f_n)| > \varepsilon \; \forall n \in \mathbb{N}$. Now $\{x_n\}$ has some $\sigma(X, X')$-accumulation point $x_0 \in X$ and [since E is $\sigma(X', X)$-compact] $\{f_n\}$ has a $\sigma(X', X)$-accumulation point $f_0 \in E$. However, taking limits of subsequences in the above inequalities yields the contradiction that $2\varepsilon/3 \leq |f_0(x_0)| \leq \varepsilon/3$. Indeed, $|\psi(f_k) - f_k(x_0)| \leq \varepsilon/3$ and $|f_k(x_0)| > 2\varepsilon/3$, $\forall k$ imply $|f_0(x_0)| \geq 2\varepsilon/3$, whereas $|f_0(x_k)| \leq \varepsilon/3 \; \forall k$ implies $|f_0(x_0)| \leq \varepsilon/3$.

If $A \subset X$ is countably compact, it is $\sigma(X, X')$-countably compact. Thus A is $\sigma(X, X')$-compact and $\sigma(X, X')$, therefore X-complete (Corollary 18.7) and relatively compact (Corollary 5.5). ■

PROBLEMS

18A. Another Definition of Polar Sets

Let $A \subset X$ for a dual pair $\langle X, Y \rangle$. In the literature (e.g., Köthe [55] and Schaeffer [88]) A° is sometimes termed the absolute polar of A and $A^\square = \{y \in Y : \mathrm{R_e}\langle a, y \rangle \leq 1 \; \forall a \in A\}$ is called the polar of A. Prove that

(a) $(A_b)^\square = A^\circ \subset A^\square$ so that $A^\circ = A^\square$ if A is balanced;
(b) $A^\square \subset Y$ is convex and $\sigma(Y, X)$-closed;
(c) properties (ii)–(iv) of Theorem 18.10 hold for "$\square$";
(d) $A^{\square\square} = \overline{(A \cup \{0\})_c}^{\sigma(Y,X)}$;
(e) $\{\bigcap_{\mathscr{A}} A_\alpha\}^\square = \overline{\{\bigcup_{\mathscr{A}} A_\alpha^\square\}_c}^{\sigma(Y,X)}$ if $\{A_\alpha : \alpha \in \mathscr{A}\}$ are convex and $\sigma(X, Y)$-closed;
(f) if A is a pointed cone in X (Problem 17G), then $A^\square = \{y \in Y : \mathrm{R_e}\langle a, y \rangle \leq 0 \; \forall a \in A\}$ is a pointed convex cone in Y.

18B. Exercise in Polar Sums

If $\langle X, Y \rangle$ is an injective dual pair and $A, B \subset X$ are abs convex, show that $\frac{1}{2}(A^\circ \cap B^\circ) \subset (A + B)^\circ \subset A^\circ + B^\circ$.

18C. Polar Topologies Via Saturated Families

The assumption that $\mathcal{S}$ satisfies (*) can be strengthened. Call $\mathcal{S}$ saturated if $\mathcal{S}_0 \cup \mathcal{S}_{\bullet} \subset \mathcal{S}$ and $\overline{(B_1 \cup B_2)}_{bc}^{\sigma(X,Y)} \subset \mathcal{S}\ \forall B_1, B_2 \in \mathcal{S}$.

(a) Every family $\mathcal{S}$ of $\sigma(X, Y)$-bounded subsets of X is contained in a smallest saturated family $\hat{\mathcal{S}}$ of $\sigma(X, Y)$-bounded sets. [*Hint*: $\mathcal{S}_B$ and the intersection of any collection of saturated families is saturated.]

(b) Show that $\mathfrak{T}_{\hat{\mathcal{S}}} = \mathfrak{T}_{\mathcal{S}}$ on Y. [*Hint*: Use Theorem 18.22 to produce a saturated family $\mathcal{S}_0 \supset \mathcal{S}$ such that $\mathfrak{T}_{\mathcal{S}} = \mathfrak{T}_{\mathcal{S}_0}$.]

(c) Given two collections $\mathcal{S}_1, \mathcal{S}_2$ of $\sigma(X, Y)$-bounded sets in X, show that $\mathfrak{T}_{\hat{\mathcal{S}}_1} = \mathfrak{T}_{\hat{\mathcal{S}}_2}$ iff $\mathfrak{T}_{\mathcal{S}_1} = \mathfrak{T}_{\mathcal{S}_2}$.

18D. $\mathfrak{T}_{\mathcal{S}}$-Total Boundedness

(a) Let $f \in \mathfrak{L}(X, Y)$ be weakly continuous for the injective dual pairs $\langle X, \tilde{X}\rangle$ and $\langle Y, \tilde{Y}\rangle$. Assume further that $\mathfrak{R}$ and $\mathcal{S}$ are families of weakly bounded sets in X and $\tilde{Y}$, respectively. Then every $f(R) \subset Y$ $(R \in \mathfrak{R})$ is $\mathfrak{T}_{\mathcal{S}}$-totally bounded iff every $f^t(S) \subset \tilde{X}$ $(S \in \mathcal{S})$ is $\mathfrak{T}_{\mathfrak{R}}$-totally bounded. [*Hint*: The argument is symmetric. Assume that every $f^t(S)$ is $\mathfrak{T}_{\mathfrak{R}}$-totally bounded and fix $R \in \mathfrak{R}$. For $S \in \mathcal{S}$ there exist $\{s_1, \ldots, s_n\} \in S$ such that $f^t(S) \subset \frac{1}{3}R^{\circ} + \bigcup_{i=1}^{n} f^t(s_i)$ and [Example 19.5(d)] corresponding $\{r_1, \ldots, r_m\} \in R$ such that $R \subset \frac{1}{3}\left\{\bigcup_{i=1}^{n} f^t(s_i)\right\}^{\circ} + \bigcup_{j=1}^{m} r_j = \bigcap_{i=1}^{n} f^{-1}(\frac{1}{3}s_i^{\circ}) + \bigcup_{j=1}^{m} r_j$. Now $(r, s) \in R \times S$ implies that $r = \alpha + r_j$ and $f^t(s) = \beta + f^t(s_i)$ for some $(\alpha, \beta) \in \bigcap_{i=1}^{n} f^{-1}(\frac{1}{3}s_i^{\circ}) \times \frac{1}{3}R^{\circ}$ and $(r_j, s_i) \in R \times S$. Show that $|\langle f(r) - f(r_j), s\rangle| = |\langle r - r_j, f^t(s)\rangle| \leq 1$ $\forall (r, s) \in R \times S$ and obtain $f(R) \subset S^{\circ} + \bigcup_{j=1}^{m} f(r_j)$.]

(b) Suppose $\tilde{X} = X'$ and $\tilde{Y} = Y'$ for LCT$_2$VSps X, Y and let $X = \bigcup_{\mathfrak{R}} R^{\circ\circ}$. Then $f(R) \subset Y$ is totally bounded $\forall R \in \mathfrak{R}$ iff $f^t(E) \subset X'$ is $\mathfrak{T}_{\mathfrak{R}}$-relatively compact $\forall E \in \mathcal{S}_e$ in Y'. [*Hint*: Take $\mathcal{S} = \mathcal{S}_e$ in (a) and use Theorem 18.26. Note, in particular, that $\overline{f^t(E)}^{\sigma(X',X)}$ is $\sigma(X', X)$-complete (Theorem 18.15); therefore $\mathfrak{T}_{\mathfrak{R}}$-complete (Corollary 18.35).]

18E. Banach-Dieudonne Theorem

Given an LCTVS X, does there exist a finest topology $\mathfrak{T}_f \in T(X')$ that agrees with $\sigma(X', X)$ on every $E \in \mathcal{S}_e$ in X'? In certain instances, yes.

(a) Show that $\mathfrak{O} = \{E \subset X' : E \cap \Phi$ is $\sigma(X', X)$-open in $\Phi\ \forall \Phi \in \mathcal{S}_e$ in $X'\}$ is the collection of all open sets for the finest T_2 topology $\mathfrak{T}_f$ on X' satisfying

$\mathfrak{T}_f/\Phi = \sigma(X', X)/\Phi \ \forall \Phi \in \mathscr{S}_e$. Show further that $\mathfrak{T}_f$ has a nbd base at $0 \in X'$ of balanced absorbing sets. [*Hint*: Consider $v(0) = \bigcup_{v(0)} \{\text{balanced sets } E \subset v\} \ \forall \mathfrak{T}_f$-nbd $v(0)$.]

Remark. In general $\mathfrak{T}_f \notin T(X')$, since Theorem 13.7(ii) may fail (Komura [54]).

Assume that X is metrizible with $\mathfrak{N}_0 = \{u_n(0) : n \in \mathbb{N}\}$ a nbd base of decreasing abs convex closed sets.

(b) For each $\mathfrak{T}_f$-open nbd $E(0) \subset X'$ there is a sequence $\{F_n \neq \varnothing : n \in \mathbb{N}\}$ of finite subsets of X satisfying $F_n \subset u_n$ and $\left(\bigcup_{k=1}^{n-1} F_k\right)^\circ \cap u_n^\circ \subset E \ \forall n \in \mathbb{N}$. [*Hint*: Use induction. If the assertion holds $\forall 1 \leq k \leq n-1$ and fails for $k = n$, then $\left(\bigcup_{k=1}^{n} F_k\right)^\circ \cap F^\circ \cap [u_{n+1}^\circ \cap (X' - E] \neq \varnothing \ \forall$ finite set $F \subset u_n$. Taking $K_n = \bigcup_{k=1}^{n} F_k$ and $v_n = u_{n+1}^\circ \cap (X' - E) \ \forall n \in \mathbb{N}$, we see that $\xi = \{K_n^\circ \cap F^\circ \cap v_n : \text{finite subsets } F \subset u_n\}$ is a filterbase of $\mathfrak{T}_f$-closed sets on the $\mathfrak{T}_f$-compact set v_n (indeed, $X' - E$ is $\mathfrak{T}_f$-closed and u_{n+1}° is $\sigma(X', X) = \mathfrak{T}_f$ compact). If $f_0 \in \bigcap_{\xi} (K_n^\circ \cap F^\circ \cap v_n)$ by Appendix t.7, then $f_0 \in K_{n-1}^\circ \cap v_{n-1}$, which is impossible, since $K_{n-1}^\circ \cap u_n^\circ \subset E$.]

(c) For $Y = \mathbb{F}$ in Example 7.8(d), write $\lambda(X', X) = \mathfrak{T}_{\mathscr{S}_\lambda}$ and prove that $\mathfrak{T}_f = \lambda(X', X) \in T(X')$. [*Hint*: $\lambda(X', X) \subset \mathfrak{T}_f$ by Example 13.27(d). If $E(0) \subset X'$ is $\mathfrak{T}_f$-open, take $K = \bigcup_{n=1}^{\infty} F_n \in \mathscr{S}_\lambda$ and conclude (from $K^\circ \cap u_n^\circ \subset E \ \forall n \in \mathbb{N}$) that $K^\circ \subset E$.]

(d) Let $\mathscr{S}_N$ denote the collection of sequences $\{x_n \cup 0 : x_n \to 0\}$. Then $\mathfrak{T}_{\mathscr{S}_N} = \lambda(X', X) = \kappa(X', X)$, where $\kappa(X', X) = \mathfrak{T}_{\mathscr{S}_\kappa} \in T(X')$ in Example 7.8(c). [*Hint*: Use Theorem 18.22(i) to get $\lambda(X', X) \subset \mathfrak{T}_{\mathscr{S}_N}$.]

(e) Prove that X is complete iff $\bar{K}_{bc}^{\sigma(X,X')}$ is compact for all compact sets $K \subset X$. [*Hint*: if $B = \{x_n\}$ is Cauchy, then $B \subset E^{\circ\circ}$ for some $E \in \mathscr{S}_N$. The sufficiency hypothesis ensures that $E^{\circ\circ}$ is compact and $\lim_n x_n \in B$.]

18F. Polar Topologies for an LCTVS

Given an LCTVS X, let $\mathscr{S}_{\kappa_{bc}}$ $\{\mathscr{S}_{\kappa_c}\}$ be the collection of abs convex, compact {convex, compact} subsets of X. Use the notation $\kappa_{bc}(X', X) = \mathfrak{T}_{\mathscr{S}_{\kappa_{bc}}} \in T(X')$ and $\kappa_c(X', X) = \mathfrak{T}_{\mathscr{S}_{\kappa_c}} \in T(X')$ and prove that

(a) $\sigma(X', X) \subset \kappa_{bc}(X', X) \subset \kappa_c(X', X) \subset \kappa(X', X) \subset \lambda(X', X) \subset \beta(X', X)$ and $\kappa_{bc}(X', X) \subset \tau(X', X)$ on X'. [*Hint*: Problem 13C can be used to prove $\mathscr{S}_{\kappa_{bc}}$ saturated.]

(b) If $\mathfrak{T} = \sigma(X', X)$, then $\kappa_{bc}(X', X) = \tau(X', X)$ and $\lambda(X', X) = \beta(X', X)$.

Remark. See Example 20.9(d) for an example in which $\kappa_{bc}(X', X) \neq \tau(X', X)$.

(c) Use Corollary 14.3⁺ to prove that $\kappa_{bc}(X', X) = \lambda(X', X)$ if X is quasi-complete.

(d) $\tau(X', X) \subsetneq \kappa(X', X)$ if X is an incomplete, metrizible LCTVS. [*Hint*: Theorem 18.22 and the fact [Problem 18E(c)] that $\bar{K}_{bc}^{\sigma(X,X')}$ is not compact for some compact set $K \subset X$.]

(e) Show that $\lambda(X', X)$ is the finest polar topology $\mathfrak{T}_{\mathcal{S}}$ on X' under which $u^{\circ} \subset X'$ is $\mathfrak{T}_{\mathcal{S}}$-compact $\forall X$-nbd $u(0)$. [*Hint*: Theorems 18.14 and 18.15 and Example 13.27(d).]

18G. β^* and κ_{bc}^* Topologies

Given an LCTVS $(X, \mathfrak{T})$, let $\mathcal{S}_{\beta^*} = \{\beta(X', X)\text{-bounded sets in } X'\}$ and $\mathcal{S}_{\kappa_{bc}^*} = \{\text{abs convex } \kappa_{bc}(X', X)\text{-compact sets in } X'\}$. Defining $\beta^*(X, X') = \mathfrak{T}_{\mathcal{S}_{\beta^*}} \in T(X)$ and $\kappa_{bc}^*(X, X') = \mathfrak{T}_{\mathcal{S}_{\kappa_{bc}^*}} \in T(X)$, prove that

(a) $\mathfrak{T} \subset \kappa_{bc}^*(X, X') \subset \tau(X, X') \subset \beta^*(X, X') \subset \beta(X, X')$. [*Hint*: $\mathfrak{T} \subset \kappa_{bc}^*(X, X')$ via Corollary 14.3, Theorem 18.11, and Problem 18F(e). For $\tau(X, X') \subset \beta^*(X, X')$ see Theorem 18.25.]

(b) Use Theorem 18.26 to show that $\mathfrak{T} = \beta^*(X, X')$ iff $\mathcal{S}_{\beta^*} = \mathcal{S}_e$ in X'. Show also that $\beta(X, X') = \beta^*(X, X')$ if X is sequentially complete.

(c) The converse of Theorem 18.25(i) fails. There exist $\sigma(X', X)$-closed, $\beta(X', X)$-bounded sets that are not $\sigma(X', X)$-compact. [*Hint*: Consider a space for which $\tau(X, X') \neq \beta^*(X, X')$. (See Problem 20E for example.)]

(d) $\mathfrak{T} = \kappa_{bc}^*(X, X')$ iff $(X, \mathfrak{T})$ is the topological isomorph of $(Y', \kappa_{bc}(Y', Y))$ for some LCT_2VS Y. [*Hint*: Take $Y = X'$.]

18H. Krein-Šmulian Theorem

(a) For a metrizible LCTVS X the following are equivalent:

(i) $E \subset X'$ is $\lambda(X', X)$-closed.

(ii) $E \cap \Phi$ is $\sigma(X', X)$-closed in X' $\forall$ abs convex $\sigma(X', X)$-closed equicontinuous set $\Phi \subset X'$. [*Hint*: For (i)$\Rightarrow$(ii) use Example 13.27(d). For (ii)$\Rightarrow$(i) use Problem 18E, after demonstrating that E is $\mathfrak{T}_f$-closed. (First show that $\Psi^{\circ\circ} \in \mathcal{S}_e$ in X' for $\Psi \in \mathcal{S}_e$ in X'.)]

(b) (Krein-Šmulian). Suppose that X above is complete and $E \subset X'$ is convex. Then E is $\sigma(X', X)$-closed iff (ii) holds. [*Hint*: Weak closure follows from (ii) by Problem 18F(a), (c) and Theorem 18.6.]

18I. Hyperplane Characterization of Completeness

An LCT_2VS X is complete iff each hyperplane in X', which intersects every abs convex $\Phi \in \mathcal{S}_e$ in a $\sigma(X', X)$-closed set, is $\sigma(X', X)$-closed. [*Hint*: Suppose that (Theorem 17.2) $H_\psi = \psi^{-1}(0) \subset X'$ for $\psi \in X'^*$. Then (Problem 13A) $H_\psi \cap E$ is $\sigma(X', X)$-closed iff ψ/E is $\sigma(X', X)$-closed $\forall$ abs convex $\Phi \in \mathcal{S}_e$. The hyperplane condition is therefore equivalent to requiring that $X = \check{X}$ (Corollary 18.36).]

18J. Additional f^t-Continuity Considerations

Let $\mathfrak{R}$, $\mathcal{S}$ be respective families of bounded sets [satisfying (*)] in LCTVSps X and Y and write $Y_{\mathcal{S}} = (Y, \mathfrak{T}_{\mathcal{S}})$ and $X_{\mathfrak{R}} = (X, \mathfrak{T}_{\mathfrak{R}})$.

(a) Suppose that in $f \in L(X, Y)$ and $\forall R \in \mathfrak{R}$ there is an $S \in \mathcal{S}$ such that $f(R) \subset S$. Then $f^t \in L(Y'_{\mathcal{S}}, X'_{\mathfrak{R}})$. [*Hint*: $f(R) \subset S \Rightarrow f^t(S^{\circ}) \subset R^{\circ}$.]

(b) If $f \in L(X, Y)$, then $f^t: (Y', \xi(Y', Y)) \to (X', \xi(X', X))$ is continuous for $\xi = \kappa_{bc}, \kappa_c, \kappa$, and β.

(c) If X, Y in (b) are LCT$_2$VSps, then $f: (X, \eta(X, X')) \to (Y, \eta(Y, Y'))$ is continuous for $\eta = \beta^*, \kappa^*_{bc}$ and each ξ topology above.

(d) Every weakly continuous $f^t \in \mathfrak{L}(Y, X)$ is strongly continuous. Can you give a condition under which the converse holds? [*Hint*: $\sigma(X', X) \subset \sigma(X', (X', \beta)')$ in Theorem 18.16. If Y is semireflexive (Theorem 20.2), then $\sigma(Y', Y) = \sigma(Y', (Y', \beta)')$.]

(e) Let $f \in L(X, Y)$, where X and Y are NLSps. Then $f^t: (Y', ||| \; |||) \to (X', ||| \; |||)$ is continuous and $|||f^t||| = |||f|||$. [*Hint*: Example 18.24(e) and Problem 17B.]

(f) (Notation as in Problem 13J). Let $f \in L_{cc}(X, Y)$, where X and Y are LCT$_2$VSps. Then $f^t: (Y', \kappa_{bc}(Y', Y)) \to (X', \kappa_{bc}(X', X))$ is completely continuous. [*Hint*: Theorem 18.18, Problem 18F.]

18K. Completion Problem

Let $(\check{X}, \check{\mathfrak{T}})$ be the completion of an LCT$_2$VS $(X, \mathfrak{T})$ and let $\bar{E}^{\vee}$ denote the $\check{\mathfrak{T}}$-closure of $E \subset X$ in $\check{X}$. Prove

(a) u° and $\bar{u}^{\vee\circ}$ coincide in $X' = \check{X}'$ $\forall \mathfrak{T}$-nbd $u(0) \subset X$. [*Hint*: Use Corollary 17.13 and extension by continuity.]

(b) $E \subset X'$ is $\mathfrak{T}$-equicontinuous iff $E \subset \check{X}'$ is $\check{\mathfrak{T}}$-equicontinuous. [*Hint*: Apply Problem 13E and (a) above to $(X, \mathfrak{T}) \subset (\check{X}, \check{\mathfrak{T}})$.]

(c) Show that $\sigma(\check{X}', \check{X})/\Phi = \sigma(X', X)/\Phi$ on every equicontinuous subset $\Phi \subset X' = \check{X}'$. Use this to obtain $\check{\mathfrak{T}} = \tau(\check{X}, \check{X}')$ if $\mathfrak{T} = \tau(X, X')$.

18L. Products and Sums of Polar Topologies

Let $X = \prod_{\mathfrak{A}} X_\alpha$ and $Y = \sum_{\mathfrak{A}} \oplus Y_\alpha$, where $\{\langle X_\alpha, Y_\alpha \rangle : \alpha \in \mathfrak{A}\}$ is a collection of dual pairs.

(a) $\langle X, Y \rangle$ under $\langle \{x_\alpha\}, \sum_{\mathfrak{A}} y_\alpha \rangle = \sum_{\mathfrak{A}} \langle x_\alpha, y_\alpha \rangle$ is X-injective $\{Y$-injective$\}$ iff each $\langle X_\alpha, Y_\alpha \rangle$ is X_α-injective $\{Y_\alpha$-injective$\}$.

(b) Show that $\pi_\alpha: (X, \sigma(X, Y)) \to (X_\alpha, \sigma(X_\alpha, Y_\alpha))$ and $j_\alpha: (Y_\alpha, \sigma(Y_\alpha, X_\sigma)) \to (Y, \sigma(Y, X))$ are continuous $\forall \alpha \in \mathfrak{A}$. [*Hint*: $\langle \pi_\beta\{x_\alpha\}, y_\beta \rangle = \langle \{x_\alpha\}, j_\beta(y_\beta) \rangle$ $\forall \beta \in \mathfrak{A}$.]

(c) If $\mathfrak{R}_\alpha$ is a family of $\sigma(X_\alpha, Y_\alpha)$-bounded sets in X_α $(\alpha \in \mathcal{Q})$, then $\mathfrak{R} = \{\prod_{\mathcal{Q}} R_\alpha : R \in \mathfrak{R}_\alpha\}$ is a family of $\sigma(X, Y)$-bounded sets in X. Furthermore, $X = \bigcup_{\mathfrak{R}} R$ if each $X_\alpha = \bigcup_{\mathfrak{R}_\alpha} R_\alpha$.

(d) If $\mathcal{S}_\alpha$ is a family of $\sigma(Y_\alpha, X_\alpha)$-bounded sets in Y_α $(\alpha \in \mathcal{Q})$, then $\mathcal{S} = \{\sum_{\mathcal{Q}_0} S_\alpha : S_\alpha \in \mathcal{S}_\alpha$ and finite index sets $\mathcal{Q}_0 \subset \mathcal{Q}\}$ is a family of $\sigma(Y, X)$-bounded sets in Y. Moreover, $Y = \bigcup_{\mathcal{S}} S$ if each $Y_\alpha = \bigcup_{\mathcal{S}_\alpha} S_\alpha$.

(e) Prove that $\prod_{\mathcal{Q}} \mathfrak{T}_{\mathcal{S}_\alpha} = \mathfrak{T}_{\mathcal{S}}$ on X. [*Hint*: $S = \sum_{\mathcal{Q}_0} S_\alpha \in \mathcal{S}$ implies $S^\bullet \subset \prod_{\mathcal{Q}_0} S_\alpha^\circ \times \prod_{\mathcal{Q}-\mathcal{Q}_0} X_\alpha \subset \aleph(\mathcal{Q}_0) S^\bullet$, where $^\circ$ denotes polars in $\langle X_\alpha, Y_\alpha \rangle$ and $^\bullet$ denotes polar in $\langle X, Y \rangle$.]

(f) Prove that $(Y, \mathfrak{T}_{\mathfrak{R}}) \overset{tc}{=} \sum_{\mathcal{Q}} \oplus (Y_\alpha, \mathfrak{T}_{\mathfrak{R}_\alpha})$. [*Hint*: If each $S_\alpha \in \mathcal{S}_\alpha$ is abs convex and closed, then (verify!) $(\prod_{\mathcal{Q}} S_\alpha)^\bullet = \overline{\{\bigcup_{\mathcal{Q}} j_\alpha(S_\alpha)^\circ\}}_{bc}$. Now use Theorem 18.22, Corollary 15.10.]

For the remainder of this problem $\langle X, Y \rangle$ is assumed to be injective.

(g) Show that $\prod_{\mathcal{Q}} \tau(X_\alpha, Y_\alpha) = \tau(X, Y)$ and $(Y, \tau(Y, X) \overset{tc}{=} \sum_{\mathcal{Q}} \oplus (Y_\alpha, \tau(Y_\alpha, X_\alpha))$. [*Hint*: The second equality follows from Corollary 18.32. For $\tau(X, Y) \subset \prod_{\mathcal{Q}} \tau(X_\alpha, Y_\alpha)$ show that $\mathcal{S} = \{\sum_{\mathcal{Q}_0} S_\alpha \in \mathcal{S}_\tau$ in Y_α, finite $\mathcal{Q}_0 \subset \mathcal{Q}\}$ is fundamental (Example 13.25) in $\mathcal{S}_\tau \subset Y$. Since every $E \in \mathcal{S}_\tau$ in Y is $\tau(Y, X) = \sum_{\mathcal{Q}} \oplus \tau(Y_\alpha, X_\alpha)$-bounded, $E \subset \sum_{\mathcal{Q}_0} \pi_\alpha(E) \in \mathcal{S}$ (Theorem 15.19).]

(h) Show that $\prod_{\mathcal{Q}} \beta(X_\alpha, Y_\alpha) = \beta(X, Y)$ on X and $(Y, \beta(Y, X)) \overset{tc}{=} \sum_{\mathcal{Q}} \oplus (Y_\alpha, \beta(Y_\alpha, X_\alpha))$. [*Hint*: $\prod_{\mathcal{Q}} \beta(X_\alpha, Y_\alpha) \subset \beta(X, Y)$ and $\sum_{\mathcal{Q}} \oplus \beta(Y_\alpha, X_\alpha) \subset \beta(Y, X)$ is straightforward. For the reverse inclusions establish that the collection of all finite products of $\sigma(Y_\alpha, X_\alpha)$-bounded sets {finite sums of $\sigma(X_\alpha, Y_\alpha)$-bounded sets} is fundamental in the family of $\sigma(Y, X)$-bounded {$\sigma(X, Y)$-bounded} sets.]

(i) Show that $\prod_{\mathcal{Q}} \sigma(X_\alpha, Y_\alpha) = \sigma(X, Y)$ and $(Y, \sigma(Y, X)) \overset{tc}{\subset} \sum_{\mathcal{Q}} \oplus (Y_\alpha, \sigma(Y_\alpha, X_\alpha))$. This last inclusion is an equality iff $\aleph(\mathcal{Q}) < \aleph_0$. [*Hint*: Use Corollary 18.32 for the first equality. Use ingenuity for the second.]

(j) Suppose each X_α is an LCT_2VS and let $Y_\alpha = X'_\alpha$. Then $\prod_{\mathcal{Q}} \xi(X_\alpha, X'_\alpha) = \xi(X, X')$ on X and $(X', \xi(X', X)) \overset{tc}{=} \sum_{\mathcal{Q}} \oplus (X'_\alpha, (\xi(X'_\alpha, X_\alpha))$ for each topology $\xi = \kappa_{bc}, \kappa_c, \kappa$ and λ. What about β^* and κ^*_{bc}?

18M. Weak Topological Direct Sums

Suppose that $\langle X, Y\rangle$ is an X-injective dual pair. If $(X, \sigma(X, Y)) \stackrel{t}{=} H \oplus K$ has projections $p_H: X \to H$ and $p_K: X \to K$, then $(Y, \sigma(Y, X)) \stackrel{t}{=} H^\circ \oplus K^\circ$ has projections $p_H^t: Y \to H^\circ$ and $p_K^t: Y \to K^\circ$.

18N. Subspaces of Polar Topologies

Let M be a subspace of X for a Y-injective dual pair $\langle X, Y\rangle$.

(a) $\langle M, Y/M^\circ\rangle$ is a Y/M°-injective dual pair under $\langle x, M^\circ + y\rangle = \langle x, y\rangle$. [*Hint*: Use Example 18.13 to verify that $\langle\, ,\rangle$ is well defined on $M \times Y/M^\circ$.]

(b) Both $\nu: (Y, \sigma(Y, X)) \to (Y/M^\circ, \sigma(Y/M^\circ, M))$ and $\mathbb{1}: (M, \sigma(M, X/M^\circ)) \to (X, \sigma(X, Y))$ are continuous.

(c) Show that $\{\nu(E)\}^\bullet = E^\circ \cap M \;\forall E \subset Y$, where $^\circ$ denotes polars in $\langle X, Y\rangle$ and $^\bullet$ denotes polars in $\langle M, Y/M^\circ\rangle$. [*Hint*: Take $\nu^t = \mathbb{1}$ in Theorem 18.19.]

Let $\mathcal{S}$ be a family of $\sigma(Y, X)$-bounded sets in Y satisfying (*), and take $\nu\mathcal{S} = \{\nu(B) = M^\circ + B : B \in \mathcal{S}\} \subset Y/M^\circ$. Then $\mathfrak{T}_{\mathcal{S}/M} = \mathfrak{T}_{\nu\mathcal{S}}$ on M. [*Hint*: $\nu\mathcal{S}$ satisfies (*) so that $(\nu S)^\bullet$ is both a $\mathfrak{T}_{\mathcal{S}/M}$ and $\mathfrak{T}_{\nu\mathcal{S}}$-nbd base at $0 \in M$.]

(e) Show that $\sigma(X, Y)/M = \sigma(M, Y/M^\circ) \subset \tau(X, Y)/M \subset \tau(X, Y/M^\circ)$ and $\beta(X, Y)/M \subset \beta(M, Y/M^\circ)$ on M.

(f) If X is an LCT_2VS, then $\sigma(X, X')/M = \sigma(M, M')$ and $\xi(X, X')/M \subset \xi(M, M')$ for each topology $\xi = \kappa_{bc}, \kappa_c, \kappa$ and λ. Can you extend this to include $\xi = \beta^*, \kappa_{bc}^*$?

18O. Quotients of Polar Topologies

Keep M and X, Y as in Problem 18N. Assume that $\mathcal{S}$ is a family of $\sigma(X, Y)$-bounded sets in X satisfying (*) and let $\mathcal{S} \cap M = \{B \cap M : B \in \mathcal{S}\}$.

(a) If $\mathfrak{T}_{\mathcal{S}} \in T(Y)$ is a topology of $\langle X, Y\rangle$, then $\mathfrak{T}_{\mathcal{S} \cap M} = \mathfrak{T}_{\mathcal{I}(\nu)}$ on Y/M°. [*Hint*: $X = (Y, \mathfrak{T}_{\mathcal{S}})'$ so that $\nu^t = \mathbb{1}$ and $\nu(B^\circ) = (B \cap M)^\bullet \;\forall B \in \mathcal{S}$. Show that $\mathcal{S} \cap M$ satisfies (*) and apply Theorem 15.12.]

(b) Show that $\sigma(Y/M^\circ, M) = \mathfrak{T}_{\mathcal{I}(\nu)}$ for $\sigma(Y, X)$ on Y and $\tau(Y/M^\circ, M) = \mathfrak{T}_{\mathcal{I}(\nu)}$ for $\tau(Y, X)$ on Y. [*Hint*: See Corollary 18.32.]

(c) If M is a closed vector subspace of an LCT_2VS X, then $\sigma(M', M) = \mathfrak{T}_{\mathcal{I}(\nu)}$ for $\sigma(X', X)$ on X'. [*Hint*: Show that $(X'/M^\circ, \mathfrak{T}_{\mathcal{I}(\nu)})$ and $(M', \sigma(M', M))$ in Example 18.12 are topological isomorphs iff M is closed.]

18P. Projective and Inductive Limits of Polar Topologies

Let $\mathfrak{F} = \{f_\alpha \in \mathfrak{L}(X_\alpha, X) : \alpha \in \mathfrak{A}\}$, where X is a vector space and each X_α $(\alpha \in \mathfrak{A})$ is an LCT_2VS whose dual X'_α carries a polar topology $\mathfrak{T}_{\mathcal{S}_\alpha}$. If $\mathfrak{T}_{\mathcal{I}\{\mathfrak{F}\}} \in T(X)$ is T_2 and

$\mathcal{S}=\left\{\bigcup_{i=1}^{n} f_{\alpha_i}(B_{\alpha_i}): B_{\alpha_i}\in\mathcal{S}_{\alpha_i} \text{ and } \alpha_i\in\mathcal{A}\right\}$, show that $\mathfrak{T}_{\mathcal{S}}=\mathfrak{T}_{\mathcal{P}\{\mathcal{F}^t\}}\in T(X')$ for $\mathcal{F}^t=\{f^t_\alpha\in\mathcal{L}(X', X'_\alpha):\alpha\in\mathcal{A}\}$. [*Hint*: Theorems 18.16 and 18.19.]

References for Further Study

On boundedness principles for bilinear mappings: Edwards [28].
On bilinear mappings and tensor products: Robertson and Robertson [84].

19. Bornological and Barreled Spaces

This section deals with two classes of LCTVSps intimately concerned with boundedness notions. Bornological spaces encompass 1° LCTVSps and are characterized by the property that bounded linear maps from such spaces to NLSps are continuous. Barreled spaces extend the Banach Steinhaus theorem 13.28 to this broader class of spaces, which generalize Catg II LCTVSps without the restriction of being 1°. Although distinct, bornological and barreled spaces both carry polar topologies and are invariant under inductive but not projective limits.

We begin by extending the terminology in Definition 13.18.

DEFINITION 19.1. A subset E of an LCTVS X is bornivorous if it absorbs every bounded subset of X. We call X bornological if every abs convex, bornivorous subset is a nbd.

THEOREM 19.2. (i) Every 1° LCTVS is bornological.
(ii) If X is bornological, $B\mathcal{L}(X, Y)=L(X, Y)\ \forall$ LCTVS Y.
(iii) An LCTVS X is bornological if $B\mathcal{L}(X, Y)=L(X, Y)\ \forall$ NLS Y.

Proof. First, assume that $\mathfrak{N}_0=\{u_1\supset u_2\supset\cdots\supset u_n\supset u_{n+1}\supset\cdots\}$ is an X-nbd base at 0 and let $E(0)$ be any abs convex, bornivorous subset of X. If E is not an X-nbd, there exist $x_n\in(1/n)u_n-E\ \forall n\in\mathbb{N}$ and $nx_n\to 0$ implies that $B=\{nx_n: n\in\mathbb{N}\}$ is bounded and absorbed by E—which is impossible (since $B\subset\lambda E$ implies $x_n\in(\lambda/n)E\subset E$ for $n\geq\lambda$). Suppose, next, that $f\in B\mathcal{L}(X, Y)$ for X, Y satisfying the hypothesis of (ii). If $B\subset X$ is bounded and $u(0)$ is an abs convex Y-nbd, then $B\subset\lambda f^{-1}(u)$ for some $\lambda>0$. The bornivorous set $f^{-1}(u)$ (being abs convex) is therefore an X-nbd. To verify (iii) assume that $E\subset X$ is abs convex and bornivorous. The Minkowski functional p_E is a seminorm on X which determines a norm on $X/p_E^{-1}(0)$ by $\|p_E^{-1}(0)+x\|_{p_E}=p_E(x)$. If $B\subset X$ is bounded, then $B\subset\lambda E$ and $\sup_B p_E(x)\leq\lambda$ for some $\lambda>0$. Therefore $\sup_{\nu(B)}\|p_E^{-1}(0)+x\|_{p_E}\leq\lambda$ and $\nu(B)$ is seen to be bounded in $X/\overset{-1}{p_E}(0)$. In essence, $\nu \in B\mathfrak{L}(X, X/\overset{-1}{p_E}(0)) = L(X, X/\overset{-1}{p_E}(0))$ and $\nu^{-1}\{S_{\frac{1}{2},\|\,\|_{p_E}}(0)\} \subset E$ ensures that E is an X-nbd. ■

THEOREM 19.3. The following are equivalent for an LCTVS X:

(i) X is bornological.

(ii) A seminorm p on X is continuous if $\sup_B p(x) < \infty \ \forall B \in \mathcal{S}_\beta$.

(iii) Let Y be a NLS. Then $\Phi \subset \mathfrak{L}(X, Y)$ is equicontinuous if $\Phi(B) \subset Y$ is bounded $\forall B \in \mathcal{S}_\beta$ in X.

(iv) $X' = B\mathfrak{L}(X, \mathbb{F})$ and X carries $\tau(X, X')$.

Proof. (i)$\Leftrightarrow$(ii). If X is bornological and $\sup_B p(x) \le \lambda_B < \infty$, then $B \subset \lambda_B \bar{S}_{1,p}(0)$. Therefore the abs convex set $\bar{S}_{1,p}(0)$ (being bornivorous) is an X-nbd and p is continuous (Theorem 14.6). If (ii) holds and $E \subset X$ is abs convex and bornivorous, then p_E (being bounded on each $B \in \mathcal{S}_\beta$) is continuous. Therefore (Theorem 14.10) $S_{1,p}(0) = E$ is an X-nbd and X is seen to be bornological. For (i)$\Rightarrow$(iii) let X be bornological and suppose that $\Phi(B) \subset Y$ is bounded $\forall B \in \mathcal{S}_\beta$. If $v(0)$ is an abs convex Y-nbd, then $u(0) = \bigcap_\Phi \phi^{-1}(v)$ (being abs convex and bornological) is an X-nbd for which $\Phi(u) \subset v$. Thus Φ is equicontinuous. Clearly (iii)$\Rightarrow$(i) follows from Theorem 19.2(ii). Also clear is $X' = B\mathfrak{L}(X, \mathbb{F})$ when $Y = \mathbb{F}$ in (iii). The second part of (iv) follows from (iii), since every $B \in \mathcal{S}_\beta$ is $\tau(X, X')$-bounded (Theorem 18.6). Indeed, $E \in \mathcal{S}_\tau$ implies $B \subset \lambda E^\circ$ and $E \subset \lambda B^\circ$ for some $\lambda > 0$. Therefore E is equicontinuous and X carries $\tau(X, X')$ by Example 18.24(d). Finally, we assume (iv) and let $\mathfrak{T}_b \in T(X)$ have $\mathfrak{N}_b = \{\text{abs convex bornivorous subsets}\}$ as its nbd base at $0 \in X$. Then (Theorem 14.11) $\tau(X, X') \subset \mathfrak{T}_b$ and $X' \subset (X, \mathfrak{T}_b)'$. Note that every $f \in (X, \mathfrak{T}_b)'$ satisfies $f \in E^\circ$ for some $E \in \mathfrak{N}_b$. If $B \in \mathcal{S}_\beta$, then $B \subset \lambda E$ and $f \in \lambda B^\circ$ for some $\lambda > 0$. This means that $X' = (X, \mathfrak{T}_b)'$ and $\tau(X, X') = \mathfrak{T}_b$. If p is a seminorm on X that is bounded on every $B \in \mathcal{S}_\beta$, then $\bar{S}_{1,p}(0) \in \mathfrak{N}_b$ is a $\mathfrak{T}_b$-nbd and p is X-continuous. This then establishes (iv)$\Rightarrow$(ii). ■

COROLLARY 19.4. If $\mathfrak{T} \in T(X)$ is a bornological topology of an X-injective dual pair $\langle X, Y \rangle$, every $\beta(Y, X)$-bounded set $\Phi \subset Y$ is $\mathfrak{T}$-equicontinuous.

Proof. If $\Phi \subset Y = (X, \mathfrak{T})'$ is $\beta(Y, X)$-bounded and $B \in \mathcal{S}_\beta$, then $\Phi \subset \lambda B^\circ$ for some $\lambda > 0$. Thus $\Phi(B) \subset \mathbb{F}$ is bounded $\forall B \in \mathcal{S}_\beta$ and Theorem 19.3(iii) applies. ■

Remark. For a NLS X, one has $\Phi \in \mathcal{S}_e$ in X' iff Φ is $\|\,\|$-bounded.

For a NLS X [Corollary 13.25, Example 18.24(e)] $(X', \beta(X', X))$ is complete. This, in fact, is a special case of the following in which [Problem 18E(d)] $\mathcal{S}_N = \{x_n \cup \{0\} : x_n \to 0\}$.

THEOREM 19.5. Let X be a bornological space and suppose that $\mathcal{S}$ is a collection of abs. convex, closed bounded subsets such that every $A \in \mathcal{S}_N$ is contained in some $B \in \mathcal{S}$. Then $L_{\mathcal{S}}(X, Y)$ is complete $\forall$ complete LCTVS Y. In particular, $(X', \mathfrak{T}_{\mathcal{S}})$ is complete.

Proof. Surely $X = \bigcup_{\mathcal{S}} A \subset \bigcup_{\mathcal{S}_N} A \subset \bigcup_{\mathcal{S}} B$, since $x = \{x, 0, 0, \ldots\}\ \forall x \in X$. Suppose that $\mathcal{G} \subset L(X, Y)$ is $\mathfrak{T}_{\mathcal{S}}$-Cauchy. Then $\mathcal{G}(x) \to y_x \in Y\ \forall x \in X$ and so (Theorem 7.9) $\mathcal{G} \xrightarrow{\mathfrak{T}_{\mathcal{S}}} \psi \in \mathfrak{L}(X, Y)$, where $\psi(x) = y_x\ \forall x \in X$. To show that $\psi \in B\mathfrak{L}(X, Y) = L(X, Y)$ we first establish that ψ is bounded on every $E \in \mathcal{S}$. Fix $E \in \mathcal{S}$ and let $v(0)$ be an abs convex Y-nbd. Let $G \in \mathcal{G}$ be $w_{E,v}(0)$-sized and let $g \in G$ so that $G \subset g + w_{E,v}$. If $g_0 \in G$ and $x \in E$, then $g_0(x) \in g(E) + v \subset (\lambda + 1)v$ for some $\lambda > 0$. Specifically, $g(E) \subset (\lambda + 1)v\ \forall g \in G$. If $g' \in G$ is chosen such that $g'(x) - y_x \in v$, there follows $y_x = (y_x - g'(x)) + g'(x) \subset (\lambda + 2)v$. Thus $\psi(E) \subset (\lambda + 2)v$ and ψ is seen to be bounded on every $E \in \mathcal{S}$. This, in turn, means that ψ is bounded. Otherwise, for some $B \in \mathcal{S}_\beta$ there is an abs convex Y-nbd $w(0)$ and sequence $\{x_n\} \subset B$ such that $\psi(x_n) \notin n^2 w\ \forall n \in \mathbb{N}$. This, however, is impossible, since (Theorem 13.22) $\{(1/n)x_n \cup 0\} \subset B$ and $\{\psi[(1/n)x_n] : n \in \mathbb{N}\} \subset n_0 w$ for $\lambda + 2 < n_0$. ■

Remark. It is easily verified that $L_{\mathcal{S}}(X, Y)$ is quasi-complete if Y is.

COROLLARY 19.6. For a bornological space X

(i) The topologies $\kappa(X', X)$, $\lambda(X', X)$, and $\beta(X', X)$ are complete.

(ii) If $\bar{K}_{bc}$ is compact $\forall$ compact set $K \subset X$, then $(X', \mathfrak{T}_{\mathcal{S}})$ is complete $\forall \mathfrak{T}_{\mathcal{S}}$ satisfying $\kappa_{bc}(X', X) = \kappa(X', X) \subset \mathfrak{T}_{\mathcal{S}} \subset \beta(X', X)$. In particular, $\tau(X', X)$ is complete.

Proof. Assertion (i) is clear, since every $\{x_n\} \cup 0 \in \mathcal{S}_N$ is X-compact. The proof of (ii) follows from Problem 13C and Theorem 18.22. ■

Remark. $\tau(X', X)$ is complete if X is sequentially complete since $\overline{\{x_n \cup 0\}_{bc}}$ is $\sigma(X, X')$-compact.

A few simple illustrations may be useful before proceeding.

Example 19.7 (a) If $\sigma(X, Y)$ is bornological for an X-injective dual pair $\langle X, Y\rangle$, then $\mathfrak{T} = \sigma(X, Y)\ \forall$ topology $\mathfrak{T}$ of $\langle X, Y\rangle$. This is true, since (Theorem 18.6) every $\mathfrak{T}$-bornivorous set is $\sigma(X, Y)$-bornivorous.

(b) Let $\langle X, Y\rangle$ be injective, where Y is a Catg II TVS. Assume further that there is a 1° topology $\mathfrak{T} \in T(X)$ of $\langle X, Y\rangle$. Then dim X is finite iff $\sigma(X, Y)$ is bornological. All this follows from Theorem 13.12, Theorem 19.2, and Example 18.5(e). Note, in particular, that a NLS X is finite dim iff $\sigma(X, X')$ is bornological (iff every $\sigma(X, X')$-bounded linear functional on X is $\sigma(X, X')$-continuous).

(c) If $(X, \mathfrak{T})$ is bornological, $\mathfrak{T} = \kappa^*_{bc}(X, X') = \tau(X, X') = \beta^*(X, X')$ by Problem 18E, Corollary 19.4.

(d) $\tau(X, X')$ cannot be both complete and bornological if $(X, \mathfrak{T})$ is an incomplete bornological space. [Otherwise, $X = (X', \tau(X', X))'$ is $\tau(X, X') = \mathfrak{T}$ complete by Corollary 19.6.] In particular, (Problem 17B) $\tau(X', X) \subsetneqq \beta(X', X) = \|\| \, \|\|$ for an incomplete seminormed space X.

Example 19.8. In contrast to Theorem 19.2(i) bornological spaces need not be 1°.

(a) $(X, \mathfrak{T}_{F_c})$ is always bornological, since (Example 18.29, Theorem 19.3) $\mathfrak{T}_{F_c} = \tau(X, X^*)$ and $(X, \mathfrak{T}_{F_c})' = X^* = B\mathfrak{L}(X; \mathbb{F})$.

(b) Both $(C_0(X; \mathbb{F}), \mathfrak{T}_{\mathcal{I}'(\mathbb{1})})$ and $(\mathfrak{D}_0^{\infty}(\mathbb{R}; \mathbb{F}), \mathfrak{T}_{\mathcal{I}'(\mathbb{1})})$ in Problem 15D are bornological and nonmetrizible. Reason: Each $(C_K(X; \mathbb{F}), \| \, \|_0)$ is Banach and each $(D_{K_n}(\mathbb{R}; \mathbb{F}), \xi_n)$ is Frechet and our next theorem prevails.

THEOREM 19.9. An inductive limit of bornological spaces is bornological.

Proof. Let $\mathfrak{F} = \{f_\alpha \in \mathfrak{L}(X_\alpha, X) : \alpha \in \mathfrak{A}\}$, where each X_α $(\alpha \in \mathfrak{A})$ is bornological. If $u(0) \subset X$ is abs convex and $\mathfrak{T}_{\mathcal{I}'\{\mathfrak{F}\}}$-bornivorous, then every $f_\alpha^{-1}(u)$ is abs convex and X_α-bornivorous. (If $B \subset X_\alpha$ is bounded, then $f_\alpha(B)$ is $\mathfrak{T}_{\mathcal{I}'\{\mathfrak{F}\}}$-bounded and $B \subset \lambda_\alpha f_\alpha^{-1}(u)$ for some $\lambda_\alpha > 0$.) Thus $f_\alpha^{-1}(u)$ is an X_α-nbd $\forall \alpha \in \mathfrak{A}$ and $u(0)$ is therefore a $\mathfrak{T}_{\mathcal{I}'\{\mathfrak{F}\}}$-nbd of $0 \in X$. ■

Remark. Each LCTVS X_α is bornological if $X = \prod_{\mathfrak{A}} X_\alpha$ is bornological [Problem 3A(b)]. However, the product of bornological spaces need not be bornological (Problem 19B) unless it is a finite product (Theorem 15.17). In the same vein a closed subspace of a bornological space X need not be bornological (Köthe [55], p. 384)—unless it has a topological supplement (Theorem 15.26) in X.

Theorem 19.9 has a converse in the following sense:

THEOREM 19.10. Every bornological $\{T_2\}$ space $(X, \mathfrak{T})$ is an inductive limit of seminormed {normed} spaces.

Proof. Let $\mathfrak{B}$ denote the collection of abs convex, closed bounded subsets of a bornological space $(X, \mathfrak{T})$. Our proof follows from Theorem

14.17 by establishing that $\mathfrak{T} = \mathfrak{T}_{\mathscr{I}'\{\mathfrak{F}\}}$ for $\mathfrak{F} = \{\mathbb{1}_B = \mathbb{1} : (E_B, p_B) \to X : B \in \mathfrak{B}\}$. Clearly, (Example 15.16) $\mathfrak{T} \subset \mathfrak{T}_{\mathscr{I}'\{\mathfrak{F}\}}$, since $\mathfrak{T}/E_B \subset \mathfrak{T}(p_B)\ \forall B \in \mathfrak{B}$. In the other direction suppose that $v(0)$ is any abs convex $\mathfrak{T}_{\mathscr{I}'\{\mathfrak{F}\}}$-nbd and let $K \subset X$ be $\mathfrak{T}$-bounded. Then $\bar{K}_{bc} = B \in \mathfrak{B}$ is $\mathfrak{T}(p_B)$-bounded, therefore $\mathfrak{T}_{\mathscr{I}'\{\mathfrak{F}\}}/E_B$-bounded, and one has $B \subset \lambda(E_B \cap v) \subset \lambda v$ for some $\lambda > 0$. The conclusion is that $v(0)$ (being $\mathfrak{T}$-bornivorous) is a $\mathfrak{T}$-nbd. ■

Corollary 19.11. Every sequentially complete bornological $\{T_2\}$ space is an inductive limit of complete {Banach} spaces.

Proof. Since every $B \in \mathfrak{B}$ is sequentially complete and sequentially closed, every (E_B, p_B) is complete {Banach}. ■

Barreled Spaces

So important is Theorem 13.28 that the class of LCTVSps for which this result holds merits special attention.

Definition 19.12. An abs convex absorbent closed subset of a TVS is called a barrel. An LCTVS X is barreled if every barrel is a nbd of $0 \in X$.

Example 19.13. (a) In a Catg II TVS X every barrel T is a nbd of 0, since (Theorem 13.6) $X = \bigcup_{n=1} nT$ implies Int $T \neq \emptyset$ and $T - T = 2T$ is a nbd of $0 \in X$. Although Catg II LCTVSps are barreled, the converse fails. $(X, \mathfrak{T}_{F_c})$ is a barreled, Catg I LCT_2VS when dim $X = \aleph_0$ (Problem 13G).

(b) An LCTVS $(X, \mathfrak{T})$ is barreled iff $\mathfrak{T}' \subset \mathfrak{T}\ \forall$ locally convex topology $\mathfrak{T}' \in T(X)$ that has a nbd base of $\mathfrak{T}$-closed sets; viz, if $\mathfrak{T}$ is barreled and $\mathfrak{T}'$ has a $\mathfrak{T}$-closed nbd base, then every $\mathfrak{T}'$-nbd $u(0)$ contains a $\mathfrak{T}$-closed, $\mathfrak{T}'$-nbd $v(0)$ and an abs convex $\mathfrak{T}'$-nbd $w(0) \subset v(0)$ such that $\bar{w} \subset v \subset u$. Since $\bar{w}$ is a $\mathfrak{T}$-barrel, it is a $\mathfrak{T}$-nbd and $\mathfrak{T}' \subset \mathfrak{T}$ follows. Suppose, conversely, that $\mathfrak{T}' \subset \mathfrak{T}\ \forall \mathfrak{T}' \in T(X)$ above. If $T \subset X$ is a $\mathfrak{T}$-barrel, then $\mathfrak{N}' = \{\alpha T : \alpha > 0\}$ is a nbd base for some $\mathfrak{T}' \in T(X)$. Since every $\alpha T \in \mathfrak{N}'$ is $\mathfrak{T}$-closed, $\mathfrak{T}' \subset \mathfrak{T}$ and T is a $\mathfrak{T}$-nbd.

Remark. If $\sigma(X, Y)$ is barreled for an X-injective dual pair $\langle X, Y\rangle$, then $\mathfrak{T} = \sigma(X, Y)\ \forall$ topology $\mathfrak{T}$ of $\langle X, Y\rangle$.

(c) Let $\langle X, Y\rangle$ be injective, where X is a 1° TVS and Y is a Catg II TVS. Then $\sigma(X, Y)$ is barreled iff dim X is finite. Clearly, $\sigma(X, Y)$ is barreled when dim $X < \aleph_0$ by Theorem 13.12 and (a) above. If $\sigma(X, Y)$ is barreled, then $\sigma(X, Y)$ is 1° and dim X is finite by (b) above and Example 19.7(b). The above also demonstrates that a NLS X is finite dim iff $\sigma(X, X')$ is barreled.

(d) Many, but not all, barreled results do not require local convexivity. Nonlocally convex (i.e., nonbarreled) spaces in which every barrel is a nbd of 0 do exist. See Example 14.22.

(e) (Theorem 10.7) Let Y be an LCTVS and suppose that X is a TVS in which every barrel is a nbd of 0. Then every $f \in \mathfrak{L}(X, Y)$ is nearly continuous and every surjection $F \in \mathfrak{L}(Y, X)$ is nearly open. Indeed, $\overline{f^{-1}(v)}$ and $\overline{F(v)}$ are abs convex and absorbent $\forall$ abs convex nbd $v(0) \subset Y$.

Boundedness is frequently characterized via polars of barrels.

THEOREM 19.14. A subset $E \subset X'$ for a TVS X is $\sigma(X', X)$-bounded $\{\beta(X', X)$-bounded$\}$ iff $E \subset T^{\circ}$ for a barrel {bornivorous barrel} $T \subset X$.

Proof. If E is $\sigma(X', X)$-bounded, then $E^{\circ} \subset X$ is a barrel (Theorems 18.10 and 18.6) and $E \subset (E^{\circ})^{\circ}$ in X. Theorem 18.10(iv) also ensures that T° is $\sigma(X', X)$-bounded $\forall$ barrel $T \subset X$. If E is $\beta(X', X)$-bounded, then $E^{\circ} \subset X$ is also bornivorous (since $B \in \mathcal{S}_{\beta}$ implies B° is a $\beta(X', X)$-nbd and $E \subset \lambda B^{\circ}$ yields $B \subset B^{\circ\circ} \subset \lambda E$ for some $\lambda > 0$). In the other direction T° is $\beta(X', X)$-bounded $\forall$ bornivorous barrel $T \subset X$. ■

THEOREM 19.15. In a TVS X a barrel absorbs every abs convex, sequentially complete, sequentially closed subset of X. Accordingly, a barrel absorbs every bounded subset of a sequentially complete LCTVS.

Proof. The first assertion is an immediate consequence of Corollary 14.18. The second follows from the first and the fact that $\bar{B}_{bc}$ is abs convex and bounded $\forall$ bounded set $B \subset X$ [Theorems 13.9, 13.20(iib)] and Corollary 14.3⁺). ■

COROLLARY 19.16. For a quasi-complete LCTVS X every $\sigma(X', X)$-bounded set $E \subset X'$ is $\beta(X', X)$-bounded.

Proof. If E is $\sigma(X', X)$-bounded, $E \subset T^{\circ}$ for some barrel $T \subset X$. This barrel is bornivorous (and so E is $\beta(X', X)$-bounded), since every $\bar{B}_{bc}$ $(B \in \mathcal{S}_{\beta})$ is complete and absorbed by T. ■

The promised extension of Theorem 13.28 is

THEOREM 19.17 (Banach Steinhaus). Let $\Phi = \{\phi_{\alpha} : \alpha \in \mathcal{Q}\} \subset L(X, Y)$, where Y is an LCTVS and X is a TVS in which every barrel is a nbd of 0. If $\Phi(X)$ is Y-bounded $\forall x \in X$, then Φ is equicontinuous.

Proof. Given any abs convex, closed nbd $u(0) \subset Y$, the set $E = \bigcap_{\mathfrak{a}} \overset{-1}{\phi_\alpha}(u)$ (being a barrel) is an X-nbd satisfying $\Phi(E) \subset u$. ■

Theorem 18.25 and Corollary 19.16 may be supplemented with

COROLLARY 19.18. The following are equivalent for $E \subset X'$ when X is a TVS in which every barrel is a nbd of 0:

(i) E is $\sigma(X', X)$-bounded;
(ii) E is $\beta(X', X)$-bounded;
(iii) E is equicontinuous;
(iv) E is $\sigma(X', X)$-relatively compact.

Proof. The first implications in each chain (iii) $\Rightarrow$ (ii) $\Rightarrow$ (i) and (iii) $\Rightarrow$ (iv) $\Rightarrow$ (i) follow from Example 13.27(e) and Theorem 18.25, respectively. Furthermore, (i) $\Rightarrow$ (iii) by Example 15.4(a) and Theorem 19.17. ■

COROLLARY 19.19. For an LCTVS $(X, \mathfrak{T})$ the following are equivalent:

(i) $(X, \mathfrak{T})$ is barreled;
(ii) $\sigma(X', X)$-boundedness implies equicontinuity;
(iii) $\mathfrak{T} = \beta(X, X')$.

Proof. If $(X, \mathfrak{T})$ is barreled, then $\mathcal{S}_e = \mathcal{S}_\beta$ in X' and (Theorem 18.26) $\mathfrak{T} = \mathfrak{T}_{\mathcal{S}_e}$ coincides with $\beta(X, X')$ on X. Thus (i) $\Rightarrow$ (iii). If $\mathfrak{T} = \beta(X, X')$, and $E \subset X'$ is $\sigma(X', X)$-bounded, then E° is a $\beta(X, X') = \mathfrak{T}$-nbd and (Theorem 18.14) $E^{\circ\circ}$, therefore E, is equicontinuous. Thus (iii) $\Rightarrow$ (ii). Clearly (ii) $\Rightarrow$ (i) since $T = (T^\circ)^\circ$ is a $\mathfrak{T}_{\mathcal{S}_e} = \mathfrak{T}$ nbd $\forall$ barrel $T \subset X$ (Theorems 18.10 and 18.11). ■

In comparison with Theorem 19.5, consider

THEOREM 19.20. Let $\mathcal{S}$ be a collection of bounded sets that covers the barreled space X. Then $L_{\mathcal{S}}(X, Y)$ is quasi-complete $\forall$ LCTVS Y. In particular, $(X', \mathfrak{T}_{\mathcal{S}})$ is quasi-complete.

Proof. Suppose $\Phi \subset L(X, Y)$ is $\mathfrak{T}_{\mathcal{S}}$-closed and bounded. If $\mathcal{G}$ is any $\mathfrak{T}_{\mathcal{S}}$-Cauchy filter in Φ, then (as in Theorem 19.5) $\mathcal{G} \xrightarrow{\mathfrak{T}_{\mathcal{S}}} \psi \in \bar{\Phi}^{\mathfrak{T}_{\mathcal{S}}} \subset \mathfrak{L}(X, Y)$. An application of Theorem 19.17 and Example 13.27(e) ensures that $\psi \in L(X, Y)$. ■

Being bornological and being barreled are generally independent concepts, yet both have some important properties in common. For example, compare Theorem 19.9 with

THEOREM 19.21. An inductive limit of barreled spaces is barreled.

Proof. Let $\mathfrak{F} = \{f_\alpha \in \mathfrak{L}(X_\alpha, X) : \alpha \in \mathcal{Q}\}$, where each X_α $(\alpha \in \mathcal{Q})$ is barreled. If $T \subset X$ is a $\mathfrak{T}_{\mathcal{G}'\{\mathfrak{F}\}}$-barrel, then $f_\alpha^{-1}(T)$ is a barrel in X_α $\forall \alpha \in \mathcal{Q}$ and so T is a $\mathfrak{T}_{\mathcal{G}'\{\mathfrak{F}\}}$-nbd of $0 \in X$. ■

Remark. Each LCTVS X_α is barreled if $X = \prod_{\mathcal{Q}} X_\alpha$ is barreled [Problem 3A(b)]. Furthermore, unlike bornological spaces, every product $X = \prod_{\mathcal{Q}} X_\alpha$ of barreled T_2 spaces is barreled. Reason: Each X_α carries $\beta(X_\alpha, X'_\alpha)$ and (Theorem 15.20) $\langle X, X'\rangle = \langle \prod_{\mathcal{Q}} X_\alpha, \sum_{\mathcal{Q}} \oplus X'_\alpha\rangle$ is injective so that $\mathfrak{T}_\pi = \beta(X, X')$ on X [Problem 18L(h)].

Remark. A closed subspace of a barreled space need not be barreled (unless it has a topological supplement in the space). For example, every complete, nonbarreled LCT_2VS is a closed, nonbarreled subspace of the barreled product of Banach spaces [Theorem 15.8, Example 19.13(a)].

Corollary 19.11 and Example 19.13(a) applied to Theorem 19.21 yield

COROLLARY 19.22. A sequentially complete bornological space is barreled.

This is a good place to embellish a few earlier results.

Example 19.23. (a) Let $X = \{x = (x_1, x_2, \ldots x_k, 0, 0, \ldots) : x_k \in \mathbb{F}\}$ carry the sup norm $\|\ \|_0$ and define $E = \{f_n \in X' : n \in \mathbb{N}\}$, where $f_n(x) = nx_n$ $\forall x \in X$. Since $|||f_n||| = n$ $\forall n \in \mathbb{N}$, the set $E \subset X'$ is $|||\ ||| = \beta(X', X)$ unbounded and nonequicontinuous. However, E is $\sigma(X', X)$-bounded because $\sup_n |\langle f_n, x\rangle| \leq \sum_{m=1}^{k} m|x_m| < \infty$ $\forall x = (x_1, x_2, \ldots x_k, 0, 0, \ldots) \in X$. This demonstrates (Theorem 19.17, Corollary 19.22) that the bornological space $(X, \|\ \|_0)$ is nonbarreled and incomplete (in fact, X is not closed in the Banach space $(c_0, \|\ \|_0)$, since $\{(1, \frac{1}{2}, \ldots, 1/n, 0, 0, \ldots) \xrightarrow{\|\ \|_0} (1, \frac{1}{2}, \frac{1}{3}, \ldots) \in c_0 - X\}$). Note also that quasi completeness cannot be omitted in Corollary 19.16.

(b) Suppose that $X = \bigcup_{n=1} K_n$ in Problem 15D(a) and define maps $\phi_n : Y = C_0(X; \mathbb{F}) \longrightarrow \mathbb{R}$, $f \rightsquigarrow \sup_{X-K_n} |f(x)|$ $\forall n \in \mathbb{N}$. Then $|||\phi_n||| = n$ $\forall n \in \mathbb{N}$ and $E =$

$\{\phi_n : n \in \mathbb{N}\}$ is $||| \, ||| = \beta(Y', Y)$ unbounded. On the other hand, $\sup_{\overline{X-K_n}} |f(x)| = 0$ for large $n = n(f)$. Therefore $\phi_n \xrightarrow{\sigma(Y', Y)} 0 \in Y'$ and E is $\sigma(Y', Y)$-bounded. The conclusion, of course, is that $(C_0(X; \mathbb{F}), \| \, \|_0)$ is nonbarreled and incomplete.

(c) Illustrations of barreled, nonbornological spaces do not abound in the literature. Such spaces are usually involved and are not given here. (See Nachbin [68], Shirota [91], Warner [103].)

(d) The complete bornological spaces in Example 19.8 are barreled. The spaces $(\mathfrak{D}^k(\mathbb{R}; \mathbb{F}), \xi)$, $(\mathfrak{D}_r^\infty(\mathbb{R}; \mathbb{F}), \zeta)$ and $\mathcal{H}_\kappa(\Omega)$ in Problem 14E (being Frechet) are bornological and barreled.

Example 19.24. Assume that Y is an LCT_2VS in Definition 16.1. Then $\mathcal{CA}(\Omega, \Sigma; Y)$ is $\mathfrak{T}_{\mathcal{P}(\mathcal{A})}$-barreled iff Σ is finite and Y is barreled.

If $\Sigma = \{E_1, \ldots, E_n\}$, then $\mathfrak{T}_{\mathcal{P}(\mathcal{A})} = \mathfrak{T}_{\mathcal{I}'(\mathbb{1}_\Omega)}$ relative to $\mathbb{1}_\Omega = \{\mathbb{1}_\omega = \mathbb{1} : (\mathfrak{D}_\omega(\Omega, \Sigma; Y), \mathfrak{T}_{\mathcal{P}(\mathcal{A})}) \to \mathcal{CA}(\Omega, \Sigma; Y); \omega \in E_i \in \Sigma\}$ (viz., if $u(0) \subset \mathcal{CA}(\Omega, \Sigma; Y)$ is a $\mathfrak{T}_{\mathcal{I}'(\mathbb{1}_\Omega)}$-nbd, then $\psi_{\omega_i}^{-1} \mathbb{1}_{\omega_i}^{-1}(u)$ is a nbd of $0 \in Y$. Therefore $j_{\omega_i} \psi_{\omega_i}^{-1} \mathbb{1}_{\omega_i}^{-1}(u) \subset \prod_{i=1}^{n} Y$ is a $\mathfrak{T}_{\mathcal{I}'(J)} = \mathfrak{T}_\pi$ nbd and $u(0) = \mathfrak{e}^{-1} j_{\omega_i} \psi_{\omega_i}^{-1} \mathbb{1}_{\omega_i}^{-1}(u)$ is a $\mathfrak{T}_{\mathcal{P}(\mathcal{A})}$-nbd). If Y is also barreled, then $(\mathfrak{D}_\omega(\Omega, \Sigma; Y), \mathfrak{T}_{\mathcal{P}(\mathcal{A})})$ is barreled [Example 16.12(a)] and so is $\mathfrak{T}_{\mathcal{P}(\mathcal{A})} = \mathfrak{T}_{\mathcal{I}'(\mathbb{1}_\Omega)}$ on $\mathcal{CA}(\Omega, \Sigma; Y)$.

If $\mathcal{CA}(\Omega, \Sigma; Y)$ is $\mathfrak{T}_{\mathcal{P}(\mathcal{A})}$-barrelcd, then [Example 16.12(b)] $(\mathfrak{D}_\omega(\Omega, \Sigma; Y), \mathfrak{T}_{\mathcal{P}(\mathcal{A})})$ and Y are barreled. Let $v(0) = \bigcap_{\Sigma} A_E^{-1}(u)$ for an abs convex, closed nbd $u(0) \subsetneqq Y$. Since $v(0)$ is a $\mathfrak{T}_{\mathcal{P}(\mathcal{A})}$-barrel (absorbency follows from Corollary 18.8), it contains a $\mathfrak{T}_{\mathcal{P}(\mathcal{A})}$-base nbd $\bigcap_{j=1}^{m} A_{F_j}^{-1}(w)$ for disjoint $\{F_1, \ldots, F_m\} \in \Sigma$. Furthermore, $v(0) \subset A_E^{-1}(u) \, \forall E \in \Sigma$ implies (Lemma 16.13, Theorem 16.3) that $\Sigma = \mathfrak{R}\{F_1, \ldots, F_m\}$ is finite.

Remark. Similar reasoning shows that Y is bornological if $\mathcal{CA}(\Omega, \Sigma; Y)$ is $\mathfrak{T}_{\mathcal{P}(\mathcal{A})}$-bornological, whereas $\mathfrak{T}_{\mathcal{P}(\mathcal{A})}$ is bornological whenever Σ is finite and Y is bornological.

PROBLEMS

19A. Bornological Space Associated with an LCTVS

Let $\mathfrak{B}$ denote the collection of all bounded subsets of an LCTVS $(X, \mathfrak{T})$. Then there is a finest bornological topology $\mathfrak{T}_{bo} \in T(X)$ which has $\mathfrak{B}$ as the collection of all $\mathfrak{T}_{bo}$-bounded subsets of X.

(a) The collection of all abs convex, $\mathfrak{T}$-bornivorous subsets of X constitutes a $\mathfrak{T}_{bo}$-nbd base at 0.

(b) Show that $\mathfrak{T}_{bo} = \mathfrak{T}_{\mathcal{G}(\mathfrak{F})}$ in Theorem 19.10. Therefore $\mathfrak{T} = \mathfrak{T}_{bo}$ iff $\mathfrak{T}$ is bornological.

(c) If $B\mathfrak{L}(X, \mathbb{F})$ is the space of $\mathfrak{T}$-bounded, linear functionals on X, then $\mathfrak{T}_{bo} = \tau(X, B\mathfrak{L}(X; \mathbb{F})$ for the dual pair $\langle X, B\mathfrak{L}(X; \mathbb{F})\rangle$.

(d) Let $T_{\mathfrak{B}}(X)$ denote the collection of locally convex, $\mathfrak{T}$-bornivorous topologies $\xi \in T(X)$. Then $\mathfrak{T}_{bo} = \mathfrak{T}_{\mathcal{P}(\mathcal{A})}$, where $\mathcal{A} = \{\mathbb{1}_\xi = \mathbb{1} : X \to (X, \xi); \xi \in T_{\mathfrak{B}}(X)\}$.

(e) Why is there also a finest bornological T_2 topology $\mathfrak{T}'_{bo} \in T(X)$ whose bounded sets coincide with $\mathfrak{B}$? [*Hint*: Consider Definition 15.13.]

19B. Products of Bornological Spaces

Let $X = \prod_{\mathfrak{A}} X_\alpha$, where each LCT_2VS X_α is bornological. Then [Theorem 19.3, Problem 18L(g)] $\mathfrak{T}_\pi = \tau(X, X')$ and X is bornological iff $X' = B\mathfrak{L}(X; \mathbb{F})$.

(a) Show that $X' = B\mathfrak{L}(X, \mathbb{F})$ whenever $\aleph(\mathfrak{A}) \leq \aleph_0$. [*Hint*: If $\aleph(\mathfrak{A}) = \aleph_0$ and $F \in B\mathfrak{L}(X; \mathbb{F})$, there is an $n_0 \in \mathbb{N}$ such that $F(x) = 0$ $\forall x = (0, \ldots, 0, x_{n_0+1}, x_{n_0+2}, \ldots) \in X$. (Otherwise there is a sequence $\{\tilde{x}_m = (0, \ldots, 0, x_{m+1}, x_{m+2}, \ldots)\} \in X$ such that $\{[m/F(\tilde{x}_m)]\tilde{x}_m : m \in \mathbb{N}\}$ is $\mathfrak{T}_\pi$-bounded but not F-bounded.) Using $Fj_i \in X'_i$ $(i = 1, 2, \ldots, n_0)$ and Theorem 15.20, obtain $F = \sum_{i=1}^{n_0} Fj_i \in X'$.]

Suppose that $\aleph(\mathfrak{A}) > \aleph_0$. Fix $F \in B\mathfrak{L}(X; \mathbb{F})$, define $F_{\mathfrak{A}_0} : \underset{\prod_{\mathfrak{A}} x_\alpha \rightsquigarrow F(\bigcup_{\mathfrak{A}_0} i_\alpha(x_\alpha))}{X \longrightarrow \mathbb{F}}$ for every subset $\mathfrak{A}_0 \subset \mathfrak{A}$, and let $\mathfrak{B} = \{\mathfrak{A}_0 \subset \mathfrak{A} : F_{\mathfrak{A}_0} \neq 0\}$.

(b) $\aleph(\mathfrak{B}_0) < \aleph_0$ for every disjoint subfamily $\mathfrak{B}_0 \subset \mathfrak{B}$. [*Hint*: Generalize the hint in (a) above.]

(c) Using (*b*), show that there is a finite disjoint cover $\{\mathfrak{B}_1, \mathfrak{B}_2, \ldots, \mathfrak{B}_{n_0}\}$ of $\mathfrak{A}$ such that $C \subset \mathfrak{B}_i$ implies $F_C = F_{\mathfrak{B}_i}$ or $F_C = 0$.

(d) Verify that $F = \sum_{i=1}^{n_0} F_{\mathfrak{B}_i}$ and $F_{\mathfrak{B}_i} \in B\mathfrak{L}(X; \mathbb{F})$ $\forall 1 \leq i \leq n_0$.

For each i $(1 \leq i \leq n_0)$ define $\mu_i : (\mathfrak{A}, \mathcal{P}(\mathfrak{A})) \to \{0, 1\}$ by $\mu_i(\mathfrak{A}_0) = 0$ if $F_{\mathfrak{B}_i \cap \mathfrak{A}_0} = 0$ and $\mu_i(\mathfrak{A}_0) = 1$ if $F_{\mathfrak{B}_i \cap \mathfrak{A}_0} = F_{\mathfrak{B}_i}$. Note that $F_{\mathfrak{B}_i \cap \mathfrak{A}_0} = \mu_i(\mathfrak{A}_0) F_{\mathfrak{B}_i}$ $\forall \mathfrak{A}_0 \subset \mathfrak{A}$ and $1 \leq i \leq n_0$.

(e) X is bornological if there does not exist a Ulam measure on $\mathfrak{A}$, that is, a nonzero measure $\mu : (\mathfrak{A}, \mathcal{P}(\mathfrak{A})) \to \{0, 1\}$ satisfying $\mu(\{\alpha\}) = 0$ $\forall \alpha \in \mathfrak{A}$. [*Hint*: Assume that no Ulam measure exists on $\mathfrak{A}$. Then $\forall 1 \leq i \leq n_0$ there is an $\alpha_i \in \mathfrak{B}_i$ such that $\mu_i(\alpha_i) = 1$. Thus $F_{\mathfrak{B}_i} = F_{\alpha_i} \in X'_{\alpha_i}$ and $F = \sum_{i=1}^{n_0} F_{\alpha_i} \in X'$.]

(f) X is bornological if there is a Ulam measure on $\mathfrak{A}$. [*Hint* (Kelley and Namioka [52]): If μ is Ulam, then $f \in L_1(\mathfrak{A}, \mu)$ $\forall f : \mathfrak{A} \to \mathbb{F}$ since $\lim_n \mu\{\alpha \in \mathfrak{A} : |f(\alpha)| > n\} = 0$. For each $\alpha \in \mathfrak{A}$, fix an $f_\alpha \neq 0 : X_\alpha \to \mathbb{F}$ and define $F : X \to \mathbb{F}$ by $F(\{x_\alpha\}) = \int_{\mathfrak{A}} f_\alpha(x_\alpha)\, d\mu(\alpha)$. Then $F \in B\mathfrak{L}(X; \mathbb{F})$, since $B \subset X$ being bounded implies $B \subset \prod_{\mathfrak{A}} B_\alpha$ for bounded $B_\alpha \subset X_\alpha$ and $\sup_B |F(\{x_\alpha\})| \leq \int_{\mathfrak{A}} \sup_{B_\alpha} |f_\alpha(x_\alpha)|\, d\mu(\alpha)$. However, $F \notin X'$.

Otherwise (Theorem 15.10) there is a finite index set $\mathcal{Q}_0 \subset \mathcal{Q}$ such that $F(x)=0$ $\forall x=\{x_\alpha\}$ having $x_\alpha=0$ for $\alpha \in \mathcal{Q}_0$. This, however, is impossible since $y=\{y_\alpha\} \in X$, defined as $y_\alpha = 0\ \forall \alpha \in \mathcal{Q}_0$ and $y_\alpha = [1/f_\alpha(x_\alpha)]^{-1} x_\alpha\ \forall \alpha \notin \mathcal{Q}_0$, yields $0=F(y)=\int_{\mathcal{Q}} f_\alpha(x_\alpha)\, d\mu(\alpha) = \mu(\mathcal{Q}-\mathcal{Q}_0)=1.$]

Remark. It remains an open question whether there exists a Ulam measure on $\mathcal{Q}$.

19C. Infrabarreled Spaces

An LCTVS $(X, \mathfrak{T})$ is infrabarreled if every bornivorous barrel is a nbd of $0 \in X$. Thus bornological and barreled spaces are infrabarreled.

(a) The following are equivalent: (i) $\mathfrak{T}$ is infrabarreled; (ii) $\beta(X', X)$-bounded sets are equicontinuous; (iii) $\mathfrak{T} = \beta^*(X, X')$. [*Hint*: Use Theorem 19.14, Theorem 18.14, Theorem 18.26, and Theorem 18.11 to get (i) $\Rightarrow$ (ii) $\Rightarrow$ (iii) $\Rightarrow$ (i), respectively.]

(b) A sequentially complete infrabarreled space is barreled.

(c) $(X, \beta(X', X))$ is quasi-complete if $(X, \mathfrak{T})$ is infrabarreled. [Compare this with Theorems 19.5 and 19.20.]

(d) An inductive limit of infrabarreled spaces is infrabarreled.

(e) Each LCTVS X_α is infrabarreled when $X = \prod_{\mathcal{Q}} X_\alpha$ is infrabarreled. If each X_α is an infrabarreled LCT_2VS, then $X = \prod_{\mathcal{Q}} X_\alpha$ is infrabarreled.

19D. Completion Problem

The completion $\check{X}$ of an LCT_2VS X is barreled {bornological and barreled} if X is infrabarreled {bornological}. [*Hint*: The key here is Problem 18K. For infrabarreled X use Problem 19C(a), (b) and the fact that $\beta(\check{X}', \check{X}) \supset \beta(X', X)$. For bornological X use Theorem 19.3 and Corollary 19.22.]

19E. An Incomplete, Quasi-Complete LCT_2VS

Verify that $(l_1, \sigma(l_1, c_0))$ is as headed above. [*Hint*: $\sigma(l_1, c_0)$-incompleteness follows from Example 18.5(b) and the fact that $l_1 = (c_0, \|\ \|_0)' \subsetneqq c_0^*$. For $\sigma(l_1, c_0)$-quasi-completeness use Theorem 19.20.]

It is frequently important to know when the strong dual of an LCTVS is bornological or barreled. The following two problems are geared to this question.

19F. $\beta(X', X)$-Bornological Property

Let X be an LCTVS with nbd base $\mathfrak{N}$ at 0 and let $\mathfrak{N}^\circ = \{u^\circ \subset X' : u \in \mathfrak{N}\}$.

(a) Show that the topology $\mathfrak{T}_{\mathcal{G}'\{\mathcal{F}\}} \in T(X')$ is bornological for $\mathcal{F} = \{\mathbb{1}_{u^\circ} = \mathbb{1} : (E_{u^\circ}, p_{u^\circ}) \to X';\ u^\circ \in \mathfrak{N}^\circ\}$.

(b) Using (a) above, prove that $\beta(X', X) \subset \mathfrak{T}_{\mathcal{G}'\{\mathcal{F}\}} \subset \beta_{bo}(X', X)$ (cf Problem 19A). [*Hint*: Theorems 18.15, 18.25, and 14.17 and Example 15.16.]

Remark. This demonstrates that $\beta(X', X)$ is bornological iff $\beta(X', X) = \mathfrak{T}_{\mathcal{G}'\{\mathcal{F}\}}$.

(c) If X is infrabarreled, then $\beta_{bo}(X', X) = \mathfrak{T}_{\mathcal{G}'\{\mathcal{F}\}}$. Furthermore, $\beta(X', X)$ is barreled if it is bornological. [*Hint*: $\beta(X', X)$-bounded sets are $\mathfrak{T}_{\mathcal{G}'\{\mathcal{F}\}}$-bounded [Problem 19C(a) and Theorems 18.14 and 14.17] so that $\beta_{bo}(X', X) \subset \{\mathfrak{T}_{\mathcal{G}'\{\mathcal{F}\}}\}_{bo}$. For the second assertion use Problem 19C(c).]

(d) Suppose that X is sequentially complete. Then $\beta(X', X)$ is barreled if it is bornological. [*Hint*: $\mathfrak{T}_{\mathcal{G}'\{\mathcal{F}\}}$ is barreled [Problem 19A(b)].]

(e) If $\beta(X', X)$ is barreled and $\mathfrak{T}_{\mathcal{G}'\{\mathcal{F}\}}$ has a nbd base of $\beta(X', X)$-closed sets, then $\beta(X', X)$ is bornological. [*Hint*: Example 19.13(b).]

19G. Distinguished Spaces

An LCTVS X is said to be distinguished if every $\sigma(X'', X')$-bounded subset of $X'' = (X', \beta(X', X))'$ is contained in the $\sigma(X'', X')$-closure of some X-bounded set. For notation $^{\circ}$ denotes polarity with respect to $\langle X, X' \rangle$ and $^{\bullet}$ denotes polarity for $\langle X', X'' \rangle$.

(a) X is distinguished iff $\forall \sigma(X'', X')$-bounded set $E \subset X''$, there is an X-bounded set B satisfying $E \subset B^{\circ\bullet}$. [*Hint*: Use Theorem 18.11.]

(b) Use (a) to prove that X is distinguished iff $\beta(X', X'') \subset \beta(X', X)$ (i.e., iff $\beta(X', X'') = \beta(X', X)$). [*Hint*: First show that $\beta(X', X) \subset \beta(X', X'')$ $\forall$ LCTVS X.]

(c) X is distinguished iff $\beta(X', X)$ is barreled. [*Hint*: If $\beta(X', X)$ is barreled and $E \subset X''$ is $\sigma(X'', X')$-bounded, then (Corollary 19.19, Theorem 19.14) $E \subset \tilde{B}^{\bullet}$ for some $\beta(X', X)$-nbd $\tilde{B} \subset X'$ and $B^{\circ} \subset \tilde{B}$ for some bounded set $B \subset X$. If X is distinguished and $T \subset X'$ is a $\beta(X', X)$-barrel, then $T^{\bullet} \subset X''$ is $\sigma(X'', X')$-bounded and $T^{\bullet} \subset B^{\circ\bullet}$ in X'' for some bounded set $B \subset X$. Therefore $T = T^{\bullet\bullet}$ contains $B^{\circ\bullet\bullet} = B^{\circ}$.]

(d) Every seminormable TVS is distinguished. [Use Problem 17B with Examples 18.24(e) and 19.13(a).]

(e) A sequentially complete, infrabarreled (therefore Frechet) space X is distinguished if $\beta(X', X)$ is bornological. [*Hint*: Use Problem 19F(c).]

(f) As a partial converse to (e) prove that the properties below are equivalent for a metrizible LCTVS X:

(i) X is distinguished;
(ii) $\beta(X', X)$ is barreled;
(iii) $\beta(X', X)$ is infrabarreled;
(iv) $\beta(X', X)$ is bornological.

[*Hint*: Clearly (iv) $\Rightarrow$ (iii) $\Leftrightarrow$ (ii) $\Leftrightarrow$ (i) (Corollary 19.6 and Problem 19C(b) and (c) above). For (ii) $\Rightarrow$ (iv) use Problem 19F(c) (Bourbaki [13]). Assume that $\mathfrak{N} = \{u_1 \supset u_2 \supset \cdots\}$ is an X-nbd base of closed abs convex sets and let $v(0)$ be any abs convex $\mathfrak{T}_{\mathcal{G}'\{\mathfrak{F}\}}$-nbd in X'. Each u_n° is $\beta(X', X)$-bounded, therefore $\mathfrak{T}_{\mathcal{G}'\{\mathfrak{F}\}}$-bounded, so that $\forall n \in \mathbb{N}$ there is an $\alpha_n > 0$ satisfying $\alpha_n u_n^{\circ} \subset \frac{1}{2}v$. If $A_n = \left\{ \bigcup_{k=1}^{n} \alpha_k u_k^{\circ} \right\}_c$, then $w = \bigcup_{n=1}^{\infty} A_n \subset \frac{1}{2}v$ is abs convex and absorbent. It remains to prove that $\bar{w}^{\beta(X',X)} \subset v$. If $x' \notin v$, then $x' \notin 2A_n$ and $\langle x_n, x' \rangle = 2$ for some $x_n \in A_n^{\circ}$ (for each $n \in \mathbb{N}$). Since $\{x_n : n \in \mathbb{N}\}$ is $\sigma(X, X')$-bounded, $\{x_n : n \in \mathbb{N}\}^{\circ}$ is a $\beta(X', X)$-nbd and $\{x_n : n \in \mathbb{N}\}^{\circ\bullet} \subset X''$ is $\sigma(X'', X')$-compact. Let $x'' \in \overline{\{x_n : n \in \mathbb{N}\}}^{\sigma(X'',X')} = \{x_n : n \in \mathbb{N}\}^{\circ\bullet}$. Then $x'' \in \bigcap_{n=1}^{\infty} A_n = w^{\bullet} \subset X''$ and $2 = \langle x'', x' \rangle = \lim \langle x_n, x' \rangle$ implies that $x' \notin w^{\bullet\bullet} = \bar{w}^{\sigma(X',X'')} \supset \bar{w}^{\beta(X',X)}$.]

(g) Use (e), (f) above to conclude that a Frechet space X is distinguished iff $\beta(X', X)$ is bornological.

(h) Prove that $\beta_{bo}(X', X) = \beta(X', X'')$ for a metrizible LCTVS X. [*Hint*: $\beta(X', X'')$ has the collection of $\beta(X', X)$-barrels as a nbd base at 0. Use Corollaries 19.6 and 14.18 to get $\beta(X', X'') \subset \beta_{bo}(X', X)$. For the reverse inclusion show (this is somewhat involved!) that every abs convex $\beta(X', X)$-bornivorous set contains a $\beta(X', X)$-barrel.]

Remark. The results of (h) and (b) also establish (f).

(i) A strict inductive limit of distinguished, metrizible spaces is distinguished.

References for Further Study

On bornological and barreled spaces: Edwards [28]; Köthe [55]; Saxon [87]; Gould and Mahowald [33].

On Frechet spaces: Treves [98].

On Schwartz spaces: Horvath [43].

On distinguished spaces: Horvath [43]; Köthe [55]; Grothendieck [35].

20. Reflexive and Montel Spaces

As is well known, $X = (X', \sigma(X', X))'$ for an LCT$_2$VS X. Here we consider the other extreme $X = (X', \beta(X', X))'$, in which case every polar topology on X' is a topology of $\langle X, X' \rangle$ and X is termed semireflexive. This algebraic equality becomes topological (and X is called reflexive) if X is also infrabarreled. Reflexive spaces are barreled and, in turn, encompass Montel spaces (i.e., barreled spaces characterized by bounded sets being relatively compact).

It is assumed throughout that $X'' = (X', \beta(X', X))'$ for an LCTVS X. Polars in $\langle X, X' \rangle$ are denoted $^{\circ}$, whereas $^{\bullet}$ denotes polarity in $\langle X', X'' \rangle$.

Semireflexive and Reflexive Spaces

Each $\langle x, \ \rangle : \underset{x' \rightsquigarrow x'(x)}{X' \longrightarrow \mathbb{F}}$ (being $\sigma(X', X)$-continuous, therefore $\beta(X', X)$-continuous) is an element of X''. If X is T_2, then X' is injective and $\underset{x \rightsquigarrow \langle x, \ \rangle}{j: X \to X''}$ is an into vector space isomorphism. In fact (Theorem 18.26), j is a topological isomorphism for $\mathfrak{T}_{\mathcal{S}_e} \in T(X'')$, the polar topology having $\mathcal{S}_e^\bullet = \{B^\bullet \subset X'' : B \in \mathcal{S}_e \text{ in } X'\}$ as its nbd base at $0 \in X'$.

If $B \in \mathcal{S}_e$, then B is $\beta(X', X)$-bounded and $\sigma(X', X'')$-bounded [Example 13.27(e), Theorem 18.6]. Thus $B^\bullet$ is a $\beta(X'', X')$-nbd and $\mathfrak{T}_{\mathcal{S}_e} \subset \beta(X'', X')$ on X'' follows. Clearly [Problem 19C(a)] $\mathfrak{F}_{\mathcal{S}_e} = \beta(X'', X')$ iff X is infrabarreled.

DEFINITION 20.1. An LCT_2VS X is semireflexive if j is surjective. A semireflexive, infrabarreled space is called reflexive. Thus X is semireflexive {reflexive} iff X and X'' are vector space {topological} isomorphs.

The most important characterizations of semireflexivity are encompassed in

THEOREM 20.2. The following are equivalent for an LCT_2VS X:

(i) X is semireflexive;
(ii) $(X', \beta(X', X))' \subset (X', \sigma(X', X))'$;
(iii) $\beta(X', X)$ is a topology of $\langle X', X \rangle$;
(iv) $\beta(X', X)$-closed, convex sets are $\sigma(X', X)$-closed;
(v) $\beta(X', X) = \tau(X', X)$;
(vi) all $B \in \mathcal{S}_\beta$ are $\sigma(X, X')$-relatively compact;
(vii) X is $\sigma(X, X')$ quasi-complete.

Proof. The first half of (i)$\Leftrightarrow$(ii)$\Rightarrow$(iii)$\Rightarrow$(iv)$\Rightarrow$(ii) holds by definition, whereas the second half follows from Theorem 18.6 and Theorem 17.2(ii), respectively. The first and third implications in (iii)$\Leftrightarrow$(v)$\Rightarrow$(vi)$\Rightarrow$(vii)$\Rightarrow$(i) are clear. If $\beta(X', X) = \tau(X', X)$ and $B \in \mathcal{S}_\beta$ (i.e., $B \subset X$ is bounded), then B° is a $\tau(X', X)$-nbd and (by Theorem 18.15) the set $B^{\circ\circ} \subset X$ is $\sigma((X', \tau(X', X))', X') = \sigma(X, X')$ compact. Consequently $B \subset B^{\circ\circ}$ is $\sigma(X, X')$-relatively compact. Finally, (vii)$\Rightarrow$(i) is based on the fact (Theorem 18.11 and Theorem 18.14 applied twice) that $X'' = \bigcup_{S_\beta} B^{\circ\bullet} = \bigcup_{\mathcal{S}_\beta} \overline{B_{bc}}^{\sigma(X'', X')}$. If $x'' \in X''$, there is a $B \in \mathcal{S}_\beta$ and filter $\mathfrak{F} \subset B_{bc}$ such that

$\mathfrak{F} \xrightarrow{\sigma(X'',X')} x''$. Since $\sigma(X'', X')/X = \sigma(X, X')$ and (by hypothesis) $\bar{B}_{bc}^{\sigma(X,X')}$ is $\sigma(X, X')$-complete, the $\sigma(X, X')$-Cauchy filter $\mathfrak{F}$ is $\sigma(X, X')$-convergent to some $x \in \bar{B}_{bc}^{\sigma(X,X')} \subset X$; but $\sigma(X'', X')$ is T_2. Therefore $x'' = x$ and $X'' = X$ follows. ■

In terms of Problem 19G we have

COROLLARY 20.3. A semireflexive space is distinguished and quasi-complete. Therefore every reflexive space is barreled.

Proof. A semireflexive space X is distinguished [Problem 19G(c)], since $\beta(X', X) = \beta(X', X'')$ implies (Corollary 19.19) that $(X', \beta(X', X''))$ is barreled. Since the equicontinuous sets in X' cover X' (Theorem 18.14), it further follows (Theorem 19.20) that $X = (X'', \beta(X'', X'))$ is quasi-complete. Reflexive spaces are therefore barreled by Problem 19C(b). ■

Remark. The converse of Corollary 20.3 fails, since the nonreflexive Banach spaces c_0, c, and l_1 are distinguished.

Although (semi-) reflexivity is not preserved under general projective or inductive limits (Problem 20C), the following hereditary properties prevail:

THEOREM 20.4. (i) If X is reflexive, so is $(X', \beta(X', X))$.

(ii) A quasi-complete, Mackey T_2 space is reflexive if $(X', \beta(X', X))$ is reflexive. Thus a Frechet space X is reflexive iff $(X', \beta(X', X))$ is reflexive.

(iii) Every product of (semi-) reflexive spaces is (semi-) reflexive.

(iv) A closed subspace of a semireflexive {reflexive Frechet} space is semireflexive {reflexive}.

(v) A convex direct sum of (semi-) reflexive spaces is (semi-) reflexive.

(vi) The quotient of a semireflexive NLS (i.e., reflexive Banach space) by a closed subspace is reflexive.

Proof. Corollary 20.3, Corollary 19.18(iv), and Theorem 20.2(vii) proves (i). If $\beta(X', X)$ is semireflexive for an LCT$_2$VS $(X, \tau(X, X'))$, then $\beta(X', X)$-bounded sets are $\sigma(X', X) = \sigma(X', X'')$ relatively compact and [Problem 19C(a)] $\tau(X, X')$ is infrabarreled. If $\tau(X, X') = \tau(X'', X')/X$ is also quasi-complete, the reasoning in Theorem 20.2 (beginning with $X'' = \bigcup_{\mathscr{S}_B} \bar{B}_{bc}^{\sigma(X'',X')} = \bigcup_{\mathscr{S}_B} \bar{B}_{bc}^{\tau(X'',X')}$ by Theorem 19.6) yields $X'' = X$. For (iii) suppose

X is the product of semireflexive spaces $\{X_\alpha : \alpha \in \mathcal{Q}\}$. Then $X' = \sum_{\mathcal{Q}} \oplus X'_\alpha$ (by Theorem 15.20), and $\beta(X', X) = \sum_{\mathcal{Q}} \oplus \beta(X'_\alpha, X_\alpha) = \sum_{\mathcal{Q}} \oplus \tau(X'_\alpha, X_\alpha) = \tau(X', X)$ [Theorem 20.2(v), Problem 18L] implies that X is semireflexive. Note, further, that X is barreled if each X_α is barreled. Assume next that M is a closed subspace of a semireflexive space X in (iv). Then every bounded set $B \subset M$ is X-bounded and $\sigma(X, X')$-relatively compact. In particular [Problem 18N(f)], B is $\sigma(X, X')/M = \sigma(M, M')$-compact and it follows that M is semireflexive. Clearly M is Frechet, therefore reflexive, if X is Frechet. (v) If X is a convex direct sum of semireflexive spaces $\{X_\alpha : \alpha \in \mathcal{Q}\}$, the references in (iii) also ensure that $\beta(X', X) = \prod_{\mathcal{Q}} \beta(X'_\alpha, X_\alpha) = \prod_{\mathcal{Q}} \tau(X'_\alpha, X_\alpha) = \tau(X', X)$ on $X' = \prod_{\mathcal{Q}} X'_\alpha$. Therefore X is semireflexive and Theorem 19.21 applies. Assume that H in (iv) is a closed subspace of a reflexive Banach space $(X, \|\ \|)$. Then (Example 15.14) $\mathfrak{T}_{\mathcal{g}(\nu)} = |\ |$ on X/H and $S_{1,|\,|}(0) = \nu\{S_{1,\|\,\|}(0)\} \subset X/H$. Since $\nu \in L(X, X/H)$ is weakly continuous (Theorem 18.16), an application of Example 20.5(c) below will confirm that $S_{1,|\,|}(0)$ is $\sigma(X/H, (X/H)')$-compact and that $(X/H, \mathfrak{T}_{\mathcal{g}(\nu)})$ is therefore reflexive. ■

Remark. A closed subspace, having a topological supplement in a reflexive space, is reflexive (Theorem 19.21).

Incomplete NLSps cannot be reflexive, whereas complete NLSps may or may not be reflexive.

Example 20.5. (a) A Hilbert space $(X, (\ ,\ ))$ is reflexive. Indeed (Riesz representation theorem), each $F \in X'$ is defined by a unique $x_F \in X$ satisfying $F(x) = (x, x_F)\ \forall x \in X$. Furthermore, $\underset{x \rightsquigarrow (\ ,\ x)}{(X, \|\ \|) \to (X', \|\|\ \|\|)}$ is an onto congruence. If $f \in X''$, then $g_f : \underset{x \rightsquigarrow f\{(\ ,\ )\}}{(X, \|\ \|) \to \mathbb{F}}$ is linear and satisfies $|g_f(x)| \leq \|\|f\|\|\ \|x\|\ \forall x \in X$. Thus $g_f \in X'$ and there is a unique $x_f \in X$ satisfying $(x_f, x) = f\{(\ ,x)\}\ \forall x \in X$. It now follows from $\{j(x_f)\}(\ ,x) = \{(\ ,x)\}(x_f) = (x_f, x) = f\{(\ ,x)\}\ \forall x \in X$ that $j(x_f) = f$ and that j is therefore surjective.

(b) l_p (and L_p) is reflexive $\forall 1 < p < \infty$, since $l'_p = l_q$ and $l'_q = l_p$. Since $c_0 \subsetneq l_\infty = l'_1 = (c'_0)'$, the Banach subspace c_0 of l_1 is nonreflexive. Thus l_1 is also nonreflexive [Theorem 20.4(iv)]. In fact, both $c = c_0 \oplus [(1, 1, 1, \ldots)]$ and l_∞ are also nonreflexive Banach spaces.

(c) The properties below are equivalent for a NLS $(X, \|\ \|)$:

(i) X is $\|\ \|$-reflexive;
(ii) $\bar{S}_{1,\|\|\,\|\|}(0)$ is $\sigma(X, X')$-compact;
(iii) $\bar{S}_{1,\|\|\,\|\|}(0)$ is $\sigma(X, X')$-complete;
(iv) X is $\sigma(X, X')$ quasi-complete.

If $(X, \| \|)$ is reflexive, then $\||\ \|| = \beta(X', X)$ is a topology of $\langle X, X' \rangle$ and (Theorem 18.15) $\bar{S}_{1,\| \|}(0) = \{\bar{S}_{1,\||\ \||}(0)\}^{\circ} \subset X'' = X$ is $\sigma(X'', X') = \sigma(X, X')$ compact. The remaining implications (ii)$\Rightarrow$(iii)$\Rightarrow$(iv)$\Rightarrow$(i) are clear.

Example 20.6. A new definition (see Problem 17H) of line integral can be made for $F \in C(\Omega; X)$ when $(X, \| \|)$ is a reflexive Banach space. Specifically, the map $\Phi_F : X' \longrightarrow \mathbb{C}$, $f \rightsquigarrow \int_\Gamma (fF)(\lambda)\, d\lambda$ belongs to X'' (with $\||\Phi_F\|| \leq \sup_\Gamma \|F(\lambda)\| \cdot l(\Gamma)$, where $l(\Gamma)$ is the length of Γ) and there is a unique $x_F \in X$ satisfying $\Phi_F(f) = f(x_F)\ \forall f \in X'$. One frequently calls x_F the weak line integral of F, which is denoted $x_F = \mathop{w\!\!\int}_\Gamma F\, d\lambda$.

Example 20.7. $\mathcal{CA}(\Omega, \Sigma; Y)$ is $\mathfrak{T}_{\mathcal{P}\{\mathcal{A}\}}$-reflexive iff Σ is finite and Y is reflexive. The conditions on Σ and Y are sufficient for $\mathfrak{T}_{\mathcal{P}\{\mathcal{A}\}}$-reflexivity by Theorems 20.4(iii) and 16.10. They are also necessary by Example 16.12(b), Theorem 20.4(iv), and Example 19.24.

Semi-Montel and Montel Spaces

In a Euclidean space (equivalently, a finite dim T_2VS) all closed bounded sets are compact. A NLS has this property iff it is finite dim (Theorems 13.12 and 13.14). From a broader perspective

DEFINITION 20.8. An LCT_2VS is semi-Montel if bounded subsets are relatively compact. An infrabarreled, semi-Montel space is called Montel.

Example 20.9. (a) (Example 14.14) $(X, \mathfrak{T}_{F_c})$ is Montel. If $B \subset X$ is $\mathfrak{T}_{F_c}$-bounded, then (Corollary 13.13) $[B]$ is a finite dim closed subspace of $(X, \mathfrak{T}_{F_c})$. Since $([B], \mathfrak{T}_{F_c})$ is semi-Montel and B is $\mathfrak{T}_{F_c}$-bounded in $[B]$, it follows that $\bar{B} \subset [B] \subset X$ is $\mathfrak{T}_{F_c}$-compact.

(b) If X is semi-Montel, then $\kappa_{bc}(X', X) = \beta(X', X)$ and [by Problem 19F(a)] $\tau(X', X) = \kappa_{bc}(X', X) = \kappa_c(X', X) = \kappa(X', X) = \lambda(X', X) = \beta(X', X)$ on X'. For any TVS Y one also has (Example 13.27) $L_\kappa(X, Y) = L_\lambda(X, Y) = L_\beta(X, Y)$ and $\Phi_\sigma = \Phi_\beta\ \forall$ equicontinuous set $\Phi \subset L(X, Y)$. Specifically, $w_{B,v}(0) = w_{\bar{B},v}(0)\ \forall B \in \mathcal{S}_\beta$ and closed nbd $v(0) \subset Y$, since $f(B) \subset v \Rightarrow f(\bar{B}) \subset \overline{f(B)} \subset \bar{v} = v\ \forall f \in L(X, Y)$.

(c) A partial converse to (b) is the fact that a quasi-complete LCT_2VS X is semi-Montel if $\kappa_{bc}(X', X) = \lambda(X', X)$ equals $\beta(X', X)$. Indeed, every $B \in \mathcal{S}_\beta$ is relatively compact in X, since $E^{\circ} \subset B^{\circ}$ in X' and $B \subset E^{\circ\circ} = E$ in X for some $E \in \mathcal{S}_{\kappa_{bc}}$.

(d) $\kappa_{bc}(X', X) \subsetneq \tau(X', X)$ for an infinite dim reflexive Banach space X, since equality implies that X is semi-Montel and so finite dim (Theorem 20.2 and (c) above.

The basic hereditary properties of being semi-Montel are given in

THEOREM 20.10. A closed subspace of a semi-Montel space is semi-Montel. Every product and convex direct sum of {semi-} Montel spaces is {semi-} Montel.

Proof. The first assertion is obvious. Assume that $\{X_\alpha : \alpha \in \mathcal{Q}\}$ are semi-Montel. If $B \subset X = \prod_{\mathcal{Q}} X_\alpha$ is bounded, then $\overline{\pi_\alpha(B)}$ is X_α-compact $\forall \alpha \in \mathcal{Q}$. Therefore $\overline{\prod_{\mathcal{Q}} \pi_\alpha(B)} = \prod_{\mathcal{Q}} \overline{\pi_\alpha(B)}$ is compact and $B \subset \prod_{\mathcal{Q}} \pi_\alpha(B)$ is relatively compact in X. This establishes that X is semi-Montel. If $B \subset \sum_{\mathcal{Q}} \oplus X_\alpha$ is $\mathfrak{T}_{\mathcal{J}'\{J\}}$-bounded, then [Theorem 15.19(i)] $B \subset \sum_{i=1}^{n} E_{\alpha_i}$ for bounded $E_{\alpha_i} \subset X_{\alpha_i}$ $(i = 1, 2, \ldots, n)$. The $\mathfrak{T}_{\mathcal{J}'\{J\}}$-relative compactness of B [therefore the semi-Montel property of $\mathfrak{T}_{\mathcal{J}'\{J\}}$) follows from Problem 13C(a), Theorem 15.19(ii) and the fact [Theorem 13.4(iii)] that $\bar{B} \subset \overline{\sum_{i=1}^{n} E_{\alpha_i}} = \sum_{i=1}^{n} \bar{E}_{\alpha_i}$.

It remains only to invoke Problem 19E for the case in which all X_α $(\alpha \in \mathcal{Q})$ are infrabarreled. ■

We already know a good deal about (semi-) Montel spaces since

THEOREM 20.11. (i) A (semi-) Montel space is (semi-) reflexive.
(ii) If $(X, \mathfrak{T})$ is semi-Montel, then $\sigma(X, X') = \mathfrak{T}$ on every $B \in \mathcal{S}_\beta$.
(iii) $(X', \beta(X', X))$ is Montel if X is Montel.

Proof. Suppose that X is (semi-) Montel and let $B \in \mathcal{S}_\beta$. Then (Theorem 18.6) $\bar{B}_c^{\sigma(X,X')} = \bar{B}_c$ is compact, therefore $\sigma(X, X')$-compact, and so X is seen to be (semi-) reflexive. Assume next that $X = (X, \mathfrak{T})$ in (ii). Since $\sigma(X, X')$ is T_2 and $\bar{B}$ is compact, every $\mathfrak{T}$-closed subset of $\bar{B}$ is $\sigma(X, X')$-closed. Therefore $\sigma(X, X') = \mathfrak{T}$ on $\bar{B}$ and on $B \in \mathcal{S}_\beta$. Finally, suppose that X in (iii) is Montel and let $E \subset X'$ be $\beta(X', X)$-bounded. Then E_c [being $\beta(X', X)$-bounded] is equicontinuous and $\sigma(X', X)$-relatively compact in X' [Problem 19E(a), Theorem 18.15, or Theorems 20.4(i) and 20.2(vi)]; but $(\bar{E}_c^{\sigma(X',X)}, \sigma(X', X)) = (\bar{E}_c^{\beta(X',X)}, \beta(X', X))$ by Theorem 20.2(iii), Theorem 18.6, and Example 20.9(b). Thus $\bar{E}^{\beta(X',X)}$ is $\beta(X', X)$-compact in X' and $(X', \beta(X', X))$ is Montel. ■

COROLLARY 20.12. The following are equivalent for an LCT_2VS X:

(i) X is semireflexive;
(ii) X is $\sigma(X, X')$ quasi-complete;
(iii) X is $\sigma(X, X')$ semi-Montel;
(iv) X is $\sigma(X, X')$ semireflexive.

Proof. (iii)$\Rightarrow$(iv) and (iv)$\Rightarrow$(ii)$\Leftrightarrow$(i)$\Leftrightarrow$(iii) by Theorem 20.2. ■

In general, weak convergence may [Example 13.27(f)] or may not [Example 15.4(f)] imply convergence.

COROLLARY 20.13. In a semi-Montel space X every $\sigma(X, X')$-convergent sequence is convergent. If X is Montel, every $\sigma(X', X)$-convergent sequence is $\beta(X', X)$-convergent.

Proof. If X is semi-Montel and $x_n \xrightarrow{\sigma(X,X')} x_0 \in X$, then $B = \{x_n : n \in \mathbb{N}_0\}$ is bounded and $(B, \sigma(X, X')) = B$ with the induced topology. Given any X-nbd $u(0)$, there is a $\sigma(X, X')$-nbd $w(0)$ such that $B \cap \{w + x_0\} = B \cap \{u + x_0\} \subset u + x_0$. This, of course, establishes that $x_n \to x_0 \in X$. If X is Montel and $x'_n \xrightarrow{\sigma(X',X)} x'_0 \in X'$, then $B'' = \{x'_n : n \in \mathbb{N}_0\}$ is equicontinuous (Corollary 19.18) and $x'_n \xrightarrow{\beta(X',X)} x'_0 \in X'$ [Example 20.9(b)]. ■

We conclude with two illustrations of varying familiarity.

Example 20.14. $\mathcal{CA}(\Omega, \Sigma; Y)$ is $\mathfrak{T}_{\mathcal{P}\{\mathcal{A}\}}$-Montel iff Σ is finite and Y is Montel. (Simply substitute Theorem 20.10 for Theorem 20.4 in Example 20.7.)

Y is semi-Montel if any A_E $(E \in \Sigma)$ is completely continuous. [If $B \subset Y$ is bounded, then $\psi(B)$ $(\omega \in E)$ is $\mathfrak{T}_{\mathcal{P}(\mathcal{A})}$-bounded in $\mathfrak{D}_\omega(\Omega, \Sigma; Y)$ and also in $\mathcal{CA}(\Omega, \Sigma; Y)$. Therefore $B = A_E \psi_\omega(B) \subset Y$ is relatively compact by Problem 13I(a).]

Example 20.15. The name "Montel space" derives from Montel's theorem in complex analysis:

Let $\Phi \subset \mathcal{H}(\Omega)$ and assume that for each compact set $K \subset \Omega$ there is a $\lambda_K > 0$ such that $\sup_{z \in K} |f(z)| \leq \lambda_K$ $\forall f \in \Phi$. Then every sequence in Φ contains a subsequence that converges uniformly on compacta to a member of $\mathcal{H}(\Omega)$.

In the context of TVSps Montel's theorem asserts that every bounded set $\Phi \subset \mathscr{H}_\kappa(\Omega)$ is relatively compact; in other words $\mathscr{H}_\kappa(\Omega)$ is a Montel space.

Proof. We already know [Problem 14H(a), Example 19.13(a)] that $\mathscr{H}_\kappa(\Omega)$ is barreled. Suppose that $\Phi \subset \mathscr{H}_\kappa(\Omega)$ is bounded. Then $\bar{\Phi}^\kappa \subset \mathscr{H}_\kappa(\Omega)$ is bounded and $\{f(z): f \in \bar{\Phi}^\kappa\}$ is compact $\forall z \in \Omega$. Given any fixed compact set $K \subset \Omega$, there is a finite cover by closed disks $\{\bar{S}_{\delta_K}(z_i): 1 \le i \le n_K\}$ in Ω having centers $z_i \in K$. If $z_0 \in K$ and $\gamma = \bar{S}_{\delta_K}(z_0) - S_{\delta_K}(z_0)$, then Cauchy's formula

$$|f'(z_0)| = \left|\frac{1}{2\pi i}\int_\gamma \frac{f(\zeta)\,d\zeta}{(\zeta - z_0)^2}\right| \le \frac{\lambda_K}{\delta_K}$$

holds $\forall f \in \bar{\Phi}^\kappa$. Therefore $z \in K \cap S_{(\varepsilon\delta_K/\lambda_K)}(z_0)$ implies $|f(x) - f(z_0)| = |\int_{z_0}^{z} f(\zeta)\,d\zeta| \le (\lambda_K/\delta_K)|z - z_0| < \varepsilon\ \forall f \in \bar{\Phi}^\kappa$. The conclusion is that $\bar{\Phi}^\kappa$ is equicontinuous on compacta and (Theorem 7.18) $\bar{\Phi}^\kappa$ is compact in $\mathscr{H}_\kappa(\Omega)$.

Problems

20A. Semireflexivity Definition Reexamined

We do not consider $j: X \to X^{**}$, $x \leadsto \langle x, \ \rangle$ in Definition 20.1 because j will be surjective iff dim X is finite. [*Hint*: X^* is injective and j is injective, therefore surjective, when dim $< \infty$. If $\mathfrak{B} = \{x_\alpha : \alpha \in \mathfrak{A}\}$ is an infinite basis for X, define $f_\alpha: X \to \mathbb{F}$ $(\alpha \in \mathfrak{A})$ by $f_\alpha\left(\sum_{i=1}^{k} \lambda_i x_{\alpha i}\right) = \sum_{i=1}^{k} \lambda_i \delta_{\alpha i,\alpha}$. Extend $\{f_\alpha : \alpha \in \mathfrak{A}\}$ to a basis $\mathfrak{B}^* = \{f_\alpha^* : \alpha \in \mathfrak{A}'\}$ for X^* and show that $g^{**} \in X^{**} - j(X)$ for $g^{**}\left(\sum_{i=1}^{k} \lambda_i f_{\alpha_i}^*\right) = \sum_{i=1}^{k} \gamma_i$.]

20B. Strong Separability

(a) A seminormable space (X, p) is separable if $(X', \beta(X', X))$ is separable. [*Hint*: $S'(0) = \{f \in X' : |||f||| = 1\}$ is $||| \ ||| = \beta(X', X)$ separable. If $\{f_n : n \in \mathbb{N}\}$ is dense in $S'(0)$, there exist $\{x_n : n \in \mathbb{N}\} \subset X$ with $p(x_n) = 1$ and $|f_n(x_n)| > \frac{1}{2}\ \forall n \in \mathbb{N}$. Show that $M = \{$finite, rational linear combination of $x_n\}$ is a countably dense subset of $[x_1, x_2, \ldots]$ and $X = \overline{[x_1, x_2, \ldots]}$.]

(b) The converse of (a) is generally false. [*Hint*: l_1 is separable but $(l_1', \beta(l_1', l_1)) = l_\infty$ is not.]

(c) Show that $(X', \beta(X', X))$ is separable if (X, p) is separable and reflexive. [*Hint*: $(X'', \beta(X'', X'))$ is separable and (a) applies.]

20C. An Incomplete, Nonreflexive Quotient Space

Let M be a proper dense subspace of a complete LCT$_2$VS X and write $M_\xi = (M, \xi)$ for a complete topology $\xi \in T(M)$ finer than the induced topology on M. Assume that $\sum_1^\infty \oplus X$ is a convex direct sum (Definition 15.8) and let $Y = \left(\sum_1^\infty \oplus X\right) \times \prod_1^\infty M_\xi$ carry the product topology.

(a) Show that $h: Y \to \sum_1^\infty \oplus X + \prod_1^\infty M \subset \prod_1^\infty X$, $\langle x, m\rangle \rightsquigarrow x+m$, is a continuous open linear surjection.

(b) Why is $h(Y)$ a proper dense subspace of $\prod_1^\infty X$? [*Hint*: $\sum_{n=1}^\infty \oplus X$ is dense in $\prod_1^\infty X$, whereas M is proper and dense in X.]

(c) Verify that Y is complete but $Y/h^{-1}(0)$ is not. [*Hint*: Use Theorem 15.19(iii) and Theorem 6.6 applied to $Y/h^{-1}(0) = h(Y) \subset \prod_1^\infty X$ in (b).]

(d) Why has $h^{-1}(0)$ no topological supplement in Y?

(e) Suppose that X and M_ξ are reflexive. Then Y is reflexive, whereas $Y/h^{-1}(0)$ is not even semireflexive. {*Hint*: To get $h(Y) \subset \prod_1^\infty X = \left(\prod_1^\infty X\right)'' = \{h(Y)\}''$ show that $\beta(M', M) = \beta(X', X)$ on $M' = X'$ [Corollary 17.13(iv)].}

Remark. This problem is due essentially to Kelley and Namioka [52].

20D. (Semi-) Reflexive Strict Inductive Limits

[Compare with Problem 20C(e)]. A strict inductive limit of (semi-) reflexive spaces $\{E_n : n \in \mathbb{N}\}$ is (semi-) reflexive if E_n is closed in E_{n+1} $\forall n \in \mathbb{N}$. [*Hint*: Use Problem 15B(a), (e) with Theorem 21.2(vi) and Theorem 19.21.]

20E. Semireflexive, Nonreflexive Mackey Spaces

(a) Every reflexive space is semireflexive and Mackey [Problems 18G(a) and 19C(a)]. A semireflexive, Mackey space need not be reflexive. {*Hint*: If X is a nonreflexive barreled LCT$_2$VS, then $(X', \tau(X', X))$ is semireflexive and nonreflexive [Corollary 19.19(iii) plus Theorems 20.2(i) and 20.4(i)].}

(b) Sharpen the above by proving that a semireflexive Mackey space $(X, \tau(X, X'))$ is nonreflexive iff $(X, \tau(X, X')) = (Y', \tau(Y', Y))$ for some nonreflexive barreled LCT$_2$VS Y. [*Hint*: If X is $\tau(X, X')$ semireflexive and nonbarreled, then $Y = (X', \tau(X', X))$ is nonreflexive and barreled.]

(c) Using Corollary 20.12, prove the following equivalent for an LCT_2VS X:

(i) $\tau(X, X')$ is reflexive;
(ii) $\sigma(X, X')$ is semireflexive;
(iii) $\tau(X, X')$ is barreled;
(iv) $\tau(X', X)$ is reflexive;
(v) $\tau(X', X)$ is barreled;
(vi) $\sigma(X', X)$ is semireflexive.

20F. Nonreflexivity and Catg I Sets

(a) A complete LCT_2VS X is $\beta(X'', X')$-closed in X''. The converse also holds for an infrabarreled LCT_2VS X having bornological dual $(X', \beta(X', X))$. [*Hint*: $j(X)$ is $\mathfrak{T}_{\mathcal{S}_e}$-closed in X'' by Theorem 6.6. The second assertion holds by Corollary 19.6, Problem 19C(a).]

Remark. Distinguished metrizible spaces fit the second situation.

(b) Use Problem 13G(a) to prove that $X \in \text{Catg}_{\text{I}}(X'')$ for a complete non-semireflexive LCT_2VS X.

20G. Quasi-Reflexive Banach Spaces

Nonreflexive Banach spaces X generally have infinite co-dim in X'' (i.e., dim X''/Y is infinite for $X''=X\oplus Y$) as is, for example, c_0 in $c_0''=l_\infty$. A nonreflexive Banach space of co-dim n in X'' is called quasi-reflexive of order n. Quasi-reflexive first order spaces frequently exhibit unusual and pathological properties missing in more standard spaces. Such spaces, for example, are Catg I hyperplanes in X'' [Problem 20F(b)]. The following example of a first-order quasi-reflexive space is due to (James [46]).

Let $\mathcal{P}$ denote the collection of all finite, increasing sequences $P=\{p_1, p_2, \ldots, p_{2n+1}\}$ of positive integers. For each sequence $x=\{x_n\}\in\prod_1^\infty \mathbb{C}$ and $P\in\mathcal{P}$, write

$$\|x\|_P = \sqrt{\sum_{i=1}^{n} (x_{p_{2i-1}} - x_{p_{2i}})^2 + (x_{p_{2n+1}})^2}$$

Define $\|x\| = \sup_{\mathcal{P}} \|x\|_P$ and let $X=\{x\in c_0 : \|x\|<\infty\}$.

(a) $(X, \|\ \|)$ is a NLS with $\|x\|_0 \le \|x\|\ \forall x\in X$. (Here $\|\ \|_0$ is the sup norm on c_0.) [*Hint*: Assume n_0 is the first coordinate of x for which $\|x\|_0=|x_n|$ and let $P_{n_0}=\{n_0-2, n_0-1, n_0\}$. Then $|x_{n_0}|\le\sqrt{(x_{n_0-2}-x_{n_0-1})^2+(x_{n_0})^2}=\|x\|_{P_{n_0}}\le\|x\|$. For the triangle inequality consider any $\varepsilon>0$ and let $P(\varepsilon)\in\mathcal{P}$ satisfy $\|x+y\|<\varepsilon+\|x+y\|_{P(\varepsilon)}$. Show that $\|x+y\|_{P(\varepsilon)}\le\|x\|_{P(\varepsilon)}+\|y\|_{P(\varepsilon)}$ so that $\|x+y\|<\varepsilon+\|x\|+\|y\|\ \forall\varepsilon>0$.]

(b) Every $x=\{x_n\}\in X$ has a representation $x=\sum_{n=1}^{\infty} x_n e_n$ in the sense that $\lim_n \left\|x-\sum_{i=1}^{n} x_i e_i\right\|=0$. Show also that $\|e_n\|=1\ \forall n\in\mathbb{N}$ and $\|x\|=\sup_{\mathbb{N}}\left\|\sum_{i=1}^{n} x_i e_i\right\|\ \forall x\in X$. [*Hint*: The first claim follows from $X\subset c_0$. If $P\in\mathcal{P}$ and $k\geq\max_P p_i$, then $\|x\|_P=\left\|\sum_{i=1}^{k} x_i e_i\right\|_P\leq\left\|\sum_{i=1}^{k} x_i e_i\right\|\leq\sup_{\mathbb{N}}\left\|\sum_{i=1}^{n} x_i e_i\right\|$ and $\|x\|\leq\sup_{\mathbb{N}}\left\|\sum_1^n x_i e_i\right\|$. The reverse inequality holds, since (verify!) $\left\|\sum_{i=1}^{n} x_i e_i\right\|\leq\|x\|\ \forall n\in\mathbb{N}$.]

(c) Show that $f_m\colon \underset{x=\{x_n\}\rightsquigarrow x_m}{X\longrightarrow\mathbb{C}}$ has $|||f_m|||=1\ \forall m\in\mathbb{N}$. Moreover, $f=\sum_{n=1}^{\infty} f(e_n)f_n\ \forall f\in X'$ and $f(x)=\sum_{n=1}^{\infty} f(e_n)x_n\ \forall x=\{x_n\}\in X$.

(d) $(X,\|\ \|)$ is not reflexive. [*Hint*: $\bar{S}_{1,\|\|}(0)$ is not weakly compact, since $\left\|\sum_{i=1}^{n} e_i\right\|=1\ \forall n\in\mathbb{N}$ and $\left\{1_n=\sum_{i=1}^{n} e_i : n\in\mathbb{N}\right\}$ has no weakly convergent subsequent (if $1_{nk}\xrightarrow{\sigma(X,X')} y=\sum_{m=1}^{\infty} y_m e_m$, then $1=\lim_k f_m(1_{nk})=f_m(y)=y_m\ \forall m\in\mathbb{N}$ and $y\in X$).]

(e) Let $g_m\colon \underset{f=\sum_{n=1}^{\infty} f(e_n)f_n\rightsquigarrow f(e_m)}{X'\longrightarrow\mathbb{C}}$. Then $||||g_m||||=1\ \forall m\in\mathbb{N}$ (where $||||\ ||||$ is the norm on X'' induced by $|||\ |||$ on X'). Moreover, $g=\sum_{n=1}^{\infty} g(f_n)g_n\ \forall g\in X''$ and $g(f)=\sum_{n=1}^{\infty} g(f_n)f(e_n)\ \forall f=\sum_{n=1}^{\infty} f(e_n)f_n\in X'$.

(f) Prove that $||||g||||=\|g\|\ \forall g=\sum_{n=1}^{\infty} g(f_n)g_n$, where $\|g\|=\left\|\sum_{n=1}^{\infty} f(g_n)e_n\right\|$. [*Hint*: $|g(f)|\leq\|g\|\cdot\|f\|$ and $||||g||||\leq\|g\|$ by (b) above. For $n\in\mathbb{N}$, define $f_{(n)}(e_k)=0\ \forall k>n$ and $f_{(n)}\left(\sum_{i=1}^{n} g(f_i)e_i\right)=\left\|\sum_{i=1}^{n} g(f_i)e_i\right\|$. If $F_{(n)}$ linearly extends $f_{(n)}$ to X, then $|||F_{(n)}|||=1$ and $F_{(n)}=\sum_{i=1}^{n} f(e_i)f_i$ so that $|g(F_{(n)})|=\left\|\sum_{i=1}^{n} g(f_i)e_i\right\|$ and $||||g||||=\sup_{|||F|||\leq 1}|g(F)|\geq\sup_{\mathbb{N}}|g(F_{(n)})|=\|g\|$.]

(g) Show that $J(X)=\left\{g=\sum_{n=1}^{\infty} g(f_n)g_n\in X'' : \lim_n g(f_n)=0\right\}$ and conclude that $J:(X,\|\ \|)\to(J(X),\|\ \|)$ is a congruence.

(h) Verify that $X''\overset{t}{\equiv} J(X)\oplus[1,1,1,\ldots)]$ and that $(X,\|\ \|)$ is a Banach space. [*Hint*: If $g\in X''=J(X)\neq\varnothing$, then $\|g\|<\infty$ and $g(f_n)|\to\alpha>0$. Thus $\tilde{g}=\sum_{n=1}^{\infty}[g(f_n)-\alpha]g_n\in J(X)$ and $g=\tilde{g}-\alpha(1,1,1,\ldots)$. The completeness of $\|\ \|$ follows from Problem 20F and the fact that $J(X)$ is closed in X''.]

20H. Uniformly Convex Spaces

A NLS $(X, \| \|)$ is uniformly convex (UC) if $\forall \varepsilon > 0$ there is a $\delta(\varepsilon) > 0$ such that $\|\frac{1}{2}(x+y)\| < 1-\delta$ whenever $x, y \in \bar{S}_{1,\| \|}(0)$ satisfy $\|x-y\| < \varepsilon$.

(a) $(X, \| \|)$ is UC iff $\lim_n \|x_n - y_n\| = 0$ whenever $\{x_n\}, \{y_n\} \in \bar{S}_{1,\| \|}(0)$ satisfy $\lim_n \|\frac{1}{2}(x_n + y_n)\| = 1$.

(b) Suppose that $(X, \| \|)$ is UC and $\|\|f\|\| = 1$ for $f \in X'$. If $\varepsilon > 0$ and $|f(x_i) - 1| < [\delta(\varepsilon)]/2$ for $x_1, x_2 \in \bar{S}_{1,\| \|}(0)$, then $\|x_1 - x_2\| < \varepsilon$. [*Hint*: $\|\frac{1}{2}(x_1+x_2)\| \geq \frac{1}{2}|f(x_1+x_2)| \geq |f(x_1)| - \frac{1}{2}|f(x_2 - x_1)| > 1 - \delta(\varepsilon)$.]

(c) Use (b) to prove that $\| \| = \sigma(X, X')$ on $B_1(0) = \{x : \|x\| = 1\}$, and $x_n \xrightarrow{\| \|} x_0$ iff $x_n \xrightarrow{\sigma(X,X')} x_0$. [*Hint*: If $\|x_0\| = 1$ and $f(x_0) = 1$, then $f^{-1}\{S_{[\delta(\varepsilon)]/2}(1)\} \cap B_1(0) \subset S_{\varepsilon,\| \|}(x_0) \cap B_1(0)$. If $x_n \xrightarrow{\sigma(X,X')} x_0$, then [Example 15.4(a), Problem 17B(f)] $\sup_{\mathbb{N}_0} \|x_k\| < \lambda_0$ for some $\lambda_0 > 0$. For $g = \lambda_0 f$ obtain $|g[(1/\lambda_0)x_0] - 1| < [\delta(\varepsilon)]/2$ and $|g((1/\lambda_0)x_n) - 1| < [\delta(\varepsilon)]/2$ $\forall n > N = N((\delta(\varepsilon))/2)$; conclude that $\|x_n - x_0\| < \lambda_0 \varepsilon$ $\forall n > N$.]

Remark. Note that this convergence property also holds in the non-UC space l_1 below.

(d) A UC Banach space $(X, \| \|)$ is reflexive. [*Hint*: Since $\bar{S}_{1,\| \|}(0) \subset \{S_{1,\| \|}(0)\}^{\circ\bullet} = \bar{S}_{1,\| \|}(0)^{\sigma(X'',X')}$ is $\sigma(X'', X')$-compact in X'' (Theorem 18.15), it suffices (Example 20.5) to show that $x'' \in \bar{S}_{1,\| \|}(0)^{\sigma(X'',X')}$ belongs to $\bar{S}_{1,\| \|}(0)$. Since $F_u = (x'' + u) \cap \bar{S}_{1,\| \|}(0) \neq \varnothing$ $\forall \sigma(X'', X')$-nbd $u(0) \subset X''$, the filter $\mathfrak{F} = \{F_u : \sigma(X'', X')\text{-nbds } u \subset X''\}$ in $\bar{S}_{1,\| \|}(0)$ is $\sigma(X'', X')$-convergent to $x'' \in \bar{S}_{1,\| \|}(0)$ is $\sigma(X'', X')$-convergent to $x'' \in \bar{S}_{1,\| \|}(0)^{\sigma(X'',X')}$; but $\mathfrak{F}$ is also $\| \|$-Cauchy [for $\varepsilon > 0$ and $x''(f) = 1$ for $f \in X'$ with $\|\|f\|\| = 1$ it follows from (b) that $F_v = (x'' + f^{-1}\{S_{\delta(\varepsilon)/2}(0)\}) \cap \bar{S}_{1,\| \|}(0)$ is ε-sized in $\mathfrak{F}$] and so $\| \|$-convergent to some $x_0 \in \bar{S}_{1,\| \|}(0)$. Finally, $x_0 = x''$, since $\sigma(X'', X')$ is T_2.]

(e) Both L_p and l_p are UC $\forall p > 1$. [*Hint*: For $p \geq 2$ and complex α, β the inequality $|\alpha + \beta|^p + |\alpha - \beta|^p \leq 2^{p-1}(|\alpha|^p + |\beta|^p)$ yields $\|f+g\|_p^p + \|f-g\|_p^p \leq 2^{p-1}(\|f\|_p^p + \|g\|_p^p)$. The case in which $1 < p < 2$ is much more involved. See Köthe [55].]

Remark. Examples of reflexive, non-UC spaces may be found in Bourbaki [13].

20I. Montel Spaces: Additional Examples

(a) Prove that $(\mathfrak{D}^\infty(\mathbb{R}; \mathbb{F}), \xi)$ in Problem 14E is Montel. [*Hint*: View $(\mathfrak{D}^\infty(\mathbb{R}; \mathbb{F}), \xi)$ as a closed subspace of $\prod_{m=1}^{\infty} C_\kappa(\mathbb{R}; \mathbb{F})$ under the embedding (verify!) $f \rightsquigarrow (f, f', f'', \ldots)$. If $\Phi \subset \mathfrak{D}^\infty(\mathbb{R}; \mathbb{F})$ is ξ-bounded, then $\pi_m(\Phi) = \{f^{(m)} : f \in \Phi\} \subset C_\kappa(\mathbb{R}; \mathbb{F})$

is equicontinuous $\forall m \in \mathbb{N}$. Use Theorem 7.18 to conclude that Φ is ξ-relatively compact.]

(b) Use (a) above to conclude that $(\mathfrak{D}^{\infty}_{K_n}(\mathbb{R}; \mathbb{F}), \xi_{K_n})$ in Problem 15C(b) is Montel $\forall n \in \mathbb{N}$. [*Hint*: $\mathfrak{D}^{\infty}_{K_n}(\mathbb{R}; \mathbb{F})$ is a closed subspace of $(\mathfrak{D}^{\infty}(\mathbb{R}; \mathbb{F}), \xi)$.]

(c) A strict inductive limit of (semi-) Montel spaces $\{E_n : n \in \mathbb{N}\}$ is (semi-) Montel if each E_n is closed in E_{n+1}.

(d) Use (b), (c) above to verify that $(\mathfrak{D}^{\infty}_0(\mathbb{R}; \mathbb{F}), \xi_0)$ in Problem 15C(b) is Montel. Show further that $(C_0(X; \mathbb{F}), \mathfrak{T}_{\mathfrak{s}(\mathfrak{1})})$ in Problem 15C(a) is Montel if X is also σ-compact.

(e) Use ingenuity to prove that $(\mathfrak{D}^{\infty}_r(\mathbb{R}; \mathbb{F}), \zeta)$ in Problem 14F is Montel.

20J. Corollary 20.13 Revisited

A separable Frechet space X is Montel iff $\sigma(X', X)$-convergence and $\beta(X', X)$-convergence are equivalent in X'. [*Hint*: Every $\beta(X', X)$-bounded set $E \subset X'$ is $\sigma(X', X)$-totally bounded and metrizible [Corollary 19.18, Examples 18.5(d) and 18.24(b)], therefore $\beta(X', X)$-totally bounded and $\beta(X', X)$-relatively compact (Theorem 19.20). Thus $(X', \beta(X', X))$ and X'' and also X, are semi-Montel.]

References for Further Study

On reflexive and Montel spaces: Edwards [28]; Fleming [30]; Köthe [55]; Wheeler [105].
On quasi-reflexive spaces; Civin and Yood [18]; James [46], [47]; Lohman and Casazza [57].

21. Full Completeness: Open Mapping and Closed Graph Theorems

For a given pair of TVSps X and Y it is generally not true that

(om) every surjection $f \in L(X, Y)$ is open;
(cg) $f \in L(X, Y)$ if $f \in \mathfrak{L}(X, Y)$ has a closed graph.

Note, however (Theorem 10.8), that both (om) and (cg) hold if X is Frechet and Y is a Catg II T_2VS. This can be extended if barreled spaces are taken as the appropriate generalization of Catg II, LCTVSps; but, first, a new completeness concept is needed.

DEFINITION 21.1. Let X be an LCT_2VS. A subset $E \subset X'$ is nearly closed if $E \cap u^{\circ}$ is $\sigma(X', X)$-closed in X' $\forall$ nbd $u(0) \subset X$. Furthermore, X is called fully complete (also a Ptak space) if every nearly closed subspace of X' is $\sigma(X', X)$-closed.

Remark. [Theorem 18.10(i)]. $E \cap u^{\circ}$ is $\sigma(X', X)$-closed in X' iff it is $\sigma(X', X)$-closed in $E \subset X'$.

Almost by definition a fully complete space X remains fully complete under any finer topology of the dual pair $\langle X, X'\rangle$. A few other illustrations pertinent to Definition 21.1 are collected below.

Example 21.2. (a) $(X^*, \sigma(X^*, X))$ is fully complete for every vector space X. Simply use Example 18.29(a) and the fact that $(X, \sigma(X, X^*)) = (Y', \sigma(Y', Y))$ for $Y = (X^*, \sigma(X^*, X))$.

(b) Every fully complete space is complete (Problem 18I) but the converse is generally false. To be sure, (Problem 15E) $\mathfrak{T}_{F_c}$ is complete and nonnormable for an infinite dim, barreled NLS $(X, \|\ \|)$. However, $\mathfrak{T}_{F_c}$ cannot be fully complete, since $\mathbb{1}: (X, \mathfrak{T}_{F_c}) \to (X, \|\ \|)$ will be a homeomorphism (Corollary 21.9).

In contrast to Problem 15B(c) strict inductive limits of fully complete spaces need not be fully complete. This follows from Problem 15D(b), Theorem 13.12, and the second assertion in

THEOREM 21.3. (i) A closed vector subspace of a fully complete space is fully complete.

(ii) If X is Frechet, then X and $(X', \mathfrak{T}_{\mathscr{S}})$ are fully complete for every $\mathfrak{T}_{\mathscr{S}}$ satisfying $\kappa(X', X) \subset \mathfrak{T}_{\mathscr{S}} \subset \tau(X', X)$.

Proof. First, observe [Problem 18O(c)] that a closed vector subspace M of a fully complete space X satisfies $(M', \sigma(M', M)) = (X'/M^{\circ}, \mathfrak{T}_{\mathscr{G}(\nu)})$. Denote polarity in $\langle M, M'\rangle$ by $^{\bullet}$ and suppose Φ is a nearly closed subspace of $M' = X'/M^{\circ}$. Then $\Phi \cap u^{\bullet}$ is $\sigma(M', M) = \mathfrak{T}_{\mathscr{G}(\nu)}$-closed $\forall$ M-nbd $u(0)$ and a simple computation reveals that $\nu^{-1}(\Phi) \cap u^{\circ} = \nu^{-1}(\Phi \cap u^{\bullet})$ is $\sigma(X', X)$-closed in X'. This ensures that $\nu^{-1}(\Phi) \cap v^{\circ} = \nu^{-1}(\Phi) \cap (M \cap v)^{\circ} \cap v^{\circ}$ is $\sigma(X', X)$-closed $\forall$ X-nbd $v(0)$. Since X is fully complete, $\nu^{-1}(\Phi)$ is $\sigma(X', X)$-closed and so Φ is $\mathfrak{T}_{\mathscr{G}(\nu)} = \sigma(M', M)$ closed in $X'/M^{\circ} = M'$. The first part of (ii) is clear, since every Frechet space is fully complete by Theorem 18.14 and Problem 18H(b). If M is a nearly closed subspace of $X = (X', \mathfrak{T}_{\mathscr{S}})'$ and $x_0 \in \bar{M}$, then $x_0 = \lim_n x_n$ for some $\{x_n\} \subset M$. Since $K = \{x_m : m \in \mathbb{N}_0\}$ is the polar of a $\kappa(X', X)$-nbd K°, it follows that $M \cap K$ is $\sigma(X, X')$-closed and $x_0 \in M$. This establishes that M is $\sigma(X, X')$-closed and $(X', \mathfrak{T}_{\mathscr{S}})$ is fully complete. ■

Remark. If X in (ii) is reflexive, $(X', \mathfrak{T}_{\mathscr{S}})$ is fully complete for $\kappa(X', X) \subset \mathfrak{T}_{\mathscr{S}} \subset \beta(X', X)$ (Theorem 20.2).

Ptak's full completeness characterization below uses

LEMMA 21.4. Let $f \in \mathfrak{L}(X, Y)$ be weakly continuous and nearly open for LCT_2VSps X, Y. Then $f^t(M)$ is nearly closed in X' for every nearly closed subspace $M \subset Y$.

Proof. It suffices to verify that $f^t(M) \cap u^{\circ}$ is $\sigma(X', X)$-closed in X' for every abs convex nbd $u(0) \subset X$. Begin by noting (Theorem 18.19) that $f^t(M) \cap u^{\circ} = f^t\{M \cap (f^t)^{-1}(u^{\circ})\} = f^t(M \cap \{f(u)\}^{\circ})$ in X'. Since $\{f(u)\}^{\circ} = \{\overline{f(u)}\}^{\circ}$ is $\sigma(Y', Y)$-compact (Theorem 18.15), it follows by hypothesis that $M \cap \{f(u)\}^{\circ} \subset Y'$ is also $\sigma(Y', Y)$-compact. Consequently (Theorem 18.18) $f^t(M) \cap u^{\bullet} = f^t(M \cap \{f(u)\}^{\circ})$ is $\sigma(X', X)$-closed in X'. ■

THEOREM 21.5. The following are equivalent for an LCT_2VS X:

(i) X is fully complete.
(ii) Every continuous, nearly open linear mapping of X into any LCT_2VS Y is open.

Proof. (i)⇒(ii) Suppose X is fully complete, $(Y, \mathfrak{T})$ is any LCT_2VS, and $f \in L(X, Y)$ is nearly open. Assume (for convenience) that f is also surjective. Then (by hypothesis and Lemma 21.4) $f^t(Y')$ is $\sigma(X', X)$-closed in X'. Evidently $\mathfrak{T} \subset \mathfrak{T}_{\mathcal{G}'(f)} \in T(Y)$ and $Y' \subset (Y, \mathfrak{T}_{\mathcal{G}'(f)})'$. To establish the reverse inclusions (and the conclusion that f is $\mathfrak{T}_{\mathcal{G}'(f)} = \mathfrak{T}$ open) we use the fact that $y' \in (Y, \mathfrak{T}_{\mathcal{G}'(f)})'$ implies $y'f \in f^t(Y')$. (Otherwise $y'f \in X' - f^t(Y')$ by Theorem 15.9(ii), and by Corollary 17.12 there is an $x_0 \in X = (X', \sigma(X', X))'$ such that $1 = \langle x_0, y'f \rangle = (y'f)(x_0)$ and $0 = \langle x_0, f^t(Y') \rangle = \langle f(x_0), Y' \rangle$. This is impossible, since Y' is injective and $f(x_0) = 0$.) Since f^t is injective (Theorem 18.19), $y' \in Y'$ and $\mathfrak{T}_{\mathcal{G}'(f)}$ is a topology of $\langle Y, Y' \rangle$. It follows, further (Theorem 18.6), that $\mathfrak{T}_{\mathcal{G}'(f)} \subset \mathfrak{T}$ since every abs convex, $\mathfrak{T}_{\mathcal{G}'(f)}$-closed nbd $v(0) \subset Y$ has an X-nbd $u(0) \subset f^{-1}(v)$ and corresponding $\mathfrak{T}$-nbd $\overline{f(u)}$ contained in $\bar{v} = v$.

For (ii)⇒(i) suppose that M is a nearly closed subspace of X' and let $^{\bullet}$ denote polarity relative to the injective dual pair $\langle M, X/M^{\circ} \rangle$. Assume that $\mathfrak{N}$ is an X-nbd base of abs convex sets and take $\mathcal{S} = \{M \cap u^{\circ} : u \in \mathfrak{N}\}$. Since u° is $\sigma(X', X)$-compact and $M \cap u^{\circ}$ is $\sigma(X', X)$-closed, every $M \cap u^{\circ} \in \mathcal{S}$ is $\sigma(M, X/M^{\circ})$-compact and so $\mathfrak{T}_{\mathcal{S}} \subset \tau(X/M^{\circ}, M)$ on X/M°. In the other direction $X' = \bigcup_{\mathfrak{N}} u^{\circ}$ ensures that $\mathcal{S}$ covers M and $(X/M^{\circ}, M) \subset \mathfrak{T}_{\mathcal{S}}$ [Example 18.24(a)]. This establishes that $M = (X/M^{\circ}, \mathfrak{T}_{\mathcal{S}})'$. Let $\underset{x' \rightsquigarrow x'}{i : M \to X'}$ so that $i^t = \nu : X \to X/M^{\circ}$. Then (using Theorem 18.19)

$\nu^{-1}\{(M\cap u^{\circ})^{\bullet}\}=(i^{t})^{-1}\{M\cap u^{\circ})^{\bullet}\}=(M\cap u^{\circ})^{\circ}$ in X and $(M\cap u^{\circ})^{\bullet}=\nu\{(M\cap u^{\circ})^{\circ}\}\supset \nu(u)\ \forall u\in\mathfrak{N}$ ensures that $\nu: X\to(X/M^{\circ},\mathfrak{T}_{\mathscr{S}})$ is continuous and $\mathfrak{T}_{\mathscr{S}}\subset\mathfrak{T}_{\mathfrak{g}(\nu)}$. Now [Problem 18N(c)] $\{\nu(u)\}^{\bullet}=M\cap u^{\circ}$ and $\overline{\nu(u)}=\{\nu(u)\}^{\bullet\bullet}=(M\cap u^{\circ})^{\bullet}$ so that ν is also $\mathfrak{T}_{\mathscr{S}}$-nearly open. By hypothesis $\mathfrak{T}_{\mathscr{S}}$ is open and $\mathfrak{T}_{\mathscr{S}}=\mathfrak{T}_{\mathfrak{g}(\nu)}$ on X/M° follows. It also follows [Example 18.13(b) slightly modified] that $M=(X/M^{\circ},\mathfrak{T}_{\mathfrak{g}(\nu)})'=M^{\circ\circ}$ in X' is $\sigma(X',X)$-closed and X is fully complete. ■

COROLLARY 21.6. Full completeness is preserved under continuous, nearly open linear surjections. In particular, the quotient of a fully complete space by a closed vector subspace is fully complete.

Proof. Suppose X is fully complete and $Y=f(X)$ for a nearly open surjection $f\in L(X,Y)$. If Z is any LCT$_2$VS and $g\in L(Y,Z)$ is nearly open, then $gf\in L(X,Z)$ is open; but $g(u)=(gf)\{f^{-1}(u)\}$ is a Z-nbd $\forall$ Y-nbd $u(0)$. Therefore g is open and Y is seen to be fully complete. ■

In order to proceed, we need

LEMMA 21.7. Let $f\in\mathfrak{L}(X,Y)$, where X and Y are LCT$_2$VSps.

(i) The graph $g_r(f)\subset X\times Y$ is closed iff $(f^t)^{-1}(X')$ is $\sigma(Y',Y)$-dense in Y'.

(ii) If f is nearly continuous, then $(f^t)^{-1}(M)$ is nearly closed in Y' for every nearly closed subspace $M\subset X'$.

Proof. (i) Begin with the observation that $\{(f^t)^{-1}(X')\}^{\circ\circ}=\overline{(f^t)^{-1}(X')}^{\sigma(Y',Y)}=Y'$ iff $\{(f^t)^{-1}(X')\}^{\circ}=\{0\}$ in Y. Suppose that $\mathfrak{N}$ is an X-nbd base of abs convex sets. Then $X'=\bigcup_{\mathfrak{N}}u^{\circ}$ for polars in X^* so that (Theorem 18.19) $(f^t)^{-1}(X')=\bigcup_{\mathfrak{N}}(f^t)^{-1}(u^{\circ})=\bigcup_{\mathfrak{N}}\{f(u)\}^{\circ}$ and $\{(f^t)^{-1}(X')\}^{\circ}=\bigcap_{\mathfrak{N}}\{f(u)\}^{\circ\circ}=\bigcap_{\mathfrak{N}}\overline{f(u)}$. Thus $(f^t)^{-1}(X')$ is $\sigma(Y',Y)$-dense in Y' iff $\bigcap_{\mathfrak{N}}\overline{f(u)}=\{0\}$. Now $y\in\bigcap_{\mathfrak{N}}\overline{f(u)}$ iff $\{y+v(0)\}\cap f(u)\neq\varnothing\ \forall u\in\mathfrak{N}$ and Y-nbd $v(0)$; that is, iff $(0,y)\in\overline{g_r(f)}\subset X\times Y$. Clearly $g_r(f)=\overline{g_r(f)}$ implies $\bigcap_{\mathfrak{N}}\overline{f(u)}=\{0\}$. On the other hand, $\bigcap_{\mathfrak{N}}\overline{f(u)}=\{0\}$ and $(x_0,y_0)\in\overline{g_r(f)}$ yield $(0,y_0-f(x_0)\in g_r(f)$; therefore $y_0-f(x_0)=0$ and $(x_0,y_0)\in g_r(f)$.

(ii) Suppose M is nearly closed in X' and let $v(0)$ be an abs convex Y-nbd. Then $\overline{f^{-1}(v)}$ is an X-nbd and $M\cap\{\overline{f^{-1}(v)}\}^{\circ}$ is $\sigma(X',X)$-closed in X'. Since $\{\overline{f^{-1}(v)}\}^{\circ}$ is $\sigma(X^*,X)$-compact (Theorem 18.15), $M\cap\{f^{-1}(v)\}^{\circ}$ is $\sigma(X^*,X)$-closed in X^* and $(f^t)^{-1}(M\cap\{\overline{f^{-1}(v)}\}^{\circ})$ is $\sigma(Y',Y)$-closed. Now $(f^t)^{-1}(M\cap\{\overline{f^{-1}(v)}\}^{\circ})=(f^t)^{-1}(M\cap\{f^{-1}(v)\}^{\circ})=(f^t)^{-1}(M)\cap(f\{f^{-1}(v)\})^{\circ}=(f^t)^{-1}(M)\cap\{v\cap f(X)\}^{\circ}$ implies $(f^t)^{-1}(M)\cap v^{\circ}=$

$(f^t)^{-1}(M) \cap \{v \cap f(X)\}^{\circ} \cap v^{\circ}$ is $\sigma(Y', Y)$-closed and $(f^t)^{-1}(M)$ is nearly closed in Y'. ■

THEOREM 21.8. For an LCT$_2$VS X and fully complete space Y:

(i) $f \in L(X, Y)$ if $f \in \mathfrak{L}(X, Y)$ is nearly continuous and $g_r(f)$ is closed.

(ii) $\tilde{f} \in \mathfrak{L}(Y, X)$ is open if $\tilde{f}$ is nearly open and surjective and $g_r(\tilde{f})$ is closed.

Proof. Suppose that $f \in \mathfrak{L}(X, Y)$ is nearly continuous and $g_r(f)$ is closed. Since X' is nearly closed and Y is fully complete, $(f^t)^{-1}(X') = Y'$ by Lemma 21.7. Accordingly, f is weakly continuous (Theorem 18.18). If $v(0)$ is any abs convex Y-closed nbd, then v is $\sigma(Y, Y')$-closed and $\overline{f^{-1}(v)} = \overline{f^{-1}(v)}^{\sigma(X,X')} = \overline{f^{-1}(v)}$ is an X-nbd. Therefore $f \in L(X, Y)$ and the proof of (i) is complete. If $\tilde{f}$ is a nearly open surjection with closed graph, then $Y/\tilde{f}^{-1}(0)$ is $\mathfrak{T}_{\mathfrak{g}(\nu)}$-fully complete (Theorem 10.3, Corollary 21.6). Furthermore (Example 9.7), $\mu_{\tilde{f}}^{-1}$ is nearly continuous. The continuity of $\mu_{\tilde{f}}^{-1}$ (i.e., the openness of $\mu_{\tilde{f}}$ and $\tilde{f}$) now follows from (i), since $g_r(\mu_{\tilde{f}})$, therefore $g_r(\mu_{\tilde{f}}^{-1})$, is closed. To be sure, ν is continuous and $g_r(\nu)$ is closed. Consequently (Lemma 21.7) $(\mu_{\tilde{f}}^t)^{-1}\{Y/\tilde{f}^{-1}(0)\}' = (\mu_{\tilde{f}}^t)^{-1}\{(\nu^t)^{-1}(Y')\} = (\tilde{f}^t)^{-1}(Y')$ is $\sigma(X', X)$-dense in X'. ■

The promised generalization of Theorem 10.8 is an easy consequence of Example 19.13(e) and Theorem 21.8.

COROLLARY 21.9. For a barreled T_2 space X and fully complete space Y

(cg) $f \in L(X, Y)$ if $f \in \mathfrak{L}(X, Y)$ has a closed graph;

(om) $\tilde{f} \in \mathfrak{L}(Y, X)$ is open if $\tilde{f}$ is surjective and $g_r(\tilde{f})$ is closed.

Remark. Note (Theorem 10.3) that $g_r(\tilde{f})$ is closed if $\tilde{f} \in L(Y, X)$.

COROLLARY 21.10. Let $(Y, \mathfrak{T})$ be fully complete. Then

(i) $\mathfrak{T} = \mathfrak{T}_0$ for every barreled T_2 topology $\mathfrak{T}_0 \in T(Y)$ weaker than $\mathfrak{T}$.

(ii) A barreled T_2 space X is fully complete if $X = \tilde{f}(Y)$ for some $\tilde{f} \in \mathfrak{L}(Y, X)$.

Proof. For (i) take $\tilde{f} = \mathbb{1}: (Y, \mathfrak{T}) \to (Y, \mathfrak{T}_0)$ in (om). For (ii) use (om) and Corollary 21.6. ■

Corollary 21.9 has been proved under the "best" conditions available if we expect reasonable (om) and (cg) results to hold. Full completeness cannot be replaced by completeness in (om) even if the barreled property is strengthened. For example (Problem 15D), the continuous surjection $\mathbb{1}:(Y, \mathfrak{T}_{F_c}) \to Y$ is not open if Y is an infinite dim Banach space. Relaxing the barreled property by taking near openness of mappings still requires (Theorem 21.5) that full completeness prevail. In a similar vein the barreled property is interwoven with full completeness for (cg) results, as the following theorem (due to Mahowald) demonstrates.

THEOREM 21.11. An LCT_2VS X is barreled if every linear mapping with closed graph of X into any Banach space is continuous.

Proof. Suppose B is a barrel in X and ν is the canonical mapping of X into $(\check{E}_B, \|\ \|)$, where $(\check{E}_B, \|\ \|)$ is the completion of $E_B = (X/p_B^{-1}(0), \|\ \|_{p_B})$. Then (Theorem 14.10) $B = \bar{S}_{1,p_B}(0)$ and $\nu(B) = \bar{S}_{1,\|\|_{p_B}}(0)$ is a $\|\ \|_{p_B}$-nbd in E_B. Moreover, $p_B^{-1}(0) \subset B$ and $\nu^{-1}\{\nu(B)\} \subset 3B$ so that it remains only to prove $g_r(\nu)$ closed in $X \times \check{E}_B$. If $(x_0, \check{y}_0) \notin g_r(\nu)$, then $\|\check{y}_0 - \nu(x_0)\| > 2\varepsilon$ for some $\varepsilon > 0$. Since E_B is dense in $\check{E}_B$, one also has $\|\check{y}_0 - \nu(x_1)\| < \varepsilon$ for some $x_1 \in X$. Thus $\|\nu(x_0) - \nu(x_1)\| > \varepsilon$ and $x_0 \notin S_{\varepsilon,p_B}(x_1) = x_1 + B$ which is closed in X. It follows from $(x_0, y_0) \in D = \{X - S_{\varepsilon,p_B}(x_1)\} \times S_{\varepsilon,\|\|}(\nu(x_1))$ and $D \cap g_r(\nu) = \varnothing$ that $(x_0, \check{y}_0) \notin \overline{g_r(\nu)}$. ■

PROBLEMS

21A. Frechet Spaces

Assume throughout that X is a Frechet space.

(a) An LCT_2VS Y is Frechet if $Y = f(X)$ for a nearly open $f \in L(X, Y)$. In particular, the quotient of a Frechet space by a closed vector subspace is Frechet. [*Hint*: Use Theorem 15.12, Corollary 21.6.]

(b) $X = H \oplus K$ implies that $X \doteq H \oplus K$ for closed subspaces $H, K \subset X$. [*Hint*: Corollary 21.9, Theorem 15.26.]

(c) X is normable (i.e., Banach) if $(X', \beta(X', X))$ is 1°. [*Hint*: Let $\{B_n : n \in \mathbb{N}\}$ be abs convex, closed bounded sets in X such that $\{B_n^{\circ} : n \in \mathbb{N}\}$ is a $\beta(X', X)$-nbd base at $0 \in X'$. If $\{u_1 \supset u_2 \supset \cdots u_n \supset u_{n+1} \supset \cdots\}$ is an X-nbd base at 0, then $u_n \subset B_n$ for some $n \in \mathbb{N}$.]

(d) Use (c) and Corollary 19.18 to confirm that a Frechet space whose strong dual is Frechet is necessarily Banach.

21B. Infrabarreled Characterization

(Compare with Theorem 21.11.) An LCT_2VS X is infrabarreled iff for any Banach space Y every $f \in B\mathfrak{L}(X, Y)$ having a closed graph is continuous. [*Hint*: If X

is infrabarreled, $f \in B\mathfrak{L}(X, Y)$ is nearly continuous and Theorem 21.8 applies. If $B \subset X$ is a bornivorous barrel, then (as in Theorems 19.2(iii) and 21.11), $\nu \in L(X, \check{E}_B)$ and B is an X-nbd.]

21C. B_r-Completeness

An LCT_2VS X is said to be B_r-complete (also called infra-Ptak) if $\Phi = X'$ for every nearly closed, weakly dense subspace $\Phi \subset X'$.

(a) The following are equivalent for an LCT_2VS $(X, \mathfrak{T})$:

(i) X is B_r-complete.
(ii) For any LCT_2VS Y every nearly open injection $f \in L(X, Y)$ is open.
(iii) $\mathfrak{T} = \mathfrak{T}_0$ $\forall T_2$ topology $\mathfrak{T}_0 \in T(X)$ weaker than $\mathfrak{T}$ such that $\bar{u}^{\mathfrak{T}_0}$ is a $\mathfrak{T}_0$-nbd $\forall \mathfrak{T}$-nbd $u(0) \subset X$.

[*Hint*: For (i)$\Rightarrow$(ii) proceed as in Theorem 21.5. To get $Y' = (Y, \mathfrak{T}_{\mathcal{G}'(f)})'$ show (using Theorems 18.18 and 18.19) that $f'(Y')$ is $\sigma(X', X)$-dense in X' and obtain $X' = f'(Y')$. For (ii)$\Rightarrow$(iii) take $f = \mathbb{1}: (X, \mathfrak{T}) \to (X, \mathfrak{T}_0)$. For (iii)$\Rightarrow$(i) assume that $\Phi \subset X'$ as above and let $\mathfrak{N}$ be a $\mathfrak{T}$-nbd base at $0 \in X$. Then $\{(\Phi \cap u^{\circ})^{\circ} : u \in \mathfrak{N}\}$ is the base for an LCT_2 topology $\mathfrak{T}_0 \in T(X)$ satisfying the hypothesis of (iii) and $\Phi = (X, \mathfrak{T}_0)'$ (since $(\Phi \cap u^{\circ})^{\circ\circ} = \Phi \cap u^{\circ}$ $\forall u \in \mathfrak{N}$).]

(b) A fully complete space is B_r-complete, which is complete. [*Hint*: See Problem 18I.]

(c) Can you find counterexamples to the reverse implications in (b) above? [*Hint*: Use (d) below to show that $(X, \mathfrak{T}_{F_c})$ is not B_r-complete.]

(d) Full completeness can be replaced by B_r-completeness in Theorem 21.8(i) and Corollary 21.9 (cg) and in Theorem 21.8(ii) if f is bijective.

(e) $X = H \oplus K$ implies $X \stackrel{\perp}{=} H \oplus K$ for closed subspaces H, K of a barreled B_r-complete space X.

21D. Hypercomplete Spaces

An LCT_2VS X is said to be hypercomplete if every nearly closed abs convex subset of X' is $\sigma(X', X)$-closed.

(a) Hypercompleteness implies full completeness. A complete metrizible LCTVS is hypercomplete. [*Hint*: Problem 19H(b).]

(b) $(X^*, \sigma(X^*, X))$ is hypercomplete for every vector space X. [*Hint*: Show that nearly closed abs convex subsets of X are $\mathfrak{T}_{F_c} = \tau(X, X^*)$ closed.]

(c) Hypercompleteness is preserved under continuous open surjections and closed subspaces.

References for Further Study

On open mapping and closed graph theorems: Ptak [81].
On B_r-completeness: Collins [21]; Edwards [28].
On hypercomplete spaces: Kelley [51].

CHAPTER IV

Topological Algebras

Perspective. A topological algebra (TVA) is a TVS in which the underlying set is an algebra with continuous multiplication. As such, TVAs combine aspects of both topological groups and TVSps. Our study of TVAs, in large part, details the effects of continuity on this combined structure. We begin with normed and Banach algebras and abstract their properties to more general TVAs in the sections that follow. An enriched version of Section 15 reveals that these general TVA properties are intimately linked to normed algebra properties by way of projective limits.

22. Algebraic Preliminaries

The notions below are basic and are in use from here on.

DEFINITION 22.1. A linear algebra (abbreviated "algebra") is a vector space X with a binary operation that satisfies $x(yz)=(xy)z$, $x(y+z)=xy+xz$, $(y+z)x=yx+zx$ and $\alpha(xy)=(\alpha x)y=x(\alpha y)$ $\forall x, y, z \in X$ and $\alpha \in \mathbb{F}$.

A homomorphism between algebras X, Y is a mapping $h \in \mathfrak{L}(X, Y)$ such that $h(x_1x_2)=h(x_1)h(x_2)$ $\forall x_1, x_2 \in X$. The collection of nonzero complex-valued homomorphisms $h \in X^*$ is denoted $\mathcal{H}$.

Remark. The theory of commutative algebras is quite different from that of noncommutative algebras. Unless specified otherwise, **all algebras are assumed to be commutative and in possession of a multiplicative identity *e*.** The latter requirement, incidentally, is not too restrictive, since every algebra X without identity may be viewed as a subalgebra of $X_e = \mathbb{F} \times X$—an algebra with identity $e=(1, 0)$ under vector addition, scalar multiplication, and vector multiplication $(\alpha x)(\beta y)=(\alpha\beta, \alpha y+\beta x+xy)$. Simply note that $\underset{x \rightsquigarrow (0\ x)}{X \Rightarrow X_e}$ is an into algebra isomorphism (i.e., an injective homomorphism).

THEOREM 22.2. $\mathcal{H}=\{h \in X^*: h(e)=1$ and $h^{-1}(0)$ is a subalgebra of $X\}$. Furthermore, distinct members of $\mathcal{H}$ are linearly independent in X^*

Proof. Every $h \in \mathcal{H}$ clearly satisfies the above conditions. If $h(e)=1$ and $h^{-1}(0)$ is a subalgebra of X for $h \in X^*$, then $x - h(x)e \in h^{-1}(0)\ \forall x \in X$ and so $xx_0 - xh(x_0)e - x_0h(x)e + h(x)\,h(x_0)e = \{x - h(x)e\}\{x_0 - h(x_0)\} \in h^{-1}(0)$. Therefore $h(xx_0) - h(x)\,h(x_0) = 0\ \forall x, x_0 \in X$ and $h \in \mathcal{H}$. The second assertion concerning linear independence is trivially true for $n=1$. If $h_2 \neq h_1$ and $\alpha_1 h_1 + \alpha_2 h_2 = 0$, then $h_2(x_0) \neq h_1(x_0)$ for some $x_0 \in X$. From $0 = \alpha_1 h_1(x)\,h_1(x_0) + \alpha_2 h_2(x)\,h_2(x_0)$ subtract $0 = h_2(x_0)\{\alpha_1 h_1(x) + \alpha_2 h_2(x)\}$ to obtain $\alpha_1\{h_1(x_0) - h_2(x_0)\}h_1(x) = 0\ \forall x \in X$. This means that α_1, α_2 are both zero and $\{h_1, h_2\}$ are linearly independent. Now assume inductively that every set of n-distinct members of h is linearly independent and let $\sum_{i=1}^{n+1} \alpha_i h_i = 0$. Then $h_{n+1}(x_0) \neq h_1(x_0)$ for some $x_0 \in X$ and $0 = \sum_{i=1}^{n+1} \alpha_i h_i(x)\,h_i(x_0) - h_{n+1}(x_0) \sum_{i=1}^{n+1} \alpha_i h_i(x) = \sum_{i=1}^{n} \alpha_i\{h_i(x_0) - h_{n+1}(x_0)\}h_i(x)\ \forall x \in X$. By inductive hypothesis $\alpha_1 = 0$ and $\sum_{i=2}^{n+1} \alpha_i h_i = 0$ implies $\alpha_i = 0\ \forall 2 \leq i \leq n+1$. ■

COROLLARY 22.3. *If $\{h_1, h_2, \ldots, h_n\} \in \mathcal{H}$ are distinct, there is a linearly independent set $\{x_1, \ldots, x_n\} \in X$ that satisfies $h_i(x_j) = \delta_{ij}\ \forall i, j = 1, 2, \ldots, n$. For each $\{\alpha_1, \alpha_2, \ldots, \alpha_n\} \in \mathbb{F}$ there is also an $x \in X$ such that $h_i(x) = \alpha_i\ \forall 1 \leq i \leq n$.*

Proof. The first assertion holds by Example 13.2(b). The second follows by induction. Specifically, $h_1(x) = \alpha_1$ if $x = \alpha_1 e$. Assuming the statement to be true for $k = n$, consider any $h_{n+1} \in \mathcal{H} - \{h_1, \ldots, h_n\}$ and $\alpha_{n+1} \in \mathbb{F}$. Since $\mathcal{H}$ is concentrating, $h_{n+1}(x_0) = 1$ for some $x_0 \in \bigcap_{i=1}^{n} h_i^{-1}(0)$. Therefore $h_i(\tilde{x}) = \alpha_i$ $(\forall 1 \leq i \leq n+1)$ for $\tilde{x} = x + \{\alpha_{n+1} - h_{n+1}(x)\}x_0$. ■

The following may be surprising in that seemingly unrelated topological and algebraic properties are proved equivalent. Simply note that $\sigma(X, \mathcal{H}) = \sigma(X, [\mathcal{H}])$ relative to $\langle X, [\mathcal{H}]\rangle$ in

COROLLARY 22.4. *If X is an n-dim algebra, then $\aleph(\mathcal{H}) \leq n$ and the statements below are equivalent:*

(i) $\aleph(\mathcal{H}) = n$; (ii) $\mathcal{H}$ *is injective;* (iii) $\sigma(X, \mathcal{H})$ *is* T_2.

Proof. Clearly X^* is n-dim and $\aleph(\mathcal{H}) \leq n$ when $\dim X = n$. If $\mathcal{H} = \{h_1, \ldots, h_n\}$, the set $\{x_1, \ldots, x_n\}$ in Corollary 22.3 is a basis for X. Since

$x=\sum_{i=1}^{n}\alpha_i x_i\in\bigcap_{j=1}^{n}h_j^{-1}(0)$ implies $0=\sum_{i=1}^{n}\alpha_i h_j(x_i)=\alpha_j\ \forall 1\le j\le n$, one has $\bigcap_{j=1}^{n}h_j^{-1}(0)=\{0\}$ and (i)$\Rightarrow$(ii). Assume, conversely, that $\aleph(\mathcal{H})=m<n$ and let $X_0=[x_1,\ldots,x_m]$, where $h_i(x_j)=\delta_{ij}\ \forall i,j=1,2,\ldots,m$. Since $m<n$, there is an $x\in X-X_0$ such that $x\neq y=\sum_{i=1}^{m}h_i(x)x_i\in X_0$. In particular, $0\neq x-y\in\bigcap_{i=1}^{m}h_i^{-1}(0)$ and $\mathcal{H}$ is not injective. Theorem 15.1(ii) assures that (ii)$\Rightarrow$(iii). ∎

DEFINITION 22.5. Members of $\mathfrak{U}=\{x\in X: x^{-1}\in X\}$ are called units in the algebra X. A scalar $\lambda\in\mathbb{C}$ is said to be a regular or singular point of $x\in X$, depending on whether or not $x=\lambda e\in\mathfrak{U}$. One calls $\rho(x)=\{\lambda\in\mathbb{C}: x-\lambda e\in\mathfrak{U}\}$ the resolvent of x; its complement $\sigma(x)=\mathbb{C}-\rho(x)$ is termed the spectrum of x.

The spectral radius of $x\in X$ is defined as $r_\sigma(x)=\sup_{\sigma(x)}|\lambda|$.

Notation. The above notions are subscripted $\mathfrak{U}_X$, ρ_X, σ_X, $r_{\sigma X}$ if any confusion is possible (e.g., when more than one algebra is involved).

THEOREM 22.6. In a complex algebra X

(i) $x_1x_2\cdots x_n\in\mathfrak{U}$ iff $x_i\in\mathfrak{U}\ \forall 1\le i\le n$.

(ii) $\sigma\{p(x)\}=p\{\sigma(x)\}$ for every polynomial $p(t)$ with complex coefficients. Therefore $r_\sigma(x^n)=\{r_\sigma(x)\}^n\ \forall x\in X$ and $n\in\mathbb{N}$.

Proof. If $\{x_i: 1\le i\le n\}\in\mathfrak{U}$, then $x=x_1x_2\cdots x_n\in\mathfrak{U}$ with $x^{-1}=x_n^{-1}\cdots x_2^{-1}x_1^{-1}$. On the other hand, $x=x_1x_2\cdots x_n\in\mathfrak{U}$ implies $(x^{-1}x_1\cdots x_{i-1}x_{i+1}\cdots x_n)x_i=e$ and $x_i\in\mathfrak{U}\ \forall 1\le i\le n$. Assume next that $p(t)$ is nth degree in (ii). Since $\mathbb{C}$ is algebraically closed, $p(t)-\lambda=\alpha\prod_{i=0}(t-\lambda_i)$. Therefore $\lambda=p(\lambda_i)\ \forall 0\le i\le n$. Take $x^{\circ}=e$ and write $p(x)-\lambda e=\alpha\prod_{i=0}^{n}(x-\lambda_i e)$. If $\lambda\in\sigma\{p(x)\}$, then $p(x)-\lambda e\notin\mathfrak{U}$ and $x-\lambda_{i_0}e\notin\mathfrak{U}$ for some $1\le i_0\le n$. Therefore $\lambda_{i_0}\in\sigma(x)$ and $\lambda\in p\{\sigma(x)\}$. Assume conversely that $\lambda=p(\gamma)$ for $\gamma\in\sigma(x)$. Then $t-\gamma$ is a factor of $p(t)-\lambda$ and $p(x)-\lambda e=\beta\prod_{i=0}^{n-1}(x-\gamma_i e)(x-\gamma e)$. Since $x-\gamma e\notin\mathfrak{U}$, it follows from (i) that $\lambda\in\sigma\{p(x)\}$. The second assertion in (ii) holds, since $\{r_\sigma(x)\}^n=\sup_{\lambda\in\sigma\{(x)\}^n}|\lambda|=\sup_{\lambda\in\sigma(x^n)}|\lambda|=r_\sigma(x^n)$. ∎

DEFINITION 22.7. A vector subspace I of an algebra X is called an ideal of X if $XI \subset I$. (Note that every algebra contains two ideals; namely, $\{0\}$ and the algebra itself.) The collection of X's proper ideals is denoted $\mathfrak{J}$.

By a maximal ideal of X is meant and $M \in \mathfrak{J}$ which is not properly contained in any other $I \in \mathfrak{J}$. The collection of X's maximal ideals is denoted $\mathfrak{M}$.

One calls $\mathfrak{R}(X) = \bigcap_{\mathfrak{M}} M$ the radical of X and terms X to be semisimple if $\mathfrak{R}(X) = \{0\}$. Note that $X/\mathfrak{R}(X)$ is always semisimple since ν maps maximal ideals of X into maximal ideals of $X/\mathfrak{R}(X)$ and so $\mathfrak{R}(X/\mathfrak{R}(X)) = \mathfrak{R}(X)$.

The ideal Xx_0 in X is improper iff $x_0 \in \mathfrak{U}$ (since $x_0 \in \mathfrak{U}$ implies every $x \in X$ can be written $x = (xx_0^{-1})x_0$, whereas $X = Xx_0$ necessitates $e = xx_0$ and $x_0 \in \mathfrak{U}$). This just about establishes the first part of

THEOREM 22.8. (i) $\mathfrak{U} = X - \bigcup_{\mathfrak{J}} I$. Accordingly, every $I \in \mathfrak{J}$ is contained in some $M \in \mathfrak{M}$.

(ii) If $x_0 \in \mathfrak{R}(X)$, then $e - x_0 \in \mathfrak{U}$ and $\sigma(x_0) = \{0\}$ or $\varnothing$.

Proof. If $x \in \mathfrak{U}$, then $x \notin I$ for any $I \in \mathfrak{J}$ (otherwise $e = x^{-1}x \in I$ and $I = X$). The reverse inclusion also holds, since $x \in Xx \in \mathfrak{J}\ \forall x \notin \mathfrak{U}$. Next, partially order $\mathfrak{J}_I = \{J \in \mathfrak{J} : J \supset I\}$ by inclusion. Since every totally ordered subset $\{J_\alpha : \alpha \in \mathcal{Q}\} \subset \mathfrak{J}_I$ has an upper bound $\bigcup_{\mathcal{Q}} J \in \mathfrak{J}_I$, there is an $M \in \mathfrak{J}_I \cap \mathfrak{M}$ as asserted (Zorn's lemma). The proof of (ii) is a consequence of (i), since $x_0 \in \mathfrak{R}(X)$ implies $e - x_0 \in \mathfrak{U}$ (otherwise $e - x_0$ and so $e = (e - x_0) + x_0$ belongs to some $M \in \mathfrak{M}$). For any $\lambda \neq 0$, one therefore has $(1/\lambda)x_0 \in \mathfrak{R}(X)$ and $e - (1/\lambda)x_0 \in \mathfrak{U}$, which means $-(1/\lambda)(x_0 - e) \in \mathfrak{U}$ and $\lambda \notin \sigma(x_0)$. ■

The following illustration also serves as a point of reference later on.

Example 22.9. The space $X = C(\Omega)$ of continuous complex-valued functions on a compact T_2 space Ω is an algebra with identity χ_Ω under pointwise multiplication $(fg)(\omega) = f(\omega)\,g(\omega)\ \forall \omega \in \Omega$. For each $\omega \in \Omega$ define $h_\omega : C(\Omega) \to \mathbb{C}$, $f \rightsquigarrow f(\omega)$ and $M_\omega = \{f \in C(\Omega) : f(\omega) = 0\}$. Then $h_\omega \in \mathcal{H}$ and $M_\omega = h^{-1}(0) \in \mathfrak{M}$.

(a) $\Omega \to \mathfrak{M}$, $\omega \rightsquigarrow M$ is a bijection: if $\omega_1 \neq \omega_2$, then $f(\omega_1) = 0 \neq 1 = f(\omega_2)$ and $f \in M_{\omega_1} - M_{\omega_2}$ for some $f \in C(\Omega)$. If $M \in \mathfrak{M}$, then $M = M_{\omega_0}$ for some $\omega_0 \in \Omega$. Otherwise there are $f_\omega \in M$ with $f_\omega(\omega) \neq 0\ \forall \omega \in \Omega$. Since $f_\omega \in C(\Omega)$, each

$f_\omega \neq 0$ on some open set $u_\omega(\omega) \subset \Omega$. Therefore $\Omega \subset \bigcup_{i=1}^{n} u_{\omega_i}(\omega_i)$ for some $\{\omega_1, \ldots, \omega_n\} \subset \Omega$ and $f = \sum_{i=1}^{n} \bar{f}_{\omega_i} f_{\omega_i} \in M \cap \mathfrak{U}$ (with inverse $1/f$) clearly contradicts Theorem 22.8(i).

(b) $\underset{\omega \rightsquigarrow h_\omega}{\Omega \to \mathcal{H}}$ and $\underset{h_\omega \rightsquigarrow M_\omega}{\mathcal{H} \to \mathfrak{M}}$ are bijections: surely $h_{\omega_1} = h_{\omega_2}$ implies $M_{\omega_1} = M_{\omega_2}$ and $\omega_1 = \omega_2$. If $h \in \mathcal{H}$, then $h^{-1}(0) \in \mathfrak{M}$ and $h^{-1}(0) = h_{\omega_0}^{-1}(0)$ for some $\omega_0 \in \Omega$. This requires that $h = h_{\omega_0}$. Otherwise $h(x_0) \neq h_{\omega_0}(x_0)$ for some $x_0 \in h^{-1}(0)$ and every $x \in X$ has a unique representation $x = x_h + x_0$ [for fixed $x_h \in h^{-1}(0)$]. Therefore $h(x) = \lambda h(x_0)$ and $h_{\omega_0}(x) = \lambda h_{\omega_0}(x_0)$ yields $h(x) = \{h(x_0)/h_{\omega_0}(x_0)\}\, h(x)\ \forall x \in X$ which contradicts Theorem 22.2.

(c) $X = C(\Omega)$ is semisimple, since $\mathfrak{R}(X) = \bigcap_{\Omega} M_\omega = \{0\}$, and $\mathcal{H}$ is injective, since $h_\omega(f) = 0\ \forall \omega \in \Omega$ implies $f \equiv 0$ on Ω.

(d) $\sigma(f) = \{\lambda \in \mathbb{C} : f - \lambda\chi_\Omega \notin \mathfrak{U}\} = \{\lambda \in \mathbb{C} : f(\omega) = \lambda$ for $\omega \in \Omega\} = f(\Omega)$ and $r_\sigma(f) = \sup_{\Omega} |f(\omega)|\ \forall f \in X$.

Remark. The results of (a) and (b) may be summarized as $\mathfrak{M} = \mathfrak{M}_\Omega$ and $\mathcal{H} = \mathcal{H}_\Omega$.

Problems

22A. Quasi-Regularity

Since units are undefined in a (not necessarily commutative) algebra X without identity, one frequently embeds X in X_e and works with $\mathfrak{U} \subset X_e$. Another approach uses the "circle operation" $x \cdot y = x + y - xy\ \forall x, y \in X$. An element $x \in X$ is said to be left {right} quasi-regular, and written $x \in \mathfrak{U}_{\odot_L}$ $\{x \in \mathfrak{U}_{\odot_R}\}$, if there is a $y \in X$ such that $y \cdot x = 0$ $\{x \cdot y = 0\}$. One calls y the left {right} quasi-inverse of $x \in X$.

(a) If $x \cdot y_1 = 0 = y_2 \cdot x$, then $y_1 = y_2$. Therefore it makes sense to call $x \in \mathfrak{U}_\odot = \mathfrak{U}_{\odot_L} \cap \mathfrak{U}_{\odot_R}$ quasi-regular, with quasi-inverse $x^\odot$ defined as the common value of its left and right quasi-inverse.

(b) $\mathfrak{U}_\odot$ is a circle multiplication group with identity $0 \in x$ and serves as the analog of $\mathfrak{U}$ in an algebra with identity. Specifically, $x \in \mathfrak{U}_{\odot_L}$ $\{\mathfrak{U}_{\odot_R}\}$ in X iff $(1, -x) = (1, 0) - (0, x) \in \mathfrak{U}_L$ $\{\mathfrak{U}_R\}$ in X_e.

(c) Use (b) to prove that $\mathfrak{U} = (1, -\mathfrak{U}_\odot)$ in X_e and $x^\odot = (0, x)(-1, x)^{-1}\ \forall x \in \mathfrak{U}_\odot$ (i.e., $x^\odot \in X$ is the preimage of $(0, x)(-1, x)^{-1} \in X_e$ under $\underset{x \rightsquigarrow (0,x)}{X \to X_e}$).

(d) Scalars in $\rho_\odot(x) = \{\lambda \neq 0 : (0, x) - \lambda(1, 0) \in \mathfrak{U}$ in $X_e\}$ are called quasi-regular points of x and elements in $\sigma_\odot(x) = \mathbb{C} - \rho_\odot(x) = \{\lambda : (1/\lambda)x \notin \mathfrak{U}_\odot\}$ are termed quasi-singular points of x. Establish that $\sigma_\odot(x) = \sigma\{(0, x)\}\ \forall x \in X$ and $\sigma_\odot(x^\odot) = \{\lambda/(\lambda - 1) : \lambda \in \sigma_\odot(x)\}\ \forall x \in \mathfrak{U}_\odot$, whereas $\sigma\{(1, x)^{-1}\} = \{1/\lambda : \lambda \in \sigma\{(1, -x)\}\}\ \forall (1, -x) \in \mathfrak{U}$. [*Hint*: First show that $[(\lambda - 1)/\lambda]x^\odot = x^\odot \cdot [(1/\lambda)x]\ \forall x \in \mathfrak{U}_\odot$.]

(e) If $x \in \mathfrak{U}_{\odot L}$ $\{\mathfrak{U}_{\odot R}\}$ and Y is an algebra with identity e_Y, then $h(x) \neq e_Y$ for every homomorphism $h: X \to Y$. [*Hint*: Observe that $h(x) + h(x^{\odot}) = h(x^{\odot})\,h(x)$ $\{h(x)\,h(x^{\odot})\}$.]

22B. Regular Maximal Ideals

Assume that X is a complex commutative algebra without identity.

(a) For each $x_0 \in X$ show that $I_{x0} = \{xx_0 - x : x \in X\} \in \mathcal{I} \cup \{X\}$, and $x_0 \in \mathfrak{U}_{\odot}$ iff $x_0 \in I_{x0}$ iff $I_{x0} \in X$.

(b) Show that $x_0 \in \mathfrak{U}_{\odot}$ iff $I_{x0} \not\subset M$ $\forall M \in \mathfrak{M}$. Use this to conclude that $\mathfrak{R}(X) \subset \mathfrak{U}_{\odot}$.

(c) An ideal $I \in \mathcal{I}$ is said to be regular if there exists an $x_0 \in X$ (called an identity mod I) such that $I = I_{x_0}$. Verify that $I = I_{x0}$ iff $I + x_0$ is an identity for X/I.

(d) If $\mathcal{I}_{\mathfrak{R}}$ denotes the set of regular ideals of X, then every $I \in \mathcal{I}_{\mathfrak{R}}$ is contained in some $M \in \mathfrak{M}_{\mathfrak{R}} = \mathcal{I}_{\mathfrak{R}} \cap \mathfrak{M}$. Thus every $M \in \mathfrak{M}_{\mathfrak{R}}$ is a maximal regular ideal (i.e., a regular ideal not properly contained in any other regular ideal of X).

The notions in Problems 22A and 22B are picked up again in Problems 23A and 23B and Problems 26A.

23. Normed and Normed *-Algebras

Normed linear algebras (NLAs) are the natural particularization of NLSps. Both serve as the prototype for the general topological structures of their categories. This comparison quickly ends when NLAs are bestowed with an involution. At this level continuity of multiplication signals where TVA theory veers from that of TVSps.

DEFINITION 23.1. A normed linear algebra (NLA) is an algebra X with a norm satisfying $\|xy\| \leq \|x\|\,\|y\|$ $\forall x, y \in X$. By a Banach algebra is meant a NLA that is also a Banach space.

Remark. It is assumed henceforth that $\|e\| = 1$. This is not especially restrictive, since $\|x\|' = \sup_{y \neq 0} \|xy\|/\|y\|$ is an equivalent norm for which $\|e\|' = 1$ and $\|xy\|' \leq \|x\|'\|y\|'$. (Specifically, $\|x\|/\|e\| \leq \|x\|' \leq \|x\|$ and $\|x_1x_2x\| \leq \|x_1\|'\|x_2\|'\|x\|$ $\forall x \in X$ implies $\|x_1x_2\|' \leq \|x_1\|'\|x_2\|'$.)

Remark. In a NLA multiplication is continuous, since $\|xy - x_0y_0\| \leq \|x - x_0\| + \|y - y_0\| + \|x_0\|\,\|y - y_0\| + \|y_0\|\,\|x - x_0\|$ $\forall x, x_0, y, y_0 \in X$.

Remark. If $(X, \|\ \|)$ is a NLA without identity, $\|(\alpha, x)\|_e = |\alpha| + \|x\|$ is a norm on X_e under which $X \to X_e$ is a congruence. Thus $(X, \|\ \|)$ is Banach iff $(X_e, \|\ \|_e)$ is Banach.

Almost all important Banach algebras fall within one of three categories: function algebras, operator algebras, or group algebras, depending on whether multiplication is defined pointwise, by composition, or by convolution. One fact worth mentioning here (cf Corollary 24.21) is that $\underset{h \rightsquigarrow h^{-1}(0)}{\mathcal{H} \to \mathfrak{M}}$ is bijective for a complex Banach algebra X with identity. Thus X is semisimple if $\mathcal{H}$ is injective.

Example 23.2. $(CB(\Omega; \mathbb{F}), \|\ \|_0)$ is a commutative Banach algebra with identity χ_Ω under pointwise multiplication. It is also semisimple, since $\mathcal{H} = \{h_\omega : \omega \in \Omega\}$ is injective.

If Ω is locally compact and T_2, then $C_\infty(\Omega; \mathbb{F})$ is a semisimple Banach subalgebra of $CB(\Omega; \mathbb{F})$ that (Appendix t.8) also contains $C_0(\Omega; \mathbb{F})$ as a $\|\ \|_0$-dense NL subalgebra. Neither $C_\infty(\Omega; \mathbb{F})$ nor $C_0(\Omega; \mathbb{F})$ need have an identity. In fact, each has an identity iff Ω is compact, in which case all algebras above coincide with $C(\Omega; \mathbb{F})$.

Lack of identity somewhat complicates matters here. We shall wait until Problem 26B to show that $C_\infty(\Omega; \mathbb{F})$ has $\mathcal{H} = \mathcal{H}_\Omega$ and $\mathfrak{M}_{\mathfrak{R}} = \mathfrak{M}_\Omega$ (Problem 22B).

Example 23.3. (a) The "disk" algebra $\mathfrak{H}(\bar{S}_1) \subset (C(\bar{S}_1; \mathbb{C}), \|\ \|_0)$ of functions analytic in the interior of $\bar{S}_1 = \{z \in \mathbb{C} : |z| \leq 1\}$ is closed in $C(\bar{S}_1; \mathbb{C})$ (by Morera's theorem) and is therefore a commutative Banach algebra with identity $\chi_{\bar{S}_1}$. To show $\mathcal{H} = \mathcal{H}_{\bar{S}_1}$ (and conclude that $\mathfrak{H}(\bar{S}_1)$ is semisimple) fix any $h \in \mathcal{H}$ and observe [Example 22.9(b)] that $h(f) = z_0 \in \bar{S}_1$ for $f(z) = z$ on $\bar{S}_1$. Therefore $h(g) = g(z_0)$ for every polynomial $g(z) = \sum_{k=0}^{n} \alpha_k z^k$ and (since polynomials are dense in $\mathfrak{H}(\bar{S}_1)$) $h(F) = F(z_0)$ $\forall F \in \mathfrak{H}(\bar{S}_1)$. In other words, $h = h_{z_0}$.

(b) The space $\mathfrak{D}^{(n)}(I)$ of complex-valued functions having continuous nth-order derivatives in $I = [0, 1]$ is a commutative Banach algebra with identity χ_I under pointwise operations and norm given by $\|f(t)\| = \sum_{k=0}^{n} (1/k!) \sup_I |f^{(k)}(t)|$. Reasoning as in Example 22.9, one obtains $\mathfrak{M} = \mathfrak{M}_I$ and $\mathcal{H} = \mathcal{H}_I$. Therefore $\mathfrak{D}^{(n)}(I)$ is semisimple.

(c) The Banach spaces L_p and l_p $(1 \leq p \leq \infty)$ are commutative Banach algebras. It is easily verified that $\|fg\|_p \leq \|f\|_p \|g\|_p$ for $p = 1$ or $p = \infty$. For $p \in (1, \infty)$ assume that $\|f\|_p = 1 = \|g\|_p$ so that $\|f\|_0 \leq 1$ and $\|g\|_0 \leq 1$ yield $|f(\omega)|^{p^2} \leq |f(\omega)|^p$ and $|g(\omega)|^{pq} \leq |g(\omega)|^p$ for $q = p/(p-1)$. Now

$$\int_\Omega |fg|^p \, d\mu \leq \left(\int_\Omega |f|^{p^2} \, d\mu \right)^{1/p} \left(\int_\Omega |g|^{pq} \, d\mu \right)^{1/q} \leq \|f\|_p \|g\|_p \text{ and } \|fg\|_p \leq 1$$

by invoking Hölder's inequality.

(d) Let $\mathcal{W}$ denote the "Wiener" algebra of abs convergent trigonometric series $x(t) = \sum_{-\infty}^{\infty} c_n \, e^{int}$ under pointwise addition, scalar multiplication, and Cauchy products. (Recall that the Cauchy product of two abs convergent series is abs convergent.) Then $(\mathcal{W}, \| \, \|)$ is a commutative Banach algebra under $\|x(t)\| = \sum_{-\infty}^{\infty} |c_n|$, since it is congruent to l_1. We show that $\mathcal{H} = \mathcal{H}_{[0,2\pi)}$ and $\mathfrak{M} = \mathfrak{M}_{[0,2\pi)}$ (and conclude therefore that $\mathcal{W}$ is semisimple). Obviously $h_{t_0} \in \mathcal{H} \; \forall t_0 \in [0, 2\pi)$. If $h \in \mathcal{H}$ and $h(e^{it}) = \alpha$, then (Corollary 23.7) $|\alpha| = |h(e^{it})| \leq \|e^{it}\| = 1$ and $|1/\alpha| = |h(e^{-it})| \leq 1$ yield $\alpha = e^{it_0}$ for some $t_0 \in [0, 2\pi)$. Since $h(e^{it}) = e^{it_0}$ and $h(e^{int}) = e^{int_0}$, there follows $h\{x(t)\} = x(t_0) \; \forall x(t) \in \mathcal{W}$ and $h = h_{t_0}$.

(e) Let $\mathcal{P}_{n+1}(\mathbb{C})$ be the NLA of polynomials $\mathcal{Q}(t) = \sum_{k=0}^{n} \alpha_k t^k \; (\alpha_k \in \mathbb{C})$ with norm $\|\mathcal{Q}(t)\| = \sum_{k=0}^{n} |\alpha_k|$ under multiplication $\left(\sum_{k=0}^{n} a_k t^k\right)\left(\sum_{k=0}^{n} \beta_k t^k\right) = \sum_{k=0}^{n} \gamma_k t^k$, where $\gamma_k = \sum_{i+j=k} \alpha_i \beta_j$. This $n+1$-dim algebra $\mathcal{P}_{n+1}(\mathbb{C})$ is Banach (Theorem 15.12) and non-semisimple (Corollary 22.4), since $\mathcal{P}_n(\mathbb{C}) \cdot t$ is the only element in $\mathfrak{M}$.

Example 23.4. If X in Corollary 13.25 is a Banach space, $(L(X, X), ||| \, |||)$ is a Banach algebra under composition $(f_2 f_1)(x) = f_2\{f_1(x)\}$. When X is finite dim, $L(X, X)$ is commutative iff dim $X = 1$. (Using a fixed-ordered basis for X, represent each $f \in L(X, X)$ by its $n \times n$ matrix. The collection of all $n \times n$ matrices is commutative iff $n = 1$.) Every NLA (Banach algebra) X may be viewed as an NL (Banach) subalgebra of $L(X, X)$, since $\underset{x \rightsquigarrow \mathfrak{m}_x}{X \to L(X, X)}$ (cf Definition 24.1) is an algebra congruence (here we use $\|e\| = 1$ to obtain $|||\mathfrak{m}_x||| \geq \|x\|$). See also Problem 23I.

Example 23.5. (a) In relation to Lebesgue measure $L_1(\mathbb{R})$ is a Banach algebra under multiplication defined by convolution $(f \star g)(t) = \int_{-\infty}^{\infty} f(t-x) \, g(x) \, dx$. This (as well as (b), (c) below) follows from the more general result (Theorem 27.1) for Haar measure on a locally compact T_2 group.

(b) If $G = \{x_1, \ldots, x_n\}$ is a group, then $(G, \mathfrak{D})$ is a topological group and $\mathcal{F}(G; \mathbb{C}) = C(G; \mathbb{C})$ is a Banach space under $\|f\| = \sum_{i=1}^{n} |f(x_i)|$. It is also a

Banach algebra under multiplication $(f_1 \star f_2)(x_i) = \sum_{j=1}^{n} f_1(x_i x_j^{-1}) f_2(x_j)$. For instance,

$$\|f_1 \star f_2\| = \sum_{i=1}^{n} |(f_1 \star f_2)(x_i)| = \sum_{i=1}^{n} \left| \sum_{j=1}^{n} f_1(x_i x_j^{-1}) f_2(x_j) \right|$$

$$\leq \sum_{i=1}^{n} \sum_{j=1}^{n} |f_1(x_i x_j^{-1})| \, |f_2(x_j)|$$

$$= \sum_{j=1}^{n} |f_2(x_j)| \sum_{i=1}^{n} |f(x_i x_j^{-1})| = \|f_2\| \, \|f_1\|.$$

(c) The Banach space $L_1(\mathbb{I})$ of complex-valued functions satisfying $\|f\| = \sum_{-\infty}^{\infty} |f(n)| < \infty$ becomes a Banach algebra under $(f \star g)(n) = \sum_{m=-\infty}^{\infty} f(n-m)\, g(m)$.

Obviously $e \in \mathfrak{U}$ in an algebra X. If X is Banach, even elements close to e are units, and this has far-reaching implications as shown below.

THEOREM 23.6. Let $(X, \|\ \|)$ be a Banach algebra (as usual commutative and with $\|e\| = 1$).

(i) If $\|e - x\| < 1$, then $x \in \mathfrak{U}$ with $x^{-1} = e + \sum_{n=1}^{\infty} (e - x)^n$.

(ii) $\mathfrak{U}$ is an open multiplicative group in X.

(iii) If $|\lambda| > \|x\|$, then $\lambda \in \rho(x)$ and $(x - e)^{-1} = -\sum_{n=1}^{\infty} \lambda^{-n} x^{n-1}$.

(iv) $r_\sigma(x) \leq \|x\|$ and $\sigma(x)$ is compact $\forall x \in X$.

Proof. Assertion (i) holds since $\sum_{n=1}^{\infty} (e - x)^n$ converges absolutely and so $x\left\{e + \sum_{n=1}^{\infty} (e-x)^n\right\} = \{e - (e - x)\}\left\{e + \sum_{n=1}^{\infty} (e-x)^n\right\} = e$. For (ii) we show that every $x \in \mathfrak{U}$ is an interior point of $\mathfrak{U}$. Clearly $xx^{-1} = e \subset S_{1,\|\ \|}(e) \subset \mathfrak{U}$ and (by continuity of multiplication) $S_{\varepsilon,\|\ \|}(x)x^{-1} \subset S_{1,\|\ \|}(e)$ for some $\varepsilon > 0$. If $y \in S_{\varepsilon,\|\ \|}(x)$, then $yx^{-1} \in \mathfrak{U}$ and $e = (yx^{-1})z = y(x^{-1}z)$ for some $z \in X$. Thus $y \in \mathfrak{U}$ and $S_{\varepsilon,\|\ \|}(x) \subset \mathfrak{U}$ follows. (iii) is based on the fact that $\lambda x \in \mathfrak{U}$ with $(\lambda x)^{-1} = \lambda^{-1} x^{-1}$ when $x \in \mathfrak{U}$ and $\lambda \neq 0$. Together with (i) above, $\|e - (e - \lambda^{-1}x)\| = \|x\|/|\lambda| \leq 1$ implies that $x - \lambda e = -\lambda(e - \lambda^{-1}x) \in \mathfrak{U}$ and $\lambda \in \rho(x)$. Taking $e = x^{\circ}$ also yields $(x - \lambda e)^{-1} = -\lambda^{-1}(e - \lambda^{-1}x)^{-1} =$

$-\lambda^{-1}\left\{e+\sum_{n=1}^{\infty}(\lambda^{-1}x)^n\right\}=-\sum_{n=1}^{\infty}\lambda^{-n}x^{n-1}$. Finally, (iv) uses (iii) to obtain $r_\sigma(x)=\sup_{\sigma(x)}|\lambda|\le\|x\|\ \forall x\in X$. If $\lambda_0\in\rho(x)$, then $x=\lambda_0 e\in\mathfrak{U}$ and $S_{\varepsilon,\|\,\|}(x-\lambda_0 e)\subset\mathfrak{U}$ for some $\varepsilon>0$. Since $\underset{\lambda\rightsquigarrow x-\lambda e}{\mathbb{C}\longrightarrow X}$ is continuous, there is a $\delta>0$ such that $x-\lambda e\in S_{\varepsilon,\|\,\|}(x-\lambda_0 e)\ \forall|\lambda-\lambda_0|<\delta$. This essentially demonstrates that $S_\delta(\lambda_0)\subset\rho(x)$; that is, $\rho(x)$ is open and (the closed, bounded set) $\sigma(x)$ is compact $\forall x\in X$. ■

Remark. If X is complex, $\sigma(x)\neq\varnothing\ \forall x\in X$ (Theorem 24.18).

COROLLARY 23.7. For a Banach algebra $(X,\|\,\|)$, every $M\in\mathfrak{M}$ is closed and every $h\in\mathcal{H}$ is continuous with $|||h|||=1$.

Proof. If $I\in\mathfrak{J}$, then $\bar{I}$ is an ideal by continuity of all operations involved. Accordingly [Theorems 23.6(i) and 22.8(i)], $\bar{I}\subset X-\mathfrak{U}$ and $e\in\mathfrak{U}$ imply that $\bar{I}\in\mathfrak{J}$. In particular, $M=\bar{M}\ \forall M\in\mathfrak{M}$. If $h\in\mathcal{H}$ and $x\in X$, then $x-h(x)e\in h^{-1}(0)$ in $\mathfrak{M}$ implies $h(x)\in\sigma(x)$ and $|h(x)|\le\|x\|$. Thus $h\in X'$ with $|||h|||\le 1$. The reverse inequality also holds, since $1=|h(e)|<|||h|||\,\|e\|=|||h|||$. ■

Theorem 22.6 can be used to characterize r_σ and sharpen Theorem 23.6(i),(iii). Since $\{r_\sigma(x)\}^n=r_\sigma(x^n)\le\|x^n\|$, one has $r_\sigma(x)\le\|x^n\|^{1/n}\ \forall n\in\mathbb{N}$. Accordingly, $r_\sigma(x)\le\liminf_n\|x^n\|^{1/n}\le\limsup_n\|x^n\|^{1/n}\quad\forall x\in X$. Although these limits are generally unequal,† they agree on a complex Banach algebra, in which context $r_\sigma(x)$ is given the common value $\lim_n\|x^n\|^{1/n}$. For notation $\underset{\lambda\rightsquigarrow(x-\lambda e)^{-1}}{\mathfrak{R}_x:\rho(x)\longrightarrow\mathfrak{U}}$ is called the resolvent of $x\in X$. We use here the fact [Theorem 24.7(i)] that $\mathfrak{R}_x$ is analytic on $\rho(x)$ in the sense of Problem 17H.

COROLLARY 23.8. Let $(X,\|\,\|)$ be a complex Banach algebra.

(i) If $|\lambda|>r_\sigma(x)$, then $\lambda\in\rho(x)$ and $(x-\lambda e)^{-1}=-\sum_{n=1}^{\infty}\lambda^{-n}x^{n-1}$. In particular, $r_\sigma(e-x)<1$ implies $x\in\mathfrak{U}$ with $x^{-1}=e+\sum_{n=1}^{\infty}(e-x)^n$.

† The lim inf (lim sup) of a sequence $\{\alpha_n\}\subset\mathbb{R}$ is defined as the smallest (largest) accumulation point of $\{\alpha_n\}$. Taking $\alpha_n=(-1)^{n-1}[(n+1)/n]$, for example, yields $\inf_n\alpha_n=-\frac{3}{2}<\liminf_n\alpha_n=-1<1=\limsup_n\alpha_n<2=\sup_n\alpha_n$. If $\liminf_n\alpha_n=\limsup_n\alpha_n$, their common value (denoted $\lim_n\alpha_n$) is called the limit of $\{\alpha_n\}$.

(ii) $r_\sigma(x) = \lim_n \|x^n\|^{1/n}$ $\forall x \in X$.

(iii) $r_\sigma = \|\ \|$ on X iff there is an $n_0 > 1$ such that $\|x^{n_0}\| = \|x\|^{n_0}$ $\forall x \in X$.

Proof. To prove (i) recall (the identity theorem in analytic function theory) that two convergent power series about z_0 in the extended complex plane coincide if their sums agree on a nbd of z_0. In other words, an analytic function has a unique power series representation within its radius of convergence. The above statements are easily seen to carry over to analytic vector-valued functions. If $|\lambda| > r_\sigma(z)$, then $\lambda \in \rho(x)$ and $\mathfrak{R}_x(\lambda) = (x - \lambda e)^{-1} \in X$ has a power series representation in $-1/\lambda$ about ∞. Since $\mathfrak{R}_x(\lambda) = -1/\lambda \sum_{n=0}^{\infty} (x/\lambda)^n$ $\forall 1/|\lambda| < 1/\|x\|$, this representation holds $\forall 1/|\lambda| < (r_\sigma(x))^{-1}$. The second assertion in (i) follows from the first by taking $\lambda = 1$ and substituting $e - x$ for x. For (ii) consider $|\lambda| > r_\sigma(x)$ and observe that $f\mathfrak{R}_x(\lambda) = -1/\lambda \lim_n f\left(\sum_{k=0}^{n} \lambda^{-k} x^k\right) = -1/\lambda \sum_{k=0}^{\infty} f(\lambda^{-n} x^n)$ implies $f(\lambda^{-n}x^n) \to 0$, $\forall f \in X'$. Therefore (Theorem 18.6 or Problem 17B) $\{\lambda^{-n}x^n : n \in \mathbb{N}\}$ is $\|\ \|$-bounded and there is a constant $M > 0$ such that $\|x^n\|^{1/n} \le M^{1/n}|\lambda|$ $\forall n \in \mathbb{N}$. In particular, $\limsup_n \|x^n\|^{1/n} \le \lim_n M^{1/n}|\lambda| = |\lambda|$ and [since this holds $\forall |\lambda| > r_\sigma(x)$] $\limsup_n \|x^n\|^{1/n} \le r_\sigma(x)$. One part of (iii) follows from Theorem 22.6(ii); namely, $r_\sigma = \|\ \|$ implies $\|x^n\| = \|x\|^n$ $\forall x \in X$ and $n \in \mathbb{N}$. The converse also holds since $\|x^{n_0}\| = \|x\|^{n_0}$ and $\|x^{2n_0}\| \le \|x^{n_0}\|^2$ implies $\|x^{n_0}\|^4 = \|x^{4n_0}\| \le \|x^{2n_0}\|^2$ and $\|x\|^{2n_0} = \|x^{2n_0}\|$. By induction $\|x^{2^k n_0}\| = \|x\|^{2^k n_0}$ $\forall k \in \mathbb{N}$ and $r_\sigma(x) = \lim_n \|x^n\|^{1/n} = \lim_k \|x^{2^k n_0}\|^{1/2^k n_0} = \|x\|$ $\forall x \in X$. ■

An algebra X may carry more than one complete compatible (i.e., algebra) norm. However,

THEOREM 23.9. All complete compatible norms are equivalent on a semisimple algebra.

Proof. Assume that X is semisimple and $X_i = (X, \|\ \|_i)$ $(i = 1, 2)$ are Banach algebras. Then $(X, \|\ \|_3)$ is a Banach algebra under $\|x\|_3 = \max\{\|x\|_1, \|x\|_2\}$, since any $\|\ \|_3$-Cauchy sequence $\{x_n\} \in X$ is $\|\ \|_i$-convergent to some $y_i \in X$ and (Corollary 23.7) $\sigma(X, \mathcal{H}) \subset \sigma(X, X_i') \subset \|\ \|_i$ implies $x_n \xrightarrow{\sigma(X,\mathcal{H})} y_i$ $(i = 1, 2)$; but X is semisimple, so that $\sigma(X, \mathcal{H})$ is T_2 (Corollary 24.21) and $x_n \xrightarrow{\|\ \|_3} y_1 = y_2 \in X$. Finally, $\|\ \|_i \subset \|\ \|_3$ and Corollary 21.9(om) ensures that each $\mathbb{1}_i : (X, \|\ \|_3) \to (X, \|\ \|_i)$ is a homeomorphism. ■

COROLLARY 23.10. An algebra isomorph between semisimple Banach algebras is a homeomorphism.

Proof. If $t:(X_1, \| \|_1) \to (X_2, \| \|_2)$ is an onto isomorphism between the semisimple Banach algebras $(X_i, \| \|_i)$, then $(X_2, \| \|_t)$ is congruent to $(X_1, \| \|_1)$ for $\|x\|_t = \|t^{-1}(x)\|_1$. Consequently, $(X_2, \| \|_t)$ is complete and $\| \|_t$ is equivalent to $\| \|_2$. ■

*-*Algebras*

A particularly important class of algebras carries an involution.

DEFINITION 23.11. An involution on an algebra X is a mapping $\underset{x \rightsquigarrow x^*}{*: X \to X}$ such that $(x+y)^* = x^* + y^*$, $(\alpha x)^* = \bar{\alpha} x^*$, $(xy)^* = y^* x^*$, and $x^{**} = x$ $\forall x, y \in X$ and $\alpha \in \mathbb{C}$. An algebra with an involution is termed a *-algebra.

A *-subalgebra of X is a subalgebra Y of X which is itself a *-algebra (independent of whether X carries an involution). This is not to be confused with a sub *-algebra Y of a *-algebra X, meaning that Y is a subalgebra of X which is closed under * (i.e., $y^* \in Y$ $\forall y \in Y$).

An element x in a (not necessarily commutative) *-algebra X is said to be self-adjoint {normal} if $x = x^*$ {if $xx^* = x^*x$}. Note that $e^* = e^*e$ implies $e^* = e$ and $x \in \mathfrak{U}$ implies $x^* \in \mathfrak{U}$ with $(x^*)^{-1} = (x^{-1})^*$.

Remark. In a complex *-algebra X every $x \in X$ has a unique representation $x = x_1 + ix_2$, where $x_1 = (x+x^*)/2$ and $x_2 = (x - x^*)/2i$ are self-adjoint.

A continuous involution * on a NLA $(X, \| \|)$ may be viewed as a congruence, since $\|x\|' = \max\{\|x\|, \|x^*\|\}$ is an equivalent norm on X which satisfies $\|x\|' = \|x^*\|'$ $\forall x \in X$.

DEFINITION 23.12. A *-normed algebra is a NLA $(X, \| \|)$ with an involution satisfying $\|x^*\| = \|x\|$ $\forall x \in X$. By a B^*-algebra we mean a Banach algebra $(X, \| \|)$ with an involution such that $\|x^*x\| = \|x\|^2$ $\forall x \in X$.

Continuity of involution for a normed *-algebra is not ensured in general, since involution continuity holds iff there is a $\lambda > 0$ such that $\|x^*\| \le \lambda \|x\|$ $\forall x \in X$. Every B^*-algebra is *-normed ($\|x\|^2 \le \|x^*\| \, \|x\| \Rightarrow \|x\| \le \|x^*\|$ and also $\|x^*\| \le \|x^{**}\| = \|x\|$) and so has continuous involution. This, in fact, also emerges from

THEOREM 23.13. The involution of a semisimple Banach *-algebra is continuous.

Proof. Suppose that $(X, \| \|)$ is a semisimple Banach *-algebra and let $X\cdot$ denote X with scalar multiplication defined by $\alpha \cdot x = \bar{\alpha}x$. Then $(X\cdot, \| \|)$ is a semisimple Banach algebra, $^*: (X, \| \|) \to (X\cdot, \| \|)$ is an isomorphism, and Corollary 22.10 applies. ■

Some key features of a B^*-algebra are given below.

THEOREM 23.14. A complex B^*-algebra $(X, \| \|)$ is semisimple and has

(i) $r_\sigma(x) = \|x\| \ \forall x \in X$;
(i) $\sigma(x) \subset \mathbb{R} \ \forall x = x^*$;
(iii) $h(x^*) = \overline{h(x)} \ \forall h \in \mathcal{H}$ and $x \in X$.

Proof. Since $\|x^2\|^2 = \|(x^2)^* x^2\| = \|(x^*x)^*(x^*x)\| = \|x^*x\|^2 = \|x\|^4$ and $\|x^2\| = \|x\|^2$, we have $r_\sigma(x) = \|x\| \ \forall x \in X$ [Corollary 23.8(iii)]. Reference to either Theorem 24.26 or Theorem 25.8 then confirms that X is semisimple. To prove (ii) assume that $x = x^*$ and let $\lambda = \alpha + i\beta \in \sigma(x)$. For any $\gamma \in \mathbb{R}$ one now has $\lambda + i\gamma \in \sigma(x + i\gamma e)$ and $\alpha^2 + 2\beta\gamma + \gamma^2 = |\lambda + i\gamma|^2 \leq \|x + i\gamma e\|^2 = \|(x + i\gamma e)^*(x + i\gamma e)\| = \|x^2 + \gamma^2 e\| \leq \|x\|^2 + \gamma^2$. Thus $\alpha^2 + 2\beta\gamma \leq \|x\|^2 \ \forall \gamma \in \mathbb{R}$ and $\beta = 0$ follows (otherwise, $\gamma = \|x\|^2/\beta - \alpha^2/2\beta$ implies $2\|x\|^2 \leq \|x\|^2$). Finally, fix $x \in X$ and $h \in \mathcal{H}$ in (iii). Write $x = x_1 + ix_2$, where $x_1 = x_1^*$ and $x_2 = x_2^*$. Then $h(x^*) = h(x_1 - ix_2) = h(x_1) - ih(x_2) = \overline{h(x)}$. ■

Definition 23.12 may ring a bell. In fact, the reader experiencing *deja vu* is not entirely wrong; there is a very close connection between B^*-algebras and algebras of operators on a Hilbert space.

Example 23.15. Let $(X, (\,,\,))$ be a Hilbert space and take $f^* \in L(X, X)$ as the adjoint of $f \in L(X, X)$, where $(f(x), y) = (x, f^*(y)) \ \forall x, y \in X$. Then $(L(X, X), \||\ \||)$ is a B^*-algebra with involution $f \rightsquigarrow f^*$. We verify only that $\||f\||^2 = \||f^*f\|| \ \forall f \in L(X, X)$. To be sure, $\|f^*(y)\|^2 = (f^*(y), f^*(y)) = (ff^*(y), y) \leq \|ff^*(y)\| \, \|y\| \leq \||f\|| \, \|f^*(y)\| \, \|y\|$ implies $\|f^*(y)\| \leq \||f\|| \, \|y\| \ \forall y \in X$. Therefore $\||f^*\|| \leq \||f\||$ and $\||f\|| = \||f^{**}\|| \leq \||f^*\||$ so that $\||f^*\|| = \||f\||$ and $\||f^*f\|| \leq \||f^*\|| \, \||f\|| = \||f\||^2$. The reverse inequality follows from $\||f\||^2 = \sup_{\|x\|=1} \|f(x)\|^2 = \sup_{\|x\|=1} (f(x), f(x)) = \sup_{\|x\|=1} (f^*f(x), x) \leq \sup_{\|x\|=1} \|f^*f(x)\| \|x\| \leq \||f^*f\||$.

A number of already familiar algebras are also *-algebras.

Example 23.16. (a) [Example 8.3(a)] $(M_n(\mathbb{F}), \| \|_0)$ is a B^*-algebra for A^* denoting the conjugate transpose of $A \in M_n(\mathbb{F})$.

(b) $CB(\Omega)$ and $C_\infty(\Omega)$ are B^*-algebras and $C_0(\Omega)$ is a *-normed algebra under complex conjugation (i.e., $f^* = \bar{f}$, where $(\bar{f})(\omega) = \overline{f(\omega)}$ on Ω).

(c) Under complex conjugation $\mathfrak{D}^{(n)}(I)$ is a *-algebra; $\mathfrak{H}(\bar{S}_1)$ is not, since $f(z) = z$ belongs to $\mathfrak{H}(\bar{S}_1)$ and $f^*(z) = \bar{z}$ does not.

(d) $\mathscr{W}$ is a B^*-algebra and $\mathscr{P}_n(\mathbb{C})$ is a *-normed, non-B^*-algebra [Example 23.3(e)] under involution defined by conjugation of coefficients.

(e) The algebras of Example 23.5 are all *-normed under $f^* = \bar{\bar{f}}$, where $f^*(x) = \Delta(x)\overline{f(x^{-1})}\ \forall x \in X$. Note, for example, that $L_1(\mathbb{R})$ is non-B^*, since $\|f^* \star f\|^2 = \frac{8}{3} < 4 = \|f\|^2$ for $f = -\chi_{[-1,0)} + \chi_{[0,1)}$.

Remark. An example of a complex B^*-algebra that does not admit an involution may be found in Rickart [82], p. 306.

PROBLEMS

23A. Algebras Without Identity

(a) In a TVA X (cf Definition 24.1) $\mathfrak{U}_\odot \subset X$ is open iff $\mathfrak{U} \subset X_e$ is open.

(b) The analog of most statements in Theorem 23.6, Corollary 23.8, hold for a complex commutative Banach algebra $(X, \|\ \|)$ without identity. Using the notation of Problem 22A, prove that

(i) $\|x\| < 1$ implies $x \in \mathfrak{U}_\odot$ with $x^\odot = -\sum_{n=1}^{\infty} x^n$;

(ii) if $|\lambda| \geq \|x\|$, then $\lambda \in \rho_\odot(x)$. Therefore $r_{\sigma_\odot}(x) = \sup_{\sigma_\odot(x)} |\lambda| \leq \|x\|$ and $\sigma_\odot(x) \neq \varnothing$ is compact $\forall x \in X$.

Remark. Note that $|\lambda| > r_{\sigma_\odot}(x)$ implies $(1/\lambda)x \in \mathfrak{U}_\odot$ (i.e., $x \in \mathfrak{U}_\odot\ \forall r_{\sigma_\odot}(x) < 1$).

(iii) $r_{\sigma_\odot}(x) = \lim_n \|x^n\|^{1/n}\ \forall x \in X$;

(iv) Every $h \in \mathfrak{H}$ is continuous with $|||h||| \leq 1$.

[*Hint*: For (i) use Problem 22A(c) after noting that $\|(1, 0) - (1, -x)\|_e = \|x\|$ and $(-1, x)^{-1} = -(1, -x)^{-1} = (-1, 0) - \sum_{n=1}^{\infty} (0, x)^n$. For (iii) observe that $r_{\sigma_\odot}(x) = r_\sigma\{(0, x)\}$ and $\|(0, x)^n)\|_e = \|x^n\|\ \forall x \in X$. If $|h(x_0)| \geq \|x_0\|$ in (iv), then $x = [h(x_0)]^{-1}x_0 \in \mathfrak{U}_\odot$ by (i) and $h(x) = 1$ contradicts Problem 22A(e).]

23B. Regular Maximal Ideals: Additional Properties

Assume that $(X, \|\ \|)$ in Problem 22B is also Banach and denote concepts in X_e with a subscript e.

(a) Every $M \in \mathfrak{M}_{\mathfrak{R}}$ is closed. [*Hint*: If $I_{x_0} = M \in \mathfrak{M}_{\mathfrak{R}}$, then $x_0 \in X - \mathfrak{U}_{\odot}$. Therefore $M + x_0 \subset X - \mathfrak{U}_{\odot}$ and $\bar{M} + x_0 \subset X - \mathfrak{U}_{\odot}$. Since $\mathfrak{U}_{\odot} \neq \varnothing$, there follows $\bar{M} \subsetneqq X$ and $\bar{M} = M$.]

(b) $(0, X) \in \mathfrak{M}_e$ and $I = \{x \in X : (0, X) \in I_e\} \in \mathfrak{I}\ \forall I_e \in \mathfrak{I}_e$.

(c) $\underset{I_e \rightsquigarrow I}{t : \mathfrak{I}_e \not\subset \{(0, X)\} \to \mathfrak{I}_{\mathfrak{R}}}$ is a bijection. [*Hint*: $I_e \not\subset (0, X)$ implies $(\alpha, y_0) \in I_e$ for some $\alpha \neq 0$ and $(-1, x_0) \in I_e$ for some $x_0 \in X$. Therefore $(0, x)(-1, x_0) \in I_e$ yields $xx_0 - x \in I\ \ \forall x \in X$. If $I \in \mathfrak{I}_{\mathfrak{R}}$ has x_0 as an identity mod I, then $I_{\bullet} = \{(\alpha, x) : (0, x_0)(\alpha, x) \in I\} \not\subset (0, X)$ belongs to $\mathfrak{I}_e$ and $t(I_{\bullet}) = I$. To prove t injective, assume $t(I_e) = t(J_e)$ for $I_e, J_e \in \mathfrak{I}_e \not\subset (0, X)$ and let $(-1, x_i) \in I_e$ and $(-1, x_j) \in J_e$. Since $(0, x_j)(-1, x_i) \in I_e$ and $(0, x_i)(-1, x_j) \in J_e$, one has $x_i x_j - x_j$ and $x_i x_j - x_i$ (therefore $x_j - x_i$) contained in $t(I_e) = t(J_e)$. Any $(\alpha, x) \in I_e$ may be written $(\alpha, x) = \alpha(1, -x_j) + \alpha(0, x_j - x_i) + (0, \alpha x_i + x)$. Therefore $(0, x_i)(\alpha, x) \in I_e$ and $(0, x)(-1, x_i) \in I_e$ imply $\alpha x_i + x \in t(J_e)$; that is $(0, \alpha x_i + x) \in J_e$ and $(\alpha, x) \in J_e$. In like manner $J_e \subset I_e$.]

(d) $\underset{M_e \rightsquigarrow M}{t : \mathfrak{M}_e - \{(0, x)\} \to \mathfrak{M}_{\mathfrak{R}}}$ is a bijection. [*Hint*: Use (c) and Problem 22B(d).]

23C. NLA Complexification

(a) If X in Problem 13D is a real algebra, then X^+ is a complex algebra with identity $(e, 0)$ under $(x_1, y_1)(x_2, y_2) = (x_1x_2 - y_1y_2, x_1x_2 + y_1y_2)$.

(b) If $(X, \| \|)$ is a real NLA, so is $(X^+, | |)$. However, $\| \|^+$ is not an algebra norm for X^+. Show that $(X^+, \sqrt{2}\| \|^+)$ is a complex NLA under which $(X, \| \|) \to (X^+, \sqrt{2}\| \|^+)$ is not a real congruence.

There is a complex algebra norm for X^+ that avoids the shortcomings of $\sqrt{2}\| \|^+$ above. This norm is unique in the sense made explicit below.

(c) Show that $\underset{(x,y) \rightsquigarrow \mathfrak{M}_x^+ + i\mathfrak{M}_y^+}{X^+ \to L(X^+, X^+)}$ is a complex isomorphism.

(d) If $\|(x, y)\|^* = \|\mathfrak{M}_x^+ + i\mathfrak{M}_y^+\|\ \forall (x, y) \in X^+$, then $(X^+, \| \|^*)$ is a complex NLA satisfying $\|(x, 0)\|^* = \|x\|\ \forall x \in X$.

(e) $(X^+, \| \|^*)$ is complete iff $(X, \| \|)$ is complete. [*Hint*: Show that $\|(x, y)\|^* \leq |(x, y)| \leq 2\sqrt{2}\|(x, y)\|^*\ \forall x, y \in X$.]

(f) If $(X, \| \|)$ is a real Banach algebra, then all complete complex algebra norms for X^+ rendering $(X, \| \|) \to X^+$ congruent are equivalent. [*Hint*: If $\| \|_0^*$ meets the above criteria, then $\|(x, y)\|_0^* = \|(x, 0) + i(y, 0)\|_0^* \leq |(x, y)|\ \forall x, y \in X$ and [Corollary 21.10(i)] the norm topologies $\| \|_0^*$ and $| |$ coincide.]

23D. Topological Divisors of Zero

An element x is a NLA $(X, \| \|)$ (not necessarily having an identity) is called a topological divisor of zero (tdz) if there exists a sequence $x_n \not\to 0$ such that $x_n x \to 0$. The set of tdz's in X is denoted $\mathscr{Z}_0(X)$.

(a) (Notation as in Definition 24.1) $\mathscr{Z}_0(X) = \{x \in X : \mathfrak{M}_x \text{ is noninjective}\}$. [*Hint*: If $\mathfrak{M}_x$ is noninjective, $x_0 x = 0$ and $(nx_0)x \to 0$ for some $x_0 \neq 0$.]

(b) Use (a) to verify that $\mathscr{Z}_0(X)\subset \bigcup_{\mathcal{I}} I$ if $\mathcal{K}$ is injective.

(c) If $xy\in\mathscr{Z}_0(X)$, either $x\in\mathscr{Z}_0(X)$ or $y\in\mathscr{Z}_0(X)$. [*Hint*: Let $\alpha(x_0)=\inf\|x_0x\|/\|x\|$. Then $x_0\in\mathscr{Z}_0(X)$ iff $\alpha(x_0)=0$. Furthermore, $\alpha(x)\alpha(y)\le\alpha(xy)\ \forall x, y\in X$.]

(d) $\mathscr{Z}_0(X)=\{x^*: x\in\mathscr{Z}_0(X)\}$ if X is *-normed.

Assume now that $(X,\|\ \|)$ is Banach and has an identity.

(e) Show that $bdy(X-\mathfrak{U})\subset\mathscr{Z}_0(X)\subset X-\mathfrak{U}$. [*Hint*: If $x\in bdy(X-\mathfrak{U})$, there is a sequence $\{y_n\}\in\mathfrak{U}$ such that $y_n\to x$ and $\|y_n^{-1}\|\to\infty$. Conclude that $x_nx=x_n(x-y_n)+e/\|y_n^{-1}\|\to 0$ for $x_n=y_n^{-1}/\|y_n^{-1}\|$.

(f) Show that $x-\lambda e\in bdy(X-\mathfrak{U})\ \forall\lambda\in bdy\,\sigma(x)$ and conclude further that $\mathfrak{R}(X)\subset bdy(X-\mathfrak{U})$.

(g) If $(X,\|\ \|)$ is a B^*-algebra, $\mathscr{Z}_0(X)=X-\mathfrak{U}$. [*Hint*: If $x\in X-\mathfrak{U}$, then $xx^*\in X-\mathfrak{U}$. Since $\sigma(xx^*)\subset\mathbb{R}$, it follows that $xx^*-(i/n)e\in\mathfrak{U}\ \forall n\in\mathbb{N}$ and so $xx^*\in\bar{\mathfrak{U}}$. Therefore $xx^*\in\bar{\mathfrak{U}}-\mathfrak{U}=bdy(X-\mathfrak{U})$.]

(h) If Ω is compact and T_2, then $X=(C(\Omega;\mathbb{C}),\|\ \|_0)$ is B^* and $\mathscr{Z}_0(X)=X-\mathfrak{U}$. Prove this by demonstrating that $X-\mathfrak{U}\subset\mathscr{Z}_0(X)$. [*Hint*: Assume that $f(\omega_0)=0$ for $f\in X-\mathfrak{U}$ and let $\mathfrak{N}_{\omega_0}$ be the Ω-nbd system at $\omega_0\in\Omega$ directed by $u<v$ iff $v\subset u$. Each $u\in\mathfrak{N}_{\omega_0}$ determines an $f_u:\Omega\to I$ with $f_u(\omega_0)=1$ and $f_u(\Omega-u)=0$. Then $f_u\not\to 0$, whereas $f_uf\to 0$.]

23E. Nested Algebras

Let Y be a subalgebra of X having the same identity as X. Denote the units in Y $\{X\}$ by $\mathfrak{U}_Y$ $\{\mathfrak{U}_X\}$ and let $\sigma_Y(y)$ $\{\sigma_X(y)\}$ be the spectrum of $y\in Y$ $\{y\in Y\subset X\}$.

(a) Verify that $\mathfrak{U}_Y\subset\mathfrak{U}_X$ and $\sigma_X(y)\subset\sigma_Y(y)\ \forall y\in Y$. If $(X,\|\ \|)$ is a NLA, then $\mathscr{Z}_0(Y)\subset\mathscr{Z}_0(X)$.

(b) Suppose X is Banach and Y is closed in X. Then $bdy\,\sigma_Y(y)\subset bdy\,\sigma_X(y)\ \forall y\in Y$. Equality holds if $\rho(y)$ is connected. [*Hint*: $\lambda\in bdy\,\sigma_Y(y)$ implies $\lambda\in\sigma_X(y)$. If $\rho(y)$ is connected and $\lambda\in\sigma_Y(y)-\sigma_X(y)$, then $\lambda\in\rho_X(y)$ and $2\|y\|\in\rho_Y(y)$ can be joined by a polygonal path $\pi\subset\rho_X(y)$. Therefore some $\delta\in\pi\cap bdy\,\sigma_Y(y)$ also belongs to $\rho_X(y)\cap\sigma_X(y)$.]

(c) The essence of (a), (b) above is that elements in $\mathscr{Z}_0(Y)$ remain permanently singular, and $bdy\,\mathfrak{U}$ $(=bdy(X-\mathfrak{U}))$ remains fixed, for every such super algebra X of Y. As this containing algebra X increases, $\sigma_X(y)$ for $y\in Y$ becomes hollowed out. By way of illustration identify $Y=\mathfrak{H}(\bar{S}_1)$ with the Banach subalgebra of $X=(C(bdy\,\bar{S}_1;\mathbb{C},\|\ \|_0)$ under $f\rightsquigarrow f/bdy\,\bar{S}_1$. If $f(z)=z$ on $\bar{S}_1$, then $\sigma_Y(f)=\bar{S}_1$ and $\sigma_X(f)=bdy\,\bar{S}_1$.

(d) If X in (b) is also B^*, then $\mathfrak{U}_Y=Y\cap\mathfrak{U}_X$ and $\sigma_Y(y)=\sigma_X(y)\ \forall y\in Y$. [*Hint*: [Problem 23D(g)] $Y-\mathfrak{U}_Y=\mathscr{Z}_0(Y)\subset\mathscr{Z}_0(X)=X-\mathfrak{U}_X$ so that $Y\cap\mathfrak{U}_X\subset\mathfrak{U}_Y$. Moreover, $\lambda\in\sigma_Y(y)\Leftrightarrow y-\lambda e\notin\mathfrak{U}_Y=Y\cap\mathfrak{U}_X\Rightarrow y-\lambda e\notin\mathfrak{U}_X\Leftrightarrow\lambda\in\sigma_X(y)$.]

(d) If $(X,\|\ \|)$ and $(Y,\|\ \|)$ are Banach, $r_{\sigma_Y}(y)=r_{\sigma_X}(y)\ \forall y\in Y$. In particular, Y is semisimple if X is [*Hint*: $r_{\sigma_Y}(y)=\lim_n\|y^n\|^{1/n}=r_{\sigma_X}(y)$.]

23F. Analytic Functions on a Banach Algebra

Suppose $(X, \|\ \|)$ is a complex Banach algebra. By analogy with $f(\lambda)=1/\lambda$ one has $f(x)=x^{-1}\in X$ only for $x\in\mathfrak{U}$. For an entire function $f(\lambda)=\sum_{n=0}^{\infty}\alpha_n\lambda^n$, however, $f(x)=\sum_{n=0}^{\infty}\alpha_n x^n\in X\ \forall x\in X$. This substitution property can be further developed via the results of Problem 17H.

(a) Let $\mathfrak{F}_x$ denote the complex-valued functions that are analytic on $\sigma(x)$. Since $\sigma(x)$ is compact, every $f\in\mathfrak{F}_x$ is analytic in some bounded region Ω with $\sigma(x)\subset \operatorname{Int}\Omega$. Moreover, a closed rectifiable curve $\Gamma\subset\Omega$ can be found that contains $\sigma(x)$ in its interior. Show that it makes sense to define $f(x)=1/2\pi i\int_\Gamma(\lambda e-x)^{-1}f(\lambda)\,d\lambda$. [*Hint*: The integral exists (since the integrand is uniformly continuous on the compact set Γ) and is independent of Γ (since the integrand is analytic on $\Omega-\sigma(x)$ and the Cauchy theorem applies).]

(b) Prove that $\underset{f\rightsquigarrow f(x)}{\mathfrak{F}_x\to X}$ is an algebra homomorphism. [*Hint*: Suppose that $f_1, f_2\in\mathfrak{F}_x$ and let $\Gamma\subset\Omega$ be as above. Pick another closed rectifiable curve $\Gamma_0\subset\Omega$ that contains Γ in its interior and justify each step of

$$
\begin{aligned}
f_1(x)f_2(x) &= \left\{\frac{1}{2\pi i}\int_\Gamma(\lambda e-x)^{-1}f_1(\lambda)\,d\lambda\right\}\left\{\frac{1}{2\pi i}\int_{\Gamma_0}(\gamma e-x)^{-1}f_2(\gamma)\,d\gamma\right\}\\
&= -\frac{1}{4\pi^2}\int_\Gamma\int_{\Gamma_0}(\lambda e-x)^{-1}(\gamma e-x)^{-1}f_1(\lambda)f_2(\gamma)\,d\lambda\,d\gamma\\
&= -\frac{1}{4\pi^2}\int_\Gamma\int_{\Gamma_0}\frac{f_1(\lambda)f_2(\gamma)}{\gamma-\lambda}\{(\lambda e-x)^{-1}(\gamma e-x)^{-1}\}\,d\lambda\,d\gamma\\
&= \frac{1}{2\pi i}\int_\Gamma(\lambda e-x)^{-1}f_1(\lambda)\left\{\frac{1}{2\pi i}\int_{\Gamma_0}\frac{f_2(\gamma)}{\gamma-\lambda}\right\}d\gamma\\
&\quad+\frac{1}{4\pi^2}\int_{\Gamma_0}(\gamma e-x)^{-1}f_2(\gamma)\left\{\int_\Gamma\frac{f_1(\lambda)}{\lambda-\gamma}\,d\lambda\right\}d\gamma
\end{aligned}
$$

The second integral equals zero (Cauchy's theorem), and $1/2\pi i\int_{\Gamma_0}f_2(\gamma)/(\gamma-\lambda)\,d\gamma=f_2(\lambda)$. Thus $f_1(x)f_2(x)=(f_1f_2)(x)$.]

(c) The following substitutions hold for $\mathfrak{F}_x$:

(i) $f(\lambda)=\lambda$ implies $f(x)=x$.

(ii) If $f_n(\lambda)\xrightarrow{\text{uniformly}}f(\lambda)$ on a region $\Omega\supset\sigma(x)$, then $f_n(x)\xrightarrow{\|\ \|}f(x)$.

[*Hint*: (i) If Γ is a circle with $\sigma(x)$ inside, then (since $1/2\pi i \int_\Gamma d\lambda/\lambda^n = \delta_{n,1}$) $f(x) = 1/2\pi i \int_\Gamma (\lambda e - x)^{-1}\lambda\, d\lambda = \sum_{n=1}^{\infty} 1/2\pi i \int_\Gamma x^{n-1}/\lambda^{n-1}\, d\lambda = x$. For (ii) us $\|f_n(x) = f(x)\| \leq \sup_\Omega |f_n(\lambda) - f(\lambda)| \cdot 1/2\pi \cdot \sup_\Gamma \|(\lambda e - x)^{-1}\| l(\Gamma)$, where $l(\Gamma)$ denotes the length of $\Gamma \subset \Omega$.]

Remark. It follows from (b) and (i) that $p(x) \in X$ for every polynomial $p(\lambda)$. From (ii) it also follows that $f(x) = \sum_{n=0}^{\infty} \alpha_n x^n \in X$ for every entire function $f(\lambda) = \sum_{n=0}^{\infty} \alpha_n \lambda^n$.

(d) Theorem 22.6 can be generalized. If Ω is an open set containing $\sigma(x)$ and f is analytic on Ω, then $\sigma\{f(x)\} = f\{\sigma(x)\}$. [*Hint*: $f \in \mathfrak{F}_x$ and $x_n \to x_0$ implies that $x_n(M) \to x_0(M)\ \forall M \in \mathfrak{M}$ (notation in Theorem 24.20). If $\{S_n\}$ is any sequence of partial sums $\|\ \|$-converging to $\int_\Gamma (\lambda e - x)^{-1} f(\lambda)\, d\lambda$, the sequence $\{S_n(M)\}$ converges to $\int_\Gamma (\lambda - x(M))^{-1} f(\lambda)\, d\lambda = \int_\Gamma f(\lambda)\, d\lambda/[\lambda - x(M)] = f\{x(M)\}$. Therefore $\{f(x)\}(M) = f\{x(M)\}\ \forall M \in \mathfrak{M}$ and Theorem 24.26, Corollary 25.4, applies.] For a fuller development of this and related material see Naimark [69].

23G. Finite Dim Spectral Theorem for Normal Operators

Let $f \in \mathfrak{L}(Y, Y)$ be normal for a finite dim, complex vector space Y. This means that $ff^* = f^*f$, where $f^* \in \mathfrak{L}(Y, Y)$ is the transpose of f relative to a (fixed basis) inner product on Y. Prove the following:

THEOREM. To the distinct eigenvalues $\{\lambda_1, \ldots, \lambda_n\}$ of f there correspond $\{E_1, \ldots, E_n\} \in \mathfrak{L}(Y, Y)$ that satisfy

(i) $E_j^2 = E_j$ and $E_j E_k = 0\ \forall 1 \leq j, k \leq n$;

(ii) $\sum_{j=1}^{n} E_j = \mathbb{1}$;

(iii) $\sum_{j=1}^{n} \lambda_j E_j = f$;

(iv) $E_j(Y) = (f - \lambda_j \mathbb{1})^{-1}(0)$.

[*Hint* (cf Problem 26C): $X = \text{alg}\{f, \mathbb{1}\}$ is a commutative, semisimple Banach algebra with identity ($f_0 \in X$ implies that f_0 is normal and $|||f_0^{2^k}||| = |||f_0|||^{2^k}\ \forall k$ yields $r_{\sigma_X}(f_0) = |||f_0|||$ and $\mathfrak{R}(X) = \{0\}$). Furthermore (e.g., see Backman and Narici [6]), $\sigma_X(f) = \{\lambda_1, \ldots, \lambda_n\}$. Pick $\varepsilon_j > 0$ such that $\{\Gamma_j = \{z : |z - \lambda_j| = \varepsilon\} : 1 \leq j \leq n\}$ are exterior to one another. Then $E_j = 1/2\pi i \int_{\Gamma_j} (\lambda \mathbb{1} - f)^{-1}\, d\lambda \in X\ \ \forall 1 \leq j \leq n$. The proofs of (i), (ii) follow similarly to those in Problem 23F (e.g., $E_j E_k = 1/2\pi i \int_{\Gamma_j} (\lambda \mathbb{1} - f)^{-1}\{1/2\pi i \int_{\Gamma_k} d\gamma/(\gamma - \lambda)\}\, d\lambda + 1/4\pi^2 \int_{\Gamma_k} (\gamma \mathbb{1} - f)^{-1}\{\int_{\Gamma_j} d\lambda/(\gamma - \lambda)\}\, d\gamma$

and the inner integral of each summond equals zero.) Since $f(M)\neq\lambda_i$ implies $E_i(M)=1/2\pi i\int_{\Gamma_i} d\lambda/[\lambda-f(M)]=0$, it follows that $(fE_i-\lambda_iE_i)(M)=0\ \forall M\in\mathfrak{M}$. Thus $fE_i=\lambda_iE_i$ and (iii) follows from (ii), whereas (iv) follows from (i)-(iii).]

23H. Derivations

A linear mapping D of a (not necessarily commutative) algebra X into itself is called a derivation if $D(xy)=xD(y)+D(x)y\ \forall x, y\in X$.

(a) Each of the following is a derivation:

(i) $D(f)=f'$ on $X=\mathfrak{D}^\infty(\Omega)$;

(ii) $D(x)=\sum_{n=0}^{\infty} n\alpha_n t^{n-1}$ for $X=\left\{\sum_{n=0}^{\infty}\alpha_n t^n:\alpha_n\in\mathbb{F}\right\}$;

(iii) $D_x(y)=xy-yx\ \forall x\in X$;

(iv) $D_1D_2-D_2D_1$ for derivations D_1, D_2 on X.

(b) The set of derivations on X is a vector subspace of $\mathfrak{L}(X, X)$. Is it an algebra under $D_1\cdot D_2=D_1D_2-D_2D_1$?

Assume now that $D\in L(X, X)$ is a derivation for a complex, commutative Banach algebra $(X, \|\ \|)$ with identity.

(c) For each $\alpha\in\mathbb{C}$ the abs convergent series $\sum_{n=0}^{\infty}\alpha^n D^n/n!$ converges to an operator $e^{\alpha D}\in L(X, X)$. If $h\in\mathcal{H}$, then $h_\alpha\in\mathcal{H}$ for $h_\alpha(x)=h\{e^{\alpha D}(x)\}$. [*Hint*: Show that $D^n(xy)=\sum_{i+j=n}[D^i(x)/i!][D^j(y)/j!]$ and use $h\in\mathcal{H}'$ (Corollary 22.16) to obtain $h_\alpha(xy)=\sum_{n=0}^{\infty}\alpha^n\sum_{i+j=n}[h\{D^i(x)\}/i!][h\{D^j(y)\}/j!]$. Then make use of $h_\alpha(x)h_\alpha(y)=\sum_{i=0}^{\infty}\alpha^i[h\{D^i(x)\}/i!]\sum_{j=0}^{\infty}\alpha^j[h\{D^j(y)\}/j!]$ and the abs convergence of each factor series to conclude that $h_\alpha(xy)=h_\alpha(x)\,h_\alpha(y)$.]

(d) Why is it true that $h_\alpha(x)=\sum_{n=0}^{\infty}\alpha^n[h\{D^n(x)\}/n!]\ \forall x\in X$? Use this to establish that $\phi_x:\underset{\alpha\rightsquigarrow h_\alpha(x)}{\mathbb{C}\to\mathbb{C}}$ is constant for fixed $x\in X$. Conclude that $h\{D(x)\}=0$ and $D(x)\in\bigcap_{\mathcal{H}}h^{-1}(0)=\mathfrak{R}(X)\ \forall x\in X$.

(e) If X is semisimple, then O is the only derivation on X.

(f) Show that $AB-BA=\mathbb{1}$ for $A, B\in L(X, X)$. [*Hint*: $D_B:\underset{C\rightsquigarrow BC-CB}{L(X, X)\to L(X, X)}$ is a continuous derivation and $\mathbb{1}\notin\mathfrak{R}\{L(X, X)\}$.]

23I. Embedding X in $L(X, X)$

Assume that $(X, \| \|)$ is a complex (not necessarily commutative) Banach algebra in Example 23.4 and let $M_X = \{\mathfrak{m}_x : x \in X\}$ be the Banach subalgebra of $(L(X, X), ||| \; |||)$ under the congruence $x \rightsquigarrow \mathfrak{m}_x$.

(a) Why is $M_X \neq L(X, X)$ if X is commutative with $\dim X > 1$?

(b) If X is a *-algebra $\{B^*$-algebra$\}$, then M_X is a *-algebra $\{B^*$-algebra$\}$ under $\mathfrak{m}_x^* = \mathfrak{m}_{x^*}$. Show further that $x = x^* \{xx^* = x^*x\}$ implies $\mathfrak{m}_x = \mathfrak{m}_x^* \{\mathfrak{m}_x \mathfrak{m}_x^* = \mathfrak{m}_x^* \mathfrak{m}_x\}$. The converse also holds if X has no tdz's (Problem 23D).

(c) In general, $\sigma(\mathfrak{m}_x) \subset \sigma(x)$ $\forall x \in X$. For $\mathfrak{S} = \{x \in X : \sigma(x) = \sigma(\mathfrak{m}_x)\}$ and terminology in Problem 25D prove

(i) if $x \in \mathfrak{S} \cap \mathfrak{U}$, then $x^{-1} \in \mathfrak{S}$;
(ii) $\mathfrak{S}$ is not bounded in $(X, \| \|)$;
(iii) $\mathfrak{S}$ is closed if $(X, \| \|)$ has spectral continuity.

(d) Show that $\mathfrak{S} = X$ if X is commutative. [*Hint*: If $\lambda \notin \sigma(\mathfrak{m}_x)$, then $\mathfrak{m}_{x-\lambda e}$ is invertible in $L(X, X)$ and $(x - \lambda e)y = \mathfrak{m}_{x-\lambda e}(y) = e$ for some $y \in X$.]

(e) The converse of (d) is false, since $\mathfrak{S} = X$ for the noncommutative Banach algebra $X = \left\{x = \begin{pmatrix} x_{11} & 0 \\ x_{12} & x_{22} \end{pmatrix} : x_{ij} \in \mathbb{C}\right\}$ under $\|x\| = \frac{1}{2}(|x_{11}| + |x_{12}| + |x_{22}|)$. [*Hint*: $\operatorname{Dim} X < \infty$ so that $\sigma(f) = \{$eigenvalues of $f\}$ $\forall f \in L(X, X)$. Therefore $\sigma(x) = \{\lambda \in \mathbb{C} : (x - \lambda e)^{-1}$ does not exist$\} = \{x_{11}, x_{12}\} = \{\lambda \in \mathbb{C} : \mathfrak{m}_{x-\lambda e}(x_0) = 0$ for some $x_0 \neq 0$ in $X\} = \sigma(\mathfrak{m}_x)$.]

Remark. A fuller development of the above notions may be found in Harris [39].

23J. Symmetric and Positive Forms

Let $S = \{x \in X : x = x^*\}$ for a complex *-algebra X. Call $f \in X^*$ symmetric $\{$positive$\}$ if $f(x) = \overline{f(x)}$ $\forall x \in S$ $\{f(xx^*) \geq 0$ $\forall x \in X\}$.

(a) Show that $f \in X^*$ is symmetric iff $f(x^*) = \overline{f(x)}$ $\forall x \in X$. Furthermore, every $F \in X^*$ has a representation $F = f_1 + if_2$, where f_1, f_2 are symmetric. $\{$*Hint*: Consider $f_1(x) = [F(x) + \overline{F(x^*)}]/2$ and $f_2(x) = [F(x) - \overline{F(x^*)}]/2i.\}$

(b) Every positive $f \in X^*$ is symmetric and satisfies $|f(y^*x)|^2 \leq f(y^*y) f(x^*x)$ $\forall x, y \in X$. [*Hint*: If $x \in S$, then $0 \leq f\{(e + x)^*(e + x)\} = f(e^*e) + 2f(x) + f(x^*x)$. For the second assertion use $0 \leq f\{(x - y)^*(x - y)\}$.]

Assume $X = (X, \| \|)$ is a Banach *-normed algebra.

(c) Every positive $f \in X^*$ is continuous with $|||f||| = f(e)$. [*Hint*: Since $\|e\| = 1 \Rightarrow |||f||| = \sup_{\|x\| \leq 1} |f(x)| \geq f(e)$, it suffices to obtain $|f(x)| \leq f(e)$ $\forall \|x\| \leq 1$. Now $x \in S$ and $\|x\| \leq 1$ imply that $e - \frac{1}{2}x - (1/2!)(1/2^2)x^2 - \cdots$ converges to some $y \in S$ and

$f(e)-f(x)=f(e-x)=f(y^*y)\geq 0$. From $f(e)+f(x)=f(e)-f(-x)\geq 0$ also follows $|f(x)|\leq f(e)$. If $\|x\|\leq 1$ for $x\in X$, then $x^*x\in S$ and $f(x^*x)\leq f(e)$. Take $y=e$ in (b) above to obtain $|f(x)|^2\leq f(e)f(x^*x)$.]

(d) Every positive $f\in X^*$ satisfies $f(xx^*)\leq f(e)r_\sigma(x^*x)$. [*Hint*: Use Corollary 23.8(ii).]

(e) Show that $E=\{\text{positive } f\in X^*: f(e)\leq 1\}$ is convex and $\sigma(X^*X)$-compact in X^*.

(f) Assuming that X does not possess an identity, verify that X_e is a *-algebra under $(a, x)^*=(\bar{\alpha}, x^*)$. Go on to prove that a positive $f\in X^*$ can be extended to a positive $f\in X_e^*$ iff there is a constant λ such that $|f(x)|^2\leq \lambda f(x^*x)\ \forall x\in X$. [*Hint*: The condition implies $F(\alpha, x)=\alpha x+f(x)$ is a positive extension of f. For positive $F\in X_e^*$, take $\lambda=F(1,0)$ and use (b) above.]

(g) Is every positive $f\in X^*$ also symmetric in (f) above?

23K. Almost B^*-Algebras

A Banach *-algebra $(X, \|\ \|)$ is almost B^* if there is another (not necessarily complete) algebra norm $\|\ \|_a$ for X and scalar $\lambda>0$ such that $\|x\|_a^2\leq\lambda\|x^*x\|_a\ \forall x\in X$. One calls $\|\ \|_a$ an auxiliary norm for X. The following problem (Rickart [83]) demonstrates that many basic B^*-related results hold in the larger class of almost B^*-algebras.

Assume throughout that $(X, \|\ \|)$ is almost B^* with auxiliary norm $\|\ \|_a$.

(a) Produce an example where $(X, \|\ \|)$ is not a B^*-algebra. [*Hint*: Consider $L_1(\mathbb{R})$ in Example 23.5(a). What about a real semisimple Banach algebra with identity mapping as involution and spectral radius as auxiliary norm?]

(b) Show that $\|x^2\|_a\leq\lambda^2 r_\sigma(x^*x)\leq\lambda^2\|x^*x\|=\lambda^2\|x\|^2\ \forall x\in X$. [*Hint*: Let r_{σ_1} denote the spectral radius in $(\check{X}_1, \|\ \|_a)$, the completion of $X_1=(X, \|\ \|_a)$. Then $r_{\sigma_1}(x)\leq r_\sigma(x)\leq\|x\|\ \forall x\in X$. If $x=x^*$ in X, then $\|x_1\|\leq\lambda^{1-1/2k}\|x^{2k}\|_a^{1/2k}\ \forall k\in\mathbb{N}$ and so $\|x\|_a\leq r_{\sigma_1}(x)$. Proceed!]

(c) Every sub *-algebra $Y\subset X$ is semisimple. [*Hint*: If $y\in\mathfrak{R}(Y)$, then $y^*y\in\mathfrak{R}(Y)$ and $r_{\sigma_Y}(y^*y)=0$ (Theorem 22.8(ii), Example 24.16(a), and Theorems 24.18(i) and 24.7). Therefore $r_\sigma(y^*y)=0$ and $y=0$ (by (b) above).]

(d) The involution of X is $\|\ \|$-continuous. [*Hint*: First establish that $y=0$ if $x_n \xrightarrow{\|\ \|} 0$ and $\|x_n^*-y^*\|\to 0$.]

(e) If $(Y, \|\ \|_0)$ is a Banach sub *-algebra of $(X, \|\ \|)$, there is a constant $\delta>0$ such that $\|y\|\leq\delta\|y\|_0\ \forall y\in Y$. [*Hint*: Since $(Y, \|\ \|_0)$ is almost B^* with auxiliary norm $\|\ \|_a$ there exist constants $\gamma, \gamma_1>0$ such that $\|x^*\|\leq\gamma^2\|x\|\ \forall x\in X$ and $\|y^*\|_a\leq\gamma_1^2\|y\|_a\ \forall y\in Y$. Let $S=\{x\in X: \|y_n-x\|\to 0$ for some $\{y_n\}\in Y$ satisfying $\|y_n\|_0\to 0\}$. Then $\sup_{\|x\|\leq 1}\|x\|_a$, $\sup_{\|y\|_0\leq 1}\|y\|_a<\infty$ imply $\|x\|_a=0$ on X. Conclude by demonstrating that (Problem 23J) $S=\{0\}$ iff there is a $\delta>0$ such that $\|y\|\leq\delta\|y\|_0\ \forall y\in Y$.]

(f) Use (e) above to confirm that all complete algebra norms are equivalent on $(X, \|\ \|)$.

(g) Let $h:(Y_1, \| \|_1) \to (Y_2, \| \|_2)$ be a homomorphism between Banach algebras $(Y_i, \| \|_i)$ such that either Y_2 is semisimple and h is onto, or Y_2 is almost B^* and $h(Y_1)$ is a sub *-algebra of Y_2. Then there is a constant $\lambda > 0$ such that $\|h(y)\|_2 \le \lambda \|y\|_1$ $\forall x \in Y_1$. [*Hint*: In either case $h(Y_1)$ is a semisimple subalgebra of Y_2. Show that $h\{h^{-1}(0)\} \subset \mathfrak{R}(Y_2)$ and $h^{-1}(0)$ is closed in $(Y_1, \| \|_1)$. Therefore $\nu: Y_1/h^{-1}(0) \to h(Y_1)$ is a congruence relative to $|||y + h^{-1}(0)||| = \inf_{z \in y + h^{-1}(0)} \|z\|_1$ on $Y_1/h^{-1}(0)$ and $\|h(y)\|_0 = |||y + h^{-1}(0)|||$ on $h(Y_1)$. Now use (e), (f), and the fact that $\|h(y)\|_0 < \|y\|_1$ $\forall y \in Y_1$.]

24. TVAs and LMC Algebras

Loosely speaking, an algebra is a vector space enriched with ring multiplication and a topological vector algebra (TVA) is a TVS in which continuity of multiplication holds. In a TVA multiplicatively related notions such as closure of ideals and continuity of inversion are considered first. Then moving somewhere between NLAs and TVAs, we focus on LMC algebras, the enriched algebraic counterparts to LCTVSps. Just as LCT_2VSps are embedded in a product of Banach spaces, so $LMCT_2$ algebras are embedded in a product of Banach algebras. This enables us to relate certain LMC algebra properties to corresponding Banach algebra properties, of which a good deal is already known.

Topological Vector Algebras

Unless specified otherwise, all algebras are commutative and have an identity element *e*.

DEFINITION 24.1. A topological vector algebra (TVA) is an algebra X with a topology $\mathfrak{T} \in T(X)$ under which $\mathfrak{m}: X \times X \to X$, $(x,y) \rightsquigarrow xy$ is continuous. This can be restated as "$\mathfrak{T}$ is compatible with X's algebraic structure." A locally convex topological algebra (LCTVA) is a TVA that is also an LCTVS.

For a TVS X define $\mathcal{H}_c = \mathcal{H} \cap X'$ and $\mathfrak{M}_c = \{\text{closed } M \in \mathfrak{M}\}$.

Remark. Each $\mathfrak{m}_{x_0}: X \to X$, $x \rightsquigarrow xx_0$ $(x_0 \in X)$ is linear and continuous. If $x_0 \in \mathfrak{U}$, then (Theorem 22.8) $\mathfrak{m}_{x_0}^{-1} = \mathfrak{m}_{x_0^{-1}} \in L(X, X)$ and $\mathfrak{m}_{x_0}$ is an onto TVS isomorphism.

$(X, \mathfrak{I})$ is an LCTVA for every algebra X, whereas $(X, \mathfrak{D})$ is a TVA iff $X = \{0\}$ (by Theorem 13.14⁺). Verification also follows from

THEOREM 24.2. X is a TVA {LCTVA} iff it has a nbd base $\mathfrak{N}_0$ as in Theorem 13.7 {Theorem 14.2} such that $\forall v \in \mathfrak{N}_0$ some $u \in \mathfrak{N}_0$ satisfies $u^2 \subset v$.

Proof. Assuming the condition on $\mathfrak{N}_0$, consider any $x_0, y_0 \in X$, and $w(0) \in \mathfrak{N}_0$. Let $v+v+v \subset w$ and $u^2 \subset v$ for $u, v \in \mathfrak{N}_0$ and choose $\alpha \in (0, 1)$ such that $\alpha x_0, \alpha y_0 \in u$. This yields $(x_0+\alpha u)(y_0+\alpha u) = x_0 y_0 + \alpha x_0 u + \alpha y_0 u + \alpha^2 u^2 \subset x_0 y_0 + w$ and the continuity of multiplication is established. Suppose, conversely, that multiplication is continuous and let $v \in \mathfrak{N}_0$. Then $u_1 u_2 \subset v$ for some $u_1, u_2 \in \mathfrak{N}_0$ and so $u^2 \subset v$ for $u = u_1 \cap u_2 \in \mathfrak{N}_0$. ■

THEOREM 24.3. For a TVA X

(i) $\mathcal{H}$ is $\sigma(X^*, X)$-closed in X^* and $\mathcal{H}_c$ is $\sigma(X', X)$-closed in X'. Furthermore, $X^* = \overline{[\mathcal{H}]}^{\sigma(X^*,X)}$ iff $\mathcal{H}$ is injective and $X' = \overline{[\mathcal{H}_c]}^{\sigma(X',X)}$ iff $\mathcal{H}_c$ is injective.

(ii) $\bar{Y}$ is a subalgebra of X for every subalgebra $Y \subset X$ and $\bar{I}$ is an ideal $\forall I \in \mathcal{J}$.

Proof. The proofs for $\mathcal{H}_c$, being similar to those for $\mathcal{H}$, are omitted. If $f \in \overline{\mathcal{H}}^{\sigma(X^*,X)}$, then $h_\varphi \xrightarrow{\sigma(X^*,X)} f$ for some net $h_\varphi \subset \mathcal{H}$ and [Example 15.4(c)] $h_\varphi(xy)$ converges to $f(xy)$ and to $f(x)f(y)$. Since $\sigma(X^*, X)$ is T_2, we have $f(xy) = f(x)f(y)$ and $f \in \mathcal{H}$. The second assertion in (i) utilizes Theorem 18.11. One always has $\{0\} \subset [\mathcal{H}]^\perp$. The reverse inclusion holds iff $\bigcap_{\mathcal{H}} h^{-1}(0) = \{0\}$. Therefore, $\mathcal{H}$ is injective iff $\{0\} = [\mathcal{H}]^\circ$ which is equivalent to $\{0\}^\circ = X^*$ being equal to $[\mathcal{H}]^{\circ\circ} = \overline{[\mathcal{H}]}^{\sigma(X^*,X)}$. Everything in (ii) follows from the continuity of multiplication. For instance, $x_0 x = \lim x_\varphi x \in \bar{I}$ for $x \in X$ and net $x_\varphi \subset I$ such that $x_\varphi \to x_0 \in \bar{I}$. ■

Remark. In general $I \in \mathcal{J}$ does not imply $\bar{I} \in \mathcal{J}$. Witness that $\bar{I} = X$ for every proper ideal I of $(X, \mathcal{J})$.

If an $\mathrm{LCT_2VA}$ X is finite dim, then (Corollary 13.13) $[\mathcal{H}_c]$ is finite dim and $\sigma(X', X)$-closed in X'. A partial converse is

COROLLARY 24.4. Suppose that X is an $\mathrm{LCT_2VA}$ such that $\mathcal{T} \in T(X')$ for some Catg II, $\mathrm{T_2VS}$ topology $\mathcal{T}$. Then dim X is finite if $[\mathcal{H}_c]$ is a $\sigma(X', X)$-closed, injective countable dim subspace of X'. In particular, a complex semisimple Banach algebra X is finite dim iff $[\mathcal{H}]$ is $\sigma(X', X)$-closed and has countable dim in X'.

Proof. The first part follows from $\dim X' = \dim [\mathcal{H}_c]$ and Problem 13G(b); the second follows from Corollaries 23.7 and 24.21. ■

A particularly important feature of TVA's is

DEFINITION 24.5. A TVA X is said to have continuous inverse if $\mathfrak{i}:\mathfrak{U}\underset{x\rightsquigarrow x^{-1}}{\longrightarrow}\mathfrak{U}$ is continuous. This is equivalent to requiring that $\mathfrak{U}$ be a topological group.

Remark. Note that $\mathfrak{i}:\mathfrak{U}\to\mathfrak{U}$ is continuous if it is continuous at $e\in\mathfrak{U}$. (If $\varphi\to x_0\in\mathfrak{U}$ for a net $\varphi\subset\mathfrak{U}$, then $\varphi x_0^{-1}\to e$ and $x_0\varphi^{-1}\to e$ yields $\varphi^{-1}\to x_0^{-1}\in\mathfrak{U}$.) One can also show that X has continuous inverse iff it has continuous quasi-inverse.

Not every TVA has continuous inverse (Problem 24G); many have.

Example 24.6. (a) $(X,\mathfrak{I})$ as well as every NLA have continuous inverse by virtue of Theorem 24.18(i) below.

(b) $\mathfrak{S}=\mathfrak{S}(X,\mu)$ in Example 13.17 is a TVA with continuous inverse under pointwise multiplication. First, $\mathfrak{S}$ is a TVA with identity χ_X, since $\{x:|f(x)g(x)|>\delta\}\subset\{x:|f(x)|>\sqrt{\delta}\}\cup\{x:|g(x)|>\sqrt{\delta}\}\ \forall f,g\in\mathfrak{S}$ implies $u^2_{\varepsilon/2,\sqrt{\delta}}\subset u_{\varepsilon,\delta}(0)\ \forall\varepsilon,\delta>0$. To prove $\mathfrak{i}$ continuous note that $f^{-1}(0)=\varnothing$ for $f\in\mathfrak{U}$ (whose inverse is denoted $1/f$). Now for any $\mathfrak{U}$-nbd $W_{\varepsilon,\delta}=\{\chi_X+u_{\varepsilon,\delta}(0)\}\cap\mathfrak{U}$ of χ_X one has $\mathfrak{i}\{W_{\varepsilon/(1+\varepsilon),\delta}\}\subset W_{\varepsilon,\delta}$ since $|f(x)-1|\le\varepsilon/(1+\varepsilon)$ implies $|1/f(x)-1|=|f(x)-1|/|f(x)|\le[\varepsilon/(1+\varepsilon)]1/|f(x)|\le\varepsilon$ and $\{x:|1/f(x)-1|>\varepsilon\}\subset\{x:|f(x)-1|>\varepsilon/(1+\varepsilon)\}$.

Remark. If $\mathfrak{S}$ is viewed as the quotient space under $f=g$ iff $f=g$ a.e., then $f\in\mathfrak{U}$ iff $f\neq0$ a.e. and $\mathfrak{i}\{W_{\varepsilon/(1+\varepsilon)}\}\subset W_{\varepsilon,\delta}$ still holds, since $(f^{-1}(0)\cup\{x:f(x)\neq0\})\cap\{x:|1/|f(x)|-1|>\varepsilon\}\subset\{x:|f(x)-1|>\varepsilon/(1+\varepsilon)\}$ and $\mu\{f^{-1}(0)\}=0$.

Our next result relies heavily on continuous inversion and the algebraic completeness of the complex field.

THEOREM 24.7. For a complex TVA X with continuous inverse

(i) $\mathfrak{R}_x:\rho(x)\underset{\lambda\rightsquigarrow(x-\lambda e)^{-1}}{\longrightarrow}\mathfrak{U}$ is analytic in $\rho(x)$ if $\rho(x)$ is open.

(ii) $\sigma(x)\neq\phi$ if X is locally convex and T_2.

Proof. For any $\lambda,\lambda_0\in\rho(x)$ use $\mathfrak{R}_x(\lambda)\mathfrak{R}_x(\lambda_0)^{-1}=\mathfrak{R}_x(\lambda)(x-\lambda_0 e)=\mathfrak{R}_x(\lambda)\{(x-\lambda e)+(\lambda e-\lambda_0 e)\}=e+(\lambda-\lambda_0)\mathfrak{R}_x(\lambda)$ to obtain $\mathfrak{R}_x(\lambda)-\mathfrak{R}_x(\lambda_0)=(\lambda-\lambda_0)\mathfrak{R}_x(\lambda)\mathfrak{R}_x(\lambda_0)$. Since $\zeta_x:\rho(x)\underset{\lambda\rightsquigarrow x-\lambda e}{\to}\mathfrak{U}$ is continuous, $\mathfrak{R}_x=\mathfrak{i}\zeta_x$ is continuous, and $\mathfrak{R}'_x(\lambda_0)=\lim_{\lambda\to\lambda_0}[\mathfrak{R}_x(\lambda)-\mathfrak{R}_x(\lambda_0)]/(\lambda-\lambda_0)=\{\mathfrak{R}_x(\lambda_0)\}^2\in X$ $\forall\lambda_0\in\rho(x)$. To prove (ii) let $\mathcal{P}$ be a family of seminorms that determines the LCT_2 topology of X and suppose that $\sigma(x)=\varnothing$. Then $\rho(x)=\mathbb{C}$ and $\mathfrak{R}_x$ is entire. Since $\mathbb{C}\times\{x\}\underset{(\lambda,x)\rightsquigarrow\lambda x}{\to}X$ is continuous, the filterbase

$\mathcal{S}=\{S_\varepsilon(0)\subset\mathbb{C}:\varepsilon>0\}$ determines a filterbase $\mathcal{S}x=\{S_\varepsilon(0)x:\varepsilon>0\}$ which converges to $0\in X$ [by Theorem 14.8(ii)] since $S_{\varepsilon/p(x)}(0)x\subset S_{\varepsilon,p}(0)\forall\varepsilon>0$ and $p\in\mathcal{P}$. Accordingly, $\mathcal{S}x-e\to-e$ and $(\mathcal{S}x-e)^{-1}\to(-e)^{-1}=-e$. This means that $\forall\varepsilon>0$ and $p\in\mathcal{P}$, there is an $\alpha=\alpha(\varepsilon,p)>0$ such that

$$(*)\qquad |p\{(\lambda x-e)^{-1}\}-p(e)|\leq p\{(\lambda x-e)^{-1}-(-e)\}<\varepsilon\quad \forall|\lambda|<\alpha$$

(i.e., $p\mathfrak{R}_x(1/\lambda)\leq p(e)+\varepsilon\ \forall|1/\lambda|>1/\alpha$}. Since $\bar{S}_{1/\alpha}(0)$ is compact and $p\mathfrak{R}_x$ is continuous, $p\mathfrak{R}_x(\bar{S}_{1/\alpha}(0))$ and so $\{p\mathfrak{R}_x(1/\lambda):|1/\lambda|\leq1/\alpha\}$ is bounded. Thus $p\{\mathfrak{R}_x(\mathbb{C})\}$ is bounded $\forall p\in\mathcal{P}$ and $\mathfrak{R}_x(\lambda)\equiv\mathfrak{R}_x(0)=x^{-1}\ \forall\lambda\in\mathbb{C}$ (Corollary 14.12, Problem 17H); but (*) also establishes that $p(x^{-1})=p\mathfrak{R}_x(1/\lambda)=p\{(x-(1/\lambda)e)^{-1}\}=|\lambda|p\{(\lambda x-e)^{-1}\}\leq|\lambda|\{p(e)+\varepsilon\}\ \forall|\lambda|<\alpha$, and this ensures that $p(x^{-1})=0\ \forall p\in\mathcal{P}$. Since X is T_2, it follows [Theorem 14.8(i)] that $x^{-1}=0$. This, being clearly impossible, negates our assumption that $\sigma(x)=\varnothing$. ■

A division algebra is an algebra X such that $\mathfrak{U}=X-\{0\}$.

COROLLARY 24.8. A complex LCT_2 division algebra with continuous inverse is isomorphic to $\mathbb{C}$.

Proof. If $x\neq0$, then $\sigma(x)\neq\varnothing$ and $x-\lambda e\in X-\mathfrak{U}=\{0\}$ for some $\lambda\in\mathbb{C}$. Therefore $X=[e]$ and $\underset{\lambda e\rightsquigarrow\lambda}{X\to\mathbb{C}}$ is an onto algebra isomorphism. Since $\dim X=1$, this is also a homeomorphism [Corollary 13.13(iii)]. ■

Remark (Arens [2]). A real LCT_2 division algebra with continuous inverse is isomorphic to either $\mathbb{C}$, $\mathbb{R}$ or the quaternions.

LMC Algebras†

LCTVSps encompass NLSps and are completely determined by their abs convex nbds or, what is equivalent, by their continuous seminorms. In like manner NLAs belong to a category of LCTVAs defined in terms of abs convex sets or continuous seminorms. This will be made clear once we refine some earlier notions to take vector multiplication into account.

DEFINITION 24.9. Let X be an algebra.

(i) A subset $u\subset X$ is multiplicative if $u^2\subset u$. If u is also convex {abs convex}, it is called m-convex {abs m-convex}.

(ii) A seminorm p on X is multiplicative if $p(xy)\leq p(x)p(y)\ \forall x,y\in X$. (Thus $p\not\equiv0$ if $p(e)\neq0$.)

† Many of the results here and in Sections 25 and 26 are due to Michael [62].

Remark. Note that $p^+(x)=\max_{1\le i\le n} p_i(x)$ is multiplicative if $\{p_1,\ldots,p_n\}$ are multiplicative, since $\max_{1\le i\le n} p_i(xy)\le(\max_{1\le i\le n} p_i(x))(\max_{1\le i\le n} p_i(y))\;\forall x, y\in X$.

THEOREM 24.10. An abs convex absorbent set u in an algebra X is multiplicative iff its Minkowski seminorm p_u is multiplicative. If $E\subset X$ is multiplicative, so is E_c and E_b (therefore E_{bc}); and $\bar{E}$ is multiplicative if X is also a TVA.

Proof. If p_u is multiplicative and $x, y\in u$, then $p_u(xy)\le p_u(x)p_u(y)\le 1$ implies $xy\in u$. On the other hand, suppose that u is multiplicative and consider $x, y\in X$. Clearly $x\in p_u(x)u$ and $y\in p_u(y)u$ imply $xy\in p_u(x)p_u(y)u$ and $p_u(xy)\le p_u(x)p_u(y)$. Next, suppose that E is multiplicative and (Theorem 13.6) consider any $x=\sum_{i=1}^{n}\alpha_i x_i$ and $y=\sum_{j=1}^{m}\beta_j y_j$, where $\{x_1,\ldots,x_n,y_1,\ldots,y_m\}\in E$ and $\alpha_i,\beta_j>0$ with $\sum_{i=1}^{n}\alpha_i=\sum_{j=1}^{m}\beta_j=1$. Then $xy=\sum_{i,j}\alpha_i\beta_j x_i y_j$, where $\alpha_i\beta_j>0$ and $\sum_{i,j}\alpha_i\beta_j=1$ ensures that $(E_c)^2\subset E_c$. From $(\lambda x)(\lambda x)=\lambda^2x^2\subset\lambda^2E\subset E_b$ for $x\in E_b$ and $|\lambda|\le 1$ also follows $(E_b)^2\subset E_b$. Finally, Corollary 8.5(v) yields $\bar{E}\bar{E}\subset\overline{E^2}\subset\bar{E}$ when X is a TVA. ■

DEFINITION 24.11. A TVA X is said to be locally m-convex (*LMC*) if it has a nbd base at 0 of abs m-convex sets. (Equivalent characterizations of being LMC are given in Problem 24B.)

Remark. An LMC algebra X is a LCTVA but the converse is false (Problem 24G) unless X is T_2 with continuous inverse (Turpin [99]).

Remark. Note that X_e is *LMC* for an algebra X without identity, since an abs m-convex nbd base $\mathfrak{N}_0$ at $0\in X$ determines such a nbd base $\{S_{1/3}(1)\times\frac{1}{3}u: u\in\mathfrak{N}_0\}$ at $(1,0)\in X_e$.

Now compare Theorems 14.7 and 14.11 with

THEOREM 24.12. A TVA $(X,\mathfrak{T})$ is LMC iff $\mathfrak{T}=\mathfrak{T}(\mathcal{P})$, where $\mathcal{P}$ is a collection of multiplicative seminorms on X.

Proof. If $\mathfrak{N}_0$ is a $\mathfrak{T}$-nbd base of abs m-convex nbds for the LMC algebra $(X,\mathfrak{T})$, then $\mathcal{P}=\{p_u: u\in\mathfrak{N}_0\}$ is a collection of multiplicative seminorms such that $\mathfrak{T}=\mathfrak{T}(\mathcal{P})$, Suppose, conversely, that $\mathfrak{T}=\mathfrak{T}(\mathcal{P})$ for a collection of multiplicative seminorms on X. Then every $p\in\mathcal{P}$ is continuous and $S_{\varepsilon,p}(0)$ is a $\mathfrak{T}$-nbd $\forall\varepsilon>0$. In particular, $\mathfrak{N}=\left\{\bigcap_{i=1}^{n} S_{\varepsilon_i,p_i}(0);\varepsilon_i>0 \text{ and } p_i\in\mathcal{P}\right\}$ is a $\mathfrak{T}$-nbd base of abs m-convex sets. ■

Example 24.13. (a) $(X, \mathfrak{I})$ is a complete LMC algebra, since X itself is an abs m-convex nbd base for $\mathfrak{I}$. Equivalently, $\mathfrak{I} = \mathfrak{T}(p_0)$ for the zero multiplication seminorm p_0 on X.

(b) If Ω is a topological space and Y is a TVA {LCTVA, LMC algebra}, then $C_\kappa(\Omega; Y)$ and $C_\sigma(\Omega; Y)$ are TVA's {LCTVA's, LMC algebras}. Clearly, $C(\Omega; Y)$ has the identity $f(x) \equiv e$ when Y has identity e. For $Y = \mathbb{C}$ (Example 14.13) $C_\sigma(\Omega; \mathbb{C})$ and $C_\kappa(\Omega; \mathbb{C})$ are determined by the collections of multiplicative seminorms $\mathfrak{P} = \{p_E(f) = \sup_E |f(x)| : E \in \mathcal{S}\}$, where $\mathcal{S} = \mathcal{S}_\sigma$ and $\mathcal{C}$, respectively.

NLAs are LMCT_2 but the converse is generally false.

Example 24.14. $(\mathfrak{D}^\infty(I), \mathfrak{T}(\mathfrak{P}))$ in Example 14.20 is a nonnormable LMCT_2 algebra with identity χ_I. Although the seminorms in $\mathfrak{P}$ are not multiplicative (e.g., $f(x) = x$ implies $p_1(f^2) = 2 \not\leq 1 = p_1(f)p_1(f)$), the multiplicative seminorms in $\mathfrak{Q} = \{q_n : 2^n \max_{1 \leq i \leq n} p_i : n \in \mathbb{N}_0\}$ do determine $\mathfrak{T}(\mathfrak{P})$. Verification is easy since $\{p_n^+ = \max_{1 \leq i \leq n} p_i : n \in \mathbb{N}_0\}$, therefore $\mathfrak{Q}$ generates $\mathfrak{T}(\mathfrak{P})$. Moreover, $p_n^+(fg) \leq 2^n p_n^+(f) p_n^+(g)$ and one obtains $q_n(fg) \leq 2^n\{2^n p_n^+(f) p_n^+(g)\} = q_n(f) q_n(g)$ $\forall f, g \in \mathfrak{D}^\infty(I)$ and $n \in \mathfrak{T}_0$.

Example 24.15. (a) Given any family $\mathfrak{F} = \{f_\alpha : X_\alpha \to X; \alpha \in \mathfrak{A}\}$ of homomorphisms from TVAs X_α to an algebra X, there is a finest compatible LMC topology under which every $f_\alpha \in \mathfrak{F}$ is continuous. Simply take (by analogy with Theorem 15.9) this to be the topology having {abs m-convex absorbent $u \subset X : f_\alpha^{-1}(u)$ is an X_α-nbd $\forall \alpha \in \mathfrak{A}$} as its nbd base at $0 \in X$.

(b) If h is an onto homomorphism from a TVA {LMC algebra} X with nbd base $\mathfrak{N}_0$ to an algebra Y, then $(Y, \mathfrak{T}_{\mathfrak{I}(h)})$ is a TVA {LMC algebra} having nbd base $n\mathfrak{N}_0$ (Theorem 15.12). In particular, $(X/I, \mathfrak{T}_{\mathfrak{I}(\nu)})$ is a TVA {LMC algebra} $\forall I \in \mathfrak{J}$ in a TVA {LMC algebra} X.

Example 24.16. (a) If $\mathfrak{F} = \{f_\alpha : X \to X_\alpha; \alpha \in \mathfrak{A}\}$ is a collection of homomorphisms from an algebra X to TVAs {LMC algebras} X_α, then $(X, \mathfrak{T}_{\mathfrak{P}\{\mathfrak{F}\}})$ is a TVA {LMC algebra}. Thus every subalgebra of a TVA {LMC algebra} is a TVA {LMC algebra} with the induced topology and a product of TVAs {LMC algebras} is a TVA {LMC algebra} under pointwise multiplication $(x_\alpha)(y_\alpha) = x_\alpha y_\alpha \in \prod_{\mathfrak{A}} X_\alpha$ $\forall (x_\alpha), (y_\alpha) \in \prod_{\mathfrak{A}} X_\alpha$.

(b) (Definition 16.1.) If Y is a TVA {LMC algebra}, so is $(\mathcal{C}\mathfrak{A}(\Omega, \Sigma; Y), \mathfrak{T}_{\mathfrak{P}\{\mathcal{A}\}})$ under $(\mu_1\mu_2)(E) = \mu_1(E)\mu_2(E)$ $\forall E \in \Sigma$.

(c) Every algebra X is LMC under $\mathfrak{T}_{\mathcal{P}\{\mathcal{H}\}}=\sigma(X, \mathcal{H})$. This also follows from the fact that $\sigma(X, \mathcal{H})$ is determined by the multiplicative seminorms in $\mathcal{P}_{\mathcal{H}}=\{p_h(x)=h(x): h\in\mathcal{H}\}$.

(d) Although $\sigma(X, \mathcal{H}_c)\subset\mathfrak{T}$ and $\sigma(X, \mathcal{H}_c)\subset\sigma(X, \mathcal{H})$ for a TVA $(X, \mathfrak{T})$, it need not be true that $\sigma(X, \mathcal{H})\subset\mathfrak{T}$. In fact, $\sigma(X, \mathcal{H})\subset\mathfrak{T}$ iff $\mathcal{H}=\mathcal{H}_c$. TVAs having $\mathcal{H}=\mathcal{H}_c$ are considered in Problem 24H and Section 25. In the interim consider a TVA X such that $\mathcal{H}_0=\mathcal{H}-h_0$ is injective for some $h_0\in\mathcal{H}$. Then $\langle X, [\mathcal{H}_o]\rangle$ is an injective dual pair and (Theorems 18.2 and 22.2) $(X, \sigma(X, \mathcal{H}_0))$ is an LMCT_2 algebra whose continuous dual $[\mathcal{H}_0]$ does not contain h_0. Thus $h_0\in\mathcal{H}$ is not $\sigma(X, \mathcal{H}_0)$ continuous (and $h_0^{-1}(0)\in\mathfrak{M}$ is not $\sigma(X, \mathcal{H}_0)$-closed).

Remark. A particular case of the above is $X=C(\Omega)$ in Example 22.9. The continuity of each $f\in X$ ensures that $\mathcal{H}_\omega=\mathcal{H}-h_\omega$ is injective and $h_\omega\not\subset[\mathcal{H}_\omega]=(X, \sigma(X, \mathcal{H}_\omega))'\ \forall\omega\in\Omega$.

If we substitute "algebra" for "space" in Theorem 15.8 and assume that the continuous seminorms there are also multiplicative, there follows

THEOREM 24.17. If $\mathcal{F}=\{f_\alpha: X\to X_\alpha; \alpha\in\mathcal{Q}\}$ is an injective family of homomorphisms from an algebra X to TVAs {LMC algebras} X_α, then $(X, \mathfrak{T}_{\mathcal{P}\{\mathcal{F}\}})$ is topologically isomorphic to a subalgebra of $\prod_{\mathcal{Q}} X_\alpha$. Therefore every LCT_2VA X is topologically isomorphic to a subalgebra of a product of NLAs (or Banach algebras).

Let p be a multiplicative seminorm on an algebra X. Then $p^{-1}(0)\in\mathcal{J}$ and $X_p=(X/p^{-1}(0), \|\ \|_p)$ in Theorem 15.8 is a NLA whose completion is still denoted $(\check{X}_p, \|\ \|_p)$. Our intent here is to show how Theorem 24.17 can be used to extend Example 24.6(b) and strength Theorem 24.7, Corollary 24.8.

THEOREM 24.18. (i) An LMC algebra has continuous inverse.

(ii) A complex, nonindiscrete LMC division algebra X has $\sigma(x)\neq\varnothing\ \forall x\in X$ and is therefore isomorphic to $\mathbb{C}$.

Proof. Assume that $\mathcal{P}$ is a collection of multiplicative seminorms that determines the topology of an LMC algebra X. If $x_\varphi\to e$ for a net $x_\varphi\subset\mathfrak{U}$, then $x_\varphi+p^{-1}(0)\subset\mathfrak{U}_p$ (the units in X_p) and $x_\varphi+p^{-1}(0)\xrightarrow{\|\ \|_p} e+p^{-1}(0)\in X_p\ \forall p\in\mathcal{P}$. Since X_p has continuous inverse, we have $x_\varphi^{-1}+p^{-1}(0)=\{x_\varphi+p^{-1}(0)\}^{-1}\xrightarrow{\|\ \|_p}\{e+p^{-1}(0)\}^{-1}=e+p^{-1}(0)$ and also $p(x_\varphi^{-1}-e)=\|x_p^{-1}-e+p^{-1}(0)\|_p=\|\{x_\varphi^{-1}+p^{-1}(0)\}-\{e+p^{-1}(0)\}\|_p\xrightarrow{\|\ \|_p}0\ \forall p\in\mathcal{P}$. Therefore

(Theorem 14.8) $x_\varphi^{-1} \to e$ and the proof of (i) is complete. Since X in (ii) is nonindiscrete, there is a nonzero $p \in \mathcal{P}$. Furthermore, $(\check{X}_p, \| \|_p)$ is Banach and its spectrum satisfies $\sigma(x+p^{-1}(0)) \neq \varnothing\ \forall x \in X$. It now follows (from $\nu_p: X \to \check{X}_p$ being an onto homomorphism) that $\sigma(x+p^{-1}(0)) \subset \sigma(x) \neq \varnothing\ \forall x \in X$. ■

Contrary to what may be conjectured, complex T_2 division algebras other than $\mathbb{C}$ do exist. The following illustration, due to Williamson [107], also demonstrates that the LMC property cannot be removed in Theorem 24.18(ii).

Example 24.19. Let $\mathbb{S}(X, \mu)$ be a complex nonlocally convex T_2VA of equivalence classes in Example 24.6(c). Then $\mathfrak{Q}(X, \mu) = \{p(x)/q(x): p, q$ are polynomials with complex coefficients on $X\}$ is a T_2 division subalgebra of $\mathbb{S}(X, \mu)$ which is distinct from $\mathbb{C}$. Since $\mathbb{S}(X, \mu)$ has continuous inverse, $\mathfrak{Q}(X, \mu)$ has continuous inverse and is therefore nonlocally convex (Corollary 24.8).

Our next result is crucial for later developments.

THEOREM 24.20. If X is an LMC algebra, then X/M is a field $\forall M \in \mathfrak{M}$. If X is also complex, every X/M $(M \in \mathfrak{M}_c)$ is isomorphic to $\mathbb{C}$ and $\underset{h \rightsquigarrow h^{-1}(0)}{\mathcal{H}_c \to \mathfrak{M}_c}$ is a bijection.

Proof. If $M \in \mathfrak{M}$ and $x \notin M$, the ideal $M + Xx$ properly contains M. Therefore $X = M + Xx$ and $e = x_m + x_0 x$ for some $x_m \in M$ and $x_0 \in X$. This yields $e + M = (x_m + M) + (x_0 x + M) = (x_0 + M)(x + M)$ and $(x+M)^{-1} = (x_0 + M) \in X/M$, which confirms that X/M is a field. If X is also complex and $M \in \mathfrak{M}_c$, then $(X/M, \mathcal{T}_{\mathcal{I}(\nu)})$ is an LMCT_2 algebra that [Theorem 24.18(ii)] is isomorphic to $\mathbb{C}$. This, in turn, means that every $(x+M) \in X/M$ has a unique representation $x + M = x(M)(e+M)$. Since the onto isomorphism $\underset{x+M \rightsquigarrow x(M)}{t_M: X/M \to \mathbb{C}}$ is continuous [Corollary 13.13(ii)], $h_M = \underset{x \rightsquigarrow x+M \rightsquigarrow x(M)}{t_M \nu: X \longrightarrow \mathbb{C}}$ is a continuous homomorphism. In fact, $h_M^{-1}(0) = M$, since $x \in h_M^{-1}(0)$ iff $0 = t_M(x+M)$ iff $x \in M$. Thus $\underset{h \rightsquigarrow h^{-1}(0)}{\mathcal{H}_c \to \mathfrak{M}_c}$ is surjective. The reasoning in Example 22.9 demonstrates that this correspondence is also injective. ■

An application of Corollary 23.7 also gives us

COROLLARY 24.21. If X is a complex Banach algebra, $\mathcal{H} \to \mathfrak{M}$ is a bijection. Therefore X is semisimple iff $\mathcal{H}$ is injective.

Theorem 24.17 offers much more than may at first be imagined. By way of illustration let $\mathcal{P}$ be a collection of multiplicative seminorms that determines the topology of an LMCT$_2$ algebra X. There is no loss in assuming that $\mathcal{P}$ is saturated (i.e., $\max_{1\leq i\leq n} p_i \in \mathcal{P}\ \forall\{p_1, \ldots, p_n\} \in \mathcal{P}$), since any family of seminorms that determines the topology on X can be extended to a saturated family determining the topology. Abbreviate $S_{\varepsilon,p}(0)$ by $S_{\varepsilon,p}$ and observe that $h \in \mathcal{H}_c \cap S^{\circ}_{1,p}$ implies $h(x)=0\ \forall x \in p^{-1}(0)$ (since $|h(x_0)|=\delta>0$ and $p(x_0)=0$ imply $nx_0 \in S_{1,p}$ and $|h(nx_0)|>1$ for large enough $n \in \mathbb{N}$). Thus each $h \in \mathcal{H}_c \cap S^{\circ}_{1,p}$ determines a homomorphism $h_p \in \mathcal{H}_{p_c}$ (the continuous homomorphism on X_p) defined by $h_p\{\nu_p(x)\}=h(x)\ \forall x \in X$. Viewing homeomorphic spaces as being equal and denoting the homomorphisms of $\check{X}_p$ by $\check{\mathcal{H}}_p$, we prove

THEOREM 24.22. For the above saturated family $\mathcal{P}$ and LMCT$_2$ algebra X

(i) $(\mathcal{H}_c \cap S^{\circ}_{1,p}, \sigma(\mathcal{H}_c \cap S^{\circ}_{1,p}, X)) = (\mathcal{H}_{p_c}, \sigma(\mathcal{H}_{p_c}, X_p)) = (\check{\mathcal{H}}_p, \sigma(\check{\mathcal{H}}_p, \check{X}))$.
(ii) $\mathcal{H}_c = \bigcup_{\mathcal{P}} \mathcal{H}_{p_c}\nu_p = \bigcup_{\mathcal{P}} \check{\mathcal{H}}_p$.

Proof. First note that $\mathcal{H}_c \cap S^{\circ}_{\varepsilon,p} \subset \mathcal{H}_c \cap S^{\circ}_{1,p}\ \forall \varepsilon>1$ and $\mathcal{H}_c \cap S^{\circ}_{\varepsilon,p} = \mathcal{H}_c \cap S^{\circ}_{1,p}\ \forall \varepsilon \in (0,1]$. (If $\varepsilon<1$, $|h(x_0)|>1$ with $x_0 \in S_{1,p}$ implies $x_0^n \in S_{\varepsilon,p}$ and $|h(x_0^n)|>1$ for large enough $n \in \mathbb{N}$.) Now $t: \mathcal{H}_c \cap S^{\circ}_{1,p} \to \mathcal{H}_{p_c}$, $h \rightsquigarrow h_p$ is injective and also surjective, since every $h_p \in \mathcal{H}_{p_c}$ satisfies $h_p = t(h)$ for $h = h_p\nu_p \in \mathcal{H}_c \cap S^{\circ}_{1/\|h_p\|,p} \subset \mathcal{H}_c \cap S^{\circ}_{1,p}$. That t is bicontinuous follows from Theorem 15.1 and the defined topologies involved. For instance, t is continuous iff $\xi_{\nu_p(x)}t$ is continuous $\forall x \in X$—and this is precisely the case, since $(\xi_{\nu_p(x)}t)(h) = h(x) = \xi_x(h)\ \forall x \in X$. The second equality in (i) merely restates that $\check{t}: \check{\mathcal{H}}_p \to \mathcal{H}_{p_c}$, $\check{h}_p \rightsquigarrow \check{h}_p/X_p$ is a homeomorphism. (Clearly t is injective, since X_p is dense in $\check{X}_p$, and $\check{t}$ is surjective since every $h_p \in \mathcal{H}_{p_c}$ has a unique extension by continuity to $\check{h}_p \in \check{\mathcal{H}}_p$.) The proof of (ii) follows from $X' = \bigcup_{\mathcal{P},\varepsilon>0} S^{\circ}_{\varepsilon,p}$ [Theorem 18.14(i)]. Specifically, $\mathcal{H}_c = \bigcup_{\mathcal{P},\varepsilon>0} \mathcal{H}_c \cap S^{\circ}_{\varepsilon,p} = \bigcup_{\mathcal{P}} \mathcal{H}_c \cap S^{\circ}_{1,p} = \bigcup_{\mathcal{P}} \mathcal{H}_{p_c}\nu_p = \bigcup_{\mathcal{P}} \check{\mathcal{H}}_p$. ■

The following consequence of Theorem 24.17 is also important in that it enables us to extend certain invertibility related questions from Banach algebras to LMC algebras. As usual, $\check{X}_p$ and $\check{X}$ are the completions of X_p and X. Let $\mathfrak{U}_p$, $\check{\mathfrak{U}}_p$, and $\check{\mathfrak{U}}$ denote the units in X_p, $\check{X}_p$, and $\check{X}$, respectively, and define $\check{\sigma}(x) = \{\lambda \in \mathbb{C} : x - \lambda e \notin \check{\mathfrak{U}}\}$ for $x \in X$. Thus $\mathfrak{U} \subset \check{\mathfrak{U}}$ and $\check{\sigma}(x) \subset \sigma(x)\ \forall x \in X$, which we now sharpen in

THEOREM 24.23. Let $\mathfrak{P}$ be a saturated family of multiplicative seminorms which determines the topology of an LMCT_2 algebra X:

(i) If $x \in \mathfrak{U}$, then $x + p^{-1}(0) \in \mathfrak{U}_p\ \forall p \in \mathfrak{P}$.

(ii) If $x + p^{-1}(0) \in \bigcup_{\mathfrak{P}} \check{\mathfrak{U}}_p$, then $x \in \check{\mathfrak{U}}$.

(iii) $\check{\sigma}(x) \subset \bigcup_{\mathfrak{P}} \check{\sigma}(x + p^{-1}(0)) \subset \bigcup_{\mathfrak{P}} \sigma(x + p^{-1}(0)) \subset \sigma(x)\ \forall x \in X$.

Proof. Part (i) is clear since each $\nu_p: X \to X_p$ is a homomorphism and $x \in \mathfrak{U}$ implies $\nu_p(x) = x + p^{-1}(0) \in \mathfrak{U}_p\ \forall p \in \mathfrak{P}$. The proof of (ii) requires much more work. Let (Problems 14A and 24C) $\check{\mathfrak{P}} = \{\check{p}: p \in \mathfrak{P}\}$ determine the topology on $\check{X}$. For each $p \in \mathfrak{P}$ assume that $\{x + p^{-1}(0)\}^{-1} = y_p \in \check{\mathfrak{U}}_p$ and let $\|\ \|_{\check{p}}$ be the norm on $\check{X}_p$ (which extends $\|\ \|_p$ on X_p). Since $(\check{X}_p, \|\ \|_{\check{p}})$ is complete, there is a sequence $\{x_n^{(p)}\} \in X$ such that $\|\{x_n^{(p)} + p^{-1}(0)\} - y_p\|_{\check{p}} < 1/n\ \forall n \in \mathbb{N}$. If $p \le q$ for $p, q \in \mathfrak{P}$, we see that $q^{-1}(0) \subset p^{-1}(0)$ and $h_{qp}: \underset{x+q^{-1}(0) \rightsquigarrow x+p^{-1}(0)}{X_q \longrightarrow X_p}$ is a continuous homomorphism with $|||h_{qp}||| \le 1$ (clearly $\|x + p^{-1}(0)\|_p = p(x) \le q(x) = \|x + q^{-1}(0)\|_q\ \forall x \in X$). The unique continuous extension $\check{h}_{qp}: \check{X}_q \to \check{X}_p$ of h_{qp} (Theorem 6.12) is also a homomorphism with $|||\check{h}_{qp}||| \le 1$ and satisfies $\check{h}_{qp}(y_q) = y_p$. Since $\mathfrak{P}$ is saturated, $\mathfrak{D} = \mathfrak{P} \times \mathbb{N}$ is directed under $(p, m) \le (q, n)$ iff $m \le n$ and $p \le q$. We claim that $\varphi = \{x_n^{(p)}: (p, n) \in \mathfrak{D}\}$ is Cauchy in $\check{X}$ and converges to $x^{-1} \in \check{X}$. First, fix $\varepsilon > 0$ and let $p \in \mathfrak{P}$. Then

$$\begin{aligned}\check{p}(x_n^{(q)} - x_m^{(n)})\\ &= p(x_n^{(q)} - x_m^{(n)}) = \|x_n^{(q)} - x_m^{(r)} + p^{-1}(0)\|_p\\ &= \|\{x_n^{(q)} + p^{-1}(0)\} - \{x_m^{(r)} + p^{-1}(0)\}\|_p\\ &\le \|\{x_n^{(q)} + p^{-1}(0)\} - y_p\|_{\check{p}} + \|y_p - \{x_m^{(n)} + p^{-1}(0)\}\|_{\check{p}}.\end{aligned}$$

If $p \le q$, $r \in \mathfrak{P}$ and $2/\varepsilon < n_0 \le n, m$, then $|||\check{h}_{qp}||| \le 1$ and $|||\check{h}_{rp}||| \le 1$ yield

$$\begin{aligned}\|\{x_n^{(q)} + p^{-1}(0)\} - y_p\|_{\check{p}} &= \|\check{h}_{qp}\{(x_n^{(q)} + q^{-1}(0)) - y_q\}\|_{\check{p}}\\ &\le \|\{x_n^{(q)} + q^{-1}(0)\} - y_q\|_{\check{q}} < 1/n < 1/n_0 < \varepsilon/2\end{aligned}$$

and

$$\begin{aligned}\|y_p - \{x_m^{(r)} + p^{-1}(0)\}\|_p &= \|\check{h}_{rp}\{y_r - (x_m^{(r)} + p_n^{-1}(0))\}\|_p\\ &\le \|y_r - (x_m^{(r)} + r^{-1}(0))\|_r < 1/m < 1/n_0 < \varepsilon/2.\end{aligned}$$

Thus $\check{p}(x_n^{(q)} - x_m^{(r)}) < \varepsilon$ for $(q, n), (r, m) > (p, n_p)$. Since this is true $\forall p \in \mathfrak{P}$, it follows (Theorem 14.8) that φ is Cauchy in $\check{X}$. It remains to show that $y = \lim \varphi \in \check{X}$ equals x^{-1}. Given any $\varepsilon > 0$, let $p \in \mathfrak{P}$ and take $q \in \mathfrak{P}$ such that $p \le q$. Then

$$\begin{aligned}\check{p}(xx_n^{(q)} - e) = p(xx_n^{(q)} - e) &= \|xx_n^{(q)} - e + p^{-1}(0)\|_p\\ &\le \|xx_n^{(q)} - e + q^{-1}(0)\|_q = \|\{xx_n^{(q)} + q^{-1}(0)\} - \{x + q^{-1}(0)\}y_q\|_{\check{q}}\\ &= \|(x + q^{-1}(0))\{x_n^{(q)} + q^{-1}(0) - y_q\}\|_{\check{q}} \le q(x) \cdot 1/n < \varepsilon\end{aligned}$$

for large $n \in \mathbb{N}$. Therefore $0 \in X$ is an $\check{X}$-accumulation point of the net $\check{\mathcal{G}} = \{\check{p}(xx_n^{(q)} - e) : (q, n) \in \mathfrak{D}\}$ in $\check{X}$. Since $\check{p}$ is continuous, $\check{\mathcal{G}}$ is $\check{X}$-Cauchy and $\check{p}(xy - e) = \lim_{\mathfrak{D}} \check{\mathcal{G}} = 0$. In particular, $\check{p}(xy - e) = 0 \ \forall \check{p} \in \check{\mathcal{P}}$ ensures (Theorem 14.8) that $xy = e$ and $y = x^{-1}$. The proof of (iii) follows quickly from (i), (ii). If $\lambda \in \check{\sigma}(x)$, then $x - \lambda e \notin \check{\mathfrak{U}}$ and $(x - p^{-1}(0)) - \lambda(e + p^{-1}(0)) \notin \mathfrak{U}_p$ so that $\lambda \in \sigma(x + p^{-1}(0))$ for some $p \in \mathcal{P}$. We also have $\bigcup_{\mathcal{P}} \sigma(x + p^{-1}(0)) \subset \sigma(x) \ \forall x \in X$. ■

Corollary 24.24. When X above is complete, we have $x \in \mathfrak{U}$ iff $(x + p^{-1}(0)) \in \check{\mathfrak{U}}_p \ \forall p \in \mathcal{P}$. Thus $\sigma(x) = \bigcup_{\mathcal{P}} \sigma(x + p^{-1}(0)) \ \forall x \in X$.

Proof. If X is complete, then $X = \check{X}$ and $\mathfrak{U} = \check{\mathfrak{U}}$ so that $\sigma(x) = \check{\sigma}(x) \ \forall x \in X$. ■

Combining Theorem 24.20 with Corollary 24.24, we obtain

Theorem 24.25. A complete, complex LMCT_2 algebra X has $\sigma(x) = \{h(x) : h \in \mathcal{H}_c\} = \{h(x) : h \in \mathcal{H}\} \ \forall x \in X$. Consequently $\bigcap_{\mathfrak{M}} M = \bigcap_{\mathfrak{M}_c} M$, and $x \in \mathfrak{R}(X)$ iff $r_\sigma(x) = 0$.

Proof. The reasoning in Corollary 23.7 establishes that $\{h(x) : h \in \mathcal{H}\} \subset \sigma(x) \ \forall x \in X$. If $\lambda \in \sigma(x)$, then (for the family $\mathcal{P}$ above) $\lambda \in \sigma\{\nu_p(x)\}$ and $\nu_p(x - \lambda e) \in \mathfrak{U}_p$ for some $p \in \mathcal{P}$. Since $\check{X}_p$ is Banach, $\nu_p(x - \lambda e) \in \check{M}_p = \check{h}_p^{-1}(0)$ for some $\check{h}_p \in \check{X}'_p$ (Theorem 22.8, Corollary 24.21). In particular, $(\check{h}_p \nu_p)(x) - \lambda = 0$ and $\lambda = (\check{h}_p \nu_p)(x) \in \{h(x) : h \in \mathcal{H}_c\}$ for $h_p \nu_p \in \mathcal{H}_c$. This proves the first part of the theorem. Clearly $\bigcap_{\mathfrak{M}} M \subset \bigcap_{\mathfrak{M}_c} M$. If $x \in \bigcap_{\mathfrak{M}_c} M = \bigcap_{\mathcal{H}_c} h^{-1}(0)$, then $1 = \{h(e - xy) : h \subset \mathcal{H}_c\} - \sigma(e - xy)$ and $0 \notin \sigma(e - xy)$ implies $e - xy \in \mathfrak{U} \ \forall y \in X$. This also requires that $x \in \bigcap_{\mathfrak{M}} M$. ■

Remark. The T_2 condition above is essential. Indeed, $\bigcap_{\mathfrak{M}} M = \{0\} \neq \varnothing = \bigcap_{\mathfrak{M}_c} M$ for $(X, \mathcal{I})$.

If Y is a subalgebra of an algebra X, it need not follow that $\mathfrak{M}_X \cap Y \subset \mathfrak{M}_Y$ (although $M \cap Y \in \mathcal{J}(Y) \ \forall M \in \mathfrak{M}_X$) nor that Y is semisimple when X is.

Corollary 24.26. Let $Y \subset X$, where both X and Y are complete, complex LMCT_2 algebras. If X is semisimple, so is Y.

Proof. $r_{\sigma_X}(y) \leq r_{\sigma_Y}(y) \ \forall y \in Y$ implies $\mathfrak{R}(Y) \subset \mathfrak{R}(X) = \{0\}$. ■

The following sharpening of Corollary 24.4 will be especially useful later on.

COROLLARY 24.27. A complete, complex, semisimple $LMCT_2$ algebra X is n dim iff $\aleph(\mathcal{H})=n$.

Proof. $\mathcal{H}$ is injective and so $\dim X=n$ implies $\aleph(\mathcal{H})=n$ (Corollary 22.4). Furthermore (Theorem 18.2), $[\mathcal{H}]=(X,\sigma(X,\mathcal{H}))'$ and $\underset{x \rightsquigarrow \langle x,\ \rangle}{X\to[\mathcal{H}]^*}=(X,\sigma(X,\mathcal{H}))'^*$ is a vector space isomorphism. If $n=\aleph(\mathcal{H})$, then $\dim X\le \dim[\mathcal{H}]^*=\dim[\mathcal{H}]=n$ and (again by Corollary 22.4) $\aleph(\mathcal{H})=\dim X$. ■

Nowhere in Theorem 24.25 did we imply that $\mathcal{H}=\mathcal{H}_c$ $\{\mathfrak{M}=\mathfrak{M}_c\}$ or that r_σ is a seminorm (i.e., $r_\sigma(x)<\infty\ \forall x\in X$). Such, in fact, is not always the case, as we illustrate below.

Example 24.28. (a) Suppose Ω in Example 24.13(c) is T_4 and either locally compact or 1°. Then $\mathcal{P}=\{p_K:K\in\mathcal{C}\}$ is saturated and determines the complete $LMCT_2$ topology on $X=C_\kappa(\Omega;\mathbb{C})$. Since every $f\in C(K;\mathbb{C})$ has a continuous extension to $C(\Omega;\mathbb{C})$ (a consequence of Ω being T_4), each homomorphism $\underset{f \rightsquigarrow f/K}{a_K:C(\Omega;\mathbb{C})\to C(K;\mathbb{C})}$ is surjective. Moreover, $\overset{-1}{a_K}(0)=\overset{-1}{p_K}(0)$ so that $\rho_K:f+\overset{-1}{p_K}(0)\rightsquigarrow f/K$ $(K\in\mathcal{C})$ is an onto congruence between $(X_{p_K},\|\ \|_{p_K})$ and $(C(K;\mathbb{C}),\|\ \|_0)$. Thus the factor algebras $\{X_{p_K}:p_K\in\mathcal{P}\}$ of X are Banach.

(b) Corollary 24.2 yields $\sigma(f)=\bigcup_{\mathcal{C}}\sigma\{f+\overset{-1}{p_K}(0)\}=\bigcup_{\mathcal{C}}\sigma(f/K)=\bigcup_{\mathcal{C}} f/K(K)=f(\Omega)$, which [in contrast to Theorem 23.6(iv)] need be neither closed nor bounded for $f\in X$. Specifically, $f(\omega)=\omega$ $\{f(\omega)=\omega/(1+|\omega|)\}$ has an unbounded {nonclosed} spectrum when $\Omega=\mathbb{C}$.

(c) Since $(X,\sigma(X,\mathcal{H}_0))$ in Example 24.16(d) need not be complete [$\sigma(X,\mathcal{H})$-completeness is considered in Section 25], the reader is entitled to another demonstration that $\mathfrak{M}\neq\mathfrak{M}_c$. Simple enough. $C_0(\mathbb{C};\mathbb{C})$ is a dense ideal in $X=C_\kappa(\mathbb{C};\mathbb{C})$, and the maximal ideal of X which contains $C_0(\mathbb{C};\mathbb{C})$ cannot be closed.

(d) If $\Omega=\mathbb{R}$, then $\{f\in X:r_\sigma(f)<\infty\}$ is a proper dense subset of $X=C_\kappa(\Omega;\mathbb{C})$.

Example 24.29. (a) Let $X=\mathcal{H}_\kappa(\Omega)$, where $\Omega=\{\omega\in\mathbb{C}:|\omega|<1\}$ in Problem 14H. If $K_n=\{\omega:|\omega|\le 1-1/n\}$ and $p_n(f)=\sup_{K_n}|f(\omega)|$, then $\mathcal{P}=\{p_n:n\in\mathbb{N}\}$ determines the LMC Frechet topology on X.

(b) By the identity theorem in analytic function theory $p_n(f)=0$ implies $f=0$. Therefore p_n is a norm and $X_{p_n}=(\mathcal{H}(\Omega)/\overset{-1}{p_n}(0),\|\ \|_{p_n})=$

$(\mathcal{H}(\Omega), p_n)\ \forall n \in \mathbb{N}$. It also follows from the maximum modulus principle that $a_{K_n}: (\mathcal{H}(\Omega), p_n) \to (C(K_n; \mathbb{C}), \|\ \|_0)$, $f \rightsquigarrow f/K_n$ is an (into) algebra congruence. In contrast to Example 24.28(a) the factor algebras X_{p_n} of X are not complete; they are not closed in X. Verification: Since $g_m(z) = \sum_{k=1}^{m} 1/k^2[\omega/(1-1/n)]^k$ is analytic on K_n and $\lim_m g_m \overset{\|\|_0}{=} f \in C(K_n; \mathbb{C})$, we have $f \in \overline{a_{K_n}\{\mathcal{H}(\Omega)\}}^{\|\|_0} \subset C(K_n; \mathbb{C})$. However, f is not analytic on K_n [therefore $f \neq F/K_n$ for any $F \in \mathcal{H}(\Omega)$], since $\sum_{k=1}^{\infty} 1/k^2[\omega/(1-1/n)]^k$ has a radius of convergence $R = 1-1/n$ as well as a singularity ω_0 satisfying $|\omega_0| = 1-1/n$ (Pringsheim's theorem).

(c) As in Example 24.28(b), we have $\sigma(f) = \bigcup_{n=1}^{\infty} \sigma(f/K_n) = \bigcup_{n=1}^{\infty} f(K_n) = f(\Omega)\ \forall f \in X$. If Ω is the complex plane, then $\{f \in X : r_\sigma(f) < \infty\}$ consists only of the constant functions (Louiville's theorem).

The pervasiveness of Theorems 24.17 and 24.23 is further illustrated by the following example, which is of interest in itself, since it threads together earlier notions (Problems 23C, D) with results to follow (Problem 24E). For a TVA X (not necessarily with an identity) generalize the definition of tdz by taking $\mathcal{Z}_0(X) = \{x \in X : xx_\varphi \to 0$ for some net $x_\varphi \not\to 0$ in $X\}$. Essentially, $\mathcal{Z}_0(X) = \{x \in X : {}_x\mathfrak{m}$ is noninjective$\}$.

Example 24.30. Let X be an LMCT$_2$ algebra whose topology is determined by the saturated family $\mathcal{P}$ and let $wk\mathcal{Z}_0(X) = \{x \in X : \nu_p(x) \in \mathcal{Z}_0(\check{X}_p)$ for some $p \in \mathcal{P}\}$.

(a) $\mathcal{Z}_0(X) \subset wk\mathcal{Z}_0(X)$ with equality holding if X is a NLA. By definition, $x \notin wk\mathcal{Z}_0(X)$ implies $\nu_p(x) \notin \mathcal{Z}_0(\check{X}_p)$ and ${}_{\nu_p(x)}\mathfrak{m}$ is a homeomorphism $\forall p \in \mathcal{P}$. If $xy_1 = xy_2$, then ${}_{\nu_p(x)}\mathfrak{m}(y_1 + p^{-1}(0)) = {}_{\nu_p(x)}\mathfrak{m}(y_2 + p^{-1}(0))$ and $p(y_1 - y_2) = 0\ \forall p \in \mathcal{P}$. Thus (Theorem 14.8(i)) $y_1 = y_2$ and ${}_x\mathfrak{m}$ is injective. The reverse inclusion holds if $X = (X, \|\ \|)$ is a NLA by simply taking $\mathcal{P} = \{\|\ \|\}$.

(b) $\bar{\mathfrak{U}} - \mathfrak{U} \subset wk\mathcal{Z}_0(X) \subset X - \mathfrak{U}$. Clearly, $x \in \bar{\mathfrak{U}} - \mathfrak{U}$ implies $\nu_p(x) \notin \check{\mathfrak{U}}_p$ for some $p \in \mathcal{P}$ and (since ν_p is continuous) $\nu_p(x) \in \overline{\check{\mathfrak{U}}_p} - \check{\mathfrak{U}}_p = bdy\ \check{\mathfrak{U}}_p = bdy\ (\check{X}_p - \check{\mathfrak{U}}_p) \subset \mathcal{Z}_0(\check{X}_p)$. Thus $x \in wk\mathcal{Z}_0(X)$. If $x_0 \in \mathcal{Z}_0(X)$, then ${}_{\nu_p(x_0)}\mathfrak{m}$ is not injective and $\nu_p(x_0) \notin \check{\mathfrak{U}}_p$ for some $p \in \mathcal{P}$. In particular, $\nu_p(x_0) \notin \mathfrak{U}_p$ and $x_0 \notin \mathfrak{U}$.

(c) Suppose that X has an identity and Y is a subalgebra of X which also has this identity. Then $wk\mathcal{Z}_0(Y) \subset wk\mathcal{Z}_0(X)$ and $\sigma_Y(y) \cap bdy\, \sigma_Y(y) \subset$

$\sigma_X(y) \cap bdy\, \sigma_X(y)$ $\forall y \in Y$. This becomes clear once we let $q_p = p/Y$ ($p \in \mathcal{P}$) and observe that each $Y_{q_p} = (Y/q_p^{-1}(0), \| \|_{q_p}) \to X_p$ is a congruence. Accordingly, $y \in wk\mathcal{Z}_0(Y)$ implies some ${}_{\nu_{q_p}(y)}\mathfrak{m}: Y_{q_p} \to Y_{q_p}$, therefore ${}_{\nu_p(y)}\mathfrak{m}: X_p \to X_p$ is noninjective and $y \in wk\mathcal{Z}_0(X)$. If $\lambda \in \sigma_Y(y) \cap bdy\, \sigma_Y(y)$, then $x - \lambda y \in \overline{\mathfrak{U}_Y} - \mathfrak{U}_Y \subset wk\mathcal{Z}_0(Y) \subset wk\mathcal{Z}_0(X) \subset X - \mathfrak{U}_X$ and $\lambda \in \sigma_X(y)$. One also has $\lambda \in bdy\, \sigma_X(y)$. Otherwise, $\lambda \in \text{Int}\, \sigma_X(y) \subset \text{Int}\, \sigma_Y(y)$ contradicts $\lambda \in bdy\, \sigma_Y(y)$.

Remark. We can easily verify that $bdy(X - \mathfrak{U}) = \bar{\mathfrak{U}} - \mathfrak{U} \subset wk\mathcal{Z}_0(X)$ if $\mathfrak{U}$ is open and $\sigma(x) \cap bdy\, \sigma(x) = bdy\, \sigma(x)$ if $\sigma(x)$ is closed. LMC algebras with the above properties are the focal point of Section 25.

PROBLEMS

24A. Continuity Considerations

(a) A (not necessarily commutative) algebra X with a compatible TVS topology is a TVA iff $\mathfrak{m}$ is continuous at $(0, 0)$ and both ${}_x\mathfrak{m}$ and $\mathfrak{m}_x$ are continuous at 0 $\forall x \in X$. [*Hint*: Use $xy - x_0y_0 = (x - x_0)(y - y_0) + (x - x_0)y_0 + x_0(y - y_0)$ and the fact that X is a TVS.]

(b) Suppose X above is 1° and Catg II. Then $\mathfrak{m}$ is continuous if ${}_x\mathfrak{m}$ and $\mathfrak{m}_x$ are continuous $\forall x \in X$. [*Hint*: It suffices to show that $(x_n, y_n) \to (0, 0)$ implies $x_ny_n \to 0$. Given any nbd $u(0)$, let $w(0)$ be a closed balanced nbd satisfying $w + w \subset u$. Since $E_n = \bigcap_{m=n}^{\infty} {}_{x_n}\mathfrak{m}^{-1}(w)$ is closed and $X = \bigcup_{n=1}^{\infty} E_n$, we have $y + v \subset E_{n_0}$ and $v \subset E_{n_0} + E_{n_1}$ for some nbd $v(0)$ and $n_0, n_1 \in \mathbb{N}$. Conclude that $x_nv \subset u$ for $n > n_0 + n_1$ and $x_ny_n \to 0$.]

Remark. This is also a special case of Problem 13A(b) applied to (a) above.

(c) For X above show [Problem 8A(a)] that $\mathfrak{m}: \mathfrak{U} \times \mathfrak{U} \to \mathfrak{U}$ is continuous iff $\mathfrak{m}$ is continuous at (e, e) and both ${}_x\mathfrak{m}$ and $\mathfrak{m}_x$ are continuous at e $\forall x \in X$.

(d) A Frechet algebra (i.e., a complete metric LCT_2VA) has continuous inverse (and is therefore $LMCT_2$) iff $\mathfrak{U}$ is a G_δ. [*Hint* (Waelbroeck [102]): If (X, d) is Frechet with continuous inverse, $w_\varepsilon = \{x \in \bar{\mathfrak{U}}$: there is a $\delta > 0$ such that $y, z \in S_\delta(x) \cap \mathfrak{U} \Rightarrow d(y^{-1}, z^{-1}) < \varepsilon\}$ is an open set in $\bar{\mathfrak{U}}$ which contains $\mathfrak{U}$. Show that $\mathfrak{U} = \bigcap_{n=1}^{\infty} W_{1/n}$. For the converse use Problem 8A(d) and the fact that $\mathfrak{U}$ is a complete metric joint topological group.]

(e) Let $\mathcal{F} = \{f_{y,y'}: X = L(Y, Y) \underset{F \rightsquigarrow y'\{F(y)\}}{\longrightarrow} \mathbb{F}\ ; (y, y') \in Y \times Y'\}$, where Y is a TVS $\{LCT_2VS\}$. Then $(X, \mathfrak{T}_{\mathcal{P}\{\mathcal{F}\}})$ is a TVS $\{LCT_2VS\}$ such that both ${}_F\mathfrak{m}$ and $\mathfrak{m}_F$ are continuous at $\mathbb{1}$ $\forall F \in X$. Show, by counterexample, that neither $\mathfrak{m}$ nor $\mathfrak{i}$ need be continuous.

(f) If X is a TVA in which $\mathfrak{U}_\odot$ is open, then $\mathcal{H} \subset B\mathcal{L}(X; \mathbb{F})$. Thus $\mathcal{H} \subset X'$ if X is bornological and $\mathfrak{U}_\odot$ is open. [*Hint*: If $B \subset X$ is bounded and $h(B)$ is not, there exists $\{x_n\} \subset B$ such that $|h(x_n)| \geq n$ $\forall n \in \mathbb{N}$. Use Theorem 13.22 to contradict problem 23A(b).]

24B. LMC Nbd Base at 0

The following are equivalent for an algebra X:

(i) X is LMC;
(ii) X is an LCTVS with a nbd base of multiplicative sets;
(iii) X is a TVS with a nbd base of abs m-convex sets.

24C. TVA Completions

Using Problem 14A, give two proofs that a TVA {LCTVA, LMC algebra} completion is a TVA {LCTVA, LMC algebra} [*Hint*: If $x_{\varphi_i} \to x_i$ for nets $\{x_{\varphi_i}\} \in X$, then $x_{\varphi_1} x_{\varphi_2} = \{x_\alpha x_\beta : (\alpha, \beta) \in \varphi_1 \times \varphi_2\}$ is a net in X which converges to $x_1 x_2$.]

24D. TVA Complexifications

The complexification of a real TVA {LCTVA, LMC algebra} with continuous inverse is a TVA {LCTVA, LMC algebra} with continuous inverse. Accordingly, the real TVS isomorphism $X \to X^+$ in Problem 13D(a) is a TVA isomorphism.

24E. Permanent Singularities

The members of $\mathfrak{P}(X) = \{x \in X : x \notin \mathfrak{U}_Y$ for any algebra $Y \supset X\}$ are called permanently singular elements of a TVA X.

(a) $\mathcal{Z}_0(X) \subset \mathfrak{P}(X) \subset X - \mathfrak{U}$ $\forall$TVA X.

(b) $\mathfrak{P}(X) \subset wk\mathcal{Z}_0(X) \subset X - \mathfrak{U}$ for $X = C_\kappa(\Omega; C)$ in Example 24.28. [*Hint*: $\mathfrak{P}(X) \subset \check{X} - \check{\mathfrak{U}}$ and (Example 24.30) $\check{X}_p - \check{\mathfrak{U}}_p = \mathcal{Z}_0(\check{X}_p)$ $\forall p \in \mathcal{P}$.]

(c) $\mathfrak{P}(X) = wk\mathcal{Z}_0(X) = X - \mathfrak{U}$ for a B^*-algebra X.

(d) (Kuczma [56] The vector space of sequences $X = \{\alpha - (\alpha_0, \alpha_1, \ldots) : \alpha_k \in \mathbb{F}\}$ is a commutative algebra with identity $e = (1, 0, 0, \ldots)$ under multiplication $(\alpha_0, \alpha_1, \ldots)(\beta_0, \beta_1, \ldots) = (\gamma_0, \gamma_1, \ldots)$ with $\gamma_k = \sum_{j=0}^{k} \alpha_j \beta_{k-j}$. Show that $\mathcal{P} = \{p_n(\alpha) = \sum_{k=0}^{n} |\alpha_k| : n \in \mathbb{N}_0\}$ defines an LMCT$_2$ Frechet topology on X. Go on to prove that $wk\mathcal{Z}_0(X) = \{0\}$ and $z = (0, 1, 0, 0, \ldots) \in \mathfrak{P}(X) \subset wk\mathcal{Z}_0(X)$. {*Hint*: $\mathfrak{T}(\mathcal{P})$-convergence means coordinate-wise convergence. Therefore $x_n \to 0$ iff $zx_n \to 0$, and $\alpha \in wk\mathcal{Z}_0(X)$ iff $z^k \alpha \in wk\mathcal{Z}_0(X)$ $\forall k \in \mathbb{N}_0$ (here $z^0 = e$). Since $\mathfrak{U} = \{\alpha \in X : \alpha_0 \neq 0\}$, every $\alpha \in X$ can be written $\alpha = z^k \beta$ for some $k \in \mathbb{N}_0$ and $\beta \in \mathfrak{U}$ [specifically, $(0, 0, \ldots 0, \alpha_{k+1}, \alpha_{k+1}, \ldots) = z^k(\alpha_{k+1}, \alpha_{k+1}, \ldots)$ for $\alpha_{k+1} \neq 0$].}

24F. Complete LCT_2VA Without Continuous Inverse

(Arens [3].) Let $L_\omega = \bigcap_{p=1}^{\infty} L_p$, where L_p is defined for $I = [0, 1]$.

(a) L_ω is an algebra and $\|f\|_1 \leq \|f\|_2 \leq \cdots \forall f \in L_\omega$. [*Hint*: (E. J. McShane, Integration, Princeton, 1944) $\|fg\|_p \leq \|f\|_q \|g\|_r$ for $1/p = 1/q + 1/r$.]

(b) The collection $\{S_{\varepsilon,p}(0) = \{f \in L_\omega : \|f\|_p < \varepsilon\} : p \in \mathbb{N}$ and $\varepsilon > 0\}$ constitutes a nbd base for a complete LCT_2VA topology $\mathcal{T}_\omega$ on L_ω. [*Hint*: $f_n \xrightarrow{\mathcal{T}_\omega} 0$ iff $f_n \xrightarrow{\|\|_p} 0 \; \forall p \geq 1$.]

(c) $(L_\omega, \mathcal{T}_\omega)$ has no continuous inverse. [*Hint*: Show that $f_n = (1/n)\chi_{[0,1/n]} + \chi_{(1/n,1]} \xrightarrow{\mathcal{T}_\omega} \chi_I$, whereas $\|(f_n)^{-1} - \chi_I\|_p = (n-1)/n^{1/p} \to \infty$ for $p \geq 2$.]

(d) We already know [Theorem 24.18(i)] that $(L_\omega, \mathcal{T}_\omega)$ is non-LMC. Sharpen this by showing that an abs m-convex nbd $u(0)$ is $\mathcal{T}_\omega$-dense in L_ω. [*Hint*: Suppose that $S_{\varepsilon,p}(0) \subset u(0)$ and consider any $b\chi_E$ with measure of E equal $a \leq 1$. Choose $k \in \mathbb{N}$ such that $a \leq (\varepsilon/2)^p k$ and let $n \in \mathbb{N}$ satisfy $bk \leq 2^n$. Now cover E by disjoint sets $\{E_1, \ldots, E_k\}$ of measure not exceeding $(\varepsilon/2)^p$ and define $f_j = (bk)^{1/n} \chi_{E_j} \in S_{\varepsilon,p}(0) \; \forall 1 \leq j \leq k$. Then $b\chi_E = \sum_{j=1}^{k} (1/k) f_j^n \in u$. Conclude that every simple function $\sum_{i=1}^{m} b_i \chi_{B_i}$ belongs to u and show that all such functions are dense in L_ω.]

(e) Show that $d_\omega(f, g) = \sum_{p=1}^{\infty} 2^{-p} \|f-g\|_p / (1 + \|f-g\|_p)$ is a complete metric for $(L_\omega, \mathcal{T}_\omega)$ [*Hint*: See Example 13.16.]

24G. Functionally Continuous Algebras

Begin where Example 24.16(d) ended by calling a TVA X functionally continuous if $\mathcal{H} = \mathcal{H}_c$.

(a) Let $X = C_\kappa(\Omega; \mathbb{C})$, where Ω is a countably compact, noncompact, completely regular space that is either locally compact or 1°. Then $\mathcal{H}_c \subsetneqq \mathcal{H} \subset B\mathcal{L}(X; \mathbb{C})$ and so X is nonbornological. [*Hint*: Since Ω is countably compact, every $f \in X$ is bounded and $\mathcal{H}_c = \{h_\omega : \omega \in \beta(\Omega)\} \subset \mathcal{H}_\Omega = \mathcal{H}$, where $\beta(\Omega)$ is the Stone–Cech compactification of Ω. If $K \subset X$ is bounded, then $F(\omega) = \sup_{f \in K} |f(\omega)|$ is bounded in $\mathbb{C}$. Show that $F: \Omega \underset{\omega \rightsquigarrow \sup_X |f(\omega)|}{\longrightarrow} \mathbb{C}$ is bounded and use the fact that $h(f) \in \sigma(f) = f(\Omega) \; \forall h \in \mathcal{H}$ and $f \in X$.]

Remark. The ordinal space $\Omega = [0, \aleph_1)$ satisfies the above requirements.

(b) A complete $LMCT_2$ algebra X (not necessarily with identity) has $\mathcal{H} = \mathcal{H}_c$ if any of the following conditions hold:

(i) $X/\mathfrak{R}(X)$ is functionally continuous;
(ii) X_e is functionally continuous;
(ii) $\mathfrak{U}$ is open.

[*Hint*: (i) For each $h \in \mathcal{H}$ there is a continuous homomorphism $\tilde{h}$ on $X/\mathfrak{R}(X)$ such that $h = \tilde{h}\nu$. A proof of (iii) is given in the next section. See what you can do here.]

24H. Aren's Theorem and Quasi-Regularity

Assuming that X in Theorem 24.23 has no identity, prove

(a) If $x \in \mathfrak{U}_\odot$ in X, then $x + p^{-1}(0) \in \mathfrak{U}_{\odot_p}$ $\forall p \in \mathcal{P}$.
(b) If $x + p^{-1}(0) \in \check{\mathfrak{U}}_{\odot_p}$ in $\check{X}_p$ $\forall p \in \mathcal{P}$, then $x \in \check{\mathfrak{U}}_\odot$ in $\check{X}$.
(c) $\check{\sigma}_\odot(x) \subset \bigcup_{\mathcal{P}} \check{\sigma}_\odot(x + p^{-1}(0)) \subset \bigcup_{\mathcal{P}} \sigma_\odot(x + p^{-1}(0)) \subset \sigma_\odot(x)$ $\forall x \in X$.
(d) X has an identity iff every $\check{X}_p$ $(p \in \mathcal{P})$ has an identity.
(e) If X is complete, $\mathfrak{U}_\odot = \bigcup_{\mathcal{P}} \check{\mathfrak{U}}_{\odot_p}$ and $\sigma_\odot(x) = \bigcup_{\mathcal{P}} \sigma_\odot(x + p^{-1}(0))$ $\forall x \in X$.

25. Q-Algebras

Certain Banach algebra properties lost in general LMC algebras are recovered in Q-algebras, that is, TVAs in which U is open. Fusing Q-algebras with complete $LMCT_2$ algebras quickly produces a synergistic effect. It turns out that r_σ is a norm and $(X, r_\sigma)'$ is easily described. What also emerges is the uniqueness of a fully complete, barreled $LMCT_2$ Q-algebra topology on a complex semisimple algebra; and, identifying (X, r_σ) with a subalgebra of $(C(\mathcal{H}_c), \| \|_0)$, we characterize complete, complex $LMCT_2$ Q-algebras as B^*-algebras.

The definition of Q-algebra above can be restated.

THEOREM 25.1. A TVA is a Q-algebra iff either

(i) $\operatorname{Int} \mathfrak{U} \neq \varnothing$;
(ii) $\operatorname{Int} \bar{S}_{1,r_\sigma}(0) \neq \varnothing$.

Proof. A Q-algebra X clearly satisfies $\operatorname{Int} \mathfrak{U} = \mathfrak{U} \neq \varnothing$. Assuming (i), let $x_0 + v \subset \mathfrak{U}$ for some balanced nbd $v(0)$. Since $_{x_0^{-1}}m$ is a homeomorphism, $x_0^{-1}v$ is also a balanced nbd of $0 \in X$. If $x \in x_0^{-1}v$ and $|\lambda| \geq 1$, then $x + \lambda e = \lambda[(1/\lambda)x + e] \subset \lambda(x_0^{-1}v + e) \subset \mathfrak{U}$ and $\lambda \notin \sigma(x)$. Thus $r_\sigma(x) < 1$ and $x_0^{-1}v \subset S_{1,r_\sigma}(0)$ ensures that $0 \in \operatorname{Int} S_{1,r_\sigma}(0) \subset \operatorname{Int} \bar{S}_{1,r_\sigma}(0)$. Conversely, suppose $x_0 +$

$w \subset \bar{S}_{1,r_\sigma}(0)$ for some nbd $w(0)$. If $y \in \frac{1}{2}x_0 + \frac{1}{2}w$, then $r_\sigma(y) \leq \frac{1}{2}$ and $e + y \in \mathfrak{U}$ yields $(e + \frac{1}{2}x_0) + \frac{1}{2}w(0) \subset \mathfrak{U}$. Thus (i)$\Leftrightarrow$(ii). If $y_0 = e + \frac{1}{2}x_0$ and $x \in \mathfrak{U}$, then (since ${}_{xy_0^{-1}}\mathfrak{m}$ is a homeomorphism) $xy_0^{-1}w$ is a nbd of 0 such that $x + \frac{1}{2}xy_0^{-1} \subset \mathfrak{U}$. In other words, $x \in \text{Int}\,\mathfrak{U}$ and $\mathfrak{U} = \text{Int}\,\mathfrak{U}$ follows. ■

Remark. Note that $S_{1,r_\sigma}(0)$ is an X-nbd of 0. If r_σ is a seminorm, its topology is weaker than X's original topology.

COROLLARY 25.2. Q-algebras are preserved under open surjective homomorphisms.

Proof. If $h: X \to Y$ is an open surjective homomorphism and $\mathfrak{U}_X \subset X$ is open, $e_Y = h(e_X) \in h(\mathfrak{U}_X) \subset \mathfrak{U}_Y$ and $e_Y \in \text{Int}\,\mathfrak{U}_Y$. ■

THEOREM 25.3. If X is a Q-algebra, then $\sigma(x)$ is compact $\forall x \in X$ and $\bar{I} \in \mathfrak{J}$ $\forall I \in \mathfrak{J}$. Therefore $\mathfrak{M} = \mathfrak{M}_c$.

Proof. Suppose that $v(0) \subset S_{1,r_\sigma}(0)$ for some X-nbd $v(0)$. Then $v(0)$ is absorbent and for $x \in X$ there is a $\delta_x > 0$ such that $x\delta_x \in v(0)$. Consequently $r_\sigma(x) < 1/\delta_x < \infty$ and $\sigma(x)$ is bounded. To prove $\sigma(x)$ compact we need only verify that $\rho(x)$ is open, and this is easily done by replacing "$S_{\varepsilon,\|\|}(x - \lambda_0 e) \subset \mathfrak{U}$" with "$x - \lambda_0 e + w \subset \mathfrak{U}$ for some nbd $w(0)$" in Theorem 23.6. The assertion about ideals is clear [Theorem 24.3(ii)] since $I \cap \mathfrak{U} = \varnothing$ implies $e \notin \bar{I}$. ■

Theorem 24.20 now yields the following extension of Corollary 23.7.

COROLLARY 25.4. If X is a complex LMC Q-algebra, then $\mathfrak{K} = \mathfrak{K}_c$ and $\mathfrak{K} \to \mathfrak{M}$ is bijective.

A few illustrations will be useful before proceeding.

Example 25.5. (a) $(X, \mathfrak{G})$ is not a Q-algebra, since $0 \in X - \mathfrak{U}$ (also compare Theorem 24.3^+ with Theorem 25.3). Neither is the LMCT_2 algebra $(X, \sigma(X, \mathfrak{K}_0))$ in Example 24.16(d) by virtue of Corollary 25.4.

(b) Banach algebras are Q-algebra [Theorem 23.6(ii)] but the converse is false. The nonnormable algebra $\mathfrak{D}^\infty(I)$ of Example 24.14 is a Q-algebra, since $\chi_I + S_{1,p_0}(0) \subset \mathfrak{U}$. (If $f \in \chi_I + S_{1,p_0}(0)$, then $1 > p_0(f - \chi_I) = \sup_I |f(\omega) - 1|$ and $f \neq 0$ on I.)

(c) The LMCT_2 Frechet algebra in Problem 24F(d) is a Q-algebra, since $e + S_{1,p_0}(0) \subset \mathfrak{U}$. However, the Frechet algebra in Problem 24G(e) is not by virtue of Problem 24A(d).

(d) $C_\kappa(\Omega;\mathbb{C})=(C(\Omega;\mathbb{C}),\|\,\|_0)$ in Example 22.9 is a Q-algebra, whereas $C_\kappa(\Omega;\mathbb{C})$ in Example 24.28 is not, since every nbd $\chi_\Omega+u_{\varepsilon,K}(0)$ contains a continuous Urysohn function $F:\Omega\to I$ [with $F(K)=1$ and $F(\omega_0)=0$ for $\omega_0\notin K$] which does not belong to $\mathfrak{U}$. Verification also follows by Theorem 25.3.

Since $X-\{0\}$ is open in a T_2VA X, we have $\mathfrak{U}\subset X$ being open iff $\mathfrak{U}$ is open in $X-\{0\}$. In this vein the following (Arens [2]) illustrates how spectacularly a TVA can fail to be a Q-algebra.

Example 25.6. $X=\mathcal{H}_\kappa(\Omega)$ in Example 24.29 is a non Q-algebra having $\mathfrak{U}$ closed in $X-\{0\}$. Verification: If $f\in X-\mathfrak{U}$ for $f\not\equiv 0$, then $f(\omega_0)=0$ for some $\omega_0\in\Omega$ and there is a circle $C_{\omega_0}=S_\varepsilon(\omega_0)\cap\Omega$ on which f attains its minimum value $m>0$. (Otherwise, $f(\omega_n)=0$ for a sequence $\omega_n\in S_{1/n}(\omega_0)\cap\Omega$ and $\omega_n\to\omega_0$ implies that $f\equiv 0$ by the identity theorem for analytic functions.) Now $u_{C_{\omega_0},m}(0)$ is a $\mathfrak{T}_\kappa$-nbd of $0\in X$ such that $f+u_{C_{\omega_0},m}\subset X-\mathfrak{U}$. Indeed, every $g\in u_{C_{\omega_0},m}$ satisfies $|g(\omega)|<|f(\omega)|$ on C_{ω_0} and (by Rouche's theorem) f and $f+g$ have the same number of zeros inside C_{ω_0}. This establishes that $\mathfrak{U}\cup\{0\}$ is closed in X. We also have $0\in\bar{\mathfrak{U}}$, since $(\varepsilon/2)\chi_\Omega\in u_{K,\varepsilon}(0)\cap\mathfrak{U}\ \forall\varepsilon>0$ and compact set $K\subset\Omega$. Thus $\mathfrak{U}\cup\{0\}=\overline{\mathfrak{U}\cup\{0\}}=\bar{\mathfrak{U}}$ and $\mathfrak{U}=\bar{\mathfrak{U}}\cap[X-\{0\}]$ is closed in $X-\{0\}$. Finally, X is not a Q-algebra (Theorem 25.3), since $\sigma(f)=\Omega$ is noncompact for $f(\omega)=\omega$ on Ω.

Remark. The above also demonstrates that X is nonnormable.

Conjecturing that Theorem 17.10 applies to homomorphisms on subalgebras is very nice—and very wrong. For instance, the homomorphism $h_0: Y=\mathfrak{H}(\bar{S}_1)\to\mathbb{C}$, $f\rightsquigarrow f(0)$ in Problem 23E(c) has no homomorphic extension $h:X\to\mathbb{C}$ (otherwise taking $f(\omega)=\omega$ on $\bar{S}_1$ and $g(\omega)=1/\omega$ on $bdy\,\bar{S}_1$ yields $1=h_0(\chi_{\bar{S}_1})=h(\chi_{bdy\,\bar{S}_1})=h_0(f)h(g)=0$). For future reference we therefore include

THEOREM 25.7. Let X be a subalgebra of a complex LMC Q-algebra Y, both of which have the same identity. Then $h\in\mathcal{H}_X$ has an extension $h_0\in\mathcal{H}_Y$ iff $h^{-1}(0)\in M$ for some $M\in\mathfrak{M}_Y$.

Proof. Certainly $h^{-1}(0)\subset h_0^{-1}(0)\in\mathfrak{M}_Y$ if $h_0\in\mathcal{H}_Y$ extends $h\in\mathcal{H}_X$. If $h^{-1}(0)\subset M$ for some $M\in\mathfrak{M}_Y$, then $M=h_0^{-1}(0)$ for some $h_0\in\mathcal{H}_Y$. Furthermore, $k=h_0/X\in\mathcal{H}_X$ and $h^{-1}(0)\subset M\cap X=k^{-1}(0)\in\mathfrak{M}_X$ implies that $k^{-1}(0)=h^{-1}(0)$; that is, $k=h$ on X. ■

Remark. Suppose $h:I\to Y$ is an onto homomorphism, where Y is an algebra and $I\in\mathcal{J}(X)$ for an algebra X (not necessarily with identity). If $h(x_0)=e_Y$, then $h(xx_0)=h(x_0xx_0)\ \forall x\in X$ and $h_0(x)=h(xx_0)$ is a homomorphism extending h to X. Moreover, $h_0\in\mathcal{H}_c$ if $h\in\mathcal{H}_c$.

If X is a complete complex LMCT_2 algebra, then $r_\sigma(x)=\sup_{\mathcal{H}}|h(x)|$ (Theorem 24.25). If X is also a Q-algebra, then (Theorem 25.3) $r_\sigma(x)<\infty\ \forall x\in X$ so that r_σ is a multiplicative seminorm on X. The converse also holds when X is barreled.

THEOREM 25.8. For a complete complex LMCT_2 algebra X, properties (a)–(g) below satisfy

(i) when (g) prevails, r_σ is a norm iff (a)–(d) hold;
(ii) the implications (e)$\Rightarrow$(f)$\Rightarrow$(g) are reversible if X is barreled.

(a) r_σ is T_2;
(b) X is semisimple;
(c) $\mathcal{H}$ is injective;
(d) $\sigma(X,\mathcal{H})$ is T_2:
(e) X is a Q-algebra;
(f) $\mathcal{H}$ is $\sigma(\mathcal{H},X)$-compact;
(g) $r_\sigma(x)<\infty\ \forall x\in X$.

Proof. *If* $r_\sigma(x)<\infty\ \forall x\in X$, then r_σ is a seminorm and the proof of (i) is complete, since (a) $\Rightarrow$ (b) $\Rightarrow$ (c) $\Rightarrow$ (d) $\Rightarrow$ (a) [Corollary 25.4 and the fact that $\sigma(X,\mathcal{H})\subset r_\sigma$]. If (e) holds, $\{\bar{S}_{1,r_\sigma}(0)\}^\circ$ is $\sigma(X',X)$-compact (Theorem 18.15) and (f) follows from $\mathcal{H}\subset\mathcal{H}^{\circ\circ}=\{\bar{S}_{\cdot,r_\sigma}(0)\}^\circ$ by Corollary 25.4, Theorem 24.3(i). Clearly, (f)$\Rightarrow$(g), since every $\underset{h\rightsquigarrow h(x)}{x:\mathcal{H}\to\mathbb{C}}$ is $\sigma(\mathcal{H},X)$-continuous. Finally, the abs convex set $\bar{S}_{1,r_\sigma}(0)=\{x\in X:\sup_{\mathcal{H}_c}|h(x)|\le 1\}=\mathcal{H}_c^\circ$ is closed in X (Theorem 18.10). If X is barreled and (g) holds, then $\bar{S}_{1,r_\sigma}(0)$ is absorbent and $\text{Int}\,\bar{S}_{1,r_\sigma}(0)\neq\varnothing$ establishes (e) (Theorem 25.1). ∎

Remark. The barreled property may be weakened to require that closed abs m-convex absorbent sets be nbds. This, however, cannot be further relaxed (Problem 25A).

This is a good place to check out the dual $X'_{r_\sigma}=(X,r_\sigma)'$ of a complete LMCT_2 Q-algebra X. Since $|h(x)|\le r_\sigma(x)\ \forall x\in X$, we have (Corollary 13.25) $h\in X'_{r_\sigma}$ with $|||h|||\le 1\ \forall h\in\mathcal{H}$. Evidently, $|h(x)|/r_\sigma(x)=|h(x)|\le|||h|||$ for $r_\sigma(x)=1$, whereas $r_\sigma(x)=\alpha>0$ implies $r_\sigma[(1/\alpha)x]=1$ and

$(1/\alpha)|h(x)| = |h[(1/\alpha)x]| \leq |||h|||$. Thus $|h(x)| \leq |||h|||r_\sigma(x)\ \forall x \in X$ and, since $r_\sigma(e) = 1$, it follows that $1 = |||h|||\ \forall h \in \mathcal{H}$. All this establishes the first part of

THEOREM 25.9. For a complete complex LMCT$_2$ Q-algebra X, we have $\mathcal{H} \subset X'_{r_\sigma}$ with $|||h||| = 1\ \forall h \in \mathcal{H}$. If X is also semisimple, then $X'_{r_\sigma} = [\overline{\mathcal{H}_{bc}}^{\sigma(X'_{r_\sigma}, X)}]$.

Proof. Observe (Corollary 25.4, Theorem 25.8) that $\mathcal{H}^\circ = \bar{S}_{1,r_\sigma}(0)$, and so $\mathcal{H}^{\circ\circ} = \overline{\mathcal{H}_{bc}}^{\sigma(X'_{r_\sigma}, X)} = \{\bar{S}_{1,r_\sigma}(0)\}^\circ$ is $\sigma(X'_{r_\sigma}, X)$-compact in X'_{r_σ}. Therefore $Y = \mathcal{H}^{\circ\circ}$ is $\sigma(X'_{r_\sigma}, X)/[Y] = \sigma([Y], X)$-compact in $[Y] \subset X'_{r_\sigma}$. Since X is semisimple, $\langle X, [Y]\rangle$ is an injective dual pair and (Theorem 18.27) $r_\sigma = \mathfrak{T}_{\mathcal{S}=\{Y\}} \subset \tau(X, [Y])$ on X. Therefore $X'_{r_\sigma} \subset [Y]$ and $X'_{r_\sigma} = [\overline{\mathcal{H}_{bc}}^{\sigma(X'_{r_\sigma}, X)}]$ follows. ■

Assume that X is a complete complex LMCT$_2$ algebra and let $\mathcal{H}_c$ carry $\sigma(\mathcal{H}_c, X)$. If $\xi_x(h) = h(x)\ \forall h \in \mathcal{H}_c$, then $\xi_X = \{\xi_x : x \in X\}$ is a separating subalgebra of $C(\mathcal{H}_c; \mathbb{C})$. It is also clear that the algebra homomorphism $\psi : X \to \xi_X$ ($x \rightsquigarrow \xi_x$) is an isomorphism iff X is semisimple. To pursue this further let $\mathcal{S}_c = \mathcal{S}_e \cap \mathcal{H}_c$ denote the equicontinuous sets in $\mathcal{H}_c$ and recall (Theorem 15.12) that $\mathfrak{T}_{\mathcal{G}(\psi)}$ has $\{\psi(u) : u(0) \text{ is an } X\text{-nbd base}\}$ as a nbd base at $0 = \xi_0 \in \xi_X$.

THEOREM 25.10. For a complete complex LMCT$_2$ algebra X

(i) $\sigma(\xi_X, \mathcal{H}_c) \subset \mathfrak{T}_{\mathcal{S}_c} \subset \mathfrak{T}_{\mathcal{G}(\psi)}$ and $\mathfrak{T}_{\mathcal{S}_c} \subset \mathfrak{T}_\kappa$ on ξ_X;

(ii) $\mathcal{H}_c$ is finite iff dim ξ_X is finite;

(iii) ξ_X is dense in $C_\kappa(\mathcal{H}_c; \mathbb{C})$ if $\forall x \in X$ there is some $y \in X$ satisfying $\xi_y(h) = \overline{\xi_x(h)}\ \forall h \in \mathcal{H}_c$;†

(iv) $\psi : X \to (\xi_X, \mathfrak{T}_{\mathcal{S}_c} = \mathfrak{T}_\kappa)$ is continuous if X is barreled.

Proof. The inclusions $\sigma(\xi_X, \mathcal{H}_c) \subset \mathfrak{T}_{\mathcal{S}_c} \subset \mathfrak{T}_{\mathcal{G}(\psi)}$ hold, since every $h \in \mathcal{H}_c$ is $\mathfrak{T}_{\mathcal{S}_c}$-continuous [specifically, $h\{W_{\{h\},\varepsilon/2}(0)\} \subset S_\varepsilon(0)\ \forall h \in \mathcal{H}_c$ and $\varepsilon > 0$] and $(\varepsilon/2)\xi_u(0) \subset W_{u^\circ \cap \mathcal{H}_c, \varepsilon}(0)\ \forall X$-nbd $u(0)$ and $\varepsilon > 0$. One also has $\mathfrak{T}_{\mathcal{S}_c} \subset \mathfrak{T}_\kappa$ because every $E \in \mathcal{S}_c$ is $\sigma(\mathcal{H}_c, X)$-relatively compact by Theorems

† This is frequently described as saying ξ_X is closed under complex conjugation. We adopt the terminology that X is an $\hat{A}$-algebra; the reasoning is made clear shortly.

18.14 and 24.3. For (ii) identify $F \in C(\mathcal{H}_c, \mathbb{C})$ with $\{F(h)\} \in \prod_{\mathcal{H}_c} \mathbb{C}$ and view $C(\mathcal{H}_c; \mathbb{C})$ as a subalgebra of $\prod_{\mathcal{H}_c} \mathbb{C}$ with product topology $\mathfrak{T}_\pi$. Evidently $\prod_{\mathcal{H}_c} \mathbb{C}$, therefore ξ_X, is finite dim when $\mathcal{H}_c$ is finite. Conversely, let $\pi_{h_0}: \beta = \{\beta_n\} \to \beta_{h_0}$ $(h_0 \in \mathcal{H}_c)$ be the h_0th projection on $\prod_{\mathcal{H}_c} \mathbb{C}$ and observe that $\bigcap_{i=1}^n h_i^{-1}\{S_\varepsilon(0)\} = \psi^{-1}\left\{\bigcap_{i=1}^n \pi_{h_i}^{-1}(S_\varepsilon(0))\right\} = \psi^{-1}\left\{\psi\left(\bigcap_{i=1}^n h_i^{-1}\{S_\varepsilon(0)\}\right)\right\}$ $\forall \sigma(X, \mathcal{H})$-nbd $\bigcap_{i=1}^n h_i^{-1}\{S_\varepsilon(0)\}$. In other words, $\psi: (X, \sigma(X, \mathcal{H})) \to (\xi_X, \mathfrak{T}_\pi)$ is continuous and open and $\mu_\psi: (X/\psi^{-1}(0), \mathfrak{T}_{\mathcal{G}(\nu)}) \to (\xi_X, \mathfrak{T}_\pi)$ is a topological isomorphism. Since $\psi^{-1}(0) = \mathfrak{R}(X) = \overline{\{0\}}^{\sigma(X, \mathcal{H}_c)}$ (via Theorem 24.25), it follows (Example 15.15) that $\dim \xi_X = \dim(X/\psi^{-1}(0), \mathfrak{T}_{\mathcal{G}(\nu)})' = \dim(X, \sigma(X, \mathcal{H}_c))'$. In particular, $\mathcal{H}_c \subset (X, \sigma(X, \mathcal{H}_c))'$ is finite when $\dim \xi_X$ is finite. To prove (iii) let $f \in C(\mathcal{H}_c; \mathbb{C})$ and consider any $\sigma(\mathcal{H}_c; X)$-compact set $K \subset \mathcal{H}_c$. We can easily verify that $A_K = \{\xi_x/K : x \in X\}$ is a separating subalgebra of $C(K; \mathbb{C})$ which also contains the identity $\xi_e/K = \chi_K$ and is closed under complex conjugation. Since $\bar{A}_K^{\|\|_0} \subset C(K; \mathbb{C})$ also shares these properties, $\bar{A}_K^{\|\|_0} = C(K; \mathbb{C})$ by the Stone Weierstrass theorem (Appendix t.9). Therefore every $\varepsilon > 0$ yields some $x \in X$ satisfying $\|\xi_x/K - f/K\|_0 < \varepsilon$. In particular, $\xi_x \in \{f + W_{K,\varepsilon}(0)\} \cap \xi_X$ and $\bar{\xi}_X^{\mathfrak{T}_k} = C(\mathcal{H}_c; \mathbb{C})$, as asserted. If X in (iv) is barreled, every $\sigma(\mathcal{H}_c, X)$-compact set is equicontinuous (Theorem 24.3, Corollary 18.15) and $\mathfrak{T}_{\mathcal{S}} = \mathfrak{T}_\kappa$ on ξ_X. Moreover (Theorem 19.14), every $\mathfrak{T}_\kappa$-nbd $W_{K,\varepsilon}(0) \subset \xi_X$ has $K \subset T^\circ$ in X' for some barrel (i.e., nbd) $T(0) \subset X$ and the continuity of ψ immediately follows from $\psi^{-1}\{W_{K,\varepsilon}(0)\} = \{x \in X : \sup_K |h(x)| \le \varepsilon\} = \varepsilon K^\circ \supset \varepsilon T(0)$. ■

If X above is also a Q-algebra, $\mathcal{H}_c = \mathcal{H}$ is $\sigma(\mathcal{H}, X)$-compact and our focus can be considerably sharpened.

THEOREM 25.11. For a complete complex LMCT$_2$ Q-algebra X

(i) $\psi: (X, r_\sigma) \to (\xi_X, \|\|_0)$ is an isometry;

(ii) $\mathfrak{T}_\pi \subset \|\|_0$ on $C(\mathcal{H}; \mathbb{C})$ and $\bar{\xi}_X^{\|\|_0} \subset C(\mathcal{H}; \mathbb{C}) \subset \prod_{\mathcal{H}} \mathbb{C} = \bar{\xi}_X^{\mathfrak{T}_\pi}$;

(iii) $\xi_X = \prod_{\mathcal{H}} \mathbb{C}$ iff $\mathcal{H}$ is finite iff $C(\mathcal{H}; \mathbb{C}) = \prod_{\mathcal{H}} \mathbb{C}$.

Proof. Theorem 25.8(ii) ensures that $\|\xi_x\|_0 = \sup_{\mathcal{H}} |h(x)| = r_\sigma(x) < \infty$ $\forall x \in X$. The proof of (ii) is also simple. First, $\mathfrak{T}_\pi \subset \|\|_0$, since $S_{\varepsilon/2, \|\|_0}(0) \subset v(0) \cap C(\mathcal{H}; \mathbb{C})$ for every $\mathfrak{T}_\pi$-nbd $v(0) = \bigcap_{i=1}^n \pi_{h_i}^{-1}\{S_\varepsilon(0)\} \subset \prod_{\mathcal{H}} \mathbb{C}$. Next,

$\bar{\xi}_X^{\mathcal{T}_\pi}=\prod_{\mathcal{H}}\mathbb{C}$, is clear since every $\mathcal{T}_\pi$-nbd $v(\beta)=\beta+v(0)=\{\{\gamma_h\}\in\prod_{\mathcal{H}}\mathbb{C}:|\gamma_{h_i}-\beta_{h_i}|<\varepsilon\ \forall 1\le i\le n\}$ contains $\xi_x\in\xi_X$, where (Corollary 22.3) $x\in X$ satisfies $h_i(x)=\gamma_{h_i}$ $(1\le i\le n)$. The preceding remark is also used to prove (iii). If $\mathcal{H}$ is finite, so is $\dim\prod_{\mathcal{H}}\mathbb{C}$. Therefore $\|\ \|_0=\mathcal{T}_\pi$ and ξ_X equals $\bar{\xi}_X^{\|\ \|_0}=\bar{\xi}_X^{\mathcal{T}_\pi}=\prod_{\mathcal{H}}\mathbb{C}$ [Theorem 13.12, Corollary 13.13(ii)]. On the other hand, $\mathcal{H}$ is finite whenever $C(\mathcal{H};\mathbb{C})=\prod_{\mathcal{H}}\mathbb{C}$. Otherwise, $\{n\delta_{h,h_n}\}\subset\prod_{\mathcal{H}}\mathbb{C}$ is the image of some $f\in C(\mathcal{H};\mathbb{C})$, contradicting the compactness of $(\mathcal{H},\sigma(\mathcal{H},X))$. ■

For X above $\mu_\psi:(X/\psi^{-1}(0),\mathcal{T}_{\mathcal{I}(\nu)})\to(\xi_X,\mathcal{T}_\pi)$ is a topological isomorphism and $\psi^{-1}(0)=\overline{\{0\}}^{\sigma(X,\mathcal{H})}$. Therefore both mappings ν and $\psi=\mu_\psi\nu$ are closed (Definition 15.13⁺). In the same vein $\mathcal{R}(X)=\bigcap_{\varepsilon>0}S_{\varepsilon,r_\sigma}(0)=\overline{\{0\}}^{r_\sigma}$ ensures that the isometry $\psi:(X,r_\sigma)\to(\xi_X,\|\ \|_0)$ is closed. These remarks are used to prove

THEOREM 25.12. The properties below are equivalent for a complete complex LMCT_2 Q-algebra X:

(i) $\mathcal{H}$ is finite;
(ii) $\sigma(X,\mathcal{H})$ is complete;
(iii) $\sigma(X,\mathcal{H})$ is barreled;
(iv) $\sigma(X,\mathcal{H})$ is semi-Montel;
(v) r_σ is semi-Montel.

Proof. First, (i)⇔(ii) by Theorem 25.11, since $\xi_X=\prod_{\mathcal{H}}\mathbb{C}$ iff $(\xi_X,\mathcal{T}_\pi)$ is complete; that is, $(X/\psi^{-1}(0),\mathcal{T}_{\mathcal{I}(\nu)})$ is complete, which is equivalent to $\sigma(X,\mathcal{H})$-completeness (Definition 15.13⁺). Next, (i)⇒(iii), since $(\xi_X,\mathcal{T}_\pi)=\prod_{\mathcal{H}}\mathbb{C}$ is barreled (Theorem 19.21⁺). If $B(0)\subset X$ is a $\sigma(X,\mathcal{H})$-barrel, then $\psi(B)\subset\xi_X=\prod_{\mathcal{H}}\mathbb{C}$ is a barrel and contains some $\mathcal{T}_\pi$-nbd $\bigcap_{i=1}^{n}\pi_{h_i}^{-1}\{S_\varepsilon(0)\}$. Therefore $\bigcap_{i=1}^{n}h_i^{-1}\{S_\varepsilon(0)\}\subset B$ and $B(0)$ is a $\sigma(X,\mathcal{H})$-nbd. Begin (iii)⇒(i) by observing that $\bar{S}_{1,r_\sigma}(0)$ is a $\sigma(X,\mathcal{H})$-barrel. (For instance, $x\in\overline{\bar{S}_{1,r_\sigma}(0)}^{\sigma(X,\mathcal{H})}$ implies that there exists an element $x_{h,\varepsilon}\in\bar{S}_{1,r_\sigma}(0)\cap(x_0+h^{-1}\{S_\varepsilon(0)\})\ \forall h\in\mathcal{H}$ and $\varepsilon>0$. Thus $|h(x_0)|\le|h(x_{h,\varepsilon})|+\varepsilon\le r_\sigma(x_{h,\varepsilon})+\varepsilon\le 1+\varepsilon\ \forall\varepsilon>0$ and $r_\sigma(x_0)=\sup_{\mathcal{H}}|h(x_0)|\le 1$.) Accordingly, $\bar{S}_{1,r_\sigma}(0)$ contains some $\sigma(X,\mathcal{H})$-nbd $\bigcap_{i=1}^{n}h_i^{-1}\{S_\varepsilon(0)\}$ and this necessitates that $\mathcal{H}$ is finite.

Otherwise [Example 13.2(a), Theorem 22.2] there is an $h_{n+1} \in \mathcal{H} - \{h_1 \cdots h_n\}$ and $x_0 \in \bigcap_{i=1}^{n} h_i^{-1}\{S_\varepsilon(0)\}$ such that $r_\sigma(x_0) \geq |h_{n+1}(x_0)| = 2$. Moving on, (i)$\Leftrightarrow$(v) by Theorem 25.12(ii), since dim ξ_X is finite iff $(\xi_X, \| \|_0)$ is Montel (Theorems 13.12 and 13.14) which is equivalent to (v). Finally, (i)$\Rightarrow$(iv), since $\sigma(X, \mathcal{H}) = r_\sigma$ [and (v) applies] when $\mathcal{H}$ is finite. In the other direction (iv) implies $\bar{S}_{1,r_\sigma}(0)$ is $\sigma(X, \mathcal{H})$-compact and $\psi\{\bar{S}_{1,r_\sigma}(0)\} = E$ is closed in $\prod_{\mathcal{H}} \mathbb{C}$. Therefore $\xi_X = \bigcup_n nE$ and $\bar{\xi}_X^{\mathfrak{T}_\pi} = \bigcup_n \overline{nE}^{\mathfrak{T}_\pi} \subset \bigcup_n n\bar{E}^{\mathfrak{T}_\pi} = \xi_X$ yields $\xi_X = \bar{\xi}_X^{\mathfrak{T}_\pi} = \prod_{\mathcal{H}} \mathbb{C}$. ■

Remark. Note (Corollary 24.27) that each of the above is equivalent to dim X being finite if X is also semisimple.

When does $\mathfrak{T}_{\mathscr{S}_c} = \mathfrak{T}_{\mathfrak{g}(\psi)}$ on ξ_X? One partial answer is

THEOREM 25.13. Let X be a complete complex $\mathrm{LMCT}_2{}^*$-algebra. Then X is semisimple and $\mathfrak{T}_{\mathscr{S}_c} = \mathfrak{T}_{\mathfrak{g}(\psi)}$ iff the topology of X is determined by a family of multiplicative norms $\mathcal{P}$, each of which satisfies $p(xx^*) = \{\mathcal{P}(x)\}^2 \; \forall x \in X$.

Proof. Assume X's topology is determined by $\mathcal{P}$ and fix $p_0 \in \mathcal{P}$. Then $p_0(x) = p_0(x^*) \; \forall x \in X$ and $*' : (X_{p_0}, \| \|_{p_0}) \to (\check{X}_{p_0}, \| \|_{p_0})$, $x + p_0^{-1}(0) \rightsquigarrow x^* + p_0^{-1}(0)$ is an involution that can be extended by continuity to $(\check{X}_{p_0}, \| \|_0)$. Since $\|x + p_0^{-1}(0)\|_{p_0}^2 = \|(x + p_0^{-1}(0))(x + p_0^{-1}(0))^{*'}\|_{p_0} \; \forall x \in X$, such is also true on $(\check{X}_{p_0}, \| \|_{p_0})$. Thus (Theorem 23.14) the spectral norm $r_{\sigma_{p_0}}$ on $\check{X}_{p_0}$ satisfies $r_{\sigma_{p_0}}(x + p_0^{-1}(0)) = \|x + p_0^{-1}(0)\|_{p_0} = p_0(x) \; \forall x \in X$. From [Theorem 24.7(ii), Corollary 24.24] $\varnothing \neq \sigma(x) = \bigcup_{\mathcal{P}} \sigma(x + p^{-1}(0))$ we further obtain $r_\sigma(x) = \sup_{\mathcal{P}} r_{\sigma_p}(x + p^{-1}(0)) = \sup_{\mathcal{P}} p(x) \; \forall x \in X$. Theorems 24.25 and 14.18(i) now confirm that X is semisimple and it remains only to verify that $\mathfrak{T}(\mathcal{P}) \subset \mathfrak{T}_{\mathscr{S}_c}$ on $\xi_X = X$. For any X-subbase nbd $\bar{S}_{1,p}(0)$ we have (Theorem 24.22) $\mathcal{H}_{p_c} = \mathcal{H}_c \cap \bar{S}_{1,p}^{\circ}(0) \in \mathscr{S}_c$. Therefore $\bar{S}_{1,p}(0) = \{x + p^{-1}(0) \in X_p : r_{\sigma_p}(x + p^{-1}(0)) \leq 1\} = \{x + p^{-1}(0) \in X_p : \sup_{\mathcal{H}_{p_c}} h(x + p^{-1}(0)) \leq 1\} = \mathcal{H}_{p_c} = (\mathcal{H}_c \cap \bar{S}_{1,p}^{\circ})$ is a $\mathfrak{T}_{\mathscr{S}_c}$-nbd in X and $\mathfrak{T}(\mathcal{P}) \subset \mathfrak{T}_{\mathscr{S}_c}$ follows. For the second half of the theorem assume that X is semisimple and let $\mathfrak{T}_{\mathscr{S}_c} = \mathfrak{T}_{\mathfrak{g}(\psi)}$. Then $\{e^{\circ} = W_{E,S_1}(0) : E \in \mathscr{S}_c\}$ is a $\mathfrak{T}_{\mathfrak{g}(\psi)}$-nbd base of abs m-convex sets at $0 \in \xi_X$ and $\mathcal{P} = \{p_E(\xi_x) = \sup_E |\xi_x(h)| : E \in \mathscr{S}_c\}$ determines $\mathfrak{T}_{\mathfrak{g}(\psi)}$. Furthermore, $p_E(\xi_x \xi_{x^*}) = \{p_E(\xi_x)\}^2 \; \forall x \in X$ and $E \in \mathscr{S}_c$. To complete our proof identify the topology on X with $\mathfrak{T}_{\mathfrak{g}(\psi)}$ and define $p_E(x) = p_E(\xi_x) \; \forall x \in X$ and $E \in \mathscr{S}_c$. ■

The norm topologies r_σ and $\|\ \|$ coincide on a B^*-algebra $(X, \|\ \|)$ (Theorem 23.14), whereas r_σ is strictly weaker than the complete LMCT$_2$ Q-algebra topology on $\mathfrak{D}^\infty(I)$ (Example 24.24). Actually the situation is not so vague as it may appear. Both r_σ and the original complete LMCT$_2$ Q-algebra topology coincide on an $\hat{A}$-algebra X iff X is a B^*-algebra. We prove even more in

THEOREM 25.14. The following are equivalent for a complete complex LMCT$_2$ Q-algebra $(X, \mathfrak{T})$:

(i) X is a *-algebra and $\mathfrak{T} = \mathfrak{T}(\mathcal{P})$, where each $p \in \mathcal{P}$ satisfies $p(xx^*) = \{p(x)\}^2\ \forall x \in X$;

(ii) X is a *-algebra and $\mathfrak{T} = \mathfrak{T}(\mathcal{P})$, where each $(\check{X}_p, \|\ \|_p)$ is a B^*-algebra;

(iii) $r_\sigma = \mathfrak{T}$ and X is an $\hat{A}$-algebra;

(iv) $(X, \mathfrak{T}) = (C(\mathcal{H}; \mathbb{C}), \|\ \|_0)$;

(v) $(X, \mathfrak{T})$ is a B^*-algebra.

Proof. Evidently (iv)$\Rightarrow$(v)$\Rightarrow$(i). Furthermore (Theorem 25.13), (i)$\Rightarrow$(ii), which implies that X is semisimple and $\mathfrak{T}_{\mathcal{S}_c} = \mathfrak{T}_{\mathcal{I}(\psi)}$. Therefore $\|\ \|_0 = \mathfrak{T}_{\mathcal{I}(\omega)}$ and $r_\sigma = \mathfrak{T}$. Also $x_0 = x_0^*$ in X implies $x_0 + p^{-1}(0) = (x_0 + p^{-1}(0))^{*\prime}$ in $\check{X}_p\ \forall p \in \mathcal{P}$ and $\sigma(x_0) = \bigcup_{\mathcal{P}} \sigma(x_0 + p^{-1}(0)) \subset \mathbb{R}$ (Corollary 24.24, Theorem 23.14). Accordingly, $h(x^*) = \overline{h(x)}\ \forall x \in X$ and $h \in \mathcal{H}$ shows that X is an $\hat{A}$-algebra. Finally, (iii)$\Rightarrow$(iv), since (iii) implies that $(\xi_X, \|\ \|_0)$ is complete and dense in $C_\kappa(\mathcal{H}_c; \mathbb{C}) = (C(\mathcal{H}; \mathbb{C}), \|\ \|_0)$ and so $(X, \mathfrak{T}) = (X, r_\sigma)$ is congruent to $(\xi_X, \|\ \|_0) = (C(\mathcal{H}; \mathbb{C}), \|\ \|_0)$. ■

Of considerable importance in Banach algebra theory is the fact that complete norms arc equivalent when X is semisimple (Theorem 23.9). This topological uniqueness can be generalized once we prove

THEOREM 25.15. Let $h: X \to Y$ be a homomorphism between complex LMCT$_2$ algebras X and Y. if X is a complete barreled Q-algebra and Y is a fully complete semisimple algebra, then $h \in L(X, Y)$.

Proof. First, $\sigma(X, \mathcal{H}) \subset r_\sigma \subset \mathfrak{T}$ (the original topology on X) and h being $\sigma(X, \mathcal{H}_X) - \sigma(Y, \mathcal{H}_Y)$ continuous [Theorem 15.1(iii)] ensures that h is $\mathfrak{T} - \sigma(Y, \mathcal{H}_Y)$ continuous. Since Y is semisimple, $\sigma(Y, \mathcal{H}_Y)$ is T_2 and (Theorem 10.3) $g_r(h)$ is closed in $X \times (Y, \sigma(Y, \mathcal{H}_Y))$. Therefore $g_r(h)$ is closed in $X \times Y$ and Corollary 21.9 (cg) applies. ■

Taking $h = \mathbb{1} \in L(X, X)$ above yields

COROLLARY 25.16. There exists at most one fully complete barreled $LMCT_2$ Q-algebra topology on a complex, semisimple algebra.

PROBLEMS

25A. Counterexamples to Theorem 25.8

(Michael [62]). Let $X = C_\kappa(\Omega_{\aleph_1}; \mathbb{C})$, where $\Omega_{\aleph_1}$ is the space of ordinals smaller than $\aleph_1$ {the first uncountable ordinal) with the order topology (e.g., cf Dujundji [15]}. Assume further that $Y = C_{\kappa_0}(I; \mathbb{C})$, where $\mathfrak{T}_{\kappa_0}$ is the polar topology determined by the compact countable subsets of $I = [0, 1]$. Prove

(a) X and Y are complete $LMCT_2$ algebras;
(b) X satisfies (g) but not (e), (f) in Theorem 25.8.
(c) Y satifies (f) and (g) but not (e).

25B. Finite Dim B^*-Algebras

The following are equivalent for a complex B^*-algebra $(X, \| \|)$: (i) Dim $X < \infty$; (ii) X is $\sigma(X, X')$-complete; (iii) X is reflexive. [*Hint*: If (Theorem 25.14) $(X, \| \|) = (C(\mathcal{H}; \mathbb{C}), \| \|_0)$ is reflexive, so is $(X', \|\| \, \|\|) = (\mathcal{C}\mathcal{a}_{\mathfrak{R}}(\mathcal{H}, \Sigma; \mathbb{C}), \| \|_v)$ by Theorem 20.4(i) and Appendix m. Accordingly, dim $\mathcal{C}\mathcal{a}_{\mathfrak{R}}(\mathcal{H}, \Sigma; \mathbb{C}) < \infty$ (Corollary 16.4, Theorem 23.14, and Corollary 24.7) and dim $X = \aleph(\mathcal{H}) < \infty$. Also, (i)$\Rightarrow$(ii)$\Rightarrow$(iii) via Theorem 13.12, Example 20.5(c).]

25C. Principal Extensions

Assume throughout that X is a complete complex $LMCT_2$ algebra. The principal extension of $\Omega \subset \mathbb{C}$ is defined as $\mathcal{E}(\Omega, X) = \{x \in X : \sigma(x) \subset \Omega\}$. [This terminology is appropriate, since $\Omega\, e \subset \mathcal{E}(\Omega, X)$.]

(a) Show that $\mathcal{E}(\Omega_1, X) \cap \mathcal{E}(\Omega_2, X) \subset \mathcal{E}(\Omega_1 \cup \Omega_2, X)$ and $\mathcal{E}(\Omega_1 \cap \Omega_2, X) = \mathcal{E}(\Omega_1, X) \cap \mathcal{E}(\Omega_2, X)\ \forall \Omega_1, \Omega_2 \subset \mathbb{C}$.

(b) If Ω is convex, so is $\mathcal{E}(\Omega, X)$. This is no longer true without commutativity. Show also that $\mathcal{E}(\Omega, X)$ need not be bounded when Ω is bounded. [*Hint*: First use Theorem 24.25. Next, take $X = M_2(\mathbb{C})$ with $\Omega = \{0\}$. Then $\frac{1}{2}\begin{pmatrix} 0 & 0 \\ 2 & 0 \end{pmatrix} + \frac{1}{2}\begin{pmatrix} 0 & 2 \\ 0 & 0 \end{pmatrix} \notin \mathcal{E}(\{0\}, \mathrm{X})$ and $n\begin{pmatrix} 0 & 2 \\ 0 & 0 \end{pmatrix} \in \mathcal{E}(\{0\}, X)\ \forall n \in \mathbb{N}$.]

(c) $\mathfrak{U} = \mathcal{E}(\mathbb{C} - \{0\}, X)$. In fact, $\mathfrak{U} = \mathcal{E}(\Omega, X)$ if Ω is open and $\mathcal{E}(\Omega, X)$ is a multiplicative group. [*Hint*: $\mathcal{E}(\Omega, X) \subset \mathfrak{U}$ implies that $0 \notin \Omega$ and $1 \in \Omega$. Therefore $\Omega = \mathbb{C} - \{0\}$.]

(d) If X is a barreled Q-algebra and Ω is open, then $\mathfrak{S}(\Omega, X)$ is open. The Q-algebra property cannot be omitted from the preceding assertion. [*Hint* (Beckenstein et al. [7]): For $x_0 \in \mathfrak{S}(\Omega, X)$, use $\sigma(x_0)$-compactness to get $d(\sigma(x_0), bdy\ \Omega) = \varepsilon > 0$ and $S_{\varepsilon/2, r_\sigma}(x_0) \subset \mathfrak{S}(\Omega, X)$. For the second assertion consider $X = C_\kappa(\mathbb{C}; \mathbb{C})$ with $\Omega = S_1(0) \subset \mathbb{C}$. Here $0 \in \mathfrak{S}(\Omega, X) - \text{Int}\,\mathfrak{S}(\Omega, X)$. (Fix $\omega_0 \notin \Omega$ and $\forall n \in \mathbb{N}$, define $f_n(\omega) = 0$ for $|\omega| \leq n$, and $f_n(\omega) = (|\omega| - n)\omega_0$ otherwise. Then each $f_n \in S_{\alpha, p_n}(0) - \mathfrak{S}(\Omega, X)\ \forall \alpha > 0$.)]

Remark. For further developments see Ackermans [1].

25D. Spectral Continuity

A complex (not necessarily commutative) TVA X has spectral continuity if $\forall \varepsilon > 0$, there is an nbd $u_\varepsilon(0)$ such that $x \in x_0 + u_\varepsilon(0)$ implies $\sigma(x) \subset \sigma(x_0) + S_\varepsilon(0)$ and $\sigma(x_0) \subset \sigma(x) + S_\varepsilon(0)\ \forall x_0 \in X$.

(a) Suppose $(X, \|\ \|)$ is Banach. For each $x_0 \in X$ and $\varepsilon > 0$ there is a $\delta > 0$ such that $\|x - x_0\| < \delta$ implies $\sigma(x) \subset \sigma(x_0) + S_\varepsilon(0)$. [*Hint*: Assume that there exist sequences $\{x_n\} \in X$ satisfying $\|x_n - x_0\| < 1/n$ and $\{\lambda_n\} \in \mathbb{C}$ satisfying $\lambda_n \in \sigma(x_n) - \{\sigma(x_0) + S_\varepsilon(0)\}\ \forall n \in \mathbb{N}$. Then $\{\lambda_n\}$ has a subsequence $\lambda_{nk} \to \lambda_0 \notin \sigma(x_0)$. Since $\mathfrak{U}$ is open and $\lambda_{nk} - \lambda_{nk}e \to x_0 - \lambda_0 e$, some $\lambda_{nk} - \lambda_{nk}e \in \mathfrak{U}$.]

(b) Use (a) to prove the following equivalent:

(i) $(X, \|\ \|)$ has spectral continuity;

(ii) $(X, \|\ \|) \underset{x \rightsquigarrow \sigma(x)}{\to} (\mathcal{K}(\mathbb{C}), \rho)$ is continuous, where $\rho(E, F) = \max\{\sup_E |e - F|, \sup_F |E - f|\}$ on the compact sets $\mathcal{K}(\mathbb{C})$ of $\mathbb{C}$;

(iii) $\mathfrak{S}(\Omega, X)$ is closed for every closed set $\Omega \subset \mathbb{C}$.

(c) $(X, \|\ \|)$ has spectral continuity if X is commutative. [*Hint*: If $(X, \|\ \|)$ does not have spectral continuity, there is a sequence $x_n \to x_0$ with $\lambda_n \in \sigma(x_0) - \{\sigma(x_n + S_\varepsilon(0)\}\ \forall n \in \mathbb{N}$ and a sequence $\lambda_{nk} \to \lambda_0 \in \sigma(x_0)$. Pick $N = N(\varepsilon/2)$ such that $|\lambda - \lambda_0| > \varepsilon/2\ \forall \lambda \in \bigcup_{nk \geq N} \sigma(x_{nk})$. Then $\lambda_0 = 0$ implies $\{r_\sigma(x_{nk}^{-1}) : k \in \mathbb{N}\}$ is bounded and $x_0 \in \mathfrak{U}$ (Corollary 23.8). Show that $\lambda_0 \neq 0$ also leads to a contradiction.]

(d) A complex barreled LMCT_2 Q-algebra X has spectral continuity if it is commutative. [*Hint*: (Beckenstein et al. [7]): If $\check{\xi}_X$ is the completion of ξ_X in $Y = (C(\mathcal{K}; \mathbb{C}), \|\ \|_0)$, then $\sigma_Y(\xi_x) \subset \sigma_{\check{\xi}_X}(\xi_x) \subset \sigma_X(x)$ and $\sigma_X(x) \subset \sigma_Y(\xi_x)$ (Example 22.9). For any $\xi_{x0} \in \xi_X$ and $\varepsilon > 0$ there is a $\check{\xi}_X$-nbd $\check{u}_\varepsilon(0)$ such that $\sigma_{\check{\xi}_X}(f) \subset \sigma_{\check{\xi}_X}(\xi_{x0}) + S_\varepsilon(0)$ and $\sigma_{\check{\xi}_X}(\xi_{x0}) \subset \sigma_{\check{\xi}_X}(f) + S_\varepsilon(0)\ \forall f \in \xi_{x0} + \check{u}_\varepsilon(0)$. Since X is barreled, ψ is continuous and $\psi^{-1}(\check{u}_\varepsilon) = u_\varepsilon(0)$ is an X-nbd satisfying $\sigma(x) \subset \sigma(x_0) + S_\varepsilon(0)$ and $\sigma(x_0) \subset \sigma(x) + S_\varepsilon(0)\ \forall x \in x_0 + u_\varepsilon(0)$.]

(e) The complex Q-algebra property in (d) cannot be omitted. [*Hint*: $C_\kappa(\mathbb{R}; \mathbb{R})$ has no spectral continuity.]

25E. Projective Limits and Invertibility

Let $\mathfrak{F}=\{\mathfrak{m}_{x_\alpha}: \underset{x \rightsquigarrow xx_\alpha}{X \to X}: \alpha \in \mathcal{A}\}$ for a TVA $(X, \mathfrak{T})$ and denote $\mathfrak{T}_{\mathcal{P}\{\mathfrak{F}\}}$ relative to $(X, \mathfrak{T})$ $\{(X, r_\sigma)\}$ by $O_{S_\mathfrak{T}}$ $\{O_{S_{r_\sigma}}\}$.

(a) $O_{S_\mathfrak{T}}=\mathfrak{T}$ if $[x_\alpha: \alpha \in \mathcal{A}] \cap \mathfrak{U} \neq \varnothing$.

(b) If $(X, \mathfrak{T})$ is a complete complex LCMT$_2$ Q-algebra, then $O_{S_{r_\sigma}} \subset O_{S_\mathfrak{T}}$ and $O_{S_{r_\sigma}} \subset r_\sigma$.

(c) Suppose $(X, \mathfrak{T})$ in (b) is also barreled. If $\{x_\alpha: \alpha \in \mathcal{A}\} \not\subset h^{-1}(0)\ \forall h \in \mathcal{K}$, then $O_{S_{r_\sigma}}=r_\sigma$. [*Hint*: $\mathcal{K}$ is covered by a finite collection $\{O_{\alpha 1, \varepsilon 1}, \ldots, O_{\alpha n, \varepsilon n}\}$ of $\sigma(\mathcal{K}, X)$-open sets $O_{\alpha,\varepsilon}=\{h \in \mathcal{K}: |h(x_\alpha)|>\varepsilon\}$. A net $x_\varphi \subset X$ satisfying $x_\varphi \xrightarrow{O_{S_{r_\sigma}}} 0$ also satisfies $r_\sigma(x_{\alpha_k} x_\varphi) \to 0$ for each fixed $k=1, 2, \ldots, n$. Since $|h(x_\varphi)| \leq (1/\varepsilon_k)|h(x_{\alpha_k} x_\varphi)| \leq (1/\varepsilon_k) r_\sigma(x_{\alpha_k} x_\varphi)$ and $x_\varphi \to 0$ uniformly on each $O_{\alpha k, \varepsilon k}$, it follows that $x_\varphi \to 0$ uniformly on $\mathcal{K}$ and $r_\sigma(x_\varphi)=\sup_{\mathcal{K}} |h(x)| \to 0$.]

(d) The converse in (c) is also true if X is semisimple and dense in $(C(\mathcal{K}; \mathbb{C}), \| \|_0)$. [*Hint*: If $\{x_\alpha: \alpha \in \mathcal{A}\} \subset h_0^{-1}(0)$ for $h_0 \in \mathcal{K}$, no $O_{S_{r_\sigma}}$-nbd $w(0)=\bigcap_{i=1}^{n} \mathfrak{m}_{x_{\alpha_i}}^{-1}\{S_{\delta, r_\sigma}(0)\}$ is contained in any $S_{\varepsilon, r_\sigma}(0)$. Fix $\gamma<\delta/4\varepsilon$, take $\lambda=1+\sum_{k=1}^{n} \sup_{\mathcal{K}} |h(x_{\alpha_k})|$ and let $v(h_0)=h_0+\bigcap_{i=1}^{n} \xi_{x_{\alpha_i}}^{-1}\{S_\gamma(0)\}$. Then there is a Urysohn function $F: \mathcal{K} \to [0, \delta/2\gamma]$ with $F\{\mathcal{K}-v(h_0)\}=0$ and $F\{v(h_0)\}=\delta/2\gamma$, and an $x_0 \in X$ satisfying $\|\xi_{x_0}-F\|_0<\delta/[4(\gamma+\lambda)]$ also satisfies $x_0 \in W(0)-S_{\varepsilon, r_\sigma}(0)$.]

(e) *Wanted*! Confirmation that $[x_\alpha: x \in \mathcal{A}] \cap \mathfrak{U}=\varnothing$ and $O_{S_\mathfrak{T}}=\mathfrak{T}$ for $X=C_\kappa(\bar{S}_1(0); \mathbb{C})$. [*Hint*: Call $F(x, y)=U(x, y)+iV(x, y)$ an extension of $f(x, y)=u(x, y)+iv(x, y)$ if U, V extend u, v. Subdivide $\bar{S}_1(0)$ into five regions by $\Gamma_k=\{z:|z|=k/5\}$ and begin with $f_1(z)=1$ and $g_1(z)=5z$ on $\bar{S}_{1/5}(0)$. Continue extending f_k and g_k to $\bar{S}_{(k+1)/5}(0)$, keeping one extension fixed while the other varies (to prevent f_k and g_k from having simultaneous zeros), and obtain $F, G \in X$ satisfying $F^{-1}(0) \cap G^{-1}(0)=\varnothing$ and $\{F(z)/G(z): z \in \bar{S}_1(0) \text{ and } G(z) \neq 0\}=\mathbb{C}$.]

Remark. See Harris [39] for a more detailed proof of (e) and a fuller development of the notions considered above.

25F. Weak and Strong Homomorphism Integrals

Let $F \in \mathfrak{F}(\Omega; X)$, where Ω is a locally compact T_2 space and X is a complete complex LMCT$_2$ Q-algebra. If $\mu \in \mathcal{CA}(\Omega, \Sigma; X)$ and $h \in \mathcal{K}$, then $hF \in \mathfrak{F}(\Omega; X)$ and $\mu_h=h\mu \in \mathcal{CA}(\Omega, \Sigma; \mathbb{C})$. Denote the ordinary Lebesgue integral of hF with respect to μ_h by $\int hF\, d\mu_h$ when it exists.

A function $F \in \mathfrak{F}(\Omega; X)$ is wh, μ-integrable {sh μ-integrable} if $\oint F\, d\mu: \underset{h \rightsquigarrow \int hF d\mu_h}{(\mathcal{K}, \sigma(\mathcal{K}, X)) \longrightarrow \mathbb{C}}$ is continuous {and $\oint F\, d_\mu=\xi_{x_F}$ for some $x_F \in X$}.

(a) [In contrast to Problem 17H(c)] $\oint F\, d\mu$ may be discontinuous on $\mathcal{K}$. To see this let Σ be the Borel subsets of $\Omega=[0, 1]$ and define $F: \underset{\lambda \rightsquigarrow F_\lambda}{\Omega \to (X=C(\Omega; \mathbb{C}), \| \|_0)}$, where $F_0 = 0$ and $F_\lambda(t) = \{(\lambda^{-2} - \lambda^{-3/2})t + \lambda^{-1/2}\}\chi_{[0,\lambda)}(t) + t^{-1}\chi_{(\lambda,1]}(t)$ for $\lambda \neq 0$. Then $\oint F\,d\mu$ is defined on $\mathcal{H} = \{h_\omega : \omega \in \Omega\}$ but discontinuous at h_0. [*Hint*: $(h_0F)(\lambda) = \lambda\chi_{\{0\}}(\lambda) + \lambda^{-1/2}\chi_{(0,1]}(\lambda)$ and $(h_tF)(\lambda) = t^{-1}\chi_{[0,t]}(\lambda) + \{(\lambda^{-2} - \lambda^{-3/2})t + \lambda^{-1/2}\}\chi_{(t,1]}(\lambda)$ yield $4 = \lim_{t\to 0} \int h_tF\,d\mu_{h_t} \neq \int h_0 t\,d\mu_{h_0} = 2$. It remains to note that $h_t \to h_0$, since X is semisimple and $x(t) = \xi_x(h_t) \to \xi_x(h_0) = 0\ \forall x \in X$.]

Let $L_{\text{wh}}(\mu)$ and $L_{\text{sh}}(\mu)$ denote the respective collections of weekly μ-integrable and strongly μ-integrable functions for $\mu \in \mathcal{CA}(\Omega, \Sigma; X)$. Prove that

(b) $L_{\text{wh}}(\mu)$ and $L_{\text{sh}}(\mu)$ are vector spaces.

(c) If $x \in X$ and $F \in L_{\text{wh}}(\mu)$ $\{F \in L_{\text{sh}}(\mu)\}$, then $xF \in L_{\text{wh}}(\mu)$ $\{xF \in L_{\text{sh}}(\mu)\}$ and $(\oint xF\,d\mu)(h) = h(x)\int hF\,d\mu_h\ \forall h \in \mathcal{H}$.

(d) The "integrable simple functions" $\sum_{i=1}^{n} \alpha_i x_i \chi_{E_i} \in L_{\text{sh}}(\mu)\ \forall x_i \in X$ and $E_i \in \Sigma$ $(i = 1, 2, \ldots, n)$. [*Hint*: Use (b), (c) and the fact that $\oint \chi_E\,d\mu = \xi_{\mu(E)}\ \forall E \in \Sigma$.]

(e) Are $L_{\text{wh}}(\mu)$ and $L_{\text{sh}}(\mu)$ algebras?

(f) $L_{\text{wh}}(\mu) = L_{\text{sh}}(\mu)$ if X is semisimple. [*Hint*: Use $X = (\mathcal{H}, \sigma(\mathcal{H}, X))'$.]

Remark. See Suffel [96] for a further development of these notions and their relation to vector-valued Lebesgue integration.

References for Further Study

On LMC algebras and Q-algebras: Michael [62].
On topological algebras: Narici [72].

26. Complete, Complex, LMCT$_2$ Q-Algebras

Unless specified otherwise, X is as above (as usual, commutative and with an identity e).† Virtually everything that follows is based on the fact (Theorem 24.20, Corollary 25.4) that X/M is isomorphic to $\mathbb{C}\ \forall M \in \mathfrak{M}_c = \mathfrak{M}$, and $\mathcal{H} = \mathcal{H}_c \underset{h \rightsquigarrow h^{-1}(0)}{\longrightarrow} \mathfrak{M}$ is a bijection with $h_M : X \underset{x \rightsquigarrow x+M = x(M)(e+M) \rightsquigarrow x(M)}{\longrightarrow X/M \longrightarrow} \mathbb{C}$, the preimage of $M \in \mathfrak{M}$.

Fixing $x \in X$ and varying $M \in \mathfrak{M}$ produces a complex-valued function $\hat{x} : \underset{M \rightsquigarrow x(M)}{\mathfrak{M} \longrightarrow \mathbb{C}}$ called the Gelfand transform of x. Since $\hat{x}(M) = 0$ iff $x \in M$, we have $x \in \mathfrak{U}$ iff $\hat{x}(M) \neq 0\ \forall M \in \mathfrak{M}$. Moreover, $h_M(x) = \hat{x}(M)$ yields $\sigma(x) = \{\hat{x}(M) : M \in \mathfrak{M}\}\ \forall x \in X$. Under pointwise operations $\hat{X} = \{\hat{x} : x \in X\}$ is a

† The reader less interested in such generalizations may consider X to be a Banach algebra with $\|e\| = 1$.

separating subalgebra with identity $\hat{e}=\chi_{\mathfrak{M}}$ of $\mathfrak{F}(\mathfrak{M};\mathbb{C})$. We call $\underset{x\rightsquigarrow\hat{x}}{\phi_g: X\to\hat{X}}$ the Gelfand mapping of X. Since $\mathfrak{R}(X)=\phi_g^{-1}(0)$, the algebraic homomorphism ϕ_g is an isomorphism iff X is semisimple. We also need

Definition 26.1. The weakest topology $\mathfrak{T}_g$ on $\mathfrak{M}$ under which every $\hat{x}\in\hat{X}$ is continuous is termed the Gelfand topology. Thus $\mathfrak{T}_g$ is the projective limit topology $\mathfrak{T}_{\mathcal{P}\{\hat{X}\}}$ having nbd base sets $v(M_0;\hat{x}_1,\ldots,\hat{x}_n;\varepsilon)=\{M\in\mathfrak{M}:|\hat{x}_i(M)-\hat{x}_i(M_0)|<\varepsilon\ \forall 1\le i\le n\}$.

It is a simple matter to compare ξ_X on $\mathcal{H}$ with $\hat{X}$ on $\mathfrak{M}$; they are the same algebraically and topologically.

Theorem 26.2. $(\mathcal{H},\sigma(\mathcal{H},X))$ and $(\mathfrak{M},\mathfrak{T}_g)$ are topological homeomorphs. Moreover, $\mathfrak{T}_g$ is the only compact T_2 topology on $\mathfrak{M}$ under which every $\hat{x}\in\hat{X}$ is continuous.

Proof. The bijection $(\mathcal{H},\sigma(\mathcal{H},X))\to(\mathfrak{M},\mathfrak{T}_g)$ is also bicontinuous by Theorem 15.1(ii) and the defined topologies. Accordingly, $(\mathfrak{M},\mathfrak{T}_g)$ is compact, T_2 (Theorem 25.8(ii)). If $(\mathfrak{M},\mathfrak{T})$ is compact, T_2 and every $\hat{x}\in\hat{X}$ is $\mathfrak{T}$-continuous, $\mathfrak{T}_g\subset\mathfrak{T}$ and $\mathbb{1}:(\mathfrak{M},\mathfrak{T})\to(\mathfrak{M},\mathfrak{T}_g)$ is closed. ■

Remark. Note that $\phi_g:(X,r_\sigma)\to(\hat{X},\|\,\|_0)$ is an isometry (congruence iff X is semisimple), since $r_\sigma(x)=\sup_{\mathfrak{M}}|\hat{x}(M)|=\|\hat{x}\|_0\ \forall x\in X$.

An especially nice extension of Example 22.9 is

Example 26.3. $(\mathfrak{M},\mathfrak{T}_g)$ is the Stone Cech compactification $\beta(\Omega)$ of a $T_{3\frac{1}{2}}$ space. Verification: Let $X=CB(\Omega)$ and (using $\hat{f}(M_\omega)=h_\omega(f)=f(\omega)\ \forall f\in X$ and $\omega\in\Omega$) observe that the bijection $\underset{\omega\rightsquigarrow M_\omega}{\eta:\Omega\to\mathfrak{M}_\Omega}=\{M_\omega:\omega\in\Omega\}$ is bicontinuous. Specifically, we see that η is continuous, since $\eta^{-1}\{v(M_{\omega_0}:\hat{f}_1,\ldots,\hat{f}_n;\varepsilon)\cap\mathfrak{M}_\Omega\}=\{\omega\in\Omega:|f_i(\omega)-f_i(\omega_0)|<\varepsilon\ \ \forall 1\le i\le n\}$ is open in $\Omega\ \forall\mathfrak{T}_g$-nbd $v(M_{\omega_0};\hat{f}_1,\ldots,\hat{f}_n;\varepsilon)$, and η is open since for any open $u\subset\Omega$ and $\omega_0\in u$, one has $f(\omega_0)=1$ and $f(\Omega-u)=0$ for some $f\in X$. Therefore, $\eta(\omega_0)\in v(M_{\omega_0};\hat{f};\frac{1}{2})\subset\eta(u)$ and $\eta(\omega_0)$ is an interior point of $\eta(u)$. Next, $\mathfrak{M}_\Omega$ is dense in $(\mathfrak{M},\mathfrak{T}_g)$. If not, $v(M_0)\cap\mathfrak{M}_\Omega=\varnothing$ for some $\mathfrak{T}_g$-nbd $v(M_0;\hat{f}_1,\ldots,\hat{f}_n;\varepsilon)$ of $M_0\in\mathfrak{M}-\mathfrak{M}_\Omega$. Let $g_i=f_i-\hat{f}_i(M_0)\chi_\Omega$ and define $f=\sum_{i=1}^n g_i\bar{g}_i\in X$. For every $\omega\in\Omega$ at least one g_i satisfies $|g_i(\omega)|=|\hat{g}_i(M_\omega)|\ge\varepsilon$. Therefore $F\ge n\varepsilon^2>0$ on Ω and $1/F\in X$. This means that $(1/F)(M)=1/\hat{F}(M)$ and $\hat{F}(M)\ne 0\ \forall M\in\mathfrak{M}$, an obvious contradiction to $\hat{F}(M_0)=0$. All this shows $(\eta; (\mathfrak{M}, \mathfrak{T}_g))$ to be a T_2 compactification of Ω. We see also that every $f \in CB(\Omega)$ has $\hat{f}$ as its unique continuous extension satisfying $f = \hat{f}\eta$. This means that $C(\mathfrak{M}; \mathbb{C}) = C(\beta(\Omega))$ are isomorphs and $(\mathfrak{M}, \mathfrak{T}_g) = \beta(\Omega)$ by Example 22.9.

$\hat{A}$ and A^*-Algebras†

Recast in this new terminology, we again ask how close X comes to filling out $C(\mathfrak{M}; \mathbb{C})$. Some additional answers can be given after introducing

DEFINITION 26.4. A complete complex LMCT$_2$ Q-algebra X is called an $\hat{A}$-algebra if $\forall x \in X$ there is a $y \in X$ such that $\hat{y} = \bar{\hat{x}}$ (i.e., $y(M) = \overline{x(M)}\ \forall M \in \mathfrak{M}$). We call a $*$-algebra X an A^*-algebra if $\widehat{x^*} = \bar{\hat{x}}\ \forall x \in X$.

A subalgebra of X which is also $\hat{A}$ $\{A^*$ relative to X's involution$\}$ is termed a sub $\hat{A}$ $\{A^*\}$ algebra of X. This is not to be confused with an $\hat{A}$ $\{A^*\}$-subalgebra of X (independent of whether X is an $\hat{A}$-algebra).

Remark. The definitions above particularize the more general situation in which X has no identity, in which case $\mathfrak{M}$ is replaced by $\mathfrak{M}_{\mathfrak{R}}$ (cf Problems 26A, B).

By way of illustration, consider

Example 26.5. $\mathfrak{S}(X) = \{x \in X : \widehat{x^*} = \bar{\hat{x}}\}$ is a subalgebra of every $*$-algebra X. If X is functionally continuous (Problem 24H) and has continuous involution, $\mathfrak{S}(X)$ is closed in X. Reason: $x_\varphi \to x_0 \in X$ for a net $\{x_\varphi\} \subset \mathfrak{S}(X)$ implies $h(x_\varphi) \to h(x_0)$ and $h(x_\varphi^*) \to h(x_0^*)\ \forall h \in \mathcal{K}$. Thus $|\overline{h(x_0)} - h(x_0^*)| \le |\overline{h(x_0)} - \overline{h(x_\varphi)}| + |\overline{h(x_\varphi)} - h(x_\varphi^*)| + |h(x_\varphi^*) - h(x_0^*)|$ can be made arbitrarily small and $x_0 \in \mathfrak{S}(X)$. This demonstrates that $\mathfrak{S}(X)$ is a closed A^*-subalgebra of every complete complex LMCT$_2$ Q-$*$algebra X.

An A^*-algebra is $\hat{A}$ and a B^*-algebra is almost B^* (Problem 23J). This nesting is complete, since every almost B^*-algebra X is an A^*-algebra. Specifically (as in Theorem 23.14), $\sigma(x) \in \mathbb{R}\ \forall x = x^* \in X$ and our next theorem prevails.

† Caution may be in order here. There is no standard terminology employed in the literature. These algebras are called by different names in different sources. $\hat{A}$-algebras are frequently termed self-adjoint; A^*-algebras are sometimes called self-adjoint algebras, symmetric algebras, or (in the case of Russian literature) completely symmetric algebras.

THEOREM 26.6. The following are equivalent for a *-algebra X:

(i) X is an A^*-algebra;
(ii) $e+xx^* \in \mathfrak{U}\ \forall x \in X$;
(iii) $\sigma(x) \in \mathbb{R}\ \forall x = x^*$;
(iv) $M = M^*\ \forall M \in \mathfrak{M}$.

Proof. If X is A^*, then $x^*(M) = \overline{x(M)}$ and $(e+xx^*)(M) = e(M) + x(M)x^*(M) = 1 + |x(M)|^2 > 0\ \forall M \in \mathfrak{M}$ yields $e + xx^* \in \bigcup_{\mathfrak{M}} M \subset \mathfrak{U}$ [Theorem 22.8(i)]. Moreover, $x^*(M) = 0$ iff $x(M) = 0$ implies $x^* \in M$ iff $x \in M$, and $M = M^*$ follows. Thus (i)$\Rightarrow$(ii), (iv). For (ii)$\Rightarrow$(iii) suppose that $x = x^*$ and let $\lambda = a + ib \in \sigma(x)$. The product $(x - \lambda e)(x - \bar{\lambda} e) = b^2[e + [(x - ae)/b][(x - ae)/b]^*]$ necessitates that $b = 0$ (otherwise $x - \lambda e \in \mathfrak{U}$ by Theorem 22.6). Since (iii)$\Rightarrow$(i) as in the proof of Theorem 25.14(iii), we establish (iv)$\Rightarrow$(i). For $M \in \mathfrak{M}$ every $x \in X$ can be written $x = x(M)e + x_M$ $(x_M \in M)$. Therefore $M = M^*$ implies that $x_M^*(M) = 0$ and $x^*(M) = \overline{x(M)} + x_M^*(M) = \overline{x(M)}$. ■

COROLLARY 26.7. Let X be a *-algebra and let Y be a complete semisimple LMCT$_2$ *-algebra. Assume that $\forall z \in Z$ $(= X$ or $Y)$ there is a $z_0 \in Z$ such that $\widehat{z_0^*} = \bar{\hat{z}}$. Then every homomorphism $h: X \to Y$ is a *-homomorphism (i.e., $h(x^*) = \{h(x)\}^*\ \forall x \in X$).

Proof. A homomorphism $k \in \mathcal{H}_Y$ satisfies $k(h(x^*) - \{h(x)\}^*) = kh(x^*) - k(\{h(x)\}^*) = \overline{kh(x)} - \overline{k\{h(x)\}} = 0$. Since Y is semisimple, $\mathcal{H}_Y$ is injective and $h(x^*) = \{h(x)\}^*$ follows. ■

Being $\hat{A}$ is equivalent to being A^* for a semisimple algebra X. Verification begins with the observation that ϕ_g is injective and $^{*_g}: \underset{x \rightsquigarrow x^{*_g} = \phi_g^{-1}(\bar{\hat{x}})}{X \longrightarrow X}$ is well defined. Moreover, ^{*_g} is conjugate linear and satisfies $x^{*_g *_g} = \{\phi_g^{-1}(\bar{\hat{x}})\}^{*_g} = x$ and (since X is commutative) $(xy)^{*_g} = y^{*_g} x^{*_g}\ \forall x, y \in X$. Accordingly, ^{*_g} is an involution and we obtain

THEOREM 26.8 (Page [74]). Every semisimple A^*-algebra under involution * has $* = *_g$. In particular, a semisimple algebra is $\hat{A}$ iff it is A^*.

Proof. If X is a semisimple A^*-algebra, ϕ_g is a *-isomorphism of X into $C(\mathfrak{M}; \mathbb{C})$ with complex conjugation involution. This ϕ_g preservation of involution means that $\phi_g(x^*) = \overline{\phi_g(x)}$ and $x^* = \phi_g^{-1}\overline{\phi_g(x)} = x^{*_g}\ \forall x \in X$. The second assertion is clear, since a semisimple $\hat{A}$-algebra X is an A^{*_g}-algebra by virtue of each $x \in X$ satisfying $\overline{x(M)} = x^{*_g}(M)\ \forall M \in \mathfrak{M}$. ■

Remark. That $x = x^*$ iff $\hat{x} \in C(\mathfrak{M}, \mathbb{C})$ is real-valued, follows from $x = x^* = x^{*g}$ iff $x = \bar{\hat{x}}$.
Remark. Not every semisimple $*$-algebra is A^* [Example 26.21(d)].

COROLLARY 26.9. A semisimple Banach $\hat{A}$-algebra $(X, \|\ \|)$ is $*^g$-congruent to $(C(\mathfrak{M}; \mathbb{C}), \|\ \|_0)$, and is therefore a B^{*g}-algebra, if any of the following equivalent conditions hold:

(i) $r_\sigma(x) = \|x\|\ \forall x \in X$;
(ii) r_σ is complete;
(iii) r_σ is barreled.

Proof. r_σ is a norm and properties (i)–(iii) are equivalent by Corollaries 19.22 and 21.9 (om). If (ii) holds, then [Theorem 25.10(iii)] $\hat{X} = C(\mathfrak{M}; \mathbb{C})$ and $\phi_g: (X, \|\ \|) \to (C(\mathfrak{M}, \mathbb{C}), \|\ \|_0)$ is an onto congruence. To complete the proof define $*^g$ on X and observe that $\|x\|^2 = \|\hat{x}\|_0^2 = \|\bar{\hat{x}}\hat{x}\|_0 = \|\phi_g(x^{*g})\phi_g(x)\|_0 = \|\phi_g(x^{*g}x)\|_0 = \|x^{*g}x\|\ \forall x \in X$. ■

Any seeming distinction between B^*-subalgebras and sub B^*-algebras of a B^*-algebra can now be removed.

COROLLARY 26.10. Every closed $\hat{A}$-subalgebra of a B^* algebra $(X, \|\ \|)$ is a sub B^*-algebra of X.

Proof. If $(Y, \|\ \|)$ is a Banach $\hat{A}$-subalgebra of X, then Y is semisimple (Corollary 24.25a) and $\|y^2\| = \|y\|^2\ \forall y \in Y$. Accordingly [Corollary 23.8(iii)], $(Y, \|\ \|)$ is a B^{*g}-algebra and Theorem 26.8 applies. ■

Completely Regular and Normal Algebras

Since $\hat{X}$ separates points in $\mathfrak{M}$, it seems logical to consider algebras in $C(\mathfrak{M}; \mathbb{C})$ that enjoy the other separating properties of continuous functions on completely regular or normal spaces.

DEFINITION 26.11. Let Ω be a topological space.

A family $\Psi \subset \mathfrak{F}(\Omega; \mathbb{C})$ is said to be completely regular if for each closed set $K \subset \Omega$ and $x \notin K$ there is a $\psi \in \Psi$ such that $\psi(x) \neq 0$ and $\psi(K) = \{0\}$. Similarly, $\Psi \subset \mathfrak{F}(\Omega; \mathbb{C})$ is termed normal if for every pair of disjoint closed sets $K_1, K_2 \in \Omega$, some $\psi \in \Psi$ satisfies $\psi(K_1) = \{1\}$ and $\psi(K_2) = \{0\}$.

A complete complex LMCT$_2$ Q-algebra X is called completely regular {normal} whenever $\hat{X} \subset C(\mathfrak{M}; \mathbb{C})$ is completely regular {normal}.

A normal algebra is completely regular. The converse, we show shortly, is also true for Frechet Q-algebras. A few simple illustrations here may prove useful.

Example 26.12. (a) $C(\mathfrak{M}, \mathbb{C})$ is normal, since $(\mathfrak{M}, \mathfrak{T}_g)$ is compact T_2. Therefore X is normal if $\hat{X}$ is dense in $C(\mathfrak{M}; \mathbb{C})$. In particular, $\hat{A}$-algebras are normal, and this extends the nesting mentioned prior to Theorem 26.6.

(b) By contrast with (a) the completely regular Banach algebra $\mathfrak{D}^{(n)}(I)$ in Example 23.3(b) is non-B^*. Complete regularity holds, since $t_0 \notin K = \bar{K} \subset I$ implies that $[t_0-\delta, t_0+\delta] \cap K = \varnothing$ for some $\delta > 0$ and $f(t_0)=1$ for $f(t)=[1-[(t-t_0)/\delta]^2]e^{-[(t-t_0)/\delta]^2}\chi_{[t_0\delta, t_0+\delta]}(t)$ in $\mathfrak{D}^{(n)}(I)$. That $\mathfrak{D}^{(n)}(I)$ is non-B^* is also clear (Theorem 23.14) by virtue of $r_\sigma(f)=1<3=\|f\|$ for $f(t)=t^2$.

(c) The disk algebra $\mathfrak{H}(\bar{S}_1)$ is not completely regular. To be sure, $\mathfrak{M}=\{M_\omega : \omega \in \Omega\}$ and (by the identity theorem in analytic function theory) $f \equiv 0$ if $f(K)=\{0\}$ for an infinite compact set $K \subset \bar{S}_1(0)$.

A normal family in $\mathfrak{F}(\Omega, \mathbb{C})$ may be viewed as a decomposition of the identity element $\chi_\Omega \in \mathfrak{F}(\Omega; \mathbb{C})$.

THEOREM 26.13. Let Ψ be a normal family of functions on a topological space Ω such that Ψ is an algebra with identity under pointwise operations. If $\{u_1, \ldots, u_n\}$ is a finite open cover of Ω, there exist $\{\psi_1, \ldots, \psi_n\} \in \Psi$ satisfying

$$(*)\ \sum_{k=1}^{n} \psi_k = \chi_\Omega \text{ and } \psi_k(\Omega - u_k)=\{0\}\ \forall 1 \le k \le n.$$

Proof. By induction; if $k=2$, then $\Omega-u_1$ and $\Omega-u_2$ are disjoint and some $\psi_1 \in \Psi$ satisfies $\psi(\Omega-u_1)=\{0\}$ and $\psi(\Omega-u_2)=\{1\}$. Thus (*) holds for $\{\psi_1, \psi_2=\chi_\Omega-\psi_1\}$. Now assume that the theorem holds for $k=n-1$ and let $\{u_1, \ldots, u_n\}$ be any open cover of Ω. Take $C=\Omega-u_n$ and define $C_k = C \cap u_k\ \forall 1 \le k \le n-1$. Surely C is closed and covered by $\{C_1, \ldots, C_{n-1}\}$. Since each C_k is open in C and the algebra Ψ/C is normal, there exist (by induction hypothesis) $\{\eta_1, \ldots, \eta_{n-1}\} \in \Psi/C$ such that $\sum_{k=1}^{n-1} \eta_k = \chi_C$ and $\eta_k(C-C_k)=\{0\}\ \forall 1 \le k \le n-1$; but C and $\Omega - \sum_{k=1}^{n-1} u_k$ are also disjoint. Therefore $\psi(C)=\{1\}$ and $\psi\left(\Omega - \bigcup_{k=1}^{n-1} u_k\right)=\{0\}$ for some $\psi \in \Psi$ and (*) is satisfied by the composite functions $\left\{\psi_k = \eta_k\psi, \psi_n = \chi_\Omega - \sum_{k=1}^{n-1} \psi_k : 1 \le k \le n-1\right\} \in \Psi$. ■

COROLLARY 26.14. Let $f \in C(\Omega; \mathbb{C})$, where Ω above is compact T_2. Suppose further that $\forall \omega \in \Omega$ there is an open set $u(\omega)$ and $\psi_\omega \in \Psi$ such that $f = \psi_\omega$ on $u(\omega)$. Then $f = \psi$ for some $\psi \in \Psi$.

Proof. Suppose Ω is compact, it has a finite open subcover $\{u(\omega_1), \ldots, u(\omega_n)\}$ with $f = \psi_{\omega_k}$ on $u(\omega_k) \forall 1 \le k \le n$. If $\{\psi_1, \ldots, \psi_n\} \in \Psi$ satisfy (*), then $f(\omega) = \sum_{k=1}^{n} f(\omega)\psi_k(\omega) = \left(\sum_{k=1}^{n} \psi_{\omega_k}\psi_k\right)(\omega) \forall \omega \in \Omega$, meaning $f = \sum_{k=1}^{n} \psi_{\omega_k}\psi_k \in \Psi$. ■

Hull Kernel Topology

It is always of interest when seemingly distinct concepts are interrelated. One case in point weaves the algebraic notion of complete regularity with topological considerations of $\mathfrak{M}$. To illustrate we introduce a new topology on $\mathfrak{M}$. Begin by noting that $k(\phi) = X$ and $k(u) = \bigcap_{M \in u} M$ define closed ideals in X for each set $u \in \mathfrak{M}$. If we also define $h(I) = \{M \in \mathfrak{M} : M \supset I\} \forall$ ideal $I \subset X$, then $\xi(u) = hk(u)$ defines a Kuratowski closure operator on $\mathfrak{M}$ (Appendix t). The simple details are left as part of Problem 26H.

DEFINITION 26.15. The ξ-induced topology on $\mathfrak{M}$ having closed sets $\{u \subset \mathfrak{M} : u = hk(u)\}$ is called the hull-kernel topology and is denoted $\mathfrak{T}_{hk}$.

The topologies $\mathfrak{T}_{hk}$ and $\mathfrak{T}_g$ on $\mathfrak{M}$ are generally distinct. For instance, compare Example 26.12(c) with

THEOREM 26.16. For a complete complex LMCT$_2$ Q-algebra X

(i) $(\mathfrak{M}, \mathfrak{T}_{hk})$ is a compact T_1 space with $\mathfrak{T}_{hk} \subset \mathfrak{T}_g$;
(ii) $\mathfrak{T}_{hk} = \mathfrak{T}_g$ iff $\mathfrak{T}_{hk}$ is T_2 iff X is completely regular.

Proof. Since $M = hk(M)$, each point $M \in \mathfrak{M}$ is $\mathfrak{T}_{hk}$-closed and $\mathfrak{T}_{hk}$ is T_1. We use Appendix t to prove $\mathfrak{T}_{hk}$ compact. If $\{C_\alpha \subset \mathfrak{M} : \alpha \in \mathcal{Q}\}$ is any collection of $\mathfrak{T}_{hk}$-closed sets having $\bigcap_{\mathcal{Q}} C_\alpha = \varnothing$, then $\varnothing = \bigcap_{\mathcal{Q}} hk(C_\alpha) = h\{\bigcup_{\mathcal{Q}} k(C_\alpha)\}$ and $\bigcup_{\mathcal{Q}} k(C_\alpha) \not\subset M$ for any $M \in \mathfrak{M}$. This, of course, is also true for the ideal $\sum_{\mathcal{Q}} k(C_\alpha) = [\bigcup_{\mathcal{Q}} k(C_\alpha)] \subset X$. Therefore [Theorem 22.8(i)]

$X=\sum_{\mathscr{A}}\mathscr{k}(C_\alpha)$ and $e=\sum_{i=1}^{n} x_{\alpha_i}$ for some $x_{\alpha_i}\in\mathscr{k}(C_{\alpha_i})$. The immediate conclusion is that $X=\sum_{i=1}^{n}\mathscr{k}(C_{\alpha_i})$ and $\varnothing=\mathscr{h}(X)=\mathscr{h}\left\{\sum_{i=1}^{n}\mathscr{k}(C_{\alpha_i})\right\}=\bigcap_{i=1}^{n}\mathscr{h}\mathscr{k}(C_{\alpha_i})=\bigcap_{i=1}^{n}C_{\alpha_i}$ (note that $\mathscr{h}(I_1+I_2)=\mathscr{h}(I_1)\cap\mathscr{h}(I_2)$ since $I_1, I_2\subset I_1+I_2$ for ideals in X). Thus $\mathfrak{T}_{\mathscr{hk}}$ is compact and we proceed to verify that $\mathfrak{T}_{\mathscr{hk}}\subset\mathfrak{T}_g$ on $\mathfrak{M}$. Since every $\hat{x}^{-1}(0)$ is $\mathfrak{T}_g$-closed, $\mathscr{h}(I)=\bigcap_I \hat{x}^{-1}(0)$ is also $\mathfrak{T}_g$-closed $\forall$ ideal $I\subset X$. In particular, $\bar{u}^{\mathfrak{T}_{\mathscr{hk}}}=\mathscr{hk}(u)$ is $\mathfrak{T}_g$-closed and $u\subset\bar{u}^{\mathfrak{T}_g}\subset\overline{\bar{u}^{\mathfrak{T}_{\mathscr{hk}}}}^{\mathfrak{T}_g}=\bar{u}^{\mathfrak{T}_{\mathscr{hk}}}$ implies every $\mathfrak{T}_{\mathscr{hk}}$-closed set is $\mathfrak{T}_g$-closed. Moving on, the first equivalence in (ii) is clear by the reasoning in Theorem 26.2. If $\mathfrak{T}_{\mathscr{hk}}=\mathfrak{T}_g$ and $M_0\notin u=\bar{u}=\mathscr{hk}(u)=\bigcap_{x\in\mathscr{k}(u)}\hat{x}^{-1}(0)$, some $x_0\in\mathscr{k}(u)$ satisfies $\hat{x}_0(M_0)\neq 0$ and $\hat{x}_0(u)=\{0\}$, thereby establishing that X is completely regular. In the other direction suppose that X is completely regular and let $M_0\notin u$ for a $\mathfrak{T}_g$-closed set $u\neq\mathfrak{M}$. If $x\in X$ satisfies $\hat{x}(M_0)\neq 0$ and $\hat{x}(u)=\{0\}$, then $x\in\mathscr{k}(u)$ and $\hat{x}(\mathscr{hk}(u))=0$ necessitates that $M_0\notin\mathscr{hk}(u)$. Thus $\mathscr{hk}(u)\subset u$ and $\mathfrak{T}_{\mathscr{hk}}=\mathfrak{T}_g$ follows. ■

The objective now is to prove that complete regularity implies normality for a complex Frechet, Q-algebra X [which is also LMCT_2 by Problem 24A(d)]. If $E\subset\mathfrak{M}$ is $\mathfrak{T}_g$-closed, then $\mathscr{k}(E)$ is a closed ideal in X and $Y=(X/\mathscr{k}(E), \mathfrak{T}_{\mathfrak{g}(\nu)})$ is a Frechet Q-algebra [Problem 21A(a), Corollary 25.2]. Taking $(\mathfrak{M}_Y, \mathfrak{T}_{g_Y})$ as the maximal ideal space of Y, we prove

THEOREM 26.17. Let X be a completely regular complex Frechet Q-algebra. Then $(E,\mathfrak{T}_g)$ and $(\mathfrak{M}_Y,\mathfrak{T}_{g_Y})$ are homeomorphs for each closed set E in $(\mathfrak{M},\mathfrak{T}_g)$.

Proof. For each $M\in E$ we have $\{\mathscr{k}(E)\}(M)=\bigcup_{x\in\mathscr{k}(E)}x(M)=0$ and $x_1+\mathscr{k}(E)=x_2+\mathscr{k}(E)$ implies $x_1(M)=x_2(M)$. Therefore $\eta_M\colon \underset{x+\mathscr{k}(E)\rightsquigarrow x(M)}{Y\longrightarrow\mathbb{C}}$ is a well-defined, nonzero homomorphism and $M+\mathscr{k}(E)=\eta_M^{-1}(0)\in\mathfrak{M}_Y$. This also establishes that $t\colon\underset{M\rightsquigarrow M+\mathscr{k}(E)}{(E,\mathfrak{T}_g)\to(\mathfrak{M}_Y,\mathfrak{T}_{g_Y})}$ is injective, meaning $M_1=M_2$ whenever $M_1+\mathscr{k}(E)=M_2+\mathscr{k}(E)$. Specifically, $x_1\in M_1$ implies $x_1+\mathscr{k}(E)=x_2+\mathscr{k}(E)$ for some $x_2\in M_2$, and $x_1(M_2)=x_2(M_2)=0$ yields $x_1\in M_2$. The same argument also gives $M_2\subset M_1$. To prove t surjective consider any $\dot{M}\in\mathfrak{M}_Y$ and let $h_M=h_{\dot{M}}\nu$. Then $M=h_M^{-1}(0)\in E$ (otherwise $x(M)=0$ and $x(E)=\{0\}$ for some $x\in\mathscr{k}(E)$ yields $0\neq x(M)=h_M(x)=h_{\dot{M}}\{x+\mathscr{k}(E)\}=h_{\dot{M}}\{\mathscr{k}(E)\}=0$) and $t(M)=\{x+\mathscr{k}(E): x\in M\}=\{x+\mathscr{k}(E): h_M(x)=0\}=\{x+\mathscr{k}(E): h_{\dot{M}}\{x+\mathscr{k}(E)\}=0\}=h_{\dot{M}}^{-1}(0)=\dot{M}$. Finally, t is bicontinuous by

Theorem 15.1 and the compact T_2 property of the involved Gelfand topologies. ■

THEOREM 26.18. Let X be a completely regular complex Frechet Q-algebra. For each ideal $I \subset X$ and $\mathfrak{T}_g$-closed set $E \subset \mathfrak{M}$ satisfying $E \cap \mathscr{h}(I) = \varnothing$ there is an $x \in I$ such that $\hat{x}(E) = \{1\}$. In particular, X is normal.

Proof. First, $\nu(I)$ equals $Y = X/\mathscr{k}(E)$ (otherwise $\nu(I) \subset M + \mathscr{k}(E) \in \mathfrak{M}_Y$ for some $M \in E$ and $I \subset M$ contradicts $E \cap \mathscr{h}(I) = \varnothing$). Therefore some $x \in I$ satisfies $x + \mathscr{k}(E) = e + \mathscr{k}(E)$ and $\hat{x}(E) = \hat{e}(E) = \{1\}$. To see that X is normal let E_1 and E_2 be $\mathfrak{T}_g = \mathfrak{T}_{\mathscr{hk}}$-closed disjoint sets in $\mathfrak{M}$ and take $I = \mathscr{k}(E_1)$. Then $\mathscr{h}(I) \cap E_2 = \mathscr{hk}(E_1) \cap E_2 = E_1 \cap E_2 = \varnothing$ and $\hat{x}(E_2) = \{1\}$ for some $x \in I = \bigcap_{M \in E_1} M$. Clearly $\hat{x}(E_1) = \{0\}$. ■

The Silov Boundary

According to the maximum modulus Principle, every member of $\mathfrak{H}(\bar{S}_1)$ attains its maximum value on *bdy* $\bar{S}_1$. In a sense elements in the complex algebras of this section also enjoy a maximum modulus type property.

THEOREM 26.19. Every complete complex LMCT$_2$ Q-algebra X has a Silov boundary; that is a unique closed minimal set $\partial_X \subset \mathfrak{M}$ satisfying

(*) $\quad r_\sigma(x) = \sup_{\partial_X} |x(M)| \;\forall x \in X.$

Proof. Partially order, by inclusion, the collection $\mathfrak{F}$ of closed sets in $\mathfrak{M}$ that satisfies (*). Surely $\mathfrak{M} \in \mathfrak{F}$, since $(\mathfrak{M}, \mathfrak{T}_g)$ is compact T_2. Every totally ordered set $\{E_\alpha : \alpha \in \mathcal{Q}\} \subset \mathfrak{F}$ (having the finite intersection property) also satisfies $\bigcap_{\mathcal{Q}} E_\alpha \neq \varnothing$. Moreover, $\{M \in \mathfrak{M} : |x(M)| = r_\sigma\} \cap \{\bigcap_{\mathcal{Q}} E_\alpha\} \neq \varnothing$ $\forall x \in X$ assures that $\bigcap_{\mathcal{Q}} E_\alpha$ satisfies (*) and that $\bigcap_{\mathcal{Q}} E_\alpha \in \mathfrak{F}$ is the glb of $\{E_\alpha : \alpha \in \mathcal{Q}\}$. Therefore (the minimal principle†) $\mathfrak{F}$ contains a minimal element ∂_X whose uniqueness will be evident once we demonstrate that $\partial_X \subset E \;\forall E \in \mathfrak{F}$. It suffices for this to show that $M_0 \in \bar{E}$ for $M_0 \in \partial_X$ and $E \in \mathfrak{F}$. For any proper $\mathfrak{T}_g$-nbd $v(M_0) = v(M_0; \hat{x}_1 \cdots \hat{x}_n; \varepsilon)$ we have $\max_{1 \le i \le n} r_\sigma(x_i) = \rho > 0$. Since $\partial_X \in \mathfrak{F}$ is minimal, $\partial_X \, v(M_0)$ does not satisfy (*).

† A partially ordered set $\mathfrak{F}$ has a minimal element if every totally ordered subset has a glb in $\mathfrak{F}$.

Choose $y \in X$ such that $\sup_{\partial_X - v(M_0)} |y(M)| < r_\sigma(y)$ and $|y(\tilde{M})| = r_\sigma(y)$ for some $\tilde{M} \in \partial_X \cap v(M_0)$. For sufficiently large $n \in \mathbb{N}$ the element $z = [y/r_\sigma(y)]^n$ satisfies $|z(M)| < \varepsilon/2\rho$ on $\partial_X - v(M_0)$. Now each $z_i = x_i z - x_i(M_0)z$ $(1 \le i \le n)$ satisfies $|z_i(M)| \le \{|x_i(M)| + |x_i(M_0)|\}|z(M)| \le 2r_\sigma(x_i)|z(M)| < \varepsilon$ on $\partial_X - v(M_0)$ and [since $r_\sigma(z) = 1$] $|z_i(M)| \le |x_i(M) - x_i(M_0)|\,|z(M)| < \varepsilon$ on $v(M_0)$. In other words, $r_\sigma(z_i) = \sup_{\partial_X} |z_i(M)| < \varepsilon$ $\forall 1 \le i \le n$. Since $E \in \mathfrak{F}$, there exists some $M_E \in E$ such that $|z(M_E)| = r_\sigma(z) = 1$. This yields $|x_i(M_E) - x_i(M_0)| = |x_i(M_E) - x_i(M_0)|\,|z(M_E)| = |z_i(M_E)| \le r_\sigma(z_i) < \varepsilon$ $\forall 1 \le i \le n$. Therefore $M_E \in v(M_0) \cap E$ and [since $v(M_0)$ was arbitrary] $M_0 \in \bar{E} = E$ as required. ■

COROLLARY 26.20. $M_0 \in \partial_X$ iff $\forall \mathfrak{T}_g$-nbd $v(M_0)$ there is an $x \in X$ satisfying $\sup_{\mathfrak{M} - v(M_0)} |x(M)| < r_\sigma(x) = \sup_{v(M_0)} |x(M)|$.

Proof. The condition is sufficient for $M_0 \in \partial_X$, since it implies that $v(M_0) \cap \partial_X \ne \varnothing$ $\forall \mathfrak{T}_g$-nbd $v(M_0)$. It is also necessary, since $\partial_X \subset \mathfrak{M} - v(M_0)$ and $M_0 \notin \partial_X$ if no such $x \in X$ exists for a $\mathfrak{T}_g$-nbd $v(M_0)$. ■

Example 26.21. (a) A closed set $\partial_{\mathfrak{M}} \subset \mathfrak{M}$ $\{\partial_{\mathcal{H}} \subset \mathcal{H}\}$ is said to be *determining* if $r_\sigma(x) = \sup_{\mathfrak{M}} |x(M)|$ $\{r_\sigma(x) = \sup_{\mathcal{H}} |h(x)|\}$ $\forall x \in X$. In many instances the use of a determining set is all that is needed. For instance, $\sigma(X, \partial_{\mathcal{H}}) = \sigma(X, \mathcal{H}) = r_\sigma$ if X has a finite determining set $\partial_{\mathcal{H}}$.

(b) $\partial_X = \mathfrak{M}$ whenever $\hat{X}$ is dense in $C(\mathfrak{M}; C)$ (in particular, whenever X is an $\hat{A}$-algebra). To see this consider any $\mathfrak{T}_g$-open set $v(M_0)$ containing $M_0 \in \mathfrak{M}$ and let $f \in C(\mathfrak{M}; \mathbb{C})$ be a Urysohn function satisfying $f(M_0) = 1$ and $f(\mathfrak{M} - v(M_0)) = \{0\}$. Since some $x \in X$ satisfies $\|\hat{x} - f\|_0 < \frac{1}{3}$ and since $\sup_{\mathfrak{M} - v(M_0)} |x(M)| < \frac{1}{3} < \frac{2}{3} = r_\sigma(x) = \sup_{v(M_0)} |x(M)|$, it follows that $M_0 \in \partial_X$.

The above reasoning also confirms that $\partial_X = \mathfrak{M}$ for a completely regular algebra X.

(c) $\mathfrak{M} = \{M_\omega : \omega \in \Omega\}$ and $\partial_X = \{M_\omega : |\omega| = 1\}$ for $X = \mathfrak{H}(\bar{S}_1)$ in Example 23.3(a). Therefore X is not an $\hat{A}$-algebra and (since X is semisimple) not A^* for any involution on X (Theorem 26.8).

Corollary 26.20 is frequently combined with Theorem 25.7 to give

THEOREM 26.22. Let $X \subset Y$, where X and Y are complete complex $LMCT_2$ Q-algebras with the same identity. If $r_{\sigma_Y}(x) = r_{\sigma_X}(x)$ $\forall x \in X$, then every $M_0 \in \partial_X$ is contained in some $M \in \mathfrak{M}_Y$. Accordingly, every $h \in \partial_X$ has an extension to $\mathcal{H}_Y$.

Proof. Assume $M_0 \in \partial_X$ is not contained in any $M \in \mathcal{M}_Y$ so that no proper ideal of Y contains M_0. Then $Y = [xy : (x, y) \in M_0 \times Y]$ and $e = \sum_{k=1}^{n} x_k y_k$ for some $(x_k, y_k) \in M_0 \times Y$. There is no loss in assuming that $r_{\sigma_X}(x_k) < 1 \; \forall 1 \leq k \leq n$ (otherwise write $e = \sum_{k=1}^{n} x'_k y'_k$ for $x'_k = x_k/[1 + r_{\sigma_X}(x_k)]$ and $y'_k = \{1 + r_{\sigma_X}(x_k)\} y_k$). Let $\lambda = \max_{1 \leq k \leq n} r_{\sigma_Y}(y_k)$. Then $v(M_0) = v(M_0; \hat{x}_1 \cdots \hat{x}_n; 1/2n\lambda)$ is $\mathcal{T}_g$-open and $\sup_{\mathcal{M} - v(M_0)} |x_0(M)| < r_{\sigma_X}(x_0) = \sup_{v(M_0)} |x_0(M)|$ for some $x_0 \in X$. Since $r_{\sigma_X}(x_0) \neq 0$, the element $x' = [x_0/r_{\sigma_X}(x_0)]^m \in X$ satisfies $\sup_{\mathcal{M} - v(M_0)} |x'(M)| < 1/2n\lambda$ for large enough $m \in \mathbb{N}$. Consequently $r_{\sigma_X}(x' x_k) \leq (1/2n\lambda) \; \forall 1 \leq k \leq n$ and there follows the contradiction

$$r_{\sigma_Y}(x') = \sup_{\mathcal{M}_Y} |x'(M)| = \sup_{\mathcal{M}_Y} |x'(M)\, e(M)| \leq \lambda \sum_{k=1}^{n} \sup_{\mathcal{M}_Y} |x'(M)\, x_k(M)|$$

$$= \lambda \sum_{k=1}^{n} r_{\sigma_Y}(x' x_k) = \lambda \sum_{k=1}^{n} r_{\sigma_X}(x' x_k) \leq \tfrac{1}{2} < 1 = r_{\sigma_X}(x').$$

The obvious conclusion is that $M_0 \subset M$ for some $M \in \mathcal{M}_Y$. ■

If $(X, \| \|)$ is a Banach subalgebra with the same identity as $(Y, \| \|)$, then $r_{\sigma_Y}(x) = \lim_n \|x^n\|^{1/n} = r_{\sigma_X}(x) \; \forall x \in X$ [Corollary 23.8(ii)] and Theorem 26.22 applies. Example 26.21(b) therefore yields

COROLLARY 26.23. Let X be a Banach $\hat{A}$-algebra and let Y be a Banach algebra with the same identity that contains X as a closed subalgebra. Then every $h \in \mathcal{H}_X$ is extendable to $\mathcal{H}_Y$.

PROBLEMS

26A. ($\mathcal{M}_{\mathcal{R}}, \mathcal{T}_g$) for Algebras without Identity

Assume that X is a complete complex LMCT$_2$ Q-algebra without identity. The notation of Problems 22B and 23B prevail.

(a) Verify that X/M is isomorphic to $\mathbb{C} \; \forall M \in \mathcal{M}_{\mathcal{R}}$. Use this to conclude that $\underset{h \rightsquigarrow h^{-1}(0)}{\mathcal{H} \rightarrow \mathcal{M}_{\mathcal{R}}}$ is bijective and $\sup_{\mathcal{M}_{\mathcal{R}}} |x(M)| = \sup_{\mathcal{H}} |(hx)| =$

$= r_{\sigma_{\odot}}(x)\ \forall x \in X$. [*Hint*: If x_0 is an identity mode M, then $x_0 + M$ is an identity for X/M. For any $x \notin M$ we have $X = [M \cup x] \neq M$ and $x_0 = x_m + yx$ for $x_m \in M$. Thus $y + M = (x + M)^{-1}$ and Corollary 24.8 applies.]

(b) $(\mathfrak{M}_{\mathfrak{R}}, \mathfrak{T}_g)$ is locally compact T_2. Furthermore, $(\mathfrak{M}_e, \mathfrak{T}_g)$ is the one point compactification of $(\mathfrak{M}_{\mathfrak{R}}, \mathfrak{T}_g)$ with $(0, X)$ as the "point at infinity." [*Hint*: $\mathfrak{M}_e - (0, X)$ is $\mathfrak{T}_g$-locally compact T_2, since $(\mathfrak{M}_e, \mathfrak{T}_g)$ is compact T_2. Show [Problem 23B(d)] that $\mathfrak{T}_g$ on $\mathfrak{M}_{\mathfrak{R}}$ coincides with the topology $\mathfrak{T} = \{u \subset \mathfrak{M}_{\mathfrak{R}} : t(u) \in \mathfrak{T}_g \text{ on } \mathfrak{M}_e - (0, X)\}$.]

(c) Let $\mathcal{H}_0 = \mathcal{H} \cup \{0\}$. Then $(\mathcal{H}, \sigma(\mathcal{H}, X)) = (\mathfrak{M}_{\mathfrak{R}}, \mathfrak{T}_g)$ and we have $(\mathcal{H}_0, \sigma(\mathcal{H}_0, X)) = (\mathfrak{M}_e, \mathfrak{T}_g)$, where equality denotes topological isomorphs. This means $(\mathcal{H}, \sigma(\mathcal{H}, X))$ is locally compact T_2 with $(\mathcal{H}_0, \sigma(\mathcal{H}_0, X))$ as its one point compactification.

(d) Show that $\xi_X \subset C_\infty(\mathcal{H}; \mathbb{C})$.

(e) If X has an identity, $\mathfrak{M}_{\mathfrak{R}} = \mathfrak{M}$ and $X = (0, X)$ is an isolated point of $\mathfrak{M}_e$ {therefore $\{0\}$ is an isolated point of $\mathcal{H}_0$} and both $(\mathfrak{M}_{\mathfrak{R}}, \mathfrak{T}_g) = (\mathfrak{M}, \mathfrak{T}_g)$ and $(\mathcal{H}, \sigma(\mathcal{H}, X))$ are compact.

26B. $\mathcal{H} = \mathcal{H}_\Omega$ and $\mathfrak{M}_{\mathfrak{R}} = \mathfrak{M}_\Omega$ for $C_\infty(\Omega; \mathbb{F})$

Let $X = C_\infty(\Omega; \mathbb{F})$, where Ω is a locally compact, noncompact T_2 space in Example 23.2. Then $\mathcal{H} = \{h_\omega : \omega \in \Omega\}$ and $\mathfrak{M}_{\mathfrak{R}} = \{M_\omega : \omega \in \Omega\}$. [*Hint*: $\mathcal{H} = \mathcal{H}_\Omega$ follows from $\mathfrak{M}_{\mathfrak{R}} = \mathfrak{M}_\Omega$ and Problem 26A(a). Identify X_e with $C(\Omega_\infty; \mathbb{F})$ under $(\alpha, f) \rightsquigarrow f + \alpha\chi_{\Omega_\infty}$ so that $\underset{\omega \rightsquigarrow M_\omega}{\Omega_\infty \to \mathcal{M}_e}$ is a bijection for $M_\infty = (0, X)$. Now apply Problem 23B(d).]

26C. System of Algebra Generators

Let $Y \subset X$, where X is a complete complex LMCT_2 Q-algebra. The smallest closed subalgebra $alg\,(Y)$ of X that contains $\{Y, e\}$ is termed the algebra generated by Y. In this context Y is said to be a system of generators of $alg\,(Y) \subset X$.

(a) $alg\,(Y)$ consists of all polynomials (with complex coefficients) in elements of Y together with limits of all such polynomials.

(b) If $X = alg\,(Y)$, then $\{V(M_0; \hat{y}_1 \cdots \hat{y}_n; \varepsilon) : y_i \in Y \text{ and } \varepsilon > 0\}$ constitutes a $\mathfrak{T}_g$-nbd base at $M_0 \in \mathfrak{M}$. [*Hint*: $r_\sigma \subset \mathfrak{T}$ implies $alg\,(Y) = \overline{alg\,(Y)}^{r_\sigma} = X$. For any $\mathfrak{T}_g$-nbd $V(M_0) = V(M_0; \hat{x}_1 \cdots \hat{x}_n; \varepsilon)$ there exist polynomials $p_i = \sum_{k=1}^{n_i} \alpha_{k_i} y_{k_i}$ with $r_\sigma(x_i - p_i) < \varepsilon/3\ \forall 1 \le i \le n$. Since p_i is $\mathfrak{T}_g$-continuous, there exist $\delta_i > 0$ such that $|p_i(M) - p_i(M_0)| < \varepsilon/3$ for $M \in V(M_0; \hat{y}_{1i}, \ldots, \hat{y}_{n_i i}; \delta_i)$. Conclude that $V(M_0; \hat{y}_{11} \cdots \hat{y}_{n_1 1}, \hat{y}_{12} \cdots \hat{y}_{n_2 2}; \cdots; \hat{y}_{1n} \cdots \hat{y}_{n_n n}; \min_{1 \le i \le n} \delta_i) \subset V(M_0)$.]

(c) If $X = alg\,(Y)$, then $(\mathfrak{M}, \mathfrak{T}_g)$ is homeomorphic to a compact subset of $\prod_Y \mathbb{C}$.

(d) $\mathfrak{T}_g$ is defined by the metric $d(M_1, M_2) = \sum_{n=1}^{\infty} 2^{-n} |x_n(M_1) - x_n(M_2)|$ whenever $X = alg\,(\{x_n : n \in \mathbb{N}\})$.

(e) If $X = \mathcal{alg}\,(x)$, then $(\mathfrak{M}, T_g)$ is homeomorphic to $\sigma(x) \subset \mathbb{C}$. Furthermore, $\rho(x)$ is connected. By way of illustration show that $\mathfrak{H}(\bar{S}_1) = \mathcal{alg}\,(\omega)$ and $(\mathfrak{M}, \mathfrak{T}_g)$ is homeomorphic to $\sigma(\omega) = \bar{S}_1(0)$.

(f) The purely topological characterization of $(\mathfrak{M}, \mathfrak{T}_g)$ in (e) no longer holds when more than one generator is involved. If $X = \mathfrak{D}^{(n)}(I)$ in Example 23.3(b), then $X = \mathcal{alg}\,(\{e^{i\theta}, e^{-i\theta} : 0 \le \theta \le 2\pi\})$ and $(\mathfrak{M}, \mathfrak{T}_g)$ is the homeomorph of $\{\omega : |\omega| = 1\} \subset \mathbb{C}$.

Remark. The notions here are further developed in Silov [92].

26D. $(\mathfrak{M}, \mathfrak{T}_g)$-Connectedness Considerations

Let X be a complete, complex, $LMCT_2$ Q-algebra.

(a) $(\mathfrak{M}, \mathfrak{T}_g)$ is discrete if $1 \le \aleph(\mathfrak{M}) < \aleph_0$.

(b) $(\mathfrak{M}, \mathfrak{T}_g)$ is disconnected if $1 < \aleph(\mathfrak{M}) \le \aleph_0$. [*Hint*: Problem 13F, Theorem 8.24.]

(c) $\aleph(\mathfrak{M}) = 1$ if X is isomorphic to $\mathbb{C}$. The converse also holds when X is semisimple. [*Hint*: $\mathfrak{M} = \{0\}$ and $\mathfrak{U} = X - \{0\}$ for semisimple X with $\aleph(\mathfrak{M}) = 1$. Apply Corollary 24.8.]

(d) Give two examples in which $\aleph(\mathfrak{M}) > \aleph_0$, one for which $(\mathfrak{M}, \mathfrak{T}_g)$ is disconnected and one for which $(\mathfrak{M}, \mathfrak{T}_g)$ is connected. [*Hint*: Consider $X = C(\Omega; \mathbb{C})$ with appropriate conditions on Ω.]

(e) If $X = I_1 \oplus I_2$ for nonzero ideals I_1, I_2 in X, then $e = e_1 + e_2$ for some $e_k \in I_k$ $(k = 1, 2)$. Show that each I_k is a complete complex $LMCT_2$ Q-algebra with identity e_k. Use this to conclude that $(\mathfrak{M}, \mathfrak{T}_g)$ is disconnected. [*Hint*: Each $e_k^2 = e_k$ and $I_k = \{x \in X : xe_k = x\}$ is closed in X. Furthermore, $\mathfrak{U} \cap I_k \subset \mathfrak{U}_k$ (the units in I_k) and $\mathfrak{M} = \hat{e}_1^{-1}(1) \cup \hat{e}_2^{-1}(1)$.]

(f) The converse of (e) also holds when X is normal. If $\mathfrak{M} = F_1 \cup F_2$ for disjoint, closed sets $F_k \in \mathfrak{M}$, then $\hat{e}_1(F_1) = \{1\}$ and $\hat{e}_1(F_2) = \{0\}$ for some $e_1 \in X$. Define $e_2 = e - e_1$ and let $I_k = \{x \in X : xe_k = x\}$. Then $X = I_1 \oplus I_2$ and each maximal ideal space $\mathfrak{M}_k$ of I_k satisfies $\mathfrak{M}_k = \{F_k\}$. [*Hint*: Establish that $\underset{M \rightsquigarrow M \cap I_k}{F_k \to \mathfrak{M}_k}$ is an onto homeomorphism.]

Remark. Silov [92] (using his theory of analytic functions of several variables) proves that the above holds without requiring normality.

26E. Measure-Induced Seminormed Topologies

(Page [75]). Assume X is a complete complex $LMCT_2$ Q-algebra. Since $(\mathfrak{M}, T_g)$ is compact, $\mathfrak{M} \in \Sigma = \mathcal{R}_\sigma(\mathcal{C})$ and (notation as in Section 16) every $\mu \in \mathcal{CA}(\mathfrak{M}, \Sigma; \mathbb{C})$ is totally finite. Define

$$\mathcal{P}_{\mathcal{R}} = \left\{ p_\mu(x) = \left| \int_{\mathfrak{M}} \hat{x}(M)\, d\mu \right| : \mu \in \mathcal{CA}_{\mathcal{R}}(\mathfrak{M}, \Sigma; \mathbb{C}) \right\}$$

$$\mathcal{P} = \left\{ p_\mu(x) = \left| \int_{\mathfrak{M}} \hat{x}(M)\, d\mu \right| : \mu \in \mathcal{CA}(\mathfrak{M}, \Sigma; \mathbb{C}) \right\}$$

$$\mathcal{P}_v = \left\{ p_{\|\mu\|_v}(x) = \int_{\mathfrak{M}} |\hat{x}(M)|\, d\|\mu\|_v : \mu \in \mathcal{CA}(\mathfrak{M}, \Sigma; \mathbb{C}) \right\}$$

$$\mathcal{P}_+ = \left\{ p_\mu(x) = \int_{\mathfrak{M}} |\hat{x}(M)|\, d\mu : \text{nonnegative } \mu \in \mathcal{CA}(\mathfrak{M}, \Sigma; \mathbb{R}) \right\}$$

(a) The seminormed topologies (also denoted $\mathcal{P}_{\mathcal{R}}$, $\mathcal{P}$, $\mathcal{P}_v$, and $\mathcal{P}_+$, respectively) satisfy $\sigma(X, \mathcal{H}) \subset \mathcal{P}_{\mathcal{R}} = \mathcal{P} \subset \mathcal{P}_v = \mathcal{P}_+ \subset \sigma(X, X'_{r_\sigma})$ on X. [*Hint*: The first inclusion in $\sigma(X, \mathcal{H}) \subset \mathcal{P}_{\mathcal{R}} \subset \mathcal{P} \subset \mathcal{P}_v \subset \mathcal{P}_+$ holds, since $h \in \mathcal{H}$ with $M_h = h^{-1}(0) \in \mathfrak{M}$ yields $\mu_{M_h,1} \in \mathcal{CA}_{\mathcal{R}}(\mathfrak{M}, \Sigma; \mathbb{C})$ and $|h(x)| = p_{\mu_{M_h,1}}(x)\ \forall x \in X$. In the other direction each $\mu \in \mathcal{CA}(\mathfrak{M}, \Sigma; \mathbb{C})$ determines an element $\Phi_\mu \in \{C(\mathfrak{M})\}'$ defined by $\Phi_\mu(f) = \int_{\mathfrak{M}} f\, d\mu\ \forall f \in C(\mathfrak{M})$. Since $\{C(\mathfrak{M})\}'$ and $\mathcal{CA}_{\mathcal{R}}(\mathfrak{M}, \Sigma; \mathbb{C})$ are isomorphs (Appendix m), $p_\mu = p_\nu$ for some $\nu \in \mathcal{CA}_{\mathcal{R}}(\mathfrak{M}, \Sigma; \mathbb{C})$ and $\mathcal{P} \subset \mathcal{P}_{\mathcal{R}}$ follows; and $\mathcal{P}_+ \subset \mathcal{P}_v$ follows from $\mu = |\mu|\ \forall$ nonnegative $\mu \in \mathcal{CA}(\mathfrak{M}, \Sigma; \mathbb{R})$. Since Φ_μ and $\psi: (X, r_\sigma) \to (\hat{X}, \|\ \|_0)$ are continuous, $p_\mu = \Phi_\mu \psi \in X'_{r_\sigma}$ and $\mathcal{P}_+ \subset \sigma(X, X'_{r_\sigma})$.]

Remark. The inclusion $\sigma(X, X'_{r_\sigma}) \subset \sigma(X, X')$ is strict for non-semisimple X.

(b) Use the reasoning in Theorem 25.12 to prove that $\mathcal{P}\{\mathcal{P}_+\} = \sigma(X, X'_{r_\sigma}) = r_\sigma$ if $\mathcal{P}\{\mathcal{P}_+\}$ is barreled.

(c) Show that $\mathcal{P} = \mathcal{P}_+ = \sigma(X, X'_{r_\sigma})$ when X is semisimple. [*Hint*: $\psi \in X'_{r_\sigma}$ implies $\psi\phi_g^{-1} \in (\hat{X}, \|\ \|_0)'$ has an extension $\Psi \in (C(\mathfrak{M}), \|\ \|_0)'$ with associated measure $\mu_\Psi \in \mathcal{CA}_{\mathcal{R}}(\mathfrak{M}, \Sigma; \mathbb{C})$ satisfying $\Psi(F) = \int_{\mathfrak{M}} F\, d\mu_\Psi\ \forall F \in C(\mathfrak{M}; \mathbb{C})$. In particular, $|\psi(x)| = \psi\phi_g^{-1}(\hat{x}) = p_{\mu_\Psi}(x)\ \forall x \in X$ and $X'_{r_\sigma} = (X, \mathcal{P}_{\mathcal{R}})'$ yields $\sigma(X, X'_{r_\sigma}) \subset \mathcal{P}_{\mathcal{R}}$.]

Remark. If X is a B^*-algebra, $\sigma(X, X'_{r_\sigma}) = \sigma(X, X')$, since $X = C(\mathfrak{M}; \mathbb{C})$ and $X' = \{\int_{\mathfrak{M}} \hat{x}\, d\mu : \mu \in \mathcal{CA}_{\mathcal{R}}(\mathfrak{M}, \Sigma; \mathbb{C})\}$ gives $\sigma(X, X') = \mathcal{P}_{\mathcal{R}}$.

(d) Give examples for which $\sigma(X, \mathcal{H}) \subsetneqq \mathcal{P}$ and $\mathcal{P} \subsetneqq \sigma(X, X')$ when X is semisimple.

(e) Let $\{\mathcal{P}\}$ denote $\mathcal{P}$ or $\mathcal{P}_+$ on X. Then X is semisimple iff $\{\mathcal{P}\}$ is T_2. In particular, X is semisimple if there is an injective collection of $\{\mathcal{P}\}$-continuous, linear mappings into a T_2VS. [*Hint*: Use $\sigma(X, X'_{r_\sigma}) \subset r_\sigma$ and Theorem 25.8. Also, $\mathcal{T}_{\mathcal{P}\{\mathcal{F}\}} \subset \mathcal{P}_+$ for any T_2VS Y and collection $\mathcal{F} \subset L((X, \mathcal{P}_+), Y)$.]

(f) Suppose that X is semisimple and $\mu \in \mathcal{CA}(\mathfrak{M}, \Sigma; \mathbb{R})$ is a nonnegative measure such that $\mu(u) > 0$ for nonempty, open sets $u \subset \mathfrak{M}$. Then p_μ is a norm on

X. [*Hint*: $p_\mu(x)=0$ implies $\hat{x}=0$ a.e. on $\mathfrak{M}$. Show that $u_x=\{M\in\mathfrak{M}:\hat{x}(M)>0\}$ is $\mathfrak{T}_g$-open and conclude that $x\in\bigcap_{\mathfrak{M}} M=\{0\}$.]

26F. Probability Measure-Induced Seminorm

Assume that X is a complete complex LMCT$_2$ Q-algebra and $\nu\in\mathcal{C}\mathcal{Q}(\mathfrak{M},\Sigma;\mathbb{C})$ is a probability measure (i.e., $\nu(\mathfrak{M})=1$) satisfying $\nu(u)>0$ for every nonempty open set $u\subset\mathfrak{M}$.

(a) If $\mathfrak{M}$ is separable with dense set $\{M_n:n\in\mathbb{N}\}$, then $\nu=\sum_{n=1}^{\infty}2^{-n}\chi_{M_n}$ meets the above requirements.

Remark. A modification of the arguments in Dunford and Schwartz ([26], p. 426) shows that $\mathfrak{M}$ is separable when X is separable or $\mathfrak{M}$ is metrizible.

(b) The seminorm $p_\alpha(x)=\{\int_{\mathfrak{M}}|\hat{x}|^\alpha\,d\nu\}^{1/\alpha}$ $(\alpha\geq 1)$ is a norm iff X is semisimple, in which case $(x,y)=\int_{\mathfrak{M}}\hat{x}\bar{\hat{g}}\,d\nu$ defines an inner product on X. [*Hint*: See Problem 26E(f).]

(c) Show that (X,p_α) is an LCTVA with $p_\alpha(e)=1$. Why is (X,p_α) not necessarily LMC? [*Hint*: $p_\alpha(xy)\leq\max\{r_\sigma(x)\,p_\alpha(y),\,r_\sigma(y)\,p_\alpha(x)\ \forall x,y\in X.$]

(d) The following are equivalent when X is semisimple:

(i) $\dim X<\infty$;
(ii) $\dim L_\alpha(\mathfrak{M},\nu)<\infty$;
(iii) $\|\;\|_\alpha$ and $\|\;\|_0$ are equivalent on $\hat{X}$;
(iv) norms p_σ and r_σ are equivalent on X;
(v) $(\hat{X}.\,\|\;\|_\alpha)$ is complete;
(vi) (X,p_α) is complete.

[*Hint*: (i)$\Leftrightarrow$(ii) via $\hat{X}\subset\overline{C(\mathfrak{M})}^{\|\|_\alpha}=L_\alpha(\mathfrak{M},\nu)$, and (i)$\Rightarrow$(iii)$\Leftrightarrow$(iv), since ϕ_g is a $p_\alpha-\|\;\|_\alpha$ congruence. For (iv)$\Rightarrow$(i) use $p_1\subset\mathcal{P}_+\subset r_\sigma$ on X. If the topologies p_α and r_σ coincide, there is a $\lambda>0$ such that $r_\sigma(x)\leq\lambda p_\alpha(x)\ \forall x\in X$. Thus $p_\alpha^\alpha(x)\leq\{r_\sigma(x)\}^{\alpha-1}p_1(x)$ and $p_\alpha(x)\leq\lambda^{(\alpha-1)/\alpha}p_1(x)$ implies $r_\sigma=p_\alpha\subset p_1$. In particular, $\mathcal{P}_+=p_1=r_\sigma$ is normed and $\dim X$ is finite [Example 19.7(b)]. Clearly (ii)$\Rightarrow$(v) $\{$(i)$\Rightarrow$(vi)$\}$ and the reverse implications follow from Theorem 23.9.]

(e) Verify that (X,p_α) is complete iff $\hat{X}$ is $\|\;\|_\alpha$-closed in $L_\alpha(\mathfrak{M},\nu)$. Furthermore, $(\bar{\hat{X}}^{\|\|_\alpha},\|\;\|_\alpha)\subset L_\alpha(\mathfrak{M},\nu)$ is the completion of (X,p_α).

(f) Extend the fact that p_α-completeness implies r_σ-completeness by proving if $\mathfrak{T}\subset r_\sigma$ on X and $x_n\xrightarrow{\mathfrak{T}}x$ implies $\hat{x}_n\xrightarrow{\text{meas}}\hat{x}$, then (X,r_σ) is complete when $(X,\mathfrak{T})$ is sequentially complete. [*Hint*: Let $f\in\bar{\hat{X}}^{\|\|_0}\subset C(\mathfrak{M};\mathbb{C})$. For each $n\in\mathbb{N}$ there is an $x_n\in X$ such that $\|\hat{x}_n-f\|_0<1/n$. Since $\{\hat{x}_n\}$ is $\|\;\|_0$-Cauchy, $\{x_n\}$ is r_σ-Cauchy and $\hat{x}_n\xrightarrow{\text{meas}}\hat{x}_0\in X$. Thus $\hat{x}_0=f$ a.e. and $f=\hat{x}_0\in\hat{X}$.]

26G. Probability-Induced Pseudometric

For X and ν in Problem 26F, define $(\hat{x} \neq \hat{y}) = \{M \in \mathfrak{M} : \hat{x}(M) \neq \hat{y}(M)\}$ $\forall x, y \in X$.

(a) Show that $d(x, y) = \nu(\hat{x} \neq \hat{y})$ is a pseudometric on X, which is a metric iff X is semisimple. [*Hint*: $(\hat{x} \neq \hat{y}) \subset (\hat{x} \neq \hat{z}) \cup (\hat{z} \neq \hat{y})\ \forall x, y \in X$.]

(b) Define $|x| = d(x, 0)\ \forall x \in X$ and obtain

(i) $|x + y| \leq |x| + |y|$;
(ii) $|\alpha x| = |x|\ \forall \alpha \neq 0$;
(iii) $|xy| \leq \min\{|x|, |y|\}$;
(iv) $|e| = 1$.

[*Hint*: $(\hat{x} + \hat{y} \neq 0) \subset (\hat{x} \neq 0) \cup (\hat{y} \neq 0)$ and $(\hat{x}\hat{y} \neq 0) = (\hat{x} \neq 0) \cap (\hat{y} \neq 0)\ \forall x, y \in X$.]

(c) Use (b) to conclude that $(X, \mathfrak{T}(d))$ is a topological ring but not a topological algebra, since scalar multiplication is not continuous.

(d) If X is semisimple, $\mathfrak{U}$ is closed in $(X, \mathfrak{T}(d))$. [*Hint*: If $\{x_n\} \in \mathfrak{U}$ is $\mathfrak{T}(d)$-convergent to $x_0 \in \bar{\mathfrak{U}}^{\mathfrak{T}(d)}$, then $x_n^{-1} \xrightarrow{\mathfrak{T}(d)} y_0 \in \bar{\mathfrak{U}}^{\mathfrak{T}(d)}$ and $|x_0 y_0 - e| \leq |x_0(y - x_n^{-1})| + |x_n^{-1}(x_0 - x_n)| \leq |y_0 - x_n^{-1}| + |x_0 - x_n|\ \forall n \in \mathbb{N}$.]

Remark. It can also be shown that $\mathfrak{U}$ is open iff $d(x, y) = \begin{cases} 0, & x = y \\ 1, & x \neq y \end{cases}\ \forall x, y \in X$.

(e) $\bar{I}^{\mathfrak{T}(d)}$ is an ideal for each ideal $I \subset X$ and there exist proper closed ideals in $(X, \mathfrak{T}(d))$. [*Hint*: If d is not as in the above remark, $|x_0| < 1$ for some nonzero $x_0 \in X - \mathfrak{U}$. Therefore $|y| \leq |x_0|\ \forall y \in Xx_0$ and $Xx_0 \subset \bar{S}_{|x_0|,d}(0)$ implies $\overline{Xx_0}^{\mathfrak{T}(d)} \subset S_{1,d}(0) \neq X$.]

(f) Maximal ideals need not be closed in $(X, \mathfrak{T}(d))$. Show that there are no closed maximal ideals in $X = C(I; \mathbb{C})$ with Lebesgue measure ν on $\mathfrak{M} = \mathfrak{M}_I$. [*Hint*: Given $M_{t_0} \in \mathfrak{M}$ and $\varepsilon > 0$, there is some $f \in C(I; \mathbb{C})$ with $f(t_0) = 0$ and $d(f, \chi_I) = \nu(\hat{f} \neq \hat{e}) < \varepsilon$. Therefore $e \in \bar{M}_{t_0}^{\mathfrak{T}(d)} - M_{t_0}$.]

(g) $M \in \mathfrak{M}$ is both open and closed in $(X, \mathfrak{T}(d))$ if M is an atom of ν. [*Hint*: Use Theorem 8.7(iv) after showing that $S_{\varepsilon,d}(x_0) \subset M$ for $x_0 \in M$ and $\nu\{M\} = \varepsilon > 0$.]

Remark. A fuller development of the above ideas and related notions are given in Cohen [19], [20].

26H. The Hull-Kernel Topology Revisited

Assume X is a complete complex LMCT$_2$ Q-algebra that is not necessarily commutative nor in possession of an identity.

(a) Show that $\xi(u) = hk(u)\ \forall u \subset \mathfrak{M}_{\mathfrak{R}}$ satisfies (K_1)–(K_4) of Appendix t.2. [*Hint*: For K_3 show that $I_1 I_2 \subset M \Rightarrow I_1 \subset M$ or $I_2 \subset M\ \forall M \in \mathfrak{M}_{\mathfrak{R}}$ and ideals $I_1, I_2 \subset X$. Then establish that $k(E_1)\, k(E_2) \subset k(E_1) \cap k(E_2) = k(E_1 \cup E_2)$ for sets $E_1, E_2 \subset \mathfrak{M}_{\mathfrak{R}}$.]

(b) Show that $(\mathfrak{M}_{\mathfrak{R}}, \mathfrak{T}_{hk})$ is T_1. If $(\mathfrak{M}_{\mathfrak{R}}, \mathfrak{T}_{hk})$ is T_2, every $M \in \mathfrak{M}_{\mathfrak{R}}$ has a $\mathfrak{T}_{hk}$-nbd $v(M_0)$ such that $k\{v(M)\} \in \mathfrak{J}_{\mathfrak{R}}$.

(c) $\mathfrak{M}_{\mathfrak{R}}$ is $\mathfrak{T}_{hk}$-compact if X has an identity. If X is also commutative, $\mathfrak{T}_{hk}$ coincides with Definition 26.15 on $\mathfrak{M} = \mathfrak{M}_{\mathfrak{R}}$.

26I. A Non-T_2 Space $(\mathfrak{M}_{\mathfrak{R}}, \mathfrak{T}_{hk})$

Let $\mathbb{M}_2(\mathbb{C})$ be the Banach algebra of complex matrices $A = \begin{pmatrix} a_{11} & a_{12} \\ a_{21} & a_{22} \end{pmatrix}$ with norm $\|A\| = \sum_{i,j=1}^{2} |a_{ij}|$ and let $X = \{x = \{A_n \in \mathbb{M}_2(\mathbb{C})\} : A_n \xrightarrow{\|\,\|} \begin{pmatrix} b_{11} & 0 \\ 0 & b_{22} \end{pmatrix}$ for some $b_{11}, b_{22} \in \mathbb{C}\}$ carry the sup norm $\|x\|_0 = \sup_n \|A_n\|$.

(a) Establish that $(X, \|\,\|_0)$ is a noncommutative B^*-algebra with identity under conjugate transpose involution.

(b) Verify that $\mathfrak{M}_{\mathfrak{R}} = \{M_{11}, M_{22}, M_m : m \in \mathbb{N}\}$, where

$$M_{11} = \{x \in X : x \xrightarrow{\|\,\|} \begin{pmatrix} 0 & 0 \\ 0 & b_{22} \end{pmatrix}, \text{ some } b_{22} \in \mathbb{C}\}$$

$$M_{22} = \{x \in X : x \xrightarrow{\|\,\|} \begin{pmatrix} b_{11} & 0 \\ 0 & 0 \end{pmatrix}, \text{ some } b_{11} \in \mathbb{C}\}$$

$$M_m = \{x \in X : A_m = 0\}$$

[*Hint*: $M_{ii} = h_i^{-1}(0)$ for $h_i : X \underset{x \rightsquigarrow \begin{pmatrix} b_{11} & 0 \\ 0 & b_{22} \end{pmatrix} \rightsquigarrow b_{ii}}{\longrightarrow} \mathbb{C}$. Furthermore, $M_m = k_m^{-1}(0)$ for $k_m : \underset{x \rightsquigarrow A_m}{X \to \mathbb{M}_2(\mathbb{C})}$ and the ideals of $\mathbb{M}_2(\mathbb{C})$ are $\{0\}$ and $\mathbb{M}_2(\mathbb{C})$ itself.]

(c) Show that M_{11}, M_{22} are accumulation points of $\{M_m : m \in \mathbb{N}\}$ and cannot be separated by disjoint $\mathfrak{T}_{hk}$-open sets.

Remark. The above result (Kaplansky [48]), due to Mackey, also demonstrates that commutativity is necessary in Theorem 26.16(ii).

References for Further Study

Gaal [31]; Naimark [69]; Rickart [82].

CHAPTER V

Abstract Harmonic Analysis

Perspective. It is assumed henceforth that G is a locally compact T_2 group and μr is a weakly regular Haar measure on G, as defined in Note 2, p. 121. We abbreviate $C_0(G; \mathbb{C})$ as $C_0(G)$ and $L_p(G, \mu r)$ as $L_p(G)$ and let $M(G)$ denote the Banach algebra of complex totally finite, regular Borel measures on G. The principal objective of Sections 27 and 28 is the study of $L_1(G)$ and $M(G)$ as Banach convolution algebras. Section 29 extends a number of classical theorems about Fourier series and Fourier integrals over $\mathbb{R}$, $\mathbb{I}$, or $S'(\mathbb{C})$ to general LCT$_2$A groups.

27. The Algebra $L_1(G)$

If $f, g \in L_1(G)$, then $F(x, y) = f(x)\, g(y) \in L_1(G \times G) = L_1(G \times G, \mu r \times \mu r)$. Since $\underset{(x,y) \rightsquigarrow (y^{-1}x, y)}{H: G \times G \to G \times G}$ is an onto homeomorphism and $\mu r \times \mu r$ is left invariant (Example 12.24), $F \circ H(x, y) = f(x^{-1}y)\, g(y) \in L_1(G \times G)$, and by Fubini's theorem

$$\iint_{G\times G} f(x^{-1}y)\, g(y)\, d(\mu r \times \mu r) = \iint_{G\times G} f(x^{-1}y)\, g(y)\, d\mu r(x)\, d\mu r(y)$$

$$= \iint_{G\times G} f(x^{-1}y)\, g(y)\, d\mu r(y)\, d\mu r(x).$$

This establishes that

$$(f \star g)(x) = \int_G f(xy)\, g(y^{-1})\, d\mu r(y) = \int_G f(y)\, g(y^{-1}x)\, d\mu r(y)$$

is measurable and exists almost everywhere.

THEOREM 27.1. $L_1(G)$ is a Banach algebra under multiplication $(f \star g)(x) = \int_G f(xy)\, g(y^{-1})\, d\mu_r(y) = \int_G f(y)\, g(y^{-1}x)\, d\mu_r(y)$.

Proof. It follows from

$$\int_G (f \star g)(x)\, d\mu_r(x) = \int_G \int_G f(xy) g(y^{-1})\, d\mu_r(y)\, d\mu_r(x)$$

$$= \int_G g(y^{-1}) \Big(\int_G f(xy)\, d\mu_r(x) \Big)\, d\mu_r(y)$$

$$= \int_G g(y^{-1}) \Big(\int_G \Delta(y) f(x)\, d\mu_r(x) \Big)\, d\mu_r(y)$$

$$= \Big(\int_G \Delta(y)\, g(y^{-1}) d\mu_r(y) \Big) \Big(\int_G f(x)\, d\mu_r(x) \Big)$$

$$= \Big(\int_G g(y)\, d\mu_r(y) \Big) \Big(\int_G f(x)\, d\mu_r(x) \Big)$$

that $\|f \star g\|_1 \le \|f\|_1 \|g\|_1$ and $f \star g \in L_1(G)\ \forall f, g \in L_1(G)$. The linearity of integration further assures that $f \star (g+h) = f \star g + f \star h$, $(g+h) \star f = g \star f + h \star f$, and $(\alpha f) \star g = \alpha (f \star g) = f \star (\alpha g)\ \forall f, g, h \in L_1(G)$ and $\alpha \in \mathbb{C}$. To prove $\star$ associative observe that ${}_x f \in L_1(G)$ and ${}_x f(z)\, g(z^{-1}y)\, h(y^{-1}) \in L_1(G \times G)\ \forall y, z \in G$ and fixed $x \in G$. Accordingly, $\{(f \star g) \star h\}(x) = \int_G (f \star g)(xy)\, h(y^{-1})\, d\mu_r(y) = \int_G \big(\int_G f(xyz)\, g(z^{-1})\, d\mu_r(z) \big)\, h(y^{-1})\, d\mu_r(y) = \int_G \big(\int_G f(xy)\, g(z^{-1}y)\, d\mu_r(z) \big)\, h(y^{-1})\, d\mu_r(y)$ (substituting $y^{-1}z$ for z) $= \int_G f(xz) \big(\int_G g(z^{-1}y)\, h(y^{-1})\, d\mu_r(y) \big)\, d\mu_r(z)$ (using Fubini's theorem) $= \int_G f(xy) \{(g \star h) z^{-1}\}\, d\mu_r(z) = \{f \star (g \star h)\}(x)$ a.e. on G. ■

Since the Borel restriction of μ_r is regular, $C_0(G)$ is dense in $L_1(G)$ (Appendix m.8). Demonstrating that $C_0(G)$ is convolution subalgebra of $L_1(G)$ therefore yields

COROLLARY 27.2. $L_1(G)$ is the completion of $(C_0(G), \|\ \|_1)$.

Proof. Consider any $f, g \in C_0(G)$ with $f(G-C)=\{0\}$ and $g(G-D)=\{0\}$ for compact sets $C, D \subset G$. If $x \in G$ is fixed, ${}_xf(y)\,g(y^{-1})$ is continuous in y and so $|(f \star g)(x)| = |\int_G f(xy)\,g(y^{-1})\,d\mu_r(y)| = |\int_{x^{-1}C \cap D^{-1}} f(xy)\,g(y^{-1})\,d\mu_r(y)| \le \sup_{x^{-1}C\cap D^{-1}} |{}_xf(y)\,g(y^{-1})|\{\mu_r(C)+\mu_r(D)\} < \infty$. Since $(f \star g)(x) \neq 0$, we have $f(xy)\,g(y^{-1}) \neq 0$ and confirmation that $f \star g$ vanishes outside the compact set $CD \subset G$. To prove $f \star g$ continuous invoke Corollary 12.3. Specifically, for any $\varepsilon > 0$ and $u(e) \in \mathfrak{N}_e$ satisfying $|f(xy)-f(\tilde{x}y)| < \varepsilon$ for $x\tilde{x}^{-1} = (xy)(\tilde{x}y)^{-1} \in u(e)$, we have $|(f \star g)(x) - (f \star g)(\tilde{x})| \le \int_G |f(xy)-f(\tilde{x}y)|\,|g(y^{-1})|\,d\mu_r(y) < \varepsilon \|\hat{g}\|_1$. ■

The reasoning above also reveals that $L_1(G) \star L_1(G) \subset L_1(G)$ and $C_0(G) \star C_0(G) \subset C_0(G)$. This is further extended in

COROLLARY 27.3. (i) $L_p(G) \star L_q(G) \subset C_\infty(G)$ with $\|f \star g\|_0 \le \|f\|_p \|g\|_q$ for $p+q=pq$ $(1<p<\infty)$.

(ii) $L_1(G) \star L_\infty(G) \subset CB(G)$ with $\|f \star g\|_0 \le \|f\|_1 \|g\|_\infty$.

Proof. Given any $f \in L_p(G)$ and $g \in L_q(G)$, there exist sequences $\{f_n\}$, $\{g_n\} \in C_0(G)$ such that $f_n \xrightarrow{\|\,\|_p} f$ and $g_n \xrightarrow{\|\,\|_q} g$. Using Hölder's inequality and Corollary 12.11, we obtain

$$|(f \star g)(x) - (f_n \star g)(x)| \le \int_G |f(xy)\,g(y^{-1}) - f_n(xy)\,g_n(y^{-1})|\,d\mu_r(y)$$

$$\le \int_G |f(xy)-f_n(xy)|\,|g(y^{-1})|\,d\mu_r(y) + \int_G |g(y^{-1})-g_n(y^{-1})|\,|f_n(xy)|\,d\mu_r(y)$$

$$\le \|{}_xf - {}_x\{f_n\}\|_p \|\hat{g}\|_q + \|\hat{g}-\hat{g}_n\|_q \|{}_xf\|_p = \|f-f_n\|_p\|g\|_q + \|g-g_n\|_q\|f_n\|_p.$$

Since $\{\|f_n\|_p : n \in \mathbb{N}\}$ is bounded, $f_n \star g_n \to f \star g$ uniformly and $f \star g \in \overline{C_0(G)}^{\|\,\|_0} = C_\infty(G)$. The above reasoning also yields $|(f \star g)(x)| \le \|{}_xf\|_p\|g\|_q = \|f\|_p\|g\|_q$ $\forall x \in G$, and so $\|f \star g\|_0 \le \|f\|_p\|g\|_q$. If $f \in L_1(G)$ and $g \in L_\infty(G)$ in (ii), then $|(f \star g)(x)| \le \|f\|_1\|g\|_\infty$ $\forall x \in G$. The continuity of $f \star g$ follows via Note 4, p. 122, applied to $|(f \star g)(x)-(f \star g)(y)| \le \|{}_xf - {}_yf\|_1\|g\|_\infty$. ■

THEOREM 27.4. (i) G, $C_0(G)$, and $L_1(G)$ are simultaneously commutative or noncommutative.

(ii) $C_0(G)$ and $L_1(G)$ have an identity iff G is discrete.

Proof. First suppose G is commutative and let $f, g \in L_1(G)$. Then $(f \star g)(x) = \int_G f(xy)\, g(y^{-1})\, d\mu_r(y) = \int_G g(xy)\, f(y^{-1})\, d\mu_r(y^{-1}x^{-1}) = \int_G g(xy)\, f(y^{-1})\, d\mu_r(y) = (g \star f)(x)\ \forall x \in G$ by Corollary 12.14 and the substitution of $y^{-1}x^{-1}$ for y. Therefore $L_1(G)$, and so $C_0(G)$, is commutative. If $C_0(G)$ is commutative and $f, g \in C_0(G)$, we use Theorem 12.10 to obtain

$$
\begin{aligned}
0 = (f \star g - g \star f)(x) &= \int_G \{f(xy)\, g(y^{-1}) - g(xy)\, f(y^{-1})\}\, d\mu_r(y) \\
&= \int_G \{f(xy)\, g(y^{-1}) - g(y)\, f(y^{-1}x)\}\, d\mu_r(y) \\
&= \int_G \{\Delta(y)\, f(xy^{-1})\, g(y) - g(y)\, f(y^{-1}x)\}\, d\mu_r(y) \\
&= \int_G \{\Delta(y)\, f(xy^{-1}) - f(y^{-1}x)\}\, g(y)\, d\mu_r(y).
\end{aligned}
$$

Since this is true $\forall g \in C_0(G)$ it follows (Corollary 12.17) that $\Delta(y)\, f(xy^{-1}) - f(y^{-1}x) \equiv 0$ for each fixed $x \in G$. Taking $x = e$ yields $\Delta(y) \equiv 1$. Thus $f(xy^{-1}) = f(y^{-1}x)\ \forall f \in C_0(G)$ and [since $C_0(G)$ separates points in G] $xy^{-1} = y^{-1}x\ \forall x, y \in G$. This completes the proof of (i). If G in (ii) is discrete (Corollary 12.16) $\mu_r(e) > 0$ and $[1/\mu_r(e)]\delta_{e,x}$ is the identity both in $C_0(G)$ and in $L_1(G)$. Suppose, conversely, that ξ is the identity in $L_1(G)$. This, we claim, necessitates the existence of an $\alpha > 0$ such that $\mu_r(u) \geq \alpha$ for every open Borel set $u \neq \varnothing$ in G. Reason: Otherwise taking $f \in C_0(G)$ with $\|\xi - f\|_1 < \varepsilon/2$ and open u with $\mu_r(u) < \varepsilon/2\|f\|_0$, we have $\int_u |\xi|\, d\mu_r = \|\chi_u \xi\|_1 \leq \|\chi_u \xi - \chi_u f\|_1 + \|\chi_u f\|_1 < \varepsilon/2 + \|f\|_0 \mu_r(u) < \varepsilon$. If $x \in v(e)$ for a symmetric nbd $v(e)$ satisfying $v^2 \subset u$, we obtain $1 = \chi_v(x) = (\xi \star \chi_v)(x) = \int_G \xi(y) \chi_v(y^{-1}x)\, d\mu_r(y) = \int_{xv} \xi(y)\, d\mu_r(y) \leq \int_u |\xi(y)|\, d\mu_r(y) < \varepsilon$, which is clearly impossible for $\varepsilon < 1$. Having established the above claim, we conclude that every relatively compact, noncompact open set w contains only finitely many elements. (As in Corollary 12.16, w can be covered by n disjoint nonempty open subsets and $\infty > \mu r(w) \geq n\alpha$.) Accordingly, $\{e\}$ is open and G is discrete. ■

Remark. Although $\|f^* \star f\|_1 \leq \|f\|_1^2 \; \forall f \in L_1(G)$, the reverse inequality need not hold. Therefore $L_1(G)$ is not necessarily B^*.

If G is nondiscrete, $C_0(G)$ and $L_1(G)$ lack an identity. One technique for overcoming this deficiency utilizes an approximate identity; that is, a directed set $\mathscr{E} \subset C_0^+(G)$ that satisfies $\lim_{\mathscr{E}} \xi_\varepsilon \star F = F = \lim_{\mathscr{E}} F \star \xi_\varepsilon \; \forall F \in L_1(G)$.

THEOREM 27.5. For fixed $F \in L_1(G)$ and each $\varepsilon > 0$ there is a G-nbd $w_\varepsilon(e)$ and a $\xi_\varepsilon \in C_0^+(G)$ with $\xi_\varepsilon(G - w_\varepsilon) = \{0\}$ and $\int_G \xi \, d\mu r = 1$ such that $\|\xi_\varepsilon \star F - F\|_1 < \varepsilon$ and $\|F \star \xi_\varepsilon - F\|_1 < \varepsilon$.

Proof. For each $\varepsilon > 0$ choose a symmetric nbd $u_\varepsilon(e)$ such that $\|_{y^{-1}}F - F\|_1 < \varepsilon \; \forall y \in u_\varepsilon(e)$. Suppose further (Corollary 8.21) that $\xi_\varepsilon \in C_0^+(G)$ is a Urysohn function satisfying $\xi_\varepsilon(G-u) = \{0\}$ and $\int_G \xi_\varepsilon \, d\mu r = 1$. Then

$$\|\xi_\varepsilon \star F - F\|_1 = \int_G \left| \int_G \xi_\varepsilon(y) F(y^{-1}x) \, d\mu r(y) - F(x) \right| d\mu r(x) =$$

$$\int_G \left| \int_G \xi_\varepsilon(y) \{_{y^{-1}}F(x) - F(x)\} \, d\mu r(y) \right| d\mu r(x)$$

$$\leq \int_G \xi_\varepsilon(y) \|_{y^{-1}}F - F\|_1 \, d\mu r(y) < \varepsilon.$$

The other inequality $\|F \star \xi_\varepsilon - F\|_1 < \varepsilon$ is somewhat complicated by the need for modularity. Letting $m_\varepsilon = \int_G \xi_\varepsilon(x^{-1}) \, d\mu r(x)$, first obtain

$$\|F \star \xi_\varepsilon - F\|_1 = \int_G \left| \int_G F(xy)\xi_\varepsilon(y^{-1}) \, d\mu r(y) - [F(x)/m_\varepsilon] \int_G \xi_\varepsilon(y^{-1}) \, d\mu r(y) \right| d\mu r(x)$$

$$= \int_G \left| \int_G \{F_y(x) - [F(x)/m_\varepsilon]\} \, d\mu r(x) \right| \xi_\varepsilon(y^{-1}) \, d\mu r(y)$$

$$\leq \int_G \|m_\varepsilon F_y - F\|_1 [\xi_\varepsilon(y^{-1})]/m_\varepsilon \, d\mu r(y) \leq \|m_\varepsilon F_y - F\|_1.$$

Now direct $\mathscr{E}=\{\xi_\varepsilon : u_\varepsilon(e)\in\mathfrak{N}_e\}$ by $\xi_{\varepsilon_1}<\xi_{\varepsilon_2}$ iff $u_{\varepsilon_2}(e)\subset u_{\varepsilon_1}(e)$ and observe that $m_\varepsilon\to 1$ as $u_\varepsilon(e)\to e$ (Theorem 12.10). In particular, $\|m_\varepsilon F_y - F_y\|_1<\varepsilon/2$ for $y\in u_\varepsilon(e)$, and $\|F_y-F\|_1<\varepsilon/2$ for $y\in v(e)$ yields $\|m_\varepsilon F_y - F\|_1\le \|m_\varepsilon F_y - F_y\|_1+\|F_y-F\|_1\ \forall y\in w_\varepsilon = u_\varepsilon(e)\cap v(e)$. ∎

The use of an approximate identity is nicely illustrated in

COROLLARY 27.6. A closed subset $I\subset L_1(G)$ is a left {right} ideal iff it is invariant under left {right} translations.

Proof. If I is a left ideal in $L_1(G)$ and $\mathscr{E}=\{\xi_\varepsilon\}$ is an approximate identity of $L_1(G)$, then $F\in I$ implies $\xi_\varepsilon \star F\in I\ \forall \xi_\varepsilon\in\mathscr{E}$ and $\lim_{\mathscr{E}} \xi_\varepsilon \star F = F$ yields $\lim_{\mathscr{E}} {}_x(\xi_\varepsilon \star F) = {}_xF\in \bar{I}=I\ \forall x\in G$. Assume, conversely, that $F\in I$, where I is left invariant. If $\psi\in C_0(G)$, then $\underset{y \rightsquigarrow \psi(y)\,{}_{y^{-1}}F}{G\to C_0(G)}$ is $\|\ \|_1$-continuous and $\{\psi \star F\}(x)=\int_G \psi(y)\,{}_{y^{-1}}F(x)\,d\mu(y)$ is the $\|\ \|_1$-limit of sums of the form $\sum_{k=1}^{n}\psi(y_k)\,{}_{y_k^{-1}}F\in\bar{I}$. Therefore, $\psi \star F\in\bar{I}=I$. If $\eta\in L_1(G)$ and $\{\psi_n\}\in C_0(G)$ satisfy $\|\psi_n-\eta\|_1<(n\|F\|_1)^{-1}\ \forall n\in\mathbb{N}$, we have $\|\psi_n \star F-\eta \star F\|_1 = \|(\psi_n-\eta)\star F\|_1<1/n$ and $\eta \star F=\lim_n \psi_n \star F\in\bar{I}=I$ as required. The same proof slightly modified holds for right ideals. ∎

Fourier Transforms

For the remainder of this section G is assumed to be abelian.

DEFINITION 27.7. The Fourier transform of $f\in L_1(G)$ is $\hat{f}(\chi)=\int_G f(x)\chi(x)\,d\mu(x)$ $(\chi\in G')$.

If $X=L_1(G)$ has no identity and $R(G)$ denotes the Banach algebra X_e formed by adjoining an identity to $L_1(G)$, then $L_1(G)=(0,L_1(G))$ is a maximal ideal in $R(G)$ and $\mathcal{H}_{R(G)}-h_{(0,L_1(G))}=\mathcal{H}_{L_1(G)}$ [the homomorphisms of $L_1(G)$]. Reason: Each arrow below denotes bijection.

$$\begin{array}{ccc} & \text{(Problem 23B)} & \\ \mathfrak{M}_{\mathcal{R}} & \longleftrightarrow & \mathfrak{M}_e-(0,X) \\ \text{(Problem 26A)}\ \updownarrow & & \updownarrow\ \text{(Problem 26A)} \\ \mathcal{H} & & \mathcal{H}_{X_e}-h_{(0,X)} \end{array}$$

We now proceed to prove $\mathcal{H}_{L_1(G)}\leftrightarrow G'$.

THEOREM 27.8. Each $\underset{f \rightsquigarrow \hat{f}(\chi)}{h_\chi : L_1(G) \to \mathbb{C}}$ $(\chi \in G')$ is a nonzero homomorphism. Furthermore, $\underset{\chi \rightsquigarrow h_\chi}{G' \to \mathcal{H}_{L_1(G)}}$ is bijective and $\mathcal{H}_{L_1(G)} = \{h_\chi : \chi \in G'\}$.

Proof. Since integration preserves linearity, h_χ is linear; and h_χ is multiplicative since

$$h_\chi(f_1 \star f_2) = \int_G (f_1 \star f_2)(x)\,\chi(x)\,d\mu_r(x)$$

$$= \int_G \left\{ \int_G f_1(xy)\,f_2(y^{-1})\,d\mu_r(y) \right\} \chi(x)\,d\mu_r(x)$$

$$= \int_G \left\{ \int_G f_1(xy)\,\chi(xy)\,f_2(y^{-1})\,\chi(y^{-1})\,d\mu_r(y) \right\} d\mu_r(x)$$

$$= \int_G \left\{ \int_G f_1(xy)\,\chi(xy)\,d\mu_r(x) \right\} f_2(y^{-1})\,\chi(y^{-1})\,d\mu_r(y) = \hat{f}_1(\chi)\,\hat{f}_2(\chi)$$

(via Corollary 12.11) and this equals $h_\chi(f_1)\,h_\chi(f_2)$. If $f \in L_1(G)$ and $\int_G f\,d\mu_r \neq 0$, then $f/\chi \in L_1(G)$ and $h_\chi(f/\chi) = \int_G f\,d\mu_r \neq 0$. This proves the theorem's first assertion and also establishes that $h_\chi \in \mathcal{H}_{L_1(G)}\ \forall \chi \in G'$. The correspondence $\chi \to h_\chi$ is injective by Corollary 12.17. To see that it is also surjective consider any $h \in \mathcal{H}_{L_1(G)}$ and (Appendix m.7) let $\xi \in L_\infty(G)$, with $\|\xi\|_\infty = |||h||| \leq 1$, satisfy $h(F) = \int_G F(x)\,\xi(x)\,d\mu_r(x)\ \forall F \in L_1(G)$. For any $f, g \in L_1(G)$ we now have

$$\int_G h(f)\,g(y)\,\xi(y)\,d\mu_r(y) = h(f) \int_G g(y)\,\xi(y)\,d\mu_r(y) = h(f)\,h(g) = h(f \star g)$$

$$= \int_G (f \star g)(x)\,\xi(x)\,d\mu_r(x)$$

$$= \int_G \left\{ \int_G f(xy)\,g(y^{-1})\,d\mu_r(y) \right\} \xi(x)\,d\mu_r(x)$$

$$= \int_G \left\{ \int_G f_y(x)\,\xi(x)\,d\mu_r(x) \right\} g(y^{-1})\,d\mu_r(y)$$

$$= \int_G h(f_y)\,g(y)\,d\mu_r(y).$$

If we fix $f \in L_1(G)$ and equate the first and last terms of the above equality, there follows $h(f)\xi(y) = h(f_y)$ a.e. on G. Note also that the continuity of $h \in \mathcal{K}_{L_1}(G)$ and $\underset{y \rightsquigarrow f_y}{G \to L_1(G)}$ ensures that of $(\tilde{h}f)(y) = h\{f_y\}$ on G. Therefore $\psi = [\tilde{h}F/h(F)]$ is a continuous homomorphism on G $\forall F \in L_1(G)$ such that $h(F) \neq 0$ (clearly $h(F)\psi(xy) = h(F_{xy}) = h\{(F_x)_y\} = h(F_x)\psi(y) = h(F)\,\psi(x)\,\psi(y)$ $\forall x, y \in G$). Since $h^{-1}(0) \neq L_1(G)$, we have $\xi = \psi$ a.e. on G and $h(f) = \int_G f\xi\,d\mu_r = \int_G f\psi\,d\mu_r$ $\forall f \in L_1(G)$. Moreover, $\|\psi\|_\infty \leq 1$ implies that $|\psi(x)| \leq 1$ a.e. and this requires that $\sup_G |\psi(x)| \leq 1$ (otherwise $\psi(x_0) > 1$ implies that $|\psi(x)| > 1$ on some open nbd $u(x_0)$ and $\mu_r(u) > 0$ by Note 2, p. 121). Actually $1 = \psi(e) = \psi(xx^{-1}) = \psi(x)\,\psi(x^{-1})$ ensures that $|\psi(x)| = 1$ $\forall x \in G$. Therefore $\psi \in G'$ and our proof is complete, since $h(f) = \hat{f}(\psi) = h_\psi(f)$ $\forall f \in L_1(G)$. ■

Remark. A simple calculation reveals that $\underset{(\alpha,f) \rightsquigarrow \alpha + \hat{f}(\chi)}{h_\chi : R(G) \longrightarrow \mathbb{C}}$ is a homomorphism that does not vanish on the maximal ideal $(0, L_1(G))$ of $R(G)$.

It is no accident that $\hat{f}$ denotes both the Gelfand and Fourier transform of $f \in L_1(G)$. Indeed $\underset{\chi \rightsquigarrow h_\chi^{-1}(0) = M_\chi}{G' \longrightarrow \mathfrak{M}_{\mathcal{R}}}$ is bijective and $\hat{f}(M_\chi) = h_\chi(f) = \hat{f}(\chi)$ $\forall \chi \in G'$. Our identification of transforms is even sharper, since $G'_\kappa = (\mathfrak{M}_{\mathcal{R}}, \mathfrak{T}_g)$, where equality denotes homeomorph. To see this let $\overset{(f)}{\widehat{L_1(G)}}$ and $\overset{(g)}{\widehat{L_1(G)}}$ denote the respective Fourier and Gelfand transforms of all $f \in L_1(G)$.† Then $(G', \mathfrak{T}_{\mathcal{P}\{\overset{(f)}{\widehat{L_1(G)}}\}}) = (\mathfrak{M}_{\mathcal{R}}, \mathfrak{T}_{\mathcal{P}\{\overset{(g)}{\widehat{L_1(G)}}\}} = \mathfrak{T}_g)$ (simply identify $v(\chi_0; \hat{f}_1 \cdots \hat{f}_n; \varepsilon) = \{\chi \in G' : |f_i(\chi) - f_i(\chi_0)| < \varepsilon \ \forall 1 \leq i \leq n\}$ with $v(M_{\chi_0}; \hat{f}_1 \cdots \hat{f}_n; \varepsilon) \subset \mathfrak{M}_{\mathcal{R}}$), and $\mathfrak{T}_{\mathcal{P}\{\overset{(f)}{\widehat{L_1(G)}}\}} = \mathfrak{T}_\kappa$ on G' will be proved shortly. First, we return to familiar surroundings (Examples 11.6 and 23.5).

Example 27.9. (a) $L_1(\mathbb{R})$ has no identity. To be sure, $\mathcal{K}_{L_1(\mathbb{R})} = \{h_{\chi_\theta} : \theta \in \mathbb{R}\}$ and $\mathfrak{M}_{\mathcal{R}} = \{M_\theta : \theta \in [0, 2\pi)\}$, where $M_\theta = \{f \in L_1(\mathbb{I}) : 0 = \hat{f}(\chi_\theta) = \int_{-\infty}^{\infty} f(t)e^{i\theta t}\,dt\}$. Since $\{\mathbb{R}\}' = \mathbb{R}$, we also write $\hat{f}(\theta) = \int_{-\infty}^{\infty} f(t)e^{i\theta t}\,dt$.

† The superscripts (f) and (g) are often omitted, since members of $\widehat{L_1(G)}$ can be deciphered by their domain of definition.

(b) $L_1(\mathbb{R}^n)$ has no identity. In this framework, $\hat{f}(\chi_\theta) = \int \cdots \int_{\mathbb{R}^n} f(t_1 \cdots t_n) e^{i(t_1\theta_1 + \cdots + t_n\theta_n)} \, d\theta \cdots d\theta \; \forall f \in L_1(\mathbb{R}^n)$ and $\chi_\theta \in \{\mathbb{R}^n\}'$.

(c) Since $\mathbb{I}$ is discrete, $L_1(\mathbb{I})$ has an identity. Here $\mathcal{H}_{L_1(\mathbb{I})} = \{h\chi_\theta : \theta \in [0, 2\pi)\}$ and $\mathfrak{M} = \{M_\theta : \theta \in [0, 2\pi]\}$, where $M_\theta = \{f \in L_1(\mathbb{I}) : 0 = \hat{f}(\chi_\theta) = \sum_{n=-\infty}^{\infty} f(n) e^{in\theta}\}$.

Remark. We can easily verify [Example 23.3(d)] that $L_1(\mathbb{I}) \underset{f \rightsquigarrow \{f(n)\}}{\longrightarrow} \mathcal{W}$ is a *-congruence under $f^* = \Delta\bar{\bar{f}}$.

(d) Identify $S'(\mathbb{C})$ with $[0, 2\pi)$ and observe that $L_1\{S'(\mathbb{C})\} = L_1([0, 2\pi)) = \{\text{measurable } f : \|f\|_1 = \int_0^{2\pi} |f(\theta)| \, d\theta < \infty\}$ has no identity. In this framework $\mathcal{H}_{L_1\{S'(\mathbb{C})\}} = \{\chi_n : n \in \mathbb{I}\}$ and $\mathfrak{M}_{\mathfrak{R}} = \{M_n : n \in \mathbb{I}\}$, where $M_n = \{f \in L_1 : 0 = \hat{f}(\chi_n) = \int_{S'(\mathbb{C})} f(e^{i\theta}) e^{in\theta} \, d\mu_r(\theta) = \int_0^{2\pi} f(\theta) e^{in\theta} \, d\theta\}$.

(e) $L_1(G)$ has an identity for G in Example 11.19(a). Furthermore, $\hat{f}(\chi_m) = \sum_{k=1}^{n} f(g^k) e^{(2mk\pi i)/n} \; \forall f \in L_1(G)$ and $\chi_m \in G'$.

Verification that $\mathfrak{T}_{\mathcal{P}\{\widehat{L_1(G)}\}} = \mathfrak{T}_\kappa$ on G' utilizes

THEOREM 27.10. $G \times G' \underset{(x,\chi) \rightsquigarrow \chi(x)}{\longrightarrow} \mathbb{C}$ is continuous when G' carries $\mathfrak{T}_{\mathcal{P}\{\widehat{L_1(G)}\}}$.

Proof. Fix any $(x_0, \chi_0) \in G \times G'$ and [since $h_{\chi_0} \not\equiv 0$ on $L_1(G)$] choose $F \in L_1(G)$ such that $1 = h_{\chi_0}(F) = \hat{F}(\chi_0)$. Clearly $\chi_0 = \psi$ in Theorem 27.8 gives $\chi_0(y) = \hat{F}_y(\chi_0) \; \forall y \in G$ and for any $(x, \chi) \in G \times G'$, we have $|\chi(x) - \chi_0(x_0)| \le |\chi(x)\hat{F}(\chi_0) - \chi(x)\hat{F}(\chi)| + |\chi(x)\hat{F}(\chi) - \chi(x_0)\hat{F}(\chi)| + |\chi(x_0)\hat{F}(\chi) - \chi_0(x_0)| = |\hat{F}(\chi_0) - \hat{F}(\chi)| + |\chi(x)\hat{F}(\chi) - \chi(x_0)\hat{F}(\chi)| + |\chi(x_0)\hat{F}(\chi) - \hat{F}_{x_0}(\chi_0)|$. If $\chi \in v(\chi_0; \hat{F}, \hat{F}_{x_0}; \varepsilon/3)$, then $h_\chi = \hat{F}(\chi) \neq 0$ and $\chi(y) = \hat{F}_y(\chi)/\hat{F}(\chi) \; \forall y \in G$ (again by taking $\chi = \psi$ in Theorem 27.8). We now have $|\chi(x) - \chi_0(x_0)| \le \varepsilon/3 + |\hat{F}_x(\chi) - \hat{F}_{x_0}(\chi)| + \varepsilon/3 \le 2\varepsilon/3 + \|F_x - F_{x_0}\|_1$ and it remains only (Note 4, p. 122) to choose $u(e) \in \mathfrak{N}_e$ such that $\|F_x - F_{x_0}\|_1 < \varepsilon/3 \; \forall x \in w(x_0) = u(e)x_0$. ■

COROLLARY 27.11. $\mathfrak{T}_{\mathcal{P}\{\widehat{L_1(G)}\}} = \mathfrak{T}_\kappa$ on G'.

Proof. It suffices to demonstrate that these topologies have internesting nbd bases at $\chi_G \in G'$. For each $f \in L_1(G)$ and $\varepsilon > 0$ there is a $\phi_f \in$

$C_0(G)$ such that $\int_G |f-\phi_f|\,d\mu_r<\varepsilon/4$. Since $\phi_f(G-K_f)=\{0\}$ for some compact set $K_f\subset G$, we also have $\int_{G-K_f} |f|\,d\mu_r = \int_{G-K_f} |f-\phi_f|\,d\mu_r<\varepsilon/4$. Given any $\mathfrak{T}_{\mathcal{P}\{\widehat{L_1(G)}\}}$-nbd $v(\chi_G)=v(\chi_G;\hat{f}_1\cdots\hat{f}_n;\varepsilon)$, take $K=\bigcup_{i=1}^n K_{f_i}$ and let $m=\max_{1\le i\le n}\|f_i\|_1$. If $\chi\in u_{K,\varepsilon/2m}(\chi_G)$, then

$$\left|\int_G f_i(x)\{\chi(x)-1\}\,d\mu_r(x)\right|$$

$$\le \int_{G-K} |f_i(x)|\,|\chi(x)-1|\,d\mu_r(x)+\int_K |f_i(x)|\,|\chi(x)-1|\,d\mu_r(x)$$

$$\le 2\cdot\varepsilon/4+\varepsilon\|f_i\|/2m<\varepsilon \quad \forall 1\le i\le n.$$

Thus $u_{K,\varepsilon/2m}(\chi_G)\subset v(\chi_G)$ and $\mathfrak{T}_{\mathcal{P}\{\widehat{L_1(G)}\}}\subset\mathfrak{T}_K$. For the reverse inclusion consider any compact set $C\subset G$ and $\varepsilon>0$. Taking $x_0=e$ and $\chi_0=\chi_G$ in Theorem 27.10 ensures that $|\chi(x)-1|<\varepsilon$ for $x\in u(e)$ and $\chi\in v(\chi_G;\hat{F};\varepsilon/3)$. Since C is compact, $C=\bigcup_{i=1}^m x_i^{-1}u(e)$ for some set $\{x_1,\ldots,x_m\}\subset G$ and it remains only to note that $v(\chi_G)= v(\chi_G;\hat{F},\hat{F}_{x_1}\cdots\hat{F}_{x_m};\varepsilon/3)$ is contained in $u_{C,\varepsilon}(\chi_G)$. Reason: $(x,\chi)\in C\times v(\chi_G)$ requires that x belongs to some $x_i^{-1}u$, and $x_i^{-1}\in v(\chi_G;\hat{F};\varepsilon/3)$ implies $|\chi(x)-1|=|_{x_i^{-1}}\chi(x_ix)-1|<\varepsilon$. ■

We show shortly (Theorem 29.8) that the topology on G can also be defined by way of G'-compact sets. A step in this direction begins with

COROLLARY 27.12. $v_{K,\varepsilon}(x_0)=\{x\in G:\sup_{x\in K}|\chi(x)-\chi(x_0)|<\varepsilon\}$ is a G-open set $\forall$ compact $K\subset G'$ and $\varepsilon>0$.

Proof. Fix $x_0=e$ and $\chi_0=\chi_G$. For each $\chi\in G'$, there exist open nbd $v_\chi(\chi_G)\subset G'$ and $w_\chi(e)\subset G$ such that $|\chi(x)-1|<\varepsilon$ for $x\in w_\chi(e)$ and $\chi\in v_\chi(\chi_G)$. Then $K=\bigcup_{i=1}^n v_{\chi_i}(\chi_G)$ and $u(e)=\bigcap_{i=1}^n w_{\chi_i}(e)\subset v_{K,\varepsilon}(e)$ for some $\{\chi_1\cdots\chi_n\}\in G'$. In particular, $v_{K,\varepsilon}(e)\subset G$ is open and so is $v_{K,\varepsilon}(x_0)= x_0v_{K,\varepsilon}(e)\ \forall x_0\in G$. ■

Theorem 27.8 can also be used to establish

THEOREM 27.13. $L_1(G)$ is a *-normed, A^*-algebra under $f\rightsquigarrow f^*=\Delta\bar{f}$.

Proof. That * is an involution on $L_1(G)$ is easily verified. For instance,

$$
\begin{aligned}
(g^* \star f^*)(x) &= \int_G g^*(xy) f^*(y^{-1})\, d\mu_r(y) \\
&= \int_G \Delta(xy)\, \overline{g(y^{-1}x^{-1})}\, \Delta(y^{-1})\, \overline{f(y)}\, d\mu_r(y) \\
&= \Delta(x) \int_G \overline{g(y^{-1}x^{-1}) f(y)}\, d\mu_r(y) \\
&= \Delta(x) \int_G \overline{f(x^{-1}y)\, g(y^{-1})}\, d\mu_r(y) \\
&= \Delta(x)\overline{(f \star g)}(x^{-1}) = (f \star g)^*(x)\ \forall x \in G.
\end{aligned}
$$

Now $L_1(G)$ is *-normed, since $C_0(G)$ is dense in $L_1(G)$ and Theorem 12.10 carries over to $L_1(G)$, giving $f^* \in L_1(G)$ and $\|f\|_1 = \|f^*\|_1\ \forall f \in L_1(G)$. Finally, we use $\chi(x^{-1}) = 1/\chi(x) = \overline{\chi(x)}$ to get

$$
\begin{aligned}
\widehat{f^*}(\chi) &= \int_G f^*(x)\chi(x)\, d\mu_r(x) = \int_G \overline{f(x^{-1})}\chi(x)\, d\mu_r(x) \\
&= \int_G \overline{f(x)}\chi(x^{-1})\, d\mu_r(x) = \overline{\hat{f}(\chi)}\ \forall \chi \in G'
\end{aligned}
$$

and then obtain $\hat{f}^*(M_\chi) = \widehat{f^*}(\chi) = \overline{\hat{f}(\chi)} = \overline{\hat{f}(M_\chi)}\ \forall M_\chi \in \mathfrak{M}_{\mathfrak{R}}$. ■

Remark. Although $\|f^* \star f\|_1 \leq \|f\|_1^2$, the reverse inequality need not hold and so $L_1(G)$ need not be B^*. In fact, $L_1(G)$ need not even be almost B^* [Problem 23J(a)] so that the ordering of Theorem 26.6 can only be assumed in one direction.

Since $|\hat{f}(\chi)| \leq \|f\|_1\ \forall \chi \in G'$, it follows that $\|\hat{f}\|_0 \leq \|f\|_1$ and $\widehat{L_1(G)} \subset CB(G')$. We can, of course, say much more, since [Problem 26A(d)] $\overset{(f)}{\widehat{L_1(G)}} \subset C_\infty(G')$ is the same as $\overset{(g)}{\widehat{L_1(G)}} \subset C_\infty(\mathfrak{M}_{\mathfrak{R}})$ and the reasoning (Appendix m.8) in Theorem 25.10(iii) also yields

COROLLARY 27.14. $\overset{(f)}{\widehat{L_1(G)}}\{\overset{(g)}{\widehat{L_1(G)}}\}$ is a $\|\ \|_0$-dense, separating subalgebra of $C_\infty(G')\ \{C_\infty(\mathfrak{M}_{\mathfrak{R}})\}$.

PROBLEMS

27A. $L_1(G)$ Convolution

Show that

$$(f\star g)(x)=\int_G \Delta(y)f(xy^{-1})g(y)\,d\mu_r(y)=\int_G \Delta(y)f(y^{-1})g(yx)\,d\mu_r(y)\,\forall f,g\in L_1(G).$$

27B. Convolution of Translates

For each $f, g\in L_1(G)$ and $x\in G$ establish that

(i) ${}_xf\star g={}_x(f\star g)$ and $f\star g_x=(f\star g)_x$;
(ii) $f_x\star g=\Delta(x^{-1})\{f\star {}_xg\}=\Delta(x)\{f\star {}_{x^{-1}}g\}$;
(iii) $f\star g_x=f\star {}_xg=f_x\star g={}_xf\star g$ if G is commutative.

27C. Convolution of Summable Functions

Let $L_p(G)=L_p(G,\mu_L)$, where μ_L is extended left Haar measure on a locally compact T_2 group G and use the superscript (r) to denote concepts relative to the right Haar measure μ_R determined by μ_L (cf remarks on p. 112). For example, $\|f\|_p^{(r)}=\{\int_G |f|^p\,d\mu_R\}^{1/p}$ for $f\in L_p^{(r)}=L_p(G,\mu_R)$. Assuming that $1/p+1/q=1$ whenever $1<p,\ q<\infty$ appear together, prove

(a) $f\star g\in L_p(G)$ with $\|f\star g\|_p\le\|f\|_1\|g\|_p$ $\forall(f,g)\in L_1(G)\times L_p(G)$ and $1\le p<\infty$;
(b) $f\star g\in L_\infty(G)$ with $\|f\star g\|_\infty\le\|f\|_\infty\|g\|_1^{(r)}$ $\forall(f,g)\in L_\infty(G)\times L_1^{(r)}(G)$;
(c) $f\star g\in L_\infty(G)$ with $\|f\star g\|_\infty\le\|f\|_p\|g\|_q^{(r)}$ $\forall(f,g)\in L_p(G)\times L_q^{(r)}(G)$;
(d) Use Corollary 27.3(i) to show that $L_1(G)$ is Banach if G is compact.

27D. Dieudonne's Convolution Theorems

G is an LCT_2A group and $\mathscr{L}_0(X)$ is as defined in Problem 23D.

(a) Prove that $L_1(G)=\mathscr{L}_0\{L_1(G)\}$. [*Hint*: G' is dense in G'_∞ and so $\chi_n\to\infty\in G'_\infty$ for some sequence $\{\chi_n\}\in G'$. Since $\widehat{L_1(G)}\subset C_\infty(G')=\{\xi\in C(G'_\infty):\xi(\infty)=0\}$ (Appendix t.8), $\lim_{n\to\infty}\hat{\xi}(\chi_n)=0\ \forall\hat{\xi}\in\widehat{L_1(G)}$. Let $g_n=\chi_nf$ for $f\in L_1(G)$ with $\|f\|_1=1$. Then $\|g_n\|_1=1\ \forall n\in\mathbb{N}$ and for any $F\in L_1(G)$, the remarks above yield $F\star g_n \xrightarrow{\|\ \|_1} 0$.]

(b) Use (a) to obtain $F \star L_1(G) \subsetneq L_1(G)\ \forall F \in L_1(G)$. {*Hint*: ${}_{x_0}\mathfrak{M}(X) \subsetneq X\ \forall x_0 \in \bigcup_{\mathfrak{I}(X)} I$ (Theorem 22.8) and $\mathfrak{T}_0(X) \subset \bigcup_{\mathfrak{I}(X)} I$ when X is semisimple [Problem 23D(b)]; but $X = L_1(G)$ is semisimple (Corollary 29.18).}

Remark. (Burnham and Goldberg [15]) $F \star L_p(G) \subsetneq L_p(G)\ \forall F \in L_1(G)$.

28. The Algebra $M(G)$ and Its Components

For $f \in C_\infty(G)$ and $\Phi \in C'_\infty(G)$ the mapping $\tilde{\Phi} f: \underset{x \rightsquigarrow \Phi({}_x f)}{G \to \mathbb{C}}$ belongs to $C_\infty(G)$. Continuity follows by Note 4 on p. 122. For verification that $\tilde{\Phi} f$ "vanishes at infinity" it suffices to assume that f and Φ are positive and (Appendix m.8) $\Phi = \Phi_\mu$ for some $\mu \in M^+(G)$. Given $\varepsilon > 0$, choose a compact set $K \subset G$ having $0 < \mu(K)$ and $\mu(G-K) < \delta = \varepsilon/(1+\|f\|_0)$ so that $(\tilde{\Phi} f)(x) = \int_K f(xy)\, d\mu(y) + \int_{G-K} f(xy)\, d\mu(y) \le \int_K f(xy)\, d\mu(y) + \delta\|f\|_0$. If $\sup_{G-C} f < \delta/\mu(K)$ for compact $C \subset G$ and $y \in K$, then $f(xy) \le \delta/\mu(K)$ and $\int_K f(xy)\, d\mu(y) < \delta\ \forall x \in G - CK^{-1}$. Therefore $\sup_{G-CK^{-1}} (\tilde{\Phi} f)(x) \le \varepsilon$ and $\tilde{\Phi} f \in C'_\infty(G)$. The preceding remarks enable us to define

$$\Psi \star \Phi = \Psi \circ \tilde{\Phi}: \underset{f \rightsquigarrow \Psi\{\tilde{\Phi} f\}}{C_\infty(G) \to \mathbb{C}} \qquad \forall \Psi, \Phi \in C'_\infty(G).$$

Taking $\bar{\bar{f}}: \underset{x \rightsquigarrow \overline{f(x^{-1})}}{G \to \mathbb{C}}$, we prove

THEOREM 28.1. $(C'_\infty(G), |\!|\!|\ |\!|\!|)$ is a *-normed Banach algebra with identity $h_e(f) = f(e)$ under multiplication $\Psi \star \Phi = \Psi \circ \tilde{\Phi}$ and involution $\Phi^*(f) = \overline{\Phi(\hat{f})}$. Furthermore, $C'_\infty(G)$ is commutative iff G is commutative.

Proof. We proceed by the following sequence of claims:

CLAIM 1. $C'_\infty(G)$ is Banach with identity h_e.

Proof. Clearly $\Psi \star \Phi$ is linear and $|(\Psi \star \Phi)(f)| \le |\!|\!|\Psi|\!|\!|\ |\!|\!|\Phi|\!|\!|\ \|f\|_0\ \forall f \in C_\infty(G)$ ensures that $\Psi \star \Phi \in C'_\infty(G)$ satisfies $|\!|\!|\Psi \star \Phi|\!|\!| \le |\!|\!|\Psi|\!|\!|\,|\!|\!|\Phi|\!|\!|\ \forall \Psi, \Phi \in C'_\infty(G)$. Since $\widetilde{\Psi + \Phi} = \tilde{\Psi} + \tilde{\Phi}$ and $\widetilde{\alpha\Phi} = \alpha\tilde{\Phi}\ \forall \alpha \in \mathbb{C}$, all algebraic distributive and scalar multiplication requirements are satisfied. We need only verify that multiplication is associative. Now $\{\tilde{\Phi}({}_x f)\}(y) = \Phi\{{}_y({}_x f)\} = \Phi\{{}_{xy} f\} = (\tilde{\Phi} f)(xy) = {}_x(\tilde{\Phi} f)(y)\ \forall y \in G$ so that $\tilde{\Phi}({}_x f) = {}_x(\tilde{\Phi} f)\ \forall x \in G$. Thus, $\{(\widetilde{\Phi \star \Psi}) f\}(x) = (\Phi \star \Psi)({}_x f) = \Phi\{\Psi({}_x f)\} = \Phi\{{}_x(\tilde{\Psi} f)\} = \{\tilde{\Phi}(\tilde{\Psi} f)\}(x)\ \forall x \in G$ yields $(\widetilde{\Phi \star \Psi}) f = \tilde{\Phi}(\tilde{\Psi} f)$, from which follows $\{\Phi \star (\Psi \star \Upsilon)\}(f) = \Phi\{(\widetilde{\Psi \star \Upsilon})(f)\} = \Phi\{\tilde{\Psi}(\tilde{\Upsilon} f)\} = (\Phi \star \Psi)(\tilde{\Upsilon} f) = \{(\Phi \star \Psi) \star \Upsilon\}(f)$ and the associativity of $\star$. Finally, $h_x \in C'_\infty(G)\ \forall x \in G$ (Problem 26B, Corollary 23.7) and since $\tilde{h}_e f = f\ \forall f \in C_\infty(G)$ we have $(h_e \star \Phi)(f) = h_e(\tilde{\Phi} f) = (\tilde{\Phi} f)(e) = \Phi({}_e f) = \Phi(f) = \Phi(\tilde{h}_e f) = (\Phi \star h_e)(f)$. Thus $\Phi \star h_e = \Phi = h_e \star \Phi\ \forall \Phi \in C'_\infty(G)$.

CLAIM 2. $C'_\infty(G)$ is *-normed under $\Phi^*(f) = \overline{\Phi(\bar{\tilde{f}})}$.

Proof. If $f \in C_\infty(G)$, then $\bar{\tilde{f}} \in C_\infty(G)$ and $\|f\|_0 = \|\bar{\tilde{f}}\|_0$. Therefore $|||\Phi^*||| = \sup_{\|f\|_0=1} |\Phi(\bar{\tilde{f}})| = |||\Phi|||$ and $\Phi^* \in C'_\infty(G)\ \forall \Phi \in C'_\infty(G)$. Verification that * is an involution on $C'(G)$ is straightforward but cumbersome. For instance, associating $\mu, \nu \in M(G)$ with Φ and Ψ, we have

$$(\Phi \star \Psi)^*(f) = \overline{(\Phi \star \Psi)(\bar{\tilde{f}})} = \overline{\int_G \int_G \bar{\tilde{f}}(xy)\, d\mu(x)\, d\nu(y)}$$

$$= \overline{\int_G \int_G \overline{f(y^{-1}x^{-1})}\, d\mu(x)\, d\nu(y)} = \overline{\int_G \left\{ \int_G \overline{{}_{y^{-1}}f(x^{-1})}\, d\mu(x) \right\} d\nu(y)}$$

$$= \overline{\int_G \left\{ \int_G \overline{{}_{y^{-1}}\tilde{f}(x)}\, d\mu(x) \right\} d\nu(y)} = \overline{\int_G \Phi({}_{y^{-1}}\bar{\tilde{f}})\, d\nu(y)}$$

$$= \overline{\int_G \overline{\overline{\Phi({}_{y^{-1}}\bar{\tilde{f}})}}\, d\nu(y)} = \overline{\int_G \overline{\Phi^*({}_{y^{-1}}f)}\, d\nu(y)} = \overline{\int_G \overline{(\tilde{\Phi}^* f)(y^{-1})}\, d\nu(y)}$$

$$= \overline{\int_G (\overline{\widetilde{\tilde{\Phi}^* f}})(y)\, d\nu(y)} = \overline{\Psi\{\overline{\widetilde{\tilde{\Phi}^* f}}\}} = \Psi^*\{\widetilde{\tilde{\Phi}^* f}\}$$

$$= \Psi^*\{\tilde{\Phi}^* f\} = (\Psi^* \star \Phi^*)(f)\ \forall f \in C_\infty(G).$$

CLAIM 3. $C'_\infty(G)$ is commutative iff G is commutative.

Proof. Since $h_x \star h_y = h_{xy} \forall x, y \in G$, noncommutativity of G implies that of $C'_\infty(G)$. In the other direction $(\Phi \star \Psi)(f) = \int_G (\tilde{\Psi} f)(x)\, d\mu(x) = \int_G \int_G {}_x f\, d\nu(y)\, d\mu(x) = \int_G \int_G f(xy) \nu(y)\, d\mu(x)$ and $f(xy) \in L_1(G \times G, \mu \times \nu)$ $\forall f \in C_\infty(G)$. Accordingly, commutativity of $C'_\infty(G)$ follows from that of G, since (together with Fubini's theorem)

$$(\Phi \star \Psi)(f) = \int_G \int_G f(xy)\,d\nu(y)\,d\mu(x) = \int_G \left\{ \int_G {}_y f(x)\,d\mu(x) \right\} d\nu(y)$$

$$= \int_G (\tilde{\Phi} f)(y)\,d\nu(y) = (\Psi \star \Phi)(f) \quad \forall f \in C_\infty(G). \blacksquare$$

Remark. Fubini's theorem is essential; without it convolution algebras may fail to be commutative even though the underlying group is commutative (Problem 28C).

If we identify $M(G)$ with $C'_\infty(G)$ under $\mu \rightsquigarrow \Phi_\mu$ (Appendix m.8), then every pair $\mu, \nu \in M(G)$ determines a unique member $\mu \star \nu \in M(G)$, namely, the preimage of $I_\mu \star I_\nu \in C'_\infty(G)$. If we similarly define $\mu^* \in M(G)$ as the preimage of $I^*_\mu \in C'_\infty(G)$, it comes as no surprise that the NLA congruence $(M(G), \| \|_v) \to (C'_\infty(G), \||\ \||)$ is also an algebra *-congruence. An adjustment or two in thinking will sharpen our focus. First, each pair $\mu, \nu \in M(G)$ determines a regular weakly Borel measure $\mu \times \nu$ on $G \times G$. Reason (Berberian [9]): The regular Borel restrictions μ', ν' of μ and ν have a unique regular Borel measure $(\mu \times \nu)'$ extending $\mu' \times \nu'$ from $\mathfrak{R}_\sigma(\mathcal{C}) \times \mathfrak{R}_\sigma(\mathcal{C})$ to $\mathfrak{R}_\sigma(\mathcal{C} \times \mathcal{C})$ and there is a unique corresponding regular weakly Borel measure extension $(\mu \times \nu)''$ of $(\mu \times \nu)'$ to $\mathfrak{R}_\sigma(\mathfrak{C} \times \mathfrak{C})$. For convenience $(\mu \times \nu)''$ is also denoted $\mu \times \nu$.

If $E \subset G$ is weakly Borel, so is $E \times G$ in $G \times G$. Since $t: \underset{(x,y) \rightsquigarrow (xy,y)}{G \times G \to G \times G}$ is a homeomorphism, $\mathfrak{m}^{-1}(E) = t^{-1}(E \times G) \subset G$ is also weakly Borel and $(\mu \times \nu)\{\mathfrak{m}^{-1}(E)\}$ is defined $\forall \mu, \nu \in M(G)$.

THEOREM 28.2. $(M(G), \| \|_v)$ is a *-normed Banach algebra with identity $\mu_{e,1} = \chi_{\{e\}}$ relative to involution $\mu^*(E) = \overline{\mu(E^{-1})}$ and multiplication $(\mu \star \nu)(E) = (\mu \times \nu)\{\mathfrak{m}^{-1}(E)\} = \int_G \nu(x^{-1}E)\,d\mu(x) = \int_G \mu(Ey^{-1})\,d\nu(y)$. Furthermore,

(i) $M(G)$ is commutative iff G is commutative;

(ii) $\int_G (f(x)\,d\mu^*(x) = \int_G f(x)\,\overline{d\mu(x^{-1})}\ \forall f \in L_1(G)$;

(iii) $\int_G f\,d(\mu \star \nu) = \int_G \int_G f(xy)\,d\nu(y)\,d\mu(x)\ \forall f \in L_1(G)$;

(iv) $|\mu \star \nu| \le |\mu| \star |\nu|$.

Proof. Almost everything in the main statement has already been established. Clearly $\Phi_{\mu_{e,1}}(f) = \int_G f\,d\mu_{e,1} = f(e) = h_e(f)\ \forall f \in C_\infty(G)$ ensures that $\Phi_{\mu_{e,1}} = h_e$ and $\mu_{e,1}$ is the identity element in $M(G)$. Furthermore, $\mu^*(E) = \int_G \chi_E \, d\mu^* = \Phi^*_\mu(\chi_E) = \overline{\Phi_\mu(\tilde{\bar{\chi}}_E)} = \overline{\Phi_\mu(\chi_{E^{-1}})} = \overline{\mu(E^{-1})}$ and (almost by definition) $\int_G \mu(Ey^{-1}) \, d\nu(y) = \int_G \mu\{\mathfrak{m}^{-1}(E)\}_y \, d\nu(y) = (\mu \times \nu)^{-1}\{\mathfrak{m}^{-1}(E)\} = \int_G \nu\{\mathfrak{m}^{-1}(E)\}_x \, d\mu(x) = \int_G \nu(x^{-1}E) \, d\mu(x) \, \forall E \in \mathfrak{R}_\sigma$. Verification of (ii) begins with the observation that $\int_G \chi_E \, d\mu^* = \mu^*(E) = \overline{\mu(E^{-1})} = \overline{\int_G \chi_{E^{-1}}(x) \, d\mu(x)} = \overline{\int_G \chi_E(x) \, d\mu(x^{-1})} = \int_G \chi_E(x) \, \overline{d\mu(x^{-1})}$. This equality extends to simple functions and (Appendix m.3) the denseness of such functions in $L_1(G)$ establishes (ii). In similar fashion (iii) follows(iii) follows from

$$\int_G \chi_E \, d(\mu \star \nu) = (\mu \star \nu)(E) = (\Phi_\mu \star I_\nu)(\chi_E) = \Phi_\mu\{\tilde{\Phi}_\nu \chi_E\}$$

$$= \int_G (\tilde{\Phi}_\nu \chi_E)(x) \, d\mu(x)$$

$$= \int_G \Phi_\nu\{{}_x(\chi_E)\} \, d\mu(x) = \int_G \left\{ \int_G ({}_x\chi_E)(y) \, d\nu(Y) \right\} d\mu(x)$$

$$= \int_G \int_G \chi_E(xy) \, d\nu(y) \, d\mu(x) \, \forall E \in \mathfrak{R}_\sigma.$$

The proof of (iv) is based on $|(\mu \star \nu)(E)| = |\int_G \nu(x^{-1}E) \, d\mu(x)| \leq \int_G |\nu(x^{-1}E)| \, d|\mu|(x) \leq \int_G |\nu|(x^{-1}E) \, d|\mu|(x) = \{|\mu| \star |\nu|\}(E) \, \forall E \in \mathfrak{R}_\sigma$. Thus

$$|\mu \star \nu|(E) = \sup \{\sum_k |(\mu \star \nu)(E_k)| : E = \bigcup_{\text{disjt}} E_k\}$$

$$\leq \sup \{\sum_k \{|\mu| \star |\nu|\}(E_k) : E = \bigcup_{\text{disjt}} E_k\} = \{|\mu| \star |\nu|\}(E) \, \forall E \in \mathfrak{R}_\sigma. \blacksquare$$

An illustration of the above in more familiar surroundings may be useful.

Example 28.3. Let m denote Lebesgue measure on $G = \mathbb{R}$ and (Appendix m.7) let $\mu = \chi_I dm$, where $I = [0, 1]$. Then $\mu \star \mu =$

$(t-1)\chi_{[1,2]}(t)\,dm(t)$. Verification:

$$(\mu \star \mu)[t,\infty) = \int_0^1 \left[\int_0^1 \chi_{[t,\infty)}(x+y)\,dm(x)\right] dm(y) = \int_0^1 \phi_t(y)\,dm(y),$$

where $\phi_t(y) = m\{[t-y,\infty)\cap I\}$. A simple calculation shows that $\phi_t(y)\equiv 0$ for $t \geq 2$ and $\phi_t(y)\equiv 1$ for $t \leq 1$, whereas $\phi_t(y) = (t-y)\chi_{[t-1,1]}(y)$ for $t\in(1,2)$. Therefore, $(\mu\star\mu)[t,\infty) = \frac{1}{2}\chi_{(-\infty,1)}(t) + (t - t^2/2)\chi_{[1,2]}(t) = \int_t^\infty (t-1)\chi_{[1,2]}(t)\,dm(t)$ and $(\mu\star\mu)[\alpha,\beta) = (\mu\star\mu)[\alpha,\infty) - (\mu\star\mu)[\beta,\infty) = \int_\alpha^\beta (t-1)\chi_{[1,2]}(t)\,dm(t)\ \forall -\infty < \alpha < \beta < \infty$. It remains only to note that sets of the form $[\alpha,\beta)$ generate the Lebesgue measurable subsets of $\mathbb{R}$.

Deeper insight into the structure of $M(G)$ follows by decomposing $M(G)$ into its components.

DEFINITION 28.4. A measure $\mu\in M(G)$ is said to be continuous if $\mu(x)=0\ \forall x\in G$, and discrete if there is a countable set $S_\mu\subset G$ such that $\mu(E)=\mu(S_\mu\cap E)\ \forall E\in\mathfrak{R}_\sigma$.

The vector subspaces of continuous and discrete measures in $M(G)$ are denoted $M_c(G)$ and $M_\mathfrak{d}(G)$, respectively.

Remark. Each point measure $\mu_x(E)=\begin{cases}1, & x\in E\\ 0, & x\notin E\end{cases}$ belongs to $M_\mathfrak{d}(G)-M_c(G)$.

If $\nu=\sum_{n=1}^\infty \alpha_n\mu_{x_n}$ and $\sum_{n=1}^\infty |\alpha_n|<\infty$, then $\nu\in M_\mathfrak{d}(G)$ with $S_\nu=\{x_n : n\in\mathbb{N}\}$. On the other hand, every $\mu\in M_\mathfrak{d}(G)$ can be written $\mu=\sum_{S_\mu}\mu(y_n)\mu_{y_n}$, in which case $|\mu|=\sum_{S_\mu}|\mu(y_n)|\mu_{y_n}$ (therefore $\|\mu\|_v=\sum_{S_\mu}|\mu(y_n)|$) and $\mu^* = \sum_{S_\mu}\overline{\mu(y_n)}\mu_{y_n^{-1}}$. We use these remarks to prove

THEOREM 28.5. (i) $M_\mathfrak{d}(G)$ is a closed subalgebra (with identity μ_e) of $M(G)$, and $M_c(G)$ is a closed two-sided ideal in $M(G)$.

(ii) $M_\mathfrak{d}(G)=M(G)$ iff $M_c(G)=\{0\}$ iff G is discrete.

(iii) If G is nondiscrete, $M(G)=M_\mathfrak{d}(G)\oplus M_c(G)$ with $|\mu|=|\mu_\mathfrak{d}|+|\mu_c|$ and $\|\mu\|_v=\|\mu_\mathfrak{d}\|_v+\|\mu_c\|_v\ \forall\mu=\mu_\mathfrak{d}+\mu_c\in M(G)$.

Proof. Both $M_\mathfrak{d}(G)$ and $M_\mathfrak{c}(G)$ are vector subspaces of $M(G)$. For any $\mu=\sum_n \alpha_n\mu_{x_n}$ and $\nu=\sum_m \beta_m\mu_{y_m}$ in $M_\mathfrak{d}(G)$ we also have

$$\{\mu\star\nu\}(E)=\int_G \nu(x^{-1}E)\,d\mu(x)=\sum_n \nu(x^{-1}E)=\sum_n \alpha_n\Big\{\sum_m \beta_m\mu_{y_m}(x_n^{-1}E)\Big\}$$

$$=\sum_{n,m}\alpha_n\beta_m\mu_{y_m}(x_n^{-1}E)=\sum_{n,m}\alpha_n\beta_m\mu_{x_ny_m}(E)$$

and so $\mu\star\nu\in M_\mathfrak{d}(G)$. To see that $M_\mathfrak{d}(G)$ is $\|\ \|_v$-closed in $M(G)$ assume that $\mu_n\xrightarrow{\|\ \|_v}\mu\in M(G)$ for $\{\mu_n\}\in M_\mathfrak{d}(G)$. Evidently $S=\bigcup_n S_{\mu_n}$ is countable and $|\mu(E)-\mu(S\cap E)|\le|\mu(E)-\mu_n(E)|+|\mu_n(E)-\mu_n(S_{\mu_n}\cap E)|+|\mu_n(S_{\mu_n}\cap E)-\mu_n(S\cap E)|+|\mu_n(S\cap E)-\mu(S\cap E)|\ \forall E\in\mathfrak{R}_\sigma$. Since the middle two terms on the right vanish, $\mu(E)=\mu(S\cap E)$ and $\mu\in M_\mathfrak{d}(G)$ as required. We show next that $M_\mathfrak{c}(G)$ is a closed two-sided ideal in $M(G)$. For any $\mu\in M_\mathfrak{c}(G)$ and $\nu\in M(G)$, we have $(\mu\star\nu)(x)=\int_G \mu(xy^{-1})\,d\nu(y)=0$ and $(\nu\star\mu)(x)=\int_G \mu(x^{-1}y)\,d\nu(x)=0\ \forall x\in G$. Thus $\mu\star\nu$ and $\nu\star\mu$ belong to $M_\mathfrak{c}(G)$. Furthermore, $\mu_n\xrightarrow{\|\ \|_v}\mu\in M(G)$ for $\{\mu_n\}\in M_\mathfrak{c}(G)$ necessitates that $|\mu(x)|=|\mu_n(x)-\mu(x)|\le\|\mu_n-\mu\|_v$ and $\mu(x)=0\ \forall x\in G$. The proof of (ii) is not difficult. If $M_\mathfrak{d}(G)=M(G)$ and $\nu\in M_\mathfrak{c}(G)$, then $\nu(S)=0$ for every countable set $S\subset G$ and $\nu(E)=\nu(E\cap S)=0\ \forall E\in\mathfrak{R}_\sigma$ yields $\nu=0$. Therefore $M_\mathfrak{c}(G)=\{0\}$. If G is nondiscrete and $F\in C(G)$ is positive, then $F\,d\mu_\mathcal{F}\neq 0$ and $(F\,d\mu_\mathcal{F})(x)=F(x)\,\mu_\mathcal{F}(x)=0\ \forall x\in G$. Consequently $M_\mathfrak{c}(G)\neq\{0\}$. Finally, if G is discrete and $\mu\in M(G)$, elements in $C_0(G)$ have the form $\sum_{i=1}^n \alpha_i\chi_{\{x_i\}}$ and $\Phi_\mu\in C_0'(G)$ has the form $\sum_{j=1}^m \beta_j\xi_{x_j}$. Since $\underset{\xi_x\rightsquigarrow\mu_x}{\xi_G\to M(G)}$ is linear, $\Phi_\mu\to\sum_{j=1}^m \beta_j\mu_{x_j}\in M_\mathfrak{d}(G)$ and so $M_\mathfrak{d}(G)=M(G)$. The results of (ii) also confirm that $M(G)=M_\mathfrak{d}(G)\oplus M_\mathfrak{c}(G)$ when G is nondiscrete. Let $S_\mu=\{x\in G:\mu(x)\neq 0\}$ for any $\mu\neq 0$ in $M(G)$. Then $S_\mu=\bigcup_{n=1}^\infty S_n$, where each $S_n=\{x\in G:|\mu|>1/n\}$ is finite. (Otherwise there exist $\{x_m:m\in\mathbb{N}\}\subset S_n$ and $|\mu|\Big(\bigcup_{m=1}^\infty x_m\Big)=\sum_{m=1}^\infty \mu(x_m)=\infty$ contradicts the finiteness of $|\mu|$.) Since S_μ is countable, $\mu_\mathfrak{d}=\sum_{S_\mu}\mu(x)\mu_x\in M_\mathfrak{d}(G)$ and $\mu_\mathfrak{c}=\mu-\mu_\mathfrak{d}\in M_\mathfrak{c}(G)$. Accordingly, $\mu=\mu_\mathfrak{d}+\mu_\mathfrak{c}$ and $|\mu|\le|\mu_\mathfrak{d}|+|\mu_\mathfrak{c}|$. The reverse inequality also holds, since $|\mu_\mathfrak{d}(E)|+|\mu_\mathfrak{c}(E)|=|\mu(E\cap S_\mu)|+|\mu(E-S_\mu)|\le|\mu(E)|\ \forall E\in\mathfrak{R}_\sigma$. ■

Theorem 28.5 can be further refined by relating $L_1(G)$ to the vector subspace $M_{\mathfrak{a}}(G)=\{\mu\in M(G):\mu\ll\mu_r\}$. Our point of departure is the Radon Nikodym theorem (Appendix m.7) and the remark preceding it.

THEOREM 28.6. $L_1(G)$ and $(M_{\mathfrak{a}}(G), \|\,\|_v)$ are algebraically congruent under $t: f \rightsquigarrow \mu_f = f\,d\mu_r$.

Proof. Surely t is linear and surjective. It is also injective, since $F\in L_1(G)$ and $\int_E F\,d\mu_r=0\ \forall E\in\mathfrak{R}_\sigma$ yield $F=0$ a.e. (relative μ_r); that is, $F=0$ in $L_1(G)$. And t is an algebra isomorphism by Fubini's theorem, left invariance of μ_r, and the substitution $x\rightsquigarrow y^{-1}x$ in the third equality of

$$
\begin{aligned}
(\mu_f\star\mu_g)(E) &= \int_G \mu_g(y^{-1}E)\,d\mu_f(y) = \int_G\int_{y^{-1}E} g(x)\,d\mu_r(x)\,d\mu_f(y)\\
&= \int_G\int_E g(y^{-1}x)\,d\mu_r(x)\,d\mu_f(y)\\
&= \int_E\int_G f(y)g(y^{-1}x)\,d\mu_r(y)\,d\mu_r(x) = \int_E (f\star g)(x)\,d\mu_r(x)\\
&= \mu_{f\star g}(E)\quad \forall E\in\mathfrak{R}_\sigma.
\end{aligned}
$$

Finally, $\|\mu_f\|_v\le\|f\|_1\ \forall f\in L_1(G)$ by Appendix m.6, and the reverse inequality follows from $|\mu_f|=|f|\,d\mu_r$ (Appendix m.5) and $\|f\|_1=\int_E|f|\,d\mu_r=\int_E d|\mu_f|\le|\mu_f|(E)\le|\mu_f|(G)=\|\mu_f\|_v$ for $E=\{x\in G: f(x)\neq 0\}\in\mathfrak{R}_\sigma$. ■

Using the notation of Appendix m.7, we write $M_{\mathfrak{s}}(G)=\{\mu\in M_{\mathfrak{c}}(G):\mu\perp\mu_r\}$ and then prove

THEOREM 28.7. (i) $M_{\mathfrak{s}}(G)$ is a closed vector subspace of $M(G)$ and $M_{\mathfrak{a}}(G)$ is a closed two-sided ideal of $M(G)$.

(ii) $M_{\mathfrak{a}}(G)=M(G)$ iff G is discrete iff $M_{\mathfrak{a}}(G)\not\subset M_{\mathfrak{c}}(G)$.

(iii) If G is nondiscrete, $M(G)=M_{\mathfrak{d}}(G)\oplus M_{\mathfrak{s}}(G)$ with $|\mu|=|\mu_{\mathfrak{d}}|+|\mu_{\mathfrak{a}}|+|\mu_{\mathfrak{s}}|$ and $\|\mu\|_v=\|\mu_{\mathfrak{d}}\|+\|\mu_{\mathfrak{a}}\|_v+\|\mu_{\mathfrak{s}}\|_v\ \forall\mu=\mu_{\mathfrak{d}}+\mu_{\mathfrak{a}}+\mu_{\mathfrak{s}}\in M(G)$.

Proof. Obviously $\alpha\mu\in M_{\mathfrak{s}}(G)\ \forall\alpha\in\mathbb{C}$ and $\mu\in M_{\mathfrak{s}}(G)$. Suppose that $\mu_i\in M_{\mathfrak{s}}(G)$ $(i=1,2)$ with $|\mu_i|(E\cap A_i)=0=\mu_r(E\cap B_i)\ \forall E\in\mathfrak{R}_\sigma$, where each union $G=A_i\cup B_i$ is disjoint. Then the union $G=$

$(A_1 \cap A_2) \cup (B_1 \cup B_2)$ is disjoint and $|\mu_1 + \mu_2|\{E \cap (A_1 \cap A_2)\} = 0 = \mu r\{E \cap (B_1 \cup B_2)\}\ \forall E \in \mathfrak{R}_\sigma$. Thus $\mu_1 + \mu_2 \in M_s(G)$ and $M_s(G)$ is a vector subspace of $M(G)$. Now let $\mu_n \xrightarrow{\|\|_v} \mu \in M_c(G)$ for $\{\mu_n : n \in \mathbb{N}\} \in M_s(G)$. Assume further that each $G = A_n \cup B_n$ is disjoint with $|\mu_n|(E \cap A_n) = 0 = \mu r(E \cap B_n)\ \forall E \in \mathfrak{R}_\sigma$. As above, G is a disjoint union of $A = \bigcap_n A_n$ and $B = \bigcup_n B_n$. For any fixed $E \in \mathfrak{R}_\sigma$, we also have $|\mu_n|(E \cap A) = 0 = \mu r(E \cap B)\ \forall n \in \mathbb{N}$. Accordingly, for any weakly Borel subset D of $E \cap A$ there follows $|\mu(D)| \leq |\mu(D) - \mu_n(D)| + |\mu_n(D)| \leq \|\mu - \mu_n\|_v$ and $\mu(D) = 0$. In other words $|\mu|(E \cap A) = 0$ and $\mu \in M_s(G)$ proves $M_s(G)$ closed in $M(G)$. The preceding theorem already establishes that $M_a(G)$ is closed in $M(G)$. Verification that $M_a(G)$ is a two-sided ideal is based on the fact that $\mu \in M_a(G)$ iff $|\mu|(C) = 0\ \forall C \in \mathcal{C}$ with $\mu r(C) = 0$. Note also (Theorem 12.13) that $\mu r(C) = 0$ implies that $\mu r(x^{-1}C) = 0 = \mu r(Cx^{-1})\ \forall x \in G$. For every pair $(\mu, \nu) \in M_a(G) \times M(G)$ we therefore have

$$|\mu \star \nu|(C) \leq \{|\mu| \star |\nu|\}(C) = \int_G |\mu|(Cy^{-1})\, d|\nu|(y) = 0$$

and

$$|\nu \star \mu|(C) \leq \{|\nu| \star |\mu|\}(C) = \int_G |\mu|(x^{-1}C)\, d|\nu|(x) = 0,$$

the conclusion being that $\mu \star \nu$ and $\nu \star \mu$ belong to $M_a(G)$. The first part of (ii) is clear [Theorems 22.8(i) and 27.4(i)], since $M_a(G) = M(G)$ iff $M_a(G)$ has an identity; that is, iff G is discrete. If G is nondiscrete, then (Corollary 12.16) $\mu r(x) = 0\ \forall x \in G$ and $M_a(G) \subset M_c(G)$. If G is discrete, $M(G) = M_a(G) \not\subset M_c(G)$ (otherwise $\mu_e \in M_c(G)$, which is impossible). Everything in (iii) follows from Theorem 28.5 and the fact that $M_c(G) = M_a(G) \oplus M_s(G)$ (Appendix m.7). ■

The following lemma is recorded for future reference; its lengthy proof (Rudin [85], pp. 104–107) is omitted, since it leads too far astray. Here $\mu^{\circ} = e$ and $\mu^k = \mu \star \mu \star \cdots \star \mu$ (k-times).

Lemma 28.8. Let G be a nondiscrete LCT_2A group. Then there is a $\mu \in M_c(G)$ with $\|\mu\|_v = 1$ such that $\mu(E) = \mu(E^{-1}) \geq 0\ \forall E \in \mathfrak{R}_\sigma$ and $\left\| \sum_{k=0}^{n} \alpha_k \mu^k \right\|_v = \sum_{k=0}^{n} |\alpha_k|\ \forall (\alpha_0, \alpha_1, \ldots, \alpha_n) \in \mathbb{C}^{n+1}$.

COROLLARY 28.9. The following are equivalent for an LCT_2A group G:

(i) G is discrete;
(ii) $M_b(G) = M(G)$;
(iii) $M_a(G) = M(G)$;
(iv) $M_c(G) = \{0\}$;
(v) $M_s(G) = \{0\}$.

Proof. Always (i)–(iv) are equivalent and (iv) $\Rightarrow$ (v). Here (v) $\Rightarrow$ (i), since $\mu \in M_s(G)$ in Lemma 28.7. ■

Fourier Stieltjes Transform

For the remainder of this section, G is assumed abelian. Since $\mu \in M_a(G)$ is $\mu = f\, d\mu_r$ for some $f \in L_1(G)$, it seems only natural to define $\hat{\mu}(\chi)$ as $\hat{f}(\chi) = \int_G f\chi\, d\mu_r = \int_G \chi\, d\mu$. Generalizing this is

DEFINITION 28.10. $\hat{\mu}: G' \to \mathbb{C}$, $\chi \rightsquigarrow \int_G \chi\, d\mu$ is called the Fourier Stieltjes transform of $\mu \in M(G)$. Here $\widehat{M(G)}^{(s)}$ denotes $\{\hat{\mu} : \mu \in M(G)\}$.

Remark. Note that $\widehat{\mu^*} = \bar{\hat{\mu}}$ since $\int_G \chi(x)\, d\mu^* = \int_G \chi(x)\, d\overline{\mu(x^{-1})} = \int_G \chi(x^{-1})\, d\overline{\mu(x)} = \int_G \overline{\chi(x)}\, d\overline{\mu(x)} = \overline{\int_G \chi(x)\, d\mu(x)}\ \forall \chi \in G'$. ■

Homomorphisms can be extended from $M_a(G)$ to all of $M(G)$ by defining $\mathcal{H}_{\Theta_a} = \{h \in \mathcal{H}_{M(G)} : h/M_a(G) \equiv 0\}$ in

THEOREM 28.11. $\mathcal{H}_{M(G)} - \mathcal{H}_{\Theta_a} = \{h_\chi : M(G) \to \mathbb{C},\ \mu \rightsquigarrow \hat{\mu}(\chi) : \chi \in G'\}$.

Proof. Since h_χ is linear (integration preserves linearity) and

$$h_\chi(\mu \star \nu) = \int_G \chi(x)\, d\{\mu \star \nu\}(x) = \int_G \left(\int_G \{_x\chi\}(y)\, d\nu(y) \right) d\mu(x)$$

$$= \int_G \int_G \chi(x)\chi(y)\, d\nu(y)\, d\mu(x) = \hat{\mu}(\chi)\, \hat{\nu}(\chi)$$

$$= h_\chi(\mu)\, h_\chi(\nu)\ \forall \mu, \nu \in M(G),$$

we have (Theorems 27.8 and 28.6) $\mathcal{H}_{M_a(G)}=\{h_\chi:\chi\in G'\}\subset\mathcal{H}_{M(G)}-\mathcal{H}_{\mathfrak{S}_a}$. If $h\in\mathcal{H}_{M(G)}-\mathcal{H}_{\mathfrak{S}_a}$, then $h/M_a(G)=h_\chi$ for some $\chi\in G'$. For $\mu\in M_a(G)$ with $h_\chi(\mu)\neq 0$ and any $\nu\in M(G)$, we have $\mu\star\nu\in M_a(G)$ and $h_\chi(\mu)h(\nu)=h(\mu\star\nu)=h_\chi(\mu\star\nu)=h_\chi(\mu)\,h_\chi(\nu)$. Therefore $h=h_\chi\in\mathcal{H}_{M_a(G)}$. ■

It makes sense that $\hat{\mu}$ denotes both the Gelfand and Fourier Stieltjes transforms of $\mu\in M(G)$.

COROLLARY 28.12. The Gelfand transform extends the Fourier Stieltjes transform of $\mu\in M(G)$.

Proof. Let $\mathfrak{M}_{\mathfrak{R}}\{M_a(G)\}$ denote the regular maximal ideals in $M_a(G)$. Then $\mathcal{H}_{M_a(G)}=\mathcal{H}_{M(G)}-\mathcal{H}_{\mathfrak{S}_a}$ implies $\mathfrak{M}_{\mathfrak{R}}\{M_a(G)\}\subset\mathfrak{M}_{M(G)}$ and $\underset{\chi\rightsquigarrow M_\chi=h_\chi^{-1}(0)}{G'\to\mathfrak{M}_{\mathfrak{R}}\{M_a(G)\}}$ is a homeomorphism. Accordingly, for any $M_\chi\in\mathfrak{M}_{\mathfrak{R}}\{M_a(G)\}$, we have $\overset{(g)}{\hat{\mu}}(M_\chi)=h_\chi(\mu)=\overset{(g)}{\hat{\mu}}(\chi)$. ■

Although $M_a(G)$ is an A^*-algebra (by Definition 28.10⁺), such need not be the case for $M(G)$. In fact, a little leapfrogging establishes

THEOREM 28.13. $M(G)$ is an A^*-algebra iff G is discrete.

Proof. Clearly $M(G)=M_a(G)$ is A^* when G is discrete. Assume that G is nondiscrete and let $\lambda=\mu^{\circ}-\mu^2$ for $\mu\in M_c(G)$ in Lemma 28.8. Then $\|\lambda^n\|_v=\left\|\sum_{k=0}^{n}\binom{n}{k}(-1)^k\mu^{2k}\right\|_v=\sum_{k=0}^{n}\binom{n}{k}=2^n$ and $r_\sigma(\lambda)=\lim_n\|\lambda^n\|^{1/n}=2$. Accordingly, (Theorem 24.25, Corollary 24.21) $2=\hat{\lambda}(M_0)=1-\{\hat{\mu}(M_0)\}^2$ and $\hat{\mu}(M_0)=\pm i$ for some $M_0\in\mathfrak{M}_{M(G)}$. At the same time $\mu^*=\mu$ on $M(G)$ and $\hat{\mu}=\widehat{\mu^*}=\bar{\hat{\mu}}$ on G', which precludes $M(G)$ from being $\hat{A}$. Specifically, $\hat{\nu}=\bar{\hat{\mu}}$ on $\mathfrak{M}_{M(G)}$ implies $\hat{\nu}=\hat{\mu}$ on G' and $\nu=\mu$ in $M(G)$ (Corollary 29.14). Therefore $\hat{\mu}=\hat{\nu}=\bar{\hat{\mu}}$ on $\mathfrak{M}_{M(G)}$ contradicts $\hat{\mu}(M_0)=\pm i$. ■

Since $G'=\mathfrak{M}_{\mathfrak{R}}\{M_a(G)\}\subset\mathfrak{M}_{M(G)}$, we include

COROLLARY 28.14. $\partial_{M(G)}\not\subset\overline{\mathfrak{M}_{\mathfrak{R}}\{M_a(G)\}}$ in $\mathfrak{M}_{M(G)}$ if G is nondiscrete.

Proof. For μ and λ above $0\le\mu\le 1$ yields $0\le\hat{\lambda}=1-\hat{\mu}^2\le 1$ on G' and $\sup_{\mathfrak{M}_{\mathfrak{R}}\{M_a(G)\}}|\hat{\lambda}(M_\chi)|=\sup_{G'}|\hat{\lambda}(\chi)|\le 1<2=\hat{\lambda}(M_0)$ implies that $\partial_{M(G)}\not\subset\mathfrak{M}_{\mathfrak{R}}\{M_a(G)\}$. ■

Remark. A stronger version of this is given in Problem 28F.

Corollary 28.9 and Theorem 28.13 derive from $\mu \in M_s(G)$ in Lemma 28.8. It is not usual for members of $M_s(G)$ to spoil things; in fact, the pathological behavior of singular measures represents the greatest obstacle to the study of $M(G)$. Some inkling why this is so is given below.

Example 28.15. (a) A nonzero $\mu \in M_c(G)$ belongs to $M_s(G)$ iff $\mu(G-B)=0$ for an uncountable set $B \in \mathfrak{R}_\sigma$ with $\mu r(B)=0$. Since Int $B = \varnothing$ (Theorem 12.15), we may view B as a μr-Cantor type perfect subset of G.

(b) The vector subspace $M_s(G)$ need not be a subalgebra of $M(G)$. For instance, (Hewitt and Ross [41]) let $G = G_1 \times G_2$ for infinite compact groups G_i $(i=1,2)$ with identity e_i and normalized Haar measure μr_i. If $F \in C(G)$, then $F_1(x) = F(x, e_2)$ belongs to $C(G_1)$ and $I_1(F) = \int_{G_1} F_1 \, d\mu r_1$ belongs to $C'(G)$. Accordingly, there is a measure $\lambda_1 \in M_s(G)$ such that $\int_G F \, d\lambda_1 = I_1(F) \; \forall F \in C(G)$. In similar fashion $F_2(y) = F(e_1, y)$ belongs to $C(G_2)$ and $I_2(F) = \int_{G_2} F_2 \, d\mu r_2$ determines $\lambda_2 \in M_s(G)$ satisfying $\int_G F \, d\lambda_2 = I_2(F) \; \forall F \in C(G)$. It is easily verified that $\lambda_1 \star \lambda_2 = \lambda_2 \star \lambda_1 = \mu r_1 \times \mu r_2$ is the normalized Haar measure on G. Since $\mu r_1 \times \mu r_2 \neq 0$ belongs to $M_a(G)$, neither $\lambda_1 \star \lambda_2$ nor $\lambda_2 \star \lambda_1$ belong to $M_s(G)$.

PROBLEMS

28A. Direct Approach to $M(G)$

For any $\mu, \nu \in M(G)$ and $E \in \mathfrak{R}_\sigma$ define $(\mu \star \nu)(E) = (\mu \times \nu)\{\mathfrak{m}^{-1}(E)\}$.

(a) Show that $\mu \star \nu \in M(G)$. [*Hint*: For inner regularity assume that $\mu, \nu \geq 0$ and let $\varepsilon > 0$ be given. Then there is a compact set $K_\varepsilon \subset \mathfrak{m}^{-1}(E)$ such that $(\mu \times \nu)(K_\varepsilon) > (\mu \times \nu)\{\mathfrak{m}^{-1}(E)\} - \varepsilon = (\mu \star \nu)(E) - \varepsilon$. Since $K_\varepsilon \subset \mathfrak{m}^{-1}\{\mathfrak{m}(K_\varepsilon)\}$ and $\mathfrak{m}(K_\varepsilon)$ is compact, $(\mu \star \nu)\{\mathfrak{m}(K_\varepsilon)\} > (\mu \star \nu)(E) - \varepsilon$ and $\sup\{(\mu \star \nu)(K) : \text{compact } K \subset E\} = (\mu \star \nu)(E)$.]

(b) Establish that $\|\mu \star \nu\|_v \leq \|\mu\|_v \|\nu\|_v \; \forall \mu, \nu \in M(G)$. [*Hint*: $\int_G \chi_E \, d(\mu \star \nu)$ $\int_{G \times G} \chi_{\mathfrak{m}^{-1}(E)}(xy) \, d(\mu \times \nu) = \int_G \int_G \chi_E(xy) \, d\mu(x) \, d\nu(y) \; \forall E \in \mathfrak{R}_\sigma$. Therefore $\|\mu \star \nu\|_v = \sup_{\|f\|_0 \leq 1} |\int_G f \, d(\mu \star \nu)| = \sup_{\|f\|_0 \leq 1} |\int_G \int_G f(xy) \, d\mu(x) \, d\nu(y)| \leq \|\mu\|_v \|\nu\|_v$.]

28B. $M(G) \star L_p(G)$ and $M(G) \star \{L_1(G) \, d\mu r\} = \{M(G) \star L_1(G)\} \, d\mu r$

(a) If $\mu \in M(G)$ and $f \in C_\infty(G)$, then $\mu \star f = \tilde{\Phi}_\mu f$ belongs to $C_\infty(G)$. Sharpen this by proving that $\mu \star F \in L_p(G)$ $(1 \leq p < \infty)$ with $\|\mu \star F\|_p \leq$

$\|\mu\|_v\|F\|_p\ \forall(\mu, F)\in M(G)\times L_p(G)$. [*Hint*: It suffices to consider $\mu\geq 0$ and $F\in C_0(G)$. (Why?) Show that $h(x, y)=f(y)F(x, y)\in C_0(G\times G)$ for $f\in C_0(G)$ and (use Hölder's inequality) obtain $\int_G |h(x, y)|\, d\mu_r(y)\leq \|_xF\|_p\|f\|_q = \|F\|_p\|f\|_q$. Therefore $\int_G\int_G |h(x, y)|\, d\mu_r(y)\, d\mu(x)\leq \|F\|_p\|f\|_q\|\mu\|_v<\infty$ and (Fubini's theorem) $(\mu\star F)(x)=\int_G F(xy)\, d\mu(y)\in L_1(G)$.]

(b) Prove that $\mu\star(f\, d\mu_r)=(\mu\star f)\, d\mu_r$ in $M_a(G)\ \forall\mu\in M(G)$ and $f\in L_1(G)$.

28C. Noncommutative Convolution Algebras

Let $(B'(G), |||\ |||)$ be the Banach space of continuous linear functionals on $(B(G), \|\ \|_0)$, the space of bounded complex-valued functions on G.

(a) Parallel the proofs in Theorem 28.1 to confirm that $B'(G)$ is a Banach algebra under $(\Psi\star\Phi)(f)=\Psi\{\tilde{\Phi}f\}\ \forall\Psi, \Phi\in B'(G)$.

(b) Suppose that M is a closed left invariant (i.e., $f_0\in M$ implies that $ff_0\in M\ \forall f\in B(G)$) subspace of $B(G)$ such that $\chi_G\in M$ and $\psi_1(\chi_G)=\psi_2(\chi_G)$ for distinct left invariant $\psi_1, \psi_2\in M'$. Then $B'(G)$ is noncommutative. [*Hint*: Extend ψ_i to $\Psi_i\in B'(G)$ and obtain $(\Psi_1\star\Psi_2)(f)=\psi_2(f)\,\psi_1(\chi_G)\neq\psi_1(f)\,\psi_2(\chi_G)=(\Psi_2\star\Psi_1)(f)$ for $f\in M-\{\psi_1-\psi_2\}^{-1}(0)$.]

(c) Although $G=\mathbb{R}$ is commutative, $B'(G)$ is not. [*Hint*: Take $M=\{f\in B(G): \psi_1(f)=\lim_{x\to\infty} f(x)$ and $\psi_2(f)=\lim_{x\to-\infty} f(x)$ exist$\}$.]

(d) $B'(G)$ is noncommutative for an infinite, abelian group G. [*Hint*: G contains an element g_0 of order $\aleph_0$. Define $M=\{f\in B(G): \psi_1(f)=\lim_{n\to\infty} g_0^n$ and $\psi_2(f)=\lim_{n\to-\infty} g_0^{\circ}$ exist$\}$.]

28D. $M(H)\subset M(G)=M(G/H)$ for $H\subset G$

Let H be a closed subgroup of an $\mathrm{LCT_2A}$ group G. Then $M(H)$ is a subalgebra of $M(G)=M(G/H)$, where equality {inclusion} denotes onto {into} algebra congruence. [*Hint*: For $M(H)\subset M(G)$ extend each $\Psi_H\in C'_\infty(H)$ to $\Psi\in C'_\infty(G)$ by $\Psi(F)=\Psi_H(F/H)$ and show that $C'_\infty(H)\to C'_\infty(G)$, $\Psi_H\rightsquigarrow\Psi$, is a congruence. If $\mu\in M(G)$ and $\nu: G\to G/H$, then $\eta: f\rightsquigarrow\int_F (f\nu)\, d\mu$ belongs to $B\mathfrak{L}(C_\infty(G/H))$ and there is a unique $\sigma_\mu\in M(G/H)$ with $\|\sigma_\mu\|_v\leq\|\mu\|_v$ such that $\int_{G/H} f\, d\sigma_\mu=\int_G (f\nu)\, d\mu\ \forall f\in C_\infty(G/H)$. Verification that $M(G)\to M(G/H)$, $\mu\rightsquigarrow\sigma_\mu$, is an onto congruence may be difficult (e.g., see Rudin [85].)]

28E. Additional Properties of $M_a(G)$

(a) Verify that $M_a(G)=\{\mu\in M(G): \underset{x\rightsquigarrow\mu(Ex^{-1})}{G\to\mathbb{C}}$ is continuous for every fixed $E\in\mathfrak{R}_\sigma\}$. [*Hint*: It suffices to consider continuity at $e\in G$. If the above continuity holds and $\varepsilon>0$, there is a G-nbd $u_\varepsilon(e)$ with $\mu_r(u_\varepsilon)<\infty$ such that $|\mu(Ex^{-1})-\mu(E)|<\varepsilon\ \forall x\in u_\varepsilon(e)$. Furthermore, $\nu=\chi_{u_\varepsilon}\,d\mu_r\in M_a(G)$ and $\mu_r(E)=0$ implies that $\nu(x^{-1}E)=0\ \forall x\in G$. Consequently $0=\{\nu\star|\mu|\}(E)=\int_{u_\varepsilon}|\mu|(Ex^{-1})\,d\mu_r(x)\geq\int_G|\mu(Ex^{-1})|\,d\mu_r(x)$ and $|\mu(E)|<\varepsilon$ ensures that $\mu\in M_a(G)$. The reverse inclusion follows by Theorem 12.10, Corollary 12.13(ii), and the regularity of $\mu_r\geq 0$ (Appendix m.8).]

(b) If G is nondiscrete, no $\mu\in M_c(G)$ satisfies $\mu\star\nu=\nu$ $\{\nu\star\mu=\nu\}$ $\forall\nu\in M_a(G)$. [*Hint* [Problem 28B(b)]: $\mu\star\nu=\nu\ \forall\nu\in M_a(G)$ implies that $\mu\star\chi_E=\chi_E$ $\forall E\in\mathfrak{R}_\sigma$. Use the fact that $|\mu|(e)=0$ and modify the argument in Theorem 27.4(ii) (replacing ξ with μ) and obtain $1\leq|\mu|(u)<1$.]

(c) Does $\{\nu_n\}$, $\nu\in M_a(G)$ with $\nu_n\to\nu$ imply $|\nu_n|\to|\nu|$? [*Hint*: If m is Lebesgue measure on $G=\mathbb{R}$ and $\nu_n=\sin nx\,dm(x)$, then $\nu_n(E)\to 0$ and $|\nu_n|(E)\not\to 0\ \forall$ measurable set $E\subset[0,2\pi)$.]

(d) (Problem 12F) Every signed measure $\mu\in M_a(G)$ is ρ-continuous on $\mathfrak{R}_\sigma$. Thus $\{E\in\mathfrak{R}_\sigma:|\mu(E)|<\varepsilon\}$ is ρ-open and $\{E\in\mathfrak{R}_\sigma:|\mu(E)|\leq\varepsilon\}$ is ρ-closed.

28F. Corollary 28.14 Strengthened

Let $\mathscr{A}=\{M\in\mathfrak{M}_{M(G)}:\widehat{\mu^*}(M)=\overline{\hat{\mu}(M)}\ \forall\mu\in M(G)\}$ for a nondiscrete LCT$_2$A group G. As usual, $\mathfrak{M}_{M(G)}$ carries $\mathfrak{T}_g$. Prove

(i) $\overline{\mathfrak{M}_{\mathfrak{R}}\{M_a(G)\}}^{\mathfrak{T}_g}\subset\bar{\mathscr{A}}^{\mathfrak{T}_g}=\mathscr{A}$ in $\mathfrak{M}_{M(G)}$;

(ii) $\widehat{M(G)}/\mathscr{A}$ is dense in $(C(\mathscr{A};\mathbb{C}),\|\,\|_0)$;

(iii) $\partial_{M(G)}\not\subset\mathscr{A}$.

[*Hint*: If $\partial_{M(G)}\subset\mathscr{A}$ then $\partial_{M(G)}=\mathfrak{M}_{M(G)}$ violates Theorem 28.13. Specifically, $X=\widehat{M(G)}/\partial_{M(G)}$ is $\|\,\|_0$-dense in $C(\partial_{M(G)};\mathbb{C})$ and $M'\in\mathfrak{M}_{M(G)}$ determines a homomorphism $h_{M'}:\underset{\hat{\mu}\rightsquigarrow\hat{\mu}(M')}{X\to\mathbb{C}}$ which is extendable to a homomorphism on $C(\partial_{M(G)};\mathbb{C})$. Go on to prove that $M'\in\partial_{M(G)}$.]

28G. Wiener-Pitt Phenomenon

Let G be a nondiscrete LCT$_2$A group. Then there is a nonregular element $\gamma\in M(G)$ satisfying $\hat{\gamma}\geq 1$ on G'. In particular, $M_0-\bigcup_{\mathfrak{M}_{\mathfrak{R}}\{M_a(G)\}}M\neq\varnothing$ for some $M_0\in\mathfrak{M}_{M(G)}$. [*Hint*: Define $\gamma=\mu_e+\mu^2$ for $\mu\in M_c(G)$ in Theorem 28.13. Then $1\leq\hat{\gamma}\leq 2$ on G' and $\hat{\gamma}(M_0)=0$.]

28H. $\widetilde{M(G)}$ Is a Semisimple Banach A^*-Algebra

Given an LCT_2A group G, define an equivalence relation $\sim$ on $M(G)$ by $\mu \sim \nu$ iff $\mu(G)=\nu(G)$ and let $M(G)$ be the quotient set of equivalence classes $\{\mu\}$ ($\mu \in M(G)$).

(a) $\widetilde{M(G)}$ is a commutative convolution algebra with identity $\{\mu_e\}$ under multiplication $\{\mu\} \star \{\nu\} = \{\mu \star \nu\}$. [*Note*: You must first show that this (and all subsequent operations) are well defined!]

(b) $M(G)$ is a *-normed Banach algebra for $\|\{\mu\}\| = |\mu(G)|$ and involution $\{\mu\}^* = \{\bar{\mu}\}$.

(c) Complete the problem by proving $M(G)$ semisimple and A^*.

28I. Symmetric Subalgebras of $M(G)$

(Notation as in Definition 26.4, Example 26.5.) A Banach sub *-algebra B of $M(G)$ is said to be symmetric {strongly symmetric} if $B \subset \mathfrak{S}\{M(G)\}$ {if $B = \mathfrak{S}\{M(G)\}$}.

(a) $B \subset M(G)$ is strongly symmetric iff it is a sub A^*-algebra of $M(G)$.

(b) A strongly symmetric subalgebra of $M(G)$ is symmetric.

(c) Every (strongly) symmetric subalgebra of $M(G)$ is contained in a maximal (strongly) symmetric subalgebra of $M(G)$. [*Hint*: Partially order $\Delta = \{$(strongly) symmetric subalgebras of $M(G)$ which contain $B\}$ by inclusion and let $K = \overline{\bigcup_{\mathcal{Q}} B_\alpha}$ for any totally ordered subset $\{B_\alpha : \alpha \in \mathcal{Q}\}$ of Δ. Then K is (strongly) symmetric and Zorn's Lemma applies.]

(d) There is only one maximal symmetric subalgebra of $M(G)$; namely, $\mathfrak{S}\{M(G)\}$.

(e) If $B \subset M(G)$ is strongly symmetric, $A = \overline{\text{alg}\,\{B \cup M_a(G)\}}^{\mathcal{T}_g} \subset M(G)$ is also strongly symmetric. [*Hint*: Every $\mu \in \text{alg}\,\{B \cup M_a(G)\}$ can be written $\mu = \lambda + \nu$ for some $\lambda \in B$ and $\nu \in M_a(G)$. If $h \in \mathcal{H}_A$, then $h \in \mathcal{H}_B$ and it remains to consider whether $h = h_\chi$ ($\chi \in G'$).]

28J. Analytic Subalgebras of $M(G)$

A Banach subalgebra $B \subset M(G)$ is termed analytic {strongly analytic} if $\overline{\{h_\chi/B : \chi \in G'\}}^{\sigma(\mathcal{H}_B, B)} = \mathcal{H}_B$ {if every $h \in \mathcal{H}_B$ can be extended to a homomorphism $h_0 \in \overline{\{h_\chi : \chi \in G'\}}^{\sigma(\mathcal{H}_{M(G)}, M(G))} \subset \mathcal{H}_{M(G)}\}$.

(a) If $B \subset M(G)$ is analytic and $\mu \in B$, then $\sigma(\mu) = \overline{\{\hat{\mu}(\chi) : \chi \in G'\}}$, meaning that $\forall h \in \mathcal{H}_B$ and $\varepsilon > 0$ there is a $\chi \in G'$ such that $|\hat{\mu}(\chi) - h(\mu)| < \varepsilon$.

(b) A strongly analytic subalgebra of $M(G)$ is analytic.

(c) An analytic sub *-algebra $B \subset M(G)$ is strongly symmetric. Thus $M(G)$ is analytic iff G is discrete. [*Hint*: Fix $\mu \in B$. For any $h \in \mathcal{H}_B$ and $\varepsilon > 0$, there is an $h_\chi \in v(h; \mu; \mu^*; \varepsilon/2)$ and so $|\overline{h(\mu)} - h(\mu^*)| \le |h(\mu^*) - h_\chi(\mu^*)| + |h_\chi(\mu^*) - \overline{h(\mu)}| < \varepsilon$.]

(d) Every strongly analytic subalgebra $B \subset M(G)$ is contained in a maximal strongly analytic subalgebra of $M(G)$. [*Hint*: Proceed as in Problem 28I(e). If $h \in \mathcal{H}_K$, then $h/B_{\alpha_0} \in \mathcal{H}_{B_{\alpha_0}}$ for some $\alpha_0 \in \mathcal{A}$ and (since $\{B_\alpha : \alpha \in \mathcal{A}\}$ is totally ordered) $h/B_\alpha \in \mathcal{H}_{B_\alpha}$ $\forall \alpha > \alpha_0$. Extend h/B_α to $h_{0_\alpha} \in \Gamma = \overline{\{h_\chi : \chi \in G'\}}^{\sigma(\mathcal{H}_{M(G)}, M(G))}$ and define $F_\alpha = \{k \in \Gamma : k/B_\alpha = h/B_\alpha\}$. Then $F_\alpha \neq \varnothing$ and $F_\alpha \subset F_{\alpha_0}$ $\forall \alpha < \alpha_0$. In particular, $\{F_\alpha : \alpha \in \mathcal{A}\}$ has the fip in the $\sigma(\mathcal{H}_{M(G)}, M(G))$-compact set $\Gamma \subset \mathcal{H}_{M(G)}$. Accordingly, $\bigcap_{\mathcal{A}} F_\alpha \cap \Gamma \neq \varnothing$ yields $K \in \Delta$ and Zorn's lemma applies.]

(e) If $B \subset M(G)$ is (strongly) analytic, then $A = \overline{\mathrm{alg}\,\{B \cup M_a(G)\}}^{\mathcal{T}_g}$ is (strongly) analytic. [*Hint*: Given any $\nu \in A$ and $\varepsilon > 0$, there is a $\lambda \in A$ such that $\|\nu - \lambda\|_v < \varepsilon/2$. Write λ as a polynomial $P(\mu, \theta)$ in $\mu \in M_a(G)$ and $\theta \in B$. If $h \in \mathcal{H}_B - \mathcal{H}_{M_a(G)}$ and $\tau \in P(0, \theta) \in B$, then $|h(\nu) - h(\tau)| = |h(\nu) - h(\lambda)| \leq \|\nu - \lambda\|_v < \varepsilon/2$ (Corollary 23.7, Theorem 28.11). Proceed!]

(f) Let $[G'] \subseteq X \subseteq CB(G')$, where X is a vector space and $[G']$ denotes the span of G' in $CB(G')$. Then $\mathscr{A}\{M(G)\} = \{\mu \in M(G) : \sigma(\mu) = \{\hat{\mu}(\chi) : \chi \in G'\}\}$ is a $\sigma(M(G), X)$-dense, analytic subalgebra of $M(G)$. [*Hint*: Use Corollary 29.19 and establish that $M_b(G) \subset \mathscr{A}\{M(G)\}$.]

28K. Normal Subalgebras of $M(G)$

A Banach subalgebra B of $M(G)$ is normal if $\{h/B \neq 0 : h \in \mathcal{H}_{M(G)}\} = \{h_\chi/B : \chi \in G'\}$ [Do not confuse this terminology with Definition 26.11!], and strongly normal if $\mathcal{H}_B = \{h_\chi/B : \chi \in G'\}$.

(a) Show that $M_a(G)$ is strongly normal in $M(G)$.

(b) Every μ in a normal subalgebra $B \subset M(G)$ satisfies $\{h(\mu) : h \in \mathcal{H}_{M(G)}\} = \sigma(\mu) \subset \{\hat{\mu}(\chi) : \chi \in G'\} \cup \{0\}$.

(c) A strongly normal subalgebra of $M(G)$ is normal.

(d) Every (strongly) normal sub $*$-algebra of $M(G)$ is (strongly) symmetric. [*Hint*: $h_\chi(\mu^*) = \overline{h_\chi(\mu)}$ $\forall \mu \in M_a(G)$ by Theorem 28.11, Definition 28.10^+.]

(e) If X is taken as in Problem 28J(f) and G is compact, $\mathfrak{N}\{M(G)\} = \{\mu \in M(G) : \sigma(\mu) = \{\hat{\mu}(\chi) : \chi \in G'\} \cup \{0\}\}$ is a $\sigma(M(G), X)$-dense, normal subalgebra of $M(G)$. [*Hint*: Show that $M_a(G) \subset \mathfrak{N}\{M(G)\}$.]

(f) If K is a finite subgroup of G, then $B = \{\mu \in M(G) : \mu(G - K) = 0\}$ is a strongly normal subalgebra of $M(G)$. [*Hint*: B consists of discrete measures only.]

Problems 28I–K can be schematically summarized by

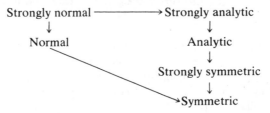

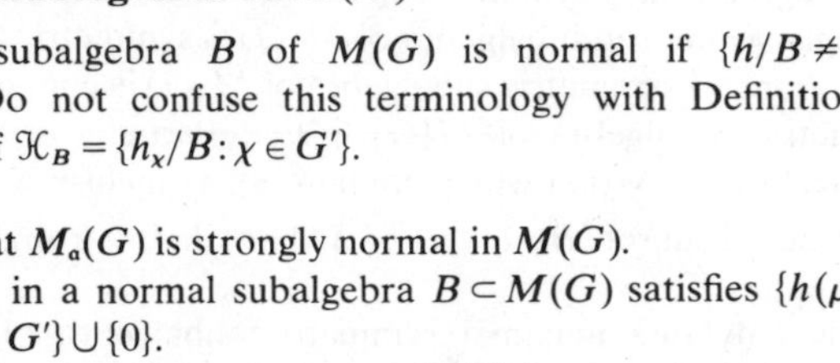

For more detailed developments, including counterexamples to reversing the above implications, see Celeste [17] and Hartman [40].

References for Further Study

On $M(G)$ and $L_1(G)$: Hewitt and Ross [41]; Rickart [83]; Rudin [85].
On $M(G)$ and its components: Hewitt and Ross [41].

29. Fourier Analysis on LCT$_2$A Groups

The theorems in this section lie at the heart of abstract harmonic analysis. Their proofs and illustrations are interwoven with the theory of earlier sections, and serve as a fitting conclusion to this text. Interested readers who correctly assume that we have not run out of subject matter—only time and space—may consult the references at the end of this section. A good starting point is Rudin [85], Chapter II. Beyond that Loomis [58], Chapter IX, and Mackey [59] offer insights into potential research problems and future developments in this area.

Unless specified otherwise G is an LCT$_2$A group.

Bochner's Theorem

We begin where Problem 11C ended by characterizing the members of $P(G)$ in terms of elements in $M^+(G')$. First, recall

DEFINITION 29.1. A complex-valued function φ on a group G is said to be positive definite if $\sum_{j,k=1}^{n} \alpha_j\bar{\alpha}_k(x_jx_k^{-1})\geq 0$ $\forall\{x_1,\ldots,x_k\}\subset G$ and $\{\alpha_1,\ldots,\alpha_n\}\in\mathbb{C}$. The continuous positive definite functions on a T_2 group G are denoted $P(G)$.

Surely, $\varphi_1+\alpha\varphi_2\in P(G)\,\forall\varphi_1,\varphi_2\in P(G)$ and $\alpha\geq 0$. That $\varphi\in P(G)$ implies $\bar{\varphi}\in P(G)$ is also clear by virtue of

THEOREM 29.2. A positive definite function φ on G satisfies

(i) $\varphi(x^{-1})=\overline{\varphi(x)}\,\forall x\in G$;
(ii) $|\varphi(x)|\leq\varphi(e)\,\forall x\in G$;
(iii) $|\varphi(x)-\varphi(x_0)|\leq 2\varphi(e)\mathcal{R}_e\{\varphi(e)-\varphi(xx_0^{-1})\}\,\forall x, x_0\in G$.

Proof. Fix $n=2$. For $\alpha_1=1$ and $\alpha_2=c$ with $x_1=e$ and $x_2=x$ we have $(1+|c|^2)\varphi(e)+\bar{c}\varphi(x^{-1})+c\varphi(x)=\sum_{j,k=1}^{2}\alpha_j\bar{\alpha}_k\varphi(x_jx_k^{-1})\geq 0$. Therefore $c=0$ yields $\varphi(e)\geq 0$. If $c=1$, then $\varphi(x^{-1})+\varphi(x)$ is real. If $c=i$, then $i\varphi(x)-i\varphi(x^{-1})$ is real and $-\varphi(x)+\varphi(x^{-1})$ is pure imaginary. Together, they give $\varphi(x^{-1})=\overline{\varphi(x)}$ in (i). Clearly (ii) holds for $\varphi\equiv 0$. Given any $x\in G$ such that

$\varphi(x)\neq 0$, substitute $c=-|\varphi(x)|/\varphi(x)$ above and obtain $2\varphi(e)-2|\varphi(x)|\geq 0$ which is property (ii). Evidently (iii) holds if $\varphi(x)=\varphi(x_0)$. For $\varphi(x)\neq\varphi(x_0)$ let $x_1=e$, $x_2=x$, $x_3=x_0$, and take $\alpha_1=1$, $\alpha_2=\lambda[|\varphi(x)-\varphi(x_0)|]/\varphi(x)-\varphi(x_0)$ (here $\lambda\in\mathbb{R}$ is arbitrary) and $\alpha_3=-\alpha_2$. Then $\sum_{j,k=1}^{3}\alpha_j\bar{\alpha}_k\varphi(x_jx_k^{-1})$ reduces to $2\lambda^2[\varphi(e)-\mathcal{R}_e\{\varphi(xx_0^{-1})\}]+2\lambda|\varphi(x)-\varphi(x_0)|+\varphi(e)$ and (since this is nonnegative $\forall\lambda\in\mathbb{R}$) the discriminant must be nonpositive; that is (iii) holds true. ■

Remark. Property (iii) ensures that φ is uniformly continuous if it is continuous at $e\in G$.

Example 29.3. (a) We have $\varphi=f\star f^*\in P(G)\,\forall f\in L_2(G)$. Clearly $\varphi\in C_\infty(G)$ by Corollary 27.3(i). Moreover (Theorem 27.1, Corollary 12.11),

$$(f\star f^*)(x_jx_k^{-1})=\int_G f(x_jx_k^{-1}y)f^*(y^{-1})\,d\mu_r(y)=\int_G f(x_jx_k^{-1}y)\overline{f(y)}\,d\mu_r(y)$$
$$=\int_G f(x_jy)\overline{f(x_ky)}\,d\mu_r(y)$$

implies that

$$\sum_{j,k=1}^{n}\alpha_j\bar{\alpha}_k\,\varphi(x_jx_k^{-1})=\int_G\sum_{j,k=1}^{n}\alpha_j\bar{\alpha}_k f(x_jy)\overline{f(x_ky)}\,d\mu_r(y)$$
$$=\int_G\left|\sum_{i=1}^{n}\alpha_i f(x_iy)\right|^2 d\mu_r(y)\geq 0$$

$\forall\{x_1,\ldots,x_n\}\in G$ and $\{\alpha_1,\ldots,\alpha_n\}\in\mathbb{C}$.

(b) If $M^+(G')$ denotes the nonnegative members of $M(G')$, then $\varphi_\mu(x)=\int_{G'}\overline{\chi(x)}\,d\mu$ belongs to $P(G)$ for each $\mu\in M^+(G')$. To be sure, $\chi(x^{-1})=\overline{\chi(x)}\,\forall x\in G$ and

$$\sum_{j,k=1}^{n}\alpha_j\bar{\alpha}_k\varphi_\mu(x_jx_k^{-1})=\int_{G'}\sum_{j,k=1}^{n}\alpha_j\bar{\alpha}_k\overline{\chi(x_jx_k^{-1})}\,d\mu(\chi)$$
$$=\int_{G'}\left|\sum_{i=1}^{n}\alpha_i\overline{\chi(x_i)}\right|^2 d\mu(\chi)\geq 0;$$

and φ_μ is continuous: for any $\varepsilon>0$ there is (by the regularity of $\mu\in M(G')$) a compact set $K\subset G'$ having $\mu(G'-K)=\mu(G)-\mu(K)<\varepsilon/4$, and (Corollary 27.12) $v = v_{K,\varepsilon/2\mu(G')}(x)$ is a G-nbd such that $y \in v$ implies

$$|\varphi_\mu(x)-\varphi_\mu(y)| \le \int_{G'} |\chi(x)-\chi(y)|\, d\mu(\chi)$$
$$= \int_K |\chi(x)-\chi(y)|\, d\mu(\chi) + \int_{G-K} |\chi(x)-\chi(y)|\, d\mu(\chi)$$
$$\le \varepsilon/2 + 2\mu(G'-K) < \varepsilon.$$

Example 29.3(b) constitutes half the proof of

THEOREM 29.4 (Bochner). $P(G)=\{\varphi \in C(G): \varphi(x)=\int_{G'} \overline{\chi(x)}\, d\mu(\chi)$ for $\mu \in M^+(G')$ with $\mu(G')=\varphi(e)\}$.

Proof. The result is trivially true for $\varphi \equiv 0$, in which case $\mu \equiv 0$ and $\varphi(e)=0$. Given any $\varphi \not\equiv 0$ in $P(G)$ and $f \in C_0(G)$ with $f(G-K)=0$ for compact $K \subset G$, we have $\xi(x, y)=f(x)\overline{f(y)}\,\varphi(xy^{-1})$ belonging to $C_0(G \times G)$ with $\xi(G \times G - K \times K)=0$. By definition of the double integral there is a partition of K by disjoint sets $\Delta(\varepsilon)=\{E_1, \ldots, E_n\} \subset \mathfrak{R}_\sigma$ and points $x_k \in E_k$ such that

$$\Big|\int_G \int_G f(x)\overline{f(y)}\,\varphi(xy^{-1})\, d\mu_r(x)\, d\mu_r(y)$$
$$- \sum_{i,j=1}^{n} f(x_i)\overline{f(x_j)}\,\varphi(x_i x_j^{-1})\mu_r(E_i)\mu_r(E_j)\Big| < \varepsilon.$$

Since φ is positive definite, each $\sum_{i,j=1}^{n} \mu_r(E_i)f(x_i)\overline{\mu_r(E_j)f(x_j)}\,\varphi(x_i x_j^{-1}) \ge 0$ and so $\int_G \int_G f(x)\overline{f(y)}\,\varphi(xy^{-1})\, d\mu_r(x)\, d\mu_r(y) \ge 0$. The denseness of $C_0(G)$ in $L_1(G)$ ensures further that $\int_G \int_G F(x)\overline{F(y)}\,\varphi(xy^{-1})\, d\mu_r(x)\, d\mu_r(y) \ge 0$ $\forall F \in L_1(G)$. Now $\Phi \in \{L_1(G)\}'$ with $|||\Phi||| \le \varphi(e)$ defined by $\Phi(F)=\int_G F(x)\,\varphi(x)\, d\mu_r(x)$ generates a sesquilinear function (Appendix a.3) on $L_1(G)$ by taking

$$((F, g)) = \Phi(F \star G^*) = \int_G \Big[\int_G F(xy)\, g^*(y^{-1})\, d\mu_r(y)\Big] d\mu_r(x)$$
$$= \int_G \Big[\int_G F(xy)\, \overline{g(y)}\, d\mu_r(y)\Big] \varphi(x)\, d\mu_r(x)$$
$$= \int_G \Big[\int_G F(xy)\, \overline{g(y)}\, \varphi(x)\, d\mu_r(x)\Big] d\mu_r(y)$$
$$= \int_G \Big[\int_G F(x)\, \overline{g(y)}\, \varphi(xy^{-1})\, d\mu_r(x)\Big] d\mu_r(y)$$

$\forall F, g \in L_1(G)$.

Therefore $|((F, g))|^2 \le ((F, F))((g, g))$ $\forall F, g \in L_1(G)$. Next, let $g_v = [1/\mu_\mathscr{T}(v)]\chi_v$ for a symmetric G-nbd $v(e)$ and obtain

$$((F, g_v)) - \Phi(F) = \int_G \left[\int_G F(x)\,\overline{g_v(y)}\,\varphi(xy^{-1})\,d\mu_\mathscr{T}(x)\right] d\mu_\mathscr{T}(y) - \int_G F(x)\,\varphi(x)\,d\mu_\mathscr{T}(x)$$

$$= \int_G \left[1/\mu_\mathscr{T}(v) \int_v F(x)\,\varphi(xy^{-1})\,d\mu_\mathscr{T}(x)\right] d\mu_\mathscr{T}(y) - \int_G F(x)\left[1/\mu_\mathscr{T}(v)\int_v \varphi(x)\,d\mu_\mathscr{T}(y)\right] d\mu_\mathscr{T}(x)$$

$$= \int_G \left[F(x)\cdot 1/\mu_\mathscr{T}(v) \int_v \{\varphi(xy^{-1}) - \varphi(x)\}\,d\mu_\mathscr{T}(y)\right] d\mu_\mathscr{T}(x)$$

and $$((g_v, g_v)) - \varphi(e) = 1/\{\mu_\mathscr{T}(v)\}^2 \int_v \left[\int_v \{\varphi(xy^{-1}) - \varphi(e)\}\,d\mu_\mathscr{T}(x)\right] d\mu_\mathscr{T}(y).$$

Since φ is uniformly continuous (Theorem 29.2), the nbd $v(e)$ can be chosen to satisfy $|((F, g_v)) - \Phi(F)| < \varepsilon$ and $|((g_v, g_v)) - \varphi(e)| < \varepsilon$ for preselected $\varepsilon > 0$. Applying this to the above Cauchy Schwartz inequality gives $|\Phi(F)|^2 \le ((F, F))\varphi(e) + \varepsilon\{((F, F)) + 2((F, g_v)) + \varepsilon\}$ and (since $\varepsilon > 0$ is arbitrary) $|\Phi(F)|^2 \le ((F, F))\,\varphi(e) = \varphi(e)\,\Phi(F \star F^*)$. Since multiplication in $L_1(G)$ is convolution, there follows $|\Phi(F)|^2 \le \varphi(e)\,\Phi(F \star F^*) \le \varphi(e)^{1+1/2}[\Phi\{(F \star F^*)^2\}]^{1/2} \le \cdots \cdots \le \varphi(e)^{1+1/2+\cdots+2^{-n}}[\Phi\{(F \star F^*)^{2^n}\}]^{-2^n}$ which, by taking the limit, yields $|\Phi(F)|^2 \le \varphi(e)^{1/2} r_{\sigma_\odot}(F \star F^*) = \varphi(e)^{1/2}\|\hat{F}\|_0^2$ [Problem 23A(b, ii), Definition 27.7]. This means that $\Phi(\hat{F}) = \Phi(F)$ is $\|\ \|_0$-continuous on $\widehat{L_1(G)}$ and so extendable with the same norm to all of $C_\infty(G')$ (Corollary 27.15). This extension is nonnegative. Reason: For any $\xi \ge 0$ in $C_\infty(G')$ and any $\varepsilon > 0$ choose $\hat{F} \in \widehat{L_1(G)}$ satisfying $\|\xi^{1/2} - \hat{F}\|_0 < \varepsilon$ and use $\|\xi - |\hat{F}|^2\|_0 \le \|\xi^{1/2} - \hat{F}\|_0 \|\xi^{1/2} - \bar{\hat{F}}\|_0 \le \varepsilon(2\|\xi^{1/2}\|_0 + \varepsilon)$ to obtain $|\Phi(\xi) - ((F, F))| = |\Phi(\xi) - \Phi(F \star F^*)| \le |||\Phi|||\ \|\xi - \widehat{F \star F^*}\|_0 = |||\Phi|||\ \|\xi - |\hat{F}|^2\|_0 \le |||\Phi|||\varepsilon(2\|\xi^{1/2}\|_0 + \varepsilon)$. The arbitrariness of $\varepsilon > 0$ necessitates that $\Phi(\xi) = ((F, F)) \ge 0$ as asserted. All this leads up to invoking the Riesz Markoff theorem (Appendix m.8). Specifically, there is a $\nu \ge 0$ in $M(G')$ with $\|\nu\|_v = |||\Phi|||$ such that

$$\Phi(\hat{F}) = \int_{G'} \hat{F}(\chi)\,d\nu(\chi) = \int_{G'} \left(\int_G F(x)\,\chi(x)\,d\mu_\mathscr{T}(x)\right) d\nu(\chi) = \int_G F(x)\left(\int_{G'} \chi(x)\,d\nu(\chi)\right) d\mu_\mathscr{T}(x)\ \forall F \in L_1(G).$$

At the same time we have $\Phi(\hat{F})=\int_G F(x)\,\varphi(x)\,d\mu_r(x)\ \forall F\in L_1(G)$ and $\int_{G'}\chi(x)\,d\nu(\chi)\in C(G)$. Therefore $\varphi(x)=\int_{G'}\chi(x)\,d\nu(\chi)$ on G by Corollary 12.17. Finally (Theorem 28.2), $\nu\in M^+(G')$ implies $\mu=\nu^*\in M^+(G')$ and $\varphi(x)=\int_{G'}\chi(x)\,d\nu(\chi)=\int_{G'}\overline{\chi(x)}\,d\nu(\chi^{-1})=\int_{G'}\overline{\chi(x)}\,d\mu(\chi)\ \forall x\in G$. ■

The Inversion Theorem

$P(G)$ is not a vector space, since it is not closed under negative scalar multiplication. Nevertheless, a good deal can be said about $[P(G)\cap L_1(G)]$—the subspace of $L_1(G)$ spanned by $P(G)\cap L_1(G)$. The main idea here is that $\{C_0(G)\}^*\star C_0(G)=\{f^*\star g: f,g\in C_0(G)\}\subset[P(G)\cap L_1(G)]$ by Example 29.3(a) and the decomposition

$$f^*\star g=\tfrac{1}{4}(f+g)^*\star(f+g)-\tfrac{1}{4}(f-g)^*\star(f-g)+(i/4)(f+ig)^*\star(f+ig)$$
$$-(i/4)(f-ig)^*\star(f-ig).$$

THEOREM 29.5. $[P(G)\cap L_1(G)]$ is dense in $L_1(G)$ and $[P(G)\cap L_1(G)\cap L_2(G)]$ is dense in $L_2(G)$.

Proof. Theorems 27.1 and 27.5 demonstrate that $\{C_0(G)\}^*\star C_0(G)$ is $\|\ \|_1$-dense in $C_0(G)$ and the denseness of $\{C_0(G)\}^*\star C_0(G)$ in $L_1(G)$ follows from that of $C_0(G)$ in $L_{p\geq 1}(G)$ (Appendix m). Similar reasoning renders $\{C_0(G)\}^*\star C_0(G)\subset[P(G)\cap L_1(G)\cap L_2(G)]$ dense in $L_2(G)$. ■

Although the linear surjection $\overset{(f)}{\widehat{L_1(G)}}\to L_1(G)$, $\hat{f}\rightsquigarrow f$, is not injective, a large number of elements in $L_1(G)$ can be recovered from their Fourier transforms. Denote $\mu\in M^+(G')$ satisfying $\varphi(x)=\int_{G'}\overline{\chi(x)}\,d\mu(\chi)$ in $P(G)$ by μ_φ.†

THEOREM 29.6 (Inversion). If $f\in[P(G)\cap L_1(G)]$, then $\hat{f}\in L_1(G')$. For fixed Haar measure μ_r on G the Haar measure μ_{r0} on G' can be normalized so that $f(x)=\int_{G'}\hat{f}(\chi)\,\overline{\chi(x)}\,d\mu_{r0}(\chi)\ \forall x\in G$.

Proof. The rather lengthy proof is best broken up into miniproofs.

CLAIM 1: $\hat{f}\,d\mu_h=\hat{h}\,d\mu_f$ on G' $\forall f,h\in[P(G)\cap L_1(G)]$.

† Do not confuse this with $\varphi\,d\mu_r$.

Proof. Integration preserves linearity. Accordingly, every $F\in[P(G)]$ satisfies $F(X)=\int_{G'}\overline{\chi(x)}\,d\mu_F(\chi)$ for some $\mu_F\in M(G')$ (Theorem 29.4) and $F(x^{-1})=\overline{F(x)}\ \forall x\in G$ [Theorem 11.6(i)]. For any $g\in L_1(G)$ we therefore have $(g\star f)(e)=\int_G g(y)f(y^{-1})\,d\mu_r(y)=\int_G g(y)[\int_{G'}\chi(y)\,d\mu_f(\chi)]\,d\mu_r(y)=\int_{G'}[\int_G g(y)\chi(y)\,d\mu_r(y)]\,d\mu_f(\chi)=\int_{G'}\hat g(\chi)\,d\mu_f(\chi)$. Now this relationship $(g\star f)(e)=\int_{G'}\hat g(\chi)\,d\mu_f(\chi)$ applied to $g\star h$ yields $\{(g\star h)\star f\}(e)=\int_{G'}\widehat{g\star h}(\chi)\,d\mu_f(\chi)=\int_{G'}\hat g(\chi)\hat h(\chi)\,d\mu_f(\chi)$. Similarly, $\{(g\star f)\star h\}(e)=\int_{G'}\hat g(\chi)\hat f(\chi)\,d\mu_h(\chi)$. Since $L_1(G)$ is commutative and $\widehat{L_1(G)}$ is dense in $C_\infty(G')$, it follows that $\int_{G'}\xi\hat f\,d\mu_h=\int_{G'}\xi\hat h\,d\mu_f\ \forall\xi\in C_\infty(G')$. In particular (Appendix m.7), the measures $\hat f\,d\mu_h$ and $\hat h\,d\mu_f$ corresponding to the linear function $\xi\rightsquigarrow\int_{G'}\xi\hat f\,d\mu_h=\int_{G'}\xi\hat h\,d\mu_f$ on $C_\infty(G')$ coincide.

CLAIM 2: For each $F\in C_0(G')$ there is an $h\in[P(G)\cap L_1(G)]$ such that $\hat h>0$ on the support $N(F)$ of F. And, the mapping $\Phi: F\rightsquigarrow\int_{G'}[1/\hat h(\chi)]F(\chi)\,d\mu_h(\chi)$ is a positive linear functional on $C_0(G')$.

Proof. For $\chi_0\in N(F)$ there is some $f_0\in L_1(G)$ such that $\hat f_0(\chi_0)\neq 0$ (Theorem 27.8). Since $C_0(G)$ is dense in $L_1(G)$, we also have (Corollary 27.15) $|\hat f_0(\chi_0)-\hat g_0(\chi_0)|\le\|\widehat{f_0-g_0}\|_0\le\|f_0-g_0\|_1<\varepsilon$ for some $g_0\in C_0(G)$. The conclusion, of course, is that $\hat g_0(\chi_0)\neq 0$. In fact, $\widehat{g_0^*\star g_0}(\chi_0)=|\hat g_0(\chi_0)|^2$ and (since $\widehat{g_0^*\star g_0}\in C(G')$) there is a G'-nbd $u_\chi(\chi_{G'})$ on which $\widehat{g_0^*\star g_0}>0$. Since $N(F)$ is compact, the same procedure yields $N(F)\subset\bigcup_{i=1}^n u_{\chi_i}(\chi_{G'})$ with $\widehat{g_i^*\star g_i}>0$ on $u_{\chi_i}(\chi_{G'})$ for some finite set $\{\chi_1,\ldots,\chi_n\}\in N(F)$. The first part of the claim now follows [Example 29.3(a)] by taking $h=\sum_{i=1}^n g_i^*\star g_i$. That Φ is well defined is clear, since for any $\psi\in[P(G)\cap L_1(G)]$ with $\hat\psi>0$ on $N(F)$, Claim 1 ensures that $\int_{G'}(1/\hat\psi)F\,d\mu_\psi=\int_{G'}(1/\hat\psi\hat h)F\hat h\,d\mu_\psi=\int_{G'}(1/\hat h)F\,d\mu_h$. Surely Φ is linear and also positive, since $\mu_h\ge 0$ implies $\Phi(F)\ge 0$ for $F\ge 0$. Moreover, $\Phi\neq 0$, since some $\xi_0\in C_0(G')$ and $f_0\in P(G)\cap L_1(G)$ satisfy $\int_{G'}\xi_0\,d\mu_f\neq 0$. Therefore $\xi_0\hat f_0\in C_0(G')$ and $\Phi(\xi_0\hat f_0)=\int_{G'}(1/\hat h)\xi_0\hat f_0\,d\mu_h=\int_{G'}\xi_0\,d\mu_{f_0}\neq 0$.

CLAIM 3: Φ is invariant on $C_0(G')$, and a Haar measure μ_{T_0} on G' can be found that satisfies all conditions of the theorem.

Proof. Fix any $F \in C_0(G')$ and $\chi_0 \in G'$. Assume that $h \in [P(G) \cap L_1(G)]$ with $\hat{h} > 0$ on $N(F)$ and define $F_0(x) = [1/\chi_0(x)]h(x)$ $\forall x \in G$. Clearly $\hat{F}_0(\chi) = \int_G (1/\chi_0)h\chi\, d\mu_T = \int_G h\chi_0^{-1}\chi\, d\mu_T = \hat{h}(\chi_0^{-1}\chi)\, \forall \chi \in G'$. It is also clear that $F_0 \in [P(G)]$. Therefore $\mu_{F_0} \in M(G')$ is defined and $F_0(x) = \int_{G'} \overline{\chi(x)}\, d\mu_{F_0}(\chi) = \int_{G'} [1/\chi_0(x)]\overline{\chi(x)}\, d\mu_h(\chi) = \int_{G'} \overline{\chi_0(x)\chi(x)}\, d\mu_h(\chi) = \int_{G'} \overline{\chi(x)}\, d\mu_h(\chi_0^{-1}\chi)$ implies that $\mu_h(\chi_0^{-1}E) = \mu_{F_0}(E)\, \forall E \in \mathfrak{R}_\sigma$ in G'. Claim 1 and our last two statements establish that

$$\begin{aligned}\Phi({}_{\chi_0}F) &= \int_{G'} [{}_{\chi_0}F(\chi)/\hat{h}(\chi)]\, d\mu_h(\chi)\\ &= \int_{G'} [F(\chi)/h(\chi_0^{-1}\chi)]\, d\mu_h(\chi_0^{-1}\chi)\\ &= \int_{G'} [F(\chi)/\hat{F}_0(\chi)]\, d_{F_0}(\chi)\\ &= \int_{G'} [F(\chi)/\hat{h}(\chi)]\, d\mu_h(\chi) = \Phi(F).\end{aligned}$$

Since the conditions in Appendix m.8 are also satisfied, there is a Haar measure $\mu_{T_0} \in M^+(G')$ such that $\Phi(F) = \int_{G'} F\, d\mu_{T_0}\, \forall F \in C_0(G')$. Furthermore, $\int_{G'} F\, d\mu_f = \int_{G'} (F\hat{f}/\hat{h})\, d\mu_h = \Phi(F\hat{f}) = \int_{G'} F\hat{f}\, d\mu_{T_0} \forall F \in C_0(G')$ establishes that such is also true $\forall F \in L_1(G')$. Thus $\mu_f = \hat{f}\, d\mu_{T_0}$ and $\hat{f} \in L_1(G')$ by virtue of $|\int_{G'} \hat{f}\, d\mu_{T_0}| = |\int_{G'} d\mu_f| = \mu_f(G') < \infty$. By definition $f(x) = \int_{G'} \overline{\chi(x)}\, d\mu_f(\chi) = \int_{G'} \hat{f}(\chi)\overline{\chi(x)}\, d\mu_{T_0}(\chi)\, \forall x \in G$. ■

Something is missing; namely, whether $f \in [P(G) \cap L_1(G)]$ can be recovered from any other $F \neq f$ in $L_1(G')$. The answer is in

THEOREM 29.7. Every $f \in [P(G) \cap L_1(G)]$ has $\hat{f} \in L_1(G')$ as its unique recovery function.

Proof. If $f \in P(G) \cap L_1(G)$, then $\hat{f} \geq 0$ in $L_1(G')$. Indeed, $\int_G (g_0 \star g_0^*)(x) f(x)\, d\mu_T(x) = ((g_0, g_0)) \geq 0\, \forall g_0 \in C_0(G)$ in Theorem 29.4.

Since $C_0(G) \star \{C_0(G)\}^*$ is dense in $L_1(G)$, we also have

$$0 \leq \int_G g(x) f(x)\, d\mu_r(x) = \int_G g(x)\Big(\int_{G'} \hat{f}(\chi)\overline{\chi(x)}\, d\mu_{r0}(\chi)\Big)\, d\mu_r(x)$$
$$= \int_{G'}\Big(\int_G g(x)\overline{\chi(x)}\, d\mu_r(x)\Big)\hat{f}(\chi)\, d\mu_{r0}(\chi)$$
$$= \int_{G'} \hat{g}(\bar{\chi})\hat{f}(\chi)\, d\mu_{r0}(\chi)\ \forall g \in L_1(G).$$

Since $\widehat{L_1(G)}$ is dense in $C_\infty(G')$, there follows $\int_{G'} \psi(\bar{\chi})\hat{f}(\chi)\, d\mu_{r0}(\chi) \geq 0$ $\forall \psi \in C_\infty(G')$. In particular, $\int_{G'} \psi(\bar{\chi})\hat{f}(\chi)\, d\mu_{r0}(\chi) \geq 0$ $\forall \psi \geq 0$ in $C_\infty(G')$ necessitates that $\hat{f} \geq 0$ on G'. Next, suppose that $f(x) = \int_{G'} F(\chi)\overline{\chi(x)}\, d\mu_{r0}(\chi)$ for $F \geq 0$ in $L_1(G')$. We claim that $F = \hat{f}$ a.e. on G'. Essentially, $\int_{G'} F(\chi)\overline{\chi(x)}\, d\mu_{r0}(\chi) = \int_{G'} \hat{f}(\chi)\overline{\chi(x)}\, d\mu_{r0}(\chi)$ $\forall x \in G$ and for $g \in L_1(G)$, we have

$$\int_{G'} \hat{g}(\bar{\chi}) F(\chi)\, d\mu_{r0}(\chi) = \int_{G'}\Big(\int_G g(x)\overline{\chi(x)}\, d\mu_r(x)\Big) F(\chi)\, d\mu_{r0}(\chi)$$
$$= \int_G g(x)\Big(\int_{G'} F(\chi)\overline{\chi(x)}\, d\mu_{r0}(\chi)\Big)\, d\mu_r(x)$$
$$= \int_G g(x)\Big(\int_{G'} \hat{f}(\chi)\overline{\chi(x)}\, d\mu_{r0}(\chi)\Big)\, d\mu_r(x)$$
$$= \int_{G'}\Big(\int_G g(x)\overline{\chi(x)}\, d\mu_r(x)\Big)\hat{f}(\chi)\, d\mu_{r0}(\chi)$$
$$= \int_{G'} \hat{g}(\bar{\chi})\hat{f}(\chi)\, d\mu_{r0}(\chi).$$

Reasoning as in Corollary 12.17 therefore ensures that $\hat{f} = F$ a.e. on G'. Extending this result to arbitrary $f \in [P(G) \cap L_1(G)]$ is straightforward; use linearity and the decomposition of $\hat{F} \in L_1(G')$ into two nonnegative functions in $L_1(G')$. ■

The inversion theorem filters out a unique Haar measure μ_{r0} on G' corresponding to each fixed Haar measure μ_r on G. This determination of μ_{r0} in terms of μ_r is sometimes a formidable task; in many instances it is not.

Example 29.8. (a) Suppose that G is compact and μ is chosen such that $\mu(G)=1$. If $f=\chi_G$, then (Example 12.18) $\hat{f}(\chi)=\int_G f(x)\chi(x)\,d\mu(x)=\begin{cases}1, & \text{if } \chi=\chi_G\\ 0, & \text{if } \chi\neq\chi_G\end{cases}$ and so $1=f(x)=\int_{G'}\hat{f}(\chi)\overline{\chi(x)}\,d\mu_0(\chi)=\int_{G'}\hat{f}(\bar{\chi})\chi(x)\,d\mu_0(\chi)=\mu_0(\chi_G)$. Thus G' is discrete with $\mu_0(\chi)=1\ \forall\chi\in G'$.

(b) Suppose G is discrete and μ is chosen such that $\mu(e)=1$. For $f=\chi_e$ we have $\hat{f}(\chi)=\int_G f(x)\overline{\chi(x)}\,d\mu(x)=\mu(e)=1\ \forall\chi\in G'$. Accordingly, $1=f(e)=\int_{G'}\hat{f}(\chi)\overline{\chi(e)}\,d\mu_0(\chi)=\mu_0(G')$.

(c) Let $\mu=\alpha\,dt$ be a fixed Haar measure on $G=\mathbb{R}$ and let $\beta\,d\theta$ be any Haar measure on $\mathbb{R}'=\mathbb{R}$ (cf Examples 11.19 and 27.9). If $f(t)=2\beta/(1+t^2)$ on $\mathbb{R}$, then $f\in P(\mathbb{R})$, since $e^{-|\theta|}\beta\,d\theta\in M^+(\mathbb{R}'=\mathbb{R})$ and $2\beta/(1+t^2)=\int_{-\infty}^{\infty}e^{-it\theta}\cdot e^{-|\theta|}\beta\,d\theta$. At the same time (Theorem 29.7) $f(t)=\int_{-\infty}^{\infty}\hat{f}(\chi_\theta)e^{-it\theta}\beta\,d\theta$ ensures that $e^{-|\theta|}=\hat{f}(\chi_\theta)=\int_{-\infty}^{\infty}2\beta/(1+t^2)e^{it\theta}\alpha\,dt=2\alpha\beta\int_{-\infty}^{\infty}e^{it\theta}/(1+t^2)\,dt$. Taking $\theta=0$ and using $\int_{-\infty}^{\infty}dt/(1+t^2)=\pi$ gives $1=2\pi\alpha\beta$. In other words, the inversion theorem for $1/(1+t^2)$ determines $\mu_0=d\theta/2\pi\alpha$ on $\mathbb{R}'$ for fixed $\mu=\alpha\,dt$ on $\mathbb{R}$.

The inversion theorem can be used to complete what was started in Corollary 27.12.

THEOREM 29.9. The collection $\{u_{K,\varepsilon}(e)\colon \text{compact } K\subset G' \text{ and } \varepsilon>0\}$ is a nbd base at $e\in G$.

Proof. Given any G-nbd $v(e)$, there is a nbd $u(e)$ and a relatively compact nbd $w(e)$ such that $\bar{w}\subset u\subset uu^{-1}\subset v$. Let $f=1/[\mu(\bar{w})]^{1/2}\chi_{\bar{w}}$ and define $F=f^*\star f\in P(G)$. Note that $F(x)=\int_G f^*(xy)f^{-1}(y)\,d\mu(y)=\int_G \overline{f(yx^{-1})}f(y)\,d\mu(y)=1/[\mu(\bar{w})]\int_G\chi_{\bar{w}}(yx^{-1})\chi_{\bar{w}}(y)\,d\mu(y)$ and $F(e)=1$. Now $\chi_{\bar{w}}(yx^{-1})\chi_{\bar{w}}(y)=0$ if either $x\notin\bar{w}y^{-1}$ or $y\notin\bar{w}$. Accordingly, $\chi_{\bar{w}}(yz^{-1})\chi_{\bar{w}}(y)\neq 0$ implies $x\in\bar{w}^{-1}y\subset\bar{w}^{-1}\bar{w}$ and F vanishes off the compact set $\bar{w}^{-1}\bar{w}\subset v$. In particular, $F\in P(G)\cap C_0(G)$ and $\hat{F}\in L_1(G')$ with $F(x)=\int_{G'}\hat{F}(\chi)\overline{\chi(x)}\,d\mu_0(\chi)\,\forall x\in G$. Furthermore, $\hat{F}=|\hat{f}|^2>0$ and $\nu=\hat{F}\,d\mu_0\in M_a^+(G')$ with $\nu(G')=F(e)=1$. The regularity of ν also ensures that for any $\varepsilon>0$ there is a compact set $K\subset G'$ such that $\nu(G'-K)<\varepsilon$ and $\nu(K)>1-\varepsilon$. We claim that $u_{K,\varepsilon}(e)\subset v(e)$ for $\varepsilon\le\frac{1}{2}$. To be sure, $x\in u_{K,\varepsilon}(e)$ implies that $\varepsilon>|\chi(x)-1|$ and $R_e\chi(x)>1-\varepsilon\ \forall\chi\in K$. Therefore (using $|z_1+z_2|\ge R_e z_1-|z_2|$ for $z_1,z_2\in\mathbb{C}$) $|F(x)|\ge R_e\int_K|\hat f(\chi)|^2\overline{\chi(x)}\,d\mu_{r0}(\chi)-|\int_{G'-K}|\hat f(\chi)|^2\overline{\chi(x)}\,d\mu_{r0}(\chi)|>(1-\varepsilon)\int_K|\hat f(\chi)|^2\,d\mu_{r0}(\chi)-\int_{G'-K}|\hat f(\chi)|^2\,d\mu_{r0}(\chi)>(1-\varepsilon)\int_K d\nu(\chi)-\int_{G'-K}d\nu(\chi)>(1-\varepsilon)^2-\varepsilon>0$ and since $F(G-\bar w^{-1}\bar w)=0)$ $x\in\bar w^{-1}\bar w\subset v$ as asserted. ■

Theorem 11.20 is now easily verified and (by analogy with Corollary 17.9) shown to have far-reaching consequences.

COROLLARY 29.10. *G' is injective for an LCT_2A group G.*

Proof. If $x_0\ne e=\bigcap_{\mathfrak{N}_e}u(e)$, then x_0 is not contained in some G-nbd $u_{K,\varepsilon}(e)$ and so $|\chi(x_0)-1|\ge\varepsilon>0$ for some $\chi\in K$ in G'. ■

Remark. Note that G' being injective is equivalent to $x\ne y$ implying $\chi(x)\ne\chi(y)$ for some $\chi\in G'$.

The gist of our next result is that every $f\in C(G)$ can be uniformly approximated on compacta by members of $[G']$.

COROLLARY 29.11. *$[G']$ is dense in $C_\kappa(G)$. If G is compact, $[G']$ is also dense in $L_p(G)$ $(1\le p<\infty)$.*

Proof. $[G']$ is separating and the Stone Weierstrass theorem applies. Regularity of μ_r and compactness of G further ensures that $C(G)$ is dense in $L_p(G)$. ■

Corollary 29.11 can be viewed from another, perhaps more familiar, perspective.

Example 29.12. (a) Every $f\in C(\mathbb{R})$ can be uniformly approximated on finite intervals by $[e^{it\theta}:\theta\in\mathbb{R}]$ and every $f\in C([0,2\pi])$ satisfying $f(0)=f(2\pi)$ can be uniformly approximated by $[e^{int}:n\in\mathbb{I}]$.

(b) A function $f=\sum_{k=1}^{n}\lambda_k\chi_k$ $(\chi_k\in G')$ is termed a trigonometric polynomial on G. Thus $[G']\subset C(G)$ denotes the class of trigonometric polynomials on G. If G is compact, G' is orthogonal (Example 12.18) and the Fourier coefficients λ_k of f can be recovered by way of $\lambda_k=1/\mu_r(G)\int_G f(x)\overline{\chi_k(x)}\,d\mu_r(x)$.

Fourier Transforms on $L_2(G)$

If G is compact, $L_p(G) \subset L_1(G)\, \forall p > 1$ and $\hat{F}$ is defined $\forall F \in L_p(G)$ $(p \geq 1)$. Such, of course, is not true in general (Appendix m.3). Nevertheless, the definition of Fourier transform can always be extended to $L_r(G)$ $(1 \leq r \leq 2)$ with the aid of the inversion theorem and our next result.

THEOREM 29.13. If $\mu \in M(G')$ and $\int_{G'} \chi(x)\, d\mu(\chi) = 0\ \forall x \in G$, then $\mu = 0$.

Proof. For any element $f \in L_1(G)$, we have $\int_{G'} \hat{f}(\chi)\, d\mu(\chi) = \int_{G'} [\int_G f(x)\chi(x)\, d\mu_{\mathcal{F}}(x)]\, d\mu(\chi) = \int_G f(x)[\int_{G'} \chi(x)\, d\mu(\chi)]\, d\mu_{\mathcal{F}}(x) = 0$. Since $\widehat{L_1(G)}$ is dense in $C_\infty(G')$, it follows further that $\int_{G'} F(x)\, d\mu(\chi) = 0\ \forall F \in C_\infty(G')$. Therefore $\Phi(F) = \int_{G'} F\, d\mu$ is the zero functional and $\mu = 0$ (Appendix m.8). ∎

THEOREM 29.14 (Plancheral). The mapping

$$L_1(G) \cap L_2(G) \to L_2(G')$$
$$f \rightsquigarrow \hat{f}$$

is an onto isometry between dense subsets of $L_2(G)$ and $L_2(G')$ and is therefore uniquely extendable to an isometry from $L_2(G)$ onto $L_2(G')$.

Proof. If $f \in L_1(G) \cap L_2(G)$, then $f \star f^* \in P(G) \cap L_1(G)$ [Example 29.3(a)] and $\|f\|_2^2 = \int_G |f|^2\, d\mu_{\mathcal{F}} = \int_G f\bar{f}\, d\mu_{\mathcal{F}} = \int_G f(x) f^*(x^{-1})\, d\mu_{\mathcal{F}}(x) = (f \star f^*)(e)$, which (Theorem 29.5) equals $\int_{G'} \widehat{f \star f}(x)\, d\mu_{\mathcal{F}_0}(\chi) = \int_{G'} |\hat{f}(x)|^2\, d\mu_{\mathcal{F}_0}(\chi) = \|\hat{f}\|_2^2$. Since $\mu_{\mathcal{F}}$ is regular, $C_0(G)$ is dense in $L_2(G)$ and $L_1(G) \cap L_2(G)$ is dense in $L_2(G)$. Accordingly, the isometry $t: L_1(G) \cap L_2(G) \to L_2(G')$ has a unique isometric extension T from $L_2(G)$ into $L_2(G')$ defined by $F \rightsquigarrow \hat{F} \overset{\|\,\|_2}{=} \lim_n \hat{f}_n \in L_2(G')$ for any sequence $\{f_n\} \in L_1(G) \cap L_2(G)$ satisfying $f_n \xrightarrow{\|\,\|_2} F \in L_2(G)$. To prove T surjective it suffices [by virtue of $L_2(G)$-completeness] to show that $T\{L_1(G) \cap L_2(G)\} = \widehat{L_1(G) \cap L_2(G)}$ is dense in $L_2(G')$. Toward this end note (Definition 18.1) that

$\langle \widehat{L_1(G)\cap L_2(G)}, L_2(G')\rangle$ is a dual pair relative to $\langle \hat{F}, \varphi\rangle = \int_{G'} \varphi(\chi)\hat{F}(\chi)\,d\mu_0(\chi)$. Clearly $\hat{F}\in L_2(G')$ and $|\langle F,\varphi\rangle| \leq \|\varphi\|_2\|\hat{F}\|_2$ (Hölder's inequality) ensures that $\varphi\hat{F}\in L_1(G')$ and $\varphi\hat{F}\,d\mu_0\in M_a(G')$. The objective is now to establish that $\widehat{L_1(G)\cap L_2(G)}^{\perp}=\{0\}$, from which the denseness of $\widehat{L_1(G)\cap L_2(G)}$ in $L_2(G')$ Example 18.13) will follow. For any $\hat{F}\in\widehat{L_1(G)\cap L_2(G)}$ and notation as in Problem 11E, we have $\hat{F}_{\chi_0} = \widehat{\langle\,,\chi_0\rangle F}\in\widehat{L_1(G)\cap L_2(G)}$ $\forall\chi_0\in G'$; specifically, $\hat{F}_{\chi_0}(\chi)=\hat{F}(\chi\chi_0)=\int_G F(x)\chi(x)\chi_0(x)\,d\mu(x)=\int_G \{\langle\,,{}_0\rangle F\}(x)\chi(x)\,d\mu(x)=\widehat{\langle\,,{}_0\rangle F}(\chi)$. In similar fashion, $\widehat{F_{x_0^{-1}}}=\langle x_0,\,\rangle\hat{F}$ $\forall x_0\in G$ since $\hat{F}_{x_0^{-1}}(\chi)=\int_G F(xx_0^{-1})\chi(x)\,d\mu(x)=\int_G F(x)\,\chi(xx_0)\,d\mu(x)=\chi(x_0)\hat{F}(\chi)=\{\langle x_0,\,\rangle\hat{F}\}(\chi)$ $\forall x\in G'$. Turning to the heart of the proof, assume that $\langle\hat{F},\varphi\rangle=0$ $\forall\hat{F}\in\widehat{L_1(G)\cap L_2(G)}$. Then $\int_{G'}\chi(x)\,\varphi(\chi)\,\hat{F}(\chi)\,d\mu_0(\chi)=\int_{G'}\varphi(\chi)\,\hat{F}_{x^{-1}}(\chi)\,d\mu_0(\chi)=0$ $\forall x\in G$ and so $\varphi\hat{F}\,d\mu_0=0$ (Theorem 12.12). In fact, $\varphi\hat{F}=0$ a.e. on G' since $0=\int_E\varphi\hat{F}\,d\mu_0$ $\forall E\in\mathfrak{R}_\sigma$ in G'. Now pick any $f\neq 0$ in $P(G)\cap L_1(G)\cap L_2(G)$ so that $\hat{f}\in\widehat{L_1(G)\cap L_2(G)}$ and $\hat{f}(\chi_0)\neq 0$ for some $\chi_0\in G'$. As above, $\hat{f}_{\chi_0}\in\widehat{L_1(G)\cap L_2(G)}$ and $\varphi\hat{f}_{\chi_0}=0$ a.e. on G'. At the same time, $\hat{f}_{\chi_0}(\chi_{G'})=\hat{f}(\chi_0)\neq 0$ and $\hat{f}_{\chi_0}\in C(G')$ (Corollary 27.11) require that $\hat{f}_{\chi_0}\neq 0$ on some G'-nbd of $\chi_{G'}$. Therefore $\varphi=0$ a.e. on G' and $\varphi=0$ in $L_2(G')$ establishes that $\widehat{L_1(G)\cap L_2(G)}^{\perp}=\{0\}$. ■

The Plancheral theorem enables us to characterize $\overset{(f)}{\widehat{L_1(G)}}$ in terms of $\overset{(f)}{\widehat{L_2(G)}}=L_2(G')$ convolutions.

THEOREM 29.15. $\overset{(f)}{\widehat{L_1(G)}}=\overset{(f)}{\widehat{L_2(G)}}\star\overset{(f)}{\widehat{L_2(G)}}=L_2(G')\star L_2(G')$.

Proof. The identity $4f\bar{g}=|f+g|^2-|f-g|^2+i|f+ig|^2-i|f-ig|^2$ $\forall f,g\in L_2(G)$ along with $\|F\|_2=\|\hat{F}\|_2$ $\forall F\in L_2(G)$ yields $\int_G f(x)\,\overline{g(x)}\,d\mu(x)=\int_{G'}\hat{f}(\chi)\,\overline{\hat{g}(\chi)}\,d\mu_0(\chi)$ $\forall f,g\in L_2(G)$. Since $\overline{g(\chi)}=\int_G\overline{g(x)\,\chi(x)\,d\mu(x)}=\int_G\overline{g(x)}\,\overline{\chi(x)}\,d\mu(x)=\hat{\bar{g}}(\bar{\chi})$, we further obtain

$\int_G f(x)\,g(x)\,d\mu_{\mathcal{T}}(x) = \int_G f(x)\,\overline{\overline{g(x)}}\,d\mu_{\mathcal{T}}(x) = \int_{G'} \hat{f}(\chi)\,\hat{g}(\bar{\chi})\,d\mu_{\mathcal{T}_0}(\chi)$. Now substitute $\langle\,,\chi_0\rangle g$ for g to get

$$\begin{aligned}\widehat{fg}(\chi_0) &= \int_G f(x)\,g(x)\,\chi_0(x)\,d\mu_{\mathcal{T}}(x)\\ &= \int_G f(x)\{\langle\,,\chi_0\rangle_g\}(x)\,d\,\mu_{\mathcal{T}}(x)\\ &= \int_{G'} \hat{f}(\chi)\,\hat{g}_{\chi_0}(\bar{\chi})\,d\mu_{\mathcal{T}_0}(\chi)\\ &= \int_{G'} \hat{f}(\chi)\hat{g}(\chi^{-1}\chi_0)\,d\mu_{\mathcal{T}_0}(\chi)\\ &= (\hat{f} \star \hat{g})(\chi_0)\ \forall \chi_0 \in G'.\end{aligned}$$

Thus $\widehat{fg} = \hat{f} \star \hat{g}\ \forall f, g \in L_2(G)$ and the conclusion $\widehat{L_1(G)}^{(f)} = L_2(G) \star L_2(G)$ follows by noting (Hölder's inequality) that $F \in L_1(G)$ iff $F = fg$ for f, g $L_2(G)$. ■

Corollary 29.16. For each open set $u \neq \varnothing$ in G' there is an $\hat{F} \neq 0$ in $\widehat{L_1(G)}$ satisfying $\hat{F}(G' - u) = 0$.

Proof. Using the regularity of $\mu_{\mathcal{T}_0}$ on G', choose a compact set $K \subset G'$ having $\mu_{\mathcal{T}_0}(K) > 0$ and (since G' is locally compact T_2) let $w = w(\chi_{G'})$ be a relatively compact open nbd such that $\bar{w}K \subset u$. Evidently $\chi_{\bar{w}}, \chi_K \in L_2(G')$ and $\hat{F} = \widehat{\chi_{\bar{w}}\chi_K} = \hat{\chi}_{\bar{w}} \star \hat{\chi}_K \in \widehat{L_1(G)}^{(f)}$. In fact, $\hat{F}(\chi_0) = \int_{G'} \chi_{\bar{w}}(\chi)\chi_K(\chi^{-1}\chi_0)\,d\mu_{\mathcal{T}_0}(\chi)$ reveals (as in Theorem 29.7) that $\hat{F}(G - \bar{w}K) = \{0\}$. The assertion that $\hat{F} \neq 0$ is clear, since $\int_{G'} \hat{F}\,d\mu_{\mathcal{T}_0} = \int_{G'} (\chi_{\bar{w}} \star \chi_K)\,d\mu_{\mathcal{T}_0} = (\int_{G'} \chi_{\bar{w}}\,d\mu_{\mathcal{T}_0})(\int_{G'} \chi_K\,d\mu_{\mathcal{T}_0}) = \mu_{\mathcal{T}_0}(\bar{w})\,\mu_{\mathcal{T}_0}(K) > 0$.

Pontryagin Duality Theorem

We now have all the machinery for proving Theorem 11.21, whose profound importance may be hinted at by recalling the influence of semireflexivity for TVSps. Notationwise, let $A \subset Y \subset X$ for a topological space X, and denote the closures of A in Y and in X by $\bar{A}^Y$ and $\bar{A}^X$, respectively. Thus $\bar{A}^Y = Y \cap \bar{A}^X$.

THEOREM 29.17. $\underset{x \rightsquigarrow \mathfrak{e}_x}{\delta: G \to G''}$ is an onto topological isomorphism.

Proof. We already know (Corollary 29.10) that δ is an into group isomorphism. If $K \subset G'$ is compact and $\varepsilon > 0$ is fixed, then $u_{K,\varepsilon}(e)$ is a G-base nbd (Theorem 29.9) and $v_{K,\varepsilon}(\chi_{G'}) = \{\eta \in G'' : \sup_K |\eta(x) - 1| < \varepsilon\}$ is a G'-base nbd. Accordingly, $\delta(u_{K,\varepsilon}(e)) = \delta(G) \cap v_{K,\varepsilon}(\chi_{G'})$ ensures that δ is continuous and open. The main task at hand is to establish that $\delta(G) = G''$. We begin by proving† $\delta(G)$ dense in G''. If $\overline{\delta(G)} \neq G''$, then $G'' - \overline{\delta(G)}$ is open and (Corollary 29.16) $\hat{F}(\overline{\delta(G)}) = 0$ for some $\hat{F} \neq 0$ in $\widehat{L_1(G')}$. However, $0 = \hat{F}(\mathfrak{e}_x) = \int_{G'} F(\chi)\,\chi(x)\,d\mu\!r(x)\;\forall x \in G$ yields $\hat{F} = 0$ (take $\varphi \equiv 1$ in Theorem 29.14), and this contradiction confirms that $\overline{\delta(G)} = G''$. Now $(\delta(G), \mathfrak{T}_\kappa)$ is locally compact T_2 and so there is a G''-nbd $v(e_e)$ such that $\overline{\delta(G) \cap v}^{\delta(G)} = \delta(G) \cap \overline{\delta(G) \cap v}^{G''}$ is compact in $\delta(G)$. In particular, $\delta(G) \cap \overline{\delta(G) \cap v}^{G''}$ is compact and closed in G''. Moreover, $\overline{\delta(G)} = G''$ implies $v = \overline{\delta(G) \cap v}^{v} \subset \overline{\delta(G) \cap v}^{G''}$ and $\delta(G) \cap v \subset \delta(G) \cap \overline{\delta(G) \cap v}^{G''}$ further yields $v \subset \overline{\delta(G) \cap v}^{G''} \subset \overline{\delta(G) \cap \overline{\delta(G) \cap v}^{G''}}^{G''} = \delta(G) \cap \overline{\delta(G) \cap v}^{G''}$. Finally, $\overline{\delta(G)} = G''$ also implies that every nbd $v\eta^{-1}$ $(\eta \in G'')$ contains an $\mathfrak{e}_x \in \delta(G)$ and $\eta \in \mathfrak{e}_{x^{-1}} v \subset \mathfrak{e}_{x^{-1}}(\delta(G) \cap \overline{\delta(G) \cap v}^{G''}) \subset \mathfrak{e}_{x^{-1}}\delta(G) = \delta(G)$. Thus $G'' \subset \delta(G)$ and our proof is complete. ■

Theorem 29.4 can now be reformulated.

COROLLARY 29.18. $P(G) = \{\varphi \in C(G) : \varphi = \hat{\overline{\mu\!r}}$ for some $\mu\!r \in M^+(G')\}$.

Proof. $\varphi(x) = \int_{G'} \overline{\chi(x)}\,d\mu\!r(\chi) = \int_{G'} \overline{\mathfrak{e}_x(x)}\,d\mu\!r(\chi) = \overline{\int_{G'} \mathfrak{e}_x(x)\,d\mu\!r(\chi)} = \overline{\hat{\mu\!r}(\mathfrak{e}_x)}\;\forall x \in G$. ■

Remark. Note also that $\overline{\hat{\mu\!r}(\mathfrak{e}_x)} = \hat{\mu\!r}(\mathfrak{e}_{x^{-1}})\;\forall x \in G$.

COROLLARY 29.19. For $\mu \in M(G)$ we have $\mu = 0$ iff $\hat{\mu} = 0$. Therefore $L_1(G)$ and $M(G)$ are semisimple.

Proof. If $\hat{\mu} = 0$, then $0 = \int_G \chi(x)\,d\mu(x) = \int_{G''} \mathfrak{e}_x(\chi)\,d\mu(\mathfrak{e}_x)\;\forall \chi \in G'$ and $\mu = 0$ by Theorem 29.13. The second assertion follows via Corollary 24.21, Theorem 28.11, since $\mathcal{H}_{L_1(G)} = \{h_\chi : \chi \in G'\}$ is injective. ■

† Communicated to me by George Bachman. The proof in a number of sources (e.g. Rudin [85] and Naimark [69]) appears to be based on the incorrect assertion that $\delta(G) \cap \overline{v}^{G''}$ is compact in $\delta(G)$.

Corollary 28.9 is easily extended by

COROLLARY 29.20. The following are equivalent:

(i) G is finite;
(ii) $\dim M(G)<\infty$;
(iii) $\dim L_1(G)<\infty$;
(iv) G' is finite;
(v) $\dim CB(G)<\infty$.

Proof. If G is finite, it is discrete and $M(G)=M_b(G)$ by Corollary 28.9. Therefore every $\mu\in M(G)$ is of the form $\mu=\sum_{i=1}^{n}\mu(x_i)\mu_{x_i}$ and $M(G)=[\mu_x : x\in G]$ is finite dim. Accordingly, $\dim L_1(G)<\infty$ and (Corollary 22.4) $\mathcal{H}_{L_1(G)}$, therefore G', is finite. From (i)$\Rightarrow$(iv) also follows (iv)$\Rightarrow$(i). Finally, (i)$\Rightarrow$(v)$\Rightarrow$(iv), since $CB(G)\subset L_1(G)$ if G is finite (Corollary 12.16) and G' is a linearly independent subset of $[G']\subset CB(G)$. ■

This interlacing of results here (other illustrations are given in the problems) with those of earlier sections also threads back to Section 18.

THEOREM 29.21. Let X be a vector space satisfying $[G']\subseteq X\subseteq CB(G)$.

(i) $\langle M(G), X\rangle$ is an injective dual pair under $\langle\mu, F\rangle=\int_G F\,d\mu$.

(ii) The LCT$_2$VS $(M(G), \sigma\{M(G), X\})$ is an LMCT$_2$ algebra if G is finite or X has a basis of homomorphisms on G.

(iii) $(X, \sigma\{X, M(G)\})$ is an LCT$_2$VS whose induced topology on G' is finer than $\mathfrak{T}_\kappa=\mathfrak{T}_{\mathcal{P}\{\widehat{L_1(G)}\}}$.

Proof. Obviously, $\langle\mu, F\rangle$ is bilinear. If $F_0\neq 0$ in X, then $F_0(x_0)\neq 0$ for some $x_0\in G$ and $\mu_{x_0}\in M(G)$ satisfies $\langle\mu_{x_0}, F_0\rangle=F_0(x_0)\neq 0$. On the other hand (Theorem 29.13), $\mu_0\neq 0$ in $M(G)$ implies $\langle\mu_0, \chi_0\rangle=\hat{\mu}_0(\chi_0)\neq 0$ for some $\chi_0\in G'$. The conclusion in (ii) holds when G is finite, since $\sigma\{M(G), X\}=\|\ \|_v$ by Corollary 29.20, Theorem 13.12. It also holds (Example 24.16) if $\mathfrak{B}$ is an X-basis of homomorphisms on G, since $\sigma\{M(G), X\}=\sigma\{M(G), \mathfrak{B}\}$ and $\langle\ ,\mathfrak{B}\rangle=\{\langle\ , h\rangle : h\in\mathfrak{B}\}$ is an injective collection of algebra homomorphisms on $M(G)$ [as in Theorem 28.11, obtain $\langle\mu\star\nu, h\rangle=\langle\mu, h\rangle\langle\nu, h\rangle\ \forall\mu, \nu\in M(G)$]. The reasoning in (iii) is that $\langle\mu, \chi\rangle=\hat{\mu}(\chi)$ and $\sigma\{X, M(G)\}/G'=\mathfrak{T}_{\mathcal{P}\{\widehat{M(G)}\}}^{(fs)}\supset\mathfrak{T}_{\mathcal{P}\{\widehat{M_a(G)}\}}^{(fs)}=\mathfrak{T}_{\mathcal{P}\{\widehat{L_1(G)}\}}$. ■

Remark. Whether X is a TVA under $\sigma\{X, M(G)\}$ depends on how multiplication is defined. For instance, consider $X \subset L_1(G)$ for compact G. If X is a convolution subalgebra of $L_1(G)$, then each $\langle \mu, \ \rangle: X \to \mathbb{C}$ is an algebra homomorphism and $(X, \sigma\{X, M(G)\})$ is an LMCT$_2$ algebra. On the other hand, $\langle \mu, \ \rangle$ are not all homomorphisms under pointwise multiplication on X.

Example 26.5, Corollary 28.9, and Theorem 28.13 are related by

COROLLARY 29.22. $\overline{M_\mathfrak{d}(G)}^{\sigma\{M(G),X\}} = \overline{S\{M(G)\}}^{\sigma\{M(G),X\}} = M(G)$ so that $M_\mathfrak{d}(G)$ $(S\{M(G)\})$ is weakly closed iff G is discrete.

If G is compact, then $\overline{M_\mathfrak{a}(G)}^{\sigma\{M(G),X\}} = M(G)$ and $M_\mathfrak{a}(G)$ is weakly closed iff G is finite.

Proof. Definition 28.4$^+$ reveals that $M_\mathfrak{d}(G) \subset S(G)$ and so only $\overline{M_\mathfrak{d}(G)}^{\sigma\{M(G),X\}} = M(G)$ needs verification. Example 18.13 and Theorem 18.11 lead to $F \in M_\mathfrak{d}(G)^\perp$ implying $0 = \int_G F\, d\mu_x = F(x)\ \forall x \in G$. Therefore $M_\mathfrak{d}(G)^\perp = \{0\} \not\ \text{in}\ X$ and $\overline{M_\mathfrak{d}(G)}^{\sigma\{M(G),X\}} = M_\mathfrak{d}(G)^{\perp\perp} = \{0\}^\perp = M(G)$. Similarly, $M_\mathfrak{a}(G)^\perp = \{0\}$ in X when G is compact. Reason: For $F \in M_\mathfrak{a}(G)^\perp$, we have $\bar{F} \in X \subset L_1(G)$ and $\bar{F}\, d\mu_\mathcal{T} \in M_\mathfrak{a}(G)$ implying $0 = \int_G F\bar{F}\, d\mu_\mathcal{T}$ and $F = 0$. It remains only to note that a compact, T_2 group is discrete iff it is finite. ■

COROLLARY 29.23. $\bar{X}^{\sigma\{CB(G),M(G)\}} = CB(G)$.

Proof. If $\mu \in X^\perp$, then $0 = \int_G F\, d\mu\ \forall F \in X$ and $\mu = 0$ (Corollary 29.19). Thus $X^\perp = \{0\}$ in $M(G)$ and $X^{\perp\perp} = CB(G)$. ■

Remark. Although $X = [G']$ is weakly dense in $CB(G)$, such need not be the case for G'. Specifically, $v(2\chi_G; \mu_e; \varepsilon) \cap G' = \varnothing$ for $\varepsilon \in (0, 1]$.

PROBLEMS

29A. Measures Concentrated on Subgroups

Let H be a closed subgroup of an LCT$_2$ abelian group G. Then $\mu \in M(G)$ is concentrated on H (i.e., $\mu(E) = \mu(E \cap H)\ \forall E \in \mathfrak{R}_\sigma$ in G) iff $\hat{\mu} = \hat{\mu}_{\chi_0}$ on $G'\ \forall \chi_0 \in H^\circ$. [*Hint*: If μ is concentrated on H and $\chi_0 \in H^\circ$, then $\int_G F\chi_0\, d\mu = \int_H F\, d\mu\ \forall F \in L_1(G)$ implies $\chi_0\, d\mu = \mu$ and $\hat{\mu}_{\chi_0} = \hat{\mu}$. Conversely, $\hat{\mu} = \hat{\mu}_{\chi_0} = \widehat{\chi_0\, d\mu}$ implies that $\mu A\chi_0 = 0$ for the open set $A_{\chi_0} = \{x \in G: \chi_0(x) \neq 1\}$. Clearly $G - H = \bigcup_{H^\circ} A\chi_0$. If B is the union of all open sets $u \subset G$ with $\mu(u) = 0$, then $\mathfrak{S}(\mu) = G - B \subset H$ and the regularity of μ yields $\mu(G - \mathfrak{S}(\mu)) = 0$. Conclude that $\mu(E) = \mu(E \cap \mathfrak{S}(\mu)) \leq \mu(E \cap H)\ \forall E \in \mathfrak{R}_\sigma$.]

29B. Positivite Definiteness on Subgroups and Quotients

For H and G, as in Problem 29A, prove

(i) $P(H^\circ)$ is $P(G')$ restricted to H°;
(ii) $P(H)$ is $P(G)$ restricted to H;
(iii) $P(G/H) = \{\varphi \in C(G/H) : \varphi = \bar{\hat{\mu}}$ for $\mu \in M^+(H^\circ)\} \subset P(G)$. [*Hint*: Use Theorem 11.18, Corollary 11.23, Corollary 29.18 and Problem 28D.]

29C. Fourier Transforms on Subgroups and Quotients

For H and G, as above, prove that $\widehat{L_1(G/H)}^{(f)}$ is $\widehat{L_1(G)}^{(f)}$ restricted to H° and $\widehat{L_1(H)}^{(f)}$ is $\widehat{L_1(G)}^{(f)}$ restricted to H'. [*Hint*: Assume that μ_H, μ_G, $\mu_{G/H}$ in Problem 12A are weakly Borel and extend the congruence $\underset{f \rightsquigarrow \xi_f}{C_0(G) \to C_0(G/H)}$ to an onto congruence $T: \underset{F d\mu_G \rightsquigarrow \xi_F d\mu_{G/H}}{L_1(G) = M_a(G) \to L_1(G/H) = M_a(G/H)}$. Then (Problem 28D) $T = \pi$ on $M_a(G)$ and $\hat{\xi}_F(\dot{\chi}) = \int_{G/H} \xi_F \dot{\chi}\, d\mu_{G/H} = \int_G F\chi\, d\mu_G = \hat{F}(\chi)\ \forall \chi \in H^\circ$. The second assertion follows from the first since $\widehat{L_1(H')}^{(f)} = \widehat{L_1(G'/H^\circ)}^{(f)}$ is $\widehat{L_1(G')}^{(f)}$ restricted to $H^{\circ\circ} = H$.]

29D. Poisson Summation Formula

Assume that H is a closed subgroup of an $\mathrm{LCT_2A}$ group G.

(a) If $F \in [P(G) \cap L_1(G)]$ and $F(xy) \in L_1(H)\ \forall x \in G$, the Haar measure μr_{H° on H° can be chosen such that $\int_H F(y)\, d\mu r_H(y) = \int_{H^\circ} \hat{F}(\chi)\, d\mu r_{H^\circ}(\chi)$. [*Hint*: Apply the inversion theorem to $\xi_F \in [P(G/H) \cap L_1(G/H)]$ and take $x = e$ in

$$\int_H F(xy)\, d\mu r_H(y) = \xi_F(Hx) = \int_{(G/H)'} \hat{\xi}_F(\dot{\chi})\, \overline{\dot{\chi}(Hx)}\, d\mu r_{(G/H)'}(\dot{\chi})$$

$$= \int_{H^\circ} \hat{F}(\chi)\, \overline{\chi(x)}\, d\mu r_{H^\circ}(\chi).]$$

(b) The theory of Fourier integrals can be linked to that of Fourier series. If $F \in L_1(\mathbb{R})$ and $\alpha > 0$, then $\sum_{n=-\infty}^{\infty} F(x + n\alpha) = \sqrt{2\pi}/\alpha \sum_{n=-\infty}^{\infty} \hat{f}(2\pi n/\alpha) e^{-2\pi i n x/\alpha}\ \forall x \in \mathbb{R}$.

[*Hint*: Let $H = \mathbb{I}\alpha$ and identify G/H with $S'(C)$ under $x + n\alpha \rightsquigarrow e^{2\pi i n x/\alpha}$. Assume that $\mu_{\mathcal{F}_{H^\circ}}(2\pi n/\alpha) = \alpha/\sqrt{2\pi}\ \forall 2\pi n/\alpha \in (2\pi/\alpha)\mathbb{I} = H^\circ$ and take $\mu_{\mathcal{F}_H}(m\alpha) = 1\ \forall m'\alpha \in H$ in (a) above.]

(c) Prove (Riemann–Lebesgue lemma) that $\hat{F}(x) \underset{|x| \rightsquigarrow \infty}{\to} 0\ \forall F \in L_1(\mathbb{R})$. [*Hint* (Corollary 27.15): Choose $g \in C_\infty(\mathbb{R})$ such that $\|\hat{F} - g\|_0 < \varepsilon$ and $\lim_{|x| \to \infty} |g(x)| = 0$. Proceed!]

29E. $\langle M(G), [\mathcal{H}_{M(G)}]\rangle$ and $\langle M(G), X\rangle$

(a) $\langle M(G), [\mathcal{H}_{M(G)}]\rangle$ is an injective dual pair under $\left\langle \mu, \sum_{i=1}^{n} \alpha_i h_i \right\rangle = \sum_{i=1}^{n} \alpha_i h_i(\mu)$. Therefore $(M(G), \sigma\{M(G), \mathcal{H}_{M(G)}\})$ is an LMCT_2 algebra.

(b) (cf Theorem 29.21). Although $([\mathcal{H}_{M(G)}], \sigma\{[\mathcal{H}_{M(G)}], M(G)\})' = (X, \sigma\{X, M(G)\})'$, it need not be true that $X = [\mathcal{H}_{M(G)}]$. [*Hint*: Let $h: M(G) \to \mathbb{C}$, $\mu \rightsquigarrow \sum_G \mu(x)$ for an infinite compact group G. Then $h \neq 0$ and $h/M_a(G) = 0$. Can $h = \langle\ , f\rangle$ for $f \neq 0$ in X?]

(c) The topologies $\sigma\{M(G), X\}$ and $\sigma\{M(G), \mathcal{H}_{M(G)}\}$ coincide iff $X = [\mathcal{H}_{M(G)}]$. [*Hint*: Both X and $[\mathcal{H}_{M(G)}]$ are injective subspaces of $\{M(G)\}^*$].

29F. $\langle \hat{Y}, M(G')\rangle$ for $Y \subset M(G)$

Let Y denote either $M_a(G)$ or $M_b(G)$ for an LCT_2 abelian group G.

(a) $\langle \hat{Y}, M(G')\rangle$ is an injective dual pair under $\langle \hat{\mu}, \nu\rangle = \int_{G'} \hat{\mu}(\chi)\, d\nu(\chi)$.

(b) If $\mathcal{S} = \{S_{\varepsilon, \|\|_v}(0) \subset M(G'): \varepsilon > 0\}$, then $\mathfrak{T}_{\mathcal{S}} = \beta(\hat{Y}, M((G'))$ coincides with the $M(G)$-norm induced topology on $\hat{Y}$. Furthermore, $\hat{Y}^{\|\|_0}$ in $CB(G')$ is $\{\xi \in \{M(G')\}^* : \xi/S_{\varepsilon, \|\|_v}(0)$ is $\sigma(M(G'), \hat{Y})$-continuous $\forall \varepsilon > 0\}$.

(c) *Wanted*! A "three-word proof" that $\langle \widehat{M_a(G')}, M(G)\rangle$ and $\langle \widehat{M_b(G')}, M(G)\rangle$ are injective dual pairs under $\langle \hat{\nu}, \mu\rangle = \int_G \hat{\nu}(\mathfrak{e}_x)\, d\mu(x)$.

References for Further Study

Harmonic Analysis on groups: Rudin [85]; Katznelson [49]; Sugivra [94].

Compact groups and almost periodic functions: Dunkl and Ramirez [27]; Edwards [29]; Loomis [58].

On $L_p(G)$-Fourier transforms: Hewitt and Ross [41].

On dual pairs: Dunkl and Ramirez [27]; Stratigos [93].

APPENDIX T

Topology

t.1. Neighborhood Systems

A topological space $(X, \mathfrak{T})$ is a set X with a family $\mathfrak{T}$ of subsets (called open) that contains $\varnothing$ and X and is closed under finite intersections and arbitrary unions. One calls $u \subset X$ a $\mathfrak{T}$-nbd of $x \in X$ if $x \in E \subset u$ for some $E \in \mathfrak{T}$. The collections $\mathfrak{N}_x = \{\mathfrak{T}\text{-nbds of } x\}$ and $\mathfrak{N} = \{\mathfrak{N}_x : x \in X\}$ are called the $\mathfrak{T}$-nbd system at $x \in X$ and the $\mathfrak{T}$-nbd system in X, respectively.

A family $\mathfrak{B}_x$ of sets containing x {subfamily of $\mathfrak{N}_x$} is a base for $\mathfrak{N}_x$ {a $\mathfrak{T}$-nbd base at x} if $\forall u \in \mathfrak{N}_x$ there is a $v \in \mathfrak{B}_x$ such that $x \in v \subset u$. Accordingly, $\mathfrak{B} = \{\mathfrak{B}_x : x \in X\}$ is called a base for $\mathfrak{N}$ {a $\mathfrak{T}$-nbd base} in X. The distribution between "base" and "nbd base" is that members of the former collection are not necessarily nbds, whereas those in the latter are.

A $\mathfrak{T}$-nbd system $\mathfrak{N}_x$ at $x \in X$ satisfies

(N_1) $x \in u \ \forall u \in \mathfrak{N}_x$;
(N_2) $u_1 \cap u_2 \in \mathfrak{N}_x \ \forall u_1, u_2 \in \mathfrak{N}_x$;
(N_3) if $u \in \mathfrak{N}_x$ and $u \subset v$, then $v \in \mathfrak{N}_x$;
(N_4) for $u \in \mathfrak{N}_x$, there is a $v \in \mathfrak{N}_x$ such that $u \in \mathfrak{N}_y \ \forall y \in v$.

Conversely, suppose each $x \in X$ determines a nonempty collection $\mathfrak{N}'_x$ of subsets of X which satisfies (N_1)–(N_3). Then there is a unique topology $\mathfrak{T}'$ on X such that $\mathfrak{N}'_x$ is a base for the $\mathfrak{T}'$-nbd system at $x \in X$. (Simply define $E \in \mathfrak{T}'$ iff $\forall x \in E$ there is a $u \in \mathfrak{N}'_x$ such that $x \in u \subset E$.) If $\mathfrak{N}'_x$ also satisfies (N_4) then $\mathfrak{N}'_x$ is precisely the $\mathfrak{T}'$-nbd system at $x \in X$.

t.2. Closure Axioms

Let $\mathfrak{P}(X)$ denote the collection of all subsets of a set X and suppose $\xi : \mathfrak{P}(X) \to \mathfrak{P}(X)$ satisfies

(K_1) $\xi(\varnothing) = \varnothing$;
(K_2) $A \subset \xi(A) \ \forall A \in \mathfrak{P}(X)$;
(K_3) $\xi(A \cup B) = \xi(A) \cup \xi(B) \ \forall A, B \in \mathfrak{P}(X)$.

Then $\mathfrak{T} = \{X - A : A = \xi(A)\}$ is the unique topology of X having $\{A \in \mathcal{P}(X) : A = \xi(A)\}$ as its collection of closed sets in X. Furthermore, $\bar{A} = \xi(A)\ \forall A \in \mathcal{P}(X)$ is equivalent to

(N_4) $\xi\{\xi(A)\} = \xi(A)\ \forall A \in \mathcal{P}(X)$.

Proof. Since $B \supset A$ implies $\xi(B) \supset \xi(A)$ and $\bar{A}$ is the intersection of all closed sets containing A, we have $\bar{A} \supset \xi(A)$. If ξ satisfies (K_4), then $\bar{A} \subset \overline{\xi(A)} = \xi(A)$. Conversely, $\bar{A} = \xi(A)\ \forall A \in \mathcal{P}(X)$ implies $\xi\{\xi(A)\} = \xi(\bar{A}) = \bar{\bar{A}} = \bar{A} = \xi(A)\ \forall A \in \mathcal{P}(X)$. ■

t.3. Category Notions

A subset E of a topological space X is nowhere dense in X if $\operatorname{Int} \bar{E} = \varnothing$ (equivalently, $X - E$ is dense in X). A set $B \subset X$ is category I in X (denoted $B \subset \mathrm{Catg_I}$ (X)) if B can be written as a countable union of nowhere dense sets in X. Otherwise, B is category II in X (written, $B \in \mathrm{Catg_{II}}(X)$). The statement $X \in \mathrm{Catg}_i(X)$ is abbreviated X is Catg i ($i = \mathrm{I, II}$). Note that a set $E \subset X$ may be of one category in X and of another category in itself as a subspace of X. For instance, $\mathbb{N}$ is Catg II and $\mathbb{N} \in \mathrm{Catg_I}(\mathbb{R})$. The following is also easily verified.

THEOREM. Let $A \subset B \subset X$, where X is a topological space:

(i) $B \in \mathrm{Catg_I}(X) \Rightarrow A \in \mathrm{Catg_I}(X)$ and $A \in \mathrm{Catg_{II}}(X) \Rightarrow B \in \mathrm{Catg_{II}}$ (X).

(ii) $A \in \mathrm{Catg_I}(B) \Rightarrow A \in \mathrm{Catg_I}(X)$.

(iii) X is Catg II iff $u \in \mathrm{Catg_{II}}(X)$ for every open set $u \in X$.

(iv) A continuous, nearly open mapping $f: X \to Y$ takes non-nowhere dense subsets of X into non-nowhere dense sets in Y and so takes $\mathrm{Catg_{II}}(X)$ sets into $\mathrm{Catg_{II}}(Y)$ sets.

THEOREM. Every locally compact regular space and every complete pseudometric space is Catg II.

t.4. Almost Open Sets. (Cech [16])

Let A be a subset of topological space X. Define $A_i = A \cap \mathbb{A}_i$ ($i = \mathrm{I, II}$), where $\mathbb{A}_\mathrm{I} = \{x \in X : A \cap u(x) \in \mathrm{Catg_I}(X)$ for some $u \in \mathfrak{N}_x\}$ and $\mathbb{A}_{\mathrm{II}} = \{x \in X : A \cap u(x) \in \mathrm{Catg_{II}}(X)\ \forall u \in \mathfrak{N}_x\}$.

DECOMPOSITION THEOREM. $X = \mathbb{A}_I \cup \mathbb{A}_{II}$ with $\mathbb{A}_I \cap \mathbb{A}_{II} = \phi$. Furthermore,

(i) $\mathbb{A}_I$ is an open, $\text{Catg}_I(X)$ set;
(ii) $\mathbb{A}_{II} \cap v(x) \in \text{Catg}_{II}(X)\ \forall v \in \mathfrak{N}_x$ of each $x \in \mathbb{A}_{II}$;
(iii) $\mathbb{A}_{II} - \overline{A_{II} \text{ Int } \mathbb{A}_{II}} = \bar{A}_{II}$.

An application of the first theorem in t.4 also yields

COROLLARY. The following are equivalent for $B \subset X$:

(i) $B = u \Delta \theta$ for open $u \subset X$ and $\theta \in \text{Catg}_I(X)$.
(ii) $B = (u - \theta_1) \cup \theta_2$ for open $u \subset X$ and $\theta_1, \theta_2 \in \text{Catg}_I(X)$.
(iii) $B \Delta u \in \text{Catg}_I(X)$ for open $u \subset X$.
(iv) $\mathbb{B}_{II} \cap ((X - B)_{II}$ is nowhere dense in X (here $((X - B)_{II}$ denotes $\mathbb{A}_{II}$ for $A = X - B$).
(v) $\mathbb{B}_{II} - B \in \text{Catg}_I(X)$.

Remark. "Open $u \subset X$" can be replaced with "closed $u \subset X$," since an open or closed set $E \subset X$ has *bdy* E nowhere dense in X and $\bar{E} = E \cup \textit{bdy}\, E$ and Int $E = E - \textit{bdy}\, E$.

A subset $B \subset X$ is said to be almost open if it satisfies any of the conditions (i)–(v) above. The σ-ring $\mathfrak{B}_{ao}(X)$ of almost open sets in X contains $\text{Catg}_I(X)$ since $E = E \,\Delta \varnothing\ \forall E \in \text{Catg}_I(X)$, and $\mathfrak{B}_{ao}(X)$ contains the σ-algebra of weakly Borel sets in X (Appendix m.8), since it contains every open (closed) set $E \subset X$ by virtue of $E \Delta E = \varnothing$.

t.5. Coverings

A family $\mathfrak{A} = \{A_\alpha\}$ of subsets of a set X is

(i) nbd finite if each $x \in X$ has a nbd intersecting at most finitely many members of $\mathfrak{A}$;
(ii) Point finite if each $x \in X$ belongs to finitely many members of $\mathfrak{A}$;
(iii) star finite if each $A_\alpha \in \mathfrak{A}$ intersects at most finitely many other members of $\mathfrak{A}$;

Remark. An open star finite cover of X is nbd finite. The converse is not true, since $\mathfrak{A} = \{\{x\} : x \in X\}$ on $X = (0, \infty)$ is both point finite and star finite but not nbd finite. The nbd finite cover $\mathfrak{B} = \{(n, \infty) : n \in \mathbb{N}_0\}$ of X is clearly not star finite.

Let $\mathfrak{A} = \{A_\alpha\}$ and $\mathfrak{B} = \{B_\beta\}$ be covers of a set X. We say that $\mathfrak{A}$ refines $\mathfrak{B}$ (denoted $\mathfrak{A} < \mathfrak{B}$) if each $A_\alpha \in \mathfrak{A}$ is contained in some $B_\beta \in \mathfrak{B}$. The cover $\mathfrak{A} \wedge \mathfrak{B} = \{A_\alpha \cap B_\beta : (A_\alpha, B_\beta) \in \mathfrak{A} \times \mathfrak{B}\}$ is termed a common refinement of $\mathfrak{A}$

and $\mathcal{B}$. Note that $\mathcal{A} \wedge \mathcal{B}$ is nbd finite {point finite} if $\mathcal{A}$, $\mathcal{B}$ are nbd finite {point finite}.

The star of $\mathcal{A} = \{A_\alpha\}$ is the cover $\mathcal{A}^* = \{\text{St}(A_\alpha, \mathcal{A}) : A_\alpha \in \mathcal{A}\}$, where $\text{St}(E, \mathcal{A}) = \bigcup\{A_\alpha \in \mathcal{A} : A_\alpha \cap E \neq \varnothing \text{ for } E \in \mathcal{P}(X)\}$. Accordingly, $\mathcal{A}$ is called a star refinement of $\mathcal{B} = \{B_\beta\}$ (denoted $\mathcal{A} \overset{*}{<} \mathcal{B}$) if $\mathcal{A}^* < \mathcal{B}$. Also, $\mathcal{A}$ is called a barycentric refinement of $\mathcal{B}$ (denoted $\mathcal{A} \overset{b}{<} \mathcal{B}$) if $\mathcal{A}^b = \{\text{St}(x, \mathcal{A}) : x \in X\}$ refines $\mathcal{B}$. We always have $\mathcal{A} \overset{*}{<} \mathcal{B} \Rightarrow \mathcal{A} \overset{b}{<} \mathcal{B} \Rightarrow \mathcal{A} < \mathcal{B}$ since $A_{\alpha_0} \subset \text{St}(x_0, \mathcal{A}) \subset \text{St}(A_{\alpha_0}, \mathcal{A})$ for $x_0 \in A_{\alpha_0}$ in $\mathcal{A}$.

THEOREM. Let $\mathcal{A}$, $\mathcal{B}$, $\mathcal{C}$ be covers of a set X. Then

(i) $\mathcal{A} < \mathcal{B} \overset{b}{<} \mathcal{C} \Rightarrow \mathcal{A} \overset{b}{<} \mathcal{C}$;
(ii) $\mathcal{A} < \mathcal{B} \overset{*}{<} \mathcal{C} \Rightarrow \mathcal{A} \overset{*}{<} \mathcal{C}$;
(iii) $\mathcal{A} \overset{b}{<} \mathcal{B} \overset{b}{<} \mathcal{C} \Rightarrow \mathcal{A} \overset{*}{<} \mathcal{C}$.

Proof. Surely (i) holds, since $\mathcal{A} < \mathcal{B}$ implies $\text{St}(x, \mathcal{A}) \subset \text{St}(x, \mathcal{B})\ \forall x \in X$. Next, if $A_0 \in \mathcal{A}$ and $B_{A_0} \in \mathcal{B}$ contains A_0, then $\text{St}(A_0, \mathcal{A}) \subset \text{St}(B_{A_0}, \mathcal{A}) \subset \text{St}(B_{A_0}, \mathcal{B})$ and $\mathcal{A} \overset{*}{<} \mathcal{B}^* < \mathcal{C}$ gives $\mathcal{A} \overset{*}{<} \mathcal{C}$. For (iii) let $A_0 \in \mathcal{A}$ and pick $x_0 \in A_0$. If $A \subset \text{St}(A_0, \mathcal{A})$ and $x_A \in A \cap A_0$, then $x_0 \in \text{St}(x_A, \mathcal{A}) \subset B_A$ for some $B_A \in \mathcal{B}$; but $B_A \subset \text{St}(x_0, \mathcal{B})$ yields $\text{St}(A_0, \mathcal{A}) \subset \text{St}(x_0, \mathcal{B})$. Thus $\mathcal{A}^* < \mathcal{B}^b < \mathcal{C}$ and $\mathcal{A} \overset{*}{<} \mathcal{C}$ follows. ■

THEOREM (Ginsberg and Isbell [32]). Let $\mathcal{A}$, $\mathcal{B}$ be covers of X.

(i) If $\mathcal{A} < \mathcal{B}$ and $\mathcal{B}$ is finite {point finite, star finite}, there is a finite {point finite, star finite} cover $\mathcal{C}$ such that $\mathcal{A} < \mathcal{C} \overset{*}{<} \mathcal{B}$.

(ii) If $\mathcal{A} \overset{**}{<} \mathcal{B}$ and $\mathcal{B}$ is countable, there is a countable cover $\mathcal{C}$ such that $\mathcal{A} < \mathcal{C} \overset{*}{<} \mathcal{B}$.

t.6. Nets and Filters

By a directed set is meant a set D with a reflexive, transitive relation $<$ satisfying: $\forall d, d' \in D$, there is a $d'' \in D$ such that $d < d''$ and $d' < d''$. A net on a set D is a function from a directed set to X. For example, a sequence is a net having domain $\mathbb{N}$.

In a topological space X a net $\{x_d : d \in D\}$

(i) converges to $x \in X$ (written $x_d \to x$) if $\forall u \in \mathcal{N}_x$ there is a $d_0 \in D$ such that $x_d \in u(x)\ \forall d > d_0$;

(ii) accumulates at $x \in X$ (written $x_d \infty x$) if $\forall u \in \mathfrak{N}_x$ and $d_0 \in D$ there is a $d > d_0$ such that $x_d \in u(x)$.

Suppose that $\{x_\alpha : \alpha \in \mathcal{A}\} \subset X$ for a normed linear space $(X, \|\ \|)$ and $D = \{$all finite subsets of $\mathcal{A}\}$ is directed by inclusion. Then $\{S_d = \sum_d x_\alpha : d \in D\}$ is a net in X and $S_d \to x \in X$ (written, $\sum_{\mathcal{A}} x_\alpha = x$) means that $\forall \varepsilon > 0$, there is a $d_0 \in D$ such that $\|S_d - x\| < \varepsilon\ \forall d \supset d_0$. Despite this notation, nowhere do we sum up an uncountably infinite number of terms.

THEOREM. If $\sum_{\mathcal{A}} x_\alpha = x$, at most countably many $x_\alpha \neq 0$.

Proof. Given $\varepsilon = 1/n$, choose $d_n \in D$ such that $\|S_{d_n} - x\| < 1/2n$ and let $d_{n.} \cap d_n = \varnothing$ for $d_{n.} \in D$. Then $\|S_{d_{n.}}\| = \|S_{d_{n.} \cup d_n} - S_{d_n}\| \leq \|x - S_{d_{n.} \cup d_n}\| + \|S_{d_n} - x\| < 1/n$. Therefore $\alpha_0 \cap \left(\bigcup_{n=1}^{\infty} d_n \right) = \varnothing$ implies that $\{\alpha_0\} \cap d_n = \varnothing$ and $\|x_{\alpha_0}\| < 1/n\ \forall n \in \mathbb{N}$ (i.e., $x_{\alpha_0} = 0$). ■

A filterbase $\mathfrak{B}$ in a set X is a collection of nonempty subsets of X such that $\forall B_1, B_2 \in \mathfrak{B}$ there is a $B_3 \in \mathfrak{B}$ with $B_3 \subset B_1 \cap B_2$. A filter in X is a filterbase with the added property that $B \subset B'$ and $B \in \mathfrak{B}$ imply $B' \in \mathfrak{B}$. Every filterbase $\mathfrak{B}$ in X generates a filter $\mathcal{F}(\mathfrak{B}) = \{F \in \mathcal{P}(X) : B \subset F$ for some $B \in \mathfrak{B}\}$.

In a topological space X a filterbase $\mathfrak{B}$

(i) converges to $x \in X$ (written, $\mathfrak{B} \to x$) if $\forall u \in \mathfrak{N}_x$ there is a set $B \in \mathfrak{B}$ such that $B \subset u(x)$.

(ii) accumulates at $x \in X$ (written $\mathfrak{B} \infty x$) if $u(x) \cap B \neq \varnothing\ \forall u \in \mathfrak{N}_x$ and $B \in \mathfrak{B}$.

Clearly $\mathfrak{B} \to x$ implies $\mathfrak{B} \infty x$. The converse also holds for maximal filters $\mathcal{M}$ (in the sense that $\mathcal{M} = \mathcal{F}$ for every filter $\mathcal{F} \supset \mathcal{M}$). Reason: A filter $\mathcal{M}$ in X is maximal iff $\forall A \in \mathcal{P}(X)$, either A or $X - A$ contains some $M \in \mathcal{M}$.

Concepts based on net convergence and on filter convergence are equivalent in that each net {filter} determines a unique filter {net}, both of which simultaneously do or do not converge to the same point.

t.7. Compactness and Compactifications

A collection $\mathcal{E}$ of sets has the finite intersection property (fip) if every finite subcollection has nonempty intersection.

THEOREM. The following are equivalent for a topological space X:

(i) X is compact.
(ii) Every collection $\mathfrak{E}$ of closed sets with the fip satisfies $\bigcap_{\mathfrak{E}} F \neq \emptyset$.
(iii) Every filter in X is contained in a convergent filter.
(iv) Every filter in X has an accumulation point.
(v) Every maximal filter in X converges.

Proof. (i)$\Leftrightarrow$(ii) by definition of compactness and de Morgan's laws of complementation. (ii)$\Rightarrow$(iii), since for any filter $\mathcal{F} = \{F_\alpha : \alpha \in \mathfrak{A}\}$ in X every $\{F_{\alpha_1}, \ldots, F_{\alpha_n}\} \subset \mathcal{F}$ has nonempty intersection and so $\bigcap_{\mathfrak{A}} \bar{F}_\alpha$ (the set of $\mathcal{F}$-accumulation points in X) is nonempty. (iii)$\Rightarrow$(iv), since $\mathcal{F} \subset \mathcal{G}$ and $\mathcal{G} \to x_0$ yield $\mathcal{F} \infty x_0$, and (iv)$\Rightarrow$(v) by the remark in t.6 for filterbases. For $\mathfrak{E}$ as in (ii) the finite intersections of members in $\mathfrak{E}$ determine a filter $\mathcal{F}$ in X and $\emptyset \neq \bigcap_{\mathcal{F}} F \subset \bigcap_{\mathfrak{E}} E$ by (iv). Evidently (v)$\Rightarrow$(iii). ∎

The Alexandroff (one-point) compactification of a topological space $(X, \mathfrak{T})$ is $X_\infty = (\mathbb{1}; (X \cup \infty, \mathfrak{T}_\infty))$, where $\infty \notin X$ and $\mathfrak{T}_\infty = \{\mathfrak{T}, E \cup \infty; X - E$ is $\mathfrak{T}$-closed and compact$\}$. Essentially, X_∞ is T_2 iff $(X, \mathfrak{T})$ is locally compact T_2 (thus demonstrating that locally compact T_2 spaces are $T_{3\frac{1}{2}}$).

For a $T_{3\frac{1}{2}}$ space $(X, \mathfrak{T})$, the evaluation mapping $\mathfrak{e}: x \rightsquigarrow \{f(x)\}$ of $(X, \mathfrak{T})$ into $\prod_{C(X;I)} I$ is a homeomorphism, and $\beta(X) = (\mathfrak{e} : (\mathfrak{e}(X), \mathfrak{T}_\pi)$ is called the Stone–Cech compactification of $(X, \mathfrak{T})$.

t.8. Function Spaces

Let $CB(\Omega)$ denote the algebra (under pointwise operations) of continuous, bounded $\mathbb{F}$-valued ($\mathbb{F} = \mathbb{R}$ or $\mathbb{C}$) functions on a topological space Ω. It is assumed that $CB(\Omega)$ carries the sup norm $\|f\|_0 = \sup_\Omega |f(\omega)|$. An $f \in CB(\Omega)$ is said to vanish at infinity if $\forall \varepsilon > 0$ there is a compact set $K = \Omega$ such that $|f(\omega)| < \varepsilon \, \forall \omega \in \Omega - K$. The subalgebra of $CB(\Omega)$ whose members vanish at infinity is denoted $C_\infty(\Omega)$. Our terminology "vanishing at infinity" is justified by

THEOREM. Let Ω be locally compact, T_2 with one point compactification Ω_∞. Then $C_\infty(\Omega) = \{^F/_\Omega : F \in C(\Omega_\infty)$ with $F(\infty) = 0\}$.

Proof. If $F(\infty)=0$ for $F\in C(\Omega_\infty)$ and $\varepsilon>0$, there is an Ω_∞-open set $u(\infty)$ such that $\sup_u |F(\omega)|<\varepsilon$. Since $u(\infty)=\Omega_\infty$ or $u(\infty)=\Omega_\infty-K$ for compact $K\subset\Omega$, we have $F/_\Omega\in C_\infty(\Omega)$. For the reverse inclusion, extend every $f\in C_\infty(\Omega)$ to $\hat f:\Omega_\infty\to\mathbb{F}$ and for $\varepsilon>0$ choose a compact set $K_f\not\subset\Omega$ $\infty \rightsquigarrow 0$ satisfying $\sup_{\Omega_\infty-K_f} |\hat f(\omega)|<\varepsilon$. Since Ω_∞ is T_2, the set $\Omega_\infty-K_f$ is open and $\hat f\in C(\Omega_\infty)$. ■

The support of $f\in CB(\Omega)$ is defined as $N(f)=\overline{\Omega-f(0)}$ and the subalgebra of $CB(\Omega)$ whose members have compact support is denoted $C_0(\Omega)$. In essence, $f\in C_0(\Omega)$ iff $f(\Omega-K)=\{0\}$ for some compact set $K\subset\Omega$ [specifically, $\Omega-N(f)\subset f^{-1}(0)$, whereas $N(f)\subset K$ when $f(\Omega-K)=\{0\}$].

The Banach algebra $(CB(\Omega),\|\;\|_0)$ contains $C_\infty(\Omega)$ as closed subalgebra. If Ω is locally compact and T_2, then $C_0(\Omega)$ is $\|\;\|_0$-dense in $C_\infty(\Omega)$. Clearly $C_0(\Omega)=C_\infty(\Omega)=C(\Omega)$ when Ω is compact; and $C_0(\Omega)\subseteq C_\infty(\Omega)\subsetneqq CB(\Omega)$ when Ω is locally compact, noncompact T_2. In fact, $C_0(G)\subsetneqq C_\infty(G)$ for a locally compact, noncompact group G (Hewitt and Ross [41]).

NOTATION. Let $\mathscr{F}(\Omega)$ denote the $\mathbb{F}$-valued functions of a set Ω. Then $\mathscr{F}(\Omega;\mathbb{R})$ and $\mathscr{F}(\Omega;\mathbb{C})$ denote $\mathscr{F}(\Omega)$ with $\mathbb{F}=\mathbb{R}$ and $\mathbb{F}=\mathbb{C}$, respectively. Similar notation applies to $CB(\Omega)$, $C_\infty(\Omega)$, and $C_0(\Omega)$.

t.9. Stone Weierstrass Theorems

There are several versions of the Stone Weierstrass Theorem, decending on $\mathbb{F}$ and the nature of Ω. A set $\mathfrak{S}\subset\mathscr{F}(\Omega)$ is separating if $\forall\omega_1\neq\omega_2$ there is a $\xi\in\mathfrak{S}$ such that $\xi(\omega_1)\neq\xi(\omega_2)$.

THEOREM. Let Ω be compact T_2.

(i) A separating subalgebra $\mathscr{A}\subset C(\Omega;\mathbb{R})$ satisfying $\bigcap_{\mathscr{A}} f^{-1}(0)=\varnothing$ is $\|\;\|_0$-dense in $C(\Omega;\mathbb{R})$.

(ii) A separating, conjugation-closed subalgebra $\mathscr{A}\subset C(\Omega;\mathbb{C})$ satisfying $\bigcap_{\mathscr{A}} f^{-1}(0)=\varnothing$ is $\|\;\|_0$-dense in $C(\Omega;\mathbb{C})$.

Remark. Statement (ii) follows from (i) by decomposing every $f\in C(\Omega;\mathbb{C})$ into $f_{\mathfrak{R}_e}+if_{\mathfrak{I}_m}$, where $f_{\mathfrak{R}_e}=(f+\bar f)/2$ and $f_{\mathfrak{I}_m}=(f-\bar f)/2i$ belong to $C(\Omega;\mathbb{R})$. The need for (ii) can be seen by observing that $\mathscr{A}=\{$holomorphic functions on $\bar S_1(0)\subset\mathbb{C}\}$ is a closed, separating subalgebra of $C(\bar S_1(0);\mathbb{C})$ satisfying $\bigcap_{\mathscr{A}} f^{-1}(0)=\varnothing$. However, $\bar f\in C(\bar S_1(0);\mathbb{C})-\mathscr{A}$ for $\bar f(\omega)=\bar\omega$.

THEOREM. Let Ω be locally compact T_2.

(i) A separating subalgebra $\mathcal{A} \subset C_\infty(\Omega; \mathbb{R})$ satisfying $\bigcap_{\mathcal{A}} f^{-1}(0) = \varnothing$ is $\| \|_0$-dense in $C(\Omega; \mathbb{R})$.

(ii) A separating, conjugation-closed subalgebra $\mathcal{A} \subset C_\infty(\Omega; \mathbb{C})$ satisfying $\bigcap_{\mathcal{A}} f^{-1}(0) = \varnothing$ is $\| \|_0$-dense in $C(\Omega; \mathbb{C})$.

Proof. As above, (ii) follows from (i) which is based on the theorem in t.8. Specifically, $\mathcal{A}_\infty = \{\hat{f} + \alpha\chi_{\Omega_\infty} : f \in \mathcal{A}$ and $\alpha \in \mathbb{R}\}$ is a separating subalgebra of $C(\Omega_\infty; \mathbb{R})$ satisfying $\bigcap_{\mathcal{A}_\infty} f^{-1}(0) = \varnothing$. Therefore, $\overline{\mathcal{A}}^{\| \|_0} = C(\Omega; \mathbb{R})$ and $\overline{\mathcal{A}}^{\| \|_0} = \{\hat{f}/_\Omega$ with $\hat{f}(\infty) = 0\} = \{\hat{f}/_\Omega : \hat{f} \in C(\Omega_\infty; \mathbb{R})$ with $\hat{f}(\infty) = 0\} = C_\infty(\Omega; \mathbb{R})$. ■

APPENDIX M

Measure and Integration

m.1. Measurable Spaces and Measure Spaces

A ring {σ-ring} in a set X is a collection $\Sigma \in \mathcal{P}(X)$ that is closed under differences and finite {countable} unions. Σ is termed an algebra {σ-algebra} if X is also a member of Σ. By a measurable space (X, Σ) is meant a set X with a σ-ring Σ such that $X = \bigcup_{\Sigma} E$. Members of Σ are called Σ-measurable sets in X.

A real-valued set function μ on a ring Σ is said to be

(i) additive {countably additive} if

$$\mu\Big(\bigcup_{k=1}^{n} E_k\Big) = \sum_{k=1}^{n} \mu(E_k) \qquad \Big\{\mu\Big(\bigcup_{k=1}^{\infty} E_k\Big) = \sum_{k=1}^{\infty} \mu(E_k) \text{ whenever } \bigcup_{k=1}^{\infty} E_k \in \Sigma\Big\}$$

for disjoint $\{E_k\} \in \Sigma$;

(ii) finite if $\mu(E) < \infty \ \forall E \in \Sigma$; σ-finite if each $E \in \Sigma$ is contained in some $\bigcup_{k=1}^{\infty} E_k$ $(E_k \in \Sigma)$ with $\mu(E_k) < \infty \ \forall k \in \mathbb{N}$;

(iii) complete if $F \in \Sigma$ whenever $F \subset E \in \Sigma$ and $\mu(E) = 0$.

"Finite" {"σ-finite"} is replaced by "totally finite" {"totally σ-finite"} when Σ is an algebra.

An extended, nonnegative, countably additive set function μ on Σ such that $\mu(\varnothing) = 0$ is called a measure. A measurable space $(X, \Sigma; \mu)$ is a measure space (X, Σ) with a measure μ on Σ.

Every collection $\mathscr{E} \subset \mathcal{P}(X)$ generates a minimal ring $\Sigma(\mathscr{E})$ {minimal σ-ring $\Sigma_\sigma(\mathscr{E})$}; namely the intersection of all rings {σ-rings} that contain $\mathscr{E}$, and every $E \in \Sigma(\mathscr{E})$ {$E \in \Sigma_\sigma(\mathscr{E})$} can be covered by a finite {countable} union of sets in $\mathscr{E}$.

UNIQUE EXTENSION THEOREM. Let $\mathfrak{R}$ be a ring and let μ, ν be measures on $\Sigma_\sigma(\mathfrak{R})$ such that $\mu(E)=\nu(E)\ \forall E\in\mathfrak{R}$. If μ, ν are σ-finite on $\mathfrak{R}$, then $\mu=\nu$ on $\Sigma_\sigma(\mathfrak{R})$.

m.2. Measurable Functions

Functions in the next two sections are assumed to be real-valued; properties of complex-valued functions are defined in terms of their real-valued components. For instance, $f=f_{\mathrm{Re}}+if_{\mathrm{Im}}$ is measurable {integrable} iff both f_{Re} and f_{Im} are measurable {integrable—in which case, $\int f\,d\mu=\int f_{\mathrm{Re}}\,d\mu+i\int f_{\mathrm{Im}}\,d\mu$}.

A function f on a measurable space (X,Σ) is measurable (more accurately, Σ-measurable) if $\{X-f^{-1}(0)\}\cap f^{-1}(B)\in\Sigma$ for each Borel set $B\subset\mathbb{R}$ (cf Section m.8). If Σ is a σ-algebra, "$\{X-f^{-1}(0)\}\cap f^{-1}(B)$" can be replaced by "$f^{-1}(B)$." For $f\in\mathscr{F}(X;\mathbb{R})$ define $f^+(x)=\max\{f(x),0\}$ and $f^-(x)=\max\{-f(x),0\}$. Then $f=f^+-f^-$ and $|f|=f^++f^-$ [as well as $f^+=(|f|+f)/2$ and $f^-=(|f|-f)/2$] are simultaneously measurable or non-measurable. The characteristic function χ_E of $E\subset X$ is measurable iff $E\in\Sigma$. Accordingly, every simple function $\varphi=\sum_{i=1}^{n}\alpha_i\chi_{E_i}$ $(E_i\in\Sigma)$ is measurable.

m.3. Integrable Functions

Let $(X,\Sigma;\mu)$ be a complete measure space (the completeness of μ ensures that measurability properties are not destroyed on subsets of measure zero). A simple function $\varphi=\sum_{i=1}^{n}\alpha_i\chi_{E_i}$ is said to be integrable (an ISF) if $\mu(E_i)<\infty\ \forall\alpha_i\neq 0$, in which case, one defines

$$\int\varphi\,d\mu=\sum_{i=1}^{n}\alpha_i\mu(E_i)<\infty.$$

Write $f\geq g$ for $f(x)\geq g(x)\ \forall x\in X$, and let $f_n\uparrow f$ denote $\lim_n f_n=f$ for $f_1\leq f_2\leq\cdots\leq f_n\leq\cdots$. If $f\geq 0$ is measurable, there exist simple functions $\{\varphi_n\}$ satisfying $0\leq\varphi_n\uparrow f$. For instance, take

$$\varphi_n=\sum_{i=1}^{n2^n}\frac{i-1}{2^n}\chi_{f^{-1}[i-1/2^n,\,1/2^n)}+n\chi_{f^{-1}[n,\infty)}.$$

Since $\lim_n \int \varphi_n \, d\mu = \sup_n \int \varphi_n \, d\mu$ is an extended real number and $\lim_n \int \varphi_n \, d\mu = \lim_n \int \psi_n \, d\mu$ for simple functions $0 \le \psi_n \uparrow f$, we call f integrable (and define $\int f \, d\mu = \lim_n \int \varphi_n \, d\mu$) if $\sup_n \int \varphi_n \, d\mu < \infty$ for ISF's $\{\varphi_n\}$ satisfying $0 \le \varphi_n \uparrow f$.

Not every measurable function f is the limit of an increasing sequence of simple functions. Nevertheless, $f = f^+ - f^-$ and f is integrable iff $f^\pm$ is integrable (iff $|f|$ is integrable). In this vein a measurable function F is termed integrable iff $F = f_1 - f_2$ for nonnegative integrable functions f_1 and f_2.

For $1 \le p < \infty$ define $\rho_p(f) = \sqrt[p]{\int |f|^p \, d\mu}$ on the vector space $L_p(X, \Sigma; \mu)$ of pth-integrable functions on X (i.e., $\int |f|^p \, d\mu < \infty$), and let $\rho_\infty(f) = \inf\{\lambda : |f| \le \lambda \text{ a.e. on } X\}$ on the vector space $L_\infty(X, \Sigma; \mu)$ of measurable, a.e. bounded functions on X. For $(f, g) \in L_p(X, \Sigma; \mu) \times L_q(X, \Sigma; \mu)$ with $1/p + 1/q = 1$ Hölders inequality $|\int fg \, d\mu| \le \rho_p(f)\rho_q(g)$ ensures that $fg \in L_1(X, \Sigma; \mu)$. An immediate byproduct is Minkowski's inequality $\rho_p(f + F) \le \rho_p(f) + \rho_p(F) \; \forall f, F \in L_p(X, \Sigma; \mu)$, which further ensures that ρ_p is a seminorm on $L_p(X, \Sigma; \mu)$. If a.e. equal functions in $L_p(X, \Sigma; \mu)$ $(1 \le p \le \infty)$ are considered equal, $L_p(X, \Sigma; \mu)$ is a Banach space for the norm ρ_p, now written $\| \, \|_p$.

If $\mu(X) < \infty$, then $L_q(X, \Sigma; \mu) \subset L_p(X, \Sigma; \mu) \; \forall 1 \le p \le q$. This need not be the case for nontotally finite μ. (Consider $[\sqrt{x}(1 - \log x)]^{-1} \in L_2(X, \Sigma; \mu) - L_1(X, \Sigma; \mu)$ for Lebesgue measure μ on $X = [0, \infty)$.) In general, $p < q$ yields $L_p(X, \Sigma; \mu) \cap L_q(X, \Sigma; \mu) \subset L_r(X, \Sigma; \mu) \; \forall 0 \le r \le q$.

The Banach space l_p $(1 \le p < \infty)$ consisting of sequences $\{x = (x_1, x_2, \ldots) : \sum_{n=1}^{\infty} |x_n|^p < \infty\}$ with norm $\|x\|_p = \sqrt[p]{\sum_{n=1}^{\infty} |x_n|^p}$ derives from $L_p(X, \Sigma; \mu)$ by taking $X = \mathbb{N}$ with $\Sigma = \mathcal{P}(\mathbb{N})$ and letting $\mu(E) = \aleph(E) \; \forall E \in \mathcal{P}(\mathbb{N})$. (Every function f on $(\mathbb{N}, \mathcal{P}(\mathbb{N}))$ is measurable and satisfies $\int_{\mathbb{N}} |f|^p \, d\mu = \sum_{n=1}^{\infty} |f(n)|^p$). Similarly, l_∞ is the Banach space of bounded sequences $\{x = (x_1, x_2, \ldots) : \sup_n |x_n| < \infty\}$ with $\|x\|_0 = \sup_n |x_n|$. Two other $\| \, \|_0$-complete spaces are of interest; namely, $c = \{x = (x_1, x_2, \ldots) : \lim_n x_n \text{ exists}\}$ and $c_0 = \{x = (x_1, x_2, \ldots) : \lim_n x_n = 0\}$.

m.4. Product Measure and Fubini's Theorem

The measurable product space of measurable spaces $(X, \mathcal{R})$ and $(Y, \mathcal{S})$ is defined as $(X \times Y, \Sigma_\sigma(\mathcal{R} \times \mathcal{S}))$. If $\mathcal{R} = \Sigma_\sigma(\mathcal{E})$ and $\mathcal{S} = \Sigma_\sigma(\mathcal{F})$, then

$\Sigma_\sigma(\mathcal{R}\times\mathcal{S})=\Sigma_\sigma(\mathcal{E}\times\mathcal{F})$. The x- and y-sections of $E\in\Sigma_\sigma(\mathcal{R}\times\mathcal{S})$ are defined as $E_x=\{y\in Y:(x,y)\in E\}\subset\mathcal{S}$ and $E_y=\{x\in X:(x,y)\in E\}\subset\mathcal{R}$.

THEOREM. Let $E\in\Sigma_\sigma(\mathcal{R}\times\mathcal{S})$, where $(X,\mathcal{R};\mu)$ and $(Y,\mathcal{S};\nu)$ are σ-finite measure spaces. Then $\nu(E_x)$ is $\mathcal{R}$-measurable, $\mu(E_y)$ is $\mathcal{S}$-measurable, and $\int_X \nu(E_x)\,d\mu=\int_Y \mu(E_y)\,d\nu$. The σ-finite set function $\pi(R\times S)=\mu(R)\nu(S)$ on $\mathcal{R}\times\mathcal{S}$ has a unique σ-finite extension to a measure $\mu\times\nu$ (called the product of μ and ν) on $\Sigma_\sigma(\mathcal{R}\times\mathcal{S})$ given by $\int_X \nu(E_x)\,d\mu=(\mu\times\nu)(E)=\int_Y \mu(E_y)\,d\nu$.

Remark. This σ-finiteness condition is essential. If μ is Lebesgue measure and ν denotes cardinality on $X=Y=[0,1]$ and E_x is the union of diagonal squares with sides of length $1/n$, then $\Delta=\bigcap_{n=1}^{\infty} E_n\in\Sigma_\sigma(X\times Y)$ and $\int_0^1 \nu(\Delta_x)\,d\mu=1\neq 0=\int_0^1 \mu(\Delta_y)\,d\nu$.

FUBINI'S THEOREM. Let $h(x,y)\in L_1\ (X\times Y,\Sigma_\sigma(\mathcal{R}\times\mathcal{S}):\mu\times\nu)$ for σ-finite measure spaces $(X,\mathcal{R};\mu)$ and $(Y,\mathcal{S};\nu)$. Then

(i) $h(x_0,y)\in L_1(Y,\mathcal{S};\nu)$ a.e. on X and $h(x,y_0)\in L_1(X,\mathcal{R};\mu)$ a.e. on Y;

(ii) $\int_Y h(x_0,y)\,d\nu\in L_1(X,\mathcal{R};\mu)$ and $\int_X h(x,y_0)\,d\mu\in L_1(Y,\mathcal{S};\nu)$;

(iii) $\int_{X\times Y} h(x,y)\,d(\mu\times\nu)$ denotes

$$\int_X\left\{\int_Y h(x,y)\,d\nu\right\}d\mu=\int_Y\left\{\int_X h(x,y)\,d\mu\right\}d\nu.$$

m.5. Signed Measures

A signed measure is an extended, real-valued, countably additive set function μ on a measurable space (X,Σ) such that $\mu(\varnothing)=0$. (Additivity necessitates that μ assume at most one of the values $+\infty$ or $-\infty$). Finiteness, total finiteness, and σ-finiteness for a signed measure μ is defined as in m.1 with $|\mu(E)|$ replacing $\mu(E)\ \forall E\in\Sigma$. Nonnegative {nonpositive} signed measures μ are denoted $\mu\leq 0$ $\{\mu\geq 0\}$.

A set $E\subset X$ is positive {negative} relative to a signed measure space $(X,\Sigma;\mu)$ if $\forall F\in\Sigma$ we have $E\cap F\in\Sigma$ and $\mu(E\cap F)\geq 0$ $\{\mu(E\cap F)\leq 0\}$. Positive {negative} sets $E\subset X$ are denoted $E\geq 0$ $\{E\leq 0\}$. Note that $\varnothing=0$, and $E\geq 0$ $\{E\leq 0\}$ implies $F\geq 0$ $\{F\leq 0\}$ $\forall F\in\Sigma$ satisfying $F\subset E$. The converse also holds whenever $E\in\Sigma$.

HAHN DECOMPOSITION THEOREM. To each signed measure μ on (X, Σ) there corresponds disjoint sets $A \geq 0$ and $B \leq 0$ such that $X = A \cup B$.

Remark. The above Hahn decomposition need not be unique. If $\{A', B'\}$ is another such decomposition of $(X, \Sigma; \mu)$, then $\mu(E \cap A') = \mu(E \cap A)$ and $\mu(E \cap B') = \mu(E \cap B)$ $\forall E \in \Sigma$.

The measures $\mu^+(E) = \mu(E \cap A)$ and $\mu^-(E) = \mu(E \cap B)$ are termed the upper and lower variations of μ, respectively. Accordingly, the measure $|\mu|(E) = \mu^+(E) + \mu^-(E)$ is called the total variation of μ. We also call $\mu = \mu^+ - \mu^-$ the Jordan decomposition of μ. Note that μ is finite $\{\sigma\text{-finite}\}$ iff $|\mu|$ is finite $\{\sigma\text{-finite}\}$, since $\mu(E) \leq |\mu(E)| \leq \max\{\mu^{\pm}(E)\} \leq |\mu|(E)\ \forall E \in \Sigma$, a consequence of which is $|\mu|(E) = \sup\{\sum_k |\mu(E_k)| : \{E_k\} \subset \Sigma$ is a finite disjoint cover of $E\}$. We also have $|\mu|(E) \leq 2 \sup_{F \in \Sigma}\{\mu(F) : F \subset E\}$ when $|\mu(E)| < \infty$.

For a measurable function f on $(X, \Sigma; \mu)$, define $\int f\,d\mu = \int f\,d\mu^+ - \int f\,d\mu^-$ and $\int f\,d|\mu| = \int f\,d\mu^+ + \int f\,d\mu^-$ so that $L_1(X, \Sigma; \mu) = L_1(X, \Sigma; \mu^+) \cap L_1(X, \Sigma; \mu^-) = L_1(X, \Sigma; \mu)$. If $f \in L_1(X, \Sigma; \mu)$, then $(f\,d\mu)(E) = \int_E f\,d\mu$ is a finite signed measure on Σ having $(f\,d\mu)^{\pm} = \int_E f^{\pm}\,d\mu$ and total variation $|f\,d\mu| = |f|\,d\mu$.

m.6. Complex Measures

A complex measure on (X, Σ) is a set function $\mu = \mu_1 + i\mu_2$, where μ_1 and μ_2 are signed measures on (X, Σ). Finiteness, total finiteness, and σ-finiteness of μ is defined as in m.1 with $|\mu(E)|$ replacing $\mu(E)\ \forall E \in \Sigma$.

Although upper and lower variations are meaningless for complex measures, the total variation of $\mu = \mu_1 + i\mu_2$ is defined as $|\mu|(E) = \sup\{\sum_k |\mu(E_k)| : \{E_k\} \subset \Sigma$ is a finite disjoint cover of $E\}$. Evidently $|\mu(E)| \leq |\mu|(E) \leq |\mu_1|(E) + |\mu_2|(E) \leq 2|\mu|(E)\ \forall E \in \Sigma$. In fact, $|\mu|$ is the smallest measure $\nu \geq 0$ satisfying $|\mu(E)| \leq \nu(E)\ \forall E \in \Sigma$. As with signed measures, $|\mu|$ is finite $\{\sigma\text{-finite}\}$ iff μ is finite $\{\sigma\text{-finite}\}$, and $L_1(X, \Sigma; \mu) = L_1(X, \Sigma; |\mu|)$. Proofs of the remarks above as well as those below may be found in Zaanen [108]; Dunford and Schwartz [26].

THEOREM. Let μ be a finite complex measure on (X, Σ). For each $E \in \Sigma$:

(i) $\left|\int_E f\,d\mu\right| \leq \int_E |f|\,d|\mu|$ for every measurable f on (X, Σ);

(ii) $|\mu|(E) = \sup\{|\int_E f\,d\mu|$: measurable f with $|f| \leq 1\}$;

(iii) $\sup_{F\in\Sigma}\{|\mu(F)| : F \subset E\} \leq |\mu|(E) \leq \sup_{F\in\Sigma}\{|\mu(F)| : F \subset E\}$.

m.7. Absolute Continuity and Mutual Singularity

A signed {complex} measure μ on (X, Σ) is absolutely continuous with respect to a signed measure ν on (X, Σ) (denoted, $\nu \ll \mu$) if $\mu(E) = 0$ implies $|\nu|(E) = 0\ \forall E \in \Sigma$. In essence, $\nu \ll \mu$ iff $\nu^{\pm} \ll \mu$ iff $|\nu| \ll \mu$ when ν is signed. For $\nu = \nu_1 + i\nu_2$ we have $\nu \ll \mu$ iff $\nu_1, \nu_2 \ll \mu$ iff $|\nu| \ll \mu$.

If ν is finite, $\nu \ll \mu$ is equivalent to requiring that for each $\varepsilon > 0$ there be a $\delta > 0$ such that $|\mu|(E) < \delta \Rightarrow |\nu|(E) < \varepsilon\ \forall E \in \Sigma$.

The antithesis of absolute continuity is mutual singularity. Signed {complex} measures μ, ν on (X, Σ) are mutually singular (denoted $\nu \perp \mu$) if there exist disjoint sets A, B with $X = A \cup B$ such that $E \cap A$, $E \cap B \in \Sigma$ and $\mu(E \cap A) = 0 = \nu(E \cap B)\ \forall E \in \Sigma$. Clearly $\nu \perp \mu$ iff $|\nu| \perp |\mu|$. For non-zero measures $\nu \perp \mu$ iff $\nu \not\ll \mu$ and $\mu \not\ll \nu$.

For each $F \in L_1(X, \Sigma; \mu)$ the signed measure $F\,d\mu$ on Σ satisfies $F\,d\mu \ll \mu$. The converse is contained in

Radon–Nikodym Theorem. If $\nu \ll \mu$ for a σ-finite measure ν on a totally σ-finite measure space $(X, \Sigma; \mu)$, then there is a finite-valued measurable function f on (X, Σ) satisfying $\nu(E) = \int_E f\,d\mu\ \forall E\Sigma$. This (not necessarily μ-integrable) function f is unique in the sense that $\int_E g\,d\mu = \int_E f\,d\mu\ \forall E \in \Sigma$ implies $f = g$ a.e. (relative μ). Furthermore, f is μ-integrable iff ν is finite, in which case the norm $\|\nu\|_v = |\nu|(X)$ has the same value as $\|f\|_1$.

If ν, μ are totally σ-finite on (X, Σ), then $|\nu| \ll |\nu| + |\mu|$ and there is a finite-valued measurable function F on (X, Σ) satisfying $|\nu|(E) = \int_E F\,d|\nu| + \int_E F\,d|\mu|\ \forall E \in \Sigma$. In fact, $0 \leq F \leq 1$ a.e. (relative $|\nu|$). Taking $A = F^{-1}(0)$ and $B = F^{-1}\{[0, 1)\}$. define $\nu_a(E) = \nu(E \cap A)$ and $\nu_s(E) = \nu(E \cap B)\ \forall E \in \Sigma$. Then $\nu_a \ll \mu$ and $\nu_s \perp \mu$. Moreover, $\nu = \nu_a + \nu_s$ and $|\nu| - |\nu_a| + |\nu_s|$; therefore $\|\nu\|_v = \|\nu_a\|_v + \|\nu_s\|_v$. (To be sure, $|\nu_a(E)| + |\nu_s(E)| \leq \{|\nu_a| + |\nu_s|\}(E)$ and $|\nu_a(E)| + |\nu_s(E)| = |\nu(E \cap A)| + |\nu(E \cap B)| \leq |\nu|(E)$ yield $|\nu_a| + |\nu_s| \leq |\nu|$. We always have $|\nu_a + \nu_s| \leq |\nu_a| + |\nu_s|$). This proves the

Lebesgue Decomposition Theorem. Let ν, μ be totally σ-finite signed measures on a σ-algebra Σ of X. Then there exist unique totally

σ-finite signed measures $\nu_a \ll \mu$ and $\nu_s \perp \mu$ such that $\nu = \nu_a + \nu_s$ and $|\nu| = |\nu_a| + |\nu_s|$.

The vector space $(X', ||| \, |||)$ of continuous linear forms on a normed linear space $(X, \| \, \|)$ is Banach (Corollary 13.25) and the Radon Nikodym theorem is frequently used to prove the following

THEOREM. Let $(X, \Sigma; \mu)$ be a totally σ-finite measure space. For each $\Phi \in L'_p(X, \Sigma; \mu)$ $(1 \le p < \infty$ and $1/p + 1/q = 1)$ there is a unique $\Psi_\Phi \in L_q(X, \Sigma; \mu)$ satisfying $\Phi(f) = \int f \Psi_\Phi \, d\mu \; \forall f \in L_p(X, \Sigma; \mu)$. In particular, $(L'_p(X, \Sigma; \mu), ||| \, |||) \underset{\Phi \rightsquigarrow \Psi_\Phi}{\to} (L_q(X, \Sigma; \mu), \| \, \|_q)$ is a congruence (i.e., a linear surjection satisfying $\|\Psi_\Phi\|_q = |||\Phi||| \; \forall \Phi \in L'_p(X, \Sigma; \mu)$).

Remark. The above theorem also yields $l'_p = l_q$ for $1 \le p < \infty$ with $1/p + 1/q = 1$. Note that $c'_0 = c' = l_1$ but $l'_\infty \ne l_1$.

m.8. Measure and Integration in Locally Compact, T_2 Spaces

Let $\mathcal{C}(\mathcal{C}_\delta\}$ denote the compact {compact G_δ} subsets of a locally compact T_2 space X and let $\mathcal{O}$ $\{\mathcal{O}_\delta\}$ be the collection of open subsets of $\mathcal{R}_\sigma(\mathcal{C})$ $\{\mathcal{R}_\sigma(\mathcal{C}_\delta)\}$, the σ-ring of Borel {Baire} subsets of X. Since every compact Baire set is a G_δ, no compact sets other than those that are G_δ's can generate $\mathcal{R}_\sigma(\mathcal{C}_\delta)$. If $\mathfrak{O}$ and $\mathfrak{C}$, respectively, denote the open and closed sets in X, the common value $\mathcal{R}_\sigma$ of $\mathcal{R}_\sigma(\mathfrak{O}) = \mathcal{R}_\sigma(\mathfrak{C})$ is termed the σ-algebra of weakly Borel sets in X.

A set $B \subset X$ is bounded {σ-bounded (σ-compact)} if $B \subset C$ $\left\{\text{if } B \subset \bigcup_{n=1}^{\infty} C_n \left(\text{if } B = \bigcup_{n=1}^{\infty} C_n\right)\right\}$ for C $\{C_n\} \in \mathcal{C}$. Thus $B \subset X$ is bounded iff B is relatively compact. It is easily verified that $\mathcal{R}_\sigma(\mathcal{C})$ coincides with the collection of σ-bounded, weakly Borel subsets of X.

A measure ν on $\mathcal{R}_\sigma(\mathcal{C}_\delta)$ is Baire if $\nu(C) < \infty \; \forall C \in \mathcal{C}$ and a measure μ on $\mathcal{R}_\sigma(\mathcal{C})$ $\{\mu_r$ on $\mathcal{R}_\sigma\}$ is Borel {weakly Borel} if $\mu(C) < \infty$ $\{\mu_r(C) < \infty\}$ $\forall C \in \mathcal{C}$. This terminology extends to signed or complex measures by replacing $\mu(C)$ with $|\mu(C)|$. Thus a signed measure μ {complex measure $\mu = \mu_1 + i\mu_2$} is Borel iff both $\mu^\pm$ {both μ_1, μ_2}; therefore $|\mu|$ is Borel.

Let $(X, \hat{\mathcal{R}}_\sigma; \hat{\mu})$ denote either $(X, \mathcal{R}_\sigma(\mathcal{C}_\delta); \nu)$, $(X, \mathcal{R}_\sigma(\mathcal{C}); \mu)$ or $(X, \mathcal{R}_\sigma; \mu_r)$ with $\hat{\mathcal{O}} = \mathcal{O}_\delta$, $\mathcal{O}$, or $\mathfrak{O}$ and $\hat{\mathcal{C}} = \mathcal{C}_\delta$ or $\mathcal{C}$ in $\hat{\mathcal{R}}_\sigma$. The measure $\hat{\mu}$ on $\hat{\mathcal{R}}_\sigma$ is said to be

(i) outer regular if $\hat{\mu}(E) = \inf\{\hat{\mu}(u) : E \subset u \in \hat{\mathcal{O}}\} \; \forall E \in \hat{\mathcal{R}}_\sigma$;

(ii) inner regular if $\hat{\mu}(E)=\sup\{\hat{\mu}(C): C\subset E \text{ and } C\in\hat{\mathcal{C}}\}\ \forall E\in\hat{\mathcal{R}}_\sigma$;

(iii) regular if $\hat{\mu}$ is both inner and outer regular.

A signed or complex measure $\hat{\mu}$ on $\hat{\mathcal{R}}_\sigma$ is regular if $\forall E\in\hat{\mathcal{R}}_\sigma$ and $\varepsilon>0$ there exist sets $C\in\hat{\mathcal{C}}$ and $u\in\hat{\mathcal{O}}$ with $C\subset E\subset u$ such that $|\hat{\mu}(A)|<\varepsilon\ \forall A\in\hat{\mathcal{R}}_\sigma$ satisfying $A\subset u-C$. This is equivalent to requiring that $|\hat{\mu}(A)-\hat{\mu}(E)|<\varepsilon\ \forall A\in\hat{\mathcal{R}}_\sigma$ satisfying $C\subset A\subset u$.

The above definitions of regularity are equivalent for a measure $\hat{\mu}\geq 0$ on $\hat{\mathcal{R}}_\sigma$; viz., $E=C\cup(E-C)\Rightarrow|\hat{\mu}(E)|\leq|\hat{\mu}(C)|+\varepsilon$ and $|\hat{\mu}(E)|=\sup\{|\hat{\mu}(C)|: C\subset E \text{ and } C\in\hat{\mathcal{C}}\}$. Similarly, $u=E\cup(u-E)\Rightarrow|\hat{\mu}(u)|\leq|\hat{\mu}(E)|+\varepsilon$ and $|\hat{\mu}(E)|=\inf\{|\hat{\mu}(u)|: E\subset u\in\hat{\mathcal{O}}\}$. The converse requires monotonicity. If $\hat{\mu}\geq 0$ satisfies (iii), then $\hat{\mu}(E)<\hat{\mu}(C)+\varepsilon/2$ and $\hat{\mu}(u)<\hat{\mu}(E)+\varepsilon/2$ for some pair $(C,u)\in\hat{\mathcal{C}}\times\hat{\mathcal{O}}$ satisfying $C\subset E\subset u$. Thus $\hat{\mu}(u)\leq\hat{\mu}(C)+\varepsilon$ and $\hat{\mu}(A)\leq\hat{\mu}(u-C)=\hat{\mu}(u)-\hat{\mu}(C)<\varepsilon\ \forall A\in u-C$.

Remark. Baire measures are regular; Borel measures need not be (if $\mathfrak{T}$ is the topology that is discrete on $\mathbb{R}-\{0\}$ and has open sets $\{E\cup\{0\}: \varnothing\neq E\subset\mathbb{R}\cup\{0\}\}$, then $\mathcal{C}=\{\text{finite subsets of } \mathbb{R}\}$ and the Borel measure $\mu(E)=\aleph(E)$ is nonregular, since $1=\mu(\{0\})=\sup_{\mathcal{C}}\{\mu(C): C\subset\{0\}\}\neq\inf_{\mathcal{O}}\{\mu(u):\{0\}\subset u\}=2$).

If μ is a regular Borel measure and $C\in\mathcal{C}$, there exists a set $D\in\mathcal{C}_\delta$ such that $C\subset D$ and $\mu(D-C)=0$ (Berberian [9]). This can be used to prove that $\forall E\in\mathcal{R}_\sigma(\mathcal{C})$ there is an $F\in\mathcal{R}_\sigma(\mathcal{C}_\delta)$ such that $\mu(E\Delta F)=0$. Accordingly, for every Borel measurable function f there is a Baire measurable function $g=f$ a.e. Since $C_0(X;\mathbb{C})$ is $\|\ \|_p$-dense in $L_p(X,\mathcal{R}_\sigma(\mathcal{C}_\delta);\nu)$ for $1\leq p<\infty$, there follows

THEOREM. $C_0(X;\mathbb{C})$ is $\|\ \|_p$-dense in $L_p(X,\mathcal{R}_\sigma(\mathcal{C});\mu)$ $(1\leq p<\infty)$ for every regular Borel measure μ on $\mathcal{R}_\sigma(\mathcal{C})$.

The vector space $M(X)$ of complex totally finite regular Borel measures is a Banach space under $\|\mu_v\|=|\mu(X)$. This is an immediate consequence of the

RIESZ REPRESENTATION THEOREM. To each $\Phi\in C'_\infty(X;\mathbb{C})$ there corresponds a unique $\mu_\Phi\in M(X)$ such that $\Phi(f)=\int f\,d\mu_\Phi\ \forall f\in C_\infty(X;\mathbb{C})$. In particular, $(C'_\infty(X;\mathbb{C}),\|\|\ \|\|)\underset{\Phi\rightsquigarrow\mu_\Phi}{\rightarrow}(M(X),\|\ \|_v)$ is an onto congruence.

Another important version of Riesz's theorem concerns positivity. Specifically, a linear functional Φ on $C_0(X;\mathbb{C})$ is positive (denoted $\Phi\geq 0$) if $\Phi(f)\geq 0\ \forall f\geq 0$ in $C_0(X;\mathbb{C})$. Thus positive linear functionals are monotone. Notationwise, $C_0^{*+}(X;\mathbb{C})$ denotes the space of positive members of $C_0^*(X;\mathbb{C})$.

RIESZ–MARKOFF THEOREM. To each $\Phi \geq 0$ in $C_0(X; \mathbb{C})$ there corresponds a unique $\mu_\Phi \geq 0$ in $M(X)$ satisfying $\Phi(f) = \int f\, d\mu_\Phi \; \forall f \in C_0(X; \mathbb{C})$. In particular, $\underset{\Phi \rightsquigarrow \mu_\Phi}{C_0^{*+}(X; \mathbb{C}) \to M^+(X)}$ is a bijection.

A locally compact T_2 space X is $T_{3\frac{1}{2}}$. Therefore for $C \subset u$ with $(C, u) \in \mathcal{C} \times \mathfrak{O}$ we have $f(C) = \{1\}$ and $f(X - u) = \{0\}$ for some $f \in C_0(X; I)$. If $C = \bigcap_{n=1}^{\infty} u_n$ $(u_n \in \mathfrak{O})$ and f_n satisfies the above for $C \subset u_n$, then $F_n = \min_{1 \leq k \leq n} f_k$ satisfies $F_n \downarrow \chi_C$. This is formally stated as

THEOREM. For every $C \in \mathcal{C}_\delta$ there is a sequence $\{F_n\} \in C_0(X; \mathbb{C})$ such that $0 \leq F_n \downarrow \chi_C$.

APPENDIX A

Linear Algebra

It is assumed here that X is an inner product space with corresponding inner product norm $\|x\|=\sqrt{(x, x)}$. We call X a Hilbert space whenever $\| \|$ is complete.

a.1. Orthogonality

The set $A \subset X$ is orthogonal if $(x, y)=0\ \forall x, y \in A$. An orthogonal set A having $\|x\|=1\ \forall x \in A$ is said to be orthonormal. Every orthogonal set not containing the zero vector is linearly independent. As is well known (Gram Schmidt), every finite dim inner product space has an orthonormal basis.

THEOREM. Let $A \subset X$ be orthonormal and let $x \in X$. Then

(i) (Bessel's inequality). $\sum_{k=1}^{n} |(x, a_k)|^2 \leq \|x\|^2$ for every finite set $\{a_1, \ldots, a_n\} \subset A$.

(ii) $\{a \in A: (x, a) \neq 0\}$ is countable.

(iii) $\sum_A |(x, a)\overline{(y, a)}| \leq \|x\|\,\|y\|\ \forall y \in X$.

(iv) For $A=\{a_1, a_2, \ldots\}$ and $\{\alpha_1, \alpha_2, \ldots\} \in \mathbb{F}$ we have $\sum_{n=1}^{\infty} \alpha_n a_n < \infty$ iff $\sum_{n=1}^{\infty} |\alpha_n|^2 < \infty$. If $\sum_{n=1}^{\infty} \alpha_n a_n = x$, then $\alpha_n = (x, a_n)\ \forall n \in \mathbb{N}$.

a.2. Completeness

An orthonormal set $A \subset X$ is complete if there is no other orthonormal set containing A (in other words, A is a maximal orthonormal set).

Every inner product space has a complete orthonormal set, since (by Zorn's lemma) every orthonormal set can be extended to a complete orthonormal set. In particular, every finite dim inner product space has a complete orthonormal basis. Such, however, is not the case for an infinite dim Hilbert space Y. Reason: If $\mathscr{B}=\{y_n : n \in \mathbb{N}\}$ is a complete orthonormal set in Y, then $\sum_{n=1}^{\infty}(1/n^4)<\infty$ and the theorem above ensures that $\sum_{n=1}^{\infty}(1/n^2)y_n$ converges to some $y \in Y$. Accordingly, $\mathscr{B}$ cannot be a basis for Y $\left(\text{otherwise } y=\sum_{i=1}^{m}\alpha_i y_{n_i}\ (y_{n_i}\in\mathscr{B}) \text{ and for } k\neq n_i \text{ we have } 1/k^2=\left(y_k, \sum_{n=1}^{\infty}(1/n^2)y_n\right)=\left(y_k, \sum_{i=1}^{m}\alpha_i y_{n_i}\right)=0\right)$.

THEOREM. For an orthonormal set $A \subset X$ the statements

(i) A is complete,
(ii) $\{x \in X : (x, a)=0\ \forall a \in A\}=\{0\}$;
(iii) $x=\sum_A (x, a)a\ \forall x \in X$,
(iv) $\|x\|^2=\sum_A |(x, a)|^2\ \forall x \in X$,
(v) $\overline{[A]}=X$

satisfy

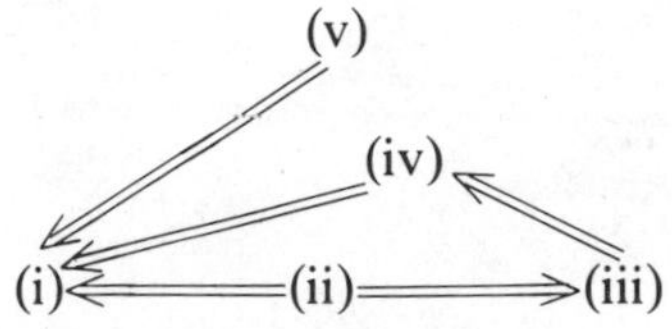

If X is Hilbert, (ii)$\Rightarrow$(iii) and (i)$\Rightarrow$(v) so that (i)–(v) are equivalent.

a.3. Sesquilinear Functionals

A sesquilinear functional on a vector space X is a mapping $f: X \times X \to \mathbb{C}$ which is linear in its first argument and conjugate linear in its second argument (i.e., $f(x, \alpha y+z)=\bar{\alpha}f(x, y)+f(x, z)\ \forall x, y, z \in X$ and $\alpha \in \mathbb{C}$). The sesquilinear functional f is

(i) symmetric if $f(x, y)=\overline{f(y, x)}\ \forall x, y \in X$;

(ii) positive {positive definite} if $f(x, x) \geq 0$ $\{f(x, x) > 0$ for $x \neq 0\}$;

(iii) bounded if there is a constant $\lambda > 0$ such that $f(x, y)| \leq \lambda \|x\| \|y\| \; \forall x, y \in X$.

A inner product is a symmetric positive definite, bounded sesquilinear functional. Boundedness holds by the Cauchy–Schwarz inequality $|(x, y)| \leq \|x\| \|y\|$, which is a special case of the following

THEOREM. A positive sesquilinear functional f on X satisfies $|f(x, y)|^2 \leq f(x, x) f(y, y) \; \forall x, y \in X$.

Proof. The result is trivially true for $f = 0$. If $f \neq 0$ and $\alpha, \beta \in \mathbb{C}$, then $0 \leq f(\alpha x + \beta y, \alpha x + \beta y) = |\alpha|^2 f(x, x) + \alpha\bar{\beta} f(x, y) + \bar{\alpha}\beta \overline{f(x, y)} + |\beta|^2 f(y, y)$. For $\beta = f(x, y)/|f(x, y)|$ and any real α one has $\bar{\beta} f(x, y) = |f(x, y)|$ and $|\beta|^2 = 1$ yielding $0 \leq \alpha^2 f(x, x) + 2\alpha |f(x, y)| + f(y, y)$. Therefore the discriminant satisfies $4|f(x, y)|^2 - 4f(x, x) f(x, x) \leq 0$. ■

Bibliography

1. S. T. M. Ackermans, "On the Principal Extensions of Complex Sets in a Banach Algebra," *Indag. Math.*, **29** (1967), 146–150.
2. R. Arens, "Linear Topological Division Algebras," *Bull. Amer. Math. Soc.*, **53** (1947), 623–630.
3. R. Arens, "The Space L^ω and Convex Topological Rings," *Bull. Amer. Math. Soc.*, **52** (1946), 931–935.
4. M. Atsuji, "Uniform Continuity of Continuous Functions on a Uniform Space," *Can. J. Math.*, **13** (1961), 657–663.
5. M. Atsuji, "Uniform Extensions of Uniformly Continuous Functions," *Proc. Jap. Acad.*, **37** (1961), 10–13.
6. G. Bachman and L. Narici, *Functional Analysis*, Academic, New York, 1968.
7. E. Beckenstein, G. Bachman, and L. Narici, "Spectral Continuity and Permanent Sets in Topological Algebras," *Linceri-Rend. Sc. fis. mat. e nat.*, Ser. VIII, Vol. L (1971), 277–283.
8. S. K. Berberian, *Measure and Integration*, Macmillan, New York, 1965.
9. S. K. Berberian, "Counterexamples in Haar Measures," *Amer. Math. Monthly*, **73**(4), Part II (1966), 135–140.
10. S. K. Berberian, "Sesquiregular Measures," *Amer. Math. Monthly*, **74**(7) (1967), 986–989.
11. G. Birkhoff, "A Note on Topological Groups, "*Composito Math.*, **3** (1936), 427–430.
12. N. Bourbaki, *General Topology*, Parts I and II, Addison-Wesley, Reading, Mass., 1966.
13. N. Bourbaki, Espaces Vectoriels Topologiques, Actualités *Sci. et Ind.*, **1189** (1953) and **1229** (1955), Hermann et Cie, Paris.
14. J. Braconnier, "Spectres d'Espaces et de Groupes Topologiques," *Portugaliae Math.*, **7** (1948), 93–111.
15. J. T. Burnham and R. R. Goldberg, "The Convolution Theorems of Diudonne," *Acta. Sci. Math. Szeged*, **36** (1974), 1–3.
16. E. Cech, *Topological Spaces*, Wiley, New York, 1966.
17. V. Celeste, "Algebras of Borel measures Over Fields With a Non Archimedean Valuation, doctoral dissertation, Polytechnic Institute of New York, 1966.
18. P. Civin and B. Yood, "Quasi Reflexive Spaces," *Proc. Amer. Math. Soc.*, **8** (1957), 906–911.
19. S. Cohen, "A Measure Induced Metric Topology for a Banach Algebra," *Colloq. Mat.*, **XXV** (2) (1972), 273–279.
20. S. Cohen, "An Inner product for a Banach Algebra," *Boll. U.M.I.*, **IV**, No. 1 (1973), 35–41.

21. H. S. Collins, "Completeness and Compactness in Linear Topological Spaces," *Trans. Amer. Math. Soc.*, **79** (1955), 256–280.

22. H. H. Corson and J. R. Isbell, "Some Properties of Strong Uniformities," *Quart. J. Math. Oxford* (2), **11** (1960), 17–33.

23. J. Dieudonne, "Sur la Completion des groupes Topologiques," *C.R. Acad. Sci. Paris*, **218** (1944), 774–776.

24. R. Doss, "On Uniform Spaces With a Unique Uniform Structure," *Amer. J. Math.*, **71** (1949), 19–23.

25. J. D. Dugundji, *Topology*, Allyn and Bacon, Boston, 1966.

26. N. Dunford and J. T. Schwartz, *Linear Operators*, Part I, Wiley, New York, 1958.

27. C. Dunkl and D. Ramirez, *Topics in Harmonic Analysis*, Appleton Century Crofts, New York, 1965.

28. R. E. Edwards, *Functional Analysis*, Holt, Rinehart and Winston, New York, 1965.

29. R. E. Edwards, *Integration and Harmonic Analysis on Compact Groups*, Cambridge University Press, 1972.

30. R. J. Flemming, "Characterizations of Semi-reflexivity and Quasi-reflexivity," *Duke Math. J.*, **36** (1969), 73–80. Correction, *Ibid.* **37** (1970), 787.

31. *S. A. Gaal, Linear Algebra and Representation Theory*, Springer-Verlag, New York, 1973.

32. S. Ginsberg and J. I. Isbell, "Some Operators on Uniform Spaces," *Trans. Amer. Math. Soc.*, **93** (1959), 145–168.

33. G. G. Gould and M. Mahowald, "Quasi-barreled Locally Convex Spaces," *Proc. Amer. Math. Soc.*, **11** (1960), 811–816.

34. M. I. Graev, "Free Topological Groups," *Amer. Math. Soc. Transl.* (1) **8** (1962), 305–364.

35. A. Grothendieck, "Sur les Espaces ($\mathfrak{F}$) et ($\mathfrak{D}\mathfrak{F}$)," *Summa Brazil Math.*, **3** (1954), 57–123.

36. A. Hager, "Projections of Zero Sets," *Trans. Amer. Math. Soc.*, **140** (1969), 87–94.

37. A. Hager, "Uniformities on a Product," *Canadian J. Math.*, **XXIV** (3) (1973), 379–389.

38. P. R. Halmos, *Measure Theory*, Van Nostrand, New York, 1950.

39. R. Harris, "Linear Multiplicative Operators on a Banach Algebra," doctoral dissertation, Polytechnic Institute of New York, New York, 1972.

40. S. Hartman, "Some Problems in the Algebra of Borel Measures," *Colloq. Math.*, **10** (1963), 73–79.

41. E. Hewitt and K. Ross, *Abstract Harmonic Analysis*, Springer-Verlag, Berlin, 1965.

42. P. Hilton, *Introduction to Homotopy Theory*, Cambridge University Press, New York, 1953.

43. J. Horvath, *Topological Vector Spaces and Distributions*, Vol. I, Addison-Wesley, Reading, Mass., 1966.

44. T. Husain, *Introduction to Topological Groups*, Saunders, Philadelphia, 1966.

45. J. R. Isbell, "Uniform Spaces," *Amer. Math. Soc. Survey 12*, Providence, Rhode Island, 1964.

46. R. C. James, "Bases and Reflexivity of Banach Spaces," *Ann. Math.*, **52** (1950), 518–527.

47. R. C. James, "A Separable Somewhat Reflexive Banach Space With Nonseparable Dual," *Bull. Amer. Math. Soc.*, **80**, No. 4 (1974), 738–743.

48. I. Kaplansky, "Normed Algebras," *Duke Math. J.*, **16** (1949), 399–418.

49. I. Katznelson, *An Introduction to Harmonic Analysis*, Wiley, New York, 1968.

50. J. L. Kelley, *General Topology*, Van Nostrand, New York, 1955.

51. J. L. Kelley, "Hypercomplete Linear Topological Spaces," *Mich. Math. J.*, **5** (1958), 235–246.

52. J. L. Kelley and I. Namioka, *Linear Topological Spaces*, Van Nostrand, Princeton, 1963.

53. V. Klee, "Invariant Metrics in Groups," *Proc. Amer. Math. Soc.*, **3** (1958), 484–487.

54. Y. Komura, "Some Examples on Linear Topological Spaces," *Math. Ann.*, **153** (1964), 150–162.

55. G. Köthe, *Topological Vector Spaces I*, Springer-Verlag, New York, 1969.

56. M. Kuczma, "On a Problem of E. Michael Concerning Topological Divisors of Zero," *Colloq. Math.*, **19** (1968), 295–299.

57. R. H. Lohman and P. G. Casazza, "A General Construction of Spaces of the Type of R. C. James," *Can. J. Math.*, **27** (1975), 1263–1270.

58. L. Loomis, *An Introduction to Abstract Harmonic Analysis*, Van Nostrand, New York, 1953.

59. G. Mackey, "Functions on Locally Compact Groups," *Bull. Amer. Math. Soc.*, **56** (1950), 385–412.

60. A. A. Markov, "On Free Topological Groups," *Amer. Math. Soc. Transl.* (1), **8** (1962), 195–272.

61. G. McCarty, *Topology: An Introduction with Applications to Topological Groups*, McGraw-Hill, New York, 1967.

62. E. A. Michael, "Locally Multiplicatively Convex Topological Algebras," *Mem. Amer. Math. Soc.*, **11** (1952), 1–79.

63. D. Montgomery, "Continuity in Topological Groups," *Bull. Amer. Math. Soc.*, **42** (1936), 879–882.

64. D. Montgomery and L. Zippin, *Topological Transformation Groups*, Wiley, New York, 1955.

65. M. G. Murdeshwar and S. A. Naimpally, *Quasi Uniform Topological Spaces*, Ser. A, Vol. 2, No. 4, Stechart Hafner, New York, 1966.

66. T. K. Murkerjee and W. H. Summers, "Continuous Linear Functionals Arising From Convergence in Measure," *Amer. Math. Monthly*, **81** (1974), 63–66.

67. L. Nachbin, *The Haar Integral*, Van Nostrand, New York, 1955.

68. L. Nachbin, "Topological Vector Spaces of Continuous Functions," *Proc. Nat. Acad. Sci. USA*, **40** (1954), 471–474.

69. M. Naimark, *Normed Rings*, Noordhoff, Groningen, 1959.

70. S. A. Naimpally, "Separation Axioms in Quasi Uniform Spaces," *Amer. Math. Monthly*, **74** (1967), 283–284.

71. S. A. Naimpally and B. D. Warrack, *Proximity Spaces*, Cambridge University Press, Great Britain, 1970.

72. L. Narici, *Topological Algebras*, American Elsevier, New York, 1978.

73. I. P. Natanson, *Theory of Functions of a Real Variable*, Vol. I, Ungar, New York, 1955.

74. W. Page, "Characterizations of B^* and Semisimple A^*-Algebras," *Rev. Rom. Mat. Pures Appl.* (8), **XVIII** (1973), 1241–1244.

75. W. Page, "Measure-Induced Seminormed Topologies on a Banach Algebra," *Rend. Mat.* (2), **6**, Ser. VI (1973), 238–255.

76. W. Page, "Projective Limits of Topological Vector Spaces," *Boll. U.M.I.* (4), **8** (1973), 35–45.

77. A. L. Peressini, "On Topologies in Ordered Vector Spaces," *Math. Ann.*, **144** (1961), 199–223.

78. B. J. Pettis, "On Continuity and Openness of Homomorphisms in Topological Groups," *Ann. Math.*, **52** (2) (1950), 293–308.

79. R. R. Phelps, "Uniqueness of Hahn Banach Extensions and Unique Best Approximations," *Trans. Amer. Math. Soc.*, **95** (1960), 238–255.

80. L. Pontryagin, *Topological Groups*, 2nd ed., Gordon and Breach, New York, 1966.

81. V. Ptak, "Completeness and the Open Mapping Theorem," *Bull. Amer. Soc. Math. France*, **86** (1958), 41–74.

82. C. Rickart, *General Theory of Banach Algebras*, Van Nostrand, New York, 1960.

83. C. Rickart, "The Uniqueness of Norm Problem in Banach Algebras," *Ann. Math.*, **51** (1950), 615–628.

84. A. P. Robertson and W. J. Robertson, *Topological Vector Spaces*, Cambridge University Press, Cambridge, 1964.

85. W. Rudin, *Fourier Analysis on Groups*, Wiley, New York, 1962.

86. S. Saxon, "Two Characterizations of Linear Baire Spaces," *Proc. Amer. Math. Soc.*, **45**(2) (1974), 204–208.

87. S. Saxon, "Some Normed Barelled Spaces which Are Not Baire," *Math. Ann. Band*, **209**, Heft 2 (1974), 153–160.

88. H. H. Schaefer, *Topological Vector Spaces*, Springer-Verlag, New York, 1971.

89. A. Schreider, "The Structure of Maximal Ideals in Rings of Measures With Convolutions," *Amer. Math. Soc. Transl.*, **8**, Ser. 1 (1953), 365–391.

90. J. Sebastiao e Silva, "Sur Certe Clasi di Spazi Localmente Convessi Importanti per le Applicazioni," *Rend. Mat. Roma.*, **V**, Ser. 14 (1955), 388–410.

91. T. Shirota, "On Locally Convex Vector Spaces of Continuous Functions," *Proc. Jap. Acad.*, **30** (1954), 294–298.

92. G. Silov, "On Decomposition of a Commutative Normed Ring in a Direct Sum of Ideals," *Amer. Math. Soc. Transl.* (2), **1** (1955), 37–48.

93. P. D. Stratigos, "Application of Duality to Spaces of Measures," doctoral dissertation, Polytechnic Institute of New York, 1972.

94. M. Sugiura, *Unitary Representations and Harmonic Analysis*, Wiley, New York, 1975.

95. C. Suffel, "On Additive Set Functions With Values in a Topological Vector Space," *Annali Mat. Pura Appl. IV*, **XCIII** (1972), 81–88.

96. C. Suffel, "Measures Taking Values in a Topological Vector Space," doctoral dissertation, Polytechnic Institute of New York, 1969.

97. O. Takenouchi, "Sur les Espaces Lineaires Localemente Convexes," *Math. J. Okayama Univ.*, **2** (1952), 57–84.

98. F. Treves, *Topological Vector Spaces, Distributions and kernels*, Academic, New York, 1967.

99. J. W. Tukey, "Convergence and Uniformity in Topology," *Ann. Math. Studies*, No. 1, Princeton University Press, 1940.

100. P. Turpin, "Une Remarque sur les Algebras a Inverse Continu," *C.R. Acad. Sci. Paris*, **270** (1970), 1686–1689.

101. V. S. Varadarajan, "Measures on Topological Spaces," *Amer. Math. Soc. Transl.*, **48**(2) (1965), 161–228.

102. L. Waelbroeck, *Topological Vector Spaces and Algebras*, Springer-Verlag, New York, 1971.

103. S. Warner, "The Topology of Compact Convergence on Continuous Function Spaces," *Duke Math. J.*, **25** (1958), 265–282.

104. A. Weil, "Sur les Espaces et Structures Uniforme et sur la Topologicique General," *Actual. Sci. Ind. fisc.*, **551**, Paris, 1937.

105. R. F. Wheeler, "The Equicontinuous Weak * Topology and Semi-Reflexivity," *Studia math.*, **XLI** (1972), 244–256.

106. J. Williams, "Locally Uniform Spaces," *Trans. Amer. Math. Soc.*, **169** (1972), 471–481.

107. J. H. Williamson, "On Topologizing the Field $C(t)$," *Proc. Amer. Math. Soc.*, **5** (1954), 729–734.

108. A. C. Zaanen, *Theory of Integration*, Wiley, New York, 1958.

INDEX

Entries designated with asterisk are for definitional purposes only.